College Algebra

About Our Team

Lead Author, Senior Content Expert

Jay Abramson has been teaching College Algebra for 33 years, the last 14 at Arizona State University, where he is a principal lecturer in the School of Mathematics and Statistics. His accomplishments at ASU include co-developing the university's first hybrid and online math courses as well as an extensive library of video lectures and tutorials. In addition, he has served as a contributing author for two of Pearson Education's math programs, NovaNet Precalculus and Trigonometry. Prior to coming to ASU, Jay taught at Texas State Technical College and Amarillo College. He received Teacher of the Year awards at both institutions.

The following faculty contributed to the development of OpenStax College Algebra, the text from which this product was updated and derived.

Contributing Authors:

Valeree Falduto, Palm Beach State College
Rachael Gross, Towson University
David Lippman, Pierce College
Melonie Rasmussen, Pierce College
Rick Norwood, East Tennessee State University
Nicholas Belloit, Florida State College Jacksonville
Jean-Marie Magnier, Springfield Technical Community College
Harold Whipple
Christina Fernandez

Honorable Mention:

Nina Alketa, Cecil College
Kiran Bhutani, Catholic University of America
Brandie Biddy, Cecil College
Lisa Blank, Lyme Central School
Bryan Blount, Kentucky Wesleyan College
Jessica Bolz, The Bryn Mawr School
Sheri Boyd, Rollins College
Sarah Brewer, Alabama School of Math and Science
Charles Buckley, St. Gregory's University
Kenneth Crane, Texarkana College
Rachel Cywinski, Alamo Colleges
Nathan Czuba
Srabasti Dutta, Ashford University
Kristy Erickson, Cecil College
Nicole Fernandez, Georgetown University / Kent State University
David French, Tidewater Community College
Douglas Furman, SUNY Ulster
Erinn Izzo, Nicaragua Christian Academy
John Jaffe
Jerry Jared, Blue Ridge School
Stan Kopec, Mount Wachusett Community College
Kathy Kovacs
Sara Lenhart, Christopher Newport University
Joanne Manville, Bunker Hill Community College
Karla McCavit, Albion College
Cynthia McGinnis, Northwest Florida State College
Lana Neal, University of Texas at Austin
Steven Purtee, Valencia College
Alice Ramos, Bethel College
Nick Reynolds, Montgomery Community College
Amanda Ross, A. A. Ross Consulting and Research, LLC
Erica Rutter, Arizona State University
Sutandra Sarkar, Georgia State University
Willy Schild, Wentworth Institute of Technology
Todd Stephen, Cleveland State University
Scott Sykes, University of West Georgia
Linda Tansil, Southeast Missouri State University
John Thomas, College of Lake County
Diane Valade, Piedmont Virginia Community College

Faculty Reviewers and Consultants:

Phil Clark, Scottsdale Community College
Michael Cohen, Hofstra University
Matthew Goodell, SUNY Ulster
Lance Hemlow, Raritan Valley Community College
Dongrin Kim, Arizona State University
Cynthia Landrigan, Erie Community College
Wendy Lightheart, Lane Community College
Carl Penziul, Tompkins-Cortland Community College
Sandra Nite, Texas A&M University
Eugenia Peterson, Richard J. Daley College
Rhonda Porter, Albany State University
Michael Price, University of Oregon
William Radulovich, Florida State College Jacksonville
Camelia Salajean, City Colleges of Chicago
Katy Shields, Oakland Community College
Nathan Schrenk, ECPI University
Pablo Suarez, Delaware State University
Allen Wolmer, Atlanta Jewish Academy

OpenStax College

Rice University
6100 Main Street MS-375
Houston, Texas 77005

To learn more about OpenStax College, visit **http://openstaxcollege.org**. Individual print copies and bulk orders can be purchased through our website.

Brief Contents

v

Contents

ix

Preface

Welcome to *College Algebra*, an OpenStax College resource. This textbook has been created with several goals in mind: accessibility, customization, and student engagement—all while encouraging students toward high levels of academic scholarship. Instructors and students alike will find that this textbook offers a strong foundation in College Algebra in an accessible format.

About OpenStax College

OpenStax College is a non-profit organization committed to improving student access to quality learning materials. Our free textbooks go through a rigorous editorial publishing process. Our texts are developed and peer-reviewed by educators to ensure they are readable, accurate, and meet the scope and sequence requirements of today's college courses. Unlike traditional textbooks, OpenStax College resources live online and are owned by the community of educators using them. Through our partnerships with companies and foundations committed to reducing costs for students, OpenStax College is working to improve access to higher education for all. OpenStax College is an initiative of Rice University and is made possible through the generous support of several philanthropic foundations. OpenStax College textbooks are used at many colleges and universities around the world. Please go to https://openstaxcollege.org/pages/adoptions to see our rapidly expanding number of adoptions.

About OpenStax College's Resources

OpenStax College resources provide quality academic instruction. Three key features set our materials apart from others: they can be customized by instructors for each class, they are a "living" resource that grows online through contributions from educators, and they are available free or for minimal cost.

Customization

OpenStax College learning resources are designed to be customized for each course. Our textbooks provide a solid foundation on which instructors can build, and our resources are conceived and written with flexibility in mind. Instructors can select the sections most relevant to their curricula and create a textbook that speaks directly to the needs of their classes and student body. Teachers are encouraged to expand on existing examples by adding unique context via geographically localized applications and topical connections.

College Algebra can be easily customized using our online platform (http://cnx.org/content/col11759/latest/). Simply select the content most relevant to your current semester and create a textbook that speaks directly to the needs of your class. *College Algebra* is organized as a collection of sections that can be rearranged, modified, and enhanced through localized examples or to incorporate a specific theme to your course. This customization feature will ensure that your textbook truly reflects the goals of your course.

Curation

To broaden access and encourage community curation, *College Algebra* is "open source" licensed under a Creative Commons Attribution (CC-BY) license. The mathematics community is invited to submit feedback to enhance and strengthen the material and keep it current and relevant for today's students. Submit your suggestions to info@openstaxcollege.org, and check in on edition status, alternate versions, errata, and news on the StaxDash at http://openstaxcollege.org.

Cost

Our textbooks are available for free online, and in low-cost print and e-book editions.

xi

About *College Algebra*

Written and reviewed by a team of highly experienced instructors, **College Algebra** provides a comprehensive and multilayered exploration of algebraic principles. The text is suitable for a typical introductory algebra course, and was developed to be used flexibly. While the breadth of topics may go beyond what an instructor would cover, the modular approach and the richness of content ensures that the book meets the needs of a variety of programs.

College Algebra guides and supports students with differing levels of preparation and experience with mathematics. Ideas are presented as clearly as possible, and progress to more complex understandings with considerable reinforcement along the way. A wealth of examples—usually several dozen per chapter—offer detailed, conceptual explanations, in order to build in students a strong, cumulative foundation in the material before asking them to apply what they've learned.

Coverage and Scope

In determining the concepts, skills, and topics to cover, we engaged dozens of highly experienced instructors with a range of student audiences. The resulting scope and sequence proceeds logically while allowing for a significant amount of flexibility in instruction.

Chapters 1 and 2 provide both a review and foundation for study of Functions that begins in Chapter 3. The authors recognize that while some institutions may find this material a prerequisite, other institutions have told us that they have a cohort that need the prerequisite skills built into the course.

> Chapter 1: Prerequisites
>
> Chapter 2: Equations and Inequalities

Chapters 3-6: The Algebraic Functions

> Chapter 3: Functions
>
> Chapter 4: Linear Functions
>
> Chapter 5: Polynomial and Rational Functions
>
> Chapter 6: Exponential and Logarithm Functions

Chapters 7-9: Further Study in College Algebra

> Chapter 7: Systems of Equations and Inequalities
>
> Chapter 8: Analytic Geometry
>
> Chapter 9: Sequences, Probability and Counting Theory

All chapters are broken down into multiple sections, the titles of which can be viewed in the Table of Contents.

Development Overview

Openstax **College Algebra** is the product of a collaborative effort by a group of dedicated authors, editors, and instructors whose collective passion for this project has resulted in a text that is remarkably unified in purpose and voice. Special thanks is due to our Lead Author, Jay Abramson of Arizona State University, who provided the overall vision for the book and oversaw the development of each and every chapter, drawing up the initial blueprint, reading numerous drafts, and assimilating field reviews into actionable revision plans for our authors and editors.

The collective experience of our author team allowed us to pinpoint the subtopics, exceptions, and individual connections that give students the most trouble. And so the textbook is replete with well-designed features and highlights, which help students overcome these barriers. As the students read and practice, they are coached in methods of thinking through problems and internalizing mathematical processes.

For example, narrative text is often followed with the "How To" feature, which summarizes the presentation into a series of distinct steps. This approach addresses varying learning styles, and models for students an important learning skill for future studies. Furthermore, the extensive graphical representations immediately connect concepts with visuals.

Accuracy of the Content

We understand that precision and accuracy are imperatives in mathematics, and undertook a dedicated accuracy program led by experienced faculty.

1. Each chapter's manuscript underwent rounds of review and revision by a panel of active instructors.
2. Then, prior to publication, a separate team of experts checked all text, examples, and graphics for mathematical accuracy; multiple reviewers were assigned to each chapter to minimize the chances of any error escaping notice.
3. A third team of experts was responsible for the accuracy of the Answer Key, dutifully re-working every solution to eradicate any lingering errors. Finally, the editorial team conducted a multi-round post-production review to ensure the integrity of the content in its final form.

The Solutions Manual, which was written and developed after the Student Edition, has also been rigorously checked for accuracy following a process similar to that described above. Incidentally, the act of writing out solutions step-by-step served as yet another round of validation for the Answer Key in the back of the Student Edition. In spite of the efforts described above, we acknowledge the possibility that—as with any textbook—some errata may have been missed. We encourage users to report errors via our Errata (https://openstaxcollege.org/errata) page.

Pedagogical Foundations and Features

Learning Objectives

Each chapter is divided into multiple sections (or modules), each of which is organized around a set of learning objectives. The learning objectives are listed explicitly at the beginning of each section, and are the focal point of every instructional element.

Narrative Text

Narrative text is used to introduce key concepts, terms, and definitions, to provide real-world context, and to provide transitions between topics and examples. Throughout this book, we rely on a few basic conventions to highlight the most important ideas:

- Key terms are boldfaced, typically when first introduced and/or when formally defined Key concepts and definitions are called out in a blue box for easy reference.
- Key concepts and definitions are called out in a blue box for easy reference.

Examples

Each learning objective is supported by one or more worked examples, which demonstrate the problem-solving approaches that students must master. The multiple Examples model different approaches to the same type of problem, or introduce similar problems of increasing complexity.

All Examples follow a simple two- or three-part format. The question clearly lays out a mathematical problem to solve. The Solution walks through the steps, usually providing context for the approach --in other words, why the instructor is solving the problem in a specific manner. Finally, the Analysis (for select examples) reflects on the broader implications of the Solution just shown. Examples are followed by a "Try It," question, as explained below.

Figures

Openstax **College Algebra** contains many figures and illustrations, the vast majority of which are graphs and diagrams. Art throughout the text adheres to a clear, understated style, drawing the eye to the most important information in each figure while minimizing visual distractions. Color contrast is employed with discretion to distinguish between the different functions or features of a graph.

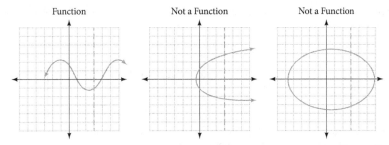

Supporting Features

Four unobtrusive but important features, each marked by a distinctive icon, contribute to and check understanding.

A "*How To*" is a list of steps necessary to solve a certain type of problem. A How To typically precedes an Example that proceeds to demonstrate the steps in action.

A "*Try It*" exercise immediately follows an Example or a set of related Examples, providing the student with an immediate opportunity to solve a similar problem. In the Web View version of the text, students can click an Answer link directly below the question to check their understanding. In the PDF, answers to the Try-It exercises are located in the Answer Key.

A "*Q & A...*" may appear at any point in the narrative, but most often follows an Example. This feature pre-empts misconceptions by posing a commonly asked yes/no question, followed by a detailed answer and explanation.

The "Media" links appear at the conclusion of each section, just prior to the Section Exercises. These are a list of links to online video tutorials that reinforce the concepts and skills introduced in the section.

Disclaimer: While we have selected tutorials that closely align to our learning objectives, we did not produce these tutorials, nor were they specifically produced or tailored to accompany Openstax *College Algebra*. We are deeply grateful to James Sousa for compiling his incredibly robust and excellent library of video tutorials, which he has made available to the public under a CC-BY-SA license at http://mathispower4u.yolasite.com/. Most or all of the videos to which we link in our "Media" feature (plus many more) are found in the Algebra 2 video library at the above site.

Section Exercises

Each section of every chapter concludes with a well-rounded set of exercises that can be assigned as homework or used selectively for guided practice. With over 4,600 exercises across the 9 chapters, instructors should have plenty to choose from[i].

Section Exercises are organized by question type, and generally appear in the following order:

Verbal questions assess conceptual understanding of key terms and concepts.

Algebraic problems require students to apply algebraic manipulations demonstrated in the section.

Graphical problems assess students' ability to interpret or produce a graph.

Numeric problems require the student perform calculations or computations.

Technology problems encourage exploration through use of a graphing utility, either to visualize or verify algebraic results or to solve problems via an alternative to the methods demonstrated in the section.

Extensions pose problems more challenging than the Examples demonstrated in the section. They require students to synthesize multiple learning objectives or apply critical thinking to solve complex problems.

Real-World Applications present realistic problem scenarios from fields such as physics, geology, biology, finance, and the social sciences.

Chapter Review Features

Each chapter concludes with a review of the most important takeaways, as well as additional practice problems that students can use to prepare for exams.

Key Terms provides a formal definition for each bold-faced term in the chapter.

Key Equations presents a compilation of formulas, theorems, and standard-form equations.

Key Concepts summarizes the most important ideas introduced in each section, linking back to the relevant Example(s) in case students need to review.

Chapter Review Exercises include 40-80 practice problems that recall the most important concepts from each section.

Practice Test includes 25-50 problems assessing the most important learning objectives from the chapter. Note that the practice test is not organized by section, and may be more heavily weighted toward cumulative objectives as opposed to the foundational objectives covered in the opening sections.

Ancillaries

OpenStax projects offer an array of ancillaries for students and instructors. Currently the following resources are available.

Instructor's Solutions Manual
Student's Solutions Manual
PowerPoint Slides

Please visit http://openstaxcollege.org to view an up-to-date list of the Learning Resources for this title and to find information on accessing these resources.

Online Homework

XYZ Homework **XYZ Homework** is built using the fastest-growing mathematics cloud platform. XYZ Homework gives instructors access to the Precalculus aligned problems, organized in the College Algebra Course Template. Instructors have access to thousands of additional algorithmically-generated questions for unparalleled course customization. For one low annual price, students can take multiple classes through XYZ Homework. Learn more at www.xyzhomework.com/openstax.

WebAssign **WebAssign** is an independent online homework and assessment solution first launched at North Carolina State University in 1997. Today, WebAssign is an employee-owned benefit corporation and participates in the education of over a million students each year. WebAssign empowers faculty to deliver fully customizable assignments and high quality content to their students in an interactive online environment. WebAssign supports College Algebra with hundreds of problems covering every concept in the course, each containing algorithmically-generated values and links directly to the eBook providing a completely integrated online learning experience.

xiv

---

i. 4,649 total exercises. Includes Chapter Reviews and Practice Tests.

1 | Prerequisites

Figure 1 Credit: Andreas Kambanls

Introduction

It's a cold day in Antarctica. In fact, it's always a cold day in Antarctica. Earth's southernmost continent, Antarctica experiences the coldest, driest, and windiest conditions known. The coldest temperature ever recorded, over one hundred degrees below zero on the Celsius scale, was recorded by remote satellite. It is no surprise then, that no native human population can survive the harsh conditions. Only explorers and scientists brave the environment for any length of time.

Measuring and recording the characteristics of weather conditions in Antarctica requires a use of different kinds of numbers. Calculating with them and using them to make predictions requires an understanding of relationships among numbers. In this chapter, we will review sets of numbers and properties of operations used to manipulate numbers. This understanding will serve as prerequisite knowledge throughout our study of algebra and trigonometry.

1

LEARNING OBJECTIVES

In this section students will:

- Classify a real number as a natural, whole, integer, rational, or irrational number.
- Perform calculations using order of operations.
- Use the following properties of real numbers: commutative, associative, distributive, inverse, and identity.
- Evaluate algebraic expressions.
- Simplify algebraic expressions.

1.1   REAL NUMBERS: ALGEBRA ESSENTIALS

It is often said that mathematics is the language of science. If this is true, then the language of mathematics is numbers. The earliest use of numbers occurred 100 centuries ago in the Middle East to count, or enumerate items. Farmers, cattlemen, and tradesmen used tokens, stones, or markers to signify a single quantity—a sheaf of grain, a head of livestock, or a fixed length of cloth, for example. Doing so made commerce possible, leading to improved communications and the spread of civilization.

Three to four thousand years ago, Egyptians introduced fractions. They first used them to show reciprocals. Later, they used them to represent the amount when a quantity was divided into equal parts.

But what if there were no cattle to trade or an entire crop of grain was lost in a flood? How could someone indicate the existence of nothing? From earliest times, people had thought of a "base state" while counting and used various symbols to represent this null condition. However, it was not until about the fifth century A.D. in India that zero was added to the number system and used as a numeral in calculations.

Clearly, there was also a need for numbers to represent loss or debt. In India, in the seventh century A.D., negative numbers were used as solutions to mathematical equations and commercial debts. The opposites of the counting numbers expanded the number system even further.

Because of the evolution of the number system, we can now perform complex calculations using these and other categories of real numbers. In this section, we will explore sets of numbers, calculations with different kinds of numbers, and the use of numbers in expressions.

Classifying a Real Number

The numbers we use for counting, or enumerating items, are the **natural numbers**: 1, 2, 3, 4, 5, and so on. We describe them in set notation as {1, 2, 3,...} where the ellipsis (...) indicates that the numbers continue to infinity. The natural numbers are, of course, also called the *counting numbers*. Any time we enumerate the members of a team, count the coins in a collection, or tally the trees in a grove, we are using the set of natural numbers. The set of **whole numbers** is the set of natural numbers plus zero: {0, 1, 2, 3,...}.

The set of **integers** adds the opposites of the natural numbers to the set of whole numbers: {..., −3, −2, −1, 0, 1, 2, 3,...}. It is useful to note that the set of integers is made up of three distinct subsets: negative integers, zero, and positive integers. In this sense, the positive integers are just the natural numbers. Another way to think about it is that the natural numbers are a subset of the integers.

$$\begin{array}{ccc} \text{negative integers} & \text{zero} & \text{positive integers} \\ \ldots, -3, -2, -1, & 0, & 1, 2, 3, \ldots \end{array}$$

The set of **rational numbers** is written as $\left\{ \frac{m}{n} \,\middle|\, m \text{ and } n \text{ are integers and } n \neq 0 \right\}$. Notice from the definition that rational numbers are fractions (or quotients) containing integers in both the numerator and the denominator, and the denominator is never 0. We can also see that every natural number, whole number, and integer is a rational number with a denominator of 1.

Because they are fractions, any rational number can also be expressed in decimal form. Any rational number can be represented as either:

1. a terminating decimal: $\frac{15}{8} = 1.875$, or

2. a repeating decimal: $\frac{4}{11} = 0.36363636\ldots = 0.\overline{36}$

We use a line drawn over the repeating block of numbers instead of writing the group multiple times.

Example 1   **Writing Integers as Rational Numbers**

Write each of the following as a rational number.

    a. 7   **b.** 0   **c.** −8

Solution   Write a fraction with the integer in the numerator and 1 in the denominator.

    a. $7 = \dfrac{7}{1}$ **b.** $0 = \dfrac{0}{1}$ **c.** $-8 = -\dfrac{8}{1}$

Try It #1

Write each of the following as a rational number.

a. 11   **b.** 3   **c.** −4

Example 2   **Identifying Rational Numbers**

Write each of the following rational numbers as either a terminating or repeating decimal.

    a. $-\dfrac{5}{7}$ **b.** $\dfrac{15}{5}$ **c.** $\dfrac{13}{25}$

Solution   Write each fraction as a decimal by dividing the numerator by the denominator.

    a. $-\dfrac{5}{7} = -0.\overline{714285}$, a repeating decimal

    b. $\dfrac{15}{5} = 3$ (or 3.0), a terminating decimal

    c. $\dfrac{13}{25} = 0.52$, a terminating decimal

Try It #2

Write each of the following rational numbers as either a terminating or repeating decimal.

a. $\dfrac{68}{17}$ **b.** $\dfrac{8}{13}$ **c.** $-\dfrac{17}{20}$

Irrational Numbers

At some point in the ancient past, someone discovered that not all numbers are rational numbers. A builder, for instance, may have found that the diagonal of a square with unit sides was not 2 or even $\dfrac{3}{2}$, but was something else. Or a garment maker might have observed that the ratio of the circumference to the diameter of a roll of cloth was a little bit more than 3, but still not a rational number. Such numbers are said to be *irrational* because they cannot be written as fractions. These numbers make up the set of **irrational numbers**. Irrational numbers cannot be expressed as a fraction of two integers. It is impossible to describe this set of numbers by a single rule except to say that a number is irrational if it is not rational. So we write this as shown.

$$\{\,h\,|\,h \text{ is not a rational number}\,\}$$

Example 3   **Differentiating Rational and Irrational Numbers**

Determine whether each of the following numbers is rational or irrational. If it is rational, determine whether it is a terminating or repeating decimal.

    a. $\sqrt{25}$ **b.** $\dfrac{33}{9}$ **c.** $\sqrt{11}$ **d.** $\dfrac{17}{34}$ **e.** 0.3033033303333...

Solution

    a. $\sqrt{25}$: This can be simplified as $\sqrt{25} = 5$. Therefore, $\sqrt{25}$ is rational.

b. $\frac{33}{9}$: Because it is a fraction, $\frac{33}{9}$ is a rational number. Next, simplify and divide.

$$\frac{33}{9} = \frac{\overset{11}{\cancel{33}}}{\underset{3}{\cancel{9}}} = \frac{11}{3} = 3.\overline{6}$$

So, $\frac{33}{9}$ is rational and a repeating decimal.

c. $\sqrt{11}$: This cannot be simplified any further. Therefore, $\sqrt{11}$ is an irrational number.

d. $\frac{17}{34}$: Because it is a fraction, $\frac{17}{34}$ is a rational number. Simplify and divide.

$$\frac{17}{34} = \frac{\overset{1}{\cancel{17}}}{\underset{2}{\cancel{34}}} = \frac{1}{2} = 0.5$$

So, $\frac{17}{34}$ is rational and a terminating decimal.

e. 0.3033033303333 … is not a terminating decimal. Also note that there is no repeating pattern because the group of 3s increases each time. Therefore it is neither a terminating nor a repeating decimal and, hence, not a rational number. It is an irrational number.

Try It #3

Determine whether each of the following numbers is rational or irrational. If it is rational, determine whether it is a terminating or repeating decimal.

a. $\frac{7}{77}$ **b.** $\sqrt{81}$ **c.** 4.27027002700027 …   **d.** $\frac{91}{13}$ **e.** $\sqrt{39}$

Real Numbers

Given any number n, we know that n is either rational or irrational. It cannot be both. The sets of rational and irrational numbers together make up the set of **real numbers**. As we saw with integers, the real numbers can be divided into three subsets: negative real numbers, zero, and positive real numbers. Each subset includes fractions, decimals, and irrational numbers according to their algebraic sign (+ or −). Zero is considered neither positive nor negative.

The real numbers can be visualized on a horizontal number line with an arbitrary point chosen as 0, with negative numbers to the left of 0 and positive numbers to the right of 0. A fixed unit distance is then used to mark off each integer (or other basic value) on either side of 0. Any real number corresponds to a unique position on the number line. The converse is also true: Each location on the number line corresponds to exactly one real number. This is known as a one-to-one correspondence. We refer to this as the **real number line** as shown in **Figure 2**.

$$-5 \;-4 \;-2 \;-1 \;\; 0 \;\;\; 1 \;\;\; 2 \;\;\; 3 \;\;\; 4 \;\;\; 5$$

Figure 2 The real number line

Example 4   **Classifying Real Numbers**

Classify each number as either positive or negative and as either rational or irrational. Does the number lie to the left or the right of 0 on the number line?

a. $-\frac{10}{3}$ **b.** $\sqrt{5}$ **c.** $-\sqrt{289}$ **d.** -6π **e.** 0.615384615384 …

Solution

a. $-\frac{10}{3}$ is negative and rational. It lies to the left of 0 on the number line.

b. $\sqrt{5}$ is positive and irrational. It lies to the right of 0.

c. $-\sqrt{289} = -\sqrt{17^2} = -17$ is negative and rational. It lies to the left of 0.

d. -6π is negative and irrational. It lies to the left of 0.

e. 0.615384615384 … is a repeating decimal so it is rational and positive. It lies to the right of 0.

Evaluating Algebraic Expressions

So far, the mathematical expressions we have seen have involved real numbers only. In mathematics, we may see expressions such as $x + 5$, $\frac{4}{3}\pi r^3$, or $\sqrt{2m^3n^2}$. In the expression $x + 5$, 5 is called a **constant** because it does not vary and x is called a **variable** because it does. (In naming the variable, ignore any exponents or radicals containing the variable.) An **algebraic expression** is a collection of constants and variables joined together by the algebraic operations of addition, subtraction, multiplication, and division.

We have already seen some real number examples of exponential notation, a shorthand method of writing products of the same factor. When variables are used, the constants and variables are treated the same way.

$$(-3)^5 = (-3) \cdot (-3) \cdot (-3) \cdot (-3) \cdot (-3) \qquad\qquad x^5 = x \cdot x \cdot x \cdot x \cdot x$$
$$(2 \cdot 7)^3 = (2 \cdot 7) \cdot (2 \cdot 7) \cdot (2 \cdot 7) \qquad\qquad (yz)^3 = (yz) \cdot (yz) \cdot (yz)$$

In each case, the exponent tells us how many factors of the base to use, whether the base consists of constants or variables.

Any variable in an algebraic expression may take on or be assigned different values. When that happens, the value of the algebraic expression changes. To evaluate an algebraic expression means to determine the value of the expression for a given value of each variable in the expression. Replace each variable in the expression with the given value, then simplify the resulting expression using the order of operations. If the algebraic expression contains more than one variable, replace each variable with its assigned value and simplify the expression as before.

Example 8   Describing Algebraic Expressions

List the constants and variables for each algebraic expression.

    a. $x + 5$ **b.** $\frac{4}{3}\pi r^3$ **c.** $\sqrt{2m^3n^2}$

Solution

	Constants	**Variables**
a. $x + 5$	5	x
b. $\frac{4}{3}\pi r^3$	$\frac{4}{3}$, π	r
c. $\sqrt{2m^3n^2}$	2	m, n

Try It #8

List the constants and variables for each algebraic expression.

a. $2\pi r(r + h)$ **b.** $2(L + W)$ **c.** $4y^3 + y$

Example 9   Evaluating an Algebraic Expression at Different Values

Evaluate the expression $2x - 7$ for each value for x.

    a. $x = 0$ **b.** $x = 1$ **c.** $x = \frac{1}{2}$ **d.** $x = -4$

Solution

    a. Substitute 0 for x.
$$\begin{aligned} 2x - 7 &= 2(0) - 7 \\ &= 0 - 7 \\ &= -7 \end{aligned}$$

    b. Substitute 1 for x.
$$\begin{aligned} 2x - 7 &= 2(1) - 7 \\ &= 2 - 7 \\ &= -5 \end{aligned}$$

c. Substitute $\frac{1}{2}$ for x.

$$2x - 7 = 2\left(\frac{1}{2}\right) - 7$$
$$= 1 - 7$$
$$= -6$$

d. Substitute -4 for x.

$$2x - 7 = 2(-4) - 7$$
$$= -8 - 7$$
$$= -15$$

Try It #9

Evaluate the expression $11 - 3y$ for each value for y.

a. $y = 2$ **b.** $y = 0$ **c.** $y = \frac{2}{3}$ **d.** $y = -5$

Example 10   **Evaluating Algebraic Expressions**

Evaluate each expression for the given values.

    a. $x + 5$ for $x = -5$ **b.** $\frac{t}{2t-1}$ for $t = 10$ **c.** $\frac{4}{3}\pi r^3$ for $r = 5$

    d. $a + ab + b$ for $a = 11$, $b = -8$ **e.** $\sqrt{2m^3n^2}$ for $m = 2$, $n = 3$

Solution

    a. Substitute -5 for x.

$$x + 5 = (-5) + 5$$
$$= 0$$

    b. Substitute 10 for t.

$$\frac{t}{2t - 1} = \frac{(10)}{2(10) - 1}$$

$$= \frac{10}{20 - 1}$$

$$= \frac{10}{19}$$

    c. Substitute 5 for r.

$$\frac{4}{3}\pi r^3 = \frac{4}{3}\pi(5)^3$$

$$= \frac{4}{3}\pi(125)$$

$$= \frac{500}{3}\pi$$

    d. Substitute 11 for a and -8 for b. $a + ab + b = (11) + (11)(-8) + (-8)$
$$= 11 - 88 - 8$$
$$= -85$$

    e. Substitute 2 for m and 3 for n. $\sqrt{2m^3n^2} = \sqrt{2(2)^3(3)^2}$
$$= \sqrt{2(8)(9)}$$
$$= \sqrt{144}$$
$$= 12$$

Try It #10

Evaluate each expression for the given values.

a. $\frac{y + 3}{y - 3}$ for $y = 5$ **b.** $7 - 2t$ for $t = -2$ **c.** $\frac{1}{3}\pi r^2$ for $r = 11$

d. $(p^2q)^3$ for $p = -2$, $q = 3$ **e.** $4(m - n) - 5(n - m)$ for $m = \frac{2}{3}$, $n = \frac{1}{3}$

Try It #9

Simplify each expression and write the answer with positive exponents only.

a. $(2uv^{-2})^{-3}$ **b.** $x^8 \cdot x^{-12} \cdot x$ **c.** $\left(\dfrac{e^2f^{-3}}{f^{-1}}\right)^2$ **d.** $(9r^{-5}s^3)(3r^6s^{-4})$ **e.** $\left(\dfrac{4}{9}tw^{-2}\right)^{-3}\left(\dfrac{4}{9}tw^{-2}\right)^3$ **f.** $\dfrac{(2h^2k)^4}{(7h^{-1}k^2)^2}$

Using Scientific Notation

Recall at the beginning of the section that we found the number 1.3×10^{13} when describing bits of information in digital images. Other extreme numbers include the width of a human hair, which is about 0.00005 m, and the radius of an electron, which is about 0.00000000000047 m. How can we effectively work read, compare, and calculate with numbers such as these?

A shorthand method of writing very small and very large numbers is called **scientific notation**, in which we express numbers in terms of exponents of 10. To write a number in scientific notation, move the decimal point to the right of the first digit in the number. Write the digits as a decimal number between 1 and 10. Count the number of places n that you moved the decimal point. Multiply the decimal number by 10 raised to a power of *n*. If you moved the decimal left as in a very large number, *n* is positive. If you moved the decimal right as in a small large number, *n* is negative.

For example, consider the number 2,780,418. Move the decimal left until it is to the right of the first nonzero digit, which is 2.

$$2{,}780418 \quad \longrightarrow \quad \overset{\text{6 places left}}{2{,}780418}$$

We obtain 2.780418 by moving the decimal point 6 places to the left. Therefore, the exponent of 10 is 6, and it is positive because we moved the decimal point to the left. This is what we should expect for a large number.

$$2.780418 \times 10^6$$

Working with small numbers is similar. Take, for example, the radius of an electron, 0.00000000000047 m. Perform the same series of steps as above, except move the decimal point to the right.

$$0.00000000000047 \quad \longrightarrow \quad \overset{\text{13 places right}}{00000000000004{.}7}$$

Be careful not to include the leading 0 in your count. We move the decimal point 13 places to the right, so the exponent of 10 is 13. The exponent is negative because we moved the decimal point to the right. This is what we should expect for a small number.

$$4.7 \times 10^{-13}$$

---

scientific notation

A number is written in scientific notation if it is written in the form $a \times 10^n$, where $1 \le |a| < 10$ and *n* is an integer.

Example 10   **Converting Standard Notation to Scientific Notation**

Write each number in scientific notation.

a. Distance to Andromeda Galaxy from Earth: 24,000,000,000,000,000,000,000 m

b. Diameter of Andromeda Galaxy: 1,300,000,000,000,000,000,000 m

c. Number of stars in Andromeda Galaxy: 1,000,000,000,000

d. Diameter of electron: 0.00000000000094 m

e. Probability of being struck by lightning in any single year: 0.00000143

Solution

a. 24,000,000,000,000,000,000,000 m
   $\leftarrow$ 22 places
2.4 $\times 10^{22}$ m

b. 1,300,000,000,000,000,000,000 m
   $\leftarrow$ 21 places
1.3 $\times 10^{21}$ m

    c. 1,000,000,000,000

               ← 12 places

      1×10^{12}

    d. 0.00000000000094 m

              → 13 places

      9.4×10^{-13} m

    e. 0.00000143

              → 6 places

      1.43×10^{-6}

Analysis *Observe that, if the given number is greater than 1, as in examples a–c, the exponent of 10 is positive; and if the number is less than 1, as in examples d–e, the exponent is negative.*

Try It #10

Write each number in scientific notation.

a. U.S. national debt per taxpayer (April 2014): $152,000

b. World population (April 2014): 7,158,000,000

c. World gross national income (April 2014): $85,500,000,000,000

d. Time for light to travel 1 m: 0.00000000334 s

e. Probability of winning lottery (match 6 of 49 possible numbers): 0.0000000715

Converting from Scientific to Standard Notation

To convert a number in **scientific notation** to standard notation, simply reverse the process. Move the decimal n places to the right if n is positive or n places to the left if n is negative and add zeros as needed. Remember, if n is positive, the value of the number is greater than 1, and if n is negative, the value of the number is less than one.

Example 11 **Converting Scientific Notation to Standard Notation**

Convert each number in scientific notation to standard notation.

    a. 3.547×10^{14} **b.** -2×10^{6} **c.** 7.91×10^{-7} **d.** -8.05×10^{-12}

Solution

    a. 3.547×10^{14}

        3.54700000000000

           → 14 places

        354,700,000,000,000

    b. -2×10^{6}

        −2.000000

           → 6 places

        −2,000,000

    c. 7.91×10^{-7}

        0000007.91

           ← 7 places

        0.000000791

    d. -8.05×10^{-12}

        −000000000008.05

           ← 12 places

        −0.00000000000805

Try It #11

Convert each number in scientific notation to standard notation.

a. 7.03×10^5 **b.** -8.16×10^{11} **c.** -3.9×10^{-13} **d.** 8×10^{-6}

Using Scientific Notation in Applications

Scientific notation, used with the rules of exponents, makes calculating with large or small numbers much easier than doing so using standard notation. For example, suppose we are asked to calculate the number of atoms in 1 L of water. Each water molecule contains 3 atoms (2 hydrogen and 1 oxygen). The average drop of water contains around 1.32×10^{21} molecules of water and 1 L of water holds about 1.22×10^4 average drops. Therefore, there are approximately $3 \times (1.32 \times 10^{21}) \times (1.22 \times 10^4) \approx 4.83 \times 10^{25}$ atoms in 1 L of water. We simply multiply the decimal terms and add the exponents. Imagine having to perform the calculation without using scientific notation!

When performing calculations with scientific notation, be sure to write the answer in proper scientific notation. For example, consider the product $(7 \times 10^4) \times (5 \times 10^6) = 35 \times 10^{10}$. The answer is not in proper scientific notation because 35 is greater than 10. Consider 35 as 3.5×10. That adds a ten to the exponent of the answer.

$$(35) \times 10^{10} = (3.5 \times 10) \times 10^{10} = 3.5 \times (10 \times 10^{10}) = 3.5 \times 10^{11}$$

Example 12   Using Scientific Notation

Perform the operations and write the answer in scientific notation.

    a. $(8.14 \times 10^{-7})(6.5 \times 10^{10})$

    b. $(4 \times 10^5) \div (-1.52 \times 10^9)$

    c. $(2.7 \times 10^5)(6.04 \times 10^{13})$

    d. $(1.2 \times 10^8) \div (9.6 \times 10^5)$

    e. $(3.33 \times 10^4)(-1.05 \times 10^7)(5.62 \times 10^5)$

Solution

    a. $(8.14 \times 10^{-7})(6.5 \times 10^{10}) = (8.14 \times 6.5)(10^{-7} \times 10^{10})$ Commutative and associative properties of multiplication

$$= (52.91)(10^3)$$ Product rule of exponents

$$= 5.291 \times 10^4$$ Scientific notation

    b. $(4 \times 10^5) \div (-1.52 \times 10^9) = \left(\dfrac{4}{-1.52}\right)\left(\dfrac{10^5}{10^9}\right)$ Commutative and associative properties of multiplication

$$\approx (-2.63)(10^{-4})$$ Quotient rule of exponents

$$= -2.63 \times 10^{-4}$$ Scientific notation

    c. $(2.7 \times 10^5)(6.04 \times 10^{13}) = (2.7 \times 6.04)(10^5 \times 10^{13})$ Commutative and associative properties of multiplication

$$= (16.308)(10^{18})$$ Product rule of exponents

$$= 1.6308 \times 10^{19}$$ Scientific notation

    d. $(1.2 \times 10^8) \div (9.6 \times 10^5) = \left(\dfrac{1.2}{9.6}\right)\left(\dfrac{10^8}{10^5}\right)$ Commutative and associative properties of multiplication

$$= (0.125)(10^3)$$ Quotient rule of exponents

$$= 1.25 \times 10^2$$ Scientific notation

    e. $(3.33 \times 10^4)(-1.05 \times 10^7)(5.62 \times 10^5) = [3.33 \times (-1.05) \times 5.62](10^4 \times 10^7 \times 10^5)$

$$\approx (-19.65)(10^{16})$$

$$= -1.965 \times 10^{17}$$

Try It #12

Perform the operations and write the answer in scientific notation.

a. $(-7.5 \times 10^8)(1.13 \times 10^{-2})$

b. $(1.24 \times 10^{11}) \div (1.55 \times 10^{18})$

c. $(3.72 \times 10^9)(8 \times 10^3)$

d. $(9.933 \times 10^{23}) \div (-2.31 \times 10^{17})$

e. $(-6.04 \times 10^9)(7.3 \times 10^2)(-2.81 \times 10^2)$

Example 13   **Applying Scientific Notation to Solve Problems**

In April 2014, the population of the United States was about 308,000,000 people. The national debt was about $17,547,000,000,000. Write each number in scientific notation, rounding figures to two decimal places, and find the amount of the debt per U.S. citizen. Write the answer in both scientific and standard notations.

Solution   The population was $308,000,000 = 3.08 \times 10^8$.

The national debt was $\$17,547,000,000,000 \approx \$1.75 \times 10^{13}$.

To find the amount of debt per citizen, divide the national debt by the number of citizens.

$$(1.75 \times 10^{13}) \div (3.08 \times 10^8) = \left(\frac{1.75}{3.08}\right) \times \left(\frac{10^{13}}{10^8}\right)$$

$$\approx 0.57 \times 10^5$$

$$= 5.7 \times 10^4$$

The debt per citizen at the time was about $\$5.7 \times 10^4$, or \$57,000.

Try It #13

An average human body contains around 30,000,000,000,000 red blood cells. Each cell measures approximately 0.000008 m long. Write each number in scientific notation and find the total length if the cells were laid end-to-end. Write the answer in both scientific and standard notations.

Access these online resources for additional instruction and practice with exponents and scientific notation.

- Exponential Notation (http://openstaxcollege.org/l/exponnot)
- Properties of Exponents (http://openstaxcollege.org/l/exponprops)
- Zero Exponent (http://openstaxcollege.org/l/zeroexponent)
- Simplify Exponent Expressions (http://openstaxcollege.org/l/exponexpres)
- Quotient Rule for Exponents (http://openstaxcollege.org/l/quotofexpon)
- Scientific Notation (http://openstaxcollege.org/l/scientificnota)
- Converting to Decimal Notation (http://openstaxcollege.org/l/decimalnota)

b. $\sqrt{81a^4b^4 \cdot 2a}$ Factor perfect square from radicand.

   $\sqrt{81a^4b^4} \cdot \sqrt{2a}$ Write radical expression as product of radical expressions.

   $9a^2b^2\sqrt{2a}$ Simplify.

Try It #2

Simplify $\sqrt{50x^2y^3z}$.

How To...

Given the product of multiple radical expressions, use the product rule to combine them into one radical expression.

1. Express the product of multiple radical expressions as a single radical expression.

2. Simplify.

Example 3   **Using the Product Rule to Simplify the Product of Multiple Square Roots**

Simplify the radical expression. $\sqrt{12} \cdot \sqrt{3}$

Solution

   $\sqrt{12 \cdot 3}$ Express the product as a single radical expression.

   $\sqrt{36}$ Simplify.

   6

Try It #3

Simplify $\sqrt{50x} \cdot \sqrt{2x}$ assuming $x > 0$.

Using the Quotient Rule to Simplify Square Roots

Just as we can rewrite the square root of a product as a product of square roots, so too can we rewrite the square root of a quotient as a quotient of square roots, using the *quotient rule for simplifying square roots*. It can be helpful to separate the numerator and denominator of a fraction under a radical so that we can take their square roots separately.

We can rewrite $\sqrt{\dfrac{5}{2}}$ as $\dfrac{\sqrt{5}}{\sqrt{2}}$.

> **the quotient rule for simplifying square roots**
>
> The square root of the quotient $\dfrac{a}{b}$ is equal to the quotient of the square roots of a and b, where $b \neq 0$.
>
> $$\sqrt{\frac{a}{b}} = \frac{\sqrt{a}}{\sqrt{b}}$$

How To...

Given a radical expression, use the quotient rule to simplify it.

1. Write the radical expression as the quotient of two radical expression.

2. Simplify the numerator and denominator.

Example 4   **Using the Quotient Rule to Simplify Square Roots**

Simplify the radical expression. $\sqrt{\dfrac{5}{36}}$

Solution

   $\dfrac{\sqrt{5}}{\sqrt{36}}$ Write as quotient of two radical expressions.

   $\dfrac{\sqrt{5}}{6}$ Simplify denominator.

Try It #4

Simplify $\sqrt{\dfrac{2x^2}{9y^4}}$.

Example 5   **Using the Quotient Rule to Simplify an Expression with Two Square Roots**

Simplify the radical expression.
$$\dfrac{\sqrt{234x^{11}y}}{\sqrt{26x^7y}}$$

Solution

$\sqrt{\dfrac{234x^{11}y}{26x^7y}}$ Combine numerator and denominator into one radical expression.

$\sqrt{9x^4}$ Simplify fraction.

$3x^2$ Simplify square root.

Try It #5

Simplify $\dfrac{\sqrt{9a^5b^{14}}}{\sqrt{3a^4b^5}}$.

Adding and Subtracting Square Roots

We can add or subtract radical expressions only when they have the same radicand and when they have the same radical type such as square roots. For example, the sum of $\sqrt{2}$ and $3\sqrt{2}$ is $4\sqrt{2}$. However, it is often possible to simplify radical expressions, and that may change the radicand. The radical expression $\sqrt{18}$ can be written with a 2 in the radicand, as $3\sqrt{2}$, so $\sqrt{2} + \sqrt{18} = \sqrt{2} + 3\sqrt{2} = 4\sqrt{2}$.

How To...

Given a radical expression requiring addition or subtraction of square roots, solve.

1. Simplify each radical expression.
2. Add or subtract expressions with equal radicands.

Example 6   **Adding Square Roots**

Add $5\sqrt{12} + 2\sqrt{3}$.

Solution

We can rewrite $5\sqrt{12}$ as $5\sqrt{4 \cdot 3}$. According the product rule, this becomes $5\sqrt{4}\,\sqrt{3}$. The square root of $\sqrt{4}$ is 2, so the expression becomes $5(2)\sqrt{3}$, which is $10\sqrt{3}$. Now the terms have the same radicand so we can add.

$10\sqrt{3} + 2\sqrt{3} = 12\sqrt{3}$

Try It #6

Add $\sqrt{5} + 6\sqrt{20}$.

Example 7   **Subtracting Square Roots**

Subtract $20\sqrt{72a^3b^4c} - 14\sqrt{8a^3b^4c}$.

Rewrite each term so they have equal radicands.

$$20\sqrt{72a^3b^4c} = 20\sqrt{9}\sqrt{4}\sqrt{2}\sqrt{a}\sqrt{a^2}\sqrt{(b^2)^2}\sqrt{c}$$

$$= 20(3)(2)|a|b^2\sqrt{2ac}$$

$$= 120|a|b^2\sqrt{2ac}$$

$$14\sqrt{8a^3b^4c} = 14\sqrt{2}\sqrt{4}\sqrt{a}\sqrt{a^2}\sqrt{(b^2)^2}\sqrt{c}$$

$$= 14(2)|a|b^2\sqrt{2ac}$$

$$= 28|a|b^2\sqrt{2ac}$$

Now the terms have the same radicand so we can subtract.

$$120|a|b^2\sqrt{2ac} - 28|a|b^2\sqrt{2ac} = 92|a|b^2\sqrt{2ac}$$

Try It #7

Subtract $3\sqrt{80x} - 4\sqrt{45x}$.

Rationalizing Denominators

When an expression involving square root radicals is written in simplest form, it will not contain a radical in the denominator. We can remove radicals from the denominators of fractions using a process called *rationalizing the denominator*.

We know that multiplying by 1 does not change the value of an expression. We use this property of multiplication to change expressions that contain radicals in the denominator. To remove radicals from the denominators of fractions, multiply by the form of 1 that will eliminate the radical.

For a denominator containing a single term, multiply by the radical in the denominator over itself. In other words, if the denominator is $b\sqrt{c}$, multiply by $\dfrac{\sqrt{c}}{\sqrt{c}}$.

For a denominator containing the sum of a rational and an irrational term, multiply the numerator and denominator by the conjugate of the denominator, which is found by changing the sign of the radical portion of the denominator. If the denominator is $a + b\sqrt{c}$, then the conjugate is $a - b\sqrt{c}$.

How To…

Given an expression with a single square root radical term in the denominator, rationalize the denominator.

1. Multiply the numerator and denominator by the radical in the denominator.
2. Simplify.

Example 8   **Rationalizing a Denominator Containing a Single Term**

Write $\dfrac{2\sqrt{3}}{3\sqrt{10}}$ in simplest form.

Solution
The radical in the denominator is $\sqrt{10}$. So multiply the fraction by $\dfrac{\sqrt{10}}{\sqrt{10}}$. Then simplify.

$$\frac{2\sqrt{3}}{3\sqrt{10}} \cdot \frac{\sqrt{10}}{\sqrt{10}}$$

$$\frac{2\sqrt{30}}{30}$$

$$\frac{\sqrt{30}}{15}$$

Try It #8

Write $\dfrac{12\sqrt{3}}{\sqrt{2}}$ in simplest form.

How To...

Given an expression with a radical term and a constant in the denominator, rationalize the denominator.

1. Find the conjugate of the denominator.
2. Multiply the numerator and denominator by the conjugate.
3. Use the distributive property.
4. Simplify.

Example 9   **Rationalizing a Denominator Containing Two Terms**

Write $\dfrac{4}{1 + \sqrt{5}}$ in simplest form.

Solution

Begin by finding the conjugate of the denominator by writing the denominator and changing the sign. So the conjugate of $1 + \sqrt{5}$ is $1 - \sqrt{5}$. Then multiply the fraction by $\dfrac{1 - \sqrt{5}}{1 - \sqrt{5}}$.

$$\dfrac{4}{1 + \sqrt{5}} \cdot \dfrac{1 - \sqrt{5}}{1 - \sqrt{5}}$$

$$\dfrac{4 - 4\sqrt{5}}{-4} \qquad \text{Use the distributive property.}$$

$$\sqrt{5} - 1 \qquad \text{Simplify.}$$

Try It #9

Write $\dfrac{7}{2 + \sqrt{3}}$ in simplest form.

Using Rational Roots

Although square roots are the most common rational roots, we can also find cuberoots, 4th roots, 5th roots, and more. Just as the square root function is the inverse of the squaring function, these roots are the inverse of their respective power functions. These functions can be useful when we need to determine the number that, when raised to a certain power, gives a certain number.

Understanding *n*th Roots

Suppose we know that $a^3 = 8$. We want to find what number raised to the 3rd power is equal to 8. Since $2^3 = 8$, we say that 2 is the cube root of 8.

The *n*th root of *a* is a number that, when raised to the *n*th power, gives *a*. For example, -3 is the 5th root of -243 because $(-3)^5 = -243$. If *a* is a real number with at least one *n*th root, then the **principal *n*th root** of *a* is the number with the same sign as *a* that, when raised to the *n*th power, equals *a*.

The principal *n*th root of *a* is written as $\sqrt[n]{a}$, where *n* is a positive integer greater than or equal to 2. In the radical expression, *n* is called the **index** of the radical.

> **principal *n*th root**
> If *a* is a real number with at least one *n*th root, then the **principal *n*th root** of *a*, written as $\sqrt[n]{a}$, is the number with the same sign as *a* that, when raised to the *n*th power, equals *a*. The **index** of the radical is *n*.

LEARNING OBJECTIVES

In this section, you will:

- Identify the degree and leading coefficient of polynomials.
- Add and subtract polynomials.
- Multiply polynomials.
- Use FOIL to multiply binomials.
- Perform operations with polynomials of several variables.

1.4 POLYNOMIALS

Earl is building a doghouse, whose front is in the shape of a square topped with a triangle. There will be a rectangular door through which the dog can enter and exit the house. Earl wants to find the area of the front of the doghouse so that he can purchase the correct amount of paint. Using the measurements of the front of the house, shown in **Figure 1**, we can create an expression that combines several variable terms, allowing us to solve this problem and others like it.

Figure 1

First find the area of the square in square feet.

$$A = s^2$$
$$= (2x)^2$$
$$= 4x^2$$

Then find the area of the triangle in square feet.

$$A = \frac{1}{2}bh$$
$$= \frac{1}{2}(2x)\left(\frac{3}{2}\right)$$
$$= \frac{3}{2}x$$

Next find the area of the rectangular door in square feet.

$$A = lw$$
$$= x \cdot 1$$
$$= x$$

The area of the front of the doghouse can be found by adding the areas of the square and the triangle, and then subtracting the area of the rectangle. When we do this, we get $4x^2 + \frac{3}{2}x - x$ ft², or $4x^2 + \frac{1}{2}x$ ft².

In this section, we will examine expressions such as this one, which combine several variable terms.

Identifying the Degree and Leading Coefficient of Polynomials

The formula just found is an example of a **polynomial**, which is a sum of or difference of terms, each consisting of a variable raised to a nonnegative integer power. A number multiplied by a variable raised to an exponent, such as 384π, is known as a **coefficient**. Coefficients can be positive, negative, or zero, and can be whole numbers, decimals, or fractions. Each product $a_i x^i$, such as $384\pi w$, is a **term of a polynomial**. If a term does not contain a variable, it is called a *constant*.

A polynomial containing only one term, such as $5x^4$, is called a **monomial**. A polynomial containing two terms, such as $2x - 9$, is called a **binomial**. A polynomial containing three terms, such as $-3x^2 + 8x - 7$, is called a **trinomial**.

We can find the **degree** of a polynomial by identifying the highest power of the variable that occurs in the polynomial. The term with the highest degree is called the **leading term** because it is usually written first. The coefficient of the leading term is called the **leading coefficient**. When a polynomial is written so that the powers are descending, we say that it is in standard form.

$$\underset{\text{Leading term}}{\underbrace{a_n x^n}} + ... + a_2 x^2 + a_1 x + a_0$$

Leading coefficient   Degree

> ***polynomials***
> A **polynomial** is an expression that can be written in the form
> $$a_n x^n + ... + a_2 x^2 + a_1 x + a_0$$
> Each real number a_i is called a **coefficient**. The number a_0 that is not multiplied by a variable is called a *constant*. Each product $a_i x^i$ is a **term of a polynomial**. The highest power of the variable that occurs in the polynomial is called the **degree** of a polynomial. The **leading term** is the term with the highest power, and its coefficient is called the **leading coefficient**.

How To...

Given a polynomial expression, identify the degree and leading coefficient.

1. Find the highest power of x to determine the degree.
2. Identify the term containing the highest power of x to find the leading term.
3. Identify the coefficient of the leading term.

Example 1   **Identifying the Degree and Leading Coefficient of a Polynomial**

For the following polynomials, identify the degree, the leading term, and the leading coefficient.

    a. $3 + 2x^2 - 4x^3$ **b.** $5t^5 - 2t^3 + 7t$ **c.** $6p - p^3 - 2$

Solution

    a. The highest power of x is 3, so the degree is 3. The leading term is the term containing that degree, $-4x^3$. The leading coefficient is the coefficient of that term, -4.

    b. The highest power of t is 5, so the degree is 5. The leading term is the term containing that degree, $5t^5$. The leading coefficient is the coefficient of that term, 5.

    c. The highest power of p is 3, so the degree is 3. The leading term is the term containing that degree, $-p^3$, The leading coefficient is the coefficient of that term, -1.

Try It #1

Identify the degree, leading term, and leading coefficient of the polynomial $4x^2 - x^6 + 2x - 6$.

Adding and Subtracting Polynomials

We can add and subtract polynomials by combining like terms, which are terms that contain the same variables raised to the same exponents. For example, $5x^2$ and $-2x^2$ are like terms, and can be added to get $3x^2$, but $3x$ and $3x^2$ are not like terms, and therefore cannot be added.

How To...

Given multiple polynomials, add or subtract them to simplify the expressions.

1. Combine like terms.
2. Simplify and write in standard form.

Example 2   **Adding Polynomials**

Find the sum.

$$(12x^2 + 9x - 21) + (4x^3 + 8x^2 - 5x + 20)$$

Solution

$$4x^3 + (12x^2 + 8x^2) + (9x - 5x) + (-21 + 20)$$ Combine like terms.

$$4x^3 + 20x^2 + 4x - 1$$ Simplify.

Analysis *We can check our answers to these types of problems using a graphing calculator. To check, graph the problem as given along with the simplified answer. The two graphs should be equivalent. Be sure to use the same window to compare the graphs. Using different windows can make the expressions seem equivalent when they are not.*

Try It #2

Find the sum.

$$(2x^3 + 5x^2 - x + 1) + (2x^2 - 3x - 4)$$

Example 3   **Subtracting Polynomials**

Find the difference.

$$(7x^4 - x^2 + 6x + 1) - (5x^3 - 2x^2 + 3x + 2)$$

Solution

$$7x^4 - 5x^3 + (-x^2 + 2x^2) + (6x - 3x) + (1 - 2)$$ Combine like terms.

$$7x^4 - 5x^3 + x^2 + 3x - 1$$ Simplify.

Analysis *Note that finding the difference between two polynomials is the same as adding the opposite of the second polynomial to the first.*

Try It #3

Find the difference.

$$(-7x^3 - 7x^2 + 6x - 2) - (4x^3 - 6x^2 - x + 7)$$

Multiplying Polynomials

Multiplying polynomials is a bit more challenging than adding and subtracting polynomials. We must use the distributive property to multiply each term in the first polynomial by each term in the second polynomial. We then combine like terms. We can also use a shortcut called the FOIL method when multiplying binomials. Certain special products follow patterns that we can memorize and use instead of multiplying the polynomials by hand each time. We will look at a variety of ways to multiply polynomials.

Multiplying Polynomials Using the Distributive Property

To multiply a number by a polynomial, we use the distributive property. The number must be distributed to each term of the polynomial. We can distribute the 2 in $2(x + 7)$ to obtain the equivalent expression $2x + 14$. When multiplying polynomials, the distributive property allows us to multiply each term of the first polynomial by each term of the second. We then add the products together and combine like terms to simplify.

Given the multiplication of two polynomials, use the distributive property to simplify the expression.

1. Multiply each term of the first polynomial by each term of the second.
2. Combine like terms.
3. Simplify.

Example 4   **Multiplying Polynomials Using the Distributive Property**

Find the product.

$(2x + 1)(3x^2 − x + 4)$

Solution

$$2x(3x^2 − x + 4) + 1(3x^2 − x + 4) \qquad \text{Use the distributive property.}$$
$$(6x^3 − 2x^2 + 8x) + (3x^2 − x + 4) \qquad \text{Multiply.}$$
$$6x^3 + (−2x^2 + 3x^2) + (8x − x) + 4 \qquad \text{Combine like terms.}$$
$$6x^3 + x^2 + 7x + 4 \qquad \text{Simplify.}$$

Analysis   We can use a table to keep track of our work, as shown in **Table 1**. Write one polynomial across the top and the other down the side. For each box in the table, multiply the term for that row by the term for that column. Then add all of the terms together, combine like terms, and simplify.

	$3x^2$	$−x$	$+4$
$2x$	$6x^3$	$−2x^2$	$8x$
$+1$	$3x^2$	$−x$	4

Table 1

Try It #4

Find the product.

$(3x + 2)(x^3 − 4x^2 + 7)$

Using FOIL to Multiply Binomials

A shortcut called FOIL is sometimes used to find the product of two binomials. It is called FOIL because we multiply the **f**irst terms, the **o**uter terms, the **i**nner terms, and then the **l**ast terms of each binomial.

$$\text{First terms} \quad \text{Last terms}$$
$$(ax + b)(cx + d) = acx^2 + adx + bcx + bd$$
$$\text{Inner terms}$$
$$\text{Outer terms}$$

The FOIL method arises out of the distributive property. We are simply multiplying each term of the first binomial by each term of the second binomial, and then combining like terms.

Given two binomials, use FOIL to simplify the expression.

1. Multiply the first terms of each binomial.
2. Multiply the outer terms of the binomials.
3. Multiply the inner terms of the binomials.
4. Multiply the last terms of each binomial.
5. Add the products.
6. Combine like terms and simplify.

In this section, you will:
- Factor the greatest common factor of a polynomial.
- Factor a trinomial.
- Factor by grouping.
- Factor a perfect square trinomial.
- Factor a difference of squares.
- Factor the sum and difference of cubes.
- Factor expressions using fractional or negative exponents.

1.5 FACTORING POLYNOMIALS

Imagine that we are trying to find the area of a lawn so that we can determine how much grass seed to purchase. The lawn is the green portion in **Figure 1**.

Figure 1

The area of the entire region can be found using the formula for the area of a rectangle.

$$A = lw$$
$$= 10x \cdot 6x$$
$$= 60x^2 \text{ units}^2$$

The areas of the portions that do not require grass seed need to be subtracted from the area of the entire region. The two square regions each have an area of $A = s^2 = 4^2 = 16 \text{ units}^2$. The other rectangular region has one side of length $10x - 8$ and one side of length 4, giving an area of $A = lw = 4(10x - 8) = 40x - 32 \text{ units}^2$. So the region that must be subtracted has an area of $2(16) + 40x - 32 = 40x \text{ units}^2$.

The area of the region that requires grass seed is found by subtracting $60x^2 - 40x \text{ units}^2$. This area can also be expressed in factored form as $20x(3x - 2) \text{ units}^2$. We can confirm that this is an equivalent expression by multiplying.

Many polynomial expressions can be written in simpler forms by factoring. In this section, we will look at a variety of methods that can be used to factor polynomial expressions.

Factoring the Greatest Common Factor of a Polynomial

When we study fractions, we learn that the **greatest common factor (GCF)** of two numbers is the largest number that divides evenly into both numbers. For instance, 4 is the GCF of 16 and 20 because it is the largest number that divides evenly into both 16 and 20 The GCF of polynomials works the same way: $4x$ is the GCF of $16x$ and $20x^2$ because it is the largest polynomial that divides evenly into both $16x$ and $20x^2$.

When factoring a polynomial expression, our first step should be to check for a GCF. Look for the GCF of the coefficients, and then look for the GCF of the variables.

> *greatest common factor*
> The **greatest common factor** (GCF) of polynomials is the largest polynomial that divides evenly into the polynomials.

How To...

Given a polynomial expression, factor out the greatest common factor.

1. Identify the GCF of the coefficients.
2. Identify the GCF of the variables.
3. Combine to find the GCF of the expression.
4. Determine what the GCF needs to be multiplied by to obtain each term in the expression.
5. Write the factored expression as the product of the GCF and the sum of the terms we need to multiply by.

Example 1   **Factoring the Greatest Common Factor**

Factor $6x^3y^3 + 45x^2y^2 + 21xy$.

Solution   First, find the GCF of the expression. The GCF of 6, 45, and 21 is 3. The GCF of x^3, x^2, and x is x. (Note that the GCF of a set of expressions in the form x^n will always be the exponent of lowest degree.) And the GCF of y^3, y^2, and y is y. Combine these to find the GCF of the polynomial, $3xy$.

Next, determine what the GCF needs to be multiplied by to obtain each term of the polynomial. We find that $3xy(2x^2y^2) = 6x^3y^3$, $3xy(15xy) = 45x^2y^2$, and $3xy(7) = 21xy$.

Finally, write the factored expression as the product of the GCF and the sum of the terms we needed to multiply by.

$$(3xy)(2x^2y^2 + 15xy + 7)$$

Analysis   After factoring, we can check our work by multiplying. Use the distributive property to confirm that $(3xy)(2x^2y^2 + 15xy + 7) = 6x^3y^3 + 45x^2y^2 + 21xy$.

Try It #1

Factor $x(b^2 - a) + 6(b^2 - a)$ by pulling out the GCF.

Factoring a Trinomial with Leading Coefficient 1

Although we should always begin by looking for a GCF, pulling out the GCF is not the only way that polynomial expressions can be factored. The polynomial $x^2 + 5x + 6$ has a GCF of 1, but it can be written as the product of the factors $(x + 2)$ and $(x + 3)$.

Trinomials of the form $x^2 + bx + c$ can be factored by finding two numbers with a product of c and a sum of b. The trinomial $x^2 + 10x + 16$, for example, can be factored using the numbers 2 and 8 because the product of those numbers is 16 and their sum is 10. The trinomial can be rewritten as the product of $(x + 2)$ and $(x + 8)$.

> *factoring a trinomial with leading coefficient* **1**
> A trinomial of the form $x^2 + bx + c$ can be written in factored form as $(x + p)(x + q)$ where $pq = c$ and $p + q = b$.

Q & A...

Can every trinomial be factored as a product of binomials?

No. Some polynomials cannot be factored. These polynomials are said to be prime.

How To…

Given a trinomial in the form $x^2 + bx + c$, factor it.

1. List factors of c.
2. Find p and q, a pair of factors of c with a sum of b.
3. Write the factored expression $(x + p)(x + q)$.

Example 2   **Factoring a Trinomial with Leading Coefficient 1**

Factor $x^2 + 2x - 15$.

Solution   We have a trinomial with leading coefficient 1, $b = 2$, and $c = -15$. We need to find two numbers with a product of -15 and a sum of 2. In **Table 1**, we list factors until we find a pair with the desired sum.

Factors of -15	Sum of Factors
1, -15	-14
$-1, 15$	14
3, -5	-2
$-3, 5$	2

Table 1

Now that we have identified p and q as -3 and 5, write the factored form as $(x - 3)(x + 5)$.

Analysis   *We can check our work by multiplying. Use FOIL to confirm that* $(x - 3)(x + 5) = x^2 + 2x - 15$.

Q & A…

Does the order of the factors matter?

No. Multiplication is commutative, so the order of the factors does not matter.

Try It #2

Factor $x^2 - 7x + 6$.

Factoring by Grouping

Trinomials with leading coefficients other than 1 are slightly more complicated to factor. For these trinomials, we can **factor by grouping** by dividing the x term into the sum of two terms, factoring each portion of the expression separately, and then factoring out the GCF of the entire expression. The trinomial $2x^2 + 5x + 3$ can be rewritten as $(2x + 3)(x + 1)$ using this process. We begin by rewriting the original expression as $2x^2 + 2x + 3x + 3$ and then factor each portion of the expression to obtain $2x(x + 1) + 3(x + 1)$. We then pull out the GCF of $(x + 1)$ to find the factored expression.

> *factor by grouping*
> To factor a trinomial in the form $ax^2 + bx + c$ by grouping, we find two numbers with a product of ac and a sum of b. We use these numbers to divide the x term into the sum of two terms and factor each portion of the expression separately, then factor out the GCF of the entire expression.

How To…

Given a trinomial in the form $ax^2 + bx + c$, factor by grouping.

1. List factors of ac.
2. Find p and q, a pair of factors of ac with a sum of b.
3. Rewrite the original expression as $ax^2 + px + qx + c$.
4. Pull out the GCF of $ax^2 + px$.
5. Pull out the GCF of $qx + c$.
6. Factor out the GCF of the expression.

Example 3   **Factoring a Trinomial by Grouping**

Factor $5x^2 + 7x - 6$ by grouping.

Solution   We have a trinomial with $a = 5$, $b = 7$, and $c = -6$. First, determine $ac = -30$. We need to find two numbers with a product of -30 and a sum of 7. In **Table 2**, we list factors until we find a pair with the desired sum.

Factors of −30	Sum of Factors
1, −30	−29
−1, 30	29
2, −15	−13
−2, 15	13
3, −10	−7
−3, 10	7

Table 2

So $p = -3$ and $q = 10$.

$$5x^2 - 3x + 10x - 6$$ Rewrite the original expression as $ax^2 + px + qx + c$.
$$x(5x - 3) + 2(5x - 3)$$ Factor out the GCF of each part.
$$(5x - 3)(x + 2)$$ Factor out the GCF of the expression.

Analysis   We can check our work by multiplying. Use FOIL to confirm that $(5x - 3)(x + 2) = 5x^2 + 7x - 6$.

Try It #3

Factor.   **a.** $2x^2 + 9x + 9$ **b.** $6x^2 + x - 1$

Factoring a Perfect Square Trinomial

A perfect square trinomial is a trinomial that can be written as the square of a binomial. Recall that when a binomial is squared, the result is the square of the first term added to twice the product of the two terms and the square of the last term.

$$a^2 + 2ab + b^2 = (a + b)^2$$

and

$$a^2 - 2ab + b^2 = (a - b)^2$$

We can use this equation to factor any perfect square trinomial.

perfect square trinomials
A perfect square trinomial can be written as the square of a binomial:
$$a^2 + 2ab + b^2 = (a + b)^2$$

How To...

Given a perfect square trinomial, factor it into the square of a binomial.

1. Confirm that the first and last term are perfect squares.
2. Confirm that the middle term is twice the product of ab.
3. Write the factored form as $(a + b)^2$.

Example 4   **Factoring a Perfect Square Trinomial**

Factor $25x^2 + 20x + 4$.

Solution   Notice that $25x^2$ and 4 are perfect squares because $25x^2 = (5x)^2$ and $4 = 2^2$. Then check to see if the middle term is twice the product of $5x$ and 2. The middle term is, indeed, twice the product: $2(5x)(2) = 20x$.

Therefore, the trinomial is a perfect square trinomial and can be written as $(5x + 2)^2$.

Try It #4

Factor $49x^2 - 14x + 1$.

Factoring a Difference of Squares

A difference of squares is a perfect square subtracted from a perfect square. Recall that a difference of squares can be rewritten as factors containing the same terms but opposite signs because the middle terms cancel each other out when the two factors are multiplied.

$$a^2 - b^2 = (a + b)(a - b)$$

We can use this equation to factor any differences of squares.

---

differences of squares
A difference of squares can be rewritten as two factors containing the same terms but opposite signs.

$$a^2 - b^2 = (a + b)(a - b)$$

How To…

Given a difference of squares, factor it into binomials.

1. Confirm that the first and last term are perfect squares.
2. Write the factored form as $(a + b)(a - b)$.

Example 5   **Factoring a Difference of Squares**

Factor $9x^2 - 25$.

Solution   Notice that $9x^2$ and 25 are perfect squares because $9x^2 = (3x)^2$ and $25 = 5^2$. The polynomial represents a difference of squares and can be rewritten as $(3x + 5)(3x - 5)$.

Try It #5

Factor $81y^2 - 100$.

Q & A…

Is there a formula to factor the sum of squares?

No. A sum of squares cannot be factored.

Factoring the Sum and Difference of Cubes

Now, we will look at two new special products: the sum and difference of cubes. Although the sum of squares cannot be factored, the sum of cubes can be factored into a binomial and a trinomial.

$$a^3 + b^3 = (a + b)(a^2 - ab + b^2)$$

Similarly, the sum of cubes can be factored into a binomial and a trinomial, but with different signs.

$$a^3 - b^3 = (a - b)(a^2 + ab + b^2)$$

We can use the acronym SOAP to remember the signs when factoring the sum or difference of cubes. The first letter of each word relates to the signs: **S**ame **O**pposite **A**lways **P**ositive. For example, consider the following example.

$$x^3 - 2^3 = (x - 2)(x^2 + 2x + 4)$$

The sign of the first 2 is the *same* as the sign between $x^3 - 2^3$. The sign of the $2x$ term is *opposite* the sign between $x^3 - 2^3$. And the sign of the last term, 4, is *always positive*.

---

sum and difference of cubes
We can factor the sum of two cubes as

$$a^3 + b^3 = (a + b)(a^2 - ab + b^2)$$

We can factor the difference of two cubes as

$$a^3 - b^3 = (a - b)(a^2 + ab + b^2)$$

How To…

Given a sum of cubes or difference of cubes, factor it.

1. Confirm that the first and last term are cubes, $a^3 + b^3$ or $a^3 - b^3$.
2. For a sum of cubes, write the factored form as $(a + b)(a^2 - ab + b^2)$. For a difference of cubes, write the factored form as $(a - b)(a^2 + ab + b^2)$.

Example 6   **Factoring a Sum of Cubes**

Factor $x^3 + 512$.

Solution   Notice that x^3 and 512 are cubes because $8^3 = 512$. Rewrite the sum of cubes as $(x + 8)(x^2 - 8x + 64)$.

Analysis   After writing the sum of cubes this way, we might think we should check to see if the trinomial portion can be factored further. However, the trinomial portion cannot be factored, so we do not need to check.

Try It #6

Factor the sum of cubes: $216a^3 + b^3$.

Example 7   **Factoring a Difference of Cubes**

Factor $8x^3 - 125$.

Solution   Notice that $8x^3$ and 125 are cubes because $8x^3 = (2x)^3$ and $125 = 5^3$. Write the difference of cubes as $(2x - 5)(4x^2 + 10x + 25)$.

Analysis   Just as with the sum of cubes, we will not be able to further factor the trinomial portion.

Try It #7

Factor the difference of cubes: $1,000x^3 - 1$.

Factoring Expressions with Fractional or Negative Exponents

Expressions with fractional or negative exponents can be factored by pulling out a GCF. Look for the variable or exponent that is common to each term of the expression and pull out that variable or exponent raised to the lowest power. These expressions follow the same factoring rules as those with integer exponents. For instance, $2x^{\frac{1}{4}} + 5x^{\frac{3}{4}}$ can be factored by pulling out $x^{\frac{1}{4}}$ and being rewritten as $x^{\frac{1}{4}}\left(2 + 5x^{\frac{1}{2}}\right)$.

Example 8 **Factoring an Expression with Fractional or Negative Exponents**

Factor $3x(x + 2)^{-\frac{1}{3}} + 4(x + 2)^{\frac{2}{3}}$.

Solution Factor out the term with the lowest value of the exponent. In this case, that would be $(x + 2)^{-\frac{1}{3}}$.

$$(x + 2)^{-\frac{1}{3}}(3x + 4(x + 2)) \qquad \text{Factor out the GCF.}$$

$$(x + 2)^{-\frac{1}{3}}(3x + 4x + 8) \qquad \text{Simplify.}$$

$$(x + 2)^{-\frac{1}{3}}(7x + 8)$$

Try It #8

Factor $2(5a - 1)^{\frac{3}{4}} + 7a(5a - 1)^{-\frac{1}{4}}$.

Access these online resources for additional instruction and practice with factoring polynomials.

- Identify GCF (http://openstaxcollege.org/l/findgcftofact)
- Factor Trinomials when *a* Equals 1 (http://openstaxcollege.org/l/facttrinom1)
- Factor Trinomials when *a* is not equal to 1 (http://openstaxcollege.org/l/facttrinom2)
- Factor Sum or Difference of Cubes (http://openstaxcollege.org/l/sumdifcube)

1.5   SECTION EXERCISES

VERBAL

1. If the terms of a polynomial do not have a GCF, does that mean it is not factorable? Explain.

2. A polynomial is factorable, but it is not a perfect square trinomial or a difference of two squares. Can you factor the polynomial without finding the GCF?

3. How do you factor by grouping?

ALGEBRAIC

For the following exercises, find the greatest common factor.

4. $14x + 4xy - 18xy^2$

5. $49mb^2 - 35m^2ba + 77ma^2$

6. $30x^3y - 45x^2y^2 + 135xy^3$

7. $200p^3m^3 - 30p^2m^3 + 40m^3$

8. $36j^4k^2 - 18j^3k^3 + 54j^2k^4$

9. $6y^4 - 2y^3 + 3y^2 - y$

For the following exercises, factor by grouping.

10. $6x^2 + 5x - 4$

11. $2a^2 + 9a - 18$

12. $6c^2 + 41c + 63$

13. $6n^2 - 19n - 11$

14. $20w^2 - 47w + 24$

15. $2p^2 - 5p - 7$

For the following exercises, factor the polynomial.

16. $7x^2 + 48x - 7$

17. $10h^2 - 9h - 9$

18. $2b^2 - 25b - 247$

19. $9d^2 - 73d + 8$

20. $90v^2 - 181v + 90$

21. $12t^2 + t - 13$

22. $2n^2 - n - 15$

23. $16x^2 - 100$

24. $25y^2 - 196$

25. $121p^2 - 169$

26. $4m^2 - 9$

27. $361d^2 - 81$

28. $324x^2 - 121$

29. $144b^2 - 25c^2$

30. $16a^2 - 8a + 1$

31. $49n^2 + 168n + 144$

32. $121x^2 - 88x + 16$

33. $225y^2 + 120y + 16$

34. $m^2 - 20m + 100$

35. $m^2 - 20m + 100$

36. $36q^2 + 60q + 25$

For the following exercises, factor the polynomials.

37. $x^3 + 216$

38. $27y^3 - 8$

39. $125a^3 + 343$

40. $b^3 - 8d^3$

41. $64x^3 - 125$

42. $729q^3 + 1331$

43. $125r^3 + 1,728s^3$

44. $4x(x - 1)^{-\frac{2}{3}} + 3(x - 1)^{\frac{1}{3}}$

45. $3c(2c + 3)^{-\frac{1}{4}} - 5(2c + 3)^{\frac{3}{4}}$

46. $3t(10t + 3)^{\frac{1}{3}} + 7(10t + 3)^{\frac{4}{3}}$

47. $14x(x + 2)^{-\frac{2}{5}} + 5(x + 2)^{\frac{3}{5}}$

48. $9y(3y - 13)^{\frac{1}{5}} - 2(3y - 13)^{\frac{6}{5}}$

49. $5z(2z - 9)^{-\frac{3}{2}} + 11(2z - 9)^{-\frac{1}{2}}$

50. $6d(2d + 3)^{-\frac{1}{6}} + 5(2d + 3)^{\frac{5}{6}}$

2

Equations and Inequalities

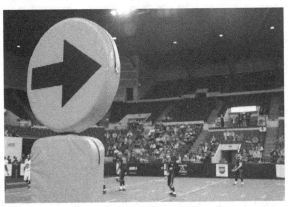

Figure 1

CHAPTER OUTLINE

Introduction

For most people, the term territorial possession indicates restrictions, usually dealing with trespassing or rite of passage and takes place in some foreign location. What most Americans do not realize is that from September through December, territorial possession dominates our lifestyles while watching the NFL. In this area, territorial possession is governed by the referees who make their decisions based on what the chains reveal. If the ball is at point $A (x_1, y_1)$, then it is up to the quarterback to decide which route to point $B (x_2, y_2)$, the end zone, is most feasible.

73

LEARNING OBJECTIVES

In this section you will:

- Plot ordered pairs in a Cartesian coordinate system.
- Graph equations by plotting points.
- Graph equations with a graphing utility.
- Find *x*-intercepts and *y*-intercepts.
- Use the distance formula.
- Use the midpoint formula.

2.1   THE RECTANGULAR COORDINATE SYSTEMS AND GRAPHS

Figure 1

Tracie set out from Elmhurst, IL, to go to Franklin Park. On the way, she made a few stops to do errands. Each stop is indicated by a red dot in **Figure 1**. Laying a rectangular coordinate grid over the map, we can see that each stop aligns with an intersection of grid lines. In this section, we will learn how to use grid lines to describe locations and changes in locations.

Plotting Ordered Pairs in the Cartesian Coordinate System

An old story describes how seventeenth-century philosopher/mathematician René Descartes invented the system that has become the foundation of algebra while sick in bed. According to the story, Descartes was staring at a fly crawling on the ceiling when he realized that he could describe the fly's location in relation to the perpendicular lines formed by the adjacent walls of his room. He viewed the perpendicular lines as horizontal and vertical axes. Further, by dividing each axis into equal unit lengths, Descartes saw that it was possible to locate any object in a two-dimensional plane using just two numbers—the displacement from the horizontal axis and the displacement from the vertical axis.

While there is evidence that ideas similar to Descartes' grid system existed centuries earlier, it was Descartes who introduced the components that comprise the **Cartesian coordinate system**, a grid system having perpendicular axes. Descartes named the horizontal axis the ***x*-axis** and the vertical axis the ***y*-axis**.

The Cartesian coordinate system, also called the rectangular coordinate system, is based on a two-dimensional plane consisting of the *x*-axis and the *y*-axis. Perpendicular to each other, the axes divide the plane into four sections. Each section is called a **quadrant**; the quadrants are numbered counterclockwise as shown in **Figure 2.**

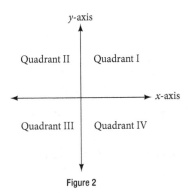

Figure 2

The center of the plane is the point at which the two axes cross. It is known as the **origin**, or point (0, 0). From the origin, each axis is further divided into equal units: increasing, positive numbers to the right on the x-axis and up the y-axis; decreasing, negative numbers to the left on the x-axis and down the y-axis. The axes extend to positive and negative infinity as shown by the arrowheads in **Figure 3**.

Figure 3

Each point in the plane is identified by its **x-coordinate**, or horizontal displacement from the origin, and its **y-coordinate**, or vertical displacement from the origin. Together, we write them as an **ordered pair** indicating the combined distance from the origin in the form (x, y). An ordered pair is also known as a coordinate pair because it consists of x- and y-coordinates. For example, we can represent the point $(3, -1)$ in the plane by moving three units to the right of the origin in the horizontal direction, and one unit down in the vertical direction. See **Figure 4**.

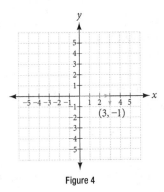

Figure 4

When dividing the axes into equally spaced increments, note that the x-axis may be considered separately from the y-axis. In other words, while the x-axis may be divided and labeled according to consecutive integers, the y-axis may be divided and labeled by increments of 2, or 10, or 100. In fact, the axes may represent other units, such as years against the balance in a savings account, or quantity against cost, and so on. Consider the rectangular coordinate system primarily as a method for showing the relationship between two quantities.

Cartesian coordinate system

A two-dimensional plane where the
- *x*-axis is the horizontal axis
- *y*-axis is the vertical axis

A point in the plane is defined as an ordered pair, (x, y), such that x is determined by its horizontal distance from the origin and y is determined by its vertical distance from the origin.

Example 1   **Plotting Points in a Rectangular Coordinate System**

Plot the points $(-2, 4)$, $(3, 3)$, and $(0, -3)$ in the plane.

Solution   To plot the point $(-2, 4)$, begin at the origin. The *x*-coordinate is -2, so move two units to the left. The *y*-coordinate is 4, so then move four units up in the positive *y* direction.

To plot the point $(3, 3)$, begin again at the origin. The *x*-coordinate is 3, so move three units to the right. The *y*-coordinate is also 3, so move three units up in the positive *y* direction.

To plot the point $(0, -3)$, begin again at the origin. The *x*-coordinate is 0. This tells us not to move in either direction along the *x*-axis. The *y*-coordinate is –3, so move three units down in the negative *y* direction. See the graph in **Figure 5**.

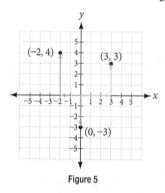

Figure 5

Analysis   *Note that when either coordinate is zero, the point must be on an axis. If the x-coordinate is zero, the point is on the y-axis. If the y-coordinate is zero, the point is on the x-axis.*

Graphing Equations by Plotting Points

We can plot a set of points to represent an equation. When such an equation contains both an *x* variable and a *y* variable, it is called an **equation in two variables**. Its graph is called a **graph in two variables**. Any graph on a two-dimensional plane is a graph in two variables.

Suppose we want to graph the equation $y = 2x - 1$. We can begin by substituting a value for *x* into the equation and determining the resulting value of *y*. Each pair of *x*- and *y*-values is an ordered pair that can be plotted. **Table 1** lists values of *x* from -3 to 3 and the resulting values for *y*.

x	$y = 2x - 1$	(x, y)
-3	$y = 2(-3) - 1 = -7$	$(-3, -7)$
-2	$y = 2(-2) - 1 = -5$	$(-2, -5)$
-1	$y = 2(-1) - 1 = -3$	$(-1, -3)$
0	$y = 2(0) - 1 = -1$	$(0, -1)$
1	$y = 2(1) - 1 = 1$	$(1, 1)$
2	$y = 2(2) - 1 = 3$	$(2, 3)$
3	$y = 2(3) - 1 = 5$	$(3, 5)$

Table 1

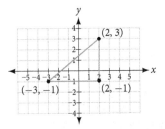

Figure 14

Then, calculate the length of d using the distance formula.

$$d = \sqrt{(x_2 - x_1)^2 + (y_2 - y_1)^2}$$
$$d = \sqrt{(2 - (-3))^2 + (3 - (-1))^2}$$
$$= \sqrt{(5)^2 + (4)^2}$$
$$= \sqrt{25 + 16}$$
$$= \sqrt{41}$$

Try It #3

Find the distance between two points: $(1, 4)$ and $(11, 9)$.

Example 6　**Finding the Distance Between Two Locations**

Let's return to the situation introduced at the beginning of this section.

Tracie set out from Elmhurst, IL, to go to Franklin Park. On the way, she made a few stops to do errands. Each stop is indicated by a red dot in **Figure 1**. Find the total distance that Tracie traveled. Compare this with the distance between her starting and final positions.

Solution　The first thing we should do is identify ordered pairs to describe each position. If we set the starting position at the origin, we can identify each of the other points by counting units east (right) and north (up) on the grid. For example, the first stop is 1 block east and 1 block north, so it is at $(1, 1)$. The next stop is 5 blocks to the east, so it is at $(5, 1)$. After that, she traveled 3 blocks east and 2 blocks north to $(8, 3)$. Lastly, she traveled 4 blocks north to $(8, 7)$. We can label these points on the grid as in **Figure 15**.

Figure 15

Next, we can calculate the distance. Note that each grid unit represents 1,000 feet.

- From her starting location to her first stop at $(1, 1)$, Tracie might have driven north 1,000 feet and then east 1,000 feet, or vice versa. Either way, she drove 2,000 feet to her first stop.
- Her second stop is at $(5, 1)$. So from $(1, 1)$ to $(5, 1)$, Tracie drove east 4,000 feet.
- Her third stop is at $(8, 3)$. There are a number of routes from $(5, 1)$ to $(8, 3)$. Whatever route Tracie decided to use, the distance is the same, as there are no angular streets between the two points. Let's say she drove east 3,000 feet and then north 2,000 feet for a total of 5,000 feet.
- Tracie's final stop is at $(8, 7)$. This is a straight drive north from $(8, 3)$ for a total of 4,000 feet.

Next, we will add the distances listed in **Table 3**.

From/To	Number of Feet Driven
$(0, 0)$ to $(1, 1)$	2,000
$(1, 1)$ to $(5, 1)$	4,000
$(5, 1)$ to $(8, 3)$	5,000
$(8, 3)$ to $(8, 7)$	4,000
Total	15,000

Table 3

The total distance Tracie drove is 15,000 feet, or 2.84 miles. This is not, however, the actual distance between her starting and ending positions. To find this distance, we can use the distance formula between the points $(0, 0)$ and $(8, 7)$.

$$\begin{aligned}
d &= \sqrt{(8-0)^2 + (7-0)^2} \\
&= \sqrt{64 + 49} \\
&= \sqrt{113} \\
&\approx 10.63 \text{ units}
\end{aligned}$$

At 1,000 feet per grid unit, the distance between Elmhurst, IL, to Franklin Park is 10,630.14 feet, or 2.01 miles. The distance formula results in a shorter calculation because it is based on the hypotenuse of a right triangle, a straight diagonal from the origin to the point $(8, 7)$. Perhaps you have heard the saying "as the crow flies," which means the shortest distance between two points because a crow can fly in a straight line even though a person on the ground has to travel a longer distance on existing roadways.

Using the Midpoint Formula

When the endpoints of a line segment are known, we can find the point midway between them. This point is known as the midpoint and the formula is known as the **midpoint formula**. Given the endpoints of a line segment, (x_1, y_1) and (x_2, y_2), the midpoint formula states how to find the coordinates of the midpoint M.

$$M = \left(\frac{x_1 + x_2}{2}, \frac{y_1 + y_2}{2} \right)$$

A graphical view of a midpoint is shown in **Figure 16**. Notice that the line segments on either side of the midpoint are congruent.

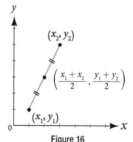

Figure 16

Example 7   **Finding the Midpoint of the Line Segment**

Find the midpoint of the line segment with the endpoints $(7, -2)$ and $(9, 5)$.

Solution   Use the formula to find the midpoint of the line segment.

$$\left(\frac{x_1 + x_2}{2}, \frac{y_1 + y_2}{2}\right) = \left(\frac{7 + 9}{2}, \frac{-2 + 5}{2}\right)$$

$$= \left(8, \frac{3}{2}\right)$$

Try It #4

Find the midpoint of the line segment with endpoints $(-2, -1)$ and $(-8, 6)$.

Example 8   **Finding the Center of a Circle**

The diameter of a circle has endpoints $(-1, -4)$ and $(5, -4)$. Find the center of the circle.

Solution   The center of a circle is the center, or midpoint, of its diameter. Thus, the midpoint formula will yield the center point.

$$\left(\frac{x_1 + x_2}{2}, \frac{y_1 + y_2}{2}\right)$$

$$\left(\frac{-1 + 5}{2}, \frac{-4 - 4}{2}\right) = \left(\frac{4}{2}, -\frac{8}{2}\right) = (2, -4)$$

Access these online resources for additional instruction and practice with the Cartesian coordinate system.

- Plotting Points on the Coordinate Plane (http://openstaxcollege.org/l/coordplotpnts)
- Find *x*- and *y*-intercepts Based on the Graph of a Line (http://openstaxcollege.org/l/xyintsgraph)

2.1   SECTION EXERCISES

VERBAL

1. Is it possible for a point plotted in the Cartesian coordinate system to not lie in one of the four quadrants? Explain.

2. Describe the process for finding the x-intercept and the y-intercept of a graph algebraically.

3. Describe in your own words what the y-intercept of a graph is.

4. When using the distance formula $d = \sqrt{(x_2 - x_1)^2 + (y_2 - y_1)^2}$, explain the correct order of operations that are to be performed to obtain the correct answer.

ALGEBRAIC

For each of the following exercises, find the x-intercept and the y-intercept without graphing. Write the coordinates of each intercept.

5. $y = -3x + 6$ **6.** $4y = 2x - 1$ **7.** $3x - 2y = 6$ **8.** $4x - 3 = 2y$

9. $3x + 8y = 9$ **10.** $2x - \dfrac{2}{3} = \dfrac{3}{4}y + 3$

For each of the following exercises, solve the equation for y in terms of x.

11. $4x + 2y = 8$ **12.** $3x - 2y = 6$ **13.** $2x = 5 - 3y$ **14.** $x - 2y = 7$

15. $5y + 4 = 10x$ **16.** $5x + 2y = 0$

For each of the following exercises, find the distance between the two points. Simplify your answers, and write the exact answer in simplest radical form for irrational answers.

17. $(-4, 1)$ and $(3, -4)$ **18.** $(2, -5)$ and $(7, 4)$ **19.** $(5, 0)$ and $(5, 6)$ **20.** $(-4, 3)$ and $(10, 3)$

21. Find the distance between the two points given using your calculator, and round your answer to the nearest hundredth. $(19, 12)$ and $(41, 71)$

For each of the following exercises, find the coordinates of the midpoint of the line segment that joins the two given points.

22. $(-5, -6)$ and $(4, 2)$ **23.** $(-1, 1)$ and $(7, -4)$ **24.** $(-5, -3)$ and $(-2, -8)$ **25.** $(0, 7)$ and $(4, -9)$

26. $(-43, 17)$ and $(23, -34)$

GRAPHICAL

For each of the following exercises, identify the information requested.

27. What are the coordinates of the origin?

28. If a point is located on the y-axis, what is the x-coordinate?

29. If a point is located on the x-axis, what is the y-coordinate?

For each of the following exercises, plot the three points on the given coordinate plane. State whether the three points you plotted appear to be collinear (on the same line).

30. $(4, 1)(-2, -3)(5, 0)$

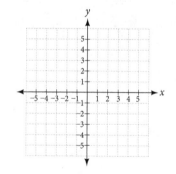

31. $(-1, 2)(0, 4)(2, 1)$

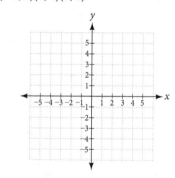

32. $(-3, 0)(-3, 4)(-3, -3)$

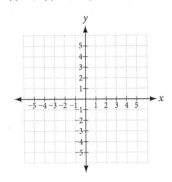

33. Name the coordinates of the points graphed.

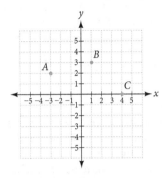

34. Name the quadrant in which the following points would be located. If the point is on an axis, name the axis.

a. $(-3, -4)$ **b.** $(-5, 0)$ **c.** $(1, -4)$ **d.** $(-2, 7)$ **e.** $(0, -3)$

For each of the following exercises, construct a table and graph the equation by plotting at least three points.

35. $y = \dfrac{1}{3}x + 2$ **36.** $y = -3x + 1$ **37.** $2y = x + 3$

NUMERIC

For each of the following exercises, find and plot the x- and y-intercepts, and graph the straight line based on those two points.

38. $4x - 3y = 12$ **39.** $x - 2y = 8$ **40.** $y - 5 = 5x$ **41.** $3y = -2x + 6$ **42.** $y = \dfrac{x - 3}{2}$

For each of the following exercises, use the graph in the figure below.

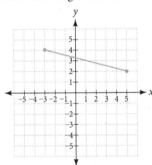

43. Find the distance between the two endpoints using the distance formula. Round to three decimal places.

44. Find the coordinates of the midpoint of the line segment connecting the two points.

45. Find the distance that $(-3, 4)$ is from the origin.

46. Find the distance that $(5, 2)$ is from the origin. Round to three decimal places.

47. Which point is closer to the origin?

TECHNOLOGY

For the following exercises, use your graphing calculator to input the linear graphs in the **Y=** graph menu.

After graphing it, use the **2ⁿᵈ CALC** button and **1:value** button, hit **ENTER**. At the lower part of the screen you will see "**x=**" and a blinking cursor. You may enter any number for x and it will display the y value for any x value you input. Use this and plug in $x = 0$, thus finding the y-intercept, for each of the following graphs.

48. $Y_1 = -2x + 5$ **49.** $Y_1 = \dfrac{3x - 8}{4}$ **50.** $Y_1 = \dfrac{x + 5}{2}$

For the following exercises, use your graphing calculator to input the linear graphs in the **Y=** graph menu.

After graphing it, use the **2ⁿᵈ CALC** button and **2:zero** button, hit **ENTER**. At the lower part of the screen you will see "**left bound?**" and a blinking cursor on the graph of the line. Move this cursor to the left of the x-intercept, hit **ENTER**. Now it says "**right bound?**" Move the cursor to the right of the x-intercept, hit **ENTER**. Now it says "**guess?**" Move your cursor to the left somewhere in between the left and right bound near the x-intercept. Hit **ENTER**. At the bottom of your screen it will display the coordinates of the x-intercept or the "zero" to the y-value. Use this to find the x-intercept.

Note: With linear/straight line functions the zero is not really a "guess," but it is necessary to enter a "guess" so it will search and find the exact x-intercept between your right and left boundaries. With other types of functions (more than one x-intercept), they may be irrational numbers so "guess" is more appropriate to give it the correct limits to find a very close approximation between the left and right boundaries.

51. $Y_1 = -8x + 6$ **52.** $Y_1 = 4x - 7$ **53.** $Y_1 = \dfrac{3x + 5}{4}$ Round your answer to the nearest thousandth.

EXTENSIONS

54. A man drove 10 mi directly east from his home, made a left turn at an intersection, and then traveled 5 mi north to his place of work. If a road was made directly from his home to his place of work, what would its distance be to the nearest tenth of a mile?

55. If the road was made in the previous exercise, how much shorter would the man's one-way trip be every day?

56. Given these four points: $A(1, 3)$, $B(-3, 5)$, $C(4, 7)$, and $D(5, -4)$, find the coordinates of the midpoint of line segments $\overline{AB}$ and $\overline{CD}$.

57. After finding the two midpoints in the previous exercise, find the distance between the two midpoints to the nearest thousandth.

58. Given the graph of the rectangle shown and the coordinates of its vertices, prove that the diagonals of the rectangle are of equal length.

59. In the previous exercise, find the coordinates of the midpoint for each diagonal.

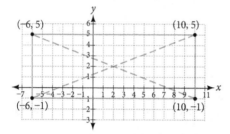

REAL-WORLD APPLICATIONS

60. The coordinates on a map for San Francisco are $(53, 17)$ and those for Sacramento are $(123, 78)$. Note that coordinates represent miles. Find the distance between the cities to the nearest mile.

61. If San Jose's coordinates are $(76, -12)$, where the coordinates represent miles, find the distance between San Jose and San Francisco to the nearest mile.

62. A small craft in Lake Ontario sends out a distress signal. The coordinates of the boat in trouble were $(49, 64)$. One rescue boat is at the coordinates $(60, 82)$ and a second Coast Guard craft is at coordinates $(58, 47)$. Assuming both rescue craft travel at the same rate, which one would get to the distressed boat the fastest?

63. A man on the top of a building wants to have a guy wire extend to a point on the ground 20 ft from the building. To the nearest foot, how long will the wire have to be if the building is 50 ft tall?

64. If we rent a truck and pay a $75/day fee plus $.20 for every mile we travel, write a linear equation that would express the total cost y, using x to represent the number of miles we travel. Graph this function on your graphing calculator and find the total cost for one day if we travel 70 mi.

LEARNING OBJECTIVES

In this section you will:

- Solve equations in one variable algebraically.
- Solve a rational equation.
- Find a linear equation.
- Given the equations of two lines, determine whether their graphs are parallel or perpendicular.
- Write the equation of a line parallel or perpendicular to a given line.

2.2 LINEAR EQUATIONS IN ONE VARIABLE

Caroline is a full-time college student planning a spring break vacation. To earn enough money for the trip, she has taken a part-time job at the local bank that pays $15.00/hr, and she opened a savings account with an initial deposit of $400 on January 15. She arranged for direct deposit of her payroll checks. If spring break begins March 20 and the trip will cost approximately $2,500, how many hours will she have to work to earn enough to pay for her vacation? If she can only work 4 hours per day, how many days per week will she have to work? How many weeks will it take? In this section, we will investigate problems like this and others, which generate graphs like the line in **Figure 1**.

Figure 1

Solving Linear Equations in One Variable

A **linear equation** is an equation of a straight line, written in one variable. The only power of the variable is 1. Linear equations in one variable may take the form $ax + b = 0$ and are solved using basic algebraic operations.

We begin by classifying linear equations in one variable as one of three types: identity, conditional, or inconsistent. An **identity equation** is true for all values of the variable. Here is an example of an identity equation.

$$3x = 2x + x$$

The **solution set** consists of all values that make the equation true. For this equation, the solution set is all real numbers because any real number substituted for x will make the equation true.

A **conditional equation** is true for only some values of the variable. For example, if we are to solve the equation $5x + 2 = 3x - 6$, we have the following:

$$5x + 2 = 3x - 6$$
$$2x = -8$$
$$x = -4$$

The solution set consists of one number: $\{-4\}$. It is the only solution and, therefore, we have solved a conditional equation.

An **inconsistent equation** results in a false statement. For example, if we are to solve $5x - 15 = 5(x - 4)$, we have the following:

$$5x - 15 = 5x - 20$$
$$5x - 15 - 5x = 5x - 20 - 5x \qquad \text{Subtract } 5x \text{ from both sides.}$$
$$-15 \neq -20 \qquad\qquad \text{False statement}$$

Indeed, $-15 \neq -20$. There is no solution because this is an inconsistent equation.

Solving linear equations in one variable involves the fundamental properties of equality and basic algebraic operations. A brief review of those operations follows.

---

linear equation in one variable

A linear equation in one variable can be written in the form
$$ax + b = 0$$
where a and b are real numbers, $a \neq 0$.

How To...

Given a linear equation in one variable, use algebra to solve it.

The following steps are used to manipulate an equation and isolate the unknown variable, so that the last line reads $x =$ _____, if x is the unknown. There is no set order, as the steps used depend on what is given:

1. We may add, subtract, multiply, or divide an equation by a number or an expression as long as we do the same thing to both sides of the equal sign. Note that we cannot divide by zero.
2. Apply the distributive property as needed: $a(b + c) = ab + ac$.
3. Isolate the variable on one side of the equation.
4. When the variable is multiplied by a coefficient in the final stage, multiply both sides of the equation by the reciprocal of the coefficient.

Example 1   **Solving an Equation in One Variable**

Solve the following equation: $2x + 7 = 19$.

Solution   This equation can be written in the form $ax + b = 0$ by subtracting 19 from both sides. However, we may proceed to solve the equation in its original form by performing algebraic operations.

$$2x + 7 = 19$$
$$2x = 12 \qquad \text{Subtract 7 from both sides.}$$
$$x = 6 \qquad \text{Multiply both sides by } \frac{1}{2} \text{ or divide by 2.}$$

The solution is 6.

Try It #1

Solve the linear equation in one variable: $2x + 1 = -9$.

Example 2   **Solving an Equation Algebraically When the Variable Appears on Both Sides**

Solve the following equation: $4(x - 3) + 12 = 15 - 5(x + 6)$.

Solution   Apply standard algebraic properties.

$$4(x - 3) + 12 = 15 - 5(x + 6)$$
$$4x - 12 + 12 = 15 - 5x - 30 \qquad \text{Apply the distributive property.}$$
$$4x = -15 - 5x \qquad \text{Combine like terms.}$$
$$9x = -15 \qquad \text{Place } x\text{- terms on one side and simplify.}$$
$$x = -\frac{15}{9} \qquad \text{Multiply both sides by } \frac{1}{9}, \text{ the reciprocal of 9.}$$
$$x = -\frac{5}{3}$$

Analysis   *This problem requires the distributive property to be applied twice, and then the properties of algebra are used to reach the final line, $x = -\frac{5}{3}$.*

We can show that two lines are perpendicular if the product of the two slopes is -1: $m_1 \cdot m_2 = -1$. For example, **Figure 5** shows the graph of two perpendicular lines. One line has a slope of 3; the other line has a slope of $-\frac{1}{3}$.

$$m_1 \cdot m_2 = -1$$

$$3 \cdot \left(-\frac{1}{3}\right) = -1$$

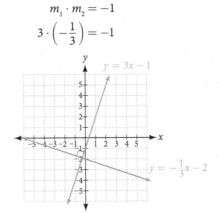

Figure 5 Perpendicular lines

Example 14 **Graphing Two Equations, and Determining Whether the Lines are Parallel, Perpendicular, or Neither**

Graph the equations of the given lines, and state whether they are parallel, perpendicular, or neither: $3y = -4x + 3$ and $3x - 4y = 8$.

Solution The first thing we want to do is rewrite the equations so that both equations are in slope-intercept form.

First equation:

$$3y = -4x + 3$$

$$y = -\frac{4}{3}x + 1$$

Second equation:

$$3x - 4y = 8$$

$$-4y = -3x + 8$$

$$y = \frac{3}{4}$$

See the graph of both lines in **Figure 6**.

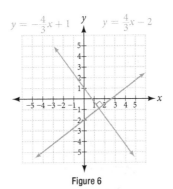

Figure 6

From the graph, we can see that the lines appear perpendicular, but we must compare the slopes.

$$m_1 = -\frac{4}{3}$$

$$m_2 = \frac{3}{4}$$

$$m_1 \cdot m_2 = \left(-\frac{4}{3}\right)\left(\frac{3}{4}\right) = -1$$

The slopes are negative reciprocals of each other, confirming that the lines are perpendicular.

Try It #11

Graph the two lines and determine whether they are parallel, perpendicular, or neither: $2y - x = 10$ and $2y = x + 4$.

Writing the Equations of Lines Parallel or Perpendicular to a Given Line

As we have learned, determining whether two lines are parallel or perpendicular is a matter of finding the slopes. To write the equation of a line parallel or perpendicular to another line, we follow the same principles as we do for finding the equation of any line. After finding the slope, use the point-slope formula to write the equation of the new line.

How To...

Given an equation for a line, write the equation of a line parallel or perpendicular to it.

1. Find the slope of the given line. The easiest way to do this is to write the equation in slope-intercept form.
2. Use the slope and the given point with the point-slope formula.
3. Simplify the line to slope-intercept form and compare the equation to the given line.

Example 15   **Writing the Equation of a Line Parallel to a Given Line Passing Through a Given Point**

Write the equation of line parallel to a $5x + 3y = 1$ and passing through the point $(3, 5)$.

Solution   First, we will write the equation in slope-intercept form to find the slope.

$$5x + 3y = 1$$

$$3y = 5x + 1$$

$$y = -\frac{5}{3}x + \frac{1}{3}$$

The slope is $m = -\frac{5}{3}$. The y-intercept is $\frac{1}{3}$, but that really does not enter into our problem, as the only thing we need for two lines to be parallel is the same slope. The one exception is that if the y-intercepts are the same, then the two lines are the same line. The next step is to use this slope and the given point with the point-slope formula.

$$y - 5 = -\frac{5}{3}(x - 3)$$

$$y - 5 = -\frac{5}{3}x + 5$$

$$y = -\frac{5}{3}x + 10$$

The equation of the line is $y = -\frac{5}{3}x + 10$. See **Figure 7**.

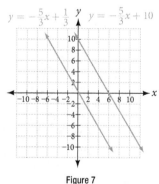

Figure 7

Try It #12

Find the equation of the line parallel to $5x = 7 + y$ and passing through the point $(-1, -2)$.

Example 16   **Finding the Equation of a Line Perpendicular to a Given Line Passing Through a Given Point**

Find the equation of the line perpendicular to $5x - 3y + 4 = 0$ and passes through the point $(-4, 1)$.

Solution   The first step is to write the equation in slope-intercept form.

$$5x - 3y + 4 = 0$$
$$-3y = -5x - 4$$
$$y = \frac{5}{3}x + \frac{4}{3}$$

We see that the slope is $m = \dfrac{5}{3}$. This means that the slope of the line perpendicular to the given line is the negative reciprocal, or $-\dfrac{3}{5}$. Next, we use the point-slope formula with this new slope and the given point.

$$y - 1 = -\frac{3}{5}(x - (-4))$$
$$y - 1 = -\frac{3}{5}x - \frac{12}{5}$$
$$y = -\frac{3}{5}x - \frac{12}{5} + \frac{5}{5}$$
$$y = -\frac{3}{5}x - \frac{7}{5}$$

Access these online resources for additional instruction and practice with linear equations.

- Solving rational equations (http://openstaxcollege.org/l/rationaleqs)
- Equation of a line given two points (http://openstaxcollege.org/l/twopointsline)
- Finding the equation of a line perpendicular to another line through a given point (http://openstaxcollege.org/l/findperpline)
- Finding the equation of a line parallel to another line through a given point (http://openstaxcollege.org/l/findparaline)

2.2   SECTION EXERCISES

VERBAL

1. What does it mean when we say that two lines are parallel?

2. What is the relationship between the slopes of perpendicular lines (assuming neither is horizontal nor vertical)?

3. How do we recognize when an equation, for example $y = 4x + 3$, will be a straight line (linear) when graphed?

4. What does it mean when we say that a linear equation is inconsistent?

5. When solving the following equation: $\dfrac{2}{x-5} = \dfrac{4}{x+1}$ explain why we must exclude $x = 5$ and $x = -1$ as possible solutions from the solution set.

ALGEBRAIC

For the following exercises, solve the equation for x.

6. $7x + 2 = 3x - 9$

7. $4x - 3 = 5$

8. $3(x + 2) - 12 = 5(x + 1)$

9. $12 - 5(x + 3) = 2x - 5$

10. $\dfrac{1}{2} - \dfrac{1}{3}x = \dfrac{4}{3}$

11. $\dfrac{x}{3} - \dfrac{3}{4} = \dfrac{2x + 3}{12}$

12. $\dfrac{2}{3}x + \dfrac{1}{2} = \dfrac{31}{6}$

13. $3(2x - 1) + x = 5x + 3$

14. $\dfrac{2x}{3} - \dfrac{3}{4} = \dfrac{x}{6} + \dfrac{21}{4}$

15. $\dfrac{x+2}{4} - \dfrac{x-1}{3} = 2$

For the following exercises, solve each rational equation for x. State all x-values that are excluded from the solution set.

16. $\dfrac{3}{x} - \dfrac{1}{3} = \dfrac{1}{6}$

17. $2 - \dfrac{3}{x+4} = \dfrac{x+2}{x+4}$

18. $\dfrac{3}{x-2} = \dfrac{1}{x-1} + \dfrac{7}{(x-1)(x-2)}$

19. $\dfrac{3x}{x-1} + 2 = \dfrac{3}{x-1}$

20. $\dfrac{5}{x+1} + \dfrac{1}{x-3} = \dfrac{-6}{x^2 - 2x - 3}$

21. $\dfrac{1}{x} = \dfrac{1}{5} + \dfrac{3}{2x}$

For the following exercises, find the equation of the line using the point-slope formula. Write all the final equations using the slope-intercept form.

22. $(0, 3)$ with a slope of $\dfrac{2}{3}$

23. $(1, 2)$ with a slope of $-\dfrac{4}{5}$

24. x-intercept is 1, and $(-2, 6)$

25. y-intercept is 2, and $(4, -1)$

26. $(-3, 10)$ and $(5, -6)$

27. $(1, 3)$ and $(5, 5)$

28. parallel to $y = 2x + 5$ and passes through the point $(4, 3)$

29. perpendicular to $3y = x - 4$ and passes through the point $(-2, 1)$.

For the following exercises, find the equation of the line using the given information.

30. $(-2, 0)$ and $(-2, 5)$

31. $(1, 7)$ and $(3, 7)$

32. The slope is undefined and it passes through the point $(2, 3)$.

33. The slope equals zero and it passes through the point $(1, -4)$.

34. The slope is $\dfrac{3}{4}$ and it passes through the point $(1,4)$.

35. $(-1, 3)$ and $(4, -5)$

Try It #1

Factor and solve the quadratic equation: $x^2 - 5x - 6 = 0$.

Example 2   **Solve the Quadratic Equation by Factoring**

Solve the quadratic equation by factoring: $x^2 + 8x + 15 = 0$.

Solution   Find two numbers whose product equals 15 and whose sum equals 8. List the factors of 15.

$$1 \cdot 15$$
$$3 \cdot 5$$
$$(-1) \cdot (-15)$$
$$(-3) \cdot (-5)$$

The numbers that add to 8 are 3 and 5. Then, write the factors, set each factor equal to zero, and solve.

$$(x + 3)(x + 5) = 0$$
$$(x + 3) = 0$$
$$x = -3$$
$$(x + 5) = 0$$
$$x = -5$$

The solutions are -3 and -5.

Try It #2

Solve the quadratic equation by factoring: $x^2 - 4x - 21 = 0$.

Example 3   **Using the Zero-Product Property to Solve a Quadratic Equation Written as the Difference of Squares**

Solve the difference of squares equation using the zero-product property: $x^2 - 9 = 0$.

Solution   Recognizing that the equation represents the difference of squares, we can write the two factors by taking the square root of each term, using a minus sign as the operator in one factor and a plus sign as the operator in the other. Solve using the zero-factor property.

$$x^2 - 9 = 0$$
$$(x - 3)(x + 3) = 0$$
$$(x - 3) = 0$$
$$x = 3$$
$$(x + 3) = 0$$
$$x = -3$$

The solutions are 3 and -3.

Try It #3

Solve by factoring: $x^2 - 25 = 0$.

Factoring and Solving a Quadratic Equation of Higher Order

When the leading coefficient is not 1, we factor a quadratic equation using the method called grouping, which requires four terms. With the equation in standard form, let's review the grouping procedures:

1. With the quadratic in standard form, $ax^2 + bx + c = 0$, multiply $a \cdot c$.
2. Find two numbers whose product equals ac and whose sum equals b.
3. Rewrite the equation replacing the bx term with two terms using the numbers found in step 1 as coefficients of x.
4. Factor the first two terms and then factor the last two terms. The expressions in parentheses must be exactly the same to use grouping.
5. Factor out the expression in parentheses.
6. Set the expressions equal to zero and solve for the variable.

Example 4   **Solving a Quadratic Equation Using Grouping**

Use grouping to factor and solve the quadratic equation: $4x^2 + 15x + 9 = 0$.

Solution   First, multiply $ac : 4(9) = 36$. Then list the factors of 36.

$$1 \cdot 36$$
$$2 \cdot 18$$
$$3 \cdot 12$$
$$4 \cdot 9$$
$$6 \cdot 6$$

The only pair of factors that sums to 15 is $3 + 12$. Rewrite the equation replacing the b term, $15x$, with two terms using 3 and 12 as coefficients of x. Factor the first two terms, and then factor the last two terms.

$$4x^2 + 3x + 12x + 9 = 0$$
$$x(4x + 3) + 3(4x + 3) = 0$$
$$(4x + 3)(x + 3) = 0$$

Solve using the zero-product property.

$$(4x + 3)(x + 3) = 0$$
$$(4x + 3) = 0$$
$$x = -\frac{3}{4}$$
$$(x + 3) = 0$$
$$x = -3$$

The solutions are $-\dfrac{3}{4}$ and -3. See **Figure 3**.

$4x^2 + 15x + 9 = 0$

Figure 3

Try It #4

Solve using factoring by grouping: $12x^2 + 11x + 2 = 0$.

Example 5   **Solving a Higher Degree Quadratic Equation by Factoring**

Solve the equation by factoring: $-3x^3 - 5x^2 - 2x = 0$.

Solution   This equation does not look like a quadratic, as the highest power is 3, not 2. Recall that the first thing we want to do when solving any equation is to factor out the GCF, if one exists. And it does here. We can factor out $-x$ from all of the terms and then proceed with grouping.

$$-3x^3 - 5x^2 - 2x = 0$$
$$-x(3x^2 + 5x + 2) = 0$$

Use grouping on the expression in parentheses.

$$-x(3x^2 + 3x + 2x + 2) = 0$$
$$-x[3x(x + 1) + 2(x + 1)] = 0$$
$$-x(3x + 2)(x + 1) = 0$$

Now, we use the zero-product property. Notice that we have three factors.

$$-x = 0$$
$$x = 0$$
$$3x + 2 = 0$$
$$x = -\frac{2}{3}$$
$$x + 1 = 0$$
$$x = -1$$

The solutions are 0, $-\frac{2}{3}$, and -1.

Try It #5

Solve by factoring: $x^3 + 11x^2 + 10x = 0$.

Using the Square Root Property

When there is no linear term in the equation, another method of solving a quadratic equation is by using the **square root property**, in which we isolate the x^2 term and take the square root of the number on the other side of the equals sign. Keep in mind that sometimes we may have to manipulate the equation to isolate the x^2 term so that the square root property can be used.

> ### the square root property
> With the x^2 term isolated, the square root property states that:
> $$\text{if } x^2 = k, \text{ then } x = \pm \sqrt{k}$$
> where k is a nonzero real number.

How To...

Given a quadratic equation with an x^2 term but no x term, use the square root property to solve it.

1. Isolate the x^2 term on one side of the equal sign.
2. Take the square root of both sides of the equation, putting a $\pm$ sign before the expression on the side opposite the squared term.
3. Simplify the numbers on the side with the $\pm$ sign.

Example 6   **Solving a Simple Quadratic Equation Using the Square Root Property**

Solve the quadratic using the square root property: $x^2 = 8$.

Solution   Take the square root of both sides, and then simplify the radical. Remember to use a $\pm$ sign before the radical symbol.

$$x^2 = 8$$
$$x = \pm \sqrt{8}$$
$$= \pm 2\sqrt{2}$$

The solutions are $2\sqrt{2}$ and $-2\sqrt{2}$.

Example 7   **Solving a Quadratic Equation Using the Square Root Property**

Solve the quadratic equation: $4x^2 + 1 = 7$.

Solution   First, isolate the x^2 term. Then take the square root of both sides.

$$4x^2 + 1 = 7$$
$$4x^2 = 6$$
$$x^2 = \frac{6}{4}$$
$$x = \pm \frac{\sqrt{6}}{2}$$

The solutions are $\frac{\sqrt{6}}{2}$ and $-\frac{\sqrt{6}}{2}$.

Try It #6

Solve the quadratic equation using the square root property: $3(x - 4)^2 = 15$.

Completing the Square

Not all quadratic equations can be factored or can be solved in their original form using the square root property. In these cases, we may use a method for solving a quadratic equation known as **completing the square**. Using this method, we add or subtract terms to both sides of the equation until we have a perfect square trinomial on one side of the equal sign. We then apply the square root property. To complete the square, the leading coefficient, a, must equal 1. If it does not, then divide the entire equation by a. Then, we can use the following procedures to solve a quadratic equation by completing the square.

We will use the example $x^2 + 4x + 1 = 0$ to illustrate each step.

1. Given a quadratic equation that cannot be factored, and with $a = 1$, first add or subtract the constant term to the right sign of the equal sign.
$$x^2 + 4x = -1$$

2. Multiply the b term by $\frac{1}{2}$ and square it.
$$\frac{1}{2}(4) = 2$$
$$2^2 = 4$$

3. Add $\left(\frac{1}{2}b\right)^2$ to both sides of the equal sign and simplify the right side. We have
$$x^2 + 4x + 4 = -1 + 4$$
$$x^2 + 4x + 4 = 3$$

4. The left side of the equation can now be factored as a perfect square.
$$x^2 + 4x + 4 = 3$$
$$(x + 2)^2 = 3$$

5. Use the square root property and solve.
$$\sqrt{(x + 2)^2} = \pm\sqrt{3}$$
$$x + 2 = \pm\sqrt{3}$$
$$x = -2 \pm\sqrt{3}$$

6. The solutions are $-2 + \sqrt{3}$ and $-2 - \sqrt{3}$.

Example 8   **Solving a Quadratic by Completing the Square**

Solve the quadratic equation by completing the square: $x^2 - 3x - 5 = 0$.

Solution   First, move the constant term to the right side of the equal sign.
$$x^2 - 3x = 5$$

Then, take $\frac{1}{2}$ of the b term and square it.
$$\frac{1}{2}(-3) = -\frac{3}{2}$$
$$\left(-\frac{3}{2}\right)^2 = \frac{9}{4}$$

Add the result to both sides of the equal sign.
$$x^2 - 3x + \left(-\frac{3}{2}\right)^2 = 5 + \left(-\frac{3}{2}\right)^2$$
$$x^2 - 3x + \frac{9}{4} = 5 + \frac{9}{4}$$

Factor the left side as a perfect square and simplify the right side.
$$\left(x - \frac{3}{2}\right)^2 = \frac{29}{4}$$

Use the square root property and solve.

$$\sqrt{\left(x - \frac{3}{2}\right)^2} = \pm\sqrt{\frac{29}{4}}$$

$$\left(x - \frac{3}{2}\right) = \pm\frac{\sqrt{29}}{2}$$

$$x = \frac{3}{2} \pm \frac{\sqrt{29}}{2}$$

The solutions are $\dfrac{3 + \sqrt{29}}{2}$ and $\dfrac{3 - \sqrt{29}}{2}$.

Try It #7

Solve by completing the square: $x^2 - 6x = 13$.

Using the Quadratic Formula

The fourth method of solving a quadratic equation is by using the quadratic formula, a formula that will solve all quadratic equations. Although the quadratic formula works on any quadratic equation in standard form, it is easy to make errors in substituting the values into the formula. Pay close attention when substituting, and use parentheses when inserting a negative number.

We can derive the quadratic formula by completing the square. We will assume that the leading coefficient is positive; if it is negative, we can multiply the equation by -1 and obtain a positive a. Given $ax^2 + bx + c = 0$, $a \neq 0$, we will complete the square as follows:

1. First, move the constant term to the right side of the equal sign:
$$ax^2 + bx = -c$$

2. As we want the leading coefficient to equal 1, divide through by a:
$$x^2 + \frac{b}{a}x = -\frac{c}{a}$$

3. Then, find $\frac{1}{2}$ of the middle term, and add $\left(\dfrac{1}{2}\dfrac{b}{a}\right)^2 = \dfrac{b^2}{4a^2}$ to both sides of the equal sign:
$$x^2 + \frac{b}{a}x + \frac{b^2}{4a^2} = \frac{b^2}{4a^2} - \frac{c}{a}$$

4. Next, write the left side as a perfect square. Find the common denominator of the right side and write it as a single fraction:
$$\left(x + \frac{b}{2a}\right)^2 = \frac{b^2 - 4ac}{4a^2}$$

5. Now, use the square root property, which gives
$$x + \frac{b}{2a} = \pm\sqrt{\frac{b^2 - 4ac}{4a^2}}$$
$$x + \frac{b}{2a} = \pm\frac{\sqrt{b^2 - 4ac}}{2a}$$

6. Finally, add $-\dfrac{b}{2a}$ to both sides of the equation and combine the terms on the right side. Thus,
$$x = \frac{-b \pm \sqrt{b^2 - 4ac}}{2a}$$

---

the quadratic formula

Written in standard form, $ax^2 + bx + c = 0$, any quadratic equation can be solved using the **quadratic formula**:

$$x = \frac{-b \pm \sqrt{b^2 - 4ac}}{2a}$$

where a, b, and c are real numbers and $a \neq 0$.

How To...

Given a quadratic equation, solve it using the quadratic formula

1. Make sure the equation is in standard form: $ax^2 + bx + c = 0$.
2. Make note of the values of the coefficients and constant term, a, b, and c.
3. Carefully substitute the values noted in step 2 into the equation. To avoid needless errors, use parentheses around each number input into the formula.
4. Calculate and solve.

Example 9   **Solve the Quadratic Equation Using the Quadratic Formula**

Solve the quadratic equation: $x^2 + 5x + 1 = 0$.

Solution   Identify the coefficients: $a = 1$, $b = 5$, $c = 1$. Then use the quadratic formula.

$$x = \frac{-(5) \pm \sqrt{(5)^2 - 4(1)(1)}}{2(1)}$$

$$= \frac{-5 \pm \sqrt{25 - 4}}{2}$$

$$= \frac{-5 \pm \sqrt{21}}{2}$$

Example 10   **Solving a Quadratic Equation with the Quadratic Formula**

Use the quadratic formula to solve $x^2 + x + 2 = 0$.

Solution   First, we identify the coefficients: $a = 1$, $b = 1$, and $c = 2$.

Substitute these values into the quadratic formula.

$$x = \frac{-b \pm \sqrt{b^2 - 4ac}}{2a}$$

$$= \frac{-(1) \pm \sqrt{(1)^2 - (4) \cdot (1) \cdot (2)}}{2 \cdot 1}$$

$$= \frac{-1 \pm \sqrt{1 - 8}}{2}$$

$$= \frac{-1 \pm \sqrt{-7}}{2}$$

$$= \frac{-1 \pm i\sqrt{7}}{2}$$

The solutions to the equation are $\dfrac{-1 + i\sqrt{7}}{2}$ and $\dfrac{-1 - i\sqrt{7}}{2}$.

Try It #8

Solve the quadratic equation using the quadratic formula: $9x^2 + 3x - 2 = 0$.

The Discriminant

The quadratic formula not only generates the solutions to a quadratic equation, it tells us about the nature of the solutions when we consider the discriminant, or the expression under the radical, $b^2 - 4ac$. The discriminant tells us whether the solutions are real numbers or complex numbers, and how many solutions of each type to expect. **Table 1** relates the value of the discriminant to the solutions of a quadratic equation.

Value of Discriminant	Results
$b^2 - 4ac = 0$	One rational solution (double solution)
$b^2 - 4ac > 0$, perfect square	Two rational solutions
$b^2 - 4ac > 0$, not a perfect square	Two irrational solutions
$b^2 - 4ac < 0$	Two complex solutions

Table 1

the discriminant

For $ax^2 + bx + c = 0$, where a, b, and c are real numbers, the **discriminant** is the expression under the radical in the quadratic formula: $b^2 - 4ac$. It tells us whether the solutions are real numbers or complex numbers and how many solutions of each type to expect.

Example 11   **Using the Discriminant to Find the Nature of the Solutions to a Quadratic Equation**

Use the discriminant to find the nature of the solutions to the following quadratic equations:

    a. $x^2 + 4x + 4 = 0$ **b.** $8x^2 + 14x + 3 = 0$ **c.** $3x^2 - 5x - 2 = 0$ **d.** $3x^2 - 10x + 15 = 0$

Solution   Calculate the discriminant $b^2 - 4ac$ for each equation and state the expected type of solutions.

    a. $x^2 + 4x + 4 = 0$
       $b^2 - 4ac = (4)^2 - 4(1)(4) = 0$. There will be one rational double solution.

    b. $8x^2 + 14x + 3 = 0$
       $b^2 - 4ac = (14)^2 - 4(8)(3) = 100$. As 100 is a perfect square, there will be two rational solutions.

    c. $3x^2 - 5x - 2 = 0$
       $b^2 - 4ac = (-5)^2 - 4(3)(-2) = 49$. As 49 is a perfect square, there will be two rational solutions.

    d. $3x^2 - 10x + 15 = 0$
       $b^2 - 4ac = (-10)^2 - 4(3)(15) = -80$. There will be two complex solutions.

Using the Pythagorean Theorem

One of the most famous formulas in mathematics is the **Pythagorean Theorem**. It is based on a right triangle, and states the relationship among the lengths of the sides as $a^2 + b^2 = c^2$, where a and b refer to the legs of a right triangle adjacent to the 90° angle, and c refers to the hypotenuse. It has immeasurable uses in architecture, engineering, the sciences, geometry, trigonometry, and algebra, and in everyday applications.

We use the Pythagorean Theorem to solve for the length of one side of a triangle when we have the lengths of the other two. Because each of the terms is squared in the theorem, when we are solving for a side of a triangle, we have a quadratic equation. We can use the methods for solving quadratic equations that we learned in this section to solve for the missing side.

The Pythagorean Theorem is given as

$$a^2 + b^2 = c^2$$

where a and b refer to the legs of a right triangle adjacent to the 90° angle, and c refers to the hypotenuse, as shown in **Figure 4**.

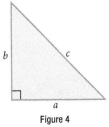

Figure 4

Example 12   **Finding the Length of the Missing Side of a Right Triangle**

Find the length of the missing side of the right triangle in **Figure 5**.

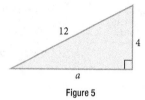

Figure 5

Solution   As we have measurements for side b and the hypotenuse, the missing side is a.

$$a^2 + b^2 = c^2$$
$$a^2 + (4)^2 = (12)^2$$
$$a^2 + 16 = 144$$
$$a^2 = 128$$
$$a = \sqrt{128}$$
$$= 8\sqrt{2}$$

Try It #9

Use the Pythagorean Theorem to solve the right triangle problem: Leg a measures 4 units, leg b measures 3 units. Find the length of the hypotenuse.

Access these online resources for additional instruction and practice with quadratic equations.

- Solving Quadratic Equations by Factoring (http://openstaxcollege.org/l/quadreqfactor)
- The Zero-Product Property (http://openstaxcollege.org/l/zeroprodprop)
- Completing the Square (http://openstaxcollege.org/l/complthesqr)
- Quadratic Formula with Two Rational Solutions (http://openstaxcollege.org/l/quadrformrat)
- Length of a Leg of a Right Triangle (http://openstaxcollege.org/l/leglengthtri)

Example 5   **Solving an Inequality Algebraically**

Solve the inequality: $13 - 7x \geq 10x - 4$.

Solution   Solving this inequality is similar to solving an equation up until the last step.

$$13 - 7x \geq 10x - 4$$

$$13 - 17x \geq -4 \qquad \text{Move variable terms to one side of the inequality.}$$

$$-17x \geq -17 \qquad \text{Isolate the variable term.}$$

$$x \leq 1 \qquad \text{Dividing both sides by } -17 \text{ reverses the inequality.}$$

The solution set is given by the interval $(-\infty, 1]$, or all real numbers less than and including 1.

Try It #5

Solve the inequality and write the answer using interval notation: $-x + 4 < \frac{1}{2}x + 1$.

Example 6   **Solving an Inequality with Fractions**

Solve the following inequality and write the answer in interval notation: $-\frac{3}{4}x \geq -\frac{5}{8} + \frac{2}{3}x$.

Solution   We begin solving in the same way we do when solving an equation.

$$-\frac{3}{4}x \geq -\frac{5}{8} + \frac{2}{3}x.$$

$$-\frac{3}{4}x - \frac{2}{3}x \geq -\frac{5}{8} \qquad \text{Put variable terms on one side.}$$

$$-\frac{9}{12}x - \frac{8}{12}x \geq -\frac{5}{8} \qquad \text{Write fractions with common denominator.}$$

$$-\frac{17}{12}x \geq -\frac{5}{8}$$

$$x \leq -\frac{5}{8}\left(-\frac{12}{17}\right) \qquad \text{Multiplying by a negative number reverses the inequality.}$$

$$x \leq \frac{15}{34}$$

The solution set is the interval $\left(-\infty, \frac{15}{34}\right]$.

Try It #6

Solve the inequality and write the answer in interval notation: $-\frac{5}{6}x \leq \frac{3}{4} + \frac{8}{3}x$.

Understanding Compound Inequalities

A **compound inequality** includes two inequalities in one statement. A statement such as $4 < x \leq 6$ means $4 < x$ and $x \leq 6$. There are two ways to solve compound inequalities: separating them into two separate inequalities or leaving the compound inequality intact and performing operations on all three parts at the same time. We will illustrate both methods.

Example 7   **Solving a Compound Inequality**

Solve the compound inequality: $3 \leq 2x + 2 < 6$.

Solution   The first method is to write two separate inequalities: $3 \leq 2x + 2$ and $2x + 2 < 6$. We solve them independently.

$$3 \leq 2x + 2 \qquad \text{and} \qquad 2x + 2 < 6$$

$$1 \leq 2x \qquad\qquad\qquad 2x < 4$$

$$\frac{1}{2} \leq x \qquad\qquad\qquad x < 2$$

Then, we can rewrite the solution as a compound inequality, the same way the problem began.

$$\frac{1}{2} \leq x < 2$$

In interval notation, the solution is written as $\left[\frac{1}{2}, 2\right)$.

The second method is to leave the compound inequality intact, and perform solving procedures on the three parts at the same time.

$$3 \leq 2x + 2 < 6$$

$$1 \leq 2x < 4 \qquad \text{Isolate the variable term, and subtract 2 from all three parts.}$$

$$\frac{1}{2} \leq x < 2 \qquad \text{Divide through all three parts by 2.}$$

We get the same solution: $\left[\frac{1}{2}, 2\right)$.

Try It #7

Solve the compound inequality: $4 < 2x - 8 \leq 10$.

Example 8   **Solving a Compound Inequality with the Variable in All Three Parts**

Solve the compound inequality with variables in all three parts: $3 + x > 7x - 2 > 5x - 10$.

Solution   Let's try the first method. Write two inequalities:

$$\begin{array}{lll}
3 + x > 7x - 2 & \text{and} & 7x - 2 > 5x - 10 \\
3 > 6x - 2 & & 2x - 2 > -10 \\
5 > 6x & & 2x > -8 \\
\frac{5}{6} > x & & x > -4 \\
x < \frac{5}{6} & & -4 < x
\end{array}$$

The solution set is $-4 < x < \frac{5}{6}$ or in interval notation $\left(-4, \frac{5}{6}\right)$. Notice that when we write the solution in interval notation, the smaller number comes first. We read intervals from left to right, as they appear on a number line. See **Figure 3**.

Figure 3

Try It #8

Solve the compound inequality: $3y < 4 - 5y < 5 + 3y$.

Solving Absolute Value Inequalities

As we know, the absolute value of a quantity is a positive number or zero. From the origin, a point located at $(-x, 0)$ has an absolute value of x, as it is x units away. Consider absolute value as the distance from one point to another point. Regardless of direction, positive or negative, the distance between the two points is represented as a positive number or zero.

An absolute value inequality is an equation of the form

$$|A| < B, |A| \leq B, |A| > B, \text{ or } |A| \geq B,$$

Where A, and sometimes B, represents an algebraic expression dependent on a variable x. Solving the inequality means finding the set of all x-values that satisfy the problem. Usually this set will be an interval or the union of two intervals and will include a range of values.

There are two basic approaches to solving absolute value inequalities: graphical and algebraic. The advantage of the graphical approach is we can read the solution by interpreting the graphs of two equations. The advantage of the algebraic approach is that solutions are exact, as precise solutions are sometimes difficult to read from a graph.

Suppose we want to know all possible returns on an investment if we could earn some amount of money within $200 of $600. We can solve algebraically for the set of x-values such that the distance between x and 600 is less than or equal to 200. We represent the distance between x and 600 as $|x - 600|$, and therefore, $|x - 600| \leq 200$ or

$$-200 \leq x - 600 \leq 200$$
$$-200 + 600 \leq x - 600 + 600 \leq 200 + 600$$
$$400 \leq x \leq 800$$

This means our returns would be between $400 and $800.

To solve absolute value inequalities, just as with absolute value equations, we write two inequalities and then solve them independently.

---

absolute value inequalities

For an algebraic expression X, and $k > 0$, an **absolute value inequality** is an inequality of the form

$$|X| < k \text{ is equivalent to } -k < X < k$$
$$|X| > k \text{ is equivalent to } X < -k \text{ or } X > k$$

These statements also apply to $|X| \leq k$ and $|X| \geq k$.

Example 9   **Determining a Number within a Prescribed Distance**

Describe all values x within a distance of 4 from the number 5.

Solution   We want the distance between x and 5 to be less than or equal to 4. We can draw a number line, such as in **Figure 4**, to represent the condition to be satisfied.

Figure 4

The distance from x to 5 can be represented using an absolute value symbol, $|x - 5|$. Write the values of x that satisfy the condition as an absolute value inequality.

$$|x - 5| \leq 4$$

We need to write two inequalities as there are always two solutions to an absolute value equation.

$$\begin{array}{ccc} x - 5 \leq 4 & \text{and} & x - 5 \geq -4 \\ x \leq 9 & & x \geq 1 \end{array}$$

If the solution set is $x \leq 9$ and $x \geq 1$, then the solution set is an interval including all real numbers between and including 1 and 9.

So $|x - 5| \leq 4$ is equivalent to $[1, 9]$ in interval notation.

Try It #9
Describe all x-values within a distance of 3 from the number 2.

Example 10   **Solving an Absolute Value Inequality**

Solve $|x - 1| \leq 3$.

Solution
$$|x - 1| \leq 3$$
$$-3 \leq x - 1 \leq 3$$
$$-2 \leq x \leq 4$$
$$[-2, 4]$$

Example 11   **Using a Graphical Approach to Solve Absolute Value Inequalities**

Given the equation $y = -\frac{1}{2}|4x - 5| + 3$, determine the x-values for which the y-values are negative.

Solution   We are trying to determine where $y < 0$, which is when $-\frac{1}{2}|4x - 5| + 3 < 0$. We begin by isolating the absolute value.

$$-\frac{1}{2}|4x - 5| < -3 \qquad \text{Multiply both sides by } -2, \text{ and reverse the inequality.}$$
$$|4x - 5| > 6$$

Next, we solve for the equality $|4x - 5| = 6$.

$$
\begin{array}{ccc}
4x - 5 = 6 & \text{or} & 4x - 5 = -6 \\
4x = 11 & & 4x = -1 \\
x = \dfrac{11}{4} & & x = -\dfrac{1}{4}
\end{array}
$$

Now, we can examine the graph to observe where the y-values are negative. We observe where the branches are below the x-axis. Notice that it is not important exactly what the graph looks like, as long as we know that it crosses the horizontal axis at $x = -\frac{1}{4}$ and $x = \frac{11}{4}$, and that the graph opens downward. See **Figure 5**.

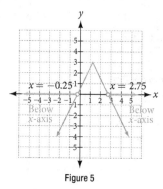

Figure 5

Try It #10

Solve $-2|k - 4| \leq -6$.

Access these online resources for additional instruction and practice with linear inequalities and absolute value inequalities.

- Interval Notation (http://openstaxcollege.org/l/intervalnotn)
- How to Solve Linear Inequalities (http://openstaxcollege.org/l/solvelinineq)
- How to Solve an Inequality (http://openstaxcollege.org/l/solveineq)
- Absolute Value Equations (http://openstaxcollege.org/l/absvaleq)
- Compound Inequalities (http://openstaxcollege.org/l/compndineqs)
- Absolute Value Inequalities (http://openstaxcollege.org/l/absvalineqs)

2.7 SECTION EXERCISES

VERBAL

1. When solving an inequality, explain what happened from Step 1 to Step 2:

Step 1 $-2x > 6$

Step 2 $x < -3$

2. When solving an inequality, we arrive at:

$x + 2 < x + 3$

$2 < 3$

Explain what our solution set is.

3. When writing our solution in interval notation, how do we represent all the real numbers?

4. When solving an inequality, we arrive at:

$x + 2 > x + 3$

$2 > 3$

Explain what our solution set is.

5. Describe how to graph $y = |x - 3|$

ALGEBRAIC

For the following exercises, solve the inequality. Write your final answer in interval notation

6. $4x - 7 \leq 9$

7. $3x + 2 \geq 7x - 1$

8. $-2x + 3 > x - 5$

9. $4(x + 3) \geq 2x - 1$

10. $-\frac{1}{2}x \leq -\frac{5}{4} + \frac{2}{5}x$

11. $-5(x - 1) + 3 > 3x - 4 - 4x$

12. $-3(2x + 1) > -2(x + 4)$

13. $\frac{x + 3}{8} - \frac{x + 5}{5} \geq \frac{3}{10}$

14. $\frac{x - 1}{3} + \frac{x + 2}{5} \leq \frac{3}{5}$

For the following exercises, solve the inequality involving absolute value. Write your final answer in interval notation.

15. $|x + 9| \geq -6$

16. $|2x + 3| < 7$

17. $|3x - 1| > 11$

18. $|2x + 1| + 1 \leq 6$

19. $|x - 2| + 4 \geq 10$

20. $|-2x + 7| \leq 13$

21. $|x - 7| < -4$

22. $|x - 20| > -1$

23. $\left|\frac{x - 3}{4}\right| < 2$

For the following exercises, describe all the x-values within or including a distance of the given values.

24. Distance of 5 units from the number 7

25. Distance of 3 units from the number 9

26. Distance of 10 units from the number 4

27. Distance of 11 units from the number 1

For the following exercises, solve the compound inequality. Express your answer using inequality signs, and then write your answer using interval notation.

28. $-4 < 3x + 2 \leq 18$

29. $3x + 1 > 2x - 5 > x - 7$

30. $3y < 5 - 2y < 7 + y$

31. $2x - 5 < -11$ or $5x + 1 \geq 6$

32. $x + 7 < x + 2$

GRAPHICAL

For the following exercises, graph the function. Observe the points of intersection and shade the *x*-axis representing the solution set to the inequality. Show your graph and write your final answer in interval notation.

33. $|x - 1| > 2$ **34.** $|x + 3| \geq 5$ **35.** $|x + 7| \leq 4$ **36.** $|x - 2| < 7$ **37.** $|x - 2| < 0$

For the following exercises, graph both straight lines (left-hand side being y_1 and right-hand side being y_2) on the same axes. Find the point of intersection and solve the inequality by observing where it is true comparing the *y*-values of the lines.

38. $x + 3 < 3x - 4$ **39.** $x - 2 > 2x + 1$ **40.** $x + 1 > x + 4$ **41.** $\frac{1}{2}x + 1 > \frac{1}{2}x - 5$

42. $4x + 1 < \frac{1}{2}x + 3$

NUMERIC

For the following exercises, write the set in interval notation.

43. $\{x | -1 < x < 3\}$ **44.** $\{x | x \geq 7\}$ **45.** $\{x | x < 4\}$ **46.** $\{x | x$ is all real numbers$\}$

For the following exercises, write the interval in set-builder notation.

47. $(-\infty, 6)$ **48.** $(4, \infty)$ **49.** $[-3, 5)$ **50.** $[-4, 1] \cup [9, \infty)$

For the following exercises, write the set of numbers represented on the number line in interval notation.

51.   **52.**   **53.**

TECHNOLOGY

For the following exercises, input the left-hand side of the inequality as a Y1 graph in your graphing utility. Enter **Y2=** the right-hand side. Entering the absolute value of an expression is found in the **MATH** menu, **Num, 1:abs(**. Find the points of intersection, recall (**2ⁿᵈ CALC 5:intersection, 1st curve, enter, 2ⁿᵈ curve, enter, guess, enter**). Copy a sketch of the graph and shade the *x*-axis for your solution set to the inequality. Write final answers in interval notation.

54. $|x + 2| - 5 < 2$ **55.** $-\frac{1}{2}|x + 2| < 4$ **56.** $|4x + 1| - 3 > 2$ **57.** $|x - 4| < 3$

58. $|x + 2| \geq 5$

EXTENSIONS

59. Solve $|3x + 1| = |2x + 3|$

60. Solve $x^2 - x > 12$

61. $\frac{x - 5}{x + 7} \leq 0, x \neq -7$

62. $p = -x^2 + 130x - 3{,}000$ is a profit formula for a small business. Find the set of *x*-values that will keep this profit positive.

REAL-WORLD APPLICATIONS

63. In chemistry the volume for a certain gas is given by $V = 20T$, where *V* is measured in cc and *T* is temperature in °C. If the temperature varies between 80°C and 120°C, find the set of volume values.

64. A basic cellular package costs $20/mo. for 60 min of calling, with an additional charge of $0.30/min beyond that time. The cost formula would be $C = \$20 + .30(x - 60)$. If you have to keep your bill lower than $50, what is the maximum calling minutes you can use?

CHAPTER 2 REVIEW

Key Terms

absolute value equation an equation in which the variable appears in absolute value bars, typically with two solutions, one accounting for the positive expression and one for the negative expression

area in square units, the area formula used in this section is used to find the area of any two-dimensional rectangular region: $A = LW$

Cartesian coordinate system a grid system designed with perpendicular axes invented by René Descartes

completing the square a process for solving quadratic equations in which terms are added to or subtracted from both sides of the equation in order to make one side a perfect square

complex conjugate a complex number containing the same terms as another complex number, but with the opposite operator. Multiplying a complex number by its conjugate yields a real number.

complex number the sum of a real number and an imaginary number; the standard form is $a + bi$, where a is the real part and b is the complex part.

complex plane the coordinate plane in which the horizontal axis represents the real component of a complex number, and the vertical axis represents the imaginary component, labeled i.

compound inequality a problem or a statement that includes two inequalities

conditional equation an equation that is true for some values of the variable

discriminant the expression under the radical in the quadratic formula that indicates the nature of the solutions, real or complex, rational or irrational, single or double roots.

distance formula a formula that can be used to find the length of a line segment if the endpoints are known

equation in two variables a mathematical statement, typically written in x and y, in which two expressions are equal

equations in quadratic form equations with a power other than 2 but with a middle term with an exponent that is one-half the exponent of the leading term

extraneous solutions any solutions obtained that are not valid in the original equation

graph in two variables the graph of an equation in two variables, which is always shown in two variables in the two-dimensional plane

identity equation an equation that is true for all values of the variable

imaginary number the square root of -1: $i = \sqrt{-1}$.

inconsistent equation an equation producing a false result

intercepts the points at which the graph of an equation crosses the x-axis and the y-axis

interval an interval describes a set of numbers within which a solution falls

interval notation a mathematical statement that describes a solution set and uses parentheses or brackets to indicate where an interval begins and ends

linear equation an algebraic equation in which each term is either a constant or the product of a constant and the first power of a variable

linear inequality similar to a linear equation except that the solutions will include sets of numbers

midpoint formula a formula to find the point that divides a line segment into two parts of equal length

ordered pair a pair of numbers indicating horizontal displacement and vertical displacement from the origin; also known as a coordinate pair, (x, y)

origin the point where the two axes cross in the center of the plane, described by the ordered pair $(0, 0)$

perimeter in linear units, the perimeter formula is used to find the linear measurement, or outside length and width, around a two-dimensional regular object; for a rectangle: $P = 2L + 2W$

polynomial equation an equation containing a string of terms including numerical coefficients and variables raised to whole-number exponents

Pythagorean Theorem a theorem that states the relationship among the lengths of the sides of a right triangle, used to solve right triangle problems

quadrant one quarter of the coordinate plane, created when the axes divide the plane into four sections

quadratic equation an equation containing a second-degree polynomial; can be solved using multiple methods

quadratic formula a formula that will solve all quadratic equations

radical equation an equation containing at least one radical term where the variable is part of the radicand

rational equation an equation consisting of a fraction of polynomials

slope the change in y-values over the change in x-values

solution set the set of all solutions to an equation

square root property one of the methods used to solve a quadratic equation, in which the x^2 term is isolated so that the square root of both sides of the equation can be taken to solve for x

volume in cubic units, the volume measurement includes length, width, and depth: $V = LWH$

x-axis the common name of the horizontal axis on a coordinate plane; a number line increasing from left to right

x-coordinate the first coordinate of an ordered pair, representing the horizontal displacement and direction from the origin

x-intercept the point where a graph intersects the x-axis; an ordered pair with a y-coordinate of zero

y-axis the common name of the vertical axis on a coordinate plane; a number line increasing from bottom to top

y-coordinate the second coordinate of an ordered pair, representing the vertical displacement and direction from the origin

y-intercept a point where a graph intercepts the y-axis; an ordered pair with an x-coordinate of zero

zero-product property the property that formally states that multiplication by zero is zero, so that each factor of a quadratic equation can be set equal to zero to solve equations

Key Equations

quadratic formula $x = \dfrac{-b \pm \sqrt{b^2 - 4ac}}{2a}$

Key Concepts

2.1 The Rectangular Coordinate Systems and Graphs

- We can locate, or plot, points in the Cartesian coordinate system using ordered pairs, which are defined as displacement from the x-axis and displacement from the y-axis. See **Example 1**.

- An equation can be graphed in the plane by creating a table of values and plotting points. See **Example 2**.

- Using a graphing calculator or a computer program makes graphing equations faster and more accurate. Equations usually have to be entered in the form $y = $ _____. See **Example 3**.

- Finding the x- and y-intercepts can define the graph of a line. These are the points where the graph crosses the axes. See **Example 4**.

- The distance formula is derived from the Pythagorean Theorem and is used to find the length of a line segment. See **Example 5** and **Example 6**.

- The midpoint formula provides a method of finding the coordinates of the midpoint dividing the sum of the x-coordinates and the sum of the y-coordinates of the endpoints by 2. See **Example 7** and **Example 8**.

2.2 Linear Equations in One Variable

- We can solve linear equations in one variable in the form $ax + b = 0$ using standard algebraic properties. See **Example 1** and **Example 2**.

- A rational expression is a quotient of two polynomials. We use the LCD to clear the fractions from an equation. See **Example 3** and **Example 4**.

- All solutions to a rational equation should be verified within the original equation to avoid an undefined term, or zero in the denominator. See **Example 5, Example 6,** and **Example 7**.

- Given two points, we can find the slope of a line using the slope formula. See **Example 8**.

- We can identify the slope and y-intercept of an equation in slope-intercept form. See **Example 9**.
- We can find the equation of a line given the slope and a point. See **Example 10**.
- We can also find the equation of a line given two points. Find the slope and use the point-slope formula. See **Example 11**.
- The standard form of a line has no fractions. See **Example 12**.
- Horizontal lines have a slope of zero and are defined as $y = c$, where c is a constant.
- Vertical lines have an undefined slope (zero in the denominator), and are defined as $x = c$, where c is a constant. See **Example 13**.
- Parallel lines have the same slope and different y-intercepts. See **Example 14** and **Example 15**.
- Perpendicular lines have slopes that are negative reciprocals of each other unless one is horizontal and the other is vertical. See **Example 16**.

2.3 Models and Applications

- A linear equation can be used to solve for an unknown in a number problem. See **Example 1**.
- Applications can be written as mathematical problems by identifying known quantities and assigning a variable to unknown quantities. See **Example 2**.
- There are many known formulas that can be used to solve applications. Distance problems, for example, are solved using the $d = rt$ formula. See **Example 3**.
- Many geometry problems are solved using the perimeter formula $P = 2L + 2W$, the area formula $A = LW$, or the volume formula $V = LWH$. See **Example 4**, **Example 5**, and **Example 6**.

2.4 Complex Numbers

- The square root of any negative number can be written as a multiple of i. See **Example 1**.
- To plot a complex number, we use two number lines, crossed to form the complex plane. The horizontal axis is the real axis, and the vertical axis is the imaginary axis. See **Example 2**.
- Complex numbers can be added and subtracted by combining the real parts and combining the imaginary parts. See **Example 3**.
- Complex numbers can be multiplied and divided.
  - To multiply complex numbers, distribute just as with polynomials. See **Example 4** and **Example 5**.
  - To divide complex numbers, multiply both numerator and denominator by the complex conjugate of the denominator to eliminate the complex number from the denominator. See **Example 6** and **Example 7**.
- The powers of i are cyclic, repeating every fourth one. See **Example 8**.

2.5 Quadratic Equations

- Many quadratic equations can be solved by factoring when the equation has a leading coefficient of 1 or if the equation is a difference of squares. The zero-product property is then used to find solutions. See **Example 1**, **Example 2**, and **Example 3**.
- Many quadratic equations with a leading coefficient other than 1 can be solved by factoring using the grouping method. See **Example 4** and **Example 5**.
- Another method for solving quadratics is the square root property. The variable is squared. We isolate the squared term and take the square root of both sides of the equation. The solution will yield a positive and negative solution. See **Example 6** and **Example 7**.
- Completing the square is a method of solving quadratic equations when the equation cannot be factored. See **Example 8**.
- A highly dependable method for solving quadratic equations is the quadratic formula, based on the coefficients and the constant term in the equation. See **Example 9** and **Example 10**.
- The discriminant is used to indicate the nature of the roots that the quadratic equation will yield: real or complex, rational or irrational, and how many of each. See **Example 11**.

- The Pythagorean Theorem, among the most famous theorems in history, is used to solve right-triangle problems and has applications in numerous fields. Solving for the length of one side of a right triangle requires solving a quadratic equation. See **Example 12**.

2.6 Other Types of Equations

- Rational exponents can be rewritten several ways depending on what is most convenient for the problem. To solve, both sides of the equation are raised to a power that will render the exponent on the variable equal to 1. See **Example 1**, **Example 2**, and **Example 3**.

- Factoring extends to higher-order polynomials when it involves factoring out the GCF or factoring by grouping. See **Example 4** and **Example 5**.

- We can solve radical equations by isolating the radical and raising both sides of the equation to a power that matches the index. See **Example 6** and **Example 7**.

- To solve absolute value equations, we need to write two equations, one for the positive value and one for the negative value. See **Example 8**.

- Equations in quadratic form are easy to spot, as the exponent on the first term is double the exponent on the second term and the third term is a constant. We may also see a binomial in place of the single variable. We use substitution to solve. See **Example 9** and **Example 10**.

- Solving a rational equation may also lead to a quadratic equation or an equation in quadratic form. See **Example 11**.

2.7 Linear Inequalities and Absolute Value Inequalities

- Interval notation is a method to indicate the solution set to an inequality. Highly applicable in calculus, it is a system of parentheses and brackets that indicate what numbers are included in a set and whether the endpoints are included as well. See **Table 1** and **Example 1** and **Example 2**.

- Solving inequalities is similar to solving equations. The same algebraic rules apply, except for one: multiplying or dividing by a negative number reverses the inequality. See **Example 3**, **Example 4**, **Example 5**, and **Example 6**.

- Compound inequalities often have three parts and can be rewritten as two independent inequalities. Solutions are given by boundary values, which are indicated as a beginning boundary or an ending boundary in the solutions to the two inequalities. See **Example 7** and **Example 8**.

- Absolute value inequalities will produce two solution sets due to the nature of absolute value. We solve by writing two equations: one equal to a positive value and one equal to a negative value. See **Example 9** and **Example 10**.

- Absolute value inequalities can also be solved by graphing. At least we can check the algebraic solutions by graphing, as we cannot depend on a visual for a precise solution. See **Example 11**.

CHAPTER 2 REVIEW EXERCISES

THE RECTANGULAR COORDINATE SYSTEMS AND GRAPHS

For the following exercises, find the x-intercept and the y-intercept without graphing.

1. $4x - 3y = 12$

2. $2y - 4 = 3x$

For the following exercises, solve for y in terms of x, putting the equation in slope–intercept form.

3. $5x = 3y - 12$

4. $2x - 5y = 7$

For the following exercises, find the distance between the two points.

5. $(-2, 5)(4, -1)$

6. $(-12, -3)(-1, 5)$

7. Find the distance between the two points $(-71,432)$ and $(511,218)$ using your calculator, and round your answer to the nearest thousandth.

For the following exercises, find the coordinates of the midpoint of the line segment that joins the two given points.

8. $(-1, 5)$ and $(4, 6)$

9. $(-13, 5)$ and $(17, 18)$

For the following exercises, construct a table and graph the equation by plotting at least three points.

10. $y = \dfrac{1}{2}x + 4$

11. $4x - 3y = 6$

LINEAR EQUATIONS IN ONE VARIABLE

For the following exercises, solve for x.

12. $5x + 2 = 7x - 8$

13. $3(x + 2) - 10 = x + 4$

14. $7x - 3 = 5$

15. $12 - 5(x + 1) = 2x - 5$

16. $\dfrac{2x}{3} - \dfrac{3}{4} = \dfrac{x}{6} + \dfrac{21}{4}$

For the following exercises, solve for x. State all x-values that are excluded from the solution set.

17. $\dfrac{x}{x^2 - 9} + \dfrac{4}{x + 3} = \dfrac{3}{x^2 - 9}$ $x \neq 3, -3$

18. $\dfrac{1}{2} + \dfrac{2}{x} = \dfrac{3}{4}$

For the following exercises, find the equation of the line using the point-slope formula.

19. Passes through these two points: $(-2, 1), (4, 2)$.

20. Passes through the point $(-3, 4)$ and has a slope of $-\dfrac{1}{3}$.

21. Passes through the point $(-3, 4)$ and is parallel to the graph $y = \dfrac{2}{3}x + 5$.

22. Passes through these two points: $(5, 1), (5, 7)$.

MODELS AND APPLICATIONS

For the following exercises, write and solve an equation to answer each question.

23. The number of males in the classroom is five more than three times the number of females. If the total number of students is 73, how many of each gender are in the class?

24. A man has 72 ft of fencing to put around a rectangular garden. If the length is 3 times the width, find the dimensions of his garden.

25. A truck rental is $25 plus $.30/mi. Find out how
many miles Ken traveled if his bill was $50.20.

COMPLEX NUMBERS

For the following exercises, use the quadratic equation to solve.

26. $x^2 - 5x + 9 = 0$ **27.** $2x^2 + 3x + 7 = 0$

For the following exercises, name the horizontal component and the vertical component.

28. $4 - 3i$ **29.** $-2 - i$

For the following exercises, perform the operations indicated.

30. $(9 - i) - (4 - 7i)$ **31.** $(2 + 3i) - (-5 - 8i)$ **32.** $2\sqrt{-75} + 3\sqrt{25}$

33. $\sqrt{-16} + 4\sqrt{-9}$ **34.** $-6i(i - 5)$ **35.** $(3 - 5i)^2$

36. $\sqrt{-4} \cdot \sqrt{-12}$ **37.** $\sqrt{-2}\left(\sqrt{-8} - \sqrt{5}\right)$ **38.** $\dfrac{2}{5 - 3i}$

39. $\dfrac{3 + 7i}{i}$

QUADRATIC EQUATIONS

For the following exercises, solve the quadratic equation by factoring.

40. $2x^2 - 7x - 4 = 0$ **41.** $3x^2 + 18x + 15 = 0$ **42.** $25x^2 - 9 = 0$

43. $7x^2 - 9x = 0$

For the following exercises, solve the quadratic equation by using the square-root property.

44. $x^2 = 49$ **45.** $(x - 4)^2 = 36$

For the following exercises, solve the quadratic equation by completing the square.

46. $x^2 + 8x - 5 = 0$ **47.** $4x^2 + 2x - 1 = 0$

For the following exercises, solve the quadratic equation by using the quadratic formula. If the solutions are not real, state *No real solution.*

48. $2x^2 - 5x + 1 = 0$ **49.** $15x^2 - x - 2 = 0$

For the following exercises, solve the quadratic equation by the method of your choice.

50. $(x - 2)^2 = 16$ **51.** $x^2 = 10x + 3$

OTHER TYPES OF EQUATIONS

For the following exercises, solve the equations.

52. $x^{\frac{3}{2}} = 27$

53. $x^{\frac{1}{2}} - 4x^{\frac{1}{4}} = 0$

54. $4x^3 + 8x^2 - 9x - 18 = 0$

55. $3x^5 - 6x^3 = 0$

56. $\sqrt{x+9} = x - 3$

57. $\sqrt{3x+7} + \sqrt{x+2} = 1$

58. $|3x - 7| = 5$

59. $|2x + 3| - 5 = 9$

LINEAR INEQUALITIES AND ABSOLUTE VALUE INEQUALITIES

For the following exercises, solve the inequality. Write your final answer in interval notation.

60. $5x - 8 \le 12$

61. $-2x + 5 > x - 7$

62. $\dfrac{x-1}{3} + \dfrac{x+2}{5} \le \dfrac{3}{5}$

63. $|3x + 2| + 1 \le 9$

64. $|5x - 1| > 14$

65. $|x - 3| < -4$

For the following exercises, solve the compound inequality. Write your answer in interval notation.

66. $-4 < 3x + 2 \le 18$

67. $3y < 1 - 2y < 5 + y$

For the following exercises, graph as described.

68. Graph the absolute value function and graph the constant function. Observe the points of intersection and shade the x-axis representing the solution set to the inequality. Show your graph and write your final answer in interval notation. $|x + 3| \ge 5$

69. Graph both straight lines (left-hand side being y_1 and right-hand side being y_2) on the same axes. Find the point of intersection and solve the inequality by observing where it is true comparing the y-values of the lines. See the interval where the inequality is true. $x + 3 < 3x - 4$

CHAPTER 2 PRACTICE TEST

1. Graph the following: $2y = 3x + 4$.

2. Find the x- and y-intercepts for the following: $2x - 5y = 6$.

3. Find the x- and y-intercepts of this equation, and sketch the graph of the line using just the intercepts plotted. $3x - 4y = 12$

4. Find the exact distance between $(5, -3)$ and $(-2, 8)$. Find the coordinates of the midpoint of the line segment joining the two points.

5. Write the interval notation for the set of numbers represented by $\{x | x \le 9\}$.

6. Solve for x: $5x + 8 = 3x - 10$.

7. Solve for x: $3(2x - 5) - 3(x - 7) = 2x - 9$.

8. Solve for x: $\dfrac{x}{2} + 1 = \dfrac{4}{x}$

9. Solve for x: $\dfrac{5}{x + 4} = 4 + \dfrac{3}{x - 2}$.

10. The perimeter of a triangle is 30 in. The longest side is 2 less than 3 times the shortest side and the other side is 2 more than twice the shortest side. Find the length of each side.

11. Solve for x. Write the answer in simplest radical form.
$\dfrac{x^2}{3} - x = -\dfrac{1}{2}$

12. Solve: $3x - 8 \le 4$.

13. Solve: $|2x + 3| < 5$.

14. Solve: $|3x - 2| \ge 4$.

For the following exercises, find the equation of the line with the given information.

15. Passes through the points $(-4, 2)$ and $(5, -3)$.

16. Has an undefined slope and passes through the point $(4, 3)$.

17. Passes through the point $(2, 1)$ and is perpendicular to $y = -\dfrac{2}{5}x + 3$.

18. Add these complex numbers: $(3 - 2i) + (4 - i)$.

19. Simplify: $\sqrt{-4} + 3\sqrt{-16}$.

20. Multiply: $5i(5 - 3i)$.

21. Divide: $\dfrac{4 - i}{2 + 3i}$.

22. Solve this quadratic equation and write the two complex roots in $a + bi$ form: $x^2 - 4x + 7 = 0$.

23. Solve: $(3x - 1)^2 - 1 = 24$.

24. Solve: $x^2 - 6x = 13$.

25. Solve: $4x^2 - 4x - 1 = 0$

26. Solve: $\sqrt{x - 7} = x - 7$

27. Solve: $2 + \sqrt{12 - 2x} = x$

28. Solve: $(x - 1)^{\frac{2}{3}} = 9$

For the following exercises, find the real solutions of each equation by factoring.

29. $2x^3 - x^2 - 8x + 4 = 0$

30. $(x + 5)^2 - 3(x + 5) - 4 = 0$

Functions

3

Figure 1 Standard and Poor's Index with dividends reinvested
(credit "bull": modification of work by Prayitno Hadinata; credit "graph": modification of work by MeasuringWorth)

CHAPTER OUTLINE

Introduction

Toward the end of the twentieth century, the values of stocks of Internet and technology companies rose dramatically. As a result, the Standard and Poor's stock market average rose as well. **Figure 1** tracks the value of that initial investment of just under $100 over the 40 years. It shows that an investment that was worth less than $500 until about 1995 skyrocketed up to about $1,100 by the beginning of 2000. That five-year period became known as the "dot-com bubble" because so many Internet startups were formed. As bubbles tend to do, though, the dot-com bubble eventually burst. Many companies grew too fast and then suddenly went out of business. The result caused the sharp decline represented on the graph beginning at the end of 2000.

Notice, as we consider this example, that there is a definite relationship between the year and stock market average. For any year we choose, we can determine the corresponding value of the stock market average. In this chapter, we will explore these kinds of relationships and their properties.

159

In this section, you will:

- Determine whether a relation represents a function.
- Find the value of a function.
- Determine whether a function is one-to-one.
- Use the vertical line test to identify functions.
- Graph the functions listed in the library of functions.

3.1   FUNCTIONS AND FUNCTION NOTATION

A jetliner changes altitude as its distance from the starting point of a flight increases. The weight of a growing child increases with time. In each case, one quantity depends on another. There is a relationship between the two quantities that we can describe, analyze, and use to make predictions. In this section, we will analyze such relationships.

Determining Whether a Relation Represents a Function

A **relation** is a set of ordered pairs. The set consisting of the first components of each ordered pair is called the **domain** and the set consisting of the second components of each ordered pair is called the **range**. Consider the following set of ordered pairs. The first numbers in each pair are the first five natural numbers. The second number in each pair is twice that of the first.

$$\{(1, 2), (2, 4), (3, 6), (4, 8), (5, 10)\}$$

The domain is $\{1, 2, 3, 4, 5\}$. The range is $\{2, 4, 6, 8, 10\}$.

Note that each value in the domain is also known as an **input** value, or **independent variable**, and is often labeled with the lowercase letter x. Each value in the range is also known as an **output** value, or **dependent variable**, and is often labeled lowercase letter y.

A function f is a relation that assigns a single element in the range to each element in the domain. In other words, no x-values are repeated. For our example that relates the first five natural numbers to numbers double their values, this relation is a function because each element in the domain, $\{1, 2, 3, 4, 5\}$, is paired with exactly one element in the range, $\{2, 4, 6, 8, 10\}$.

Now let's consider the set of ordered pairs that relates the terms "even" and "odd" to the first five natural numbers. It would appear as

$$\{(\text{odd}, 1), (\text{even}, 2), (\text{odd}, 3), (\text{even}, 4), (\text{odd}, 5)\}$$

Notice that each element in the domain, $\{\text{even}, \text{odd}\}$ is *not* paired with exactly one element in the range, $\{1, 2, 3, 4, 5\}$. For example, the term "odd" corresponds to three values from the domain, $\{1, 3, 5\}$ and the term "even" corresponds to two values from the range, $\{2, 4\}$. This violates the definition of a function, so this relation is not a function. **Figure 1** compares relations that are functions and not functions.

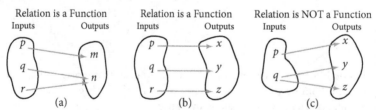

Figure 1 (a) This relationship is a function because each input is associated with a single output. Note that input *q* and *r* both give output *n*.
(b) This relationship is also a function. In this case, each input is associated with a single output.
(c) This relationship is not a function because input *q* is associated with two different outputs.

---

function

A **function** is a relation in which each possible input value leads to exactly one output value. We say "the output is a function of the input."

The **input** values make up the **domain**, and the **output** values make up the **range**.

How To...

Given a relationship between two quantities, determine whether the relationship is a function.

1. Identify the input values.
2. Identify the output values.
3. If each input value leads to only one output value, classify the relationship as a function. If any input value leads to two or more outputs, do not classify the relationship as a function.

Example 1   **Determining If Menu Price Lists Are Functions**

The coffee shop menu, shown in **Figure 2** consists of items and their prices.

> **a.** Is price a function of the item?   **b.** Is the item a function of the price?

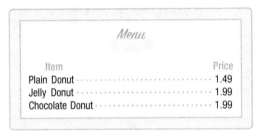

Figure 2

Solution

a. Let's begin by considering the input as the items on the menu. The output values are then the prices. See **Figure 2**.

Each item on the menu has only one price, so the price is a function of the item.

b. Two items on the menu have the same price. If we consider the prices to be the input values and the items to be the output, then the same input value could have more than one output associated with it. See **Figure 3**.

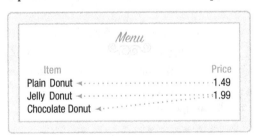

Figure 3

Therefore, the item is a not a function of price.

Example 2   **Determining If Class Grade Rules Are Functions**

In a particular math class, the overall percent grade corresponds to a grade-point average. Is grade-point average a function of the percent grade? Is the percent grade a function of the grade-point average? **Table 1** shows a possible rule for assigning grade points.

Percent grade	0-56	57-61	62-66	67-71	72-77	78-86	87-91	92-100
Grade-point average	0.0	1.0	1.5	2.0	2.5	3.0	3.5	4.0

Table 1

Solution   For any percent grade earned, there is an associated grade-point average, so the grade-point average is a function of the percent grade. In other words, if we input the percent grade, the output is a specific grade-point average.

In the grading system given, there is a range of percent grades that correspond to the same grade-point average. For example, students who receive a grade-point average of 3.0 could have a variety of percent grades ranging from 78 all the way to 86. Thus, percent grade is not a function of grade-point average

Try It #1

Table 2[1] lists the five greatest baseball players of all time in order of rank.

Player	Rank
Babe Ruth	1
Willie Mays	2
Ty Cobb	3
Walter Johnson	4
Hank Aaron	5

Table 2

a. Is the rank a function of the player name?
b. Is the player name a function of the rank?

Using Function Notation

Once we determine that a relationship is a function, we need to display and define the functional relationships so that we can understand and use them, and sometimes also so that we can program them into graphing calculators and computers. There are various ways of representing functions. A standard function notation is one representation that facilitates working with functions.

To represent "height is a function of age," we start by identifying the descriptive variables h for height and a for age. The letters f, g, and h are often used to represent functions just as we use x, y, and z to represent numbers and A, B, and C to represent sets.

h is f of a	We name the function f; height is a function of age.
$h = f(a)$	We use parentheses to indicate the function input.
$f(a)$	We name the function f; the expression is read as "f of a."

Remember, we can use any letter to name the function; the notation $h(a)$ shows us that h depends on a. The value a must be put into the function h to get a result. The parentheses indicate that age is input into the function; they do not indicate multiplication.

We can also give an algebraic expression as the input to a function. For example $f(a + b)$ means "first add a and b, and the result is the input for the function f." The operations must be performed in this order to obtain the correct result.

> **function notation**
>
> The notation $y = f(x)$ defines a function named f. This is read as "y is a function of x." The letter x represents the input value, or independent variable. The letter y, or $f(x)$, represents the output value, or dependent variable.

---

1 http://www.baseball-almanac.com/legendary/lisn100.shtml. Accessed 3/24/2014.

Example 3   **Using Function Notation for Days in a Month**

Use function notation to represent a function whose input is the name of a month and output is the number of days in that month.

Solution   The number of days in a month is a function of the name of the month, so if we name the function f, we write days $= f$(month) or $d = f(m)$. The name of the month is the input to a "rule" that associates a specific number (the output) with each input.

$$31 = f(\text{January})$$

output | input

rule

Figure 4

For example, $f(\text{March}) = 31$, because March has 31 days. The notation $d = f(m)$ reminds us that the number of days, d (the output), is dependent on the name of the month, m (the input).

Analysis   Note that the inputs to a function do not have to be numbers; function inputs can be names of people, labels of geometric objects, or any other element that determines some kind of output. However, most of the functions we will work with in this book will have numbers as inputs and outputs.

Example 4   **Interpreting Function Notation**

A function $N = f(y)$ gives the number of police officers, N, in a town in year y. What does $f(2005) = 300$ represent?

Solution   When we read $f(2005) = 300$, we see that the input year is 2005. The value for the output, the number of police officers (N), is 300. Remember $N = f(y)$. The statement $f(2005) = 300$ tells us that in the year 2005 there were 300 police officers in the town.

Try It #2

Use function notation to express the weight of a pig in pounds as a function of its age in days d.

Q & A…

Instead of a notation such as $y = f(x)$, could we use the same symbol for the output as for the function, such as $y = y(x)$, meaning "y is a function of x?"

Yes, this is often done, especially in applied subjects that use higher math, such as physics and engineering. However, in exploring math itself we like to maintain a distinction between a function such as f, which is a rule or procedure, and the output y we get by applying f to a particular input x. This is why we usually use notation such as $y = f(x)$, $P = W(d)$, and so on.

Representing Functions Using Tables

A common method of representing functions is in the form of a table. The table rows or columns display the corresponding input and output values. In some cases, these values represent all we know about the relationship; other times, the table provides a few select examples from a more complete relationship.

Table 3 lists the input number of each month (January $= 1$, February $= 2$, and so on) and the output value of the number of days in that month. This information represents all we know about the months and days for a given year (that is not a leap year). Note that, in this table, we define a days-in-a-month function f where $D = f(m)$ identifies months by an integer rather than by name.

Month number, m (input)	1	2	3	4	5	6	7	8	9	10	11	12
Days in month, D (output)	31	28	31	30	31	30	31	31	30	31	30	31

Table 3

Table 4 defines a function $Q = g(n)$. Remember, this notation tells us that g is the name of the function that takes the input n and gives the output Q.

n	1	2	3	4	5
Q	8	6	7	6	8

Table 4

Table 5 below displays the age of children in years and their corresponding heights. This table displays just some of the data available for the heights and ages of children. We can see right away that this table does not represent a function because the same input value, 5 years, has two different output values, 40 in. and 42 in.

Age in years, a (input)	5	5	6	7	8	9	10
Height in inches, h (output)	40	42	44	47	50	52	54

Table 5

How To...

Given a table of input and output values, determine whether the table represents a function.

1. Identify the input and output values.
2. Check to see if each input value is paired with only one output value. If so, the table represents a function.

Example 5   **Identifying Tables that Represent Functions**

Which table, **Table 6**, **Table 7**, or **Table 8**, represents a function (if any)?

Input	Output
2	1
5	3
8	6

Table 6

Input	Output
−3	5
0	1
4	5

Table 7

Input	Output
1	0
5	2
5	4

Table 8

Solution   **Table 6** and **Table 7** define functions. In both, each input value corresponds to exactly one output value. **Table 8** does not define a function because the input value of 5 corresponds to two different output values.

When a table represents a function, corresponding input and output values can also be specified using function notation.

The function represented by **Table 6** can be represented by writing

$$f(2) = 1, f(5) = 3, \text{and } f(8) = 6$$

Similarly, the statements

$$g(-3) = 5, g(0) = 1, \text{and } g(4) = 5$$

represent the function in table **Table 7.**

Table 8 cannot be expressed in a similar way because it does not represent a function.

Try It #3

Does **Table 9** represent a function?

Input	Output
1	10
2	100
3	1000

Table 9

Finding Input and Output Values of a Function

When we know an input value and want to determine the corresponding output value for a function, we evaluate the function. Evaluating will always produce one result because each input value of a function corresponds to exactly one output value.

When we know an output value and want to determine the input values that would produce that output value, we set the output equal to the function's formula and solve for the input. Solving can produce more than one solution because different input values can produce the same output value.

Evaluation of Functions in Algebraic Forms

When we have a function in formula form, it is usually a simple matter to evaluate the function. For example, the function $f(x) = 5 - 3x^2$ can be evaluated by squaring the input value, multiplying by 3, and then subtracting the product from 5.

How To…

Given the formula for a function, evaluate.

1. Replace the input variable in the formula with the value provided.
2. Calculate the result.

Example 6   **Evaluating Functions at Specific Values**

Evaluate $f(x) = x^2 + 3x - 4$ at:

  a. 2   **b.** a **c.** $a + h$ **d.** $\dfrac{f(a + h) - f(a)}{h}$

Solution   Replace the x in the function with each specified value.

  a. Because the input value is a number, 2, we can use simple algebra to simplify.

$$f(2) = 2^2 + 3(2) - 4$$
$$= 4 + 6 - 4$$
$$= 6$$

  b. In this case, the input value is a letter so we cannot simplify the answer any further.

$$f(a) = a^2 + 3a - 4$$

  c. With an input value of $a + h$, we must use the distributive property.

$$f(a + h) = (a + h)^2 + 3(a + h) - 4$$
$$= a^2 + 2ah + h^2 + 3a + 3h - 4$$

  d. In this case, we apply the input values to the function more than once, and then perform algebraic operations on the result. We already found that

$$f(a + h) = a^2 + 2ah + h^2 + 3a + 3h - 4$$

and we know that

$$f(a) = a^2 + 3a - 4$$

Now we combine the results and simplify.

$$\frac{f(a + h) - f(a)}{h} = \frac{(a^2 + 2ah + h^2 + 3a + 3h - 4) - (a^2 + 3a - 4)}{h}$$
$$= \frac{2ah + h^2 + 3h}{h}$$
$$= \frac{h(2a + h + 3)}{h} \qquad \text{Factor out } h.$$
$$= 2a + h + 3 \qquad \text{Simplify.}$$

Example 7   **Evaluating Functions**

Given the function $h(p) = p^2 + 2p$, evaluate $h(4)$.

Solution   To evaluate $h(4)$, we substitute the value 4 for the input variable p in the given function.

$$h(p) = p^2 + 2p$$
$$h(4) = (4)^2 + 2\,(4)$$
$$= 16 + 8$$
$$= 24$$

Therefore, for an input of 4, we have an output of 24.

Try It #4

Given the function $g(m) = \sqrt{m - 4}$. Evaluate $g(5)$.

Example 8   **Solving Functions**

Given the function $h(p) = p^2 + 2p$, solve for $h(p) = 3$.

Solution

$$h(p) = 3$$
$$p^2 + 2p = 3 \qquad \text{Substitute the original function } h(p) = p^2 + 2p.$$
$$p^2 + 2p - 3 = 0 \qquad \text{Subtract 3 from each side.}$$
$$(p + 3)(p - 1) = 0 \qquad \text{Factor.}$$

If $(p + 3)(p - 1) = 0$, either $(p + 3) = 0$ or $(p - 1) = 0$ (or both of them equal 0). We will set each factor equal to 0 and solve for p in each case.

$$(p + 3) = 0, \quad p = -3$$
$$(p - 1) = 0, \quad p = 1$$

This gives us two solutions. The output $h(p) = 3$ when the input is either $p = 1$ or $p = -3$. We can also verify by graphing as in **Figure 5**. The graph verifies that $h(1) = h(-3) = 3$ and $h(4) = 24$.

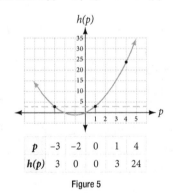

p	–3	–2	0	1	4
$h(p)$	3	0	0	3	24

Figure 5

Try It #5

Given the function $g(m) = \sqrt{m - 4}$, solve $g(m) = 2$.

Evaluating Functions Expressed in Formulas

Some functions are defined by mathematical rules or procedures expressed in equation form. If it is possible to express the function output with a formula involving the input quantity, then we can define a function in algebraic form. For example, the equation $2n + 6p = 12$ expresses a functional relationship between n and p. We can rewrite it to decide if p is a function of n.

How To...

Given a function in equation form, write its algebraic formula.

1. Solve the equation to isolate the output variable on one side of the equal sign, with the other side as an expression that involves *only* the input variable.
2. Use all the usual algebraic methods for solving equations, such as adding or subtracting the same quantity to or from both sides, or multiplying or dividing both sides of the equation by the same quantity.

Example 9 Finding an Equation of a Function

Express the relationship $2n + 6p = 12$ as a function $p = f(n)$, if possible.

Solution To express the relationship in this form, we need to be able to write the relationship where p is a function of n, which means writing it as $p = $ [expression involving n].

$$2n + 6p = 12$$

$$6p = 12 - 2n \qquad \text{Subtract } 2n \text{ from both sides.}$$

$$p = \frac{12 - 2n}{6} \qquad \text{Divide both sides by 6 and simplify.}$$

$$p = \frac{12}{6} - \frac{2n}{6}$$

$$p = 2 - \frac{1}{3}n$$

Therefore, p as a function of n is written as

$$p = f(n) = 2 - \frac{1}{3}n$$

Example 10 Expressing the Equation of a Circle as a Function

Does the equation $x^2 + y^2 = 1$ represent a function with x as input and y as output? If so, express the relationship as a function $y = f(x)$.

Solution First we subtract x^2 from both sides.

$$y^2 = 1 - x^2$$

We now try to solve for y in this equation.

$$y = \pm\sqrt{1 - x^2}$$

$$= +\sqrt{1 - x^2} \quad \text{and} \quad -\sqrt{1 - x^2}$$

We get two outputs corresponding to the same input, so this relationship cannot be represented as a single function $y = f(x)$. If we graph both functions on a graphing calculator, we will get the upper and lower semicircles.

Try It #6

If $x - 8y^3 = 0$, express y as a function of x.

Q & A...

Are there relationships expressed by an equation that do represent a function but that still cannot be represented by an algebraic formula?

Yes, this can happen. For example, given the equation $x = y + 2^y$, if we want to express y as a function of x, there is no simple algebraic formula involving only x that equals y. However, each x does determine a unique value for y, and there are mathematical procedures by which y can be found to any desired accuracy. In this case, we say that the equation gives an implicit (implied) rule for y as a function of x, even though the formula cannot be written explicitly.

Evaluating a Function Given in Tabular Form

As we saw above, we can represent functions in tables. Conversely, we can use information in tables to write functions, and we can evaluate functions using the tables. For example, how well do our pets recall the fond memories we share with them? There is an urban legend that a goldfish has a memory of 3 seconds, but this is just a myth. Goldfish can remember up to 3 months, while the beta fish has a memory of up to 5 months. And while a puppy's memory span is no longer than 30 seconds, the adult dog can remember for 5 minutes. This is meager compared to a cat, whose memory span lasts for 16 hours.

The function that relates the type of pet to the duration of its memory span is more easily visualized with the use of a table. See **Table 10**.[2]

Pet	Memory span in hours
Puppy	0.008
Adult dog	0.083
Cat	16
Goldfish	2160
Beta fish	3600

Table 10

At times, evaluating a function in table form may be more useful than using equations. Here let us call the function P. The domain of the function is the type of pet and the range is a real number representing the number of hours the pet's memory span lasts. We can evaluate the function P at the input value of "goldfish." We would write P(goldfish) = 2160. Notice that, to evaluate the function in table form, we identify the input value and the corresponding output value from the pertinent row of the table. The tabular form for function P seems ideally suited to this function, more so than writing it in paragraph or function form.

How To...

Given a function represented by a table, identify specific output and input values.

1. Find the given input in the row (or column) of input values.
2. Identify the corresponding output value paired with that input value.
3. Find the given output values in the row (or column) of output values, noting every time that output value appears.
4. Identify the input value(s) corresponding to the given output value.

Example 11   **Evaluating and Solving a Tabular Function**

Using **Table 11**,

      a. Evaluate $g(3)$ **b.** Solve $g(n) = 6$.

n	1	2	3	4	5
$g(n)$	8	6	7	6	8

Table 11

Solution

      a. Evaluating $g(3)$ means determining the output value of the function g for the input value of $n = 3$. The table output value corresponding to $n = 3$ is 7, so $g(3) = 7$.

      b. Solving $g(n) = 6$ means identifying the input values, n, that produce an output value of 6. **Table 11** shows two solutions: 2 and 4. When we input 2 into the function g, our output is 6. When we input 4 into the function g, our output is also 6.

---

2 http://www.kgbanswers.com/how-long-is-a-dogs-memory-span/4221590. Accessed 3/24/2014.

Try It #7

Using **Table 11**, evaluate $g(1)$.

Finding Function Values from a Graph

Evaluating a function using a graph also requires finding the corresponding output value for a given input value, only in this case, we find the output value by looking at the graph. Solving a function equation using a graph requires finding all instances of the given output value on the graph and observing the corresponding input value(s).

Example 12   **Reading Function Values from a Graph**

Given the graph in **Figure 6**,

    a. Evaluate $f(2)$.   **b.** Solve $f(x) = 4$.

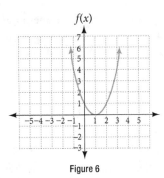

Figure 6

Solution

    a. To evaluate $f(2)$, locate the point on the curve where $x = 2$, then read the y-coordinate of that point. The point has coordinates $(2, 1)$, so $f(2) = 1$. See **Figure 7**.

Figure 7

    b. To solve $f(x) = 4$, we find the output value 4 on the vertical axis. Moving horizontally along the line $y = 4$, we locate two points of the curve with output value 4: $(-1, 4)$ and $(3, 4)$. These points represent the two solutions to $f(x) = 4$: -1 or 3. This means $f(-1) = 4$ and $f(3) = 4$, or when the input is -1 or 3, the output is 4. See **Figure 8**.

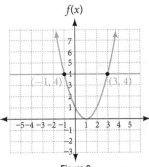

Figure 8

Try It #8

Using **Figure 7**, solve $f(x) = 1$.

Determining Whether a Function is One-to-One

Some functions have a given output value that corresponds to two or more input values. For example, in the stock chart shown in **Figure 1** at the beginning of this chapter, the stock price was $1,000 on five different dates, meaning that there were five different input values that all resulted in the same output value of $1,000.

However, some functions have only one input value for each output value, as well as having only one output for each input. We call these functions one-to-one functions. As an example, consider a school that uses only letter grades and decimal equivalents, as listed in **Table 12**.

Letter grade	Grade-point average
A	4.0
B	3.0
C	2.0
D	1.0

Table 12

This grading system represents a one-to-one function, because each letter input yields one particular grade-point average output and each grade-point average corresponds to one input letter.

To visualize this concept, let's look again at the two simple functions sketched in **Figure 1(a)** and **Figure 1(b)**. The function in part (a) shows a relationship that is not a one-to-one function because inputs q and r both give output n. The function in part (b) shows a relationship that is a one-to-one function because each input is associated with a single output.

> **one-to-one function**
> A **one-to-one function** is a function in which each output value corresponds to exactly one input value. There are no repeated x- or y-values.

Example 13 Determining Whether a Relationship Is a One-to-One Function

Is the area of a circle a function of its radius? If yes, is the function one-to-one?

Solution   A circle of radius r has a unique area measure given by $A = \pi r^2$, so for any input, r, there is only one output, A. The area is a function of radius r.

If the function is one-to-one, the output value, the area, must correspond to a unique input value, the radius. Any area measure A is given by the formula $A = \pi r^2$. Because areas and radii are positive numbers, there is exactly one solution: $r = \sqrt{\dfrac{A}{\pi}}$ So the area of a circle is a one-to-one function of the circle's radius.

Try It #9

a. Is a balance a function of the bank account number?
b. Is a bank account number a function of the balance?
c. Is a balance a one-to-one function of the bank account number?

Try It #10

a. If each percent grade earned in a course translates to one letter grade, is the letter grade a function of the percent grade?
b. If so, is the function one-to-one?

Using the Vertical Line Test

As we have seen in some examples above, we can represent a function using a graph. Graphs display a great many input-output pairs in a small space. The visual information they provide often makes relationships easier to understand. By convention, graphs are typically constructed with the input values along the horizontal axis and the output values along the vertical axis.

The most common graphs name the input value x and the output value y, and we say y is a function of x, or $y = f(x)$ when the function is named f. The graph of the function is the set of all points (x, y) in the plane that satisfies the equation $y = f(x)$. If the function is defined for only a few input values, then the graph of the function consists of only a few points, where the x-coordinate of each point is an input value and the y-coordinate of each point is the corresponding output value. For example, the black dots on the graph in **Figure 9** tell us that $f(0) = 2$ and $f(6) = 1$. However, the set of all points (x, y) satisfying $y = f(x)$ is a curve. The curve shown includes $(0, 2)$ and $(6, 1)$ because the curve passes through those points.

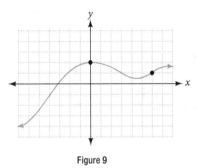

Figure 9

The **vertical line test** can be used to determine whether a graph represents a function. If we can draw any vertical line that intersects a graph more than once, then the graph does not define a function because a function has only one output value for each input value. See **Figure 10**.

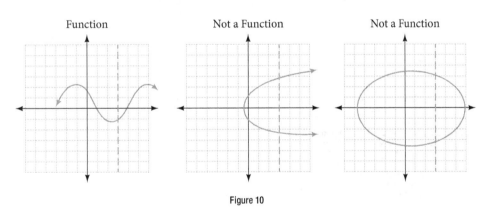

Figure 10

How To…

Given a graph, use the vertical line test to determine if the graph represents a function.

1. Inspect the graph to see if any vertical line drawn would intersect the curve more than once.
2. If there is any such line, determine that the graph does not represent a function.

Example 14   **Applying the Vertical Line Test**

Which of the graphs in **Figure 11** represent(s) a function $y = f(x)$?

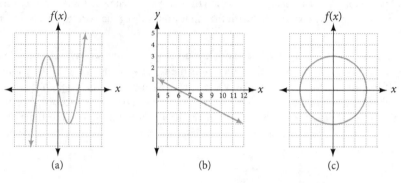

Figure 11

Solution   If any vertical line intersects a graph more than once, the relation represented by the graph is not a function. Notice that any vertical line would pass through only one point of the two graphs shown in parts (a) and (b) of **Figure 11**. From this we can conclude that these two graphs represent functions. The third graph does not represent a function because, at most x-values, a vertical line would intersect the graph at more than one point, as shown in **Figure 12**.

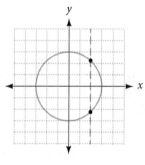

Figure 12

Try It #11

Does the graph in **Figure 13** represent a function?

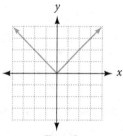

Figure 13

Using the Horizontal Line Test

Once we have determined that a graph defines a function, an easy way to determine if it is a one-to-one function is to use the **horizontal line test**. Draw horizontal lines through the graph. If any horizontal line intersects the graph more than once, then the graph does not represent a one-to-one function.

How To...

Given a graph of a function, use the horizontal line test to determine if the graph represents a one-to-one function.

1. Inspect the graph to see if any horizontal line drawn would intersect the curve more than once.
2. If there is any such line, determine that the function is not one-to-one.

Example 15   **Applying the Horizontal Line Test**

Consider the functions shown in **Figure 11(a)** and **Figure 11(b)**. Are either of the functions one-to-one?

Solution   The function in **Figure 11(a)** is not one-to-one. The horizontal line shown in **Figure 14** intersects the graph of the function at two points (and we can even find horizontal lines that intersect it at three points.)

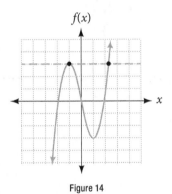

Figure 14

The function in **Figure 11(b)** is one-to-one. Any horizontal line will intersect a diagonal line at most once.

Try It #12

Is the graph shown here one-to-one?

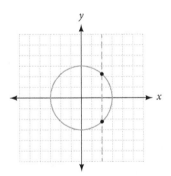

Identifying Basic Toolkit Functions

In this text, we will be exploring functions—the shapes of their graphs, their unique characteristics, their algebraic formulas, and how to solve problems with them. When learning to read, we start with the alphabet. When learning to do arithmetic, we start with numbers. When working with functions, it is similarly helpful to have a base set of building-block elements. We call these our "toolkit functions," which form a set of basic named functions for which we know the graph, formula, and special properties. Some of these functions are programmed to individual buttons on many calculators. For these definitions we will use x as the input variable and $y = f(x)$ as the output variable.

We will see these toolkit functions, combinations of toolkit functions, their graphs, and their transformations frequently throughout this book. It will be very helpful if we can recognize these toolkit functions and their features quickly by name, formula, graph, and basic table properties. The graphs and sample table values are included with each function shown in **Table 13**.

Toolkit Functions		
Name	**Function**	**Graph**
Constant	$f(x) = c$, where c is a constant	
Identity	$f(x) = x$	
Absolute value	$f(x) = \|x\|$	
Quadratic	$f(x) = x^2$	
Cubic	$f(x) = x^3$	

Reciprocal	$f(x) = \dfrac{1}{x}$	graph	x / $f(x)$: -2 / -0.5, -1 / -1, -0.5 / -2, 0.5 / 2, 1 / 1, 2 / 0.5
Reciprocal squared	$f(x) = \dfrac{1}{x^2}$	graph	x / $f(x)$: -2 / 0.25, -1 / 1, -0.5 / 4, 0.5 / 4, 1 / 1, 2 / 0.25
Square root	$f(x) = \sqrt{x}$	graph	x / $f(x)$: 0 / 0, 1 / 1, 4 / 2
Cube root	$f(x) = \sqrt[3]{x}$	graph	x / $f(x)$: -1 / -1, -0.125 / -0.5, 0 / 0, 0.125 / 0.5, 1 / 1

Table 13

Access the following online resources for additional instruction and practice with functions.

- Determine if a Relation is a Function (http://openstaxcollege.org/l/relationfunction)
- Vertical Line Test (http://openstaxcollege.org/l/vertlinetest)
- Introduction to Functions (http://openstaxcollege.org/l/introtofunction)
- Vertical Line Test of Graph (http://openstaxcollege.org/l/vertlinegraph)
- One-to-one Functions (http://openstaxcollege.org/l/onetoone)
- Graphs as One-to-one Functions (http://openstaxcollege.org/l/graphonetoone)

3.1   SECTION EXERCISES

VERBAL

1. What is the difference between a relation and a function?

2. What is the difference between the input and the output of a function?

3. Why does the vertical line test tell us whether the graph of a relation represents a function?

4. How can you determine if a relation is a one-to-one function?

5. Why does the horizontal line test tell us whether the graph of a function is one-to-one?

ALGEBRAIC

For the following exercises, determine whether the relation represents a function.

6. $\{(a, b), (c, d), (a, c)\}$

7. $\{(a, b),(b, c),(c, c)\}$

For the following exercises, determine whether the relation represents y as a function of x.

8. $5x + 2y = 10$

9. $y = x^2$

10. $x = y^2$

11. $3x^2 + y = 14$

12. $2x + y^2 = 6$

13. $y = -2x^2 + 40x$

14. $y = \dfrac{1}{x}$

15. $x = \dfrac{3y + 5}{7y - 1}$

16. $x = \sqrt{1 - y^2}$

17. $y = \dfrac{3x + 5}{7x - 1}$

18. $x^2 + y^2 = 9$

19. $2xy = 1$

20. $x = y^3$

21. $y = x^3$

22. $y = \sqrt{1 - x^2}$

23. $x = \pm\sqrt{1 - y}$

24. $y = \pm\sqrt{1 - x}$

25. $y^2 = x^2$

26. $y^3 = x^2$

For the following exercises, evaluate the function f at the indicated values $f(-3), f(2), f(-a), -f(a), f(a + h)$.

27. $f(x) = 2x - 5$

28. $f(x) = -5x^2 + 2x - 1$

29. $f(x) = \sqrt{2 - x} + 5$

30. $f(x) = \dfrac{6x - 1}{5x + 2}$

31. $f(x) = |x - 1| - |x + 1|$

32. Given the function $g(x) = 5 - x^2$, simplify $\dfrac{g(x + h) - g(x)}{h}, h \neq 0$

33. Given the function $g(x) = x^2 + 2x$, simplify $\dfrac{g(x) - g(a)}{x - a}, x \neq a$

34. Given the function $k(t) = 2t - 1$:
    a. Evaluate $k(2)$.
    b. Solve $k(t) = 7$.

35. Given the function $f(x) = 8 - 3x$:
    a. Evaluate $f(-2)$.
    b. Solve $f(x) = -1$.

36. Given the function $p(c) = c^2 + c$:
    a. Evaluate $p(-3)$.
    b. Solve $p(c) = 2$.

37. Given the function $f(x) = x^2 - 3x$
    a. Evaluate $f(5)$.
    b. Solve $f(x) = 4$

38. Given the function $f(x) = \sqrt{x + 2}$:
    a. Evaluate $f(7)$.
    b. Solve $f(x) = 4$

39. Consider the relationship $3r + 2t = 18$.
    a. Write the relationship as a function $r = f(t)$.
    b. Evaluate $f(-3)$.
    c. Solve $f(t) = 2$.

GRAPHICAL

For the following exercises, use the vertical line test to determine which graphs show relations that are functions.

40.

41.

42.

43.

44.

45.

46.

47.

48.

49.

50.

51.
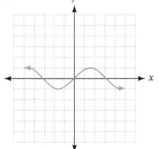

52. Given the following graph
   a. Evaluate $f(-1)$.
   b. Solve for $f(x) = 3$.

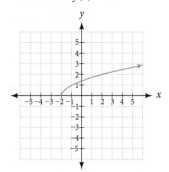

53. Given the following graph
   a. Evaluate $f(0)$.
   b. Solve for $f(x) = -3$.

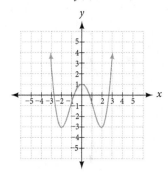

54. Given the following graph
   a. Evaluate $f(4)$.
   b. Solve for $f(x) = 1$.

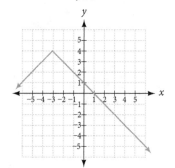

For the following exercises, determine if the given graph is a one-to-one function.

55.

56.

57.

58.

59.

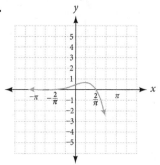

NUMERIC

For the following exercises, determine whether the relation represents a function.

60. $\{(-1, -1),(-2, -2),(-3, -3)\}$ **61.** $\{(3, 4),(4, 5),(5, 6)\}$ **62.** $\{(2, 5),(7, 11),(15, 8),(7, 9)\}$

For the following exercises, determine if the relation represented in table form represents y as a function of x.

63.

x	5	10	15
y	3	8	14

64.

x	5	10	15
y	3	8	8

65.

x	5	10	10
y	3	8	14

For the following exercises, use the function f represented in **Table 14** below.

x	0	1	2	3	4	5	6	7	8	9
f(x)	74	28	1	53	56	3	36	45	14	47

Table 14

66. Evaluate $f(3)$.

67. Solve $f(x) = 1$

For the following exercises, evaluate the function f at the values $f(-2), f(-1), f(0), f(1)$, and $f(2)$.

68. $f(x) = 4 - 2x$

69. $f(x) = 8 - 3x$

70. $f(x) = 8x^2 - 7x + 3$

71. $f(x) = 3 + \sqrt{x + 3}$

72. $f(x) = \dfrac{x - 2}{x + 3}$

73. $f(x) = 3^x$

For the following exercises, evaluate the expressions, given functions f, g, and h:

$$f(x) = 3x - 2 \qquad g(x) = 5 - x^2 \qquad h(x) = -2x^2 + 3x - 1$$

74. $3f(1) - 4g(-2)$

75. $f\left(\dfrac{7}{3}\right) - h(-2)$

TECHNOLOGY

For the following exercises, graph $y = x^2$ on the given viewing window. Determine the corresponding range for each viewing window. Show each graph.

76. $[-0.1, 0.1]$

77. $[-10, 10]$

78. $[-100, 100]$

For the following exercises, graph $y = x^3$ on the given viewing window. Determine the corresponding range for each viewing window. Show each graph.

79. $[-0.1, 0.1]$

80. $[-10, 10]$

81. $[-100, 100]$

For the following exercises, graph $y = \sqrt{x}$ on the given viewing window. Determine the corresponding range for each viewing window. Show each graph.

82. $[0, 0.01]$

83. $[0, 100]$

84. $[0, 10,000]$

For the following exercises, graph $y = \sqrt[3]{x}$ on the given viewing window. Determine the corresponding range for each viewing window. Show each graph.

85. $[-0.001, 0.001]$

86. $[-1,000, 1,000]$

87. $[-1,000,000, 1,000,000]$

REAL-WORLD APPLICATIONS

88. The amount of garbage, G, produced by a city with population p is given by $G = f(p)$. G is measured in tons per week, and p is measured in thousands of people.

   a. The town of Tola has a population of 40,000 and produces 13 tons of garbage each week. Express this information in terms of the function f.

   b. Explain the meaning of the statement $f(5) = 2$.

89. The number of cubic yards of dirt, D, needed to cover a garden with area a square feet is given by $D = g(a)$.

   a. A garden with area 5,000 ft² requires 50 yd³ of dirt. Express this information in terms of the function g.

   b. Explain the meaning of the statement $g(100) = 1$.

90. Let $f(t)$ be the number of ducks in a lake t years after 1990. Explain the meaning of each statement:

   a. $f(5) = 30$

   b. $f(10) = 40$

91. Let $h(t)$ be the height above ground, in feet, of a rocket t seconds after launching. Explain the meaning of each statement:

   a. $h(1) = 200$

   b. $h(2) = 350$

92. Show that the function $f(x) = 3(x - 5)^2 + 7$ is <u>not</u> one-to-one.

LEARNING OBJECTIVES

In this section, you will:

- Find the domain of a function defined by an equation.
- Graph piecewise-defined functions.

3.2   DOMAIN AND RANGE

If you're in the mood for a scary movie, you may want to check out one of the five most popular horror movies of all time—*I am Legend, Hannibal, The Ring, The Grudge,* and *The Conjuring.* **Figure 1** shows the amount, in dollars, each of those movies grossed when they were released as well as the ticket sales for horror movies in general by year. Notice that we can use the data to create a function of the amount each movie earned or the total ticket sales for all horror movies by year. In creating various functions using the data, we can identify different independent and dependent variables, and we can analyze the data and the functions to determine the domain and range. In this section, we will investigate methods for determining the domain and range of functions such as these.

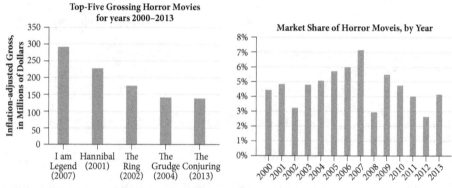

Figure 1   Based on data compiled by www.the-numbers.com.[3]

Finding the Domain of a Function Defined by an Equation

In **Functions and Function Notation**, we were introduced to the concepts of domain and range. In this section, we will practice determining domains and ranges for specific functions. Keep in mind that, in determining domains and ranges, we need to consider what is physically possible or meaningful in real-world examples, such as tickets sales and year in the horror movie example above. We also need to consider what is mathematically permitted. For example, we cannot include any input value that leads us to take an even root of a negative number if the domain and range consist of real numbers. Or in a function expressed as a formula, we cannot include any input value in the domain that would lead us to divide by 0.

We can visualize the domain as a "holding area" that contains "raw materials" for a "function machine" and the range as another "holding area" for the machine's products. See **Figure 2**.

Figure 2

We can write the domain and range in **interval notation**, which uses values within brackets to describe a set of numbers. In interval notation, we use a square bracket [when the set includes the endpoint and a parenthesis (to indicate that the endpoint is either not included or the interval is unbounded. For example, if a person has $100 to spend, he or she would need to express the interval that is more than 0 and less than or equal to 100 and write (0, 100]. We will discuss interval notation in greater detail later.

3 The Numbers: Where Data and the Movie Business Meet. "Box Office History for Horror Movies." http://www.the-numbers.com/market/genre/Horror. Accessed 3/24/2014

Let's turn our attention to finding the domain of a function whose equation is provided. Oftentimes, finding the domain of such functions involves remembering three different forms. First, if the function has no denominator or an even root, consider whether the domain could be all real numbers. Second, if there is a denominator in the function's equation, exclude values in the domain that force the denominator to be zero. Third, if there is an even root, consider excluding values that would make the radicand negative.

Before we begin, let us review the conventions of interval notation:

- The smallest number from the interval is written first.

- The largest number in the interval is written second, following a comma.

- Parentheses, (or), are used to signify that an endpoint value is not included, called exclusive.

- Brackets, [or], are used to indicate that an endpoint value is included, called inclusive.

See **Figure 3** for a summary of interval notation.

Inequality	Interval Notation	Graph on Number Line	Description
$x > a$	(a, ∞)		x is greater than a
$x < a$	$(-\infty, a)$		x is less than a
$x \geq a$	$[a, \infty)$		x is greater than or equal to a
$x \leq a$	$(-\infty, a]$		x is less than or equal to a
$a < x < b$	(a, b)		x is strictly between a and b
$a \leq x < b$	$[a, b)$		x is between a and b, to include a
$a < x \leq b$	$(a, b]$		x is between a and b, to include b
$a \leq x \leq b$	$[a, b]$		x is between a and b, to include a and b

Figure 3

Example 1   **Finding the Domain of a Function as a Set of Ordered Pairs**

Find the domain of the following function: {(2, 10), (3, 10), (4, 20), (5, 30), (6, 40)}.

Solution   First identify the input values. The input value is the first coordinate in an ordered pair. There are no restrictions, as the ordered pairs are simply listed. The domain is the set of the first coordinates of the ordered pairs.

$$\{2, 3, 4, 5, 6\}$$

Try It #1

Find the domain of the function: {(−5, 4), (0, 0), (5, −4), (10, −8), (15, −12)}

How To...

Given a function written in equation form, find the domain.

1. Identify the input values.
2. Identify any restrictions on the input and exclude those values from the domain.
3. Write the domain in interval form, if possible.

Example 2   **Finding the Domain of a Function**

Find the domain of the function $f(x) = x^2 - 1$.

Solution   The input value, shown by the variable x in the equation, is squared and then the result is lowered by one. Any real number may be squared and then be lowered by one, so there are no restrictions on the domain of this function. The domain is the set of real numbers.

In interval form, the domain of f is $(-\infty, \infty)$.

Try It #2

Find the domain of the function: $f(x) = 5 - x + x^3$.

How To...

Given a function written in an equation form that includes a fraction, find the domain.

1. Identify the input values.
2. Identify any restrictions on the input. If there is a denominator in the function's formula, set the denominator equal to zero and solve for x. If the function's formula contains an even root, set the radicand greater than or equal to 0, and then solve.
3. Write the domain in interval form, making sure to exclude any restricted values from the domain.

Example 3   **Finding the Domain of a Function Involving a Denominator**

Find the domain of the function $f(x) = \dfrac{x+1}{2-x}$.

Solution   When there is a denominator, we want to include only values of the input that do not force the denominator to be zero. So, we will set the denominator equal to 0 and solve for x.

$$2 - x = 0$$
$$-x = -2$$
$$x = 2$$

Now, we will exclude 2 from the domain. The answers are all real numbers where $x < 2$ or $x > 2$ as shown in **Figure 4**. We can use a symbol known as the union, $\cup$, to combine the two sets. In interval notation, we write the solution: $(-\infty, 2) \cup (2, \infty)$.

$$x < 2 \text{ or } x > 2$$
$$\downarrow \qquad \downarrow$$
$$(-\infty, 2) \cup (2, \infty)$$

Figure 4

Try It #3

Find the domain of the function: $f(x) = \dfrac{1+4x}{2x-1}$.

How To...

Given a function written in equation form including an even root, find the domain.

1. Identify the input values.
2. Since there is an even root, exclude any real numbers that result in a negative number in the radicand. Set the radicand greater than or equal to zero and solve for x.
3. The solution(s) are the domain of the function. If possible, write the answer in interval form.

Example 4 **Finding the Domain of a Function with an Even Root**

Find the domain of the function $f(x) = \sqrt{7 - x}$.

Solution   When there is an even root in the formula, we exclude any real numbers that result in a negative number in the radicand.

Set the radicand greater than or equal to zero and solve for x.

$$7 - x \geq 0$$
$$-x \geq -7$$
$$x \leq 7$$

Now, we will exclude any number greater than 7 from the domain. The answers are all real numbers less than or equal to 7, or $(-\infty, 7]$.

Try It #4

Find the domain of the function $f(x) = \sqrt{5 + 2x}$.

Q & A...

Can there be functions in which the domain and range do not intersect at all?

Yes. For example, the function $f(x) = -\dfrac{1}{\sqrt{x}}$ has the set of all positive real numbers as its domain but the set of all negative real numbers as its range. As a more extreme example, a function's inputs and outputs can be completely different categories (for example, names of weekdays as inputs and numbers as outputs, as on an attendance chart), in such cases the domain and range have no elements in common.

Using Notations to Specify Domain and Range

In the previous examples, we used inequalities and lists to describe the domain of functions. We can also use inequalities, or other statements that might define sets of values or data, to describe the behavior of the variable in **set-builder notation**. For example, $\{x \mid 10 \leq x < 30\}$ describes the behavior of x in set-builder notation. The braces { } are read as "the set of," and the vertical bar | is read as "such that," so we would read $\{x \mid 10 \leq x < 30\}$ as "the set of x-values such that 10 is less than or equal to x, and x is less than 30."

Figure 5 compares inequality notation, set-builder notation, and interval notation.

	Inequality Notation	Set-builder Notation	Interval Notation
	$5 < h \leq 10$	$\{h \mid 5 < h \leq 10\}$	$(5, 10]$
	$5 \leq h < 10$	$\{h \mid 5 \leq h < 10\}$	$[5, 10)$
	$5 < h < 10$	$\{h \mid 5 < h < 10\}$	$(5, 10)$
	$h < 10$	$\{h \mid h < 10\}$	$(-\infty, 10)$
	$h \geq 10$	$\{h \mid h \geq 10\}$	$[10, \infty)$
	All real numbers	$\mathbb{R}$	$(-\infty, \infty)$

Figure 5

To combine two intervals using inequality notation or set-builder notation, we use the word "or." As we saw in earlier examples, we use the union symbol, $\cup$, to combine two unconnected intervals. For example, the union of the sets $\{2, 3, 5\}$ and $\{4, 6\}$ is the set $\{2, 3, 4, 5, 6\}$. It is the set of all elements that belong to one *or* the other (or both) of the original two sets. For sets with a finite number of elements like these, the elements do not have to be listed in ascending order of numerical value. If the original two sets have some elements in common, those elements should be listed only once in the union set. For sets of real numbers on intervals, another example of a union is

$$\{x \mid |x| \geq 3\} = (-\infty, -3] \cup [3, \infty)$$

---

set-builder notation and **interval notation**

Set-builder notation is a method of specifying a set of elements that satisfy a certain condition. It takes the form $\{x \mid \text{statement about } x\}$ which is read as, "the set of all x such that the statement about x is true." For example,

$$\{x \mid 4 < x \leq 12\}$$

Interval notation is a way of describing sets that include all real numbers between a lower limit that may or may not be included and an upper limit that may or may not be included. The endpoint values are listed between brackets or parentheses. A square bracket indicates inclusion in the set, and a parenthesis indicates exclusion from the set. For example,

$$(4, 12]$$

How To...

Given a line graph, describe the set of values using interval notation.

1. Identify the intervals to be included in the set by determining where the heavy line overlays the real line.
2. At the left end of each interval, use [with each end value to be included in the set (solid dot) or (for each excluded end value (open dot).
3. At the right end of each interval, use] with each end value to be included in the set (filled dot) or) for each excluded end value (open dot).
4. Use the union symbol $\cup$ to combine all intervals into one set.

Example 5   **Describing Sets on the Real-Number Line**

Describe the intervals of values shown in **Figure 6** using inequality notation, set-builder notation, and interval notation.

Figure 6

Solution   To describe the values, x, included in the intervals shown, we would say, "x is a real number greater than or equal to 1 and less than or equal to 3, or a real number greater than 5."

Inequality	$1 \leq x \leq 3 \text{ or } x > 5$
Set-builder notation	$\{x \mid 1 \leq x \leq 3 \text{ or } x > 5\}$
Interval notation	$[1, 3] \cup (5, \infty)$

Remember that, when writing or reading interval notation, using a square bracket means the boundary is included in the set. Using a parenthesis means the boundary is not included in the set.

Try It #5

Given this figure, specify the graphed set in

a. words

b. set-builder notation

c. interval notation

Figure 7

Finding Domain and Range from Graphs

Another way to identify the domain and range of functions is by using graphs. Because the domain refers to the set of possible input values, the domain of a graph consists of all the input values shown on the *x*-axis. The range is the set of possible output values, which are shown on the *y*-axis. Keep in mind that if the graph continues beyond the portion of the graph we can see, the domain and range may be greater than the visible values. See **Figure 8**.

Figure 8

We can observe that the graph extends horizontally from -5 to the right without bound, so the domain is $[-5, \infty)$. The vertical extent of the graph is all range values 5 and below, so the range is $(-\infty, 5]$. Note that the domain and range are always written from smaller to larger values, or from left to right for domain, and from the bottom of the graph to the top of the graph for range.

Example 6   **Finding Domain and Range from a Graph**

Find the domain and range of the function *f* whose graph is shown in **Figure 9**.

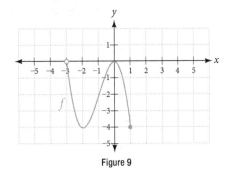

Figure 9

Solution   We can observe that the horizontal extent of the graph is −3 to 1, so the domain of *f* is (−3, 1].

The vertical extent of the graph is 0 to −4, so the range is [−4, 0]. See **Figure 10**.

Figure 10

Example 7   **Finding Domain and Range from a Graph of Oil Production**

Find the domain and range of the function *f* whose graph is shown in **Figure 11**.

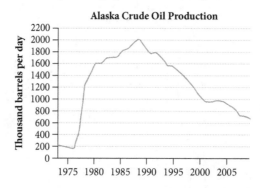

Figure 11   (credit: modification of work by the U.S. Energy Information Administration) [4]

Solution   The input quantity along the horizontal axis is "years," which we represent with the variable *t* for time. The output quantity is "thousands of barrels of oil per day," which we represent with the variable *b* for barrels. The graph may continue to the left and right beyond what is viewed, but based on the portion of the graph that is visible, we can determine the domain as $1973 \leq t \leq 2008$ and the range as approximately $180 \leq b \leq 2010$.

In interval notation, the domain is [1973, 2008], and the range is about [180, 2010]. For the domain and the range, we approximate the smallest and largest values since they do not fall exactly on the grid lines.

Try It #6

Given **Figure 12**, identify the domain and range using interval notation.

Figure 12

Q & A...

Can a function's domain and range be the same?

Yes. For example, the domain and range of the cube root function are both the set of all real numbers.

4 http://www.eia.gov/dnav/pet/hist/LeafHandler.ashx?n=PET&s=MCRFPAK2&f=A.

Finding Domains and Ranges of the Toolkit Functions

We will now return to our set of toolkit functions to determine the domain and range of each.

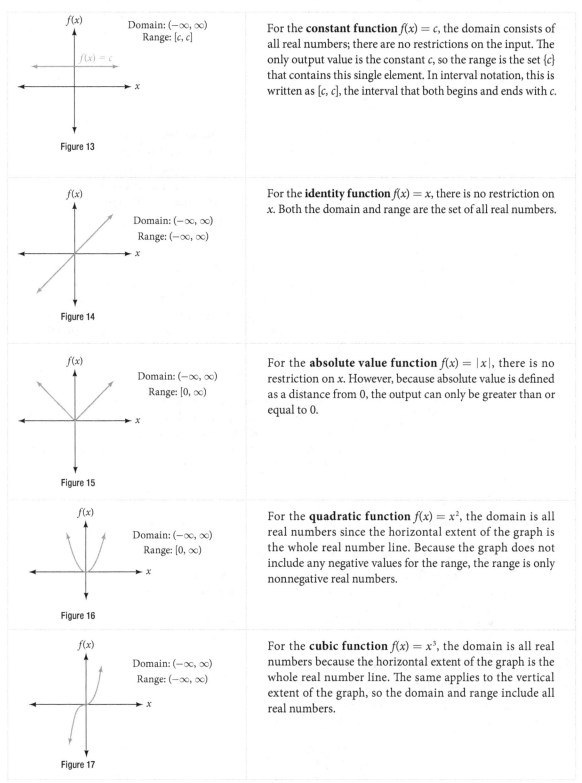

Domain: $(-\infty, \infty)$
Range: $[c, c]$

For the **constant function** $f(x) = c$, the domain consists of all real numbers; there are no restrictions on the input. The only output value is the constant c, so the range is the set $\{c\}$ that contains this single element. In interval notation, this is written as $[c, c]$, the interval that both begins and ends with c.

Figure 13

Domain: $(-\infty, \infty)$
Range: $(-\infty, \infty)$

For the **identity function** $f(x) = x$, there is no restriction on x. Both the domain and range are the set of all real numbers.

Figure 14

Domain: $(-\infty, \infty)$
Range: $[0, \infty)$

For the **absolute value function** $f(x) = |x|$, there is no restriction on x. However, because absolute value is defined as a distance from 0, the output can only be greater than or equal to 0.

Figure 15

Domain: $(-\infty, \infty)$
Range: $[0, \infty)$

For the **quadratic function** $f(x) = x^2$, the domain is all real numbers since the horizontal extent of the graph is the whole real number line. Because the graph does not include any negative values for the range, the range is only nonnegative real numbers.

Figure 16

Domain: $(-\infty, \infty)$
Range: $(-\infty, \infty)$

For the **cubic function** $f(x) = x^3$, the domain is all real numbers because the horizontal extent of the graph is the whole real number line. The same applies to the vertical extent of the graph, so the domain and range include all real numbers.

Figure 17

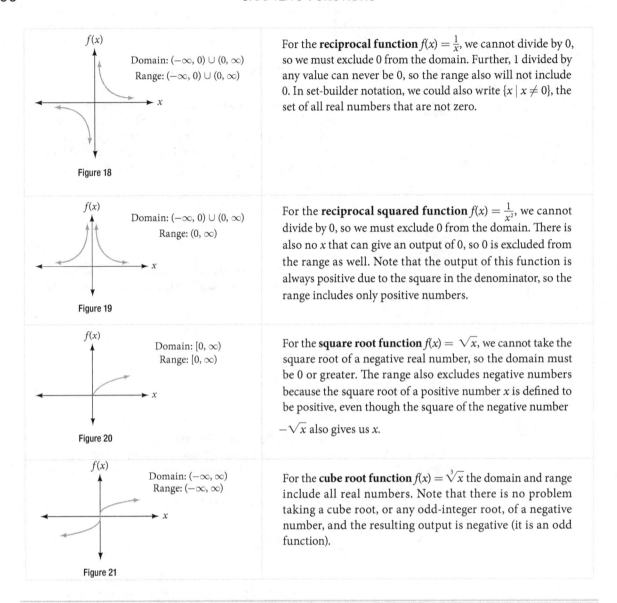

For the **reciprocal function** $f(x) = \frac{1}{x}$, we cannot divide by 0, so we must exclude 0 from the domain. Further, 1 divided by any value can never be 0, so the range also will not include 0. In set-builder notation, we could also write $\{x \mid x \neq 0\}$, the set of all real numbers that are not zero.

For the **reciprocal squared function** $f(x) = \frac{1}{x^2}$, we cannot divide by 0, so we must exclude 0 from the domain. There is also no x that can give an output of 0, so 0 is excluded from the range as well. Note that the output of this function is always positive due to the square in the denominator, so the range includes only positive numbers.

For the **square root function** $f(x) = \sqrt{x}$, we cannot take the square root of a negative real number, so the domain must be 0 or greater. The range also excludes negative numbers because the square root of a positive number x is defined to be positive, even though the square of the negative number $-\sqrt{x}$ also gives us x.

For the **cube root function** $f(x) = \sqrt[3]{x}$ the domain and range include all real numbers. Note that there is no problem taking a cube root, or any odd-integer root, of a negative number, and the resulting output is negative (it is an odd function).

How To...

Given the formula for a function, determine the domain and range.

1. Exclude from the domain any input values that result in division by zero.
2. Exclude from the domain any input values that have nonreal (or undefined) number outputs.
3. Use the valid input values to determine the range of the output values.
4. Look at the function graph and table values to confirm the actual function behavior.

Example 8   **Finding the Domain and Range Using Toolkit Functions**

Find the domain and range of $f(x) = 2x^3 - x$.

Solution   There are no restrictions on the domain, as any real number may be cubed and then subtracted from the result. The domain is $(-\infty, \infty)$ and the range is also $(-\infty, \infty)$.

Example 9 **Finding the Domain and Range**

Find the domain and range of $f(x) = \dfrac{2}{x+1}$.

Solution   We cannot evaluate the function at -1 because division by zero is undefined. The domain is $(-\infty, -1) \cup (-1, \infty)$. Because the function is never zero, we exclude 0 from the range. The range is $(-\infty, 0) \cup (0, \infty)$.

Example 10 **Finding the Domain and Range**

Find the domain and range of $f(x) = 2\sqrt{x+4}$.

Solution   We cannot take the square root of a negative number, so the value inside the radical must be nonnegative.

$$x + 4 \geq 0 \text{ when } x \geq -4$$

The domain of $f(x)$ is $[-4, \infty)$.

We then find the range. We know that $f(-4) = 0$, and the function value increases as x increases without any upper limit. We conclude that the range of f is $[0, \infty)$.

Analysis   **Figure 22** *represents the function f.*

Figure 22

Try It #7

Find the domain and range of $f(x) = -\sqrt{2-x}$.

Graphing Piecewise-Defined Functions

Sometimes, we come across a function that requires more than one formula in order to obtain the given output. For example, in the toolkit functions, we introduced the absolute value function $f(x) = |x|$. With a domain of all real numbers and a range of values greater than or equal to 0, absolute value can be defined as the magnitude, or modulus, of a real number value regardless of sign. It is the distance from 0 on the number line. All of these definitions require the output to be greater than or equal to 0.

If we input 0, or a positive value, the output is the same as the input.

$$f(x) = x \text{ if } x \geq 0$$

If we input a negative value, the output is the opposite of the input.

$$f(x) = -x \text{ if } x < 0$$

Because this requires two different processes or pieces, the absolute value function is an example of a piecewise function. A **piecewise function** is a function in which more than one formula is used to define the output over different pieces of the domain.

We use piecewise functions to describe situations in which a rule or relationship changes as the input value crosses certain "boundaries." For example, we often encounter situations in business for which the cost per piece of a certain item is discounted once the number ordered exceeds a certain value. Tax brackets are another real-world example of piecewise functions. For example, consider a simple tax system in which incomes up to $10,000 are taxed at 10%, and any additional income is taxed at 20%. The tax on a total income S would be $0.1S$ if $S \leq \$10,000$ and $\$1000 + 0.2(S - \$10,000)$ if $S > \$10,000$.

---

piecewise function

A **piecewise function** is a function in which more than one formula is used to define the output. Each formula has its own domain, and the domain of the function is the union of all these smaller domains. We notate this idea like this:

$$f(x) = \begin{cases} \text{formula 1} & \text{if } x \text{ is in domain 1} \\ \text{formula 2} & \text{if } x \text{ is in domain 2} \\ \text{formula 3} & \text{if } x \text{ is in domain 3} \end{cases}$$

In piecewise notation, the absolute value function is

$$|x| = \begin{cases} x & \text{if } x \geq 0 \\ -x & \text{if } x < 0 \end{cases}$$

How To...

Given a piecewise function, write the formula and identify the domain for each interval.

1. Identify the intervals for which different rules apply.
2. Determine formulas that describe how to calculate an output from an input in each interval.
3. Use braces and if-statements to write the function.

Example 11 **Writing a Piecewise Function**

A museum charges \$5 per person for a guided tour with a group of 1 to 9 people or a fixed \$50 fee for a group of 10 or more people. Write a function relating the number of people, *n*, to the cost, *C*.

Solution   Two different formulas will be needed. For *n*-values under 10, $C = 5n$. For values of *n* that are 10 or greater, $C = 50$.

$$C(n) = \begin{cases} 5n & \text{if } 0 < n < 10 \\ 50 & \text{if } n \geq 10 \end{cases}$$

Analysis The function is represented in **Figure 23**. The graph is a diagonal line from $n = 0$ to $n = 10$ and a constant after that. In this example, the two formulas agree at the meeting point where $n = 10$, but not all piecewise functions have this property.

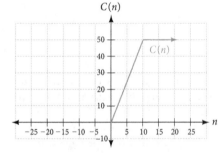

Figure 23

Example 12 **Working with a Piecewise Function**

A cell phone company uses the function below to determine the cost, *C*, in dollars for *g* gigabytes of data transfer.

$$C(g) = \begin{cases} 25 & \text{if } 0 < g < 2 \\ 25 + 10(g - 2) & \text{if } g \geq 2 \end{cases}$$

Find the cost of using 1.5 gigabytes of data and the cost of using 4 gigabytes of data.

Solution   To find the cost of using 1.5 gigabytes of data, $C(1.5)$, we first look to see which part of the domain our input falls in. Because 1.5 is less than 2, we use the first formula.

$$C(1.5) = \$25$$

To find the cost of using 4 gigabytes of data, $C(4)$, we see that our input of 4 is greater than 2, so we use the second formula.

$$C(4) = 25 + 10(4 - 2) = \$45$$

Analysis The function is represented in **Figure 24**. We can see where the function changes from a constant to a shifted and stretched identity at $g = 2$. We plot the graphs for the different formulas on a common set of axes, making sure each formula is applied on its proper domain.

Figure 24

How To...

Given a piecewise function, sketch a graph.

1. Indicate on the x-axis the boundaries defined by the intervals on each piece of the domain.
2. For each piece of the domain, graph on that interval using the corresponding equation pertaining to that piece. Do not graph two functions over one interval because it would violate the criteria of a function.

Example 13 **Graphing a Piecewise Function**

Sketch a graph of the function.

$$f(x) = \begin{cases} x^2 & \text{if} \quad x \leq 1 \\ 3 & \text{if} \ 1 < x \leq 2 \\ x & \text{if} \quad x > 2 \end{cases}$$

Solution Each of the component functions is from our library of toolkit functions, so we know their shapes. We can imagine graphing each function and then limiting the graph to the indicated domain. At the endpoints of the domain, we draw open circles to indicate where the endpoint is not included because of a less-than or greater-than inequality; we draw a closed circle where the endpoint is included because of a less-than-or-equal-to or greater-than-or-equal-to inequality.

Figure 25 shows the three components of the piecewise function graphed on separate coordinate systems.

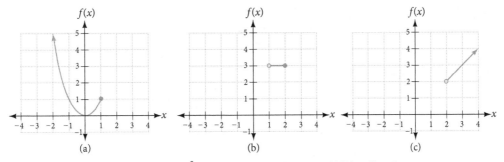

Figure 25 (a) $f(x) = x^2$ if $x \leq 1$; (b) $f(x) = 3$ if $1 < x \leq 2$; (c) $f(x) = x$ if $x > 2$

Now that we have sketched each piece individually, we combine them in the same coordinate plane. See **Figure 26**.

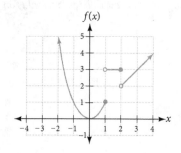

Figure 26

Analysis Note that the graph does pass the vertical line test even at $x = 1$ and $x = 2$ because the points $(1, 3)$ and $(2, 2)$ are not part of the graph of the function, though $(1, 1)$ and $(2, 3)$ are.

Try It #8

Graph the following piecewise function.

$$f(x) = \begin{cases} x^3 & \text{if} \quad x < -1 \\ -2 & \text{if} \ -1 < x < 4 \\ \sqrt{x} & \text{if} \quad x > 4 \end{cases}$$

Q & A...

Can more than one formula from a piecewise function be applied to a value in the domain?

No. Each value corresponds to one equation in a piecewise formula.

Access these online resources for additional instruction and practice with domain and range.

- Domain and Range of Square Root Functions (http://openstaxcollege.org/l/domainsqroot)
- Determining Domain and Range (http://openstaxcollege.org/l/determinedomain)
- Find Domain and Range Given the Graph (http://openstaxcollege.org/l/drgraph)
- Find Domain and Range Given a Table (http://openstaxcollege.org/l/drtable)
- Find Domain and Range Given Points on a Coordinate Plane (http://openstaxcollege.org/l/drcoordinate)

3.2 SECTION EXERCISES

VERBAL

1. Why does the domain differ for different functions?

2. How do we determine the domain of a function defined by an equation?

3. Explain why the domain of $f(x) = \sqrt[3]{x}$ is different from the domain of $f(x) = \sqrt{x}$.

4. When describing sets of numbers using interval notation, when do you use a parenthesis and when do you use a bracket?

5. How do you graph a piecewise function?

ALGEBRAIC

For the following exercises, find the domain of each function using interval notation.

6. $f(x) = -2x(x-1)(x-2)$

7. $f(x) = 5 - 2x^2$

8. $f(x) = 3\sqrt{x-2}$

9. $f(x) = 3 - \sqrt{6-2x}$

10. $f(x) = \sqrt{4-3x}$

11. $f(x) = \sqrt{x^2+4}$

12. $f(x) = \sqrt[3]{1-2x}$

13. $f(x) = \sqrt[3]{x-1}$

14. $f(x) = \dfrac{9}{x-6}$

15. $f(x) = \dfrac{3x+1}{4x+2}$

16. $f(x) = \dfrac{\sqrt{x+4}}{x-4}$

17. $f(x) = \dfrac{x-3}{x^2+9x-22}$

18. $f(x) = \dfrac{1}{x^2-x-6}$

19. $f(x) = \dfrac{2x^3-250}{x^2-2x-15}$

20. $f(x) = \dfrac{5}{\sqrt{x-3}}$

21. $f(x) = \dfrac{2x+1}{\sqrt{5-x}}$

22. $f(x) = \dfrac{\sqrt{x-4}}{\sqrt{x-6}}$

23. $f(x) = \dfrac{\sqrt{x-6}}{\sqrt{x-4}}$

24. $f(x) = \dfrac{x}{x}$

25. $f(x) = \dfrac{x^2-9x}{x^2-81}$

26. Find the domain of the function $f(x) = \sqrt{2x^3 - 50x}$ by:

    a. using algebra.

    b. graphing the function in the radicand and determining intervals on the x-axis for which the radicand is nonnegative.

GRAPHICAL

For the following exercises, write the domain and range of each function using interval notation.

27.

28.

29.

30.

31.

32.

33.

34.

35.

36.

37.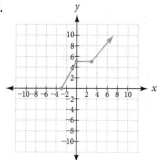

For the following exercises, sketch a graph of the piecewise function. Write the domain in interval notation.

38. $f(x) = \begin{cases} x + 1 & \text{if } x < -2 \\ -2x - 3 & \text{if } x \geq -2 \end{cases}$

39. $f(x) = \begin{cases} 2x - 1 & \text{if } x < 1 \\ 1 + x & \text{if } x \geq 1 \end{cases}$

40. $f(x) = \begin{cases} x + 1 & \text{if } x < 0 \\ x - 1 & \text{if } x > 0 \end{cases}$

41. $f(x) = \begin{cases} 3 & \text{if } x < 0 \\ \sqrt{x} & \text{if } x \geq 0 \end{cases}$

42. $f(x) = \begin{cases} x^2 & \text{if } x < 0 \\ 1 - x & \text{if } x > 0 \end{cases}$

43. $f(x) = \begin{cases} x^2 & \text{if } x < 0 \\ x + 2 & \text{if } x \geq 0 \end{cases}$

44. $f(x) = \begin{cases} x + 1 & \text{if } x < 1 \\ x^3 & \text{if } x \geq 1 \end{cases}$

45. $f(x) = \begin{cases} |x| & \text{if } x < 2 \\ 1 & \text{if } x \geq 2 \end{cases}$

NUMERIC

For the following exercises, given each function f, evaluate $f(-3)$, $f(-2)$, $f(-1)$, and $f(0)$.

46. $f(x) = \begin{cases} x+1 & \text{if } x < -2 \\ -2x - 3 & \text{if } x \geq -2 \end{cases}$

47. $f(x) = \begin{cases} 1 & \text{if } x \leq -3 \\ 0 & \text{if } x > -3 \end{cases}$

48. $f(x) = \begin{cases} -2x^2 + 3 & \text{if } x \leq -1 \\ 5x - 7 & \text{if } x > -1 \end{cases}$

For the following exercises, given each function f, evaluate $f(-1)$, $f(0)$, $f(2)$, and $f(4)$.

49. $f(x) = \begin{cases} 7x + 3 & \text{if } x < 0 \\ 7x + 6 & \text{if } x \geq 0 \end{cases}$

50. $f(x) = \begin{cases} x^2 - 2 & \text{if } x < 2 \\ 4 + |x - 5| & \text{if } x \geq 2 \end{cases}$

51. $f(x) = \begin{cases} 5x & \text{if } x < 0 \\ 3 & \text{if } 0 \leq x \leq 3 \\ x^2 & \text{if } x > 3 \end{cases}$

For the following exercises, write the domain for the piecewise function in interval notation.

52. $f(x) = \begin{cases} x+1 & \text{if } x < -2 \\ -2x - 3 & \text{if } x \geq -2 \end{cases}$

53. $f(x) = \begin{cases} x^2 - 2 & \text{if } x < 1 \\ -x^2 + 2 & \text{if } x > 1 \end{cases}$

54. $f(x) = \begin{cases} 2x - 3 & \text{if } x < 0 \\ -3x^2 & \text{if } x \geq 2 \end{cases}$

TECHNOLOGY

55. Graph $y = \dfrac{1}{x^2}$ on the viewing window $[-0.5, -0.1]$ and $[0.1, 0.5]$. Determine the corresponding range for the viewing window. Show the graphs.

56. Graph $y = \dfrac{1}{x}$ on the viewing window $[-0.5, -0.1]$ and $[0.1, 0.5]$. Determine the corresponding range for the viewing window. Show the graphs.

EXTENSION

57. Suppose the range of a function f is $[-5, 8]$. What is the range of $|f(x)|$?

58. Create a function in which the range is all nonnegative real numbers.

59. Create a function in which the domain is $x > 2$.

REAL-WORLD APPLICATIONS

60. The height h of a projectile is a function of the time t it is in the air. The height in feet for t seconds is given by the function $h(t) = -16t^2 + 96t$. What is the domain of the function? What does the domain mean in the context of the problem?

61. The cost in dollars of making x items is given by the function $C(x) = 10x + 500$.

   a. The fixed cost is determined when zero items are produced. Find the fixed cost for this item.

   b. What is the cost of making 25 items?

   c. Suppose the maximum cost allowed is $1500. What are the domain and range of the cost function, $C(x)$?

LEARNING OBJECTIVES

In this section, you will:

- Find the average rate of change of a function.
- Use a graph to determine where a function is increasing, decreasing, or constant.
- Use a graph to locate local maxima and local minima.
- Use a graph to locate the absolute maximum and absolute minimum.

3.3   RATES OF CHANGE AND BEHAVIOR OF GRAPHS

Gasoline costs have experienced some wild fluctuations over the last several decades. **Table 1**[5] lists the average cost, in dollars, of a gallon of gasoline for the years 2005–2012. The cost of gasoline can be considered as a function of year.

y	2005	2006	2007	2008	2009	2010	2011	2012
$C(y)$	2.31	2.62	2.84	3.30	2.41	2.84	3.58	3.68

Table 1

If we were interested only in how the gasoline prices changed between 2005 and 2012, we could compute that the cost per gallon had increased from $2.31 to $3.68, an increase of $1.37. While this is interesting, it might be more useful to look at how much the price changed *per year*. In this section, we will investigate changes such as these.

Finding the Average Rate of Change of a Function

The price change per year is a **rate of change** because it describes how an output quantity changes relative to the change in the input quantity. We can see that the price of gasoline in **Table 1** did not change by the same amount each year, so the rate of change was not constant. If we use only the beginning and ending data, we would be finding the **average rate of change** over the specified period of time. To find the average rate of change, we divide the change in the output value by the change in the input value.

$$\text{Average rate of change} = \frac{\text{Change in output}}{\text{Change in input}}$$

$$= \frac{\Delta y}{\Delta x}$$

$$= \frac{y_2 - y_1}{x_2 - x_1}$$

$$= \frac{f(x_2) - f(x_1)}{x_2 - x_1}$$

The Greek letter Δ (delta) signifies the change in a quantity; we read the ratio as "delta-y over delta-x" or "the change in y divided by the change in x." Occasionally we write Δf instead of Δy, which still represents the change in the function's output value resulting from a change to its input value. It does not mean we are changing the function into some other function.

In our example, the gasoline price increased by $1.37 from 2005 to 2012. Over 7 years, the average rate of change was

$$\frac{\Delta y}{\Delta x} = \frac{\$1.37}{7 \text{ years}} \approx 0.196 \text{ dollars per year}$$

On average, the price of gas increased by about 19.6¢ each year.

Other examples of rates of change include:

- A population of rats increasing by 40 rats per week
- A car traveling 68 miles per hour (distance traveled changes by 68 miles each hour as time passes)
- A car driving 27 miles per gallon (distance traveled changes by 27 miles for each gallon)
- The current through an electrical circuit increasing by 0.125 amperes for every volt of increased voltage
- The amount of money in a college account decreasing by $4,000 per quarter

5 http://www.eia.gov/totalenergy/data/annual/showtext.cfm?t=ptb0524. Accessed 3/5/2014.

---

rate of change

A rate of change describes how an output quantity changes relative to the change in the input quantity. The units on a rate of change are "output units per input units."

The average rate of change between two input values is the total change of the function values (output values) divided by the change in the input values.

$$\frac{\Delta y}{\Delta x} = \frac{f(x_2) - f(x_1)}{x_2 - x_1}$$

How To...

Given the value of a function at different points, calculate the average rate of change of a function for the interval between two values x_1 and x_2.

1. Calculate the difference $y_2 - y_1 = \Delta y$.

2. Calculate the difference $x_2 - x_1 = \Delta x$.

3. Find the ratio $\dfrac{\Delta y}{\Delta x}$.

Example 1   **Computing an Average Rate of Change**

Using the data in **Table 1**, find the average rate of change of the price of gasoline between 2007 and 2009.

Solution   In 2007, the price of gasoline was $2.84. In 2009, the cost was $2.41. The average rate of change is

$$\frac{\Delta y}{\Delta x} = \frac{y_2 - y_1}{x_2 - x_1}$$

$$= \frac{\$2.41 - \$2.84}{2009 - 2007}$$

$$= \frac{-\$0.43}{2 \text{ years}}$$

$$= -\$0.22 \text{ per year}$$

Analysis   *Note that a decrease is expressed by a negative change or "negative increase." A rate of change is negative when the output decreases as the input increases or when the output increases as the input decreases.*

Try It #1

Using the data in **Table 1** at the beginning of this section, find the average rate of change between 2005 and 2010.

Example 2   **Computing Average Rate of Change from a Graph**

Given the function $g(t)$ shown in **Figure 1**, find the average rate of change on the interval $[-1, 2]$.

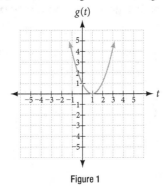

Figure 1

Solution   At $t = -1$, **Figure 2** shows $g(-1) = 4$. At $t = 2$, the graph shows $g(2) = 1$.

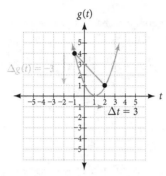

Figure 2

The horizontal change $\Delta t = 3$ is shown by the red arrow, and the vertical change $\Delta g(t) = -3$ is shown by the turquoise arrow. The average rate of change is shown by the slope of the red line segment. The output changes by -3 while the input changes by 3, giving an average rate of change of

$$\frac{1 - 4}{2 - (-1)} = \frac{-3}{3} = -1$$

Analysis   *Note that the order we choose is very important. If, for example, we use $\frac{y_2 - y_1}{x_1 - x_2}$, we will not get the correct answer. Decide which point will be 1 and which point will be 2, and keep the coordinates fixed as (x_1, y_1) and (x_2, y_2).*

Example 3   **Computing Average Rate of Change from a Table**

After picking up a friend who lives 10 miles away and leaving on a trip, Anna records her distance from home over time. The values are shown in **Table 2**. Find her average speed over the first 6 hours.

t (hours)	0	1	2	3	4	5	6	7
$D(t)$ (miles)	10	55	90	153	214	240	282	300

Table 2

Solution   Here, the average speed is the average rate of change. She traveled 282 miles in 6 hours.

$$\frac{292 - 10}{6 - 0} = \frac{282}{6}$$
$$= 47$$

The average speed is 47 miles per hour.

Analysis   *Because the speed is not constant, the average speed depends on the interval chosen. For the interval [2, 3], the average speed is 63 miles per hour.*

Example 4   **Computing Average Rate of Change for a Function Expressed as a Formula**

Compute the average rate of change of $f(x) = x^2 - \frac{1}{x}$ on the interval [2, 4].

Solution   We can start by computing the function values at each endpoint of the interval.

$$f(2) = 2^2 - \frac{1}{2} \qquad\qquad f(4) = 4^2 - \frac{1}{4}$$

$$= 4 - \frac{1}{2} \qquad\qquad\quad = 16 - \frac{1}{4}$$

$$= \frac{7}{2} \qquad\qquad\qquad\quad = \frac{63}{4}$$

Now we compute the average rate of change.

$$\text{Average rate of change} = \frac{f(4) - f(2)}{4 - 2}$$

$$= \frac{\frac{63}{4} - \frac{7}{2}}{4 - 2}$$

$$= \frac{\frac{49}{4}}{2}$$

$$= \frac{49}{8}$$

Try It #2

Find the average rate of change of $f(x) = x - 2\sqrt{x}$ on the interval $[1, 9]$.

Example 5   **Finding the Average Rate of Change of a Force**

The electrostatic force F, measured in newtons, between two charged particles can be related to the distance between the particles d, in centimeters, by the formula $F(d) = \frac{2}{d^2}$. Find the average rate of change of force if the distance between the particles is increased from 2 cm to 6 cm.

Solution   We are computing the average rate of change of $F(d) = \frac{2}{d^2}$ on the interval $[2, 6]$.

$$\text{Average rate of change} = \frac{F(6) - F(2)}{6 - 2}$$

$$= \frac{\frac{2}{6^2} - \frac{2}{2^2}}{6 - 2} \qquad \text{Simplify.}$$

$$= \frac{\frac{2}{36} - \frac{2}{4}}{4}$$

$$= \frac{-\frac{16}{36}}{4} \qquad \text{Combine numerator terms.}$$

$$= -\frac{1}{9} \qquad \text{Simplify.}$$

The average rate of change is $-\frac{1}{9}$ newton per centimeter.

Example 6   **Finding an Average Rate of Change as an Expression**

Find the average rate of change of $g(t) = t^2 + 3t + 1$ on the interval $[0, a]$. The answer will be an expression involving a in simplest form.

Solution   We use the average rate of change formula.

$$\text{Average rate of change} = \frac{g(a) - g(0)}{a - 0} \qquad \text{Evaluate.}$$

$$= \frac{(a^2 + 3a + 1) - (0^2 + 3(0) + 1)}{a - 0} \qquad \text{Simplify.}$$

$$= \frac{a^2 + 3a + 1 - 1}{a} \qquad \text{Simplify and factor.}$$

$$= \frac{a(a + 3)}{a} \qquad \text{Divide by the common factor } a.$$

$$= a + 3$$

This result tells us the average rate of change in terms of a between $t = 0$ and any other point $t = a$. For example, on the interval $[0, 5]$, the average rate of change would be $5 + 3 = 8$.

Try It #3

Find the average rate of change of $f(x) = x^2 + 2x - 8$ on the interval $[5, a]$ in simplest forms in terms of a.

Using a Graph to Determine Where a Function is Increasing, Decreasing, or Constant

As part of exploring how functions change, we can identify intervals over which the function is changing in specific ways. We say that a function is increasing on an interval if the function values increase as the input values increase within that interval. Similarly, a function is decreasing on an interval if the function values decrease as the input values increase over that interval. The average rate of change of an increasing function is positive, and the average rate of change of a decreasing function is negative. **Figure 3** shows examples of increasing and decreasing intervals on a function.

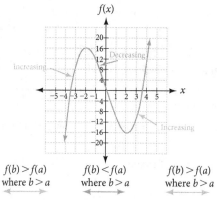

$f(b) > f(a)$ $f(b) < f(a)$ $f(b) > f(a)$
where $b > a$ where $b > a$ where $b > a$

Figure 3 The function $f(x) = x^3 - 12x$ is increasing on $(-\infty, -2) \cup (2, \infty)$ and is decreasing on $(-2, 2)$.

While some functions are increasing (or decreasing) over their entire domain, many others are not. A value of the input where a function changes from increasing to decreasing (as we go from left to right, that is, as the input variable increases) is called a **local maximum**. If a function has more than one, we say it has local maxima. Similarly, a value of the input where a function changes from decreasing to increasing as the input variable increases is called a **local minimum**. The plural form is "local minima." Together, local maxima and minima are called **local extrema**, or local extreme values, of the function. (The singular form is "extremum.") Often, the term *local* is replaced by the term *relative*. In this text, we will use the term *local*.

Clearly, a function is neither increasing nor decreasing on an interval where it is constant. A function is also neither increasing nor decreasing at extrema. Note that we have to speak of *local* extrema, because any given local extremum as defined here is not necessarily the highest maximum or lowest minimum in the function's entire domain.

For the function whose graph is shown in **Figure 4**, the local maximum is 16, and it occurs at $x = -2$. The local minimum is -16 and it occurs at $x = 2$.

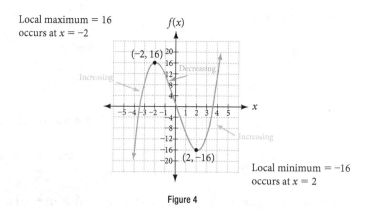

Figure 4

To locate the local maxima and minima from a graph, we need to observe the graph to determine where the graph attains its highest and lowest points, respectively, within an open interval. Like the summit of a roller coaster, the graph of a function is higher at a local maximum than at nearby points on both sides. The graph will also be lower at a local minimum than at neighboring points. **Figure 5** illustrates these ideas for a local maximum.

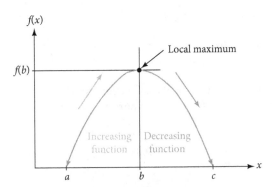

Figure 5 Definition of a local maximum

These observations lead us to a formal definition of local extrema.

---

local minima and local maxima

A function f is an **increasing function** on an open interval if $f(b) > f(a)$ for any two input values a and b in the given interval where $b > a$.

A function f is a **decreasing function** on an open interval if $f(b) < f(a)$ for any two input values a and b in the given interval where $b > a$.

A function f has a local maximum at $x = b$ if there exists an interval (a, c) with $a < b < c$ such that, for any x in the interval $(a, c), f(x) \leq f(b)$. Likewise, f has a local minimum at $x = b$ if there exists an interval (a, c) with $a < b < c$ such that, for any x in the interval $(a, c), f(x) \geq f(b)$.

Example 7   **Finding Increasing and Decreasing Intervals on a Graph**

Given the function $p(t)$ in **Figure 6**, identify the intervals on which the function appears to be increasing.

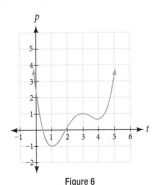

Figure 6

Solution   We see that the function is not constant on any interval. The function is increasing where it slants upward as we move to the right and decreasing where it slants downward as we move to the right. The function appears to be increasing from $t = 1$ to $t = 3$ and from $t = 4$ on.

In interval notation, we would say the function appears to be increasing on the interval $(1, 3)$ and the interval $(4, \infty)$.

Analysis Notice in this example that we used open intervals (intervals that do not include the endpoints), because the function is neither increasing nor decreasing at $t = 1$, $t = 3$, and $t = 4$. These points are the local extrema (two minima and a maximum).

Example 8   **Finding Local Extrema from a Graph**

Graph the function $f(x) = \dfrac{2}{x} + \dfrac{x}{3}$. Then use the graph to estimate the local extrema of the function and to determine the intervals on which the function is increasing.

Solution   Using technology, we find that the graph of the function looks like that in **Figure 7**. It appears there is a low point, or local minimum, between $x = 2$ and $x = 3$, and a mirror-image high point, or local maximum, somewhere between $x = -3$ and $x = -2$.

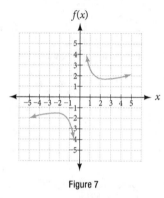

Figure 7

Analysis Most graphing calculators and graphing utilities can estimate the location of maxima and minima. **Figure 8** provides screen images from two different technologies, showing the estimate for the local maximum and minimum.

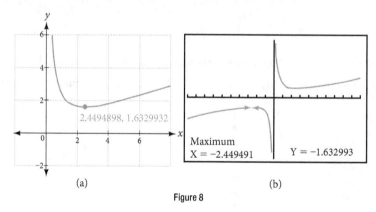

(a)   (b)

Figure 8

Based on these estimates, the function is increasing on the interval $(-\infty, -2.449)$ and $(2.449, \infty)$. Notice that, while we expect the extrema to be symmetric, the two different technologies agree only up to four decimals due to the differing approximation algorithms used by each. (The exact location of the extrema is at $\pm\sqrt{6}$, but determining this requires calculus.)

Try It #4

Graph the function $f(x) = x^3 - 6x^2 - 15x + 20$ to estimate the local extrema of the function. Use these to determine the intervals on which the function is increasing and decreasing.

Example 9   **Finding Local Maxima and Minima from a Graph**

For the function f whose graph is shown in **Figure 9**, find all local maxima and minima.

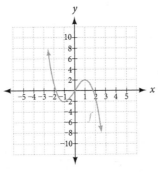

Figure 9

Solution   Observe the graph of f. The graph attains a local maximum at $x = 1$ because it is the highest point in an open interval around $x = 1$. The local maximum is the y-coordinate at $x = 1$, which is 2.

The graph attains a local minimum at $x = -1$ because it is the lowest point in an open interval around $x = -1$. The local minimum is the y-coordinate at $x = -1$, which is -2.

Analyzing the Toolkit Functions for Increasing or Decreasing Intervals

We will now return to our toolkit functions and discuss their graphical behavior in **Figure 10**, **Figure 11**, and **Figure 12**.

Function	Increasing/Decreasing	Example
Constant Function $f(x) = c$	Neither increasing nor decreasing	
Identity Function $f(x) = x$	Increasing	
Quadratic Function $f(x) = x^2$	Increasing on $(0, \infty)$ Decreasing on $(-\infty, 0)$ Minimum at $x = 0$	

Figure 10

Function	Increasing/Decreasing	Example
Cubic Function $f(x) = x^3$	Increasing	
Reciprocal $f(x) = \dfrac{1}{x}$	Decreasing $(-\infty, 0) \cup (0, \infty)$	
Reciprocal Squared $f(x) = \dfrac{1}{x^2}$	Increasing on $(-\infty, 0)$ Decreasing on $(0, \infty)$	

Figure 11

Function	Increasing/Decreasing	Example
Cube Root $f(x) = \sqrt[3]{x}$	Increasing	
Square Root $f(x) = \sqrt{x}$	Increasing on $(0, \infty)$	
Absolute Value $f(x) = |x|$	Increasing on $(0, \infty)$ Decreasing on $(-\infty, 0)$	

Figure 12

Use A Graph to Locate the Absolute Maximum and Absolute Minimum

There is a difference between locating the highest and lowest points on a graph in a region around an open interval (locally) and locating the highest and lowest points on the graph for the entire domain. The y-coordinates (output) at the highest and lowest points are called the **absolute maximum** and **absolute minimum**, respectively.

To locate absolute maxima and minima from a graph, we need to observe the graph to determine where the graph attains it highest and lowest points on the domain of the function. See **Figure 13**.

Figure 13

Not every function has an absolute maximum or minimum value. The toolkit function $f(x) = x^3$ is one such function.

---

absolute maxima and minima

The **absolute maximum** of f at $x = c$ is $f(c)$ where $f(c) \geq f(x)$ for all x in the domain of f.

The **absolute minimum** of f at $x = d$ is $f(d)$ where $f(d) \leq f(x)$ for all x in the domain of f.

Example 10 **Finding Absolute Maxima and Minima from a Graph**

For the function f shown in **Figure 14**, find all absolute maxima and minima.

Figure 14

Solution   Observe the graph of f. The graph attains an absolute maximum in two locations, $x = -2$ and $x = 2$, because at these locations, the graph attains its highest point on the domain of the function. The absolute maximum is the y-coordinate at $x = -2$ and $x = 2$, which is 16.

The graph attains an absolute minimum at $x = 3$, because it is the lowest point on the domain of the function's graph. The absolute minimum is the y-coordinate at $x = 3$, which is -10.

Access this online resource for additional instruction and practice with rates of change.

- Average Rate of Change (http://openstaxcollege.org/l/aroc)

3.3   SECTION EXERCISES

VERBAL

1. Can the average rate of change of a function be constant?

2. If a function f is increasing on (a, b) and decreasing on (b, c), then what can be said about the local extremum of f on (a, c)?

3. How are the absolute maximum and minimum similar to and different from the local extrema?

4. How does the graph of the absolute value function compare to the graph of the quadratic function, $y = x^2$, in terms of increasing and decreasing intervals?

ALGEBRAIC

For the following exercises, find the average rate of change of each function on the interval specified for real numbers b or h in simplest form.

5. $f(x) = 4x^2 - 7$ on $[1, b]$

6. $g(x) = 2x^2 - 9$ on $[4, b]$

7. $p(x) = 3x + 4$ on $[2, 2 + h]$

8. $k(x) = 4x - 2$ on $[3, 3 + h]$

9. $f(x) = 2x^2 + 1$ on $[x, x + h]$

10. $g(x) = 3x^2 - 2$ on $[x, x + h]$

11. $a(t) = \dfrac{1}{t + 4}$ on $[9, 9 + h]$

12. $b(x) = \dfrac{1}{x + 3}$ on $[1, 1 + h]$

13. $j(x) = 3x^3$ on $[1, 1 + h]$

14. $r(t) = 4t^3$ on $[2, 2 + h]$

15. $\dfrac{f(x + h) - f(x)}{h}$ given $f(x) = 2x^2 - 3x$ on $[x, x + h]$

GRAPHICAL

For the following exercises, consider the graph of f shown in **Figure 15**.

16. Estimate the average rate of change from $x = 1$ to $x = 4$.

17. Estimate the average rate of change from $x = 2$ to $x = 5$.

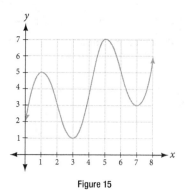

Figure 15

For the following exercises, use the graph of each function to estimate the intervals on which the function is increasing or decreasing.

18.

19.   20.

21.

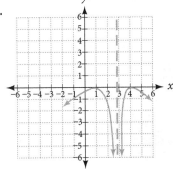

For the following exercises, consider the graph shown in **Figure 16**.

22. Estimate the intervals where the function is increasing or decreasing.

23. Estimate the point(s) at which the graph of *f* has a local maximum or a local minimum.

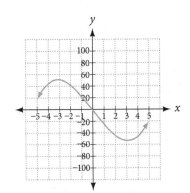

Figure 16

For the following exercises, consider the graph in **Figure 17.**

24. If the complete graph of the function is shown, estimate the intervals where the function is increasing or decreasing.

25. If the complete graph of the function is shown, estimate the absolute maximum and absolute minimum.

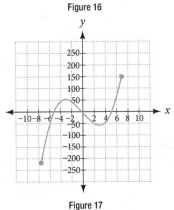

Figure 17

NUMERIC

26. Table 3 gives the annual sales (in millions of dollars) of a product from 1998 to 2006. What was the average rate of change of annual sales **(a)** between 2001 and 2002, and **(b)** between 2001 and 2004?

Year	1998	1999	2000	2001	2002	2003	2004	2005	2006
Sales (millions of dollars)	201	219	233	243	249	251	249	243	233

Table 3

27. Table 4 gives the population of a town (in thousands) from 2000 to 2008. What was the average rate of change of population **(a)** between 2002 and 2004, and **(b)** between 2002 and 2006?

Year	2000	2001	2002	2003	2004	2005	2006	2007	2008
Population (thousands)	87	84	83	80	77	76	78	81	85

Table 4

For the following exercises, find the average rate of change of each function on the interval specified.

28. $f(x) = x^2$ on $[1, 5]$

29. $h(x) = 5 - 2x^2$ on $[-2, 4]$

30. $q(x) = x^3$ on $[-4, 2]$

31. $g(x) = 3x^3 - 1$ on $[-3, 3]$

32. $y = \dfrac{1}{x}$ on $[1, 3]$

33. $p(t) = \dfrac{(t^2 - 4)(t + 1)}{t^2 + 3}$ on $[-3, 1]$

34. $k(t) = 6t^2 + \dfrac{4}{t^3}$ on $[-1, 3]$

TECHNOLOGY

For the following exercises, use a graphing utility to estimate the local extrema of each function and to estimate the intervals on which the function is increasing and decreasing.

35. $f(x) = x^4 - 4x^3 + 5$

36. $h(x) = x^5 + 5x^4 + 10x^3 + 10x^2 - 1$

37. $g(t) = t\sqrt{t + 3}$

38. $k(t) = 3t^{\frac{2}{3}} - t$

39. $m(x) = x^4 + 2x^3 - 12x^2 - 10x + 4$

40. $n(x) = x^4 - 8x^3 + 18x^2 - 6x + 2$

EXTENSION

41. The graph of the function f is shown in **Figure 18**.

Maximum
X = 1.3333324   Y = 5.1851852

Figure 18

Based on the calculator screen shot, the point (1.333, 5.185) is which of the following?

a. a relative (local) maximum of the function

b. the vertex of the function

c. the absolute maximum of the function

d. a zero of the function

42. Let $f(x) = \dfrac{1}{x}$. Find a number c such that the average rate of change of the function f on the interval $(1, c)$ is $-\dfrac{1}{4}$

43. Let $f(x) = \dfrac{1}{x}$. Find the number b such that the average rate of change of f on the interval $(2, b)$ is $-\dfrac{1}{10}$.

REAL-WORLD APPLICATIONS

44. At the start of a trip, the odometer on a car read 21,395. At the end of the trip, 13.5 hours later, the odometer read 22,125. Assume the scale on the odometer is in miles. What is the average speed the car traveled during this trip?

45. A driver of a car stopped at a gas station to fill up his gas tank. He looked at his watch, and the time read exactly 3:40 p.m. At this time, he started pumping gas into the tank. At exactly 3:44, the tank was full and he noticed that he had pumped 10.7 gallons. What is the average rate of flow of the gasoline into the gas tank?

46. Near the surface of the moon, the distance that an object falls is a function of time. It is given by $d(t) = 2.6667t^2$, where t is in seconds and $d(t)$ is in feet. If an object is dropped from a certain height, find the average velocity of the object from $t = 1$ to $t = 2$.

47. The graph in **Figure 19** illustrates the decay of a radioactive substance over t days.

Figure 19

Use the graph to estimate the average decay rate from $t = 5$ to $t = 15$.

In this section, you will:

- Combine functions using algebraic operations.
- Create a new function by composition of functions.
- Evaluate composite functions.
- Find the domain of a composite function.
- Decompose a composite function into its component functions.

3.4 COMPOSITION OF FUNCTIONS

Suppose we want to calculate how much it costs to heat a house on a particular day of the year. The cost to heat a house will depend on the average daily temperature, and in turn, the average daily temperature depends on the particular day of the year. Notice how we have just defined two relationships: The cost depends on the temperature, and the temperature depends on the day.

Using descriptive variables, we can notate these two functions. The function $C(T)$ gives the cost C of heating a house for a given average daily temperature in T degrees Celsius. The function $T(d)$ gives the average daily temperature on day d of the year. For any given day, Cost $= C(T(d))$ means that the cost depends on the temperature, which in turns depends on the day of the year. Thus, we can evaluate the cost function at the temperature $T(d)$. For example, we could evaluate $T(5)$ to determine the average daily temperature on the 5th day of the year. Then, we could evaluate the cost function at that temperature. We would write $C(T(5))$.

Cost for the temperature

$C(T(5))$

Temperature on day 5

By combining these two relationships into one function, we have performed function composition, which is the focus of this section.

Combining Functions Using Algebraic Operations

Function composition is only one way to combine existing functions. Another way is to carry out the usual algebraic operations on functions, such as addition, subtraction, multiplication and division. We do this by performing the operations with the function outputs, defining the result as the output of our new function.

Suppose we need to add two columns of numbers that represent a husband and wife's separate annual incomes over a period of years, with the result being their total household income. We want to do this for every year, adding only that year's incomes and then collecting all the data in a new column. If $w(y)$ is the wife's income and $h(y)$ is the husband's income in year y, and we want T to represent the total income, then we can define a new function.

$$T(y) = h(y) + w(y)$$

If this holds true for every year, then we can focus on the relation between the functions without reference to a year and write

$$T = h + w$$

Just as for this sum of two functions, we can define difference, product, and ratio functions for any pair of functions that have the same kinds of inputs (not necessarily numbers) and also the same kinds of outputs (which do have to be numbers so that the usual operations of algebra can apply to them, and which also must have the same units or no units when we add and subtract). In this way, we can think of adding, subtracting, multiplying, and dividing functions.

For two functions $f(x)$ and $g(x)$ with real number outputs, we define new functions $f + g$, $f - g$, fg, and $\frac{f}{g}$ by the relations

$$(f + g)(x) = f(x) + g(x)$$

$$(f - g)(x) = f(x) - g(x)$$

$$(fg)(x) = f(x)g(x)$$

$$\left(\frac{f}{g}\right)(x) = \frac{f(x)}{g(x)} \qquad \text{where } g(x) \neq 0$$

Example 1   **Performing Algebraic Operations on Functions**

Find and simplify the functions $(g - f)(x)$ and $\left(\frac{g}{f}\right)(x)$, given $f(x) = x - 1$ and $g(x) = x^2 - 1$. Are they the same function?

Solution   Begin by writing the general form, and then substitute the given functions.

$$(g - f)(x) = g(x) - f(x)$$

$$(g - f)(x) = x^2 - 1 - (x - 1)$$

$$(g - f)(x) = x^2 - x$$

$$(g - f)(x) = x(x - 1)$$

$$\left(\frac{g}{f}\right)(x) = \frac{g(x)}{f(x)}$$

$$\left(\frac{g}{f}\right)(x) = \frac{x^2 - 1}{x - 1} \qquad \text{where } x \neq 1$$

$$\left(\frac{g}{f}\right)(x) = \frac{(x + 1)(x - 1)}{x - 1} \qquad \text{where } x \neq 1$$

$$\left(\frac{g}{f}\right)(x) = x + 1 \qquad \text{where } x \neq 1$$

No, the functions are not the same.

Note: For $\left(\frac{g}{f}\right)(x)$, the condition $x \neq 1$ is necessary because when $x = 1$, the denominator is equal to 0, which makes the function undefined.

Try It #1

Find and simplify the functions $(fg)(x)$ and $(f - g)(x)$.

$$f(x) = x - 1 \text{ and } g(x) = x^2 - 1$$

Are they the same function?

Create a Function by Composition of Functions

Performing algebraic operations on functions combines them into a new function, but we can also create functions by composing functions. When we wanted to compute a heating cost from a day of the year, we created a new function that takes a day as input and yields a cost as output. The process of combining functions so that the output of one function becomes the input of another is known as a composition of functions. The resulting function is known as a **composite function**. We represent this combination by the following notation:

$$(f \circ g)(x) = f(g(x))$$

We read the left-hand side as "f composed with g at x," and the right-hand side as "f of g of x." The two sides of the equation have the same mathematical meaning and are equal. The open circle symbol ∘ is called the composition operator. We use this operator mainly when we wish to emphasize the relationship between the functions themselves without referring to any particular input value. Composition is a binary operation that takes two functions and forms a new function, much as addition or multiplication takes two numbers and gives a new number. However, it is important not to confuse function composition with multiplication because, as we learned above, in most cases $f(g(x)) \neq f(x)g(x)$.

It is also important to understand the order of operations in evaluating a composite function. We follow the usual convention with parentheses by starting with the innermost parentheses first, and then working to the outside. In the equation above, the function g takes the input x first and yields an output $g(x)$. Then the function f takes $g(x)$ as an input and yields an output $f(g(x))$.

$$g(x), \text{ the output of } g$$
$$\text{is the input of } f$$
$$\downarrow$$
$$(f \circ g)(x) = f(\overline{g(x)})$$
$$\uparrow$$
$$x \text{ is the input of } g$$

In general, $f \circ g$ and $g \circ f$ are different functions. In other words, in many cases $f(g(x)) \neq g(f(x))$ for all x. We will also see that sometimes two functions can be composed only in one specific order.

For example, if $f(x) = x^2$ and $g(x) = x + 2$, then

$$f(g(x)) = f(x + 2)$$
$$= (x + 2)^2$$
$$= x^2 + 4x + 4$$

but

$$g(f(x)) = g(x^2)$$
$$= x^2 + 2$$

These expressions are not equal for all values of x, so the two functions are not equal. It is irrelevant that the expressions happen to be equal for the single input value $x = -\dfrac{1}{2}$.

Note that the range of the inside function (the first function to be evaluated) needs to be within the domain of the outside function. Less formally, the composition has to make sense in terms of inputs and outputs.

---

composition of functions

When the output of one function is used as the input of another, we call the entire operation a composition of functions. For any input x and functions f and g, this action defines a **composite function**, which we write as $f \circ g$ such that

$$(f \circ g)(x) = f(g(x))$$

The domain of the composite function $f \circ g$ is all x such that x is in the domain of g and $g(x)$ is in the domain of f. It is important to realize that the product of functions fg is not the same as the function composition $f(g(x))$, because, in general, $f(x)g(x) \neq f(g(x))$.

Example 2   **Determining whether Composition of Functions is Commutative**

Using the functions provided, find $f(g(x))$ and $g(f(x))$. Determine whether the composition of the functions is commutative.

$$f(x) = 2x + 1 \qquad g(x) = 3 - x$$

Solution   Let's begin by substituting $g(x)$ into $f(x)$.

$$f(g(x)) = 2(3 - x) + 1$$
$$= 6 - 2x + 1$$
$$= 7 - 2x$$

Now we can substitute $f(x)$ into $g(x)$.

$$g(f(x)) = 3 - (2x + 1)$$
$$= 3 - 2x - 1$$
$$= -2x + 2$$

We find that $g(f(x)) \neq f(g(x))$, so the operation of function composition is not commutative.

Example 3   Interpreting Composite Functions

The function $c(s)$ gives the number of calories burned completing s sit-ups, and $s(t)$ gives the number of sit-ups a person can complete in t minutes. Interpret $c(s(3))$.

Solution   The inside expression in the composition is $s(3)$. Because the input to the s-function is time, $t = 3$ represents 3 minutes, and $s(3)$ is the number of sit-ups completed in 3 minutes.

Using $s(3)$ as the input to the function $c(s)$ gives us the number of calories burned during the number of sit-ups that can be completed in 3 minutes, or simply the number of calories burned in 3 minutes (by doing sit-ups).

Example 4   Investigating the Order of Function Composition

Suppose $f(x)$ gives miles that can be driven in x hours and $g(y)$ gives the gallons of gas used in driving y miles. Which of these expressions is meaningful: $f(g(y))$ or $g(f(x))$?

Solution   The function $y = f(x)$ is a function whose output is the number of miles driven corresponding to the number of hours driven.

$$\text{number of miles} = f(\text{number of hours})$$

The function $g(y)$ is a function whose output is the number of gallons used corresponding to the number of miles driven. This means:

$$\text{number of gallons} = g(\text{number of miles})$$

The expression $g(y)$ takes miles as the input and a number of gallons as the output. The function $f(x)$ requires a number of hours as the input. Trying to input a number of gallons does not make sense. The expression $f(g(y))$ is meaningless.

The expression $f(x)$ takes hours as input and a number of miles driven as the output. The function $g(y)$ requires a number of miles as the input. Using $f(x)$ (miles driven) as an input value for $g(y)$, where gallons of gas depends on miles driven, does make sense. The expression $g(f(x))$ makes sense, and will yield the number of gallons of gas used, g, driving a certain number of miles, $f(x)$, in x hours.

Q & A...

Are there any situations where $f(g(y))$ and $g(f(x))$ would both be meaningful or useful expressions?

Yes. For many pure mathematical functions, both compositions make sense, even though they usually produce different new functions. In real-world problems, functions whose inputs and outputs have the same units also may give compositions that are meaningful in either order.

Try It #2

The gravitational force on a planet a distance r from the sun is given by the function $G(r)$. The acceleration of a planet subjected to any force F is given by the function $a(F)$. Form a meaningful composition of these two functions, and explain what it means.

Evaluating Composite Functions

Once we compose a new function from two existing functions, we need to be able to evaluate it for any input in its domain. We will do this with specific numerical inputs for functions expressed as tables, graphs, and formulas and with variables as inputs to functions expressed as formulas. In each case, we evaluate the inner function using the starting input and then use the inner function's output as the input for the outer function.

Evaluating Composite Functions Using Tables

When working with functions given as tables, we read input and output values from the table entries and always work from the inside to the outside. We evaluate the inside function first and then use the output of the inside function as the input to the outside function.

Example 5   **Using a Table to Evaluate a Composite Function**

Using **Table 1**, evaluate $f(g(3))$ and $g(f(3))$.

x	$f(x)$	$g(x)$
1	6	3
2	8	5
3	3	2
4	1	7

Table 1

Solution   To evaluate $f(g(3))$, we start from the inside with the input value 3. We then evaluate the inside expression $g(3)$ using the table that defines the function g: $g(3) = 2$. We can then use that result as the input to the function f, so $g(3)$ is replaced by 2 and we get $f(2)$. Then, using the table that defines the function f, we find that $f(2) = 8$.

$$g(3) = 2$$

$$f(g(3)) = f(2) = 8$$

To evaluate $g(f(3))$, we first evaluate the inside expression $f(3)$ using the first table: $f(3) = 3$. Then, using the table for g, we can evaluate

$$g(f(3)) = g(3) = 2$$

Table 2 shows the composite functions $f \circ g$ and $g \circ f$ as tables.

x	$g(x)$	$f(g(x))$	$f(x)$	$g(f(x))$
3	2	8	3	2

Table 2

Try It #3

Using **Table 1**, evaluate $f(g(1))$ and $g(f(4))$.

Evaluating Composite Functions Using Graphs

When we are given individual functions as graphs, the procedure for evaluating composite functions is similar to the process we use for evaluating tables. We read the input and output values, but this time, from the x- and y-axes of the graphs.

How To...

Given a composite function and graphs of its individual functions, evaluate it using the information provided by the graphs.

1. Locate the given input to the inner function on the *x*-axis of its graph.
2. Read off the output of the inner function from the *y*-axis of its graph.
3. Locate the inner function output on the *x*-axis of the graph of the outer function.
4. Read the output of the outer function from the *y*-axis of its graph. This is the output of the composite function.

Example 6   **Using a Graph to Evaluate a Composite Function**

Using **Figure 1**, evaluate $f(g(1))$.

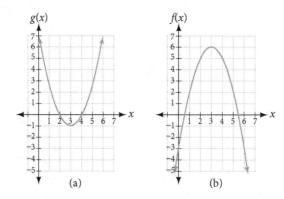

Figure 1

Solution   To evaluate $f(g(1))$, we start with the inside evaluation. See **Figure 2**.

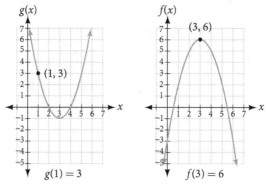

Figure 2

We evaluate $g(1)$ using the graph of $g(x)$, finding the input of 1 on the *x*-axis and finding the output value of the graph at that input. Here, $g(1) = 3$. We use this value as the input to the function *f*.

$$f(g(1)) = f(3)$$

We can then evaluate the composite function by looking to the graph of $f(x)$, finding the input of 3 on the *x*-axis and reading the output value of the graph at this input. Here, $f(3) = 6$, so $f(g(1)) = 6$.

Analysis **Figure 3** *shows how we can mark the graphs with arrows to trace the path from the input value to the output value.*

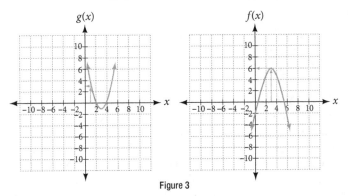

Figure 3

Try It #4

Using **Figure 1**, evaluate $g(f(2))$.

Evaluating Composite Functions Using Formulas

When evaluating a composite function where we have either created or been given formulas, the rule of working from the inside out remains the same. The input value to the outer function will be the output of the inner function, which may be a numerical value, a variable name, or a more complicated expression.

While we can compose the functions for each individual input value, it is sometimes helpful to find a single formula that will calculate the result of a composition $f(g(x))$. To do this, we will extend our idea of function evaluation. Recall that, when we evaluate a function like $f(t) = t^2 - t$, we substitute the value inside the parentheses into the formula wherever we see the input variable.

How To...

Given a formula for a composite function, evaluate the function.

1. Evaluate the inside function using the input value or variable provided.
2. Use the resulting output as the input to the outside function.

Example 7 **Evaluating a Composition of Functions Expressed as Formulas with a Numerical Input**

Given $f(t) = t^2 - t$ and $h(x) = 3x + 2$, evaluate $f(h(1))$.

Solution Because the inside expression is $h(1)$, we start by evaluating $h(x)$ at 1.

$$h(1) = 3(1) + 2$$

$$h(1) = 5$$

Then $f(h(1)) = f(5)$, so we evaluate $f(t)$ at an input of 5.

$$f(h(1)) = f(5)$$

$$f(h(1)) = 5^2 - 5$$

$$f(h(1)) = 20$$

Analysis *It makes no difference what the input variables t and x were called in this problem because we evaluated for specific numerical values.*

Try It #5

Given $f(t) = t^2 - t$ and $h(x) = 3x + 2$, evaluate

a. $h(f(2))$ **b.** $h(f(-2))$

Finding the Domain of a Composite Function

As we discussed previously, the domain of a composite function such as $f \circ g$ is dependent on the domain of g and the domain of f. It is important to know when we can apply a composite function and when we cannot, that is, to know the domain of a function such as $f \circ g$. Let us assume we know the domains of the functions f and g separately. If we write the composite function for an input x as $f(g(x))$, we can see right away that x must be a member of the domain of g in order for the expression to be meaningful, because otherwise we cannot complete the inner function evaluation. However, we also see that $g(x)$ must be a member of the domain of f, otherwise the second function evaluation in $f(g(x))$ cannot be completed, and the expression is still undefined. Thus the domain of $f \circ g$ consists of only those inputs in the domain of g that produce outputs from g belonging to the domain of f. Note that the domain of f composed with g is the set of all x such that x is in the domain of g and $g(x)$ is in the domain of f.

---

domain of a composite function

The domain of a composite function $f(g(x))$ is the set of those inputs x in the domain of g for which $g(x)$ is in the domain of f.

How To...

Given a function composition $f(g(x))$, determine its domain.

1. Find the domain of g.
2. Find the domain of f.
3. Find those inputs x in the domain of g for which $g(x)$ is in the domain of f. That is, exclude those inputs x from the domain of g for which $g(x)$ is not in the domain of f. The resulting set is the domain of $f \circ g$.

Example 8   **Finding the Domain of a Composite Function**

Find the domain of

$$(f \circ g)(x) \text{ where } f(x) = \frac{5}{x-1} \text{ and } g(x) = \frac{4}{3x-2}$$

Solution   The domain of $g(x)$ consists of all real numbers except $x = \frac{2}{3}$, since that input value would cause us to divide by 0. Likewise, the domain of f consists of all real numbers except 1. So we need to exclude from the domain of $g(x)$ that value of x for which $g(x) = 1$.

$$\frac{4}{3x-2} = 1$$

$$4 = 3x - 2$$

$$6 = 3x$$

$$x = 2$$

So the domain of $f \circ g$ is the set of all real numbers except $\frac{2}{3}$ and 2. This means that

$$x \neq \frac{2}{3} \text{ or } x \neq 2$$

We can write this in interval notation as

$$\left(-\infty, \frac{2}{3}\right) \cup \left(\frac{2}{3}, 2\right) \cup (2, \infty)$$

Example 9 **Finding the Domain of a Composite Function Involving Radicals**

Find the domain of

$$(f \circ g)(x) \text{ where } f(x) = \sqrt{x+2} \text{ and } g(x) = \sqrt{3-x}$$

Solution Because we cannot take the square root of a negative number, the domain of g is $(-\infty, 3]$. Now we check the domain of the composite function

$$(f \circ g)(x) = \sqrt{3-x+2} \text{ or } (f \circ g)(x) = \sqrt{5-x}$$

The domain of this function is $(-\infty, 5]$. To find the domain of $f \circ g$, we ask ourselves if there are any further restrictions offered by the domain of the composite function. The answer is no, since $(-\infty, 3]$ is a proper subset of the domain of $f \circ g$. This means the domain of $f \circ g$ is the same as the domain of g, namely, $(-\infty, 3]$.

Analysis This example shows that knowledge of the range of functions (specifically the inner function) can also be helpful in finding the domain of a composite function. It also shows that the domain of $f \circ g$ can contain values that are not in the domain of f, though they must be in the domain of g.

Try It #6

Find the domain of $(f \circ g)(x)$ where $f(x) = \dfrac{1}{x-2}$ and $g(x) = \sqrt{x+4}$

Decomposing a Composite Function into its Component Functions

In some cases, it is necessary to decompose a complicated function. In other words, we can write it as a composition of two simpler functions. There may be more than one way to decompose a composite function, so we may choose the decomposition that appears to be most expedient.

Example 10 **Decomposing a Function**

Write $f(x) = \sqrt{5-x^2}$ as the composition of two functions.

Solution We are looking for two functions, g and h, so $f(x) = g(h(x))$. To do this, we look for a function inside a function in the formula for $f(x)$. As one possibility, we might notice that the expression $5-x^2$ is the inside of the square root. We could then decompose the function as

$$h(x) = 5 - x^2 \text{ and } g(x) = \sqrt{x}$$

We can check our answer by recomposing the functions.

$$g(h(x)) = g(5-x^2) = \sqrt{5-x^2}$$

Try It #7

Write $f(x) = \dfrac{4}{3-\sqrt{4+x^2}}$ as the composition of two functions.

Access these online resources for additional instruction and practice with composite functions.

- Composite Functions (http://openstaxcollege.org/l/compfunction)
- Composite Function Notation Application (http://openstaxcollege.org/l/compfuncnot)
- Composite Functions Using Graphs (http://openstaxcollege.org/l/compfuncgraph)
- Decompose Functions (http://openstaxcollege.org/l/decompfunction)
- Composite Function Values (http://openstaxcollege.org/l/compfuncvalue)

3.4   SECTION EXERCISES

VERBAL

1. How does one find the domain of the quotient of two functions, $\dfrac{f}{g}$?

2. What is the composition of two functions, $f \circ g$?

3. If the order is reversed when composing two functions, can the result ever be the same as the answer in the original order of the composition? If yes, give an example. If no, explain why not.

4. How do you find the domain for the composition of two functions, $f \circ g$?

ALGEBRAIC

For the following exercises, determine the domain for each function in interval notation.

5. Given $f(x) = x^2 + 2x$ and $g(x) = 6 - x^2$, find $f + g$, $f - g$, fg, and $\dfrac{f}{g}$.

6. Given $f(x) = -3x^2 + x$ and $g(x) = 5$, find $f + g$, $f - g$, fg, and $\dfrac{f}{g}$.

7. Given $f(x) = 2x^2 + 4x$ and $g(x) = \dfrac{1}{2x}$, find $f + g$, $f - g$, fg, and $\dfrac{f}{g}$.

8. Given $f(x) = \dfrac{1}{x-4}$ and $g(x) = \dfrac{1}{6-x}$, find $f + g, f - g, fg$, and $\dfrac{f}{g}$.

9. Given $f(x) = 3x^2$ and $g(x) = \sqrt{x-5}$, find $f + g$, $f - g$, fg, and $\dfrac{f}{g}$.

10. Given $f(x) = \sqrt{x}$ and $g(x) = |x - 3|$, find $\dfrac{g}{f}$.

11. For the following exercise, find the indicated function given $f(x) = 2x^2 + 1$ and $g(x) = 3x - 5$.
   a. $f(g(2))$ **b.** $f(g(x))$ **c.** $g(f(x))$ **d.** $(g \circ g)(x)$ **e.** $(f \circ f)(-2)$

For the following exercises, use each pair of functions to find $f(g(x))$ and $g(f(x))$. Simplify your answers.

12. $f(x) = x^2 + 1, g(x) = \sqrt{x + 2}$

13. $f(x) = \sqrt{x} + 2, g(x) = x^2 + 3$

14. $f(x) = |x|, g(x) = 5x + 1$

15. $f(x) = \sqrt[3]{x}, g(x) = \dfrac{x+1}{x^3}$

16. $f(x) = \dfrac{1}{x-6}, g(x) = \dfrac{7}{x} + 6$

17. $f(x) = \dfrac{1}{x-4}, g(x) = \dfrac{2}{x} + 4$

For the following exercises, use each set of functions to find $f(g(h(x)))$. Simplify your answers.

18. $f(x) = x^4 + 6, g(x) = x - 6$, and $h(x) = \sqrt{x}$

19. $f(x) = x^2 + 1, g(x) = \dfrac{1}{x}$, and $h(x) = x + 3$

20. Given $f(x) = \dfrac{1}{x}$, and $g(x) = x - 3$, find the following:
   a. $(f \circ g)(x)$
   b. the domain of $(f \circ g)(x)$ in interval notation
   c. $(g \circ f)(x)$
   d. the domain of $(g \circ f)(x)$
   e. $\left(\dfrac{f}{g} \right) x$

21. Given $f(x) = \sqrt{2 - 4x}$ and $g(x) = -\dfrac{3}{x}$, find the following:
   a. $(g \circ f)(x)$
   b. the domain of $(g \circ f)(x)$ in interval notation

22. Given the functions $f(x) = \dfrac{1-x}{x}$ and $g(x) = \dfrac{1}{1+x^2}$, find the following:

 a. $(g \circ f)(x)$

 b. $(g \circ f)(2)$

23. Given functions $p(x) = \dfrac{1}{\sqrt{x}}$ and $m(x) = x^2 - 4$, state the domain of each of the following functions using interval notation:

 a. $\dfrac{p(x)}{m(x)}$

 b. $p(m(x))$

 c. $m(p(x))$

24. Given functions $q(x) = \dfrac{1}{\sqrt{x}}$ and $h(x) = x^2 - 9$, state the domain of each of the following functions using interval notation.

 a. $\dfrac{q(x)}{h(x)}$

 b. $q(h(x))$

 c. $h(q(x))$

25. For $f(x) = \dfrac{1}{x}$ and $g(x) = \sqrt{x-1}$, write the domain of $(f \circ g)(x)$ in interval notation.

For the following exercises, find functions $f(x)$ and $g(x)$ so the given function can be expressed as $h(x) = f(g(x))$.

26. $h(x) = (x+2)^2$

27. $h(x) = (x-5)^3$

28. $h(x) = \dfrac{3}{x-5}$

29. $h(x) = \dfrac{4}{(x+2)^2}$

30. $h(x) = 4 + \sqrt[3]{x}$

31. $h(x) = \sqrt[3]{\dfrac{1}{2x-3}}$

32. $h(x) = \dfrac{1}{(3x^2-4)^{-3}}$

33. $h(x) = \sqrt[4]{\dfrac{3x-2}{x+5}}$

34. $h(x) = \left(\dfrac{8+x^3}{8-x^3}\right)^4$

35. $h(x) = \sqrt{2x+6}$

36. $h(x) = (5x-1)^3$

37. $h(x) = \sqrt[3]{x-1}$

38. $h(x) = |x^2 + 7|$

39. $h(x) = \dfrac{1}{(x-2)^3}$

40. $h(x) = \left(\dfrac{1}{2x-3}\right)^2$

41. $h(x) = \sqrt{\dfrac{2x-1}{3x+4}}$

GRAPHICAL

For the following exercises, use the graphs of f, shown in **Figure 4**, and g, shown in **Figure 5**, to evaluate the expressions.

Figure 4

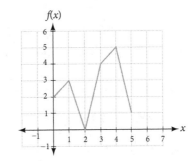

Figure 5

42. $f(g(3))$

43. $f(g(1))$

44. $g(f(1))$

45. $g(f(0))$

46. $f(f(5))$

47. $f(f(4))$

48. $g(g(2))$

49. $g(g(0))$

For the following exercises, use graphs of $f(x)$, shown in **Figure 6**, $g(x)$, shown in **Figure 7**, and $h(x)$, shown in **Figure 8**, to evaluate the expressions.

Figure 6

Figure 7

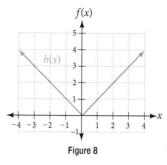

Figure 8

50. $g(f(1))$ **51.** $g(f(2))$ **52.** $f(g(4))$ **53.** $f(g(1))$

54. $f(h(2))$ **55.** $h(f(2))$ **56.** $f(g(h(4)))$ **57.** $f(g(f(-2)))$

NUMERIC

For the following exercises, use the function values for f and g shown in **Table 3** to evaluate each expression.

x	0	1	2	3	4	5	6	7	8	9
$f(x)$	7	6	5	8	4	0	2	1	9	3
$g(x)$	9	5	6	2	1	8	7	3	4	0

Table 3

58. $f(g(8))$ **59.** $f(g(5))$ **60.** $g(f(5))$ **61.** $g(f(3))$
62. $f(f(4))$ **63.** $f(f(1))$ **64.** $g(g(2))$ **65.** $g(g(6))$

For the following exercises, use the function values for f and g shown in **Table 4** to evaluate the expressions.

x	-3	-2	-1	0	1	2	3
$f(x)$	11	9	7	5	3	1	-1
$g(x)$	-8	-3	0	1	0	-3	-8

Table 4

66. $(f \circ g)(1)$ **67.** $(f \circ g)(2)$ **68.** $(g \circ f)(2)$
69. $(g \circ f)(3)$ **70.** $(g \circ g)(1)$ **71.** $(f \circ f)(3)$

For the following exercises, use each pair of functions to find $f(g(0))$ and $g(f(0))$.

72. $f(x) = 4x + 8, g(x) = 7 - x^2$ **73.** $f(x) = 5x + 7, g(x) = 4 - 2x^2$

74. $f(x) = \sqrt{x + 4}, g(x) = 12 - x^3$ **75.** $f(x) = \dfrac{1}{x + 2}, g(x) = 4x + 3$

For the following exercises, use the functions $f(x) = 2x^2 + 1$ and $g(x) = 3x + 5$ to evaluate or find the composite function as indicated.

76. $f(g(2))$ **77.** $f(g(x))$ **78.** $g(f(-3))$ **79.** $(g \circ g)(x)$

EXTENSIONS

For the following exercises, use $f(x) = x^3 + 1$ and $g(x) = \sqrt[3]{x-1}$.

80. Find $(f \circ g)(x)$ and $(g \circ f)(x)$. Compare the two answers.

81. Find $(f \circ g)(2)$ and $(g \circ f)(2)$.

82. What is the domain of $(g \circ f)(x)$?

83. What is the domain of $(f \circ g)(x)$?

84. Let $f(x) = \dfrac{1}{x}$.

   a. Find $(f \circ f)(x)$.

   b. Is $(f \circ f)(x)$ for any function f the same result as the answer to part (a) for any function? Explain.

For the following exercises, let $F(x) = (x+1)^5$, $f(x) = x^5$, and $g(x) = x + 1$.

85. True or False: $(g \circ f)(x) = F(x)$.

86. True or False: $(f \circ g)(x) = F(x)$.

For the following exercises, find the composition when $f(x) = x^2 + 2$ for all $x \geq 0$ and $g(x) = \sqrt{x-2}$.

87. $(f \circ g)(6)$; $(g \circ f)(6)$

88. $(g \circ f)(a)$; $(f \circ g)(a)$

89. $(f \circ g)(11)$; $(g \circ f)(11)$

REAL-WORLD APPLICATIONS

90. The function $D(p)$ gives the number of items that will be demanded when the price is p. The production cost $C(x)$ is the cost of producing x items. To determine the cost of production when the price is $6, you would do which of the following?

   a. Evaluate $D(C(6))$.

   b. Evaluate $C(D(6))$.

   c. Solve $D(C(x)) = 6$.

   d. Solve $C(D(p)) = 6$.

91. The function $A(d)$ gives the pain level on a scale of 0 to 10 experienced by a patient with d milligrams of a pain-reducing drug in her system. The milligrams of the drug in the patient's system after t minutes is modeled by $m(t)$. Which of the following would you do in order to determine when the patient will be at a pain level of 4?

   a. Evaluate $A(m(4))$.

   b. Evaluate $m(A(4))$.

   c. Solve $A(m(t)) = 4$.

   d. Solve $m(A(d)) = 4$.

92. A store offers customers a 30% discount on the price x of selected items. Then, the store takes off an additional 15% at the cash register. Write a price function $P(x)$ that computes the final price of the item in terms of the original price x. (Hint: Use function composition to find your answer.)

93. A rain drop hitting a lake makes a circular ripple. If the radius, in inches, grows as a function of time in minutes according to $r(t) = 25\sqrt{t+2}$, find the area of the ripple as a function of time. Find the area of the ripple at $t = 2$.

94. A forest fire leaves behind an area of grass burned in an expanding circular pattern. If the radius of the circle of burning grass is increasing with time according to the formula $r(t) = 2t + 1$, express the area burned as a function of time, t (minutes).

95. Use the function you found in the previous exercise to find the total area burned after 5 minutes.

96. The radius r, in inches, of a spherical balloon is related to the volume, V, by $r(V) = \sqrt[3]{\dfrac{3V}{4\pi}}$. Air is pumped into the balloon, so the volume after t seconds is given by $V(t) = 10 + 20t$.

   a. Find the composite function $r(V(t))$.

   b. Find the *exact* time when the radius reaches 10 inches.

97. The number of bacteria in a refrigerated food product is given by

$$N(T) = 23T^2 - 56T + 1, \, 3 < T < 33,$$

where T is the temperature of the food. When the food is removed from the refrigerator, the temperature is given by $T(t) = 5t + 1.5$, where t is the time in hours.

   a. Find the composite function $N(T(t))$.

   b. Find the time (round to two decimal places) when the bacteria count reaches 6,752.

LEARNING OBJECTIVES

In this section, you will:
- Graph functions using vertical and horizontal shifts.
- Graph functions using reflections about the *x*-axis and the *y*-axis.
- Determine whether a function is even, odd, or neither from its graph.
- Graph functions using compressions and stretches.
- Combine transformations.

3.5   TRANSFORMATION OF FUNCTIONS

Figure 1 (credit: "Misko"/Flickr)

We all know that a flat mirror enables us to see an accurate image of ourselves and whatever is behind us. When we tilt the mirror, the images we see may shift horizontally or vertically. But what happens when we bend a flexible mirror? Like a carnival funhouse mirror, it presents us with a distorted image of ourselves, stretched or compressed horizontally or vertically. In a similar way, we can distort or transform mathematical functions to better adapt them to describing objects or processes in the real world. In this section, we will take a look at several kinds of transformations.

Graphing Functions Using Vertical and Horizontal Shifts

Often when given a problem, we try to model the scenario using mathematics in the form of words, tables, graphs, and equations. One method we can employ is to adapt the basic graphs of the toolkit functions to build new models for a given scenario. There are systematic ways to alter functions to construct appropriate models for the problems we are trying to solve.

Identifying Vertical Shifts

One simple kind of transformation involves shifting the entire graph of a function up, down, right, or left. The simplest shift is a **vertical shift**, moving the graph up or down, because this transformation involves adding a positive or negative constant to the function. In other words, we add the same constant to the output value of the function regardless of the input. For a function $g(x) = f(x) + k$, the function $f(x)$ is shifted vertically k units. See **Figure 2** for an example.

Figure 2 Vertical shift by $k = 1$ of the cube root function $f(x) = \sqrt[3]{x}$.

To help you visualize the concept of a vertical shift, consider that $y = f(x)$. Therefore, $f(x) + k$ is equivalent to $y + k$. Every unit of y is replaced by $y + k$, so the y-value increases or decreases depending on the value of k. The result is a shift upward or downward.

---

vertical shift

Given a function $f(x)$, a new function $g(x) = f(x) + k$, where k is a constant, is a **vertical shift** of the function $f(x)$. All the output values change by k units. If k is positive, the graph will shift up. If k is negative, the graph will shift down.

Example 1   **Adding a Constant to a Function**

To regulate temperature in a green building, airflow vents near the roof open and close throughout the day. **Figure 3** shows the area of open vents V (in square feet) throughout the day in hours after midnight, t. During the summer, the facilities manager decides to try to better regulate temperature by increasing the amount of open vents by 20 square feet throughout the day and night. Sketch a graph of this new function.

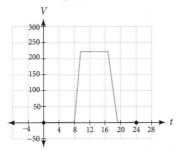

Figure 3

Solution   We can sketch a graph of this new function by adding 20 to each of the output values of the original function. This will have the effect of shifting the graph vertically up, as shown in **Figure 4**.

Figure 4

Notice that in **Figure 4**, for each input value, the output value has increased by 20, so if we call the new function $S(t)$, we could write

$$S(t) = V(t) + 20$$

This notation tells us that, for any value of t, $S(t)$ can be found by evaluating the function V at the same input and then adding 20 to the result. This defines S as a transformation of the function V, in this case a vertical shift up 20 units. Notice that, with a vertical shift, the input values stay the same and only the output values change. See **Table 1**.

t	0	8	10	17	19	24
$V(t)$	0	0	220	220	0	0
$S(t)$	20	20	240	240	20	20

Table 1

How To...

Given a tabular function, create a new row to represent a vertical shift.

1. Identify the output row or column.
2. Determine the magnitude of the shift.
3. Add the shift to the value in each output cell. Add a positive value for up or a negative value for down.

Example 2   **Shifting a Tabular Function Vertically**

A function $f(x)$ is given in **Table 2**. Create a table for the function $g(x) = f(x) - 3$.

x	2	4	6	8
$f(x)$	1	3	7	11

Table 2

Solution   The formula $g(x) = f(x) - 3$ tells us that we can find the output values of g by subtracting 3 from the output values of f. For example:

$$f(2) = 1 \qquad\qquad \text{Given}$$
$$g(x) = f(x) - 3 \qquad\qquad \text{Given transformation}$$
$$g(2) = f(2) - 3$$
$$= 1 - 3$$
$$= -2$$

Subtracting 3 from each $f(x)$ value, we can complete a table of values for $g(x)$ as shown in **Table 3**.

x	2	4	6	8
$f(x)$	1	3	7	11
$g(x)$	−2	0	4	8

Table 3

Analysis   As with the earlier vertical shift, notice the input values stay the same and only the output values change.

Try It #1

The function $h(t) = -4.9t^2 + 30t$ gives the height h of a ball (in meters) thrown upward from the ground after t seconds. Suppose the ball was instead thrown from the top of a 10-m building. Relate this new height function $b(t)$ to $h(t)$, and then find a formula for $b(t)$.

Identifying Horizontal Shifts

We just saw that the vertical shift is a change to the output, or outside, of the function. We will now look at how changes to input, on the inside of the function, change its graph and meaning. A shift to the input results in a movement of the graph of the function left or right in what is known as a **horizontal shift**, shown in **Figure 5**.

Figure 5 Horizontal shift of the function $f(x) = \sqrt[3]{x}$. Note that $h = +1$ shifts the graph to the left, that is, towards negative values of x.

For example, if $f(x) = x^2$, then $g(x) = (x - 2)^2$ is a new function. Each input is reduced by 2 prior to squaring the function. The result is that the graph is shifted 2 units to the right, because we would need to increase the prior input by 2 units to yield the same output value as given in f.

> **horizontal shift**
>
> Given a function f, a new function $g(x) = f(x - h)$, where h is a constant, is a **horizontal shift** of the function f. If h is positive, the graph will shift right. If h is negative, the graph will shift left.

Example 3 **Adding a Constant to an Input**

Returning to our building airflow example from **Figure 3**, suppose that in autumn the facilities manager decides that the original venting plan starts too late, and wants to begin the entire venting program 2 hours earlier. Sketch a graph of the new function.

Solution We can set $V(t)$ to be the original program and $F(t)$ to be the revised program.

$$V(t) = \text{the original venting plan}$$

$$F(t) = \text{starting 2 hrs sooner}$$

In the new graph, at each time, the airflow is the same as the original function V was 2 hours later. For example, in the original function V, the airflow starts to change at 8 a.m., whereas for the function F, the airflow starts to change at 6 a.m. The comparable function values are $V(8) = F(6)$. See **Figure 6**. Notice also that the vents first opened to 220 ft² at 10 a.m. under the original plan, while under the new plan the vents reach 220 ft² at 8 a.m., so $V(10) = F(8)$.

In both cases, we see that, because $F(t)$ starts 2 hours sooner, $h = -2$. That means that the same output values are reached when $F(t) = V(t - (-2)) = V(t + 2)$.

Figure 6

Analysis   Note that $V(t + 2)$ *has the effect of shifting the graph to the* left.

Horizontal changes or "inside changes" affect the domain of a function (the input) instead of the range and often seem counterintuitive. The new function F(t) uses the same outputs as V(t), but matches those outputs to inputs 2 hours earlier than those of V(t). Said another way, we must add 2 hours to the input of V to find the corresponding output for F : F(t) = V(t + 2).

How To...

Given a tabular function, create a new row to represent a horizontal shift.

1. Identify the input row or column.
2. Determine the magnitude of the shift.
3. Add the shift to the value in each input cell.

Example 4   **Shifting a Tabular Function Horizontally**

A function $f(x)$ is given in **Table 4**. Create a table for the function $g(x) = f(x - 3)$.

x	2	4	6	8
$f(x)$	1	3	7	11

Table 4

Solution   The formula $g(x) = f(x - 3)$ tells us that the output values of g are the same as the output value of f when the input value is 3 less than the original value. For example, we know that $f(2) = 1$. To get the same output from the function g, we will need an input value that is 3 *larger*. We input a value that is 3 larger for $g(x)$ because the function takes 3 away before evaluating the function f.

$$g(5) = f(5 - 3)$$
$$= f(2)$$
$$= 1$$

We continue with the other values to create **Table 5**.

x	5	7	9	11
$x - 3$	2	4	6	8
$f(x)$	1	3	7	11
$g(x)$	1	3	7	11

Table 5

The result is that the function $g(x)$ has been shifted to the right by 3. Notice the output values for $g(x)$ remain the same as the output values for $f(x)$, but the corresponding input values, x, have shifted to the right by 3. Specifically, 2 shifted to 5, 4 shifted to 7, 6 shifted to 9, and 8 shifted to 11.

Analysis   **Figure 7** *represents both of the functions. We can see the horizontal shift in each point.*

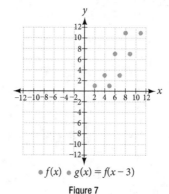

$\bullet f(x)$ $\bullet g(x) = f(x - 3)$

Figure 7

Example 5   **Identifying a Horizontal Shift of a Toolkit Function**

Figure 8 represents a transformation of the toolkit function $f(x) = x^2$. Relate this new function $g(x)$ to $f(x)$, and then find a formula for $g(x)$.

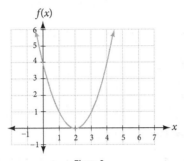

Figure 8

Solution   Notice that the graph is identical in shape to the $f(x) = x^2$ function, but the x-values are shifted to the right 2 units. The vertex used to be at (0,0), but now the vertex is at (2,0). The graph is the basic quadratic function shifted 2 units to the right, so

$$g(x) = f(x - 2)$$

Notice how we must input the value $x = 2$ to get the output value $y = 0$; the x-values must be 2 units larger because of the shift to the right by 2 units. We can then use the definition of the $f(x)$ function to write a formula for $g(x)$ by evaluating $f(x - 2)$.

$$f(x) = x^2$$
$$g(x) = f(x - 2)$$
$$g(x) = f(x - 2) = (x - 2)^2$$

Analysis *To determine whether the shift is +2 or −2, consider a single reference point on the graph. For a quadratic, looking at the vertex point is convenient. In the original function, $f(0) = 0$. In our shifted function, $g(2) = 0$. To obtain the output value of 0 from the function f, we need to decide whether a plus or a minus sign will work to satisfy $g(2) = f(x - 2) = f(0) = 0$. For this to work, we will need to subtract 2 units from our input values.*

Example 6   **Interpreting Horizontal versus Vertical Shifts**

The function $G(m)$ gives the number of gallons of gas required to drive m miles. Interpret $G(m) + 10$ and $G(m + 10)$.

Solution $G(m) + 10$ can be interpreted as adding 10 to the output, gallons. This is the gas required to drive m miles, plus another 10 gallons of gas. The graph would indicate a vertical shift.

$G(m + 10)$ can be interpreted as adding 10 to the input, miles. So this is the number of gallons of gas required to drive 10 miles more than m miles. The graph would indicate a horizontal shift.

Try It #2

Given the function $f(x) = \sqrt{x}$, graph the original function $f(x)$ and the transformation $g(x) = f(x + 2)$ on the same axes. Is this a horizontal or a vertical shift? Which way is the graph shifted and by how many units?

Combining Vertical and Horizontal Shifts

Now that we have two transformations, we can combine them. Vertical shifts are outside changes that affect the output (y-) axis values and shift the function up or down. Horizontal shifts are inside changes that affect the input (x-) axis values and shift the function left or right. Combining the two types of shifts will cause the graph of a function to shift up or down *and* right or left.

Given a function and both a vertical and a horizontal shift, sketch the graph.

1. Identify the vertical and horizontal shifts from the formula.

2. The vertical shift results from a constant added to the output. Move the graph up for a positive constant and down for a negative constant.

3. The horizontal shift results from a constant added to the input. Move the graph left for a positive constant and right for a negative constant.

4. Apply the shifts to the graph in either order.

Example 7   **Graphing Combined Vertical and Horizontal Shifts**

Given $f(x) = |x|$, sketch a graph of $h(x) = f(x + 1) - 3$.

Solution   The function f is our toolkit absolute value function. We know that this graph has a V shape, with the point at the origin. The graph of h has transformed f in two ways: $f(x + 1)$ is a change on the inside of the function, giving a horizontal shift left by 1, and the subtraction by 3 in $f(x + 1) - 3$ is a change to the outside of the function, giving a vertical shift down by 3. The transformation of the graph is illustrated in **Figure 9**.

Let us follow one point of the graph of $f(x) = |x|$.

• The point $(0, 0)$ is transformed first by shifting left 1 unit: $(0, 0) \rightarrow (-1, 0)$

• The point $(-1, 0)$ is transformed next by shifting down 3 units: $(-1, 0) \rightarrow (-1, -3)$

Figure 9

Figure 10 shows the graph of h.

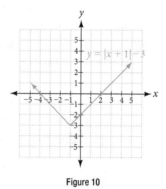

Figure 10

Given $f(x) = |x|$, sketch a graph of $h(x) = f(x - 2) + 4$.

Example 8   **Identifying Combined Vertical and Horizontal Shifts**

Write a formula for the graph shown in **Figure 11**, which is a transformation of the toolkit square root function.

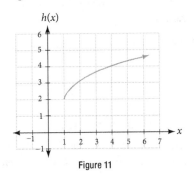

Figure 11

Solution   The graph of the toolkit function starts at the origin, so this graph has been shifted 1 to the right and up 2. In function notation, we could write that as

$$h(x) = f(x - 1) + 2$$

Using the formula for the square root function, we can write

$$h(x) = \sqrt{x - 1} + 2$$

Analysis   Note that this transformation has changed the domain and range of the function. This new graph has domain $[1, \infty)$ and range $[2, \infty)$.

Try It #4

Write a formula for a transformation of the toolkit reciprocal function $f(x) = \dfrac{1}{x}$ that shifts the function's graph one unit to the right and one unit up.

Graphing Functions Using Reflections about the Axes

Another transformation that can be applied to a function is a reflection over the x- or y-axis. A **vertical reflection** reflects a graph vertically across the x-axis, while a **horizontal reflection** reflects a graph horizontally across the y-axis. The reflections are shown in **Figure 12**.

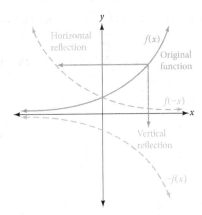

Figure 12 Vertical and horizontal reflections of a function.

Notice that the vertical reflection produces a new graph that is a mirror image of the base or original graph about the x-axis. The horizontal reflection produces a new graph that is a mirror image of the base or original graph about the y-axis.

> **reflections**
>
> Given a function $f(x)$, a new function $g(x) = -f(x)$ is a **vertical reflection** of the function $f(x)$, sometimes called a reflection about (or over, or through) the x-axis.
>
> Given a function $f(x)$, a new function $g(x) = f(-x)$ is a **horizontal reflection** of the function $f(x)$, sometimes called a reflection about the y-axis.

How To...

Given a function, reflect the graph both vertically and horizontally.

1. Multiply all outputs by –1 for a vertical reflection. The new graph is a reflection of the original graph about the x-axis.
2. Multiply all inputs by –1 for a horizontal reflection. The new graph is a reflection of the original graph about the y-axis.

Example 9   **Reflecting a Graph Horizontally and Vertically**

Reflect the graph of $s(t) = \sqrt{t}$ **a.** vertically and **b.** horizontally.

Solution

    a. Reflecting the graph vertically means that each output value will be reflected over the horizontal t-axis as shown in **Figure 13**.

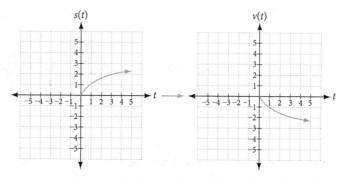

Figure 13 Vertical reflection of the square root function

Because each output value is the opposite of the original output value, we can write

$$V(t) = -s(t) \text{ or } V(t) = -\sqrt{t}$$

Notice that this is an outside change, or vertical shift, that affects the output $s(t)$ values, so the negative sign belongs outside of the function.

    b. Reflecting horizontally means that each input value will be reflected over the vertical axis as shown in **Figure 14**.

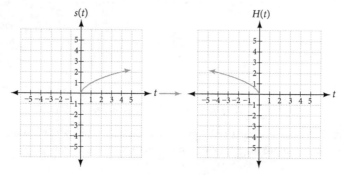

Figure 14 Horizontal reflection of the square root function

Because each input value is the opposite of the original input value, we can write

$$H(t) = s(-t) \text{ or } H(t) = \sqrt{-t}$$

Notice that this is an inside change or horizontal change that affects the input values, so the negative sign is on the inside of the function.

Note that these transformations can affect the domain and range of the functions. While the original square root function has domain $[0, \infty)$ and range $[0, \infty)$, the vertical reflection gives the $V(t)$ function the range $(-\infty, 0]$ and the horizontal reflection gives the $H(t)$ function the domain $(-\infty, 0]$.

Try It #5

Reflect the graph of $f(x) = |x - 1|$ **a.** vertically and **b.** horizontally.

Example 10   **Reflecting a Tabular Function Horizontally and Vertically**

A function $f(x)$ is given as **Table 6**. Create a table for the functions below.

a. $g(x) = -f(x)$ **b.** $h(x) = f(-x)$

x	2	4	6	8
$f(x)$	1	3	7	11

Table 6

Solution

a. For $g(x)$, the negative sign outside the function indicates a vertical reflection, so the x-values stay the same and each output value will be the opposite of the original output value. See **Table 7**.

x	2	4	6	8
$g(x)$	−1	−3	−7	−11

Table 7

b. For $h(x)$, the negative sign inside the function indicates a horizontal reflection, so each input value will be the opposite of the original input value and the $h(x)$ values stay the same as the $f(x)$ values. See **Table 8**.

x	−2	−4	−6	−8
$h(x)$	1	3	7	11

Table 8

Try It #6

A function $f(x)$ is given as **Table 9**. Create a table for the functions below.

x	−2	0	2	4
$f(x)$	5	10	15	20

Table 9

a. $g(x) = -f(x)$

b. $h(x) = f(-x)$

Example 11 **Applying a Learning Model Equation**

A common model for learning has an equation similar to $k(t) = -2^{-t} + 1$, where k is the percentage of mastery that can be achieved after t practice sessions. This is a transformation of the function $f(t) = 2^t$ shown in **Figure 15**. Sketch a graph of $k(t)$.

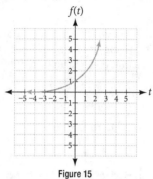

Figure 15

Solution This equation combines three transformations into one equation.

- A horizontal reflection: $f(-t) = 2^{-t}$
- A vertical reflection: $-f(-t) = -2^{-t}$
- A vertical shift: $-f(-t) + 1 = -2^{-t} + 1$

We can sketch a graph by applying these transformations one at a time to the original function. Let us follow two points through each of the three transformations. We will choose the points $(0, 1)$ and $(1, 2)$.

1. First, we apply a horizontal reflection: $(0, 1)$ $(-1, 2)$.
2. Then, we apply a vertical reflection: $(0, -1)$ $(1, -2)$.
3. Finally, we apply a vertical shift: $(0, 0)$ $(1, 1)$.

This means that the original points, $(0,1)$ and $(1,2)$ become $(0,0)$ and $(1,1)$ after we apply the transformations.

In **Figure 16**, the first graph results from a horizontal reflection. The second results from a vertical reflection. The third results from a vertical shift up 1 unit.

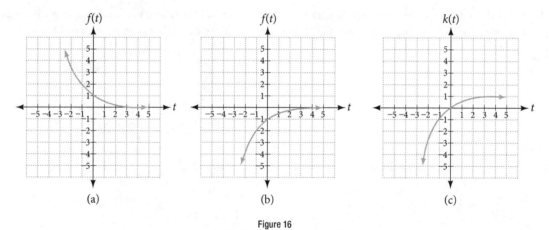

(a)   (b)   (c)

Figure 16

Analysis *As a model for learning, this function would be limited to a domain of $t \geq 0$, with corresponding range $[0, 1)$.*

Try It #7

Given the toolkit function $f(x) = x^2$, graph $g(x) = -f(x)$ and $h(x) = f(-x)$. Take note of any surprising behavior for these functions.

Determining Even and Odd Functions

Some functions exhibit symmetry so that reflections result in the original graph. For example, horizontally reflecting the toolkit functions $f(x) = x^2$ or $f(x) = |x|$ will result in the original graph. We say that these types of graphs are symmetric about the y-axis. A function whose graph is symmetric about the y-axis is called an **even function**.

If the graphs of $f(x) = x^3$ or $f(x) = \frac{1}{x}$ were reflected over *both* axes, the result would be the original graph, as shown in **Figure 17**.

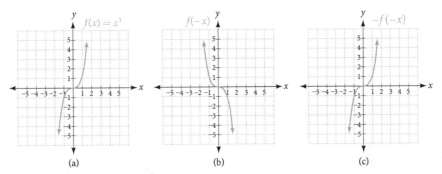

Figure 17 (a) The cubic toolkit function (b) Horizontal reflection of the cubic toolkit function
(c) Horizontal and vertical reflections reproduce the original cubic function.

We say that these graphs are symmetric about the origin. A function with a graph that is symmetric about the origin is called an **odd function**.

Note: A function can be neither even nor odd if it does not exhibit either symmetry. For example, $f(x) = 2^x$ is neither even nor odd. Also, the only function that is both even and odd is the constant function $f(x) = 0$.

---

even and odd functions

A function is called an **even function** if for every input x: $f(x) = f(-x)$

The graph of an even function is symmetric about the y-axis.

A function is called an **odd function** if for every input x: $f(x) = -f(-x)$

The graph of an odd function is symmetric about the origin.

How To…

Given the formula for a function, determine if the function is even, odd, or neither.

1. Determine whether the function satisfies $f(x) = f(-x)$. If it does, it is even.
2. Determine whether the function satisfies $f(x) = -f(-x)$. If it does, it is odd.
3. If the function does not satisfy either rule, it is neither even nor odd.

Example 12   **Determining whether a Function Is Even, Odd, or Neither**

Is the function $f(x) = x^3 + 2x$ even, odd, or neither?

Solution   Without looking at a graph, we can determine whether the function is even or odd by finding formulas for the reflections and determining if they return us to the original function. Let's begin with the rule for even functions.

$$f(-x) = (-x)^3 + 2(-x) = -x^3 - 2x$$

This does not return us to the original function, so this function is not even. We can now test the rule for odd functions.

$$-f(-x) = -(-x^3 - 2x) = x^3 + 2x$$

Because $-f(-x) = f(x)$, this is an odd function.

Analysis Consider the graph of *f* in **Figure 18**. Notice that the graph is symmetric about the origin. For every point (*x*, *y*) on the graph, the corresponding point (−*x*, −*y*) is also on the graph. For example, (1, 3) is on the graph of *f*, and the corresponding point (−1, −3) is also on the graph.

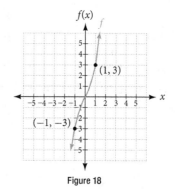

Figure 18

Try It #8

Is the function $f(s) = s^4 + 3s^2 + 7$ even, odd, or neither?

Graphing Functions Using Stretches and Compressions

Adding a constant to the inputs or outputs of a function changed the position of a graph with respect to the axes, but it did not affect the shape of a graph. We now explore the effects of multiplying the inputs or outputs by some quantity.

We can transform the inside (input values) of a function or we can transform the outside (output values) of a function. Each change has a specific effect that can be seen graphically.

Vertical Stretches and Compressions

When we multiply a function by a positive constant, we get a function whose graph is stretched or compressed vertically in relation to the graph of the original function. If the constant is greater than 1, we get a **vertical stretch**; if the constant is between 0 and 1, we get a **vertical compression**. **Figure 19** shows a function multiplied by constant factors 2 and 0.5 and the resulting vertical stretch and compression.

Figure 19 Vertical stretch and compression

---

vertical stretches and compressions

Given a function $f(x)$, a new function $g(x) = af(x)$, where *a* is a constant, is a **vertical stretch** or **vertical compression** of the function $f(x)$.

- If $a > 1$, then the graph will be stretched.

- If $0 < a < 1$, then the graph will be compressed.

- If $a < 0$, then there will be combination of a vertical stretch or compression with a vertical reflection.

Given a function, graph its vertical stretch.

1. Identify the value of *a*.

2. Multiply all range values by *a*.

3. If $a > 1$, the graph is stretched by a factor of *a*.
If $0 < a < 1$, the graph is compressed by a factor of *a*.
If $a < 0$, the graph is either stretched or compressed and also reflected about the *x*-axis.

Example 13   **Graphing a Vertical Stretch**

A function $P(t)$ models the population of fruit flies. The graph is shown in **Figure 20**.

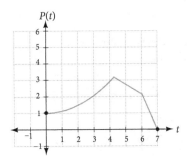

Figure 20

A scientist is comparing this population to another population, *Q*, whose growth follows the same pattern, but is twice as large. Sketch a graph of this population.

Solution   Because the population is always twice as large, the new population's output values are always twice the original function's output values. Graphically, this is shown in **Figure 21**.

If we choose four reference points, (0, 1), (3, 3), (6, 2) and (7, 0) we will multiply all of the outputs by 2.

The following shows where the new points for the new graph will be located.

$$(0, 1) \rightarrow (0, 2)$$
$$(3, 3) \rightarrow (3, 6)$$
$$(6, 2) \rightarrow (6, 4)$$
$$(7, 0) \rightarrow (7, 0)$$

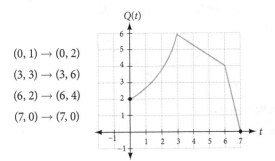

Figure 21

Symbolically, the relationship is written as

$$Q(t) = 2P(t)$$

This means that for any input *t*, the value of the function *Q* is twice the value of the function *P*. Notice that the effect on the graph is a vertical stretching of the graph, where every point doubles its distance from the horizontal axis. The input values, *t*, stay the same while the output values are twice as large as before.

How To...

Given a tabular function and assuming that the transformation is a vertical stretch or compression, create a table for a vertical compression.

1. Determine the value of a.
2. Multiply all of the output values by a.

Example 14   **Finding a Vertical Compression of a Tabular Function**

A function f is given as **Table 10**. Create a table for the function $g(x) = \frac{1}{2}f(x)$.

x	2	4	6	8
$f(x)$	1	3	7	11

Table 10

Solution   The formula $g(x) = \frac{1}{2}f(x)$ tells us that the output values of g are half of the output values of f with the same inputs. For example, we know that $f(4) = 3$. Then

$$g(4) = \frac{1}{2}f(4) = \frac{1}{2}(3) = \frac{3}{2}$$

We do the same for the other values to produce **Table 11**.

x	2	4	6	8
$g(x)$	$\frac{1}{2}$	$\frac{3}{2}$	$\frac{7}{2}$	$\frac{11}{2}$

Table 11

Analysis   The result is that the function $g(x)$ has been compressed vertically by $\frac{1}{2}$. Each output value is divided in half, so the graph is half the original height.

Try It #9

A function f is given as **Table 12**. Create a table for the function $g(x) = \frac{3}{4}f(x)$.

x	2	4	6	8
$f(x)$	12	16	20	0

Table 12

Example 15   **Recognizing a Vertical Stretch**

The graph in **Figure 22** is a transformation of the toolkit function $f(x) = x^3$. Relate this new function $g(x)$ to $f(x)$, and then find a formula for $g(x)$.

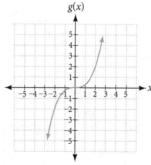

Figure 22

Solution   When trying to determine a vertical stretch or shift, it is helpful to look for a point on the graph that is relatively clear. In this graph, it appears that $g(2) = 2$. With the basic cubic function at the same input, $f(2) = 2^3 = 8$. Based on that, it appears that the outputs of g are $\frac{1}{4}$ the outputs of the function f because $g(2) = \frac{1}{4} f(2)$. From this we can fairly safely conclude that $g(x) = \frac{1}{4} f(x)$.

We can write a formula for g by using the definition of the function f.

$$g(x) = \frac{1}{4} f(x) = \frac{1}{4} x^3$$

Try It #10

Write the formula for the function that we get when we stretch the identity toolkit function by a factor of 3, and then shift it down by 2 units.

Horizontal Stretches and Compressions

Now we consider changes to the inside of a function. When we multiply a function's input by a positive constant, we get a function whose graph is stretched or compressed horizontally in relation to the graph of the original function. If the constant is between 0 and 1, we get a **horizontal stretch**; if the constant is greater than 1, we get a **horizontal compression** of the function.

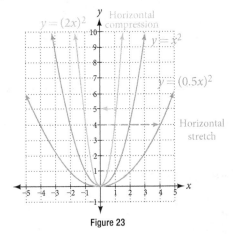

Figure 23

Given a function $y = f(x)$, the form $y = f(bx)$ results in a horizontal stretch or compression. Consider the function $y = x^2$. Observe **Figure 23**. The graph of $y = (0.5x)^2$ is a horizontal stretch of the graph of the function $y = x^2$ by a factor of 2. The graph of $y = (2x)^2$ is a horizontal compression of the graph of the function $y = x^2$ by a factor of 2.

---

horizontal stretches and compressions

Given a function $f(x)$, a new function $g(x) = f(bx)$, where b is a constant, is a **horizontal stretch** or **horizontal compression** of the function $f(x)$.

- If $b > 1$, then the graph will be compressed by $\frac{1}{b}$.

- If $0 < b < 1$, then the graph will be stretched by $\frac{1}{b}$.

- If $b < 0$, then there will be combination of a horizontal stretch or compression with a horizontal reflection.

How To...

Given a description of a function, sketch a horizontal compression or stretch.

1. Write a formula to represent the function.

2. Set $g(x) = f(bx)$ where $b > 1$ for a compression or $0 < b < 1$ for a stretch.

Example 16   **Graphing a Horizontal Compression**

Suppose a scientist is comparing a population of fruit flies to a population that progresses through its lifespan twice as fast as the original population. In other words, this new population, R, will progress in 1 hour the same amount as the original population does in 2 hours, and in 2 hours, it will progress as much as the original population does in 4 hours. Sketch a graph of this population.

Solution   Symbolically, we could write

$$R(1) = P(2),$$
$$R(2) = P(4), \text{ and in general,}$$
$$R(t) = P(2t).$$

See **Figure 24** for a graphical comparison of the original population and the compressed population.

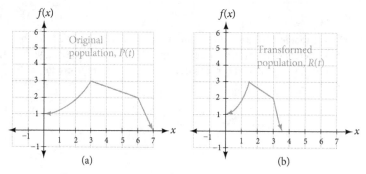

Figure 24 (a) Original population graph (b) Compressed population graph

Analysis   Note that the effect on the graph is a horizontal compression where all input values are half of their original distance from the vertical axis.

Example 17   **Finding a Horizontal Stretch for a Tabular Function**

A function $f(x)$ is given as **Table 13**. Create a table for the function $g(x) = f\left(\frac{1}{2}x\right)$.

x	2	4	6	8
$f(x)$	1	3	7	11

Table 13

Solution   The formula $g(x) = f\left(\frac{1}{2}x\right)$ tells us that the output values for g are the same as the output values for the function f at an input half the size. Notice that we do not have enough information to determine $g(2)$ because $g(2) = f\left(\frac{1}{2} \cdot 2\right) = f(1)$, and we do not have a value for $f(1)$ in our table. Our input values to g will need to be twice as large to get inputs for f that we can evaluate. For example, we can determine $g(4)$.

$$g(4) = f\left(\frac{1}{2} \cdot 4\right) = f(2) = 1$$

We do the same for the other values to produce **Table 14**.

x	4	8	12	16
$g(x)$	1	3	7	11

Table 14

Figure 25 shows the graphs of both of these sets of points.

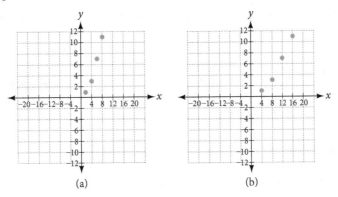

(a)   (b)

Figure 25

Analysis   *Because each input value has been doubled, the result is that the function g(x) has been stretched horizontally by a factor of 2.*

Example 18   **Recognizing a Horizontal Compression on a Graph**

Relate the function $g(x)$ to $f(x)$ in **Figure 26**.

Figure 26

Solution   The graph of $g(x)$ looks like the graph of $f(x)$ horizontally compressed. Because $f(x)$ ends at $(6, 4)$ and $g(x)$ ends at $(2, 4)$, we can see that the x-values have been compressed by $\frac{1}{3}$, because $6\left(\frac{1}{3}\right) = 2$. We might also notice that $g(2) = f(6)$ and $g(1) = f(3)$. Either way, we can describe this relationship as $g(x) = f(3x)$. This is a horizontal compression by $\frac{1}{3}$.

Analysis   *Notice that the coefficient needed for a horizontal stretch or compression is the reciprocal of the stretch or compression. So to stretch the graph horizontally by a scale factor of 4, we need a coefficient of $\frac{1}{4}$ in our function: $f\left(\frac{1}{4}x\right)$. This means that the input values must be four times larger to produce the same result, requiring the input to be larger, causing the horizontal stretching.*

Try It #11

Write a formula for the toolkit square root function horizontally stretched by a factor of 3.

Performing a Sequence of Transformations

When combining transformations, it is very important to consider the order of the transformations. For example, vertically shifting by 3 and then vertically stretching by 2 does not create the same graph as vertically stretching by 2 and then vertically shifting by 3, because when we shift first, both the original function and the shift get stretched, while only the original function gets stretched when we stretch first.

When we see an expression such as $2f(x) + 3$, which transformation should we start with? The answer here follows nicely from the order of operations. Given the output value of $f(x)$, we first multiply by 2, causing the vertical stretch, and then add 3, causing the vertical shift. In other words, multiplication before addition.

Horizontal transformations are a little trickier to think about. When we write $g(x) = f(2x + 3)$, for example, we have to think about how the inputs to the function g relate to the inputs to the function f. Suppose we know $f(7) = 12$. What input to g would produce that output? In other words, what value of x will allow $g(x) = f(2x + 3) = 12$? We would need $2x + 3 = 7$. To solve for x, we would first subtract 3, resulting in a horizontal shift, and then divide by 2, causing a horizontal compression.

This format ends up being very difficult to work with, because it is usually much easier to horizontally stretch a graph before shifting. We can work around this by factoring inside the function.

$$f(bx + p) = f\left(b\left(x + \frac{p}{b}\right)\right)$$

Let's work through an example.

$$f(x) = (2x + 4)^2$$

We can factor out a 2.

$$f(x) = (2(x + 2))^2$$

Now we can more clearly observe a horizontal shift to the left 2 units and a horizontal compression. Factoring in this way allows us to horizontally stretch first and then shift horizontally.

---

combining transformations

When combining vertical transformations written in the form $af(x) + k$, first vertically stretch by a and then vertically shift by k.

When combining horizontal transformations written in the form $f(bx + h)$, first horizontally shift by h and then horizontally stretch by $\frac{1}{b}$.

When combining horizontal transformations written in the form $f(b(x + h))$, first horizontally stretch by $\frac{1}{b}$ and then horizontally shift by h.

Horizontal and vertical transformations are independent. It does not matter whether horizontal or vertical transformations are performed first.

Example 19   **Finding a Triple Transformation of a Tabular Function**

Given **Table 15** for the function $f(x)$, create a table of values for the function $g(x) = 2f(3x) + 1$.

x	6	12	18	24
$f(x)$	10	14	15	17

Table 15

Solution   There are three steps to this transformation, and we will work from the inside out. Starting with the horizontal transformations, $f(3x)$ is a horizontal compression by $\frac{1}{3}$, which means we multiply each x-value by $\frac{1}{3}$. See **Table 16**.

x	2	4	6	8
$f(3x)$	10	14	15	17

Table 16

Looking now to the vertical transformations, we start with the vertical stretch, which will multiply the output values by 2. We apply this to the previous transformation. See **Table 17**.

x	2	4	6	8
$2f(3x)$	20	28	30	34

Table 17

Finally, we can apply the vertical shift, which will add 1 to all the output values. See **Table 18**.

x	2	4	6	8
$g(x) = 2f(3x) + 1$	21	29	31	35

Table 18

Example 20 **Finding a Triple Transformation of a Graph**

Use the graph of $f(x)$ in **Figure 27** to sketch a graph of $k(x) = f\left(\frac{1}{2}x + 1\right) - 3$.

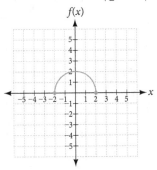

Figure 27

Solution To simplify, let's start by factoring out the inside of the function.

$$f\left(\frac{1}{2}x + 1\right) - 3 = f\left(\frac{1}{2}(x + 2)\right) - 3$$

By factoring the inside, we can first horizontally stretch by 2, as indicated by the $\frac{1}{2}$ on the inside of the function. Remember that twice the size of 0 is still 0, so the point (0, 2) remains at (0, 2) while the point (2, 0) will stretch to (4, 0). See **Figure 28**.

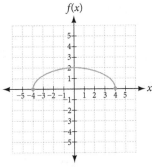

Figure 28

Next, we horizontally shift left by 2 units, as indicated by $x + 2$. See **Figure 29**.

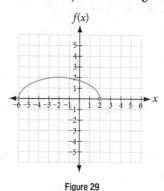

Figure 29

Last, we vertically shift down by 3 to complete our sketch, as indicated by the −3 on the outside of the function. See **Figure 30**.

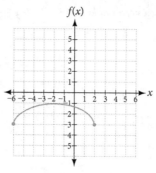

Figure 30

Access this online resource for additional instruction and practice with transformation of functions.

• Function Transformations (http://openstaxcollege.org/l/functrans)

3.5   SECTION EXERCISES

VERBAL

1. When examining the formula of a function that is the result of multiple transformations, how can you tell a horizontal shift from a vertical shift?

2. When examining the formula of a function that is the result of multiple transformations, how can you tell a horizontal stretch from a vertical stretch?

3. When examining the formula of a function that is the result of multiple transformations, how can you tell a horizontal compression from a vertical compression?

4. When examining the formula of a function that is the result of multiple transformations, how can you tell a reflection with respect to the x-axis from a reflection with respect to the y-axis?

5. How can you determine whether a function is odd or even from the formula of the function?

ALGEBRAIC

For the following exercises, write a formula for the function obtained when the graph is shifted as described.

6. $f(x) = \sqrt{x}$ is shifted up 1 unit and to the left 2 units.

7. $f(x) = |x|$ is shifted down 3 units and to the right 1 unit.

8. $f(x) = \dfrac{1}{x}$ is shifted down 4 units and to the right 3 units.

9. $f(x) = \dfrac{1}{x^2}$ is shifted up 2 units and to the left 4 units.

For the following exercises, describe how the graph of the function is a transformation of the graph of the original function f.

10. $y = f(x - 49)$

11. $y = f(x + 43)$

12. $y = f(x + 3)$

13. $y = f(x - 4)$

14. $y = f(x) + 5$

15. $y = f(x) + 8$

16. $y = f(x) - 2$

17. $y = f(x) - 7$

18. $y = f(x - 2) + 3$

19. $y = f(x + 4) - 1$

For the following exercises, determine the interval(s) on which the function is increasing and decreasing.

20. $f(x) = 4(x + 1)^2 - 5$

21. $g(x) = 5(x + 3)^2 - 2$

22. $a(x) = \sqrt{-x + 4}$

23. $k(x) = -3\sqrt{x} - 1$

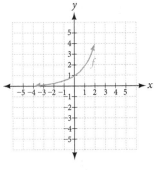

Figure 31

GRAPHICAL

For the following exercises, use the graph of $f(x) = 2^x$ shown in **Figure 31** to sketch a graph of each transformation of $f(x)$.

24. $g(x) = 2^x + 1$

25. $h(x) = 2^x - 3$

26. $w(x) = 2^{x-1}$

For the following exercises, sketch a graph of the function as a transformation of the graph of one of the toolkit functions.

27. $f(t) = (t + 1)^2 - 3$

28. $h(x) = |x - 1| + 4$

29. $k(x) = (x - 2)^3 - 1$

30. $m(t) = 3 + \sqrt{t + 2}$

NUMERIC

31. Tabular representations for the functions f, g, and h are given below. Write $g(x)$ and $h(x)$ as transformations of $f(x)$.

x	−2	−1	0	1	2
$f(x)$	−2	−1	−3	1	2

x	−1	0	1	2	3
$g(x)$	−2	−1	−3	1	2

x	−2	−1	0	1	2
$h(x)$	−1	0	−2	2	3

32. Tabular representations for the functions f, g, and h are given below. Write $g(x)$ and $h(x)$ as transformations of $f(x)$.

x	−2	−1	0	1	2
$f(x)$	−1	−3	4	2	1

x	−3	−2	−1	0	1
$g(x)$	−1	−3	4	2	1

x	−2	−1	0	1	2
$h(x)$	−2	−4	3	1	0

For the following exercises, write an equation for each graphed function by using transformations of the graphs of one of the toolkit functions.

33.

34.

35.

36.

37.

38.

39.

40.

For the following exercises, use the graphs of transformations of the square root function to find a formula for each of the functions.

41.

42.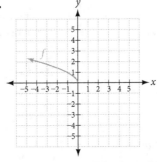

For the following exercises, use the graphs of the transformed toolkit functions to write a formula for each of the resulting functions.

43.

44.

45.

46.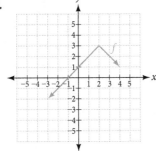

For the following exercises, determine whether the function is odd, even, or neither.

47. $f(x) = 3x^4$

48. $g(x) = \sqrt{x}$

49. $h(x) = \dfrac{1}{x} + 3x$

50. $f(x) = (x - 2)^2$

51. $g(x) = 2x^4$

52. $h(x) = 2x - x^3$

For the following exercises, describe how the graph of each function is a transformation of the graph of the original function f.

53. $g(x) = -f(x)$

54. $g(x) = f(-x)$

55. $g(x) = 4f(x)$

56. $g(x) = 6f(x)$

57. $g(x) = f(5x)$

58. $g(x) = f(2x)$

59. $g(x) = f\left(\dfrac{1}{3}x\right)$

60. $g(x) = f\left(\dfrac{1}{5}x\right)$

61. $g(x) = 3f(-x)$

62. $g(x) = -f(3x)$

For the following exercises, write a formula for the function g that results when the graph of a given toolkit function is transformed as described.

63. The graph of $f(x) = |x|$ is reflected over the y-axis and horizontally compressed by a factor of $\dfrac{1}{4}$.

64. The graph of $f(x) = \sqrt{x}$ is reflected over the x-axis and horizontally stretched by a factor of 2.

65. The graph of $f(x) = \dfrac{1}{x^2}$ is vertically compressed by a factor of $\dfrac{1}{3}$, then shifted to the left 2 units and down 3 units.

66. The graph of $f(x) = \dfrac{1}{x}$ is vertically stretched by a factor of 8, then shifted to the right 4 units and up 2 units.

67. The graph of $f(x) = x^2$ is vertically compressed by a factor of $\dfrac{1}{2}$, then shifted to the right 5 units and up 1 unit.

68. The graph of $f(x) = x^2$ is horizontally stretched by a factor of 3, then shifted to the left 4 units and down 3 units.

For the following exercises, describe how the formula is a transformation of a toolkit function. Then sketch a graph of the transformation.

69. $g(x) = 4(x + 1)^2 - 5$

70. $g(x) = 5(x + 3)^2 - 2$

71. $h(x) = -2|x - 4| + 3$

72. $k(x) = -3\sqrt{x} - 1$

73. $m(x) = \dfrac{1}{2}x^3$

74. $n(x) = \dfrac{1}{3}|x - 2|$

75. $p(x) = \left(\dfrac{1}{3}x\right)^3 - 3$

76. $q(x) = \left(\dfrac{1}{4}x\right)^3 + 1$

77. $a(x) = \sqrt{-x + 4}$

For the following exercises, use the graph in **Figure 32** to sketch the given transformations.

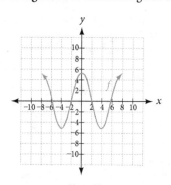

Figure 32

78. $g(x) = f(x) - 2$

79. $g(x) = -f(x)$

80. $g(x) = f(x + 1)$

81. $g(x) = f(x - 2)$

LEARNING OBJECTIVES

In this section you will:

- Graph an absolute value function.
- Solve an absolute value equation.

3.6 ABSOLUTE VALUE FUNCTIONS

Figure 1 Distances in deep space can be measured in all directions. As such, it is useful to consider distance in terms of absolute values. (credit: "s58y"/Flickr)

Until the 1920s, the so-called spiral nebulae were believed to be clouds of dust and gas in our own galaxy, some tens of thousands of light years away. Then, astronomer Edwin Hubble proved that these objects are galaxies in their own right, at distances of millions of light years. Today, astronomers can detect galaxies that are billions of light years away. Distances in the universe can be measured in all directions. As such, it is useful to consider distance as an absolute value function. In this section, we will continue our investigation of absolute value functions.

Understanding Absolute Value

Recall that in its basic form $f(x) = |x|$, the absolute value function, is one of our toolkit functions. The absolute value function is commonly thought of as providing the distance the number is from zero on a number line. Algebraically, for whatever the input value is, the output is the value without regard to sign. Knowing this, we can use absolute value functions to solve some kinds of real-world problems.

---

absolute value function

The absolute value function can be defined as a piecewise function

$$f(x) = |x| = \begin{cases} x & \text{if } x \geq 0 \\ -x & \text{if } x < 0 \end{cases}$$

Example 1   **Using Absolute Value to Determine Resistance**

Electrical parts, such as resistors and capacitors, come with specified values of their operating parameters: resistance, capacitance, etc. However, due to imprecision in manufacturing, the actual values of these parameters vary somewhat from piece to piece, even when they are supposed to be the same. The best that manufacturers can do is to try to guarantee that the variations will stay within a specified range, often $\pm 1\%$, $\pm 5\%$, or $\pm 10\%$.

Suppose we have a resistor rated at 680 ohms, $\pm 5\%$. Use the absolute value function to express the range of possible values of the actual resistance.

Solution   We can find that 5% of 680 ohms is 34 ohms. The absolute value of the difference between the actual and nominal resistance should not exceed the stated variability, so, with the resistance R in ohms,

$$|R - 680| \leq 34$$

Try It #1
Students who score within 20 points of 80 will pass a test. Write this as a distance from 80 using absolute value notation.

Graphing an Absolute Value Function

The most significant feature of the absolute value graph is the corner point at which the graph changes direction. This point is shown at the origin in **Figure 2**.

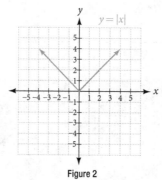

Figure 2

Figure 3 shows the graph of $y = 2|x - 3| + 4$. The graph of $y = |x|$ has been shifted right 3 units, vertically stretched by a factor of 2, and shifted up 4 units. This means that the corner point is located at $(3, 4)$ for this transformed function.

Figure 3

Writing an Equation for an Absolute Value Function Given a Graph

Write an equation for the function graphed in **Figure 4**.

Figure 4

Solution   The basic absolute value function changes direction at the origin, so this graph has been shifted to the right 3 units and down 2 units from the basic toolkit function. See **Figure 5**.

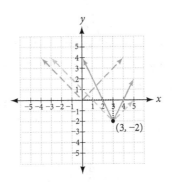

Figure 5

We also notice that the graph appears vertically stretched, because the width of the final graph on a horizontal line is not equal to 2 times the vertical distance from the corner to this line, as it would be for an unstretched absolute value function. Instead, the width is equal to 1 times the vertical distance as shown in **Figure 6**.

Figure 6

From this information we can write the equation

$$f(x) = 2|x - 3| - 2, \text{ treating the stretch as a vertical stretch, or}$$

$$f(x) = |2(x - 3)| - 2, \text{ treating the stretch as a horizontal compression.}$$

Analysis   Note that these equations are algebraically equivalent—the stretch for an absolute value function can be written interchangeably as a vertical or horizontal stretch or compression.

Q & A...

If we couldn't observe the stretch of the function from the graphs, could we algebraically determine it?

Yes. If we are unable to determine the stretch based on the width of the graph, we can solve for the stretch factor by putting in a known pair of values for x and $f(x)$.

$$f(x) = a|x - 3| - 2$$

Now substituting in the point (1, 2)

$$2 = a|1 - 3| - 2$$
$$4 = 2a$$
$$a = 2$$

Try It #2

Write the equation for the absolute value function that is horizontally shifted left 2 units, is vertically flipped, and vertically shifted up 3 units.

Do the graphs of absolute value functions always intersect the vertical axis? The horizontal axis?

Yes, they always intersect the vertical axis. The graph of an absolute value function will intersect the vertical axis when the input is zero.

No, they do not always intersect the horizontal axis. The graph may or may not intersect the horizontal axis, depending on how the graph has been shifted and reflected. It is possible for the absolute value function to intersect the horizontal axis at zero, one, or two points (see **Figure 7**).

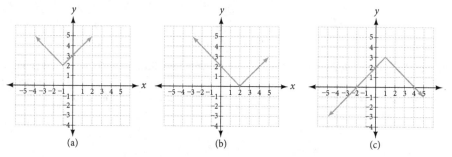

(a)   (b)   (c)

Figure 7 (a) The absolute value function does not intersect the horizontal axis. (b) The absolute value function intersects the horizontal axis at one point. (c) The absolute value function intersects the horizontal axis at two points.

Solving an Absolute Value Equation

In **Other Types of Equations**, we touched on the concepts of absolute value equations. Now that we understand a little more about their graphs, we can take another look at these types of equations. Now that we can graph an absolute value function, we will learn how to solve an absolute value equation. To solve an equation such as $8 = |2x - 6|$, we notice that the absolute value will be equal to 8 if the quantity inside the absolute value is 8 or -8. This leads to two different equations we can solve independently.

$$2x - 6 = 8 \ \text{ or } \ 2x - 6 = -8$$
$$2x = 14 \qquad\quad 2x = -2$$
$$x = 7 \qquad\qquad x = -1$$

Knowing how to solve problems involving absolute value functions is useful. For example, we may need to identify numbers or points on a line that are at a specified distance from a given reference point.

An absolute value equation is an equation in which the unknown variable appears in absolute value bars. For example,

$$|x| = 4,$$
$$|2x - 1| = 3$$
$$|5x + 2| - 4 = 9$$

---

solutions to absolute value equations

For real numbers A and B, an equation of the form $|A| = B$, with $B \geq 0$, will have solutions when $A = B$ or $A = -B$. If $B < 0$, the equation $|A| = B$ has no solution.

Given the formula for an absolute value function, find the horizontal intercepts of its graph.

1. Isolate the absolute value term.
2. Use $|A| = B$ to write $A = B$ or $-A = B$, assuming $B > 0$.
3. Solve for x.

Example 3   **Finding the Zeros of an Absolute Value Function**

For the function $f(x) = |4x + 1| - 7$, find the values of x such that $f(x) = 0$.

Solution

$0 = |4x + 1| - 7$	Substitute 0 for $f(x)$.
$7 = |4x + 1|$	Isolate the absolute value on one side of the equation.
$7 = 4x + 1$ or $-7 = 4x + 1$	Break into two separate equations and solve.
$6 = 4x \qquad -8 = 4x$	
$x = \dfrac{6}{4} = 1.5 \quad x = \dfrac{-8}{4} = -2$	

The function outputs 0 when $x = 1.5$ or $x = -2$. See **Figure 8**.

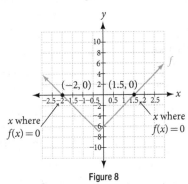

Figure 8

Try It #3

For the function $f(x) = |2x - 1| - 3$, find the values of x such that $f(x) = 0$.

Q & A...

Should we always expect two answers when solving $|A| = B$?

No. We may find one, two, or even no answers. For example, there is no solution to $2 + |3x - 5| = 1$.

Access these online resources for additional instruction and practice with absolute value.

- Graphing Absolute Value Functions (http://openstaxcollege.org/l/graphabsvalue)
- Graphing Absolute Value Functions 2 (http://openstaxcollege.org/l/graphabsvalue2)

3.6   SECTION EXERCISES

VERBAL

1. How do you solve an absolute value equation?

2. How can you tell whether an absolute value function has two x-intercepts without graphing the function?

3. When solving an absolute value function, the isolated absolute value term is equal to a negative number. What does that tell you about the graph of the absolute value function?

4. How can you use the graph of an absolute value function to determine the x-values for which the function values are negative?

ALGEBRAIC

5. Describe all numbers x that are at a distance of 4 from the number 8. Express this using absolute value notation.

6. Describe all numbers x that are at a distance of $\frac{1}{2}$ from the number -4. Express this using absolute value notation.

7. Describe the situation in which the distance that point x is from 10 is at least 15 units. Express this using absolute value notation.

8. Find all function values $f(x)$ such that the distance from $f(x)$ to the value 8 is less than 0.03 units. Express this using absolute value notation.

For the following exercises, find the x- and y-intercepts of the graphs of each function.

9. $f(x) = 4|x - 3| + 4$

10. $f(x) = -3|x - 2| - 1$

11. $f(x) = -2|x + 1| + 6$

12. $f(x) = -5|x + 2| + 15$

13. $f(x) = 2|x - 1| - 6$

14. $f(x) = |-2x + 1| - 13$

15. $f(x) = -|x - 9| + 16$

GRAPHICAL

For the following exercises, graph the absolute value function. Plot at least five points by hand for each graph.

16. $y = |x - 1|$

17. $y = |x + 1|$

18. $y = |x| + 1$

For the following exercises, graph the given functions by hand.

19. $y = |x| - 2$

20. $y = -|x|$

21. $y = -|x| - 2$

22. $y = -|x - 3| - 2$

23. $f(x) = -|x - 1| - 2$

24. $f(x) = -|x + 3| + 4$

25. $f(x) = 2|x + 3| + 1$

26. $f(x) = 3|x - 2| + 3$

27. $f(x) = |2x - 4| - 3$

28. $f(x) = |3x + 9| + 2$

29. $f(x) = -|x - 1| - 3$

30. $f(x) = -|x + 4| - 3$

31. $f(x) = \frac{1}{2}|x + 4| - 3$

TECHNOLOGY

32. Use a graphing utility to graph $f(x) = 10|x - 2|$ on the viewing window $[0, 4]$. Identify the corresponding range. Show the graph.

33. Use a graphing utility to graph $f(x) = -100|x| + 100$ on the viewing window $[-5, 5]$. Identify the corresponding range. Show the graph.

For the following exercises, graph each function using a graphing utility. Specify the viewing window.

34. $f(x) = -0.1|0.1(0.2 - x)| + 0.3$

35. $f(x) = 4 \times 10^9 |x - (5 \times 10^9)| + 2 \times 10^9$

EXTENSIONS

For the following exercises, solve the inequality.

36. If possible, find all values of a such that there are no x-intercepts for $f(x) = 2|x + 1| + a$.

37. If possible, find all values of a such that there are no y-intercepts for $f(x) = 2|x + 1| + a$.

REAL-WORLD APPLICATIONS

38. Cities A and B are on the same east-west line. Assume that city A is located at the origin. If the distance from city A to city B is at least 100 miles and x represents the distance from city B to city A, express this using absolute value notation.

39. The true proportion p of people who give a favorable rating to Congress is 8% with a margin of error of 1.5%. Describe this statement using an absolute value equation.

40. Students who score within 18 points of the number 82 will pass a particular test. Write this statement using absolute value notation and use the variable x for the score.

41. A machinist must produce a bearing that is within 0.01 inches of the correct diameter of 5.0 inches. Using x as the diameter of the bearing, write this statement using absolute value notation.

42. The tolerance for a ball bearing is 0.01. If the true diameter of the bearing is to be 2.0 inches and the measured value of the diameter is x inches, express the tolerance using absolute value notation.

LEARNING OBJECTIVES

In this section, you will:

- Verify inverse functions.
- Determine the domain and range of an inverse function, and restrict the domain of a function to make it one-to-one.
- Find or evaluate the inverse of a function.
- Use the graph of a one-to-one function to graph its inverse function on the same axes.

3.7 INVERSE FUNCTIONS

A reversible heat pump is a climate-control system that is an air conditioner and a heater in a single device. Operated in one direction, it pumps heat out of a house to provide cooling. Operating in reverse, it pumps heat into the building from the outside, even in cool weather, to provide heating. As a heater, a heat pump is several times more efficient than conventional electrical resistance heating.

If some physical machines can run in two directions, we might ask whether some of the function "machines" we have been studying can also run backwards. **Figure 1** provides a visual representation of this question. In this section, we will consider the reverse nature of functions.

Figure 1 Can a function "machine" operate in reverse?

Verifying That Two Functions Are Inverse Functions

Suppose a fashion designer traveling to Milan for a fashion show wants to know what the temperature will be. He is not familiar with the Celsius scale. To get an idea of how temperature measurements are related, he asks his assistant, Betty, to convert 75 degrees Fahrenheit to degrees Celsius. She finds the formula

$$C = \frac{5}{9}(F - 32)$$

and substitutes 75 for F to calculate

$$\frac{5}{9}(75 - 32) \approx 24°C.$$

Knowing that a comfortable 75 degrees Fahrenheit is about 24 degrees Celsius, he sends his assistant the week's weather forecast from **Figure 2** for Milan, and asks her to convert all of the temperatures to degrees Fahrenheit.

Mon	Tue	Web	Thu
26°C | 19°C	29°C | 19°C	30°C | 20°C	26°C | 18°C

Figure 2

Example 3 **Determining Inverse Relationships for Power Functions**

If $f(x) = x^3$ (the cube function) and $g(x) = \frac{1}{3}x$, is $g = f^{-1}$?

Solution $f(g(x)) = \frac{x^3}{27} \neq x$

No, the functions are not inverses.

Analysis *The correct inverse to the cube is, of course, the cube root $\sqrt[3]{x} = x^{1/3}$ that is, the one-third is an exponent, not a multiplier.*

Try It #3

If $f(x) = (x - 1)^3$ and $g(x) = \sqrt[3]{x} + 1$, is $g = f^{-1}$?

Finding Domain and Range of Inverse Functions

The outputs of the function f are the inputs to f^{-1}, so the range of f is also the domain of f^{-1}. Likewise, because the inputs to f are the outputs of f^{-1}, the domain of f is the range of f^{-1}. We can visualize the situation as in **Figure 3**.

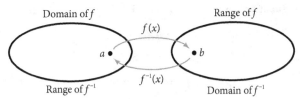

Figure 3 Domain and range of a function and its inverse

When a function has no inverse function, it is possible to create a new function where that new function on a limited domain does have an inverse function. For example, the inverse of $f(x) = \sqrt{x}$ is $f^{-1}(x) = x^2$, because a square "undoes" a square root; but the square is only the inverse of the square root on the domain $[0, \infty)$, since that is the range of $f(x) = \sqrt{x}$.

We can look at this problem from the other side, starting with the square (toolkit quadratic) function $f(x) = x^2$. If we want to construct an inverse to this function, we run into a problem, because for every given output of the quadratic function, there are two corresponding inputs (except when the input is 0). For example, the output 9 from the quadratic function corresponds to the inputs 3 and -3. But an output from a function is an input to its inverse; if this inverse input corresponds to more than one inverse output (input of the original function), then the "inverse" is not a function at all! To put it differently, the quadratic function is not a one-to-one function; it fails the horizontal line test, so it does not have an inverse function. In order for a function to have an inverse, it must be a one-to-one function.

In many cases, if a function is not one-to-one, we can still restrict the function to a part of its domain on which it is one-to-one. For example, we can make a restricted version of the square function $f(x) = x^2$ with its range limited to $[0, \infty)$, which is a one-to-one function (it passes the horizontal line test) and which has an inverse (the square-root function).

If $f(x) = (x - 1)^2$ on $[1, \infty)$, then the inverse function is $f^{-1}(x) = \sqrt{x} + 1$.

- The domain of $f =$ range of $f^{-1} = [1, \infty)$.
- The domain of $f^{-1} =$ range of $f = [0, \infty)$.

Q & A...

Is it possible for a function to have more than one inverse?

No. If two supposedly different functions, say, g and h, both meet the definition of being inverses of another function f, then you can prove that $g = h$. We have just seen that some functions only have inverses if we restrict the domain of the original function. In these cases, there may be more than one way to restrict the domain, leading to different inverses. However, on any one domain, the original function still has only one unique inverse.

> **domain and range of inverse functions**
>
> The range of a function $f(x)$ is the domain of the inverse function $f^{-1}(x)$. The domain of $f(x)$ is the range of $f^{-1}(x)$.

How To...

Given a function, find the domain and range of its inverse.

1. If the function is one-to-one, write the range of the original function as the domain of the inverse, and write the domain of the original function as the range of the inverse.
2. If the domain of the original function needs to be restricted to make it one-to-one, then this restricted domain becomes the range of the inverse function.

Example 4   **Finding the Inverses of Toolkit Functions**

Identify which of the toolkit functions besides the quadratic function are not one-to-one, and find a restricted domain on which each function is one-to-one, if any. The toolkit functions are reviewed in **Table 2**. We restrict the domain in such a fashion that the function assumes all y-values exactly once.

Constant	Identity	Quadratic	Cubic	Reciprocal
$f(x) = c$	$f(x) = x$	$f(x) = x^2$	$f(x) = x^3$	$f(x) = \frac{1}{x}$

Reciprocal squared	Cube root	Square root	Absolute value	
$f(x) = \frac{1}{x^2}$	$f(x) = \sqrt[3]{x}$	$f(x) = \sqrt{x}$	$f(x) = |x|$	

Table 2

Solution   The constant function is not one-to-one, and there is no domain (except a single point) on which it could be one-to-one, so the constant function has no meaningful inverse.

The absolute value function can be restricted to the domain $[0, \infty)$, where it is equal to the identity function.

The reciprocal-squared function can be restricted to the domain $(0, \infty)$.

Analysis   *We can see that these functions (if unrestricted) are not one-to-one by looking at their graphs, shown in* **Figure 4**. *They both would fail the horizontal line test. However, if a function is restricted to a certain domain so that it passes the horizontal line test, then in that restricted domain, it can have an inverse.*

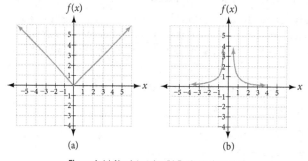

Figure 4 (a) Absolute value (b) Reciprocal squared

Try It #4

The domain of function f is $(1, \infty)$ and the range of function f is $(-\infty, -2)$. Find the domain and range of the inverse function.

Finding and Evaluating Inverse Functions

Once we have a one-to-one function, we can evaluate its inverse at specific inverse function inputs or construct a complete representation of the inverse function in many cases.

Inverting Tabular Functions

Suppose we want to find the inverse of a function represented in table form. Remember that the domain of a function is the range of the inverse and the range of the function is the domain of the inverse. So we need to interchange the domain and range.

Each row (or column) of inputs becomes the row (or column) of outputs for the inverse function. Similarly, each row (or column) of outputs becomes the row (or column) of inputs for the inverse function.

Example 5   **Interpreting the Inverse of a Tabular Function**

A function $f(t)$ is given in **Table 3**, showing distance in miles that a car has traveled in t minutes. Find and interpret $f^{-1}(70)$.

t (minutes)	30	50	70	90
$f(t)$ (miles)	20	40	60	70

Table 3

Solution   The inverse function takes an output of f and returns an input for f. So in the expression $f^{-1}(70)$, 70 is an output value of the original function, representing 70 miles. The inverse will return the corresponding input of the original function f, 90 minutes, so $f^{-1}(70) = 90$. The interpretation of this is that, to drive 70 miles, it took 90 minutes.

Alternatively, recall that the definition of the inverse was that if $f(a) = b$, then $f^{-1}(b) = a$. By this definition, if we are given $f^{-1}(70) = a$, then we are looking for a value a so that $f(a) = 70$. In this case, we are looking for a t so that $f(t) = 70$, which is when $t = 90$.

Try It #5

Using **Table 4**, find and interpret **a.** $f(60)$, and **b.** $f^{-1}(60)$.

t (minutes)	30	50	60	70	90
$f(t)$ (miles)	20	40	50	60	70

Table 4

Evaluating the Inverse of a Function, Given a Graph of the Original Function

We saw in **Functions and Function Notation** that the domain of a function can be read by observing the horizontal extent of its graph. We find the domain of the inverse function by observing the *vertical* extent of the graph of the original function, because this corresponds to the horizontal extent of the inverse function. Similarly, we find the range of the inverse function by observing the *horizontal* extent of the graph of the original function, as this is the vertical extent of the inverse function. If we want to evaluate an inverse function, we find its input within its domain, which is all or part of the vertical axis of the original function's graph.

How To…

Given the graph of a function, evaluate its inverse at specific points.

1. Find the desired input on the *y*-axis of the given graph.
2. Read the inverse function's output from the *x*-axis of the given graph.

Example 6   **Evaluating a Function and Its Inverse from a Graph at Specific Points**

A function $g(x)$ is given in **Figure 5**. Find $g(3)$ and $g^{-1}(3)$.

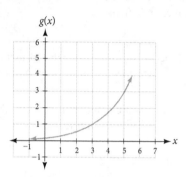

Figure 5

Solution   To evaluate $g(3)$, we find 3 on the x-axis and find the corresponding output value on the y-axis. The point $(3, 1)$ tells us that $g(3) = 1$.

To evaluate $g^{-1}(3)$, recall that by definition $g^{-1}(3)$ means the value of x for which $g(x) = 3$. By looking for the output value 3 on the vertical axis, we find the point $(5, 3)$ on the graph, which means $g(5) = 3$, so by definition, $g^{-1}(3) = 5$. See **Figure 6**.

Figure 6

Try It #6

Using the graph in **Figure 6**, **a.** find $g^{-1}(1)$, and **b.** estimate $g^{-1}(4)$.

Finding Inverses of Functions Represented by Formulas

Sometimes we will need to know an inverse function for all elements of its domain, not just a few. If the original function is given as a formula—for example, y as a function of x—we can often find the inverse function by solving to obtain x as a function of y.

How To...

Given a function represented by a formula, find the inverse.

1. Make sure f is a one-to-one function.
2. Solve for x.
3. Interchange x and y.

Example 7   **Inverting the Fahrenheit-to-Celsius Function**

Find a formula for the inverse function that gives Fahrenheit temperature as a function of Celsius temperature.

$$C = \frac{5}{9}(F - 32)$$

Solution

$$C = \frac{5}{9}(F - 32)$$

$$C \cdot \frac{9}{5} = F - 32$$

$$F = \frac{9}{5}C + 32$$

By solving in general, we have uncovered the inverse function. If

$$C = h(F) = \frac{5}{9}(F - 32),$$

then

$$F = h^{-1}(C) = \frac{9}{5}C + 32.$$

In this case, we introduced a function h to represent the conversion because the input and output variables are descriptive, and writing C^{-1} could get confusing.

Try It #7

Solve for x in terms of y given $y = \frac{1}{3}(x - 5)$

Example 8   **Solving to Find an Inverse Function**

Find the inverse of the function $f(x) = \frac{2}{x - 3} + 4.$

Solution

$$y = \frac{2}{x - 3} + 4 \quad \text{Set up an equation.}$$

$$y - 4 = \frac{2}{x - 3} \quad \text{Subtract 4 from both sides.}$$

$$x - 3 = \frac{2}{y - 4} \quad \text{Multiply both sides by } x - 3 \text{ and divide by } y - 4.$$

$$x = \frac{2}{y - 4} + 3 \quad \text{Add 3 to both sides.}$$

So $f^{-1}(y) = \frac{2}{y - 4} + 3$ or $f^{-1}(x) = \frac{2}{x - 4} + 3.$

*Analysis The domain and range of f exclude the values 3 and 4, respectively. f and f^{-1} are equal at two points but are not the same function, as we can see by creating **Table 5**.*

x	1	2	5	$f^{-1}(y)$
$f(x)$	3	2	5	y

Table 5

Example 9   **Solving to Find an Inverse with Radicals**

Find the inverse of the function $f(x) = 2 + \sqrt{x - 4}.$

Solution

$$y = 2 + \sqrt{x - 4}$$

$$(y - 2)^2 = x - 4$$

$$x = (y - 2)^2 + 4$$

So $f^{-1}(x) = (x - 2)^2 + 4.$

The domain of f is $[4, \infty)$. Notice that the range of f is $[2, \infty)$, so this means that the domain of the inverse function f^{-1} is also $[2, \infty)$.

Analysis   The formula we found for $f^{-1}(x)$ looks like it would be valid for all real x. However, f^{-1} itself must have an inverse (namely, f) so we have to restrict the domain of f^{-1} to $[2, \infty)$ in order to make f^{-1} a one-to-one function. This domain of f^{-1} is exactly the range of f.

Try It #8

What is the inverse of the function $f(x) = 2 - \sqrt{x}$? State the domains of both the function and the inverse function.

Finding Inverse Functions and Their Graphs

Now that we can find the inverse of a function, we will explore the graphs of functions and their inverses. Let us return to the quadratic function $f(x) = x^2$ restricted to the domain $[0, \infty)$, on which this function is one-to-one, and graph it as in **Figure 7**.

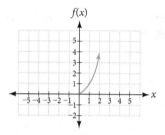

Figure 7 Quadratic function with domain restricted to [0, ∞).

Restricting the domain to $[0, \infty)$ makes the function one-to-one (it will obviously pass the horizontal line test), so it has an inverse on this restricted domain.

We already know that the inverse of the toolkit quadratic function is the square root function, that is, $f^{-1}(x) = \sqrt{x}$. What happens if we graph both f and f^{-1} on the same set of axes, using the x-axis for the input to both f and f^{-1}?

We notice a distinct relationship: The graph of $f^{-1}(x)$ is the graph of $f(x)$ reflected about the diagonal line $y = x$, which we will call the identity line, shown in **Figure 8**.

Figure 8 Square and square-root functions on the non-negative domain

This relationship will be observed for all one-to-one functions, because it is a result of the function and its inverse swapping inputs and outputs. This is equivalent to interchanging the roles of the vertical and horizontal axes.

Example 10   **Finding the Inverse of a Function Using Reflection about the Identity Line**

Given the graph of $f(x)$ in **Figure 9**, sketch a graph of $f^{-1}(x)$.

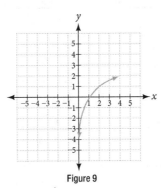

Figure 9

Solution   This is a one-to-one function, so we will be able to sketch an inverse. Note that the graph shown has an apparent domain of $(0, \infty)$ and range of $(-\infty, \infty)$, so the inverse will have a domain of $(-\infty, \infty)$ and range of $(0, \infty)$.

If we reflect this graph over the line $y = x$, the point $(1, 0)$ reflects to $(0, 1)$ and the point $(4, 2)$ reflects to $(2, 4)$. Sketching the inverse on the same axes as the original graph gives **Figure 10**.

Figure 10 The function and its inverse, showing reflection about the identity line

Try It #9

Draw graphs of the functions f and f^{-1} from **Example 8**.

Q & A...

Is there any function that is equal to its own inverse?

Yes. If $f = f^{-1}$, then $f(f(x)) = x$, and we can think of several functions that have this property. The identity function does, and so does the reciprocal function, because

$$\frac{1}{\frac{1}{x}} = x$$

Any function $f(x) = c - x$, where c is a constant, is also equal to its own inverse.

Access these online resources for additional instruction and practice with inverse functions.

- Inverse Functions (http://openstaxcollege.org/l/inversefunction)
- One-to-one Functions (http://openstaxcollege.org/l/onetoone)
- Inverse Function Values Using Graph (http://openstaxcollege.org/l/inversfuncgraph)
- Restricting the Domain and Finding the Inverse (http://openstaxcollege.org/l/restrictdomain)

3.7   SECTION EXERCISES

VERBAL

1. Describe why the horizontal line test is an effective way to determine whether a function is one-to-one?

2. Why do we restrict the domain of the function $f(x) = x^2$ to find the function's inverse?

3. Can a function be its own inverse? Explain.

4. Are one-to-one functions either always increasing or always decreasing? Why or why not?

5. How do you find the inverse of a function algebraically?

ALGEBRAIC

6. Show that the function $f(x) = a - x$ is its own inverse for all real numbers a.

For the following exercises, find $f^{-1}(x)$ for each function.

7. $f(x) = x + 3$

8. $f(x) = x + 5$

9. $f(x) = 2 - x$

10. $f(x) = 3 - x$

11. $f(x) = \dfrac{x}{x + 2}$

12. $f(x) = \dfrac{2x + 3}{5x + 4}$

For the following exercises, find a domain on which each function f is one-to-one and non-decreasing. Write the domain in interval notation. Then find the inverse of f restricted to that domain.

13. $f(x) = (x + 7)^2$

14. $f(x) = (x - 6)^2$

15. $f(x) = x^2 - 5$

16. Given $f(x) = x^3 - 5$ and $g(x) = \dfrac{2x}{1 - x}$:

   a. Find $f(g(x))$ and $g(f(x))$.

   b. What does the answer tell us about the relationship between $f(x)$ and $g(x)$?

For the following exercises, use function composition to verify that $f(x)$ and $g(x)$ are inverse functions.

17. $f(x) = \sqrt[3]{x - 1}$ and $g(x) = x^3 + 1$

18. $f(x) = -3x + 5$ and $g(x) = \dfrac{x - 5}{-3}$

GRAPHICAL

For the following exercises, use a graphing utility to determine whether each function is one-to-one.

19. $f(x) = \sqrt{x}$

20. $f(x) = \sqrt[3]{3x + 1}$

21. $f(x) = -5x + 1$

22. $f(x) = x^3 - 27$

For the following exercises, determine whether the graph represents a one-to-one function.

23.

24.

For the following exercises, use the graph of *f* shown in **Figure 11**.

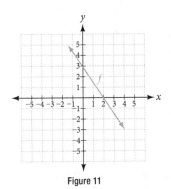

Figure 11

25. Find $f(0)$.

26. Solve $f(x) = 0$.

27. Find $f^{-1}(0)$.

28. Solve $f^{-1}(x) = 0$.

For the following exercises, use the graph of the one-to-one function shown in **Figure 12**.

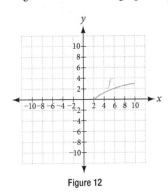

Figure 12

29. Sketch the graph of f^{-1}.

30. Find $f(6)$ and $f^{-1}(2)$.

31. If the complete graph of *f* is shown, find the domain of *f*.

32. If the complete graph of *f* is shown, find the range of *f*.

NUMERIC

For the following exercises, evaluate or solve, assuming that the function *f* is one-to-one.

33. If $f(6) = 7$, find $f^{-1}(7)$.

34. If $f(3) = 2$, find $f^{-1}(2)$.

35. If $f^{-1}(-4) = -8$, find $f(-8)$.

36. If $f^{-1}(-2) = -1$, find $f(-1)$.

For the following exercises, use the values listed in **Table 6** to evaluate or solve.

x	0	1	2	3	4	5	6	7	8	9
$f(x)$	8	0	7	4	2	6	5	3	9	1

Table 6

37. Find $f(1)$.

38. Solve $f(x) = 3$.

39. Find $f^{-1}(0)$.

40. Solve $f^{-1}(x) = 7$.

41. Use the tabular representation of *f* in **Table 7** to create a table for $f^{-1}(x)$.

x	3	6	9	13	14
$f(x)$	1	4	7	12	16

Table 7

TECHNOLOGY

For the following exercises, find the inverse function. Then, graph the function and its inverse.

42. $f(x) = \dfrac{3}{x-2}$

43. $f(x) = x^3 - 1$

44. Find the inverse function of $f(x) = \dfrac{1}{x-1}$. Use a graphing utility to find its domain and range. Write the domain and range in interval notation.

REAL-WORLD APPLICATIONS

45. To convert from x degrees Celsius to y degrees Fahrenheit, we use the formula $f(x) = \dfrac{9}{5}x + 32$. Find the inverse function, if it exists, and explain its meaning.

46. The circumference C of a circle is a function of its radius given by $C(r) = 2\pi r$. Express the radius of a circle as a function of its circumference. Call this function $r(C)$. Find $r(36\pi)$ and interpret its meaning.

47. A car travels at a constant speed of 50 miles per hour. The distance the car travels in miles is a function of time, t, in hours given by $d(t) = 50t$. Find the inverse function by expressing the time of travel in terms of the distance traveled. Call this function $t(d)$. Find $t(180)$ and interpret its meaning.

- The order in which different transformations are applied does affect the final function. Both vertical and horizontal transformations must be applied in the order given. However, a vertical transformation may be combined with a horizontal transformation in any order. See **Example 19** and **Example 20**.

3.6 Absolute Value Functions

- Applied problems, such as ranges of possible values, can also be solved using the absolute value function. See **Example 1**.

- The graph of the absolute value function resembles a letter V. It has a corner point at which the graph changes direction. See **Example 2**.

- In an absolute value equation, an unknown variable is the input of an absolute value function.

- If the absolute value of an expression is set equal to a positive number, expect two solutions for the unknown variable. See **Example 3**.

3.7 Inverse Functions

- If $g(x)$ is the inverse of $f(x)$, then $g(f(x)) = f(g(x)) = x$. See **Example 1**, **Example 2**, and **Example 3**.

- Each of the toolkit functions has an inverse. See **Example 4**.

- For a function to have an inverse, it must be one-to-one (pass the horizontal line test).

- A function that is not one-to-one over its entire domain may be one-to-one on part of its domain.

- For a tabular function, exchange the input and output rows to obtain the inverse. See **Example 5**.

- The inverse of a function can be determined at specific points on its graph. See **Example 6**.

- To find the inverse of a formula, solve the equation $y = f(x)$ for x as a function of y. Then exchange the labels x and y. See **Example 7**, **Example 8**, and **Example 9**.

- The graph of an inverse function is the reflection of the graph of the original function across the line $y = x$. See **Example 10**.

CHAPTER 3 REVIEW EXERCISES

FUNCTIONS AND FUNCTION NOTATION

For the following exercises, determine whether the relation is a function.

1. $\{(a, b), (c, d), (e, d)\}$

2. $\{(5, 2), (6, 1), (6, 2), (4, 8)\}$

3. $y^2 + 4 = x$, for x the independent variable and y the dependent variable

4. Is the graph in **Figure 1** a function?

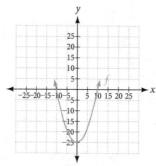

Figure 1

For the following exercises, evaluate the function at the indicated values: $f(-3); f(2); f(-a); -f(a); f(a + h)$.

5. $f(x) = -2x^2 + 3x$ **6.** $f(x) = 2|3x - 1|$

For the following exercises, determine whether the functions are one-to-one.

7. $f(x) = -3x + 5$ **8.** $f(x) = |x - 3|$

For the following exercises, use the vertical line test to determine if the relation whose graph is provided is a function.

9.   **10.**   **11.**

      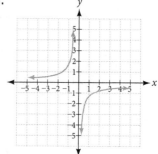

For the following exercises, graph the functions.

12. $f(x) = |x + 1|$ **13.** $f(x) = x^2 - 2$

For the following exercises, use **Figure 2** to approximate the values.

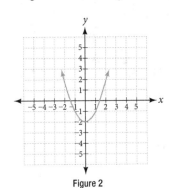

Figure 2

14. $f(2)$

15. $f(-2)$

16. If $f(x) = -2$, then solve for x.

17. If $f(x) = 1$, then solve for x.

For the following exercises, use the function $h(t) = -16t^2 + 80t$ to find the values.

18. $\dfrac{h(2) - h(1)}{2 - 1}$

19. $\dfrac{h(a) - h(1)}{a - 1}$

DOMAIN AND RANGE

For the following exercises, find the domain of each function, expressing answers using interval notation.

20. $f(x) = \dfrac{2}{3x + 2}$

21. $f(x) = \dfrac{x - 3}{x^2 - 4x - 12}$

22. $f(x) = \dfrac{\sqrt{x - 6}}{\sqrt{x - 4}}$

23. Graph this piecewise function: $f(x) = \begin{cases} x + 1 & x < -2 \\ -2x - 3 & x \geq -2 \end{cases}$

RATES OF CHANGE AND BEHAVIOR OF GRAPHS

For the following exercises, find the average rate of change of the functions from $x = 1$ to $x = 2$.

24. $f(x) = 4x - 3$

25. $f(x) = 10x^2 + x$

26. $f(x) = -\dfrac{2}{x^2}$

For the following exercises, use the graphs to determine the intervals on which the functions are increasing, decreasing, or constant.

27.

28.

29.

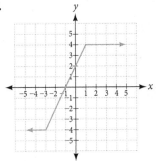

30. Find the local minimum of the function graphed in **Exercise 27**.

31. Find the local extrema for the function graphed in **Exercise 28**.

32. For the graph in **Figure 3**, the domain of the
function is $[-3, 3]$. The range is $[-10, 10]$. Find the
absolute minimum of the function on this interval.

33. Find the absolute maximum of the function graphed
in **Figure 3**.

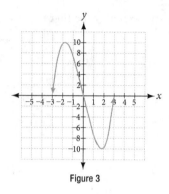

Figure 3

COMPOSITION OF FUNCTIONS

For the following exercises, find $(f \circ g)(x)$ and $(g \circ f)(x)$ for each pair of functions.

34. $f(x) = 4 - x$, $g(x) = -4x$

35. $f(x) = 3x + 2$, $g(x) = 5 - 6x$

36. $f(x) = x^2 + 2x$, $g(x) = 5x + 1$

37. $f(x) = \sqrt{x + 2}$, $g(x) = \frac{1}{x}$

38. $f(x) = \frac{x + 3}{2}$, $g(x) = \sqrt{1 - x}$

For the following exercises, find $(f \circ g)$ and the domain for $(f \circ g)(x)$ for each pair of functions.

39. $f(x) = \frac{x + 1}{x + 4}$, $g(x) = \frac{1}{x}$

40. $f(x) = \frac{1}{x + 3}$, $g(x) = \frac{1}{x - 9}$

41. $f(x) = \frac{1}{x}$, $g(x) = \sqrt{x}$

42. $f(x) = \frac{1}{x^2 - 1}$, $g(x) = \sqrt{x + 1}$

For the following exercises, express each function H as a composition of two functions f and g where $H(x) = (f \circ g)(x)$.

43. $H(x) = \sqrt{\dfrac{2x - 1}{3x + 4}}$

44. $H(x) = \dfrac{1}{(3x^2 - 4)^{-3}}$

TRANSFORMATION OF FUNCTIONS

For the following exercises, sketch a graph of the given function.

45. $f(x) = (x - 3)^2$

46. $f(x) = (x + 4)^3$

47. $f(x) = \sqrt{x} + 5$

48. $f(x) = -x^3$

49. $f(x) = \sqrt[3]{-x}$

50. $f(x) = 5\sqrt{-x} - 4$

51. $f(x) = 4[|x - 2| - 6]$

52. $f(x) = -(x + 2)^2 - 1$

For the following exercises, sketch the graph of the function g if the graph of the function f is shown in **Figure 4**.

Figure 4

53. $g(x) = f(x - 1)$

54. $g(x) = 3f(x)$

For the following exercises, write the equation for the standard function represented by each of the graphs below.

55.

56.

For the following exercises, determine whether each function below is even, odd, or neither.

57. $f(x) = 3x^4$

58. $g(x) = \sqrt{x}$

59. $h(x) = \dfrac{1}{x} + 3x$

For the following exercises, analyze the graph and determine whether the graphed function is even, odd, or neither.

60.

61.

62.

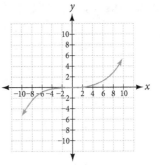

ABSOLUTE VALUE FUNCTIONS

For the following exercises, write an equation for the transformation of $f(x) = |x|$.

63.

64.

65.

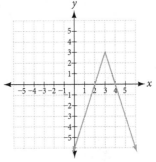

For the following exercises, graph the absolute value function.

66. $f(x) = |x - 5|$

67. $f(x) = -|x - 3|$

68. $f(x) = |2x - 4|$

INVERSE FUNCTIONS

For the following exercises, find $f^{-1}(x)$ for each function.

69. $f(x) = 9 + 10x$

70. $f(x) = \dfrac{x}{x + 2}$

For the following exercise, find a domain on which the function f is one-to-one and non-decreasing. Write the domain in interval notation. Then find the inverse of f restricted to that domain.

71. $f(x) = x^2 + 1$

72. Given $f(x) = x^3 - 5$ and $g(x) = \sqrt[3]{x + 5}$:
   a. Find $f(g(x))$ and $g(f(x))$.
   b. What does the answer tell us about the relationship between $f(x)$ and $g(x)$?

For the following exercises, use a graphing utility to determine whether each function is one-to-one.

73. $f(x) = \dfrac{1}{x}$

74. $f(x) = -3x^2 + x$

75. If $f(5) = 2$, find $f^{-1}(2)$.

76. If $f(1) = 4$, find $f^{-1}(4)$.

CHAPTER 3 PRACTICE TEST

For the following exercises, determine whether each of the following relations is a function.

1. $y = 2x + 8$

2. $\{(2, 1), (3, 2), (-1, 1), (0, -2)\}$

For the following exercises, evaluate the function $f(x) = -3x^2 + 2x$ at the given input.

3. $f(-2)$

4. $f(a)$

5. Show that the function $f(x) = -2(x - 1)^2 + 3$ is not one-to-one.

6. Write the domain of the function $f(x) = \sqrt{3 - x}$ in interval notation.

7. Given $f(x) = 2x^2 - 5x$, find $f(a + 1) - f(1)$.

8. Graph the function $f(x) = \begin{cases} x + 1 \text{ if } -2 < x < 3 \\ -x \quad \text{if} \qquad x \geq 3 \end{cases}$

9. Find the average rate of change of the function $f(x) = 3 - 2x^2 + x$ by finding $\dfrac{f(b) - f(a)}{b - a}$.

For the following exercises, use the functions $f(x) = 3 - 2x^2 + x$ and $g(x) = \sqrt{x}$ to find the composite functions.

10. $(g \circ f)(x)$

11. $(g \circ f)(1)$

12. Express $H(x) = \sqrt[3]{5x^2 - 3x}$ as a composition of two functions, f and g, where $(f \circ g)(x) = H(x)$.

For the following exercises, graph the functions by translating, stretching, and/or compressing a toolkit function.

13. $f(x) = \sqrt{x + 6} - 1$

14. $f(x) = \dfrac{1}{x + 2} - 1$

For the following exercises, determine whether the functions are even, odd, or neither.

15. $f(x) = -\dfrac{5}{x^2} + 9x^6$

16. $f(x) = -\dfrac{5}{x^3} + 9x^5$

17. $f(x) = \dfrac{1}{x}$

18. Graph the absolute value function $f(x) = -2|x - 1| + 3$.

For the following exercises, find the inverse of the function.

19. $f(x) = 3x - 5$

20. $f(x) = \dfrac{4}{x + 7}$

For the following exercises, use the graph of g shown in **Figure 1**.

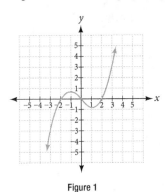

Figure 1

21. On what intervals is the function increasing?

22. On what intervals is the function decreasing?

23. Approximate the local minimum of the function. Express the answer as an ordered pair.

24. Approximate the local maximum of the function. Express the answer as an ordered pair.

For the following exercises, use the graph of the piecewise function shown in **Figure 2**.

Figure 2

25. Find $f(2)$.

26. Find $f(-2)$.

27. Write an equation for the piecewise function.

For the following exercises, use the values listed in **Table 1**.

x	0	1	2	3	4	5	6	7	8
$F(x)$	1	3	5	7	9	11	13	15	17

Table 1

28. Find $F(6)$.

29. Solve the equation $F(x) = 5$.

30. Is the graph increasing or decreasing on its domain?

31. Is the function represented by the graph one-to-one?

32. Find $F^{-1}(15)$.

33. Given $f(x) = -2x + 11$, find $f^{-1}(x)$.

4

Linear Functions

Figure 1 A bamboo forest in China (credit: "JFXie"/Flickr)

CHAPTER OUTLINE

Introduction

Imagine placing a plant in the ground one day and finding that it has doubled its height just a few days later. Although it may seem incredible, this can happen with certain types of bamboo species. These members of the grass family are the fastest-growing plants in the world. One species of bamboo has been observed to grow nearly 1.5 inches every hour.[6] In a twenty-four hour period, this bamboo plant grows about 36 inches, or an incredible 3 feet! A constant rate of change, such as the growth cycle of this bamboo plant, is a linear function.

Recall from **Functions and Function Notation** that a function is a relation that assigns to every element in the domain exactly one element in the range. Linear functions are a specific type of function that can be used to model many real-world applications, such as plant growth over time. In this chapter, we will explore linear functions, their graphs, and how to relate them to data.

6 http://www.guinnessworldrecords.com/records-3000/fastest-growing-plant/

279

In this section, you will:

- Represent a linear function.
- Determine whether a linear function is increasing, decreasing, or constant.
- Interpret slope as a rate of change.
- Write and interpret an equation for a linear function.
- Graph linear functions.
- Determine whether lines are parallel or perpendicular.
- Write the equation of a line parallel or perpendicular to a given line.

4.1 LINEAR FUNCTIONS

Figure 1 Shanghai MagLev Train (credit: "kanegen"/Flickr)

Just as with the growth of a bamboo plant, there are many situations that involve constant change over time. Consider, for example, the first commercial maglev train in the world, the Shanghai MagLev Train (**Figure 1**). It carries passengers comfortably for a 30-kilometer trip from the airport to the subway station in only eight minutes.[7]

Suppose a maglev train travels a long distance, and that the train maintains a constant speed of 83 meters per second for a period of time once it is 250 meters from the station. How can we analyze the train's distance from the station as a function of time? In this section, we will investigate a kind of function that is useful for this purpose, and use it to investigate real-world situations such as the train's distance from the station at a given point in time.

Representing Linear Functions

The function describing the train's motion is a linear function, which is defined as a function with a constant rate of change, that is, a polynomial of degree 1. There are several ways to represent a linear function, including word form, function notation, tabular form, and graphical form. We will describe the train's motion as a function using each method.

Representing a Linear Function in Word Form

Let's begin by describing the linear function in words. For the train problem we just considered, the following word sentence may be used to describe the function relationship.

- *The train's distance from the station is a function of the time during which the train moves at a constant speed plus its original distance from the station when it began moving at constant speed.*

The speed is the rate of change. Recall that a rate of change is a measure of how quickly the dependent variable changes with respect to the independent variable. The rate of change for this example is constant, which means that it is the same for each input value. As the time (input) increases by 1 second, the corresponding distance (output) increases by 83 meters. The train began moving at this constant speed at a distance of 250 meters from the station.

7 http://www.chinahighlights.com/shanghai/transportation/maglev-train.htm

Representing a Linear Function in Function Notation

Another approach to representing linear functions is by using function notation. One example of function notation is an equation written in the slope-intercept form of a line, where x is the input value, m is the rate of change, and b is the initial value of the dependent variable.

Equation form $\qquad y = mx + b$

Function notation $\qquad f(x) = mx + b$

In the example of the train, we might use the notation $D(t)$ in which the total distance D is a function of the time t. The rate, m, is 83 meters per second. The initial value of the dependent variable b is the original distance from the station, 250 meters. We can write a generalized equation to represent the motion of the train.

$$D(t) = 83t + 250$$

Representing a Linear Function in Tabular Form

A third method of representing a linear function is through the use of a table. The relationship between the distance from the station and the time is represented in **Figure 2**. From the table, we can see that the distance changes by 83 meters for every 1 second increase in time.

t	0	1	2	3
$D(t)$	250	333	416	499

Figure 2 Tabular representation of the function D showing selected input and output values

Q & A...

Can the input in the previous example be any real number?

No. The input represents time, so while nonnegative rational and irrational numbers are possible, negative real numbers are not possible for this example. The input consists of non-negative real numbers.

Representing a Linear Function in Graphical Form

Another way to represent linear functions is visually, using a graph. We can use the function relationship from above, $D(t) = 83t + 250$, to draw a graph, represented in **Figure 3**. Notice the graph is a line. When we plot a linear function, the graph is always a line.

The rate of change, which is constant, determines the slant, or slope of the line. The point at which the input value is zero is the vertical intercept, or y-intercept, of the line. We can see from the graph that the y-intercept in the train example we just saw is (0, 250) and represents the distance of the train from the station when it began moving at a constant speed.

Figure 3 The graph of $D(t) = 83t + 250$. Graphs of linear functions are lines because the rate of change is constant.

Notice that the graph of the train example is restricted, but this is not always the case. Consider the graph of the line $f(x) = 2x + 1$. Ask yourself what numbers can be input to the function. In other words, what is the domain of the function? The domain is comprised of all real numbers because any number may be doubled, and then have one added to the product.

> ### linear function
>
> A **linear function** is a function whose graph is a line. Linear functions can be written in the **slope-intercept form** of a line
>
> $$f(x) = mx + b$$
>
> where b is the initial or starting value of the function (when input, $x = 0$), and m is the constant rate of change, or slope of the function. The y-intercept is at $(0, b)$.

Example 1   **Using a Linear Function to Find the Pressure on a Diver**

The pressure, P, in pounds per square inch (PSI) on the diver in **Figure 4** depends upon her depth below the water surface, d, in feet. This relationship may be modeled by the equation, $P(d) = 0.434d + 14.696$. Restate this function in words.

Figure 4 (credit: Ilse Reijs and Jan-Noud Hutten)

Solution   To restate the function in words, we need to describe each part of the equation. The pressure as a function of depth equals four hundred thirty-four thousandths times depth plus fourteen and six hundred ninety-six thousandths.

Analysis   The initial value, 14.696, is the pressure in PSI on the diver at a depth of 0 feet, which is the surface of the water. The rate of change, or slope, is 0.434 PSI per foot. This tells us that the pressure on the diver increases 0.434 PSI for each foot her depth increases.

Determining Whether a Linear Function Is Increasing, Decreasing, or Constant

The linear functions we used in the two previous examples increased over time, but not every linear function does. A linear function may be increasing, decreasing, or constant. For an increasing function, as with the train example, the output values increase as the input values increase. The graph of an increasing function has a positive slope. A line with a positive slope slants upward from left to right as in **Figure 5(a)**. For a decreasing function, the slope is negative. The output values decrease as the input values increase. A line with a negative slope slants downward from left to right as in **Figure 5(b)**. If the function is constant, the output values are the same for all input values so the slope is zero. A line with a slope of zero is horizontal as in **Figure 5(c)**.

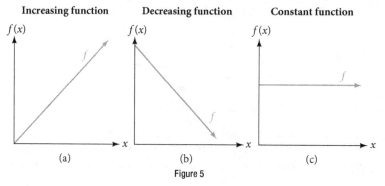

Figure 5

---

increasing and decreasing functions

The slope determines if the function is an **increasing linear function**, a **decreasing linear function**, or a constant function.

- $f(x) = mx + b$ is an increasing function if $m > 0$.
- $f(x) = mx + b$ is an decreasing function if $m < 0$.
- $f(x) = mx + b$ is a constant function if $m = 0$.

Example 2   **Deciding Whether a Function Is Increasing, Decreasing, or Constant**

Some recent studies suggest that a teenager sends an average of 60 texts per day.[8] For each of the following scenarios, find the linear function that describes the relationship between the input value and the output value. Then, determine whether the graph of the function is increasing, decreasing, or constant.

    a. The total number of texts a teen sends is considered a function of time in days. The input is the number of days, and output is the total number of texts sent.

    b. A teen has a limit of 500 texts per month in his or her data plan. The input is the number of days, and output is the total number of texts remaining for the month.

    c. A teen has an unlimited number of texts in his or her data plan for a cost of $50 per month. The input is the number of days, and output is the total cost of texting each month.

Solution   Analyze each function.

    a. The function can be represented as $f(x) = 60x$ where x is the number of days. The slope, 60, is positive so the function is increasing. This makes sense because the total number of texts increases with each day.

    b. The function can be represented as $f(x) = 500 - 60x$ where x is the number of days. In this case, the slope is negative so the function is decreasing. This makes sense because the number of texts remaining decreases each day and this function represents the number of texts remaining in the data plan after x days.

    c. The cost function can be represented as $f(x) = 50$ because the number of days does not affect the total cost. The slope is 0 so the function is constant.

Interpreting Slope as a Rate of Change

In the examples we have seen so far, we have had the slope provided for us. However, we often need to calculate the slope given input and output values. Recall that given two values for the input, x_1 and x_2, and two corresponding values for the output, y_1 and y_2—which can be represented by a set of points, (x_1, y_1) and (x_2, y_2)—we can calculate the slope m.

$$m = \frac{\text{change in output (rise)}}{\text{change in input (run)}} = \frac{\Delta y}{\Delta x} = \frac{y_2 - y_1}{x_2 - x_1}$$

Note in function notation two corresponding values for the output y_1 and y_2 for the function f, $y_1 = f(x_1)$ and $y_2 = f(x_2)$, so we could equivalently write

$$m = \frac{f(x_2) - f(x_1)}{x_2 - x_1}$$

Figure 6 indicates how the slope of the line between the points, (x_1, y_1) and (x_2, y_2), is calculated. Recall that the slope measures steepness, or slant. The greater the absolute value of the slope, the steeper the line is.

---

8 http://www.cbsnews.com/8301-501465_162-57400228-501465/teens-are-sending-60-texts-a-day-study-says/

1.

Figure 6 The slope of a function is calculated by the change in y divided by the change in x. It does not matter which coordinate is used as the (x_2, y_2) and which is the (x_1, y_1), as long as each calculation is started with the elements from the same coordinate pair.

Q & A...

Are the units for slope always $\dfrac{\textbf{units for the output}}{\textbf{units for the input}}$ **?**

Yes. Think of the units as the change of output value for each unit of change in input value. An example of slope could be miles per hour or dollars per day. Notice the units appear as a ratio of units for the output per units for the input.

---

calculate slope

The slope, or rate of change, of a function m can be calculated according to the following:

$$m = \frac{\text{change in output (rise)}}{\text{change in input (run)}} = \frac{\Delta y}{\Delta x} = \frac{y_2 - y_1}{x_2 - x_1}$$

where x_1 and x_2 are input values, y_1 and y_2 are output values.

How To...

Given two points from a linear function, calculate and interpret the slope.

1. Determine the units for output and input values.
2. Calculate the change of output values and change of input values.
3. Interpret the slope as the change in output values per unit of the input value.

Example 3   **Finding the Slope of a Linear Function**

If $f(x)$ is a linear function, and $(3, -2)$ and $(8, 1)$ are points on the line, find the slope. Is this function increasing or decreasing?

Solution   The coordinate pairs are $(3, -2)$ and $(8, 1)$. To find the rate of change, we divide the change in output by the change in input.

$$m = \frac{\text{change in output}}{\text{change in input}} = \frac{1 - (-2)}{8 - 3} = \frac{3}{5}$$

We could also write the slope as $m = 0.6$. The function is increasing because $m > 0$.

Analysis   *As noted earlier, the order in which we write the points does not matter when we compute the slope of the line as long as the first output value, or y-coordinate, used corresponds with the first input value, or x-coordinate, used. Note that if we had reversed them, we would have obtained the same slope.*

$$m = \frac{(-2) - (1)}{3 - 8} = \frac{-3}{-5} = \frac{3}{5}$$

Try It #1

If $f(x)$ is a linear function, and (2, 3) and (0, 4) are points on the line, find the slope. Is this function increasing or decreasing?

Example 4   **Finding the Population Change from a Linear Function**

The population of a city increased from 23,400 to 27,800 between 2008 and 2012. Find the change of population per year if we assume the change was constant from 2008 to 2012.

Solution   The rate of change relates the change in population to the change in time. The population increased by $27,800 - 23,400 = 4,400$ people over the four-year time interval. To find the rate of change, divide the change in the number of people by the number of years.

$$\frac{4,400 \text{ people}}{4 \text{ years}} = 1,100 \frac{\text{people}}{\text{year}}$$

So the population increased by 1,100 people per year.

Analysis   Because we are told that the population increased, we would expect the slope to be positive. This positive slope we calculated is therefore reasonable.

Try It #2

The population of a small town increased from 1,442 to 1,868 between 2009 and 2012. Find the change of population per year if we assume the change was constant from 2009 to 2012.

Writing and Interpreting an Equation for a Linear Function

Recall from **Equations and Inequalities** that we wrote equations in both the slope-intercept form and the point-slope form. Now we can choose which method to use to write equations for linear functions based on the information we are given. That information may be provided in the form of a graph, a point and a slope, two points, and so on. Look at the graph of the function f in **Figure 7**.

Figure 7

We are not given the slope of the line, but we can choose any two points on the line to find the slope. Let's choose (0, 7) and (4, 4). We can use these points to calculate the slope.

$$m = \frac{y_2 - y_1}{x_2 - x_1}$$

$$= \frac{4 - 7}{4 - 0}$$

$$= -\frac{3}{4}$$

Now we can substitute the slope and the coordinates of one of the points into the point-slope form.

$$y - y_1 = m(x - x_1)$$

$$y - 4 = -\frac{3}{4}(x - 4)$$

If we want to rewrite the equation in the slope-intercept form, we would find

$$y - 4 = -\frac{3}{4}(x - 4)$$

$$y - 4 = -\frac{3}{4}x + 3$$

$$y = -\frac{3}{4}x + 7$$

If we wanted to find the slope-intercept form without first writing the point-slope form, we could have recognized that the line crosses the y-axis when the output value is 7. Therefore, $b = 7$. We now have the initial value b and the slope m so we can substitute m and b into the slope-intercept form of a line.

$$f(x) = mx + b$$
$$\uparrow \quad \uparrow$$
$$-\frac{3}{4} \quad 7$$

$$f(x) = -\frac{3}{4}x + 7$$

So the function is $f(x) = -\frac{3}{4}x + 7$, and the linear equation would be $y = -\frac{3}{4}x + 7$.

How To...

Given the graph of a linear function, write an equation to represent the function.

1. Identify two points on the line.
2. Use the two points to calculate the slope.
3. Determine where the line crosses the y-axis to identify the y-intercept by visual inspection.
4. Substitute the slope and y-intercept into the slope-intercept form of a line equation.

Example 5   **Writing an Equation for a Linear Function**

Write an equation for a linear function given a graph of f shown in **Figure 8**.

Figure 8

Solution   Identify two points on the line, such as $(0, 2)$ and $(-2, -4)$. Use the points to calculate the slope.

$$m = \frac{y_2 - y_1}{x_2 - x_1}$$

$$= \frac{-4 - 2}{-2 - 0}$$

$$= \frac{-6}{-2}$$

$$= 3$$

Substitute the slope and the coordinates of one of the points into the point-slope form.

$$y - y_1 = m(x - x_1)$$
$$y - (-4) = 3(x - (-2))$$
$$y + 4 = 3(x + 2)$$

We can use algebra to rewrite the equation in the slope-intercept form.

$$y + 4 = 3(x + 2)$$
$$y + 4 = 3x + 6$$
$$y = 3x + 2$$

Analysis *This makes sense because we can see from* **Figure 9** *that the line crosses the y-axis at the point (0, 2), which is the y-intercept, so b* = 2.

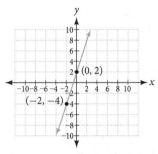

Figure 9

Example 6   **Writing an Equation for a Linear Cost Function**

Suppose Ben starts a company in which he incurs a fixed cost of $1,250 per month for the overhead, which includes his office rent. His production costs are $37.50 per item. Write a linear function C where $C(x)$ is the cost for x items produced in a given month.

Solution   The fixed cost is present every month, $1,250. The costs that can vary include the cost to produce each item, which is $37.50 for Ben. The variable cost, called the marginal cost, is represented by 37.5. The cost Ben incurs is the sum of these two costs, represented by $C(x) = 1250 + 37.5x$.

Analysis *If Ben produces* 100 *items in a month, his monthly cost is found by substitution* 100 *for x.*

$$C(100) = 1,250 + 37.5(100)$$
$$= 5,000$$

So his monthly cost would be $5,000.

Example 7   **Writing an Equation for a Linear Function Given Two Points**

If f is a linear function, with $f(3) = -2$, and $f(8) = 1$, find an equation for the function in slope-intercept form.

Solution   We can write the given points using coordinates.

$$f(3) = -2 \rightarrow (3, -2)$$
$$f(8) = 1 \rightarrow (8, 1)$$

We can then use the points to calculate the slope.

$$m = \frac{y_2 - y_1}{x_2 - x_1}$$
$$= \frac{1 - (-2)}{8 - 3}$$
$$= \frac{3}{5}$$

Substitute the slope and the coordinates of one of the points into the point-slope form.

$$y - y_1 = m(x - x_1)$$
$$y - (-2) = \frac{3}{5}(x - 3)$$

We can use algebra to rewrite the equation in the slope-intercept form.

$$y + 2 = \frac{3}{5}(x - 3)$$

$$y + 2 = \frac{3}{5}x - \frac{9}{5}$$

$$y = \frac{3}{5}x - \frac{19}{5}$$

Try It #3

If $f(x)$ is a linear function, with $f(2) = -11$, and $f(4) = -25$, find an equation for the function in slope-intercept form.

Modeling Real-World Problems with Linear Functions

In the real world, problems are not always explicitly stated in terms of a function or represented with a graph. Fortunately, we can analyze the problem by first representing it as a linear function and then interpreting the components of the function. As long as we know, or can figure out, the initial value and the rate of change of a linear function, we can solve many different kinds of real-world problems.

How To...

Given a linear function f and the initial value and rate of change, evaluate $f(c)$.

1. Determine the initial value and the rate of change (slope).
2. Substitute the values into $f(x) = mx + b$.
3. Evaluate the function at $x = c$.

Example 8   Using a Linear Function to Determine the Number of Songs in a Music Collection

Marcus currently has 200 songs in his music collection. Every month, he adds 15 new songs. Write a formula for the number of songs, N, in his collection as a function of time, t, the number of months. How many songs will he own in a year?

Solution   The initial value for this function is 200 because he currently owns 200 songs, so $N(0) = 200$, which means that $b = 200$.

The number of songs increases by 15 songs per month, so the rate of change is 15 songs per month. Therefore we know that $m = 15$. We can substitute the initial value and the rate of change into the slope-intercept form of a line.

$$f(x) = mx + b$$
$$\uparrow \quad \uparrow$$
$$15 \quad 200$$
$$N(t) = 15t + 200$$

Figure 10

We can write the formula $N(t) = 15t + 200$.

With this formula, we can then predict how many songs Marcus will have in 1 year (12 months). In other words, we can evaluate the function at $t = 12$.

$$N(12) = 15(12) + 200$$
$$= 180 + 200$$
$$= 380$$

Marcus will have 380 songs in 12 months.

Analysis *Notice that N is an increasing linear function. As the input (the number of months) increases, the output (number of songs) increases as well.*

Example 9 **Using a Linear Function to Calculate Salary Based on Commission**

Working as an insurance salesperson, Ilya earns a base salary plus a commission on each new policy. Therefore, Ilya's weekly income, I, depends on the number of new policies, n, he sells during the week. Last week he sold 3 new policies, and earned \$760 for the week. The week before, he sold 5 new policies and earned \$920. Find an equation for $I(n)$, and interpret the meaning of the components of the equation.

Solution The given information gives us two input-output pairs: (3,760) and (5,920). We start by finding the rate of change.

$$m = \frac{920 - 760}{5 - 3}$$
$$= \frac{\$160}{2 \text{ policies}}$$
$$= \$80 \text{ per policy}$$

Keeping track of units can help us interpret this quantity. Income increased by \$160 when the number of policies increased by 2, so the rate of change is \$80 per policy. Therefore, Ilya earns a commission of \$80 for each policy sold during the week.

We can then solve for the initial value.

$$I(n) = 80n + b$$
$$760 = 80(3) + b \quad \text{when } n = 3, I(3) = 760$$
$$760 - 80(3) = b$$
$$520 = b$$

The value of b is the starting value for the function and represents Ilya's income when $n = 0$, or when no new policies are sold. We can interpret this as Ilya's base salary for the week, which does not depend upon the number of policies sold.

We can now write the final equation.

$$I(n) = 80n + 520$$

Our final interpretation is that Ilya's base salary is \$520 per week and he earns an additional \$80 commission for each policy sold.

Example 10 **Using Tabular Form to Write an Equation for a Linear Function**

Table 1 relates the number of rats in a population to time, in weeks. Use the table to write a linear equation.

Number of weeks, w	0	2	4	6
Number of rats, $P(w)$	1,000	1,080	1,160	1,240

Table 1

Solution We can see from the table that the initial value for the number of rats is 1,000, so $b = 1,000$.

Rather than solving for m, we can tell from looking at the table that the population increases by 80 for every 2 weeks that pass. This means that the rate of change is 80 rats per 2 weeks, which can be simplified to 40 rats per week.

$$P(w) = 40w + 1000$$

If we did not notice the rate of change from the table we could still solve for the slope using any two points from the table. For example, using (2, 1080) and (6, 1240)

$$m = \frac{1240 - 1080}{6 - 2}$$
$$= \frac{160}{4}$$
$$= 40$$

Q & A...

Is the initial value always provided in a table of values like Table 1?

No. Sometimes the initial value is provided in a table of values, but sometimes it is not. If you see an input of 0, then the initial value would be the corresponding output. If the initial value is not provided because there is no value of input on the table equal to 0, find the slope, substitute one coordinate pair and the slope into $f(x) = mx + b$, and solve for b.

Try It #4

A new plant food was introduced to a young tree to test its effect on the height of the tree. **Table 2** shows the height of the tree, in feet, x months since the measurements began. Write a linear function, $H(x)$, where x is the number of months since the start of the experiment.

x	0	2	4	8	12
$H(x)$	12.5	13.5	14.5	16.5	18.5

Table 2

Graphing Linear Functions

Now that we've seen and interpreted graphs of linear functions, let's take a look at how to create the graphs. There are three basic methods of graphing linear functions. The first is by plotting points and then drawing a line through the points. The second is by using the y-intercept and slope. And the third method is by using transformations of the identity function $f(x) = x$.

Graphing a Function by Plotting Points

To find points of a function, we can choose input values, evaluate the function at these input values, and calculate output values. The input values and corresponding output values form coordinate pairs. We then plot the coordinate pairs on a grid. In general, we should evaluate the function at a minimum of two inputs in order to find at least two points on the graph. For example, given the function, $f(x) = 2x$, we might use the input values 1 and 2. Evaluating the function for an input value of 1 yields an output value of 2, which is represented by the point (1, 2). Evaluating the function for an input value of 2 yields an output value of 4, which is represented by the point (2, 4). Choosing three points is often advisable because if all three points do not fall on the same line, we know we made an error.

How To...

Given a linear function, graph by plotting points.

1. Choose a minimum of two input values.
2. Evaluate the function at each input value.
3. Use the resulting output values to identify coordinate pairs.
4. Plot the coordinate pairs on a grid.
5. Draw a line through the points.

Example 11   **Graphing by Plotting Points**

Graph $f(x) = -\frac{2}{3}x + 5$ by plotting points.

Solution   Begin by choosing input values. This function includes a fraction with a denominator of 3, so let's choose multiples of 3 as input values. We will choose 0, 3, and 6.

Evaluate the function at each input value, and use the output value to identify coordinate pairs.

$$x = 0 \qquad f(0) = -\frac{2}{3}(0) + 5 = 5 \Rightarrow (0, 5)$$

$$x = 3 \qquad f(3) = -\frac{2}{3}(3) + 5 = 3 \Rightarrow (3, 3)$$

$$x = 6 \qquad f(6) = -\frac{2}{3}(6) + 5 = 1 \Rightarrow (6, 1)$$

Plot the coordinate pairs and draw a line through the points. **Figure 11** represents the graph of the function $f(x) = -\frac{2}{3}x + 5$.

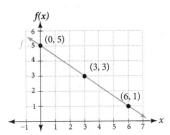

Figure 11 The graph of the linear function $f(x) = -\frac{2}{3}x + 5$.

Analysis   *The graph of the function is a line as expected for a linear function. In addition, the graph has a downward slant, which indicates a negative slope. This is also expected from the negative, constant rate of change in the equation for the function.*

Try It #5

Graph $f(x) = -\frac{3}{4}x + 6$ by plotting points.

Graphing a Function Using *y*-intercept and Slope

Another way to graph linear functions is by using specific characteristics of the function rather than plotting points. The first characteristic is its *y*-intercept, which is the point at which the input value is zero. To find the *y*-intercept, we can set $x = 0$ in the equation.

The other characteristic of the linear function is its slope.

Let's consider the following function.

$$f(x) = \frac{1}{2}x + 1$$

The slope is $\frac{1}{2}$. Because the slope is positive, we know the graph will slant upward from left to right. The *y*-intercept is the point on the graph when $x = 0$. The graph crosses the *y*-axis at $(0, 1)$. Now we know the slope and the *y*-intercept. We can begin graphing by plotting the point $(0, 1)$. We know that the slope is rise over run, $m = \frac{\text{rise}}{\text{run}}$. From our example, we have $m = \frac{1}{2}$, which means that the rise is 1 and the run is 2. So starting from our *y*-intercept $(0, 1)$, we can rise 1 and then run 2, or run 2 and then rise 1. We repeat until we have a few points, and then we draw a line through the points as shown in **Figure 12**.

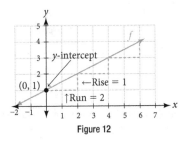

Figure 12

---

graphical interpretation of a linear function

In the equation $f(x) = mx + b$

- b is the *y*-intercept of the graph and indicates the point $(0, b)$ at which the graph crosses the *y*-axis.

- m is the slope of the line and indicates the vertical displacement (rise) and horizontal displacement (run) between each successive pair of points. Recall the formula for the slope:

$$m = \frac{\text{change in output (rise)}}{\text{change in input (run)}} = \frac{\Delta y}{\Delta x} = \frac{y_2 - y_1}{x_2 - x_1}$$

Q & A...

Do all linear functions have *y*-intercepts?

Yes. All linear functions cross the *y*-axis and therefore have *y*-intercepts. (*Note:* A vertical line parallel to the *y*-axis does not have a *y*-intercept, but it is not a function.)

How To...

Given the equation for a linear function, graph the function using the *y*-intercept and slope.

1. Evaluate the function at an input value of zero to find the *y*-intercept.
2. Identify the slope as the rate of change of the input value.
3. Plot the point represented by the *y*-intercept.
4. Use $\dfrac{\text{rise}}{\text{run}}$ to determine at least two more points on the line.
5. Sketch the line that passes through the points.

Example 12   **Graphing by Using the *y*-intercept and Slope**

Graph $f(x) = -\dfrac{2}{3}x + 5$ using the *y*-intercept and slope.

Solution   Evaluate the function at $x = 0$ to find the *y*-intercept. The output value when $x = 0$ is 5, so the graph will cross the *y*-axis at (0, 5).

According to the equation for the function, the slope of the line is $-\dfrac{2}{3}$. This tells us that for each vertical decrease in the "rise" of -2 units, the "run" increases by 3 units in the horizontal direction. We can now graph the function by first plotting the *y*-intercept on the graph in **Figure 13**. From the initial value (0, 5) we move down 2 units and to the right 3 units. We can extend the line to the left and right by repeating, and then draw a line through the points.

Figure 13   Graph of $f(x) = -\dfrac{2}{3}x + 5$ and shows how to calculate the rise over run for the slope.

Analysis   *The graph slants downward from left to right, which means it has a negative slope as expected.*

Try It #6

Find a point on the graph we drew in **Example 12** that has a negative *x*-value.

Graphing a Function Using Transformations

Another option for graphing is to use a transformation of the identity function $f(x) = x$. A function may be transformed by a shift up, down, left, or right. A function may also be transformed using a reflection, stretch, or compression.

Vertical Stretch or Compression

In the equation $f(x) = mx$, the m is acting as the vertical stretch or compression of the identity function. When m is negative, there is also a vertical reflection of the graph. Notice in **Figure 14** that multiplying the equation of $f(x) = x$ by m stretches the graph of f by a factor of m units if $m > 1$ and compresses the graph of f by a factor of m units if $0 < m < 1$. This means the larger the absolute value of m, the steeper the slope.

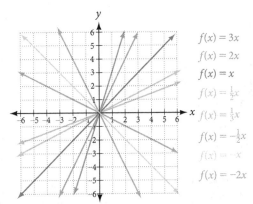

Figure 14 Vertical stretches and compressions and reflections on the function $f(x) = x$.

Vertical Shift

In $f(x) = mx + b$, the b acts as the vertical shift, moving the graph up and down without affecting the slope of the line. Notice in **Figure 15** that adding a value of b to the equation of $f(x) = x$ shifts the graph of f a total of b units up if b is positive and $|b|$ units down if b is negative.

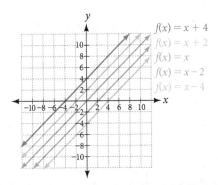

Figure 15 This graph illustrates vertical shifts of the function $f(x) = x$.

Using vertical stretches or compressions along with vertical shifts is another way to look at identifying different types of linear functions. Although this may not be the easiest way to graph this type of function, it is still important to practice each method.

How To...

Given the equation of a linear function, use transformations to graph the linear function in the form $f(x) = mx + b$.

1. Graph $f(x) = x$.
2. Vertically stretch or compress the graph by a factor m.
3. Shift the graph up or down b units.

Example 13 **Graphing by Using Transformations**

Graph $f(x) = \dfrac{1}{2}x - 3$ using transformations.

Solution The equation for the function shows that $m = \dfrac{1}{2}$ so the identity function is vertically compressed by $\dfrac{1}{2}$. The equation for the function also shows that $b = -3$ so the identity function is vertically shifted down 3 units. First, graph the identity function, and show the vertical compression as in **Figure 16**.

Then show the vertical shift as in **Figure 17**.

1.

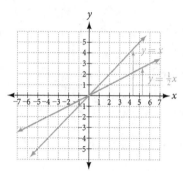

Figure 16 The function, $y = x$, compressed by a factor of $\frac{1}{2}$.

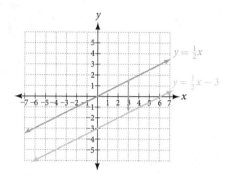

Figure 17 The function $y = \frac{1}{2}x$, shifted down 3 units.

Try It #7

Graph $f(x) = 4 + 2x$, using transformations.

Q & A...

In Example 15, could we have sketched the graph by reversing the order of the transformations?

No. The order of the transformations follows the order of operations. When the function is evaluated at a given input, the corresponding output is calculated by following the order of operations. This is why we performed the compression first. For example, following the order: Let the input be 2.

$$f(2) = \frac{1}{2}(2) - 3$$
$$= 1 - 3$$
$$= -2$$

Writing the Equation for a Function from the Graph of a Line

Earlier, we wrote the equation for a linear function from a graph. Now we can extend what we know about graphing linear functions to analyze graphs a little more closely. Begin by taking a look at **Figure 18**. We can see right away that the graph crosses the y-axis at the point $(0, 4)$ so this is the y-intercept.

Figure 18

Then we can calculate the slope by finding the rise and run. We can choose any two points, but let's look at the point $(-2, 0)$. To get from this point to the y-intercept, we must move up 4 units (rise) and to the right 2 units (run). So the slope must be

$$m = \frac{\text{rise}}{\text{run}} = \frac{4}{2} = 2$$

Substituting the slope and y-intercept into the slope-intercept form of a line gives

$$y = 2x + 4$$

Given a graph of linear function, find the equation to describe the function.

1. Identify the y-intercept of an equation.
2. Choose two points to determine the slope.
3. Substitute the y-intercept and slope into the slope-intercept form of a line.

Example 14   **Matching Linear Functions to Their Graphs**

Match each equation of the linear functions with one of the lines in **Figure 19**.

$\quad$ **a.** $f(x) = 2x + 3$ $\qquad$ **b.** $g(x) = 2x - 3$ $\qquad$ **c.** $h(x) = -2x + 3$ $\qquad$ **d.** $j(x) = \frac{1}{2}x + 3$

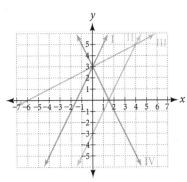

Figure 19

Solution   Analyze the information for each function.

$\quad$ **a.** This function has a slope of 2 and a y-intercept of 3. It must pass through the point (0, 3) and slant upward from left to right. We can use two points to find the slope, or we can compare it with the other functions listed. Function g has the same slope, but a different y-intercept. Lines I and III have the same slant because they have the same slope. Line III does not pass through (0, 3) so f must be represented by line I.

$\quad$ **b.** This function also has a slope of 2, but a y-intercept of -3. It must pass through the point (0, -3) and slant upward from left to right. It must be represented by line III.

$\quad$ **c.** This function has a slope of -2 and a y-intercept of 3. This is the only function listed with a negative slope, so it must be represented by line IV because it slants downward from left to right.

$\quad$ **d.** This function has a slope of $\frac{1}{2}$ and a y-intercept of 3. It must pass through the point (0, 3) and slant upward from left to right. Lines I and II pass through (0, 3), but the slope of j is less than the slope of f so the line for j must be flatter. This function is represented by Line II.

Now we can re-label the lines as in **Figure 20**.

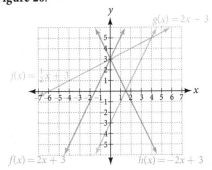

Figure 20

Finding the x-intercept of a Line

So far, we have been finding the y-intercept of a function: the point at which the graph of the function crosses the y-axis. Recall that a function may also have an x-intercept, which is the x-coordinate of the point where the graph of the function crosses the x-axis. In other words, it is the input value when the output value is zero.

To find the x-intercept, set a function $f(x)$ equal to zero and solve for the value of x. For example, consider the function shown.

$$f(x) = 3x - 6$$

Set the function equal to 0 and solve for x.

$$0 = 3x - 6$$
$$6 = 3x$$
$$2 = x$$
$$x = 2$$

The graph of the function crosses the x-axis at the point $(2, 0)$.

Q & A...

Do all linear functions have x-intercepts?

No. However, linear functions of the form $y = c$, where c is a nonzero real number, are the only examples of linear functions with no x-intercept. For example, $y = 5$ is a horizontal line 5 units above the x-axis. This function has no x-intercepts, as shown in **Figure 21**.

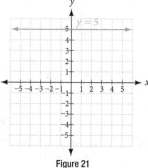

Figure 21

---

x-intercept

The **x-intercept** of the function is value of x when $f(x) = 0$. It can be solved by the equation $0 = mx + b$.

Example 15   **Finding an x-intercept**

Find the x-intercept of $f(x) = \dfrac{1}{2}x - 3$.

Solution   Set the function equal to zero to solve for x.

$$0 = \frac{1}{2}x - 3$$
$$3 = \frac{1}{2}x$$
$$6 = x$$
$$x = 6$$

The graph crosses the x-axis at the point $(6, 0)$.

*Analysis   A graph of the function is shown in **Figure 22**. We can see that the x-intercept is $(6, 0)$ as we expected.*

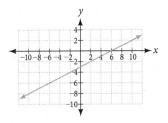

Figure 22

Try It #8

Find the *x*-intercept of $f(x) = \frac{1}{4}x - 4$.

Describing Horizontal and Vertical Lines

There are two special cases of lines on a graph—horizontal and vertical lines. A horizontal line indicates a constant output, or *y*-value. In **Figure 23**, we see that the output has a value of 2 for every input value. The change in outputs between any two points, therefore, is 0. In the slope formula, the numerator is 0, so the slope is 0. If we use $m = 0$ in the equation $f(x) = mx + b$, the equation simplifies to $f(x) = b$. In other words, the value of the function is a constant. This graph represents the function $f(x) = 2$.

x	−4	−2	0	2	4
y	2	2	2	2	2

Figure 23 A horizontal line representing the function *f(x) = 2*.

A vertical line indicates a constant input, or *x*-value. We can see that the input value for every point on the line is 2, but the output value varies. Because this input value is mapped to more than one output value, a vertical line does not represent a function. Notice that between any two points, the change in the input values is zero. In the slope formula, the denominator will be zero, so the slope of a vertical line is undefined.

$$m = \frac{\text{change of output}}{\text{change of input}} \quad \begin{array}{l} \leftarrow \text{Non-zero real number} \\ \leftarrow 0 \end{array}$$

Figure 24 Example of how a line has a vertical slope. 0 in the denominator of the slope.

Notice that a vertical line, such as the one in **Figure 25**, has an *x*-intercept, but no *y*-intercept unless it's the line $x = 0$. This graph represents the line $x = 2$.

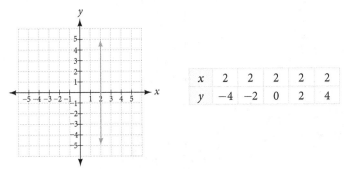

x	2	2	2	2	2
y	−4	−2	0	2	4

Figure 25 The vertical line, *x = 2*, which does not represent a function.

> *horizontal and vertical lines*
>
> Lines can be horizontal or vertical.
>
> A **horizontal line** is a line defined by an equation in the form $f(x) = b$.
>
> A **vertical line** is a line defined by an equation in the form $x = a$.

Example 16   **Writing the Equation of a Horizontal Line**

Write the equation of the line graphed in **Figure 26**.

Figure 26

Solution   For any x-value, the y-value is -4, so the equation is $y = -4$.

Example 17   **Writing the Equation of a Vertical Line**

Write the equation of the line graphed in **Figure 27**.

Figure 27

Solution   The constant x-value is 7, so the equation is $x = 7$.

Determining Whether Lines are Parallel or Perpendicular

The two lines in **Figure 28** are parallel lines: they will never intersect. They have exactly the same steepness, which means their slopes are identical. The only difference between the two lines is the y-intercept. If we shifted one line vertically toward the y-intercept of the other, they would become coincident.

Figure 28 Parallel lines

We can determine from their equations whether two lines are parallel by comparing their slopes. If the slopes are the same and the y-intercepts are different, the lines are parallel. If the slopes are different, the lines are not parallel.

$$f(x) = -2x + 6 \atop f(x) = -2x - 4 \Bigg\} \text{ parallel} \qquad f(x) = 3x + 2 \atop f(x) = 2x + 2 \Bigg\} \text{ not parallel}$$

Unlike parallel lines, perpendicular lines do intersect. Their intersection forms a right, or 90-degree, angle. The two lines in **Figure 29** are perpendicular.

Figure 29 Perpendicular lines

Perpendicular lines do not have the same slope. The slopes of perpendicular lines are different from one another in a specific way. The slope of one line is the negative reciprocal of the slope of the other line. The product of a number and its reciprocal is 1. So, if m_1 and m_2 are negative reciprocals of one another, they can be multiplied together to yield -1.

$$m_1 m_2 = -1$$

To find the reciprocal of a number, divide 1 by the number. So the reciprocal of 8 is $\frac{1}{8}$, and the reciprocal of $\frac{1}{8}$ is 8. To find the negative reciprocal, first find the reciprocal and then change the sign.

As with parallel lines, we can determine whether two lines are perpendicular by comparing their slopes, assuming that the lines are neither horizontal nor vertical. The slope of each line below is the negative reciprocal of the other so the lines are perpendicular.

$$f(x) = \frac{1}{4}x + 2 \quad \text{negative reciprocal of } \frac{1}{4} \text{ is } -4$$

$$f(x) = -4x + 3 \quad \text{negative reciprocal of } -4 \text{ is } \frac{1}{4}$$

The product of the slopes is -1.

$$-4\left(\frac{1}{4}\right) = -1$$

---

parallel and perpendicular lines

Two lines are **parallel lines** if they do not intersect. The slopes of the lines are the same.

$$f(x) = m_1 x + b_1 \text{ and } g(x) = m_2 x + b_2 \text{ are parallel if } m_1 = m_2.$$

If and only if $b_1 = b_2$ and $m_1 = m_2$, we say the lines coincide. Coincident lines are the same line.

Two lines are **perpendicular lines** if they intersect at right angles.

$$f(x) = m_1 x + b_1 \text{ and } g(x) = m_2 x + b_2 \text{ are perpendicular if and only if } m_1 m_2 = -1, \text{ and so } m_2 = -\frac{1}{m_1}$$

Example 18 **Identifying Parallel and Perpendicular Lines**

Given the functions below, identify the functions whose graphs are a pair of parallel lines and a pair of perpendicular lines.

$$f(x) = 2x + 3 \qquad\qquad h(x) = -2x + 2$$

$$g(x) = \frac{1}{2}x - 4 \qquad\qquad j(x) = 2x - 6$$

Solution   Parallel lines have the same slope. Because the functions $f(x) = 2x + 3$ and $j(x) = 2x - 6$ each have a slope of 2, they represent parallel lines. Perpendicular lines have negative reciprocal slopes. Because -2 and $\frac{1}{2}$ are negative reciprocals, the equations, $g(x) = \frac{1}{2}x - 4$ and $h(x) = -2x + 2$ represent perpendicular lines.

Analysis   *A graph of the lines is shown in* **Figure 30**.

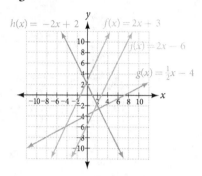

Figure 30

The graph shows that the lines $f(x) = 2x + 3$ and $j(x) = 2x - 6$ are parallel, and the lines $g(x) = \frac{1}{2}x - 4$ and $h(x) = -2x + 2$ are perpendicular.

Writing the Equation of a Line Parallel or Perpendicular to a Given Line

If we know the equation of a line, we can use what we know about slope to write the equation of a line that is either parallel or perpendicular to the given line.

Writing Equations of Parallel Lines

Suppose for example, we are given the equation shown.

$$f(x) = 3x + 1$$

We know that the slope of the line formed by the function is 3. We also know that the y-intercept is $(0, 1)$. Any other line with a slope of 3 will be parallel to $f(x)$. So the lines formed by all of the following functions will be parallel to $f(x)$.

$$g(x) = 3x + 6 \qquad h(x) = 3x + 1 \qquad p(x) = 3x + \frac{2}{3}$$

Suppose then we want to write the equation of a line that is parallel to f and passes through the point $(1, 7)$. This type of problem is often described as a point-slope problem because we have a point and a slope. In our example, we know that the slope is 3. We need to determine which value for b will give the correct line. We can begin with the point-slope form of an equation for a line, and then rewrite it in the slope-intercept form.

$$y - y_1 = m(x - x_1)$$
$$y - 7 = 3(x - 1)$$
$$y - 7 = 3x - 3$$
$$y = 3x + 4$$

So $g(x) = 3x + 4$ is parallel to $f(x) = 3x + 1$ and passes through the point $(1, 7)$.

How To...

Given the equation of a function and a point through which its graph passes, write the equation of a line parallel to the given line that passes through the given point.

1. Find the slope of the function.
2. Substitute the given values into either the general point-slope equation or the slope-intercept equation for a line.
3. Simplify.

Example 19   **Finding a Line Parallel to a Given Line**

Find a line parallel to the graph of $f(x) = 3x + 6$ that passes through the point $(3, 0)$.

Solution   The slope of the given line is 3. If we choose the slope-intercept form, we can substitute $m = 3$, $x = 3$, and $f(x) = 0$ into the slope-intercept form to find the y-intercept.

$$g(x) = 3x + b$$
$$0 = 3(3) + b$$
$$b = -9$$

The line parallel to $f(x)$ that passes through $(3, 0)$ is $g(x) = 3x - 9$.

Analysis   We can confirm that the two lines are parallel by graphing them. **Figure 31** *shows that the two lines will never intersect.*

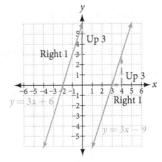

Figure 31

Writing Equations of Perpendicular Lines

We can use a very similar process to write the equation for a line perpendicular to a given line. Instead of using the same slope, however, we use the negative reciprocal of the given slope. Suppose we are given the following function:

$$f(x) = 2x + 4$$

The slope of the line is 2, and its negative reciprocal is $-\frac{1}{2}$. Any function with a slope of $-\frac{1}{2}$ will be perpendicular to $f(x)$. So the lines formed by all of the following functions will be perpendicular to $f(x)$.

$$g(x) = -\frac{1}{2}x + 4 \qquad h(x) = -\frac{1}{2}x + 2 \qquad p(x) = -\frac{1}{2}x - \frac{1}{2}$$

As before, we can narrow down our choices for a particular perpendicular line if we know that it passes through a given point. Suppose then we want to write the equation of a line that is perpendicular to $f(x)$ and passes through the point $(4, 0)$. We already know that the slope is $-\frac{1}{2}$. Now we can use the point to find the y-intercept by substituting the given values into the slope-intercept form of a line and solving for b.

$$g(x) = mx + b$$
$$0 = -\frac{1}{2}(4) + b$$
$$0 = -2 + b$$
$$2 = b$$
$$b = 2$$

The equation for the function with a slope of $-\frac{1}{2}$ and a y-intercept of 2 is

$$g(x) = -\frac{1}{2}x + 2.$$

So $g(x) = -\frac{1}{2}x + 2$ is perpendicular to $f(x) = 2x + 4$ and passes through the point $(4, 0)$. Be aware that perpendicular lines may not look obviously perpendicular on a graphing calculator unless we use the square zoom feature.

Q & A...

A horizontal line has a slope of zero and a vertical line has an undefined slope. These two lines are perpendicular, but the product of their slopes is not -1. Doesn't this fact contradict the definition of perpendicular lines?

No. For two perpendicular linear functions, the product of their slopes is -1. However, a vertical line is not a function so the definition is not contradicted.

Given the equation of a function and a point through which its graph passes, write the equation of a line perpendicular to the given line.

1. Find the slope of the function.
2. Determine the negative reciprocal of the slope.
3. Substitute the new slope and the values for x and y from the coordinate pair provided into $g(x) = mx + b$.
4. Solve for b.
5. Write the equation for the line.

Example 20   **Finding the Equation of a Perpendicular Line**

Find the equation of a line perpendicular to $f(x) = 3x + 3$ that passes through the point $(3, 0)$.

Solution   The original line has slope $m = 3$, so the slope of the perpendicular line will be its negative reciprocal, or $-\frac{1}{3}$. Using this slope and the given point, we can find the equation for the line.

$$g(x) = -\frac{1}{3}x + b$$
$$0 = -\frac{1}{3}(3) + b$$
$$1 = b$$
$$b = 1$$

The line perpendicular to $f(x)$ that passes through $(3, 0)$ is $g(x) = -\frac{1}{3}x + 1$.

Analysis   *A graph of the two lines is shown in **Figure 32**.*

Figure 32

Note that that if we graph perpendicular lines on a graphing calculator using standard zoom, the lines may not appear to be perpendicular. Adjusting the window will make it possible to zoom in further to see the intersection more closely.

Try It #9

Given the function $h(x) = 2x - 4$, write an equation for the line passing through $(0, 0)$ that is

a. parallel to $h(x)$ b. perpendicular to $h(x)$

Given two points on a line and a third point, write the equation of the perpendicular line that passes through the point.

1. Determine the slope of the line passing through the points.
2. Find the negative reciprocal of the slope.
3. Use the slope-intercept form or point-slope form to write the equation by substituting the known values.
4. Simplify.

Example 21 **Finding the Equation of a Line Perpendicular to a Given Line Passing through a Point**

A line passes through the points $(-2, 6)$ and $(4, 5)$. Find the equation of a perpendicular line that passes through the point $(4, 5)$.

Solution From the two points of the given line, we can calculate the slope of that line.

$$m_1 = \frac{5 - 6}{4 - (-2)}$$

$$= \frac{-1}{6}$$

$$= -\frac{1}{6}$$

Find the negative reciprocal of the slope.

$$m_2 = \frac{-1}{-\frac{1}{6}}$$

$$= -1\left(-\frac{6}{1}\right)$$

$$= 6$$

We can then solve for the y-intercept of the line passing through the point $(4, 5)$.

$$g(x) = 6x + b$$

$$5 = 6(4) + b$$

$$5 = 24 + b$$

$$-19 = b$$

$$b = -19$$

The equation for the line that is perpendicular to the line passing through the two given points and also passes through point $(4, 5)$ is

$$y = 6x - 19$$

Try It #10

A line passes through the points, $(-2, -15)$ and $(2, -3)$. Find the equation of a perpendicular line that passes through the point, $(6, 4)$.

Access this online resource for additional instruction and practice with linear functions.

- Linear Functions (http://openstaxcollege.org/l/linearfunctions)
- Finding Input of Function from the Output and Graph (http://openstaxcollege.org/l/findinginput)
- Graphing Functions Using Tables (http://openstaxcollege.org/l/graphwithtable)

4.1   SECTION EXERCISES

VERBAL

1. Terry is skiing down a steep hill. Terry's elevation, $E(t)$, in feet after t seconds is given by $E(t) = 3000 - 70t$. Write a complete sentence describing Terry's starting elevation and how it is changing over time.

2. Jessica is walking home from a friend's house. After 2 minutes she is 1.4 miles from home. Twelve minutes after leaving, she is 0.9 miles from home. What is her rate in miles per hour?

3. A boat is 100 miles away from the marina, sailing directly toward it at 10 miles per hour. Write an equation for the distance of the boat from the marina after t hours.

4. If the graphs of two linear functions are perpendicular, describe the relationship between the slopes and the y-intercepts.

5. If a horizontal line has the equation $f(x) = a$ and a vertical line has the equation $x = a$, what is the point of intersection? Explain why what you found is the point of intersection.

ALGEBRAIC

For the following exercises, determine whether the equation of the curve can be written as a linear function.

6. $y = \frac{1}{4}x + 6$

7. $y = 3x - 5$

8. $y = 3x^2 - 2$

9. $3x + 5y = 15$

10. $3x^2 + 5y = 15$

11. $3x + 5y^2 = 15$

12. $-2x^2 + 3y^2 = 6$

13. $-\frac{x - 3}{5} = 2y$

For the following exercises, determine whether each function is increasing or decreasing.

14. $f(x) = 4x + 3$

15. $g(x) = 5x + 6$

16. $a(x) = 5 - 2x$

17. $b(x) = 8 - 3x$

18. $h(x) = -2x + 4$

19. $k(x) = -4x + 1$

20. $j(x) = \frac{1}{2}x - 3$

21. $p(x) = \frac{1}{4}x - 5$

22. $n(x) = -\frac{1}{3}x - 2$

23. $m(x) = -\frac{3}{8}x + 3$

For the following exercises, find the slope of the line that passes through the two given points.

24. $(2, 4)$ and $(4, 10)$

25. $(1, 5)$ and $(4, 11)$

26. $(-1, 4)$ and $(5, 2)$

27. $(8, -2)$ and $(4, 6)$

28. $(6, 11)$ and $(-4, 3)$

For the following exercises, given each set of information, find a linear equation satisfying the conditions, if possible.

29. $f(-5) = -4$, and $f(5) = 2$

30. $f(-1) = 4$ and $f(5) = 1$

31. Passes through $(2, 4)$ and $(4, 10)$

32. Passes through $(1, 5)$ and $(4, 11)$

33. Passes through $(-1, 4)$ and $(5, 2)$

34. Passes through $(-2, 8)$ and $(4, 6)$

35. x-intercept at $(-2, 0)$ and y-intercept at $(0, -3)$

36. x-intercept at $(-5, 0)$ and y-intercept at $(0, 4)$

For the following exercises, determine whether the lines given by the equations below are parallel, perpendicular, or neither.

37. $4x - 7y = 10$
    $7x + 4y = 1$

38. $3y + x = 12$
    $-y = 8x + 1$

39. $3y + 4x = 12$
    $-6y = 8x + 1$

40. $6x - 9y = 10$
    $3x + 2y = 1$

For the following exercises, find the x- and y-intercepts of each equation

41. $f(x) = -x + 2$ **42.** $g(x) = 2x + 4$ **43.** $h(x) = 3x - 5$

44. $k(x) = -5x + 1$ **45.** $-2x + 5y = 20$ **46.** $7x + 2y = 56$

For the following exercises, use the descriptions of each pair of lines given below to find the slopes of Line 1 and Line 2. Is each pair of lines parallel, perpendicular, or neither?

47. Line 1: Passes through $(0, 6)$ and $(3, -24)$
Line 2: Passes through $(-1, 19)$ and $(8, -71)$

48. Line 1: Passes through $(-8, -55)$ and $(10, 89)$
Line 2: Passes through $(9, -44)$ and $(4, -14)$

49. Line 1: Passes through $(2, 3)$ and $(4, -1)$
Line 2: Passes through $(6, 3)$ and $(8, 5)$

50. Line 1: Passes through $(1, 7)$ and $(5, 5)$
Line 2: Passes through $(-1, -3)$ and $(1, 1)$

51. Line 1: Passes through $(2, 5)$ and $(5, -1)$
Line 2: Passes through $(-3, 7)$ and $(3, -5)$

For the following exercises, write an equation for the line described.

52. Write an equation for a line parallel to $f(x) = -5x - 3$ and passing through the point $(2, -12)$.

53. Write an equation for a line parallel to $g(x) = 3x - 1$ and passing through the point $(4, 9)$.

54. Write an equation for a line perpendicular to $h(t) = -2t + 4$ and passing through the point $(-4, -1)$.

55. Write an equation for a line perpendicular to $p(t) = 3t + 4$ and passing through the point $(3, 1)$.

GRAPHICAL

For the following exercises, find the slope of the lines graphed.

56.

57.
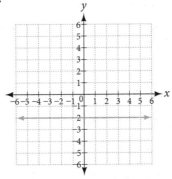

For the following exercises, write an equation for the lines graphed.

58.

59.

60.

61.

62.

63.

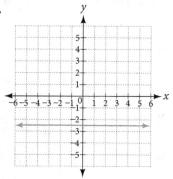

For the following exercises, match the given linear equation with its graph in **Figure 33**.

Figure 33

64. $f(x) = -x - 1$

65. $f(x) = -2x - 1$

66. $f(x) = -\frac{1}{2}x - 1$

67. $f(x) = 2$

68. $f(x) = 2 + x$

69. $f(x) = 3x + 2$

For the following exercises, sketch a line with the given features.

70. An x-intercept of $(-4, 0)$ and y-intercept of $(0, -2)$

71. An x-intercept of $(-2, 0)$ and y-intercept of $(0, 4)$

72. A y-intercept of $(0, 7)$ and slope $-\frac{3}{2}$

73. A y-intercept of $(0, 3)$ and slope $\frac{2}{5}$

74. Passing through the points $(-6, -2)$ and $(6, -6)$

75. Passing through the points $(-3, -4)$ and $(3, 0)$

For the following exercises, sketch the graph of each equation.

76. $f(x) = -2x - 1$

77. $g(x) = -3x + 2$

78. $h(x) = \frac{1}{3}x + 2$

79. $k(x) = \frac{2}{3}x - 3$

80. $f(t) = 3 + 2t$

81. $p(t) = -2 + 3t$

82. $x = 3$

83. $x = -2$

84. $r(x) = 4$

For the following exercises, write the equation of the line shown in the graph.

85.

86.

87.

88.

NUMERIC

For the following exercises, which of the tables could represent a linear function? For each that could be linear, find a linear equation that models the data.

89.

x	0	5	10	15
$g(x)$	5	−10	−25	−40

90.

x	0	5	10	15
$h(x)$	5	30	105	230

91.

x	0	5	10	15
$f(x)$	−5	20	45	70

92.

x	5	10	20	25
$k(x)$	13	28	58	73

93.

x	0	2	4	6
$g(x)$	6	−19	−44	−69

94.

x	2	4	8	10
$h(x)$	13	23	43	53

95.

x	2	4	6	8
$f(x)$	−4	16	36	56

96.

x	0	2	6	8
$k(x)$	6	31	106	231

TECHNOLOGY

For the following exercises, use a calculator or graphing technology to complete the task.

97. If f is a linear function, $f(0.1) = 11.5$, and $f(0.4) = -5.9$, find an equation for the function.

98. Graph the function f on a domain of $[-10, 10] : f(x) = 0.02x - 0.01$. Enter the function in a graphing utility. For the viewing window, set the minimum value of x to be -10 and the maximum value of x to be 10.

99. Graph the function f on a domain of $[-10, 10] : f(x) = 2{,}500x + 4{,}000$

100. Table 3 shows the input, w, and output, k, for a linear function k. **a.** Fill in the missing values of the table. **b.** Write the linear function k, round to 3 decimal places.

w	−10	5.5	67.5	b
k	30	−26	a	−44

Table 3

101. Table 4 shows the input, p, and output, q, for a linear function q. **a.** Fill in the missing values of the table. **b.** Write the linear function k.

p	0.5	0.8	12	b
q	400	700	a	1,000,000

Table 4

102. Graph the linear function f on a domain of $[-10, 10]$ for the function whose slope is $\frac{1}{8}$ and y-intercept is $\frac{31}{16}$. Label the points for the input values of -10 and 10.

103. Graph the linear function f on a domain of $[-0.1, 0.1]$ for the function whose slope is 75 and y-intercept is -22.5. Label the points for the input values of -0.1 and 0.1.

104. Graph the linear function f where $f(x) = ax + b$ on the same set of axes on a domain of $[-4, 4]$ for the following values of a and b.

    a. $a = 2; b = 3$ **b.** $a = 2; b = 4$ **c.** $a = 2; b = -4$ **d.** $a = 2; b = -5$

EXTENSIONS

105. Find the value of x if a linear function goes through the following points and has the following slope: $(x, 2), (-4, 6), m = 3$

106. Find the value of y if a linear function goes through the following points and has the following slope: $(10, y), (25, 100), m = -5$

107. Find the equation of the line that passes through the following points: (a, b) and $(a, b + 1)$

108. Find the equation of the line that passes through the following points: $(2a, b)$ and $(a, b + 1)$

109. Find the equation of the line that passes through the following points: $(a, 0)$ and (c, d)

110. Find the equation of the line parallel to the line $g(x) = -0.01x + 2.01$ through the point $(1, 2)$.

111. Find the equation of the line perpendicular to the line $g(x) = -0.01x + 2.01$ through the point $(1, 2)$.

For the following exercises, use the functions $f(x) = -0.1x + 200$ and $g(x) = 20x + 0.1$.

112. Find the point of intersection of the lines f and g.

113. Where is $f(x)$ greater than $g(x)$? Where is $g(x)$ greater than $f(x)$?

REAL-WORLD APPLICATIONS

114. At noon, a barista notices that she has $20 in her tip jar. If she makes an average of $0.50 from each customer, how much will she have in her tip jar if she serves n more customers during her shift?

115. A gym membership with two personal training sessions costs $125, while gym membership with five personal training sessions costs $260. What is cost per session?

116. A clothing business finds there is a linear relationship between the number of shirts, n, it can sell and the price, p, it can charge per shirt. In particular, historical data shows that 1,000 shirts can be sold at a price of $30, while 3,000 shirts can be sold at a price of $22. Find a linear equation in the form $p(n) = mn + b$ that gives the price p they can charge for n shirts.

117. A phone company charges for service according to the formula: $C(n) = 24 + 0.1n$, where n is the number of minutes talked, and $C(n)$ is the monthly charge, in dollars. Find and interpret the rate of change and initial value.

118. A farmer finds there is a linear relationship between the number of bean stalks, n, she plants and the yield, y, each plant produces. When she plants 30 stalks, each plant yields 30 oz of beans. When she plants 34 stalks, each plant produces 28 oz of beans. Find a linear relationship in the form $y = mn + b$ that gives the yield when n stalks are planted.

119. A city's population in the year 1960 was 287,500. In 1989 the population was 275,900. Compute the rate of growth of the population and make a statement about the population rate of change in people per year.

120. A town's population has been growing linearly. In 2003, the population was 45,000, and the population has been growing by 1,700 people each year. Write an equation, $P(t)$, for the population t years after 2003.

121. Suppose that average annual income (in dollars) for the years 1990 through 1999 is given by the linear function: $I(x) = 1{,}054x + 23{,}286$, where x is the number of years after 1990. Which of the following interprets the slope in the context of the problem?

  a. As of 1990, average annual income was $23,286.

  b. In the ten-year period from 1990–1999, average annual income increased by a total of $1,054.

  c. Each year in the decade of the 1990s, average annual income increased by $1,054.

  d. Average annual income rose to a level of $23,286 by the end of 1999.

122. When temperature is 0 degrees Celsius, the Fahrenheit temperature is 32. When the Celsius temperature is 100, the corresponding Fahrenheit temperature is 212. Express the Fahrenheit temperature as a linear function of C, the Celsius temperature, $F(C)$.

  a. Find the rate of change of Fahrenheit temperature for each unit change temperature of Celsius.

  b. Find and interpret $F(28)$.

  c. Find and interpret $F(-40)$.

LEARNING OBJECTIVES

In this section, you will:

- Build linear models from verbal descriptions.
- Model a set of data with a linear function.

4.2   MODELING WITH LINEAR FUNCTIONS

Figure 1 (credit: EEK Photography/Flickr)

Emily is a college student who plans to spend a summer in Seattle. She has saved $3,500 for her trip and anticipates spending $400 each week on rent, food, and activities. How can we write a linear model to represent her situation? What would be the *x*-intercept, and what can she learn from it? To answer these and related questions, we can create a model using a linear function. Models such as this one can be extremely useful for analyzing relationships and making predictions based on those relationships. In this section, we will explore examples of linear function models.

Building Linear Models from Verbal Descriptions

When building linear models to solve problems involving quantities with a constant rate of change, we typically follow the same problem strategies that we would use for any type of function. Let's briefly review them:

1. Identify changing quantities, and then define descriptive variables to represent those quantities. When appropriate, sketch a picture or define a coordinate system.
2. Carefully read the problem to identify important information. Look for information that provides values for the variables or values for parts of the functional model, such as slope and initial value.
3. Carefully read the problem to determine what we are trying to find, identify, solve, or interpret.
4. Identify a solution pathway from the provided information to what we are trying to find. Often this will involve checking and tracking units, building a table, or even finding a formula for the function being used to model the problem.
5. When needed, write a formula for the function.
6. Solve or evaluate the function using the formula.
7. Reflect on whether your answer is reasonable for the given situation and whether it makes sense mathematically.
8. Clearly convey your result using appropriate units, and answer in full sentences when necessary.

Now let's take a look at the student in Seattle. In her situation, there are two changing quantities: time and money. The amount of money she has remaining while on vacation depends on how long she stays. We can use this information to define our variables, including units.

Output: M, money remaining, in dollars

Input: t, time, in weeks

So, the amount of money remaining depends on the number of weeks: $M(t)$

We can also identify the initial value and the rate of change.

Initial Value: She saved $3,500, so $3,500 is the initial value for M.

Rate of Change: She anticipates spending $400 each week, so $-$$400 per week is the rate of change, or slope.

Notice that the unit of dollars per week matches the unit of our output variable divided by our input variable. Also, because the slope is negative, the linear function is decreasing. This should make sense because she is spending money each week.

The rate of change is constant, so we can start with the linear model $M(t) = mt + b$. Then we can substitute the intercept and slope provided.

$$M(t) = mt + b$$

$$-400 \quad 3500$$

$$M(t) = -400t + 3500$$

To find the x-intercept, we set the output to zero, and solve for the input.

$$0 = -400t + 3500$$

$$t = \frac{3500}{400}$$

$$= 8.75$$

The x-intercept is 8.75 weeks. Because this represents the input value when the output will be zero, we could say that Emily will have no money left after 8.75 weeks.

When modeling any real-life scenario with functions, there is typically a limited domain over which that model will be valid—almost no trend continues indefinitely. Here the domain refers to the number of weeks. In this case, it doesn't make sense to talk about input values less than zero. A negative input value could refer to a number of weeks before she saved $3,500, but the scenario discussed poses the question once she saved $3,500 because this is when her trip and subsequent spending starts. It is also likely that this model is not valid after the x-intercept, unless Emily will use a credit card and goes into debt. The domain represents the set of input values, so the reasonable domain for this function is $0 \le t \le 8.75$.

In the above example, we were given a written description of the situation. We followed the steps of modeling a problem to analyze the information. However, the information provided may not always be the same. Sometimes we might be provided with an intercept. Other times we might be provided with an output value. We must be careful to analyze the information we are given, and use it appropriately to build a linear model.

Using a Given Intercept to Build a Model

Some real-world problems provide the y-intercept, which is the constant or initial value. Once the y-intercept is known, the x-intercept can be calculated. Suppose, for example, that Hannah plans to pay off a no-interest loan from her parents. Her loan balance is $1,000. She plans to pay $250 per month until her balance is $0. The y-intercept is the initial amount of her debt, or $1,000. The rate of change, or slope, is $-$$250 per month. We can then use the slope-intercept form and the given information to develop a linear model.

$$f(x) = mx + b$$

$$= -250x + 1000$$

Now we can set the function equal to 0, and solve for x to find the x-intercept.

$$0 = -250x + 1000$$
$$1000 = 250x$$
$$4 = x$$
$$x = 4$$

The x-intercept is the number of months it takes her to reach a balance of $0. The x-intercept is 4 months, so it will take Hannah four months to pay off her loan.

Using a Given Input and Output to Build a Model

Many real-world applications are not as direct as the ones we just considered. Instead they require us to identify some aspect of a linear function. We might sometimes instead be asked to evaluate the linear model at a given input or set the equation of the linear model equal to a specified output.

How To…

Given a word problem that includes two pairs of input and output values, use the linear function to solve a problem.

1. Identify the input and output values.
2. Convert the data to two coordinate pairs.
3. Find the slope.
4. Write the linear model.
5. Use the model to make a prediction by evaluating the function at a given x-value.
6. Use the model to identify an x-value that results in a given y-value.
7. Answer the question posed.

Example   1   **Using a Linear Model to Investigate a Town's Population**

A town's population has been growing linearly. In 2004 the population was 6,200. By 2009 the population had grown to 8,100. Assume this trend continues.

   a. Predict the population in 2013.

   b. Identify the year in which the population will reach 15,000.

Solution   The two changing quantities are the population size and time. While we could use the actual year value as the input quantity, doing so tends to lead to very cumbersome equations because the y-intercept would correspond to the year 0, more than 2,000 years ago!

To make computation a little nicer, we will define our input as the number of years since 2004:

Input: t, years since 2004
Output: $P(t)$, the town's population

To predict the population in 2013 ($t = 9$), we would first need an equation for the population. Likewise, to find when the population would reach 15,000, we would need to solve for the input that would provide an output of 15,000. To write an equation, we need the initial value and the rate of change, or slope.

To determine the rate of change, we will use the change in output per change in input.

$$m = \frac{\text{change in output}}{\text{change in input}}$$

The problem gives us two input-output pairs. Converting them to match our defined variables, the year 2004 would correspond to $t = 0$, giving the point (0, 6200). Notice that through our clever choice of variable definition, we have "given" ourselves the y-intercept of the function. The year 2009 would correspond to $t = 5$, giving the point (5, 8100).

The two coordinate pairs are (0, 6200) and (5, 8100). Recall that we encountered examples in which we were provided two points earlier in the chapter. We can use these values to calculate the slope.

$$m = \frac{8100 - 6200}{5 - 0}$$

$$= \frac{1900}{5}$$

$$= 380 \text{ people per year}$$

We already know the y-intercept of the line, so we can immediately write the equation:

$$P(t) = 380t + 6200$$

To predict the population in 2013, we evaluate our function at $t = 9$.

$$P(9) = 380(9) + 6{,}200$$

$$= 9{,}620$$

If the trend continues, our model predicts a population of 9,620 in 2013.

To find when the population will reach 15,000, we can set $P(t) = 15000$ and solve for t.

$$15000 = 380t + 6200$$

$$8800 = 380t$$

$$t \approx 23.158$$

Our model predicts the population will reach 15,000 in a little more than 23 years after 2004, or somewhere around the year 2027.

Try It #1

A company sells doughnuts. They incur a fixed cost of $25,000 for rent, insurance, and other expenses. It costs $0.25 to produce each doughnut.

a. Write a linear model to represent the cost C of the company as a function of x, the number of doughnuts produced.
b. Find and interpret the y-intercept.

Try It #2

A city's population has been growing linearly. In 2008, the population was 28,200. By 2012, the population was 36,800. Assume this trend continues.

a. Predict the population in 2014.

b. Identify the year in which the population will reach 54,000.

Using a Diagram to Model a Problem

It is useful for many real-world applications to draw a picture to gain a sense of how the variables representing the input and output may be used to answer a question. To draw the picture, first consider what the problem is asking for. Then, determine the input and the output. The diagram should relate the variables. Often, geometrical shapes or figures are drawn. Distances are often traced out. If a right triangle is sketched, the Pythagorean Theorem relates the sides. If a rectangle is sketched, labeling width and height is helpful.

Example 2   **Using a Diagram to Model Distance Walked**

Anna and Emanuel start at the same intersection. Anna walks east at 4 miles per hour while Emanuel walks south at 3 miles per hour. They are communicating with a two-way radio that has a range of 2 miles. How long after they start walking will they fall out of radio contact?

Solution   In essence, we can partially answer this question by saying they will fall out of radio contact when they are 2 miles apart, which leads us to ask a new question:

"How long will it take them to be 2 miles apart?"

In this problem, our changing quantities are time and position, but ultimately we need to know how long will it take for them to be 2 miles apart. We can see that time will be our input variable, so we'll define our input and output variables.

Input: t, time in hours.

Output: $A(t)$, distance in miles, and $E(t)$, distance in miles

Because it is not obvious how to define our output variable, we'll start by drawing a picture such as **Figure 2**.

Figure 2

Initial Value: They both start at the same intersection so when $t = 0$, the distance traveled by each person should also be 0. Thus the initial value for each is 0.

Rate of Change: Anna is walking 4 miles per hour and Emanuel is walking 3 miles per hour, which are both rates of change. The slope for A is 4 and the slope for E is 3.

Using those values, we can write formulas for the distance each person has walked.

$$A(t) = 4t$$

$$E(t) = 3t$$

For this problem, the distances from the starting point are important. To notate these, we can define a coordinate system, identifying the "starting point" at the intersection where they both started. Then we can use the variable, A, which we introduced above, to represent Anna's position, and define it to be a measurement from the starting point in the eastward direction. Likewise, can use the variable, E, to represent Emanuel's position, measured from the starting point in the southward direction. Note that in defining the coordinate system, we specified both the starting point of the measurement and the direction of measure.

We can then define a third variable, D, to be the measurement of the distance between Anna and Emanuel.

Showing the variables on the diagram is often helpful, as we can see from **Figure 3**.

Recall that we need to know how long it takes for D, the distance between them, to equal 2 miles. Notice that for any given input t, the outputs $A(t)$, $E(t)$, and $D(t)$ represent distances.

Figure 3

Figure 3 shows us that we can use the Pythagorean Theorem because we have drawn a right angle.

Using the Pythagorean Theorem, we get:

$$D(t)^2 = A(t)^2 + E(t)^2$$
$$= (4t)^2 + (3t)^2$$
$$= 16t^2 + 9t^2$$
$$= 25t^2$$
$$D(t) = \pm\sqrt{25t^2} \qquad \text{Solve for } D(t) \text{ using the square root.}$$
$$= \pm 5|t|$$

In this scenario we are considering only positive values of t, so our distance $D(t)$ will always be positive. We can simplify this answer to $D(t) = 5t$. This means that the distance between Anna and Emanuel is also a linear function. Because D is a linear function, we can now answer the question of when the distance between them will reach 2 miles. We will set the output $D(t) = 2$ and solve for t.

$$D(t) = 2$$
$$5t = 2$$
$$t = \frac{2}{5} = 0.4$$

They will fall out of radio contact in 0.4 hours, or 24 minutes.

Q & A...

Should I draw diagrams when given information based on a geometric shape?

Yes. Sketch the figure and label the quantities and unknowns on the sketch.

Example 3   **Using a Diagram to Model Distance Between Cities**

There is a straight road leading from the town of Westborough to Agritown 30 miles east and 10 miles north. Partway down this road, it junctions with a second road, perpendicular to the first, leading to the town of Eastborough. If the town of Eastborough is located 20 miles directly east of the town of Westborough, how far is the road junction from Westborough?

Solution   It might help here to draw a picture of the situation. See **Figure 4**. It would then be helpful to introduce a coordinate system. While we could place the origin anywhere, placing it at Westborough seems convenient. This puts Agritown at coordinates (30, 10), and Eastborough at (20, 0).

Figure 4

Using this point along with the origin, we can find the slope of the line from Westborough to Agritown:

$$m = \frac{10 - 0}{30 - 0} = \frac{1}{3}$$

Now we can write an equation to describe the road from Westborough to Agritown.

$$W(x) = \frac{1}{3}x$$

From this, we can determine the perpendicular road to Eastborough will have slope $m = -3$. Because the town of Eastborough is at the point (20, 0), we can find the equation:

$$E(x) = -3x + b$$
$$0 = -3(20) + b \qquad \text{Substitute } (20, 0) \text{ into the equation.}$$
$$b = 60$$
$$E(x) = -3x + 60$$

We can now find the coordinates of the junction of the roads by finding the intersection of these lines. Setting them equal,

$$\frac{1}{3}x = -3x + 60$$

$$\frac{10}{3}x = 60$$

$$10x = 180$$

$$x = 18 \qquad \text{Substitute this back into } W(x).$$

$$y = W(18)$$

$$= \frac{1}{3}(18)$$

$$= 6$$

The roads intersect at the point (18, 6). Using the distance formula, we can now find the distance from Westborough to the junction.

$$\text{distance} = \sqrt{(x_2 - x_1)^2 + (y_2 - y_1)^2}$$

$$= \sqrt{(18 - 0)^2 + (6 - 0)^2}$$

$$\approx 18.974 \text{ miles}$$

Analysis   One nice use of linear models is to take advantage of the fact that the graphs of these functions are lines. This means real-world applications discussing maps need linear functions to model the distances between reference points.

Try It #3

There is a straight road leading from the town of Timpson to Ashburn 60 miles east and 12 miles north. Partway down the road, it junctions with a second road, perpendicular to the first, leading to the town of Garrison. If the town of Garrison is located 22 miles directly east of the town of Timpson, how far is the road junction from Timpson?

Modeling a Set of Data with Linear Functions

Real-world situations including two or more linear functions may be modeled with a system of linear equations. Remember, when solving a system of linear equations, we are looking for points the two lines have in common. Typically, there are three types of answers possible, as shown in **Figure 5**.

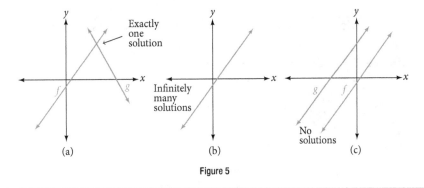

Figure 5

How To…

Given a situation that represents a system of linear equations, write the system of equations and identify the solution.

1. Identify the input and output of each linear model.
2. Identify the slope and y-intercept of each linear model.
3. Find the solution by setting the two linear functions equal to one another and solving for x, or find the point of intersection on a graph.

Example 4   **Building a System of Linear Models to Choose a Truck Rental Company**

Jamal is choosing between two truck-rental companies. The first, Keep on Trucking, Inc., charges an up-front fee of $20, then 59 cents a mile. The second, Move It Your Way, charges an up-front fee of $16, then 63 cents a mile[9]. When will Keep on Trucking, Inc. be the better choice for Jamal?

Solution   The two important quantities in this problem are the cost and the number of miles driven. Because we have two companies to consider, we will define two functions in **Table 1**.

Input	d, distance driven in miles
Outputs	$K(d)$: cost, in dollars, for renting from Keep on Trucking $M(d)$ cost, in dollars, for renting from Move It Your Way
Initial Value	Up-front fee: $K(0) = 20$ and $M(0) = 16$
Rate of Change	$K(d) = \$0.59$/mile and $P(d) = \$0.63$/mile

Table 1

A linear function is of the form $f(x) = mx + b$. Using the rates of change and initial charges, we can write the equations

$$K(d) = 0.59d + 20$$

$$M(d) = 0.63d + 16$$

Using these equations, we can determine when Keep on Trucking, Inc., will be the better choice. Because all we have to make that decision from is the costs, we are looking for when Move It Your Way, will cost less, or when $K(d) < M(d)$. The solution pathway will lead us to find the equations for the two functions, find the intersection, and then see where the $K(d)$ function is smaller.

These graphs are sketched in **Figure 6**, with $K(d)$ in red.

Figure 6

To find the intersection, we set the equations equal and solve:

$$K(d) = M(d)$$

$$0.59d + 20 = 0.63d + 16$$

$$4 = 0.04d$$

$$100 = d$$

$$d = 100$$

This tells us that the cost from the two companies will be the same if 100 miles are driven. Either by looking at the graph, or noting that $K(d)$ is growing at a slower rate, we can conclude that Keep on Trucking, Inc. will be the cheaper price when more than 100 miles are driven, that is $d > 100$.

Access this online resources for additional instruction and practice with linear function models.

• Interpreting a Linear Function (http://openstaxcollege.org/l/interpretlinear)

9 Rates retrieved Aug 2, 2010 from http://www.budgettruck.com and http://www.uhaul.com/

4.2 SECTION EXERCISES

VERBAL

1. Explain how to find the input variable in a word problem that uses a linear function.

2. Explain how to find the output variable in a word problem that uses a linear function.

3. Explain how to interpret the initial value in a word problem that uses a linear function.

4. Explain how to determine the slope in a word problem that uses a linear function.

ALGEBRAIC

5. Find the area of a parallelogram bounded by the y-axis, the line $x = 3$, the line $f(x) = 1 + 2x$, and the line parallel to $f(x)$ passing through $(2, 7)$.

6. Find the area of a triangle bounded by the x-axis, the line $f(x) = 12 - \frac{1}{3}x$, and the line perpendicular to $f(x)$ that passes through the origin.

7. Find the area of a triangle bounded by the y-axis, the line $f(x) = 9 - \frac{6}{7}x$, and the line perpendicular to $f(x)$ that passes through the origin.

8. Find the area of a parallelogram bounded by the x-axis, the line $g(x) = 2$, the line $f(x) = 3x$, and the line parallel to $f(x)$ passing through $(6, 1)$.

For the following exercises, consider this scenario: A town's population has been decreasing at a constant rate. In 2010 the population was 5,900. By 2012 the population had dropped 4,700. Assume this trend continues.

9. Predict the population in 2016.

10. Identify the year in which the population will reach 0.

For the following exercises, consider this scenario: A town's population has been increased at a constant rate. In 2010 the population was 46,020. By 2012 the population had increased to 52,070. Assume this trend continues.

11. Predict the population in 2016.

12. Identify the year in which the population will reach 75,000.

For the following exercises, consider this scenario: A town has an initial population of 75,000. It grows at a constant rate of 2,500 per year for 5 years.

13. Find the linear function that models the town's population P as a function of the year, t, where t is the number of years since the model began.

14. Find a reasonable domain and range for the function P.

15. If the function P is graphed, find and interpret the x-and y-intercepts.

16. If the function P is graphed, find and interpret the slope of the function.

17. When will the output reached 100,000?

18. What is the output in the year 12 years from the onset of the model?

For the following exercises, consider this scenario: The weight of a newborn is 7.5 pounds. The baby gained one-half pound a month for its first year.

19. Find the linear function that models the baby's weight, W, as a function of the age of the baby, in months, t.

20. Find a reasonable domain and range for the function W.

21. If the function W is graphed, find and interpret the x- and y-intercepts.

22. If the function W is graphed, find and interpret the slope of the function.

23. When did the baby weight 10.4 pounds?

24. What is the output when the input is 6.2? Interpret your answer.

For the following exercises, consider this scenario: The number of people afflicted with the common cold in the winter months steadily decreased by 205 each year from 2005 until 2010. In 2005, 12,025 people were afflicted.

25. Find the linear function that models the number of people inflicted with the common cold, C, as a function of the year, t.

26. Find a reasonable domain and range for the function C.

27. If the function C is graphed, find and interpret the x-and y-intercepts.

28. If the function C is graphed, find and interpret the slope of the function.

29. When will the output reach 0?

30. In what year will the number of people be 9,700?

GRAPHICAL

For the following exercises, use the graph in **Figure 7**, which shows the profit, y, in thousands of dollars, of a company in a given year, t, where t represents the number of years since 1980.

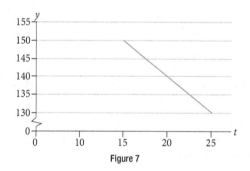

Figure 7

31. Find the linear function y, where y depends on t, the number of years since 1980.

32. Find and interpret the y-intercept.

33. Find and interpret the x-intercept.

34. Find and interpret the slope.

For the following exercises, use the graph in **Figure 8**, which shows the profit, y, in thousands of dollars, of a company in a given year, t, where t represents the number of years since 1980.

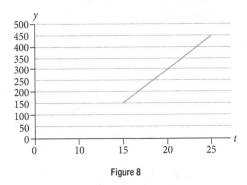

Figure 8

35. Find the linear function y, where y depends on t, the number of years since 1980.

36. Find and interpret the y-intercept.

37. Find and interpret the x-intercept.

38. Find and interpret the slope.

NUMERIC

For the following exercises, use the median home values in Mississippi and Hawaii (adjusted for inflation) shown in **Table 2**. Assume that the house values are changing linearly.

Year	Mississippi	Hawaii
1950	$25,200	$74,400
2000	$71,400	$272,700

Table 2

39. In which state have home values increased at a higher rate?

40. If these trends were to continue, what would be the median home value in Mississippi in 2010?

41. If we assume the linear trend existed before 1950 and continues after 2000, the two states' median house values will be (or were) equal in what year? (The answer might be absurd.)

For the following exercises, use the median home values in Indiana and Alabama (adjusted for inflation) shown in **Table 3**. Assume that the house values are changing linearly.

Year	Indiana	Alabama
1950	$37,700	$27,100
2000	$94,300	$85,100

Table 3

42. In which state have home values increased at a higher rate?

43. If these trends were to continue, what would be the median home value in Indiana in 2010?

44. If we assume the linear trend existed before 1950 and continues after 2000, the two states' median house values will be (or were) equal in what year? (The answer might be absurd.)

REAL-WORLD APPLICATIONS

45. In 2004, a school population was 1,001. By 2008 the population had grown to 1,697. Assume the population is changing linearly.

   a. How much did the population grow between the year 2004 and 2008?

   b. How long did it take the population to grow from 1,001 students to 1,697 students?

   c. What is the average population growth per year?

   d. What was the population in the year 2000?

   e. Find an equation for the population, P, of the school t years after 2000.

   f. Using your equation, predict the population of the school in 2011.

46. In 2003, a town's population was 1,431. By 2007 the population had grown to 2,134. Assume the population is changing linearly.

   a. How much did the population grow between the year 2003 and 2007?

   b. How long did it take the population to grow from 1,431 people to 2,134 people?

   c. What is the average population growth per year?

   d. What was the population in the year 2000?

   e. Find an equation for the population, P of the town t years after 2000.

   f. Using your equation, predict the population of the town in 2014.

47. A phone company has a monthly cellular plan where a customer pays a flat monthly fee and then a certain amount of money per minute used on the phone. If a customer uses 410 minutes, the monthly cost will be $71.50. If the customer uses 720 minutes, the monthly cost will be $118.

   a. Find a linear equation for the monthly cost of the cell plan as a function of x, the number of monthly minutes used.

   b. Interpret the slope and y-intercept of the equation.

   c. Use your equation to find the total monthly cost if 687 minutes are used.

48. A phone company has a monthly cellular data plan where a customer pays a flat monthly fee of $10 and then a certain amount of money per megabyte (MB) of data used on the phone. If a customer uses 20 MB, the monthly cost will be $11.20. If the customer uses 130 MB, the monthly cost will be $17.80.

   a. Find a linear equation for the monthly cost of the data plan as a function of x, the number of MB used.

   b. Interpret the slope and y-intercept of the equation.

   c. Use your equation to find the total monthly cost if 250 MB are used.

49. In 1991, the moose population in a park was measured to be 4,360. By 1999, the population was measured again to be 5,880. Assume the population continues to change linearly.

   a. Find a formula for the moose population, P since 1990.

   b. What does your model predict the moose population to be in 2003?

50. In 2003, the owl population in a park was measured to be 340. By 2007, the population was measured again to be 285. The population changes linearly. Let the input be years since 1990.

   a. Find a formula for the owl population, P. Let the input be years since 2003.

   b. What does your model predict the owl population to be in 2012?

51. The Federal Helium Reserve held about 16 billion cubic feet of helium in 2010 and is being depleted by about 2.1 billion cubic feet each year.

   a. Give a linear equation for the remaining federal helium reserves, R, in terms of t, the number of years since 2010.

   b. In 2015, what will the helium reserves be?

   c. If the rate of depletion doesn't change, in what year will the Federal Helium Reserve be depleted?

52. Suppose the world's oil reserves in 2014 are 1,820 billion barrels. If, on average, the total reserves are decreasing by 25 billion barrels of oil each year:

   a. Give a linear equation for the remaining oil reserves, R, in terms of t, the number of years since now.

   b. Seven years from now, what will the oil reserves be?

   c. If the rate at which the reserves are decreasing is constant, when will the world's oil reserves be depleted?

53. You are choosing between two different prepaid cell phone plans. The first plan charges a rate of 26 cents per minute. The second plan charges a monthly fee of $19.95 *plus* 11 cents per minute. How many minutes would you have to use in a month in order for the second plan to be preferable?

54. You are choosing between two different window washing companies. The first charges $5 per window. The second charges a base fee of $40 plus $3 per window. How many windows would you need to have for the second company to be preferable?

55. When hired at a new job selling jewelry, you are given two pay options:
- Option A: Base salary of $17,000 a year with a commission of 12% of your sales
- Option B: Base salary of $20,000 a year with a commission of 5% of your sales

How much jewelry would you need to sell for option A to produce a larger income?

56. When hired at a new job selling electronics, you are given two pay options:
- Option A: Base salary of $14,000 a year with a commission of 10% of your sales
- Option B: Base salary of $19,000 a year with a commission of 4% of your sales

How much electronics would you need to sell for option A to produce a larger income?

57. When hired at a new job selling electronics, you are given two pay options:
- Option A: Base salary of $20,000 a year with a commission of 12% of your sales
- Option B: Base salary of $26,000 a year with a commission of 3% of your sales

How much electronics would you need to sell for option A to produce a larger income?

58. When hired at a new job selling electronics, you are given two pay options:
- Option A: Base salary of $10,000 a year with a commission of 9% of your sales
- Option B: Base salary of $20,000 a year with a commission of 4% of your sales

How much electronics would you need to sell for option A to produce a larger income?

LEARNING OBJECTIVES

In this section, you will:

- Draw and interpret scatter plots.
- Use a graphing utility to find the line of best fit.
- Distinguish between linear and nonlinear relations.
- Fit a regression line to a set of data and use the linear model to make predictions.

4.3   FITTING LINEAR MODELS TO DATA

A professor is attempting to identify trends among final exam scores. His class has a mixture of students, so he wonders if there is any relationship between age and final exam scores. One way for him to analyze the scores is by creating a diagram that relates the age of each student to the exam score received. In this section, we will examine one such diagram known as a scatter plot.

Drawing and Interpreting Scatter Plots

A scatter plot is a graph of plotted points that may show a relationship between two sets of data. If the relationship is from a linear model, or a model that is nearly linear, the professor can draw conclusions using his knowledge of linear functions. **Figure 1** shows a sample scatter plot.

Figure 1 A scatter plot of age and final exam score variables.

Notice this scatter plot does *not* indicate a linear relationship. The points do not appear to follow a trend. In other words, there does not appear to be a relationship between the age of the student and the score on the final exam.

Example 1   **Using a Scatter Plot to Investigate Cricket Chirps**

Table 1 shows the number of cricket chirps in 15 seconds, for several different air temperatures, in degrees Fahrenheit[10]. Plot this data, and determine whether the data appears to be linearly related.

Chirps	44	35	20.4	33	31	35	18.5	37	26
Temperature	80.5	70.5	57	66	68	72	52	73.5	53

Table 1 Cricket Chirps vs Air Temperature

Solution   Plotting this data, as depicted in **Figure 2** suggests that there may be a trend. We can see from the trend in the data that the number of chirps increases as the temperature increases. The trend appears to be roughly linear, though certainly not perfectly so.

---

10 Selected data from http://classic.globe.gov/fsl/scientistsblog/2007/10/. Retrieved Aug 3, 2010.

Figure 2

Finding the Line of Best Fit

Once we recognize a need for a linear function to model that data, the natural follow-up question is "what is that linear function?" One way to approximate our linear function is to sketch the line that seems to best fit the data. Then we can extend the line until we can verify the y-intercept. We can approximate the slope of the line by extending it until we can estimate the $\frac{\text{rise}}{\text{run}}$.

Example 2   Finding a Line of Best Fit

Find a linear function that fits the data in **Table 1** by "eyeballing" a line that seems to fit.

Solution   On a graph, we could try sketching a line. Using the starting and ending points of our hand drawn line, points (0, 30) and (50, 90), this graph has a slope of

$$m = \frac{60}{50} = 1.2$$

and a y-intercept at 30. This gives an equation of

$$T(c) = 1.2c + 30$$

where c is the number of chirps in 15 seconds, and $T(c)$ is the temperature in degrees Fahrenheit. The resulting equation is represented in **Figure 3**.

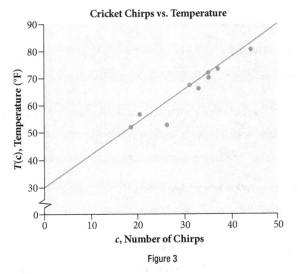

Figure 3

Analysis   *This linear equation can then be used to approximate answers to various questions we might ask about the trend.*

While the data for most examples does not fall perfectly on the line, the equation is our best guess as to how the relationship will behave outside of the values for which we have data. We use a process known as **interpolation** when we predict a value inside the domain and range of the data. The process of **extrapolation** is used when we predict a value outside the domain and range of the data.

Figure 4 compares the two processes for the cricket-chirp data addressed in **Example 2**. We can see that interpolation would occur if we used our model to predict temperature when the values for chirps are between 18.5 and 44. Extrapolation would occur if we used our model to predict temperature when the values for chirps are less than 18.5 or greater than 44.

There is a difference between making predictions inside the domain and range of values for which we have data and outside that domain and range. Predicting a value outside of the domain and range has its limitations. When our model no longer applies after a certain point, it is sometimes called **model breakdown**. For example, predicting a cost function for a period of two years may involve examining the data where the input is the time in years and the output is the cost. But if we try to extrapolate a cost when $x = 50$, that is in 50 years, the model would not apply because we could not account for factors fifty years in the future.

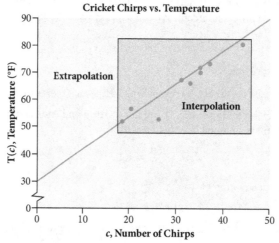

Figure 4 Interpolation occurs within the domain and range of the provided data whereas extrapolation occurs outside.

---

interpolation and extrapolation

Different methods of making predictions are used to analyze data.

The method of **interpolation** involves predicting a value inside the domain and/or range of the data.

The method of **extrapolation** involves predicting a value outside the domain and/or range of the data.

Model breakdown occurs at the point when the model no longer applies.

Example 3   **Understanding Interpolation and Extrapolation**

Use the cricket data from **Table 1** to answer the following questions:

   a. Would predicting the temperature when crickets are chirping 30 times in 15 seconds be interpolation or extrapolation? Make the prediction, and discuss whether it is reasonable.

   b. Would predicting the number of chirps crickets will make at 40 degrees be interpolation or extrapolation? Make the prediction, and discuss whether it is reasonable.

a. The number of chirps in the data provided varied from 18.5 to 44. A prediction at 30 chirps per 15 seconds is inside the domain of our data, so would be interpolation. Using our model:

$$T(30) = 30 + 1.2(30)$$
$$= 66 \text{ degrees}$$

Based on the data we have, this value seems reasonable.

b. The temperature values varied from 52 to 80.5. Predicting the number of chirps at 40 degrees is extrapolation because 40 is outside the range of our data. Using our model:

$$40 = 30 + 1.2c$$
$$10 = 1.2c$$
$$c \approx 8.33$$

We can compare the regions of interpolation and extrapolation using **Figure 5**.

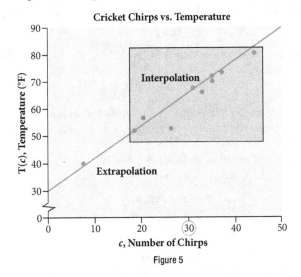

Figure 5

Analysis   Our model predicts the crickets would chirp 8.33 times in 15 seconds. While this might be possible, we have no reason to believe our model is valid outside the domain and range. In fact, generally crickets stop chirping altogether below around 50 degrees.

Try It #1

According to the data from **Table 1**, what temperature can we predict it is if we counted 20 chirps in 15 seconds?

Finding the Line of Best Fit Using a Graphing Utility

While eyeballing a line works reasonably well, there are statistical techniques for fitting a line to data that minimize the differences between the line and data values[11]. One such technique is called **least squares regression** and can be computed by many graphing calculators, spreadsheet software, statistical software, and many web-based calculators[12]. Least squares regression is one means to determine the line that best fits the data, and here we will refer to this method as linear regression.

11 Technically, the method minimizes the sum of the squared differences in the vertical direction between the line and the data values.
12 For example, http://www.shodor.org/unchem/math/lls/leastsq.html

How To…

Given data of input and corresponding outputs from a linear function, find the best fit line using linear regression.

1. Enter the input in **List 1 (L1)**.
2. Enter the output in **List 2 (L2)**.
3. On a graphing utility, select **Linear Regression (LinReg)**.

Example 4   **Finding a Least Squares Regression Line**

Find the least squares regression line using the cricket-chirp data in **Table 2**.

Solution

1. Enter the input (chirps) in **List 1 (L1)**.
2. Enter the output (temperature) in **List 2 (L2)**. See **Table 2**.

L1	44	35	20.4	33	31	35	18.5	37	26
L2	80.5	70.5	57	66	68	72	52	73.5	53

Table 2

3. On a graphing utility, select **Linear Regression (LinReg)**. Using the cricket chirp data from earlier, with technology we obtain the equation:

$$T(c) = 30.281 + 1.143c$$

Analysis *Notice that this line is quite similar to the equation we "eyeballed" but should fit the data better. Notice also that using this equation would change our prediction for the temperature when hearing 30 chirps in 15 seconds from 66 degrees to:*

$$T(30) = 30.281 + 1.143(30)$$

$$= 64.571$$

$$\approx 64.6 \text{ degrees}$$

The graph of the scatter plot with the least squares regression line is shown in **Figure 6**.

**Number of Cricket Chirps
vs. Temperature**

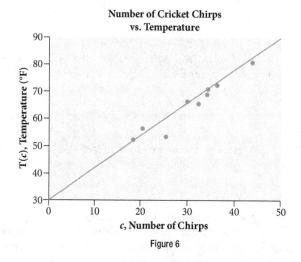

Figure 6

Q & A…

Will there ever be a case where two different lines will serve as the best fit for the data?

No. There is only one best fit line.

Distinguishing Between Linear and Non-Linear Models

As we saw above with the cricket-chirp model, some data exhibit strong linear trends, but other data, like the final exam scores plotted by age, are clearly nonlinear. Most calculators and computer software can also provide us with the correlation coefficient, which is a measure of how closely the line fits the data. Many graphing calculators require the user to turn a "diagnostic on" selection to find the correlation coefficient, which mathematicians label as *r*. The correlation coefficient provides an easy way to get an idea of how close to a line the data falls.

We should compute the correlation coefficient only for data that follows a linear pattern or to determine the degree to which a data set is linear. If the data exhibits a nonlinear pattern, the correlation coefficient for a linear regression is meaningless. To get a sense for the relationship between the value of *r* and the graph of the data, **Figure 7** shows some large data sets with their correlation coefficients. Remember, for all plots, the horizontal axis shows the input and the vertical axis shows the output.

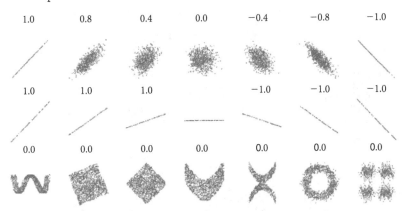

Figure 7 Plotted data and related correlation coefficients. (credit: "DenisBoigelot," Wikimedia Commons)

---

correlation coefficient

The **correlation coefficient** is a value, *r*, between −1 and 1.

- $r > 0$ suggests a positive (increasing) relationship
- $r < 0$ suggests a negative (decreasing) relationship
- The closer the value is to 0, the more scattered the data.
- The closer the value is to 1 or −1, the less scattered the data is.

Example 5   **Finding a Correlation Coefficient**

Calculate the correlation coefficient for cricket-chirp data in **Table 1**.

Solution   Because the data appear to follow a linear pattern, we can use technology to calculate *r*. Enter the inputs and corresponding outputs and select the Linear Regression. The calculator will also provide you with the correlation coefficient, $r = 0.9509$. This value is very close to 1, which suggests a strong increasing linear relationship.

Note: For some calculators, the Diagnostics must be turned "on" in order to get the correlation coefficient when linear regression is performed: [**2nd**]>[**0**]>[**alpha**][$x - 1$], then scroll to **DIAGNOSTICSON**.

Fitting a Regression Line to a Set of Data

Once we determine that a set of data is linear using the correlation coefficient, we can use the regression line to make predictions. As we learned above, a regression line is a line that is closest to the data in the scatter plot, which means that only one such line is a best fit for the data.

Example 6   **Using a Regression Line to Make Predictions**

Gasoline consumption in the United States has been steadily increasing. Consumption data from 1994 to 2004 is shown in **Table 3**[13]. Determine whether the trend is linear, and if so, find a model for the data. Use the model to predict the consumption in 2008.

Year	'94	'95	'96	'97	'98	'99	'00	'01	'02	'03	'04
Consumption (billions of gallons)	113	116	118	119	123	125	126	128	131	133	136

Table 3

The scatter plot of the data, including the least squares regression line, is shown in **Figure 8**.

Figure 8

Solution   We can introduce a new input variable, t, representing years since 1994.

The least squares regression equation is:

$$C(t) = 113.318 + 2.209t$$

Using technology, the correlation coefficient was calculated to be 0.9965, suggesting a very strong increasing linear trend.

Using this to predict consumption in 2008 ($t = 14$),

$$C(14) = 113.318 + 2.209(14)$$

$$= 144.244$$

The model predicts 144.244 billion gallons of gasoline consumption in 2008.

Try It #2

Use the model we created using technology in **Example 6** to predict the gas consumption in 2011. Is this an interpolation or an extrapolation?

Access these online resources for additional instruction and practice with fitting linear models to data.

- Introduction to Regression Analysis (http://openstaxcollege.org/l/introregress)
- Linear Regression (http://openstaxcollege.org/l/linearregress)

13 http://www.bts.gov/publications/national_transportation_statistics/2005/html/table_04_10.html

4.3 SECTION EXERCISES

VERBAL

1. Describe what it means if there is a model breakdown when using a linear model.

2. What is interpolation when using a linear model?

3. What is extrapolation when using a linear model?

4. Explain the difference between a positive and a negative correlation coefficient.

5. Explain how to interpret the absolute value of a correlation coefficient.

ALGEBRAIC

6. A regression was run to determine whether there is a relationship between hours of TV watched per day (x) and number of sit-ups a person can do (y). The results of the regression are given below. Use this to predict the number of situps a person who watches 11 hours of TV can do.

$$y = ax + b$$
$$a = -1.341$$
$$b = 32.234$$
$$r = -0.896$$

7. A regression was run to determine whether there is a relationship between the diameter of a tree (x, in inches) and the tree's age (y, in years). The results of the regression are given below. Use this to predict the age of a tree with diameter 10 inches.

$$y = ax + b$$
$$a = 6.301$$
$$b = -1.044$$
$$r = -0.970$$

For the following exercises, draw a scatter plot for the data provided. Does the data appear to be linearly related?

8.

0	2	4	6	8	10
−22	−19	−15	−11	−6	−2

9.

1	2	3	4	5	6
46	50	59	75	100	136

10.

100	250	300	450	600	750
12	12.6	13.1	14	14.5	15.2

11.

1	3	5	7	9	11
1	9	28	65	125	216

12. For the following data, draw a scatter plot. If we wanted to know when the population would reach 15,000, would the answer involve interpolation or extrapolation? Eyeball the line, and estimate the answer.

Year	1990	1995	2000	2005	2010
Population	11,500	12,100	12,700	13,000	13,750

13. For the following data, draw a scatter plot. If we wanted to know when the temperature would reach 28°F, would the answer involve interpolation or extrapolation? Eyeball the line and estimate the answer.

Temperature, °F	16	18	20	25	30
Time, seconds	46	50	54	55	62

GRAPHICAL

For the following exercises, match each scatterplot with one of the four specified correlations in **Figure 9** and **Figure 10**.

(a)   (b)

Figure 9

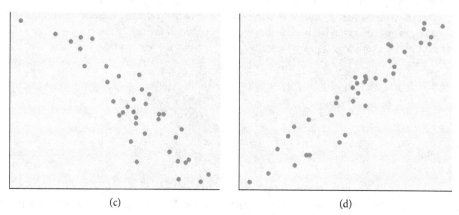

(c)   (d)

Figure 10

14. $r = 0.95$ **15.** $r = -0.89$

16. $r = -0.26$ **17.** $r = -0.39$

For the following exercises, draw a best-fit line for the plotted data.

18.

19.

20.

21.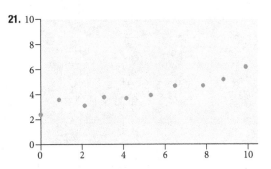

NUMERIC

22. The U.S. Census tracks the percentage of persons 25 years or older who are college graduates. That data for several years is given in **Table 4**[14]. Determine whether the trend appears linear. If so, and assuming the trend continues, in what year will the percentage exceed 35%?

Year	1990	1992	1994	1996	1998	2000	2002	2004	2006	2008
Percent Graduates	21.3	21.4	22.2	23.6	24.4	25.6	26.7	27.7	28	29.4

Table 4

23. The U.S. import of wine (in hectoliters) for several years is given in **Table 5**. Determine whether the trend appears linear. If so, and assuming the trend continues, in what year will imports exceed 12,000 hectoliters?

Year	1992	1994	1996	1998	2000	2002	2004	2006	2008	2009
Imports	2665	2688	3565	4129	4584	5655	6549	7950	8487	9462

Table 5

24. **Table 6** shows the year and the number of people unemployed in a particular city for several years. Determine whether the trend appears linear. If so, and assuming the trend continues, in what year will the number of unemployed reach 5 people?

Year	1990	1992	1994	1996	1998	2000	2002	2004	2006	2008
Number Unemployed	750	670	650	605	550	510	460	420	380	320

Table 6

TECHNOLOGY

For the following exercises, use each set of data to calculate the regression line using a calculator or other technology tool, and determine the correlation coefficient to 3 decimal places of accuracy.

25.

x	8	15	26	31	56
y	23	41	53	72	103

26.

x	5	7	10	12	15
y	4	12	17	22	24

14 http://www.census.gov/hhes/socdemo/education/data/cps/historical/index.html. Accessed 5/1/2014.

27.

x	3	4	5	6	7	8	9
y	21.9	22.22	22.74	22.26	20.78	17.6	16.52

x	10	11	12	13	14	15	16
y	18.54	15.76	13.68	14.1	14.02	11.94	12.76

28.

x	4	5	6	7	8	9	10	11	12	13
y	44.8	43.1	38.8	39	38	32.7	30.1	29.3	27	25.8

29.

x	21	25	30	31	40	50
y	17	11	2	−1	−18	−40

30.

x	100	80	60	55	40	20
y	2000	1798	1589	1580	1390	1202

31.

x	900	988	1000	1010	1200	1205
y	70	80	82	84	105	108

EXTENSIONS

32. Graph $f(x) = 0.5x + 10$. Pick a set of 5 ordered pairs using inputs $x = -2, 1, 5, 6, 9$ and use linear regression to verify that the function is a good fit for the data.

33. Graph $f(x) = -2x - 10$. Pick a set of 5 ordered pairs using inputs $x = -2, 1, 5, 6, 9$ and use linear regression to verify the function.

For the following exercises, consider this scenario: The profit of a company decreased steadily over a ten-year span. The following ordered pairs shows dollars and the number of units sold in hundreds and the profit in thousands of over the ten-year span, (number of units sold, profit) for specific recorded years:

(46, 600), (48, 550), (50, 505), (52, 540), (54, 495).

34. Use linear regression to determine a function P where the profit in thousands of dollars depends on the number of units sold in hundreds.

35. Find to the nearest tenth and interpret the x-intercept.

36. Find to the nearest tenth and interpret the y-intercept.

REAL-WORLD APPLICATIONS

For the following exercises, consider this scenario: The population of a city increased steadily over a ten-year span. The following ordered pairs shows the population and the year over the ten-year span, (population, year) for specific recorded years:

(2500, 2000), (2650, 2001), (3000, 2003), (3500, 2006), (4200, 2010)

37. Use linear regression to determine a function y, where the year depends on the population. Round to three decimal places of accuracy.

38. Predict when the population will hit 8,000.

For the following exercises, consider this scenario: The profit of a company increased steadily over a ten-year span. The following ordered pairs show the number of units sold in hundreds and the profit in thousands of over the ten-year span, (number of units sold, profit) for specific recorded years:

(46, 250), (48, 305), (50, 350), (52, 390), (54, 410).

39. Use linear regression to determine a function y, where the profit in thousands of dollars depends on the number of units sold in hundreds.

40. Predict when the profit will exceed one million dollars.

For the following exercises, consider this scenario: The profit of a company decreased steadily over a ten-year span. The following ordered pairs show dollars and the number of units sold in hundreds and the profit in thousands of over the ten-year span (number of units sold, profit) for specific recorded years:

(46, 250), (48, 225), (50, 205), (52, 180), (54, 165).

41. Use linear regression to determine a function y, where the profit in thousands of dollars depends on the number of units sold in hundreds.

42. Predict when the profit will dip below the $25,000 threshold.

CHAPTER 4 REVIEW

Key Terms

correlation coefficient a value, r, between -1 and 1 that indicates the degree of linear correlation of variables, or how closely a regression line fits a data set.

decreasing linear function a function with a negative slope: If $f(x) = mx + b$, then $m < 0$.

extrapolation predicting a value outside the domain and range of the data

horizontal line a line defined by $f(x) = b$, where b is a real number. The slope of a horizontal line is 0.

increasing linear function a function with a positive slope: If $f(x) = mx + b$, then $m > 0$.

interpolation predicting a value inside the domain and range of the data

least squares regression a statistical technique for fitting a line to data in a way that minimizes the differences between the line and data values

linear function a function with a constant rate of change that is a polynomial of degree 1, and whose graph is a straight line

model breakdown when a model no longer applies after a certain point

parallel lines two or more lines with the same slope

perpendicular lines two lines that intersect at right angles and have slopes that are negative reciprocals of each other

point-slope form the equation for a line that represents a linear function of the form $y - y_1 = m(x - x_1)$

slope the ratio of the change in output values to the change in input values; a measure of the steepness of a line

slope-intercept form the equation for a line that represents a linear function in the form $f(x) = mx + b$

vertical line a line defined by $x = a$, where a is a real number. The slope of a vertical line is undefined.

Key Concepts

4.1 Linear Functions

- Linear functions can be represented in words, function notation, tabular form, and graphical form. See **Example 1**.

- An increasing linear function results in a graph that slants upward from left to right and has a positive slope. A decreasing linear function results in a graph that slants downward from left to right and has a negative slope. A constant linear function results in a graph that is a horizontal line. See **Example 2**.

- Slope is a rate of change. The slope of a linear function can be calculated by dividing the difference between y-values by the difference in corresponding x-values of any two points on the line. See **Example 3** and **Example 4**.

- An equation for a linear function can be written from a graph. See **Example 5**.

- The equation for a linear function can be written if the slope m and initial value b are known. See **Example 6** and **Example 7**.

- A linear function can be used to solve real-world problems given information in different forms. See **Example 8**, **Example 9**, and **Example 10**.

- Linear functions can be graphed by plotting points or by using the y-intercept and slope. See **Example 11** and **Example 12**.

- Graphs of linear functions may be transformed by using shifts up, down, left, or right, as well as through stretches, compressions, and reflections. See **Example 13**.

- The equation for a linear function can be written by interpreting the graph. See **Example 14**.

- The x-intercept is the point at which the graph of a linear function crosses the x-axis. See **Example 15**.

- Horizontal lines are written in the form, $f(x) = b$. See **Example 16**.

- Vertical lines are written in the form, $x = b$. See **Example 17**.

- Parallel lines have the same slope. Perpendicular lines have negative reciprocal slopes, assuming neither is vertical. See **Example 18**.

- A line parallel to another line, passing through a given point, may be found by substituting the slope value of the line and the x- and y-values of the given point into the equation, $f(x) = mx + b$, and using the b that results. Similarly, the point-slope form of an equation can also be used. See **Example 19**.

- A line perpendicular to another line, passing through a given point, may be found in the same manner, with the exception of using the negative reciprocal slope. See **Example 20** and **Example 21**.

4.2 Modeling with Linear Functions

- We can use the same problem strategies that we would use for any type of function.

- When modeling and solving a problem, identify the variables and look for key values, including the slope and y-intercept. See **Example 1**.

- Draw a diagram, where appropriate. See **Example 2** and **Example 3**.

- Check for reasonableness of the answer.

- Linear models may be built by identifying or calculating the slope and using the y-intercept.

  ○ The x-intercept may be found by setting $y = 0$, which is setting the expression $mx + b$ equal to 0.

  ○ The point of intersection of a system of linear equations is the point where the x- and y-values are the same. See **Example 4**.

  ○ A graph of the system may be used to identify the points where one line falls below (or above) the other line.

4.3 Fitting Linear Models to Data

- Scatter plots show the relationship between two sets of data. See **Example 1**.

- Scatter plots may represent linear or non-linear models.

- The line of best fit may be estimated or calculated, using a calculator or statistical software. See **Example 2**.

- Interpolation can be used to predict values inside the domain and range of the data, whereas extrapolation can be used to predict values outside the domain and range of the data. See **Example 3**.

- The correlation coefficient, r, indicates the degree of linear relationship between data. See **Example 4**.

- A regression line best fits the data. See **Example 5**.

- The least squares regression line is found by minimizing the squares of the distances of points from a line passing through the data and may be used to make predictions regarding either of the variables. See **Example 6**.

CHAPTER 4   REVIEW EXERCISES

LINEAR FUNCTIONS

1. Determine whether the algebraic equation is linear. $2x + 3y = 7$

2. Determine whether the algebraic equation is linear. $6x^2 - y = 5$

3. Determine whether the function is increasing or decreasing. $f(x) = 7x - 2$

4. Determine whether the function is increasing or decreasing. $g(x) = -x + 2$

5. Given each set of information, find a linear equation that satisfies the given conditions, if possible. Passes through $(7, 5)$ and $(3, 17)$

6. Given each set of information, find a linear equation that satisfies the given conditions, if possible. x-intercept at $(6, 0)$ and y-intercept at $(0, 10)$

7. Find the slope of the line shown in the graph.

8. Find the slope of the line shown in the graph.

9. Write an equation in slope-intercept form for the line shown.

10. Does the following table represent a linear function? If so, find the linear equation that models the data.

x	−4	0	2	10
$g(x)$	18	−2	−12	−52

11. Does the following table represent a linear function? If so, find the linear equation that models the data.

x	6	8	12	26
$g(x)$	−8	−12	−18	−46

12. On June 1ˢᵗ, a company has $4,000,000 profit. If the company then loses 150,000 dollars per day thereafter in the month of June, what is the company's profit nᵗʰ day after June 1ˢᵗ?

For the following exercises, determine whether the lines given by the equations below are parallel, perpendicular, or neither parallel nor perpendicular:

13. $2x - 6y = 12$
$-x + 3y = 1$

14. $y = \frac{1}{3}x - 2$
$3x + y = -9$

For the following exercises, find the x- and y-intercepts of the given equation

15. $7x + 9y = -63$

16. $f(x) = 2x - 1$

For the following exercises, use the descriptions of the pairs of lines to find the slopes of Line 1 and Line 2. Is each pair of lines parallel, perpendicular, or neither?

17. Line 1: Passes through (5, 11) and (10, 1)
Line 2: Passes through (−1, 3) and (−5, 11)

18. Line 1: Passes through (8, −10) and (0, −26)
Line 2: Passes through (2, 5) and (4, 4)

19. Write an equation for a line perpendicular to $f(x) = 5x − 1$ and passing through the point (5, 20).

20. Find the equation of a line with a y-intercept of (0, 2) and slope $-\dfrac{1}{2}$.

21. Sketch a graph of the linear function $f(t) = 2t − 5$.

22. Find the point of intersection for the 2 linear functions: $\quad x = y + 6$
$$2x − y = 13$$

23. A car rental company offers two plans for renting a car.
Plan A: 25 dollars per day and 10 cents per mile
Plan B: 50 dollars per day with free unlimited mileage
How many miles would you need to drive for plan B to save you money?

MODELING WITH LINEAR FUNCTIONS

24. Find the area of a triangle bounded by the y-axis, the line $f(x) = 10 − 2x$, and the line perpendicular to f that passes through the origin.

25. A town's population increases at a constant rate. In 2010 the population was 55,000. By 2012 the population had increased to 76,000. If this trend continues, predict the population in 2016.

26. The number of people afflicted with the common cold in the winter months dropped steadily by 50 each year since 2004 until 2010. In 2004, 875 people were inflicted.

Find the linear function that models the number of people afflicted with the common cold C as a function of the year, t. When will no one be afflicted?

For the following exercises, use the graph in **Figure 1** showing the profit, y, in thousands of dollars, of a company in a given year, x, where x represents years since 1980.

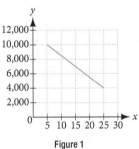

Figure 1

27. Find the linear function y, where y depends on x, the number of years since 1980.

28. Find and interpret the y-intercept.

For the following exercise, consider this scenario: In 2004, a school population was 1,700. By 2012 the population had grown to 2,500.

29. Assume the population is changing linearly.
   a. How much did the population grow between the year 2004 and 2012?
   b. What is the average population growth per year?
   c. Find an equation for the population, P, of the school t years after 2004.

For the following exercises, consider this scenario: In 2000, the moose population in a park was measured to be 6,500. By 2010, the population was measured to be 12,500. Assume the population continues to change linearly.

30. Find a formula for the moose population, *P*.

31. What does your model predict the moose population to be in 2020?

For the following exercises, consider this scenario: The median home values in subdivisions Pima Central and East Valley (adjusted for inflation) are shown in **Table 1**. Assume that the house values are changing linearly.

Year	Pima Central	East Valley
1970	32,000	120,250
2010	85,000	150,000

Table 1

32. In which subdivision have home values increased at a higher rate?

33. If these trends were to continue, what would be the median home value in Pima Central in 2015?

FITTING LINEAR MODELS TO DATA

34. Draw a scatter plot for the data in **Table 2**. Then determine whether the data appears to be linearly related.

0	2	4	6	8	10
−105	−50	1	55	105	160

Table 2

35. Draw a scatter plot for the data in **Table 3**. If we wanted to know when the population would reach 15,000, would the answer involve interpolation or extrapolation?

Year	1990	1995	2000	2005	2010
Population	5,600	5,950	6,300	6,600	6,900

Table 3

36. Eight students were asked to estimate their score on a 10-point quiz. Their estimated and actual scores are given in **Table 4**. Plot the points, then sketch a line that fits the data.

Predicted	6	7	7	8	7	9	10	10
Actual	6	7	8	8	9	10	10	9

Table 4

37. Draw a best-fit line for the plotted data.

For the following exercises, consider the data in **Table 5**, which shows the percent of unemployed in a city of people 25 years or older who are college graduates is given below, by year.

Year	2000	2002	2005	2007	2010
Percent Graduates	6.5	7.0	7.4	8.2	9.0

Table 5

38. Determine whether the trend appears to be linear. If so, and assuming the trend continues, find a linear regression model to predict the percent of unemployed in a given year to three decimal places.

39. In what year will the percentage exceed 12%?

40. Based on the set of data given in **Table 6**, calculate the regression line using a calculator or other technology tool, and determine the correlation coefficient to three decimal places.

x	17	20	23	26	29
y	15	25	31	37	40

Table 6

41. Based on the set of data given in **Table 7**, calculate the regression line using a calculator or other technology tool, and determine the correlation coefficient to three decimal places.

x	10	12	15	18	20
y	36	34	30	28	22

Table 7

For the following exercises, consider this scenario: The population of a city increased steadily over a ten-year span. The following ordered pairs show the population and the year over the ten-year span (population, year) for specific recorded years:

(3,600, 2000); (4,000, 2001); (4,700, 2003); (6,000, 2006)

42. Use linear regression to determine a function y, where the year depends on the population, to three decimal places of accuracy.

43. Predict when the population will hit 12,000.

44. What is the correlation coefficient for this model to three decimal places of accuracy?

45. According to the model, what is the population in 2014?

CHAPTER 4   PRACTICE TEST

1. Determine whether the following algebraic equation can be written as a linear function.
   $2x + 3y = 7$

2. Determine whether the following function is increasing or decreasing. $f(x) = -2x + 5$

3. Determine whether the following function is increasing or decreasing. $f(x) = 7x + 9$

4. Given the following set of information, find a linear equation satisfying the conditions, if possible. Passes through $(5, 1)$ and $(3, -9)$

5. Given the following set of information, find a linear equation satisfying the conditions, if possible. x-intercept at $(-4, 0)$ and y-intercept at $(0, -6)$

6. Find the slope of the line in **Figure 1**.

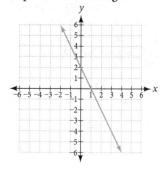

Figure 1

7. Write an equation for line in **Figure 2**.

Figure 2

8. Does **Table 1** represent a linear function? If so, find a linear equation that models the data.

x	-6	0	2	4
$g(x)$	14	32	38	44

Table 1

9. Does **Table 2** represent a linear function? If so, find a linear equation that models the data.

x	1	3	7	11
$g(x)$	4	9	19	12

Table 2

10. At 6 am, an online company has sold 120 items that day. If the company sells an average of 30 items per hour for the remainder of the day, write an expression to represent the number of items that were sold n after 6 am.

For the following exercises, determine whether the lines given by the equations below are parallel, perpendicular, or neither parallel nor perpendicular:

11. $y = \dfrac{3}{4}x - 9$
    $-4x - 3y = 8$

12. $-2x + y = 3$
    $3x + \dfrac{3}{2}y = 5$

13. Find the x- and y-intercepts of the equation $2x + 7y = -14$.

14. Given below are descriptions of two lines. Find the slopes of Line 1 and Line 2. Is the pair of lines parallel, perpendicular, or neither?
    Line 1: Passes through $(-2, -6)$ and $(3, 14)$
    Line 2: Passes through $(2, 6)$ and $(4, 14)$

15. Write an equation for a line perpendicular to $f(x) = 4x + 3$ and passing through the point (8, 10).

16. Sketch a line with a y-intercept of (0, 5) and slope $-\dfrac{5}{2}$.

17. Graph of the linear function $f(x) = -x + 6$.

18. For the two linear functions, find the point of intersection:
$$x = y + 2$$
$$2x - 3y = -1$$

19. A car rental company offers two plans for renting a car.

   Plan A: $25 per day and $0.10 per mile
   Plan B: $40 per day with free unlimited mileage

How many miles would you need to drive for plan B to save you money?

20. Find the area of a triangle bounded by the y-axis, the line $f(x) = 12 - 4x$, and the line perpendicular to f that passes through the origin.

21. A town's population increases at a constant rate. In 2010 the population was 65,000. By 2012 the population had increased to 90,000. Assuming this trend continues, predict the population in 2018.

22. The number of people afflicted with the common cold in the winter months dropped steadily by 25 each year since 2002 until 2012. In 2002, 8,040 people were inflicted. Find the linear function that models the number of people afflicted with the common cold C as a function of the year, t. When will less than 6,000 people be afflicted?

For the following exercises, use the graph in **Figure 3**, showing the profit, y, in thousands of dollars, of a company in a given year, x, where x represents years since 1980.

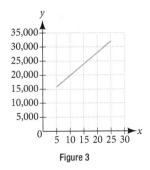

Figure 3

23. Find the linear function y, where y depends on x, the number of years since 1980.

24. Find and interpret the y-intercept.

25. In 2004, a school population was 1250. By 2012 the population had dropped to 875. Assume the population is changing linearly.
   a. How much did the population drop between the year 2004 and 2012?
   b. What is the average population decline per year?
   c. Find an equation for the population, P, of the school t years after 2004.

26. Draw a scatter plot for the data provided in **Table 3**. Then determine whether the data appears to be linearly related.

0	2	4	6	8	10
−450	−200	10	265	500	755

Table 3

27. Draw a best-fit line for the plotted data.

For the following exercises, use **Table 4** which shows the percent of unemployed persons 25 years or older who are college graduates in a particular city, by year.

Year	2000	2002	2005	2007	2010
Percent Graduates	8.5	8.0	7.2	6.7	6.4

Table 4

28. Determine whether the trend appears linear. If so, and assuming the trend continues, find a linear regression model to predict the percent of unemployed in a given year to three decimal places.

29. In what year will the percentage drop below 4%?

30. Based on the set of data given in **Table 5**, calculate the regression line using a calculator or other technology tool, and determine the correlation coefficient. Round to three decimal places of accuracy.

x	16	18	20	24	26
y	106	110	115	120	125

Table 5

For the following exercises, consider this scenario: The population of a city increased steadily over a ten-year span. The following ordered pairs shows the population (in hundreds) and the year over the ten-year span, (population, year) for specific recorded years:

(4,500, 2000); (4,700, 2001); (5,200, 2003); (5,800, 2006)

31. Use linear regression to determine a function y, where the year depends on the population. Round to three decimal places of accuracy.

32. Predict when the population will hit 20,000.

33. What is the correlation coefficient for this model?

Polynomial and Rational Functions

Figure 1 35-mm film, once the standard for capturing photographic images, has been made largely obsolete by digital photography. (credit "film": modification of work by Horia Varlan; credit "memory cards": modification of work by Paul Hudson)

CHAPTER OUTLINE

Introduction

Digital photography has dramatically changed the nature of photography. No longer is an image etched in the emulsion on a roll of film. Instead, nearly every aspect of recording and manipulating images is now governed by mathematics. An image becomes a series of numbers, representing the characteristics of light striking an image sensor. When we open an image file, software on a camera or computer interprets the numbers and converts them to a visual image. Photo editing software uses complex polynomials to transform images, allowing us to manipulate the image in order to crop details, change the color palette, and add special effects. Inverse functions make it possible to convert from one file format to another. In this chapter, we will learn about these concepts and discover how mathematics can be used in such applications.

343

LEARNING OBJECTIVES

In this section, you will:

- Recognize characteristics of parabolas.
- Understand how the graph of a parabola is related to its quadratic function.
- Determine a quadratic function's minimum or maximum value.
- Solve problems involving a quadratic function's minimum or maximum value.

5.1   QUADRATIC FUNCTIONS

Figure 1 An array of satellite dishes. (credit: Matthew Colvin de Valle, Flickr)

Curved antennas, such as the ones shown in **Figure 1** are commonly used to focus microwaves and radio waves to transmit television and telephone signals, as well as satellite and spacecraft communication. The cross-section of the antenna is in the shape of a parabola, which can be described by a quadratic function.

In this section, we will investigate quadratic functions, which frequently model problems involving area and projectile motion. Working with quadratic functions can be less complex than working with higher degree functions, so they provide a good opportunity for a detailed study of function behavior.

Recognizing Characteristics of Parabolas

The graph of a quadratic function is a U-shaped curve called a parabola. One important feature of the graph is that it has an extreme point, called the **vertex**. If the parabola opens up, the vertex represents the lowest point on the graph, or the minimum value of the quadratic function. If the parabola opens down, the vertex represents the highest point on the graph, or the maximum value. In either case, the vertex is a turning point on the graph. The graph is also symmetric with a vertical line drawn through the vertex, called the **axis of symmetry**. These features are illustrated in **Figure 2**.

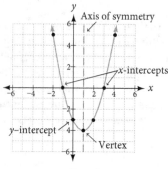

Figure 2

The y-intercept is the point at which the parabola crosses the y-axis. The x-intercepts are the points at which the parabola crosses the x-axis. If they exist, the x-intercepts represent the **zeros**, or **roots**, of the quadratic function, the values of x at which $y = 0$.

Example 1   **Identifying the Characteristics of a Parabola**

Determine the vertex, axis of symmetry, zeros, and y-intercept of the parabola shown in **Figure 3**.

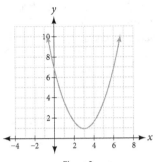

Figure 3

Solution   The vertex is the turning point of the graph. We can see that the vertex is at (3, 1). Because this parabola opens upward, the axis of symmetry is the vertical line that intersects the parabola at the vertex. So the axis of symmetry is $x = 3$. This parabola does not cross the x-axis, so it has no zeros. It crosses the y-axis at (0, 7) so this is the y-intercept.

Understanding How the Graphs of Parabolas are Related to Their Quadratic Functions

The **general form of a quadratic function** presents the function in the form

$$f(x) = ax^2 + bx + c$$

where a, b, and c are real numbers and $a \neq 0$. If $a > 0$, the parabola opens upward. If $a < 0$, the parabola opens downward. We can use the general form of a parabola to find the equation for the axis of symmetry.

The axis of symmetry is defined by $x = -\dfrac{b}{2a}$. If we use the quadratic formula, $x = \dfrac{-b \pm \sqrt{b^2 - 4ac}}{2a}$, to solve $ax^2 + bx + c = 0$ for the x-intercepts, or zeros, we find the value of x halfway between them is always $x = -\dfrac{b}{2a}$, the equation for the axis of symmetry.

Figure 4 represents the graph of the quadratic function written in general form as $y = x^2 + 4x + 3$. In this form, $a = 1$, $b = 4$, and $c = 3$. Because $a > 0$, the parabola opens upward. The axis of symmetry is $x = -\dfrac{4}{2(1)} = -2$. This also makes sense because we can see from the graph that the vertical line $x = -2$ divides the graph in half. The vertex always occurs along the axis of symmetry. For a parabola that opens upward, the vertex occurs at the lowest point on the graph, in this instance, $(-2, -1)$. The x-intercepts, those points where the parabola crosses the x-axis, occur at $(-3, 0)$ and $(-1, 0)$.

Figure 4

The **standard form of a quadratic function** presents the function in the form

$$f(x) = a(x - h)^2 + k$$

where (h, k) is the vertex. Because the vertex appears in the standard form of the quadratic function, this form is also known as the **vertex form of a quadratic function**.

As with the general form, if $a > 0$, the parabola opens upward and the vertex is a minimum. If $a < 0$, the parabola opens downward, and the vertex is a maximum. **Figure 5** represents the graph of the quadratic function written in standard form as $y = -3(x + 2)^2 + 4$. Since $x - h = x + 2$ in this example, $h = -2$. In this form, $a = -3, h = -2$, and $k = 4$. Because $a < 0$, the parabola opens downward. The vertex is at $(-2, 4)$.

Figure 5

The standard form is useful for determining how the graph is transformed from the graph of $y = x^2$. **Figure 6** is the graph of this basic function.

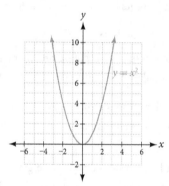

Figure 6

If $k > 0$, the graph shifts upward, whereas if $k < 0$, the graph shifts downward. In **Figure 5**, $k > 0$, so the graph is shifted 4 units upward. If $h > 0$, the graph shifts toward the right and if $h < 0$, the graph shifts to the left. In **Figure 5**, $h < 0$, so the graph is shifted 2 units to the left. The magnitude of a indicates the stretch of the graph. If $|a| > 1$, the point associated with a particular x-value shifts farther from the x-axis, so the graph appears to become narrower, and there is a vertical stretch. But if $|a| < 1$, the point associated with a particular x-value shifts closer to the x-axis, so the graph appears to become wider, but in fact there is a vertical compression. In **Figure 5**, $|a| > 1$, so the graph becomes narrower.

The standard form and the general form are equivalent methods of describing the same function. We can see this by expanding out the general form and setting it equal to the standard form.

$$a(x - h)^2 + k = ax^2 + bx + c$$

$$ax^2 - 2ahx + (ah^2 + k) = ax^2 + bx + c$$

For the linear terms to be equal, the coefficients must be equal.

$$-2ah = b, \text{ so } h = -\frac{b}{2a}.$$

This is the axis of symmetry we defined earlier. Setting the constant terms equal:

$$ah^2 + k = c$$

$$k = c - ah^2$$

$$= c - a\left(\frac{b}{2a}\right)^2$$

$$= c - \frac{b^2}{4a}$$

In practice, though, it is usually easier to remember that k is the output value of the function when the input is h, so $f(h) = k$.

---

forms of quadratic functions

A quadratic function is a polynomial function of degree two. The graph of a quadratic function is a parabola.

The **general form of a quadratic function** is $f(x) = ax^2 + bx + c$ where a, b, and c are real numbers and $a \neq 0$.

The **standard form of a quadratic function** is $f(x) = a(x - h)^2 + k$ where $a \neq 0$.

The vertex (h, k) is located at

$$h = -\frac{b}{2a}, k = f(h) = f\left(\frac{-b}{2a}\right).$$

How To...

Given a graph of a quadratic function, write the equation of the function in general form.

1. Identify the horizontal shift of the parabola; this value is h. Identify the vertical shift of the parabola; this value is k.
2. Substitute the values of the horizontal and vertical shift for h and k. in the function $f(x) = a(x - h)^2 + k$.
3. Substitute the values of any point, other than the vertex, on the graph of the parabola for x and $f(x)$.
4. Solve for the stretch factor, $|a|$.
5. If the parabola opens up, $a > 0$. If the parabola opens down, $a < 0$ since this means the graph was reflected about the x-axis.
6. Expand and simplify to write in general form.

Example 2   **Writing the Equation of a Quadratic Function from the Graph**

Write an equation for the quadratic function g in **Figure 7** as a transformation of $f(x) = x^2$, and then expand the formula, and simplify terms to write the equation in general form.

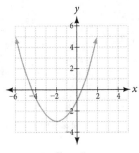

Figure 7

Solution   We can see the graph of g is the graph of $f(x) = x^2$ shifted to the left 2 and down 3, giving a formula in the form $g(x) = a(x - (-2))^2 - 3 = a(x + 2)^2 - 3$.

Substituting the coordinates of a point on the curve, such as $(0, -1)$, we can solve for the stretch factor.

$$-1 = a(0 + 2)^2 - 3$$
$$2 = 4a$$
$$a = \frac{1}{2}$$

In standard form, the algebraic model for this graph is $g(x) = \frac{1}{2}(x + 2)^2 - 3$.

To write this in general polynomial form, we can expand the formula and simplify terms.

$$\begin{aligned} g(x) &= \frac{1}{2}(x + 2)^2 - 3 \\ &= \frac{1}{2}(x + 2)(x + 2) - 3 \\ &= \frac{1}{2}(x^2 + 4x + 4) - 3 \\ &= \frac{1}{2}x^2 + 2x + 2 - 3 \\ &= \frac{1}{2}x^2 + 2x - 1 \end{aligned}$$

Notice that the horizontal and vertical shifts of the basic graph of the quadratic function determine the location of the vertex of the parabola; the vertex is unaffected by stretches and compressions.

Analysis *We can check our work using the table feature on a graphing utility. First enter* $Y1 = \frac{1}{2}(x + 2)^2 - 3$. *Next, select* **TBLSET**, *then use* **TblStart** $= -6$ *and* **ΔTbl** $= 2$, *and select* **TABLE**. *See* **Table 1**.

x	-6	-4	-2	0	2
y	5	-1	-3	-1	5

Table 1

The ordered pairs in the table correspond to points on the graph.

Try It #1

A coordinate grid has been superimposed over the quadratic path of a basketball in **Figure 8** Find an equation for the path of the ball. Does the shooter make the basket?

Figure 8 (credit: modification of work by Dan Meyer)

How To…

Given a quadratic function in general form, find the vertex of the parabola.

1. Identify a, b, and c.
2. Find h, the x-coordinate of the vertex, by substituting a and b into $h = -\frac{b}{2a}$.
3. Find k, the y-coordinate of the vertex, by evaluating $k = f(h) = f\left(-\frac{b}{2a}\right)$.

Example 3 **Finding the Vertex of a Quadratic Function**

Find the vertex of the quadratic function $f(x) = 2x^2 - 6x + 7$. Rewrite the quadratic in standard form (vertex form).

Solution The horizontal coordinate of the vertex will be at

$$h = -\frac{b}{2a}$$

$$= -\frac{-6}{2(2)}$$

$$= \frac{6}{4}$$

$$= \frac{3}{2}$$

The vertical coordinate of the vertex will be at

$$k = f(h)$$

$$= f\left(\frac{3}{2}\right)$$

$$= 2\left(\frac{3}{2}\right)^2 - 6\left(\frac{3}{2}\right) + 7$$

$$= \frac{5}{2}$$

Rewriting into standard form, the stretch factor will be the same as the a in the original quadratic. First, find the horizontal coordinate of the vertex. Then find the vertical coordinate of the vertex. Substitute the values into standard form, using the "a" from the general form.

$$f(x) = ax^2 + bx + c$$

$$f(x) = 2x^2 - 6x + 7$$

The standard form of a quadratic function prior to writing the function then becomes the following:

$$f(x) = 2\left(x - \frac{3}{2}\right)^2 + \frac{5}{2}$$

Analysis *One reason we may want to identify the vertex of the parabola is that this point will inform us where the maximum or minimum value of the output occurs, k, and where it occurs, x.*

Try It #2

Given the equation $g(x) = 13 + x^2 - 6x$, write the equation in general form and then in standard form.

Finding the Domain and Range of a Quadratic Function

Any number can be the input value of a quadratic function. Therefore, the domain of any quadratic function is all real numbers. Because parabolas have a maximum or a minimum point, the range is restricted. Since the vertex of a parabola will be either a maximum or a minimum, the range will consist of all y-values greater than or equal to the y-coordinate at the turning point or less than or equal to the y-coordinate at the turning point, depending on whether the parabola opens up or down.

---

domain and range of a quadratic function

The domain of any quadratic function is all real numbers unless the context of the function presents some restrictions. The range of a quadratic function written in general form $f(x) = ax^2 + bx + c$ with a positive a value is

$f(x) \geq f\left(-\frac{b}{2a}\right)$, or $\left[f\left(-\frac{b}{2a}\right), \infty\right)$.

The range of a quadratic function written in general form with a negative a value is $f(x) \leq f\left(-\frac{b}{2a}\right)$, or $\left(-\infty, f\left(-\frac{b}{2a}\right)\right]$.

The range of a quadratic function written in standard form $f(x) = a(x - h)^2 + k$ with a positive a value is $f(x) \geq k$; the range of a quadratic function written in standard form with a negative a value is $f(x) \leq k$.

How To…

Given a quadratic function, find the domain and range.

1. Identify the domain of any quadratic function as all real numbers.
2. Determine whether a is positive or negative. If a is positive, the parabola has a minimum. If a is negative, the parabola has a maximum.
3. Determine the maximum or minimum value of the parabola, k.
4. If the parabola has a minimum, the range is given by $f(x) \geq k$, or $[k, \infty)$. If the parabola has a maximum, the range is given by $f(x) \leq k$, or $(-\infty, k]$.

Example 4   **Finding the Domain and Range of a Quadratic Function**

Find the domain and range of $f(x) = -5x^2 + 9x - 1$.

Solution   As with any quadratic function, the domain is all real numbers.

Because a is negative, the parabola opens downward and has a maximum value. We need to determine the maximum value. We can begin by finding the x-value of the vertex.

$$h = -\frac{b}{2a}$$

$$= -\frac{9}{2(-5)}$$

$$= \frac{9}{10}$$

The maximum value is given by $f(h)$.

$$f\left(\frac{9}{10}\right) = -5\left(\frac{9}{10}\right)^2 + 9\left(\frac{9}{10}\right) - 1$$

$$= \frac{61}{20}$$

The range is $f(x) \leq \dfrac{61}{20}$, or $\left(-\infty, \dfrac{61}{20}\right]$.

Try It #3

Find the domain and range of $f(x) = 2\left(x - \dfrac{4}{7}\right)^2 + \dfrac{8}{11}$.

Determining the Maximum and Minimum Values of Quadratic Functions

The output of the quadratic function at the vertex is the maximum or minimum value of the function, depending on the orientation of the parabola. We can see the maximum and minimum values in **Figure 9**.

Figure 9

There are many real-world scenarios that involve finding the maximum or minimum value of a quadratic function, such as applications involving area and revenue.

Example 5   Finding the Maximum Value of a Quadratic Function

A backyard farmer wants to enclose a rectangular space for a new garden within her fenced backyard. She has purchased 80 feet of wire fencing to enclose three sides, and she will use a section of the backyard fence as the fourth side.

    a. Find a formula for the area enclosed by the fence if the sides of fencing perpendicular to the existing fence have length L.

    b. What dimensions should she make her garden to maximize the enclosed area?

Solution   Let's use a diagram such as **Figure 10** to record the given information. It is also helpful to introduce a temporary variable, W, to represent the width of the garden and the length of the fence section parallel to the backyard fence.

Figure 10

    a. We know we have only 80 feet of fence available, and $L + W + L = 80$, or more simply, $2L + W = 80$. This allows us to represent the width, W, in terms of L.

$$W = 80 - 2L$$

Now we are ready to write an equation for the area the fence encloses. We know the area of a rectangle is length multiplied by width, so

$$A = LW = L(80 - 2L)$$
$$A(L) = 80L - 2L^2$$

This formula represents the area of the fence in terms of the variable length L. The function, written in general form, is

$$A(L) = -2L^2 + 80L.$$

    b. The quadratic has a negative leading coefficient, so the graph will open downward, and the vertex will be the maximum value for the area. In finding the vertex, we must be careful because the equation is not written in standard polynomial form with decreasing powers. This is why we rewrote the function in general form above. Since a is the coefficient of the squared term, $a = -2$, $b = 80$, and $c = 0$.

To find the vertex:

$$h = -\frac{b}{2a} \qquad\qquad k = A(20)$$
$$h = -\frac{80}{2(-2)} \qquad\qquad = 80(20) - 2(20)^2$$
$$= 20 \qquad \text{and} \qquad = 800$$

The maximum value of the function is an area of 800 square feet, which occurs when $L = 20$ feet. When the shorter sides are 20 feet, there is 40 feet of fencing left for the longer side. To maximize the area, she should enclose the garden so the two shorter sides have length 20 feet and the longer side parallel to the existing fence has length 40 feet.

Analysis   This problem also could be solved by graphing the quadratic function. We can see where the maximum area occurs on a graph of the quadratic function in **Figure 11**.

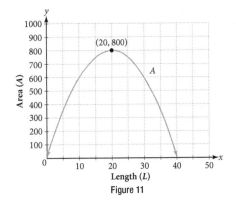

Figure 11

How To...

Given an application involving revenue, use a quadratic equation to find the maximum.

1. Write a quadratic equation for a revenue function.
2. Find the vertex of the quadratic equation.
3. Determine the y-value of the vertex.

Example 6   **Finding Maximum Revenue**

The unit price of an item affects its supply and demand. That is, if the unit price goes up, the demand for the item will usually decrease. For example, a local newspaper currently has 84,000 subscribers at a quarterly charge of $30. Market research has suggested that if the owners raise the price to $32, they would lose 5,000 subscribers. Assuming that subscriptions are linearly related to the price, what price should the newspaper charge for a quarterly subscription to maximize their revenue?

Solution   Revenue is the amount of money a company brings in. In this case, the revenue can be found by multiplying the price per subscription times the number of subscribers, or quantity. We can introduce variables, p for price per subscription and Q for quantity, giving us the equation Revenue $= pQ$.

Because the number of subscribers changes with the price, we need to find a relationship between the variables. We know that currently $p = 30$ and $Q = 84{,}000$. We also know that if the price rises to $32, the newspaper would lose 5,000 subscribers, giving a second pair of values, $p = 32$ and $Q = 79{,}000$. From this we can find a linear equation relating the two quantities. The slope will be

$$m = \frac{79{,}000 - 84{,}000}{32 - 30}$$

$$= \frac{-5{,}000}{2}$$

$$= -2{,}500$$

This tells us the paper will lose 2,500 subscribers for each dollar they raise the price. We can then solve for the y-intercept.

$$Q = -2{,}500p + b \qquad \text{Substitute in the point } Q = 84{,}000 \text{ and } p = 30$$
$$84{,}000 = -2{,}500(30) + b \qquad \text{Solve for } b$$
$$b = 159{,}000$$

This gives us the linear equation $Q = -2{,}500p + 159{,}000$ relating cost and subscribers. We now return to our revenue equation.

$$\text{Revenue} = pQ$$
$$\text{Revenue} = p(-2{,}500p + 159{,}000)$$
$$\text{Revenue} = -2{,}500p^2 + 159{,}000p$$

We now have a quadratic function for revenue as a function of the subscription charge. To find the price that will maximize revenue for the newspaper, we can find the vertex.

$$h = -\frac{159{,}000}{2(-2{,}500)}$$

$$= 31.8$$

The model tells us that the maximum revenue will occur if the newspaper charges $31.80 for a subscription. To find what the maximum revenue is, we evaluate the revenue function.

$$\text{maximum revenue} = -2{,}500(31.8)^2 + 159{,}000(31.8)$$

$$= 2{,}528{,}100$$

Analysis   This could also be solved by graphing the quadratic as in **Figure 12**. *We can see the maximum revenue on a graph of the quadratic function.*

Figure 12

Finding the *x*- and *y*-Intercepts of a Quadratic Function

Much as we did in the application problems above, we also need to find intercepts of quadratic equations for graphing parabolas. Recall that we find the *y*-intercept of a quadratic by evaluating the function at an input of zero, and we find the *x*-intercepts at locations where the output is zero. Notice in **Figure 13** that the number of *x*-intercepts can vary depending upon the location of the graph.

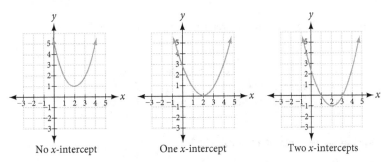

No *x*-intercept One *x*-intercept Two *x*-intercepts

Figure 13 Number of *x*-intercepts of a parabola

How To…

Given a quadratic function $f(x)$, find the *y*- and *x*-intercepts.

1. Evaluate $f(0)$ to find the *y*-intercept.
2. Solve the quadratic equation $f(x) = 0$ to find the *x*-intercepts.

Example 7 **Finding the *y*- and *x*-Intercepts of a Parabola**

Find the *y*- and *x*-intercepts of the quadratic $f(x) = 3x^2 + 5x - 2$.

Solution We find the *y*-intercept by evaluating $f(0)$.

$$f(0) = 3(0)^2 + 5(0) - 2$$
$$= -2$$

So the *y*-intercept is at $(0, -2)$.

For the *x*-intercepts, we find all solutions of $f(x) = 0$.

$$0 = 3x^2 + 5x - 2$$

In this case, the quadratic can be factored easily, providing the simplest method for solution.

$$0 = (3x - 1)(x + 2)$$
$$0 = 3x - 1 \qquad 0 = x + 2$$
$$x = \frac{1}{3} \qquad \text{or} \qquad x = -2$$

So the *x*-intercepts are at $\left(\frac{1}{3}, 0\right)$ and $(-2, 0)$.

Analysis By graphing the function, we can confirm that the graph crosses the y-axis at $(0, -2)$. We can also confirm that the graph crosses the x-axis at $\left(\frac{1}{3}, 0\right)$ and $(-2, 0)$. See **Figure 14**.

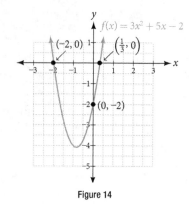

Figure 14

Rewriting Quadratics in Standard Form

In **Example 7**, the quadratic was easily solved by factoring. However, there are many quadratics that cannot be factored. We can solve these quadratics by first rewriting them in standard form.

How To...

Given a quadratic function, find the x-intercepts by rewriting in standard form.

1. Substitute a and b into $h = -\dfrac{b}{2a}$.

2. Substitute $x = h$ into the general form of the quadratic function to find k.

3. Rewrite the quadratic in standard form using h and k.

4. Solve for when the output of the function will be zero to find the x-intercepts.

Example 8   **Finding the x-Intercepts of a Parabola**

Find the x-intercepts of the quadratic function $f(x) = 2x^2 + 4x - 4$.

Solution We begin by solving for when the output will be zero.

$$0 = 2x^2 + 4x - 4$$

Because the quadratic is not easily factorable in this case, we solve for the intercepts by first rewriting the quadratic in standard form.

$$f(x) = a(x - h)^2 + k$$

We know that $a = 2$. Then we solve for h and k.

$$h = -\frac{b}{2a} \qquad\qquad k = f(-1)$$

$$= -\frac{4}{2(2)} \qquad\qquad = 2(-1)^2 + 4(-1) - 4$$

$$= -1 \qquad\qquad\qquad = -6$$

So now we can rewrite in standard form.

$$f(x) = 2(x + 1)^2 - 6$$

We can now solve for when the output will be zero.

$$0 = 2(x + 1)^2 - 6$$
$$6 = 2(x + 1)^2$$
$$3 = (x + 1)^2$$
$$x + 1 = \pm\sqrt{3}$$
$$x = -1 \pm\sqrt{3}$$

The graph has x-intercepts at $(-1 - \sqrt{3}, 0)$ and $(-1 + \sqrt{3}, 0)$.

We can check our work by graphing the given function on a graphing utility and observing the x-intercepts. See **Figure 15**.

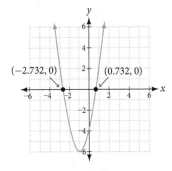

Figure 15

Analysis *We could have achieved the same results using the quadratic formula. Identify $a = 2$, $b = 4$, and $c = -4$.*

$$x = \frac{-b \pm \sqrt{b^2 - 4ac}}{2a}$$

$$x = \frac{-4 \pm \sqrt{4^2 - 4(2)(-4)}}{2(2)}$$

$$x = \frac{-4 \pm \sqrt{48}}{4}$$

$$x = \frac{-4 \pm \sqrt{3(16)}}{4}$$

$$x = -1 \pm \sqrt{3}$$

So the x-intercepts occur at $(-1 - \sqrt{3}, 0)$ and $(-1 + \sqrt{3}, 0)$.

Try It #4

In a separate **Try It**, we found the standard and general form for the function $g(x) = 13 + x^2 - 6x$. Now find the y- and x-intercepts (if any).

Example 9   **Applying the Vertex and *x*-Intercepts of a Parabola**

A ball is thrown upward from the top of a 40 foot high building at a speed of 80 feet per second. The ball's height above ground can be modeled by the equation $H(t) = -16t^2 + 80t + 40$.

    a. When does the ball reach the maximum height?
    b. What is the maximum height of the ball?
    c. When does the ball hit the ground?

Solution

    a. The ball reaches the maximum height at the vertex of the parabola.

$$h = -\frac{80}{2(-16)}$$

$$= \frac{80}{32}$$

$$= \frac{5}{2}$$

$$= 2.5$$

    The ball reaches a maximum height after 2.5 seconds.

b. To find the maximum height, find the *y*-coordinate of the vertex of the parabola.

$$k = H\left(-\frac{b}{2a}\right)$$
$$= H(2.5)$$
$$= -16(2.5)^2 + 80(2.5) + 40$$
$$= 140$$

The ball reaches a maximum height of 140 feet.

c. To find when the ball hits the ground, we need to determine when the height is zero, $H(t) = 0$. We use the quadratic formula.

$$t = \frac{-80 \pm \sqrt{80^2 - 4(-16)(40)}}{2(-16)}$$

$$= \frac{-80 \pm \sqrt{8960}}{-32}$$

Because the square root does not simplify nicely, we can use a calculator to approximate the values of the solutions.

$$t = \frac{-80 - \sqrt{8960}}{-32} \approx 5.458 \text{ or } t = \frac{-80 + \sqrt{8960}}{-32} \approx -0.458$$

The second answer is outside the reasonable domain of our model, so we conclude the ball will hit the ground after about 5.458 seconds. See **Figure 16.**

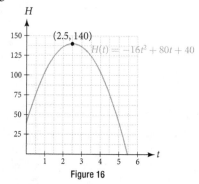

Figure 16

Notice that the graph does not represent the physical path of the ball upward and downward. Keep quantities on each axis in mind while interpreting the graph.

Try It #5

A rock is thrown upward from the top of a 112-foot high cliff overlooking the ocean at a speed of 96 feet per second. The rock's height above ocean can be modeled by the equation
$H(t) = -16t^2 + 96t + 112$.

a. When does the rock reach the maximum height?

b. What is the maximum height of the rock?

c. When does the rock hit the ocean?

Access these online resources for additional instruction and practice with quadratic equations.

- Graphing Quadratic Functions in General Form (http://openstaxcollege.org/l/graphquadgen)
- Graphing Quadratic Functions in Standard Form (http://openstaxcollege.org/l/graphquadstan)
- Quadratic Function Review (http://openstaxcollege.org/l/quadfuncrev)
- Characteristics of a Quadratic Function (http://openstaxcollege.org/l/characterquad)

5.1 SECTION EXERCISES

VERBAL

1. Explain the advantage of writing a quadratic function in standard form.

2. How can the vertex of a parabola be used in solving real-world problems?

3. Explain why the condition of $a \neq 0$ is imposed in the definition of the quadratic function.

4. What is another name for the standard form of a quadratic function?

5. What two algebraic methods can be used to find the horizontal intercepts of a quadratic function?

ALGEBRAIC

For the following exercises, rewrite the quadratic functions in standard form and give the vertex.

6. $f(x) = x^2 - 12x + 32$

7. $g(x) = x^2 + 2x - 3$

8. $f(x) = x^2 - x$

9. $f(x) = x^2 + 5x - 2$

10. $h(x) = 2x^2 + 8x - 10$

11. $k(x) = 3x^2 - 6x - 9$

12. $f(x) = 2x^2 - 6x$

13. $f(x) = 3x^2 - 5x - 1$

For the following exercises, determine whether there is a minimum or maximum value to each quadratic function. Find the value and the axis of symmetry.

14. $y(x) = 2x^2 + 10x + 12$

15. $f(x) = 2x^2 - 10x + 4$

16. $f(x) = -x^2 + 4x + 3$

17. $f(x) = 4x^2 + x - 1$

18. $h(t) = -4t^2 + 6t - 1$

19. $f(x) = \frac{1}{2}x^2 + 3x + 1$

20. $f(x) = -\frac{1}{3}x^2 - 2x + 3$

For the following exercises, determine the domain and range of the quadratic function.

21. $f(x) = (x - 3)^2 + 2$

22. $f(x) = -2(x + 3)^2 - 6$

23. $f(x) = x^2 + 6x + 4$

24. $f(x) = 2x^2 - 4x + 2$

25. $k(x) = 3x^2 - 6x - 9$

For the following exercises, use the vertex (h, k) and a point on the graph (x, y) to find the general form of the equation of the quadratic function.

26. $(h, k) = (2, 0), (x, y) = (4, 4)$

27. $(h, k) = (-2, -1), (x, y) = (-4, 3)$

28. $(h, k) = (0, 1), (x, y) = (2, 5)$

29. $(h, k) = (2, 3), (x, y) = (5, 12)$

30. $(h, k) = (-5, 3), (x, y) = (2, 9)$

31. $(h, k) = (3, 2), (x, y) = (10, 1)$

32. $(h, k) = (0, 1), (x, y) = (1, 0)$

33. $(h, k) = (1, 0), (x, y) = (0, 1)$

GRAPHICAL

For the following exercises, sketch a graph of the quadratic function and give the vertex, axis of symmetry, and intercepts.

34. $f(x) = x^2 - 2x$

35. $f(x) = x^2 - 6x - 1$

36. $f(x) = x^2 - 5x - 6$

37. $f(x) = x^2 - 7x + 3$

38. $f(x) = -2x^2 + 5x - 8$

39. $f(x) = 4x^2 - 12x - 3$

For the following exercises, write the equation for the graphed function.

40.

41.

42.

43.

44.

45.

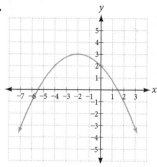

NUMERIC

For the following exercises, use the table of values that represent points on the graph of a quadratic function. By determining the vertex and axis of symmetry, find the general form of the equation of the quadratic function.

46.

x	−2	−1	0	1	2
y	5	2	1	2	5

47.

x	−2	−1	0	1	2
y	1	0	1	4	9

48.

x	−2	−1	0	1	2
y	−2	1	2	1	−2

49.

x	−2	−1	0	1	2
y	−8	−3	0	1	0

50.

x	−2	−1	0	1	2
y	8	2	0	2	8

TECHNOLOGY

For the following exercises, use a calculator to find the answer.

51. Graph on the same set of axes the functions $f(x) = x^2$, $f(x) = 2x^2$, and $f(x) = \frac{1}{3}x^2$. What appears to be the effect of changing the coefficient?

52. Graph on the same set of axes $f(x) = x^2$, $f(x) = x^2 + 2$ and $f(x) = x^2$, $f(x) = x^2 + 5$ and $f(x) = x^2 - 3$. What appears to be the effect of adding a constant?

53. Graph on the same set of axes $f(x) = x^2$, $f(x) = (x - 2)^2$, $f(x - 3)^2$, and $f(x) = (x + 4)^2$. What appears to be the effect of adding or subtracting those numbers?

54. The path of an object projected at a 45 degree angle with initial velocity of 80 feet per second is given by the function $h(x) = \frac{-32}{(80)^2}x^2 + x$ where x is the horizontal distance traveled and $h(x)$ is the height in feet. Use the [**TRACE**] feature of your calculator to determine the height of the object when it has traveled 100 feet away horizontally.

55. A suspension bridge can be modeled by the quadratic function $h(x) = 0.0001x^2$ with $-2000 \le x \le 2000$ where $|x|$ is the number of feet from the center and $h(x)$ is height in feet. Use the [**TRACE**] feature of your calculator to estimate how far from the center does the bridge have a height of 100 feet.

EXTENSIONS

For the following exercises, use the vertex of the graph of the quadratic function and the direction the graph opens to find the domain and range of the function.

56. Vertex $(1, -2)$, opens up.

57. Vertex $(-1, 2)$ opens down.

58. Vertex $(-5, 11)$, opens down.

59. Vertex $(-100, 100)$, opens up.

For the following exercises, write the equation of the quadratic function that contains the given point and has the same shape as the given function.

60. Contains $(1, 1)$ and has shape of $f(x) = 2x^2$. Vertex is on the y-axis.

61. Contains $(-1, 4)$ and has the shape of $f(x) = 2x^2$. Vertex is on the y-axis.

62. Contains $(2, 3)$ and has the shape of $f(x) = 3x^2$. Vertex is on the y-axis.

63. Contains $(1, -3)$ and has the shape of $f(x) = -x^2$. Vertex is on the y-axis.

64. Contains $(4, 3)$ and has the shape of $f(x) = 5x^2$. Vertex is on the y-axis.

65. Contains $(1, -6)$ has the shape of $f(x) = 3x^2$. Vertex has x-coordinate of -1.

REAL-WORLD APPLICATIONS

66. Find the dimensions of the rectangular corral producing the greatest enclosed area given 200 feet of fencing.

67. Find the dimensions of the rectangular corral split into 2 pens of the same size producing the greatest possible enclosed area given 300 feet of fencing.

68. Find the dimensions of the rectangular corral producing the greatest enclosed area split into 3 pens of the same size given 500 feet of fencing.

69. Among all of the pairs of numbers whose sum is 6, find the pair with the largest product. What is the product?

70. Among all of the pairs of numbers whose difference is 12, find the pair with the smallest product. What is the product?

71. Suppose that the price per unit in dollars of a cell phone production is modeled by $p = \$45 - 0.0125x$, where x is in thousands of phones produced, and the revenue represented by thousands of dollars is $R = x \cdot p$. Find the production level that will maximize revenue.

72. A rocket is launched in the air. Its height, in meters above sea level, as a function of time, in seconds, is given by $h(t) = -4.9t^2 + 229t + 234$. Find the maximum height the rocket attains.

73. A ball is thrown in the air from the top of a building. Its height, in meters above ground, as a function of time, in seconds, is given by $h(t) = -4.9t^2 + 24t + 8$. How long does it take to reach maximum height?

74. A soccer stadium holds 62,000 spectators. With a ticket price of $11, the average attendance has been 26,000. When the price dropped to $9, the average attendance rose to 31,000. Assuming that attendance is linearly related to ticket price, what ticket price would maximize revenue?

75. A farmer finds that if she plants 75 trees per acre, each tree will yield 20 bushels of fruit. She estimates that for each additional tree planted per acre, the yield of each tree will decrease by 3 bushels. How many trees should she plant per acre to maximize her harvest?

LEARNING OBJECTIVES

In this section, you will:

- Identify power functions.
- Identify end behavior of power functions.
- Identify polynomial functions.
- Identify the degree and leading coefficient of polynomial functions.

5.2   POWER FUNCTIONS AND POLYNOMIAL FUNCTIONS

Figure 1 (credit: Jason Bay, Flickr)

Suppose a certain species of bird thrives on a small island. Its population over the last few years is shown in **Table 1**.

Year	2009	2010	2011	2012	2013
Bird Population	800	897	992	1,083	1,169

Table 1

The population can be estimated using the function $P(t) = -0.3t^3 + 97t + 800$, where $P(t)$ represents the bird population on the island t years after 2009. We can use this model to estimate the maximum bird population and when it will occur. We can also use this model to predict when the bird population will disappear from the island. In this section, we will examine functions that we can use to estimate and predict these types of changes.

Identifying Power Functions

Before we can understand the bird problem, it will be helpful to understand a different type of function. A **power function** is a function with a single term that is the product of a real number, a **coefficient**, and a variable raised to a fixed real number. (A number that multiplies a variable raised to an exponent is known as a coefficient.)

As an example, consider functions for area or volume. The function for the area of a circle with radius r is

$$A(r) = \pi r^2$$

and the function for the volume of a sphere with radius r is

$$V(r) = \frac{4}{3}\pi r^3$$

Both of these are examples of power functions because they consist of a coefficient, π or $\frac{4}{3}\pi$, multiplied by a variable r raised to a power.

---

power function

A **power function** is a function that can be represented in the form

$$f(x) = kx^p$$

where k and p are real numbers, and k is known as the **coefficient**.

Q & A...

Is $f(x) = 2^x$ a power function?

No. A power function contains a variable base raised to a fixed power. This function has a constant base raised to a variable power. This is called an exponential function, not a power function.

Example 1   **Identifying Power Functions**

Which of the following functions are power functions?

$$f(x) = 1 \qquad\qquad \text{Constant function}$$
$$f(x) = x \qquad\qquad \text{Identify function}$$
$$f(x) = x^2 \qquad\qquad \text{Quadratic function}$$
$$f(x) = x^3 \qquad\qquad \text{Cubic function}$$
$$f(x) = \frac{1}{x} \qquad\qquad \text{Reciprocal function}$$
$$f(x) = \frac{1}{x^2} \qquad\qquad \text{Reciprocal squared function}$$
$$f(x) = \sqrt{x} \qquad\qquad \text{Square root function}$$
$$f(x) = \sqrt[3]{x} \qquad\qquad \text{Cube root function}$$

Solution   All of the listed functions are power functions.

The constant and identity functions are power functions because they can be written as $f(x) = x^0$ and $f(x) = x^1$ respectively.

The quadratic and cubic functions are power functions with whole number powers $f(x) = x^2$ and $f(x) = x^3$.

The reciprocal and reciprocal squared functions are power functions with negative whole number powers because they can be written as $f(x) = x^{-1}$ and $f(x) = x^{-2}$.

The square and cube root functions are power functions with fractional powers because they can be written as $f(x) = x^{1/2}$ or $f(x) = x^{1/3}$.

Try It #1

Which functions are power functions?

$$f(x) = 2x^2 \cdot 4x^3 \qquad\qquad g(x) = -x^5 + 5x^3 \qquad\qquad h(x) = \frac{2x^5 - 1}{3x^2 + 4}$$

Identifying End Behavior of Power Functions

Figure 2 shows the graphs of $f(x) = x^2$, $g(x) = x^4$ and $h(x) = x^6$, which are all power functions with even, whole-number powers. Notice that these graphs have similar shapes, very much like that of the quadratic function in the toolkit. However, as the power increases, the graphs flatten somewhat near the origin and become steeper away from the origin.

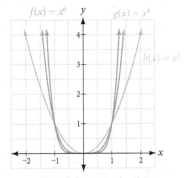

Figure 2 Even-power functions

To describe the behavior as numbers become larger and larger, we use the idea of infinity. We use the symbol ∞ for positive infinity and $-\infty$ for negative infinity. When we say that "x approaches infinity," which can be symbolically written as $x \rightarrow \infty$, we are describing a behavior; we are saying that x is increasing without bound.

With the positive even-power function, as the input increases or decreases without bound, the output values become very large, positive numbers. Equivalently, we could describe this behavior by saying that as x approaches positive or negative infinity, the $f(x)$ values increase without bound. In symbolic form, we could write

$$\text{as } x \rightarrow \pm\infty, \quad f(x) \rightarrow \infty$$

Figure 3 shows the graphs of $f(x) = x^3$, $g(x) = x^5$, and $h(x) = x^7$, which are all power functions with odd, whole-number powers. Notice that these graphs look similar to the cubic function in the toolkit. Again, as the power increases, the graphs flatten near the origin and become steeper away from the origin.

Figure 3 Odd-power functions

These examples illustrate that functions of the form $f(x) = x^n$ reveal symmetry of one kind or another. First, in **Figure 2** we see that even functions of the form $f(x) = x^n$, n even, are symmetric about the y-axis. In **Figure 3** we see that odd functions of the form $f(x) = x^n$, n odd, are symmetric about the origin.

For these odd power functions, as x approaches negative infinity, $f(x)$ decreases without bound. As x approaches positive infinity, $f(x)$ increases without bound. In symbolic form we write

$$\text{as } x \rightarrow -\infty, \quad f(x) \rightarrow -\infty \qquad \text{as } x \rightarrow \infty, \quad f(x) \rightarrow \infty$$

The behavior of the graph of a function as the input values get very small ($x \rightarrow -\infty$) and get very large ($x \rightarrow \infty$) is referred to as the **end behavior** of the function. We can use words or symbols to describe end behavior.

Figure 4 shows the end behavior of power functions in the form $f(x) = kx^n$ where n is a non-negative integer depending on the power and the constant.

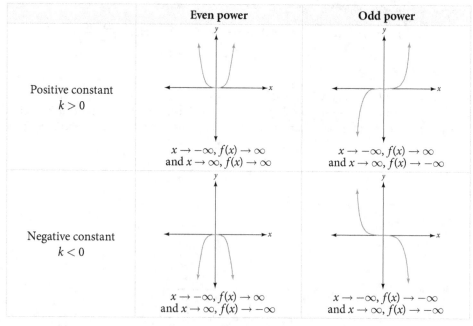

Figure 4

How To...
Given a power function $f(x) = kx^n$ where n is a non-negative integer, identify the end behavior.

1. Determine whether the power is even or odd.
2. Determine whether the constant is positive or negative.
3. Use **Figure 4** to identify the end behavior.

Example 2   Identifying the End Behavior of a Power Function

Describe the end behavior of the graph of $f(x) = x^8$.

Solution   The coefficient is 1 (positive) and the exponent of the power function is 8 (an even number). As x approaches infinity, the output (value of $f(x)$) increases without bound. We write as $x \to \infty$, $f(x) \to \infty$. As x approaches negative infinity, the output increases without bound. In symbolic form, as $x \to -\infty$, $f(x) \to \infty$. We can graphically represent the function as shown in **Figure 5**.

Figure 5

Example 3   Identifying the End Behavior of a Power Function

Describe the end behavior of the graph of $f(x) = -x^9$.

Solution   The exponent of the power function is 9 (an odd number). Because the coefficient is -1 (negative), the graph is the reflection about the x-axis of the graph of $f(x) = x^9$. **Figure 6** shows that as x approaches infinity, the output decreases without bound. As x approaches negative infinity, the output increases without bound. In symbolic form, we would write

$$\text{as } x \to -\infty, \quad f(x) \to \infty$$
$$\text{as } x \to \infty, \quad f(x) \to -\infty$$

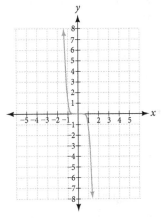

Figure 6

Analysis We can check our work by using the table feature on a graphing utility.

x	$f(x)$
−10	1,000,000,000
−5	1,953,125
0	0
5	−1,953,125
10	−1,000,000,000

Table 2

We can see from **Table 2** that, when we substitute very small values for x, the output is very large, and when we substitute very large values for x, the output is very small (meaning that it is a very large negative value).

Try It #2

Describe in words and symbols the end behavior of $f(x) = -5x^4$.

Identifying Polynomial Functions

An oil pipeline bursts in the Gulf of Mexico, causing an oil slick in a roughly circular shape. The slick is currently 24 miles in radius, but that radius is increasing by 8 miles each week. We want to write a formula for the area covered by the oil slick by combining two functions. The radius r of the spill depends on the number of weeks w that have passed. This relationship is linear.

$$r(w) = 24 + 8w$$

We can combine this with the formula for the area A of a circle.

$$A(r) = \pi r^2$$

Composing these functions gives a formula for the area in terms of weeks.

$$A(w) = A(r(w))$$
$$= A(24 + 8w)$$
$$= \pi(24 + 8w)^2$$

Multiplying gives the formula.

$$A(w) = 576\pi + 384\pi w + 64\pi w^2$$

This formula is an example of a **polynomial function**. A polynomial function consists of either zero or the sum of a finite number of non-zero terms, each of which is a product of a number, called the coefficient of the term, and a variable raised to a non-negative integer power.

---

polynomial functions

Let n be a non-negative integer. A **polynomial function** is a function that can be written in the form
$$f(x) = a_n x^n + \ldots + a_2 x^2 + a_1 x + a_0$$

This is called the general form of a polynomial function. Each a_i is a coefficient and can be any real number other than zero. Each expression $a_i x^i$ is a **term of a polynomial function**.

Example 4   **Identifying Polynomial Functions**

Which of the following are polynomial functions?

$$f(x) = 2x^3 \cdot 3x + 4 \qquad g(x) = -x(x^2 - 4) \qquad h(x) = 5\sqrt{x + 2}$$

Solution   The first two functions are examples of polynomial functions because they can be written in the form $f(x) = a_n x^n + \ldots + a_2 x^2 + a_1 x + a_0$, where the powers are non-negative integers and the coefficients are real numbers.

- $f(x)$ can be written as $f(x) = 6x^4 + 4$.
- $g(x)$ can be written as $g(x) = -x^3 + 4x$.
- $h(x)$ cannot be written in this form and is therefore not a polynomial function.

Identifying the Degree and Leading Coefficient of a Polynomial Function

Because of the form of a polynomial function, we can see an infinite variety in the number of terms and the power of the variable. Although the order of the terms in the polynomial function is not important for performing operations, we typically arrange the terms in descending order of power, or in general form. The **degree** of the polynomial is the highest power of the variable that occurs in the polynomial; it is the power of the first variable if the function is in general form. The **leading term** is the term containing the highest power of the variable, or the term with the highest degree. The **leading coefficient** is the coefficient of the leading term.

---

terminology of polynomial functions

We often rearrange polynomials so that the powers are descending.

Leading coefficient   Degree

$$f(x) = a_n x^n + \dots + a_2 x^2 + a_1 x + a_0$$

Leading term

When a polynomial is written in this way, we say that it is in general form.

How To...

Given a polynomial function, identify the degree and leading coefficient.

1. Find the highest power of x to determine the degree function.
2. Identify the term containing the highest power of x to find the leading term.
3. Identify the coefficient of the leading term.

Example 5   **Identifying the Degree and Leading Coefficient of a Polynomial Function**

Identify the degree, leading term, and leading coefficient of the following polynomial functions.

$$f(x) = 3 + 2x^2 - 4x^3$$
$$g(t) = 5t^5 - 2t^3 + 7t$$
$$h(p) = 6p - p^3 - 2$$

Solution   For the function $f(x)$, the highest power of x is 3, so the degree is 3. The leading term is the term containing that degree, $-4x^3$. The leading coefficient is the coefficient of that term, -4.

For the function $g(t)$, the highest power of t is 5, so the degree is 5. The leading term is the term containing that degree, $5t^5$. The leading coefficient is the coefficient of that term, 5.

For the function $h(p)$, the highest power of p is 3, so the degree is 3. The leading term is the term containing that degree, $-p^3$; the leading coefficient is the coefficient of that term, -1.

Try It #3

Identify the degree, leading term, and leading coefficient of the polynomial $f(x) = 4x^2 - x^6 + 2x - 6$.

Identifying End Behavior of Polynomial Functions

Knowing the degree of a polynomial function is useful in helping us predict its end behavior. To determine its end behavior, look at the leading term of the polynomial function. Because the power of the leading term is the highest, that term will grow significantly faster than the other terms as x gets very large or very small, so its behavior will dominate the graph. For any polynomial, the end behavior of the polynomial will match the end behavior of the power function consisting of the leading term. See **Table 3**.

Polynomial Function	Leading Term	Graph of Polynomial Function
$f(x) = 5x^4 + 2x^3 - x - 4$	$5x^4$	
$f(x) = -2x^6 - x^5 + 3x^4 + x^3$	$-2x^6$	
$f(x) = 3x^5 - 4x^4 + 2x^2 + 1$	$3x^5$	
$f(x) = -6x^3 + 7x^2 + 3x + 1$	$-6x^3$	

Table 3

Example 6   **Identifying End Behavior and Degree of a Polynomial Function**

Describe the end behavior and determine a possible degree of the polynomial function in **Figure 7**.

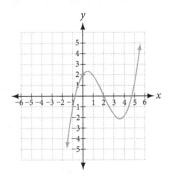

Figure 7

Solution   As the input values x get very large, the output values $f(x)$ increase without bound. As the input values x get very small, the output values $f(x)$ decrease without bound. We can describe the end behavior symbolically by writing

$$\text{as } x \to -\infty, \quad f(x) \to -\infty$$
$$\text{as } x \to \infty, \quad f(x) \to \infty$$

In words, we could say that as x values approach infinity, the function values approach infinity, and as x values approach negative infinity, the function values approach negative infinity.

We can tell this graph has the shape of an odd degree power function that has not been reflected, so the degree of the polynomial creating this graph must be odd and the leading coefficient must be positive.

Try It #4

Describe the end behavior, and determine a possible degree of the polynomial function in **Figure 8**.

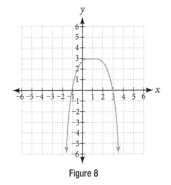

Figure 8

Example 7   **Identifying End Behavior and Degree of a Polynomial Function**

Given the function $f(x) = -3x^2(x - 1)(x + 4)$, express the function as a polynomial in general form, and determine the leading term, degree, and end behavior of the function.

Solution   Obtain the general form by expanding the given expression for $f(x)$.

$$f(x) = -3x^2(x - 1)(x + 4)$$
$$= -3x^2(x^2 + 3x - 4)$$
$$= -3x^4 - 9x^3 + 12x^2$$

The general form is $f(x) = -3x^4 - 9x^3 + 12x^2$. The leading term is $-3x^4$; therefore, the degree of the polynomial is 4. The degree is even (4) and the leading coefficient is negative (-3), so the end behavior is

$$\text{as } x \to -\infty, \quad f(x) \to -\infty$$
$$\text{as } x \to \infty, \quad f(x) \to -\infty$$

Try It #5

Given the function $f(x) = 0.2(x - 2)(x + 1)(x - 5)$, express the function as a polynomial in general form and determine the leading term, degree, and end behavior of the function.

Identifying Local Behavior of Polynomial Functions

In addition to the end behavior of polynomial functions, we are also interested in what happens in the "middle" of the function. In particular, we are interested in locations where graph behavior changes. A **turning point** is a point at which the function values change from increasing to decreasing or decreasing to increasing.

We are also interested in the intercepts. As with all functions, the *y*-intercept is the point at which the graph intersects the vertical axis. The point corresponds to the coordinate pair in which the input value is zero. Because a polynomial is a function, only one output value corresponds to each input value so there can be only one *y*-intercept $(0, a_0)$. The *x*-intercepts occur at the input values that correspond to an output value of zero. It is possible to have more than one *x*-intercept. See **Figure 9**.

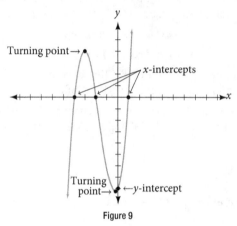

Figure 9

> *intercepts and turning points of polynomial functions*
>
> A **turning point** of a graph is a point at which the graph changes direction from increasing to decreasing or decreasing to increasing. The *y*-intercept is the point at which the function has an input value of zero. The *x*-intercepts are the points at which the output value is zero.

How To…

Given a polynomial function, determine the intercepts.

1. Determine the *y*-intercept by setting $x = 0$ and finding the corresponding output value.
2. Determine the *x*-intercepts by solving for the input values that yield an output value of zero.

Example 8   **Determining the Intercepts of a Polynomial Function**

Given the polynomial function $f(x) = (x - 2)(x + 1)(x - 4)$, written in factored form for your convenience, determine the *y*- and *x*-intercepts.

Solution   The *y*-intercept occurs when the input is zero so substitute 0 for *x*.

$$f(0) = (0 - 2)(0 + 1)(0 - 4)$$
$$= (-2)(1)(-4)$$
$$= 8$$

The *y*-intercept is (0, 8).

The x-intercepts occur when the output is zero.

$$0 = (x - 2)(x + 1)(x - 4)$$
$$x - 2 = 0 \quad \text{or} \quad x + 1 = 0 \quad \text{or} \quad x - 4 = 0$$
$$x = 2 \quad \text{or} \quad x = -1 \quad \text{or} \quad x = 4$$

The x-intercepts are $(2, 0)$, $(-1, 0)$, and $(4, 0)$.

We can see these intercepts on the graph of the function shown in **Figure 10**.

Figure 10

Example 9   **Determining the Intercepts of a Polynomial Function with Factoring**

Given the polynomial function $f(x) = x^4 - 4x^2 - 45$, determine the y- and x-intercepts.

Solution   The y-intercept occurs when the input is zero.

$$f(0) = (0)^4 - 4(0)^2 - 45$$
$$= -45$$

The y-intercept is $(0, -45)$.

The x-intercepts occur when the output is zero. To determine when the output is zero, we will need to factor the polynomial.

$$f(x) = x^4 - 4x^2 - 45$$
$$= (x^2 - 9)(x^2 + 5)$$
$$= (x - 3)(x + 3)(x^2 + 5)$$
$$0 = (x - 3)(x + 3)(x^2 + 5)$$
$$x - 3 = 0 \quad \text{or} \quad x + 3 = 0 \quad \text{or} \quad x^2 + 5 = 0$$
$$x = 3 \quad \text{or} \quad x = -3 \text{ or} \quad \text{(no real solution)}$$

The x-intercepts are $(3, 0)$ and $(-3, 0)$.

We can see these intercepts on the graph of the function shown in **Figure 11**. We can see that the function is even because $f(x) = f(-x)$.

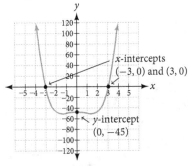

Figure 11

Try It #6

Given the polynomial function $f(x) = 2x^3 - 6x^2 - 20x$, determine the y- and x-intercepts.

Comparing Smooth and Continuous Graphs

The degree of a polynomial function helps us to determine the number of x-intercepts and the number of turning points. A polynomial function of nth degree is the product of n factors, so it will have at most n roots or zeros, or x-intercepts. The graph of the polynomial function of degree n must have at most $n - 1$ turning points. This means the graph has at most one fewer turning point than the degree of the polynomial or one fewer than the number of factors.

A **continuous function** has no breaks in its graph: the graph can be drawn without lifting the pen from the paper. A **smooth curve** is a graph that has no sharp corners. The turning points of a smooth graph must always occur at rounded curves. The graphs of polynomial functions are both continuous and smooth.

---

intercepts and turning points of polynomials

A polynomial of degree n will have, at most, n x-intercepts and $n - 1$ turning points.

Example 10   **Determining the Number of Intercepts and Turning Points of a Polynomial**

Without graphing the function, determine the local behavior of the function by finding the maximum number of x-intercepts and turning points for $f(x) = -3x^{10} + 4x^7 - x^4 + 2x^3$.

Solution   The polynomial has a degree of 10, so there are at most 10 x-intercepts and at most 9 turning points.

Try It #7

Without graphing the function, determine the maximum number of x-intercepts and turning points for
$f(x) = 108 - 13x^9 - 8x^4 + 14x^{12} + 2x^3$

Example 11   **Drawing Conclusions about a Polynomial Function from the Graph**

What can we conclude about the polynomial represented by the graph shown in **Figure 12** based on its intercepts and turning points?

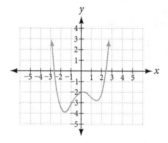

Figure 12

Solution   The end behavior of the graph tells us this is the graph of an even-degree polynomial. See **Figure 13**.

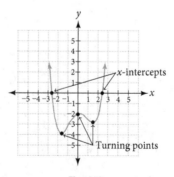

Figure 13

The graph has 2 x-intercepts, suggesting a degree of 2 or greater, and 3 turning points, suggesting a degree of 4 or greater. Based on this, it would be reasonable to conclude that the degree is even and at least 4.

Try It #8

What can we conclude about the polynomial represented by the graph shown in **Figure 14** based on its intercepts and turning points?

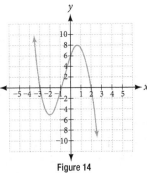

Figure 14

Example 12   **Drawing Conclusions about a Polynomial Function from the Factors**

Given the function $f(x) = -4x(x + 3)(x - 4)$, determine the local behavior.

Solution   The y-intercept is found by evaluating $f(0)$.

$$f(0) = -4(0)(0 + 3)(0 - 4)$$
$$= 0$$

The y-intercept is $(0, 0)$.

The x-intercepts are found by determining the zeros of the function.

$$0 = -4x(x + 3)(x - 4)$$
$$x = 0 \quad \text{or} \quad x + 3 = 0 \quad \text{or } x - 4 = 0$$
$$x = 0 \quad \text{or} \quad x = -3 \quad \text{or} \quad x = 4$$

The x-intercepts are $(0, 0)$, $(-3, 0)$, and $(4, 0)$.

The degree is 3 so the graph has at most 2 turning points.

Try It #9

Given the function $f(x) = 0.2(x - 2)(x + 1)(x - 5)$, determine the local behavior.

Access these online resources for additional instruction and practice with power and polynomial functions.

- Find Key Information About a Given Polynomial Function (http://openstaxcollege.org/l/keyinfopoly)
- End Behavior of a Polynomial Function (http://openstaxcollege.org/l/endbehavior)
- Turning Points and x-intercepts of Polynomial Functions (http://openstaxcollege.org/l/turningpoints)
- Least Possible Degree of a Polynomial Function (http://openstaxcollege.org/l/leastposdegree)

5.2 SECTION EXERCISES

VERBAL

1. Explain the difference between the coefficient of a power function and its degree.

2. If a polynomial function is in factored form, what would be a good first step in order to determine the degree of the function?

3. In general, explain the end behavior of a power function with odd degree if the leading coefficient is positive.

4. What is the relationship between the degree of a polynomial function and the maximum number of turning points in its graph?

5. What can we conclude if, in general, the graph of a polynomial function exhibits the following end behavior? As $x \to -\infty$, $f(x) \to -\infty$ and as $x \to \infty$, $f(x) \to -\infty$.

ALGEBRAIC

For the following exercises, identify the function as a power function, a polynomial function, or neither.

6. $f(x) = x^5$

7. $f(x) = (x^2)^3$

8. $f(x) = x - x^4$

9. $f(x) = \dfrac{x^2}{x^2 - 1}$

10. $f(x) = 2x(x + 2)(x - 1)^2$

11. $f(x) = 3^{x+1}$

For the following exercises, find the degree and leading coefficient for the given polynomial.

12. $-3x$

13. $7 - 2x^2$

14. $-2x^2 - 3x^5 + x - 6$

15. $x(4 - x^2)(2x + 1)$

16. $x^2(2x - 3)^2$

For the following exercises, determine the end behavior of the functions.

17. $f(x) = x^4$

18. $f(x) = x^3$

19. $f(x) = -x^4$

20. $f(x) = -x^9$

21. $f(x) = -2x^4 - 3x^2 + x - 1$

22. $f(x) = 3x^2 + x - 2$

23. $f(x) = x^2(2x^3 - x + 1)$

24. $f(x) = (2 - x)^7$

For the following exercises, find the intercepts of the functions.

25. $f(t) = 2(t - 1)(t + 2)(t - 3)$

26. $g(n) = -2(3n - 1)(2n + 1)$

27. $f(x) = x^4 - 16$

28. $f(x) = x^3 + 27$

29. $f(x) = x(x^2 - 2x - 8)$

30. $f(x) = (x + 3)(4x^2 - 1)$

GRAPHICAL

For the following exercises, determine the least possible degree of the polynomial function shown.

31.

32.

33.

34.

35.

36.

37.

38.

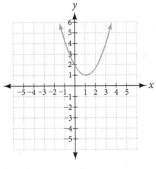

For the following exercises, determine whether the graph of the function provided is a graph of a polynomial function. If so, determine the number of turning points and the least possible degree for the function.

39.

40.

41.

42.

43.

44.

45.

NUMERIC

For the following exercises, make a table to confirm the end behavior of the function.

46. $f(x) = -x^3$

47. $f(x) = x^4 - 5x^2$

48. $f(x) = x^2(1 - x)^2$

49. $f(x) = (x - 1)(x - 2)(3 - x)$

50. $f(x) = \dfrac{x^5}{10} - x^4$

TECHNOLOGY

For the following exercises, graph the polynomial functions using a calculator. Based on the graph, determine the intercepts and the end behavior.

51. $f(x) = x^3(x - 2)$

52. $f(x) = x(x - 3)(x + 3)$

53. $f(x) = x(14 - 2x)(10 - 2x)$

54. $f(x) = x(14 - 2x)(10 - 2x)^2$

55. $f(x) = x^3 - 16x$

56. $f(x) = x^3 - 27$

57. $f(x) = x^4 - 81$

58. $f(x) = -x^3 + x^2 + 2x$

59. $f(x) = x^3 - 2x^2 - 15x$

60. $f(x) = x^3 - 0.01x$

EXTENSIONS

For the following exercises, use the information about the graph of a polynomial function to determine the function. Assume the leading coefficient is 1 or −1. There may be more than one correct answer.

61. The y-intercept is $(0, -4)$. The x-intercepts are $(-2, 0)$, $(2, 0)$. Degree is 2. End behavior: as $x \to -\infty$, $f(x) \to \infty$, as $x \to \infty, f(x) \to \infty$.

62. The y-intercept is $(0, 9)$. The x-intercepts are $(-3, 0)$, $(3, 0)$. Degree is 2. End behavior: as $x \to -\infty$, $f(x) \to -\infty$, as $x \to \infty, f(x) \to -\infty$.

63. The y-intercept is $(0, 0)$. The x-intercepts are $(0, 0)$, $(2, 0)$. Degree is 3. End behavior: as $x \to -\infty$, $f(x) \to -\infty$, as $x \to \infty, f(x) \to \infty$.

64. The y-intercept is $(0, 1)$. The x-intercept is $(1, 0)$. Degree is 3. End behavior: as $x \to -\infty, f(x) \to \infty$, as $x \to \infty, f(x) \to -\infty$.

65. The y-intercept is $(0, 1)$. There is no x-intercept. Degree is 4. End behavior: as $x \to -\infty, f(x) \to \infty$, as $x \to \infty, f(x) \to \infty$.

REAL-WORLD APPLICATIONS

For the following exercises, use the written statements to construct a polynomial function that represents the required information.

66. An oil slick is expanding as a circle. The radius of the circle is increasing at the rate of 20 meters per day. Express the area of the circle as a function of d, the number of days elapsed.

67. A cube has an edge of 3 feet. The edge is increasing at the rate of 2 feet per minute. Express the volume of the cube as a function of m, the number of minutes elapsed.

68. A rectangle has a length of 10 inches and a width of 6 inches. If the length is increased by x inches and the width increased by twice that amount, express the area of the rectangle as a function of x.

69. An open box is to be constructed by cutting out square corners of x-inch sides from a piece of cardboard 8 inches by 8 inches and then folding up the sides. Express the volume of the box as a function of x.

70. A rectangle is twice as long as it is wide. Squares of side 2 feet are cut out from each corner. Then the sides are folded up to make an open box. Express the volume of the box as a function of the width (x).

LEARNING OBJECTIVES

In this section, you will:

- Recognize characteristics of graphs of polynomial functions.
- Use factoring to find zeros of polynomial functions.
- Identify zeros and their multiplicities.
- Determine end behavior.
- Understand the relationship between degree and turning points.
- Graph polynomial functions.
- Use the Intermediate Value Theorem.

5.3 GRAPHS OF POLYNOMIAL FUNCTIONS

The revenue in millions of dollars for a fictional cable company from 2006 through 2013 is shown in **Table 1**.

Year	2006	2007	2008	2009	2010	2011	2012	2013
Revenues	52.4	52.8	51.2	49.5	48.6	48.6	48.7	47.1

Table 1

The revenue can be modeled by the polynomial function

$$R(t) = -0.037t^4 + 1.414t^3 - 19.777t^2 + 118.696t - 205.332$$

where R represents the revenue in millions of dollars and t represents the year, with $t = 6$ corresponding to 2006. Over which intervals is the revenue for the company increasing? Over which intervals is the revenue for the company decreasing? These questions, along with many others, can be answered by examining the graph of the polynomial function. We have already explored the local behavior of quadratics, a special case of polynomials. In this section we will explore the local behavior of polynomials in general.

Recognizing Characteristics of Graphs of Polynomial Functions

Polynomial functions of degree 2 or more have graphs that do not have sharp corners; recall that these types of graphs are called smooth curves. Polynomial functions also display graphs that have no breaks. Curves with no breaks are called continuous. **Figure 1** shows a graph that represents a polynomial function and a graph that represents a function that is not a polynomial.

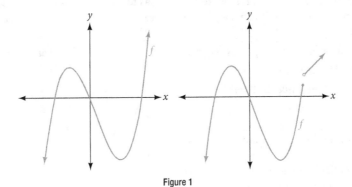

Figure 1

Example 1   **Recognizing Polynomial Functions**

Which of the graphs in **Figure 2** represents a polynomial function?

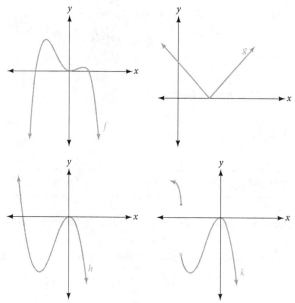

Figure 2

Solution   The graphs of f and h are graphs of polynomial functions. They are smooth and continuous.

The graphs of g and k are graphs of functions that are not polynomials. The graph of function g has a sharp corner. The graph of function k is not continuous.

─────────────────

Q & A...

Do all polynomial functions have as their domain all real numbers?

Yes. Any real number is a valid input for a polynomial function.

─────────────────

Using Factoring to Find Zeros of Polynomial Functions

Recall that if f is a polynomial function, the values of x for which $f(x) = 0$ are called zeros of f. If the equation of the polynomial function can be factored, we can set each factor equal to zero and solve for the zeros.

We can use this method to find x-intercepts because at the x-intercepts we find the input values when the output value is zero. For general polynomials, this can be a challenging prospect. While quadratics can be solved using the relatively simple quadratic formula, the corresponding formulas for cubic and fourth-degree polynomials are not simple enough to remember, and formulas do not exist for general higher-degree polynomials. Consequently, we will limit ourselves to three cases:

1. The polynomial can be factored using known methods: greatest common factor and trinomial factoring.
2. The polynomial is given in factored form.
3. Technology is used to determine the intercepts.

─────────────────

How To...

Given a polynomial function f, find the x-intercepts by factoring.

1. Set $f(x) = 0$.
2. If the polynomial function is not given in factored form:
    a. Factor out any common monomial factors.
    b. Factor any factorable binomials or trinomials.
3. Set each factor equal to zero and solve to find the x-intercepts.

─────────────────

Example 2   **Finding the x-Intercepts of a Polynomial Function by Factoring**

Find the x-intercepts of $f(x) = x^6 - 3x^4 + 2x^2$.

Solution   We can attempt to factor this polynomial to find solutions for $f(x) = 0$.

$$x^6 - 3x^4 + 2x^2 = 0 \qquad \text{Factor out the greatest common factor.}$$
$$x^2(x^4 - 3x^2 + 2) = 0 \qquad \text{Factor the trinomial.}$$
$$x^2(x^2 - 1)(x^2 - 2) = 0 \qquad \text{Set each factor equal to zero.}$$
$$(x^2 - 1) = 0 \qquad (x^2 - 2) = 0$$
$$x^2 = 0 \quad \text{or} \quad x^2 = 1 \quad \text{or } x^2 = 2$$
$$x = 0 \qquad x = \pm 1 \qquad x = \pm\sqrt{2}$$

This gives us five x-intercepts: $(0, 0)$, $(1, 0)$, $(-1, 0)$, $(\sqrt{2}, 0)$, and $(-\sqrt{2}, 0)$. See **Figure 3**. We can see that this is an even function because it is symmetric about the y-axis.

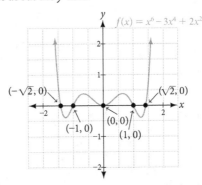

Figure 3

Example 3   **Finding the x-Intercepts of a Polynomial Function by Factoring**

Find the x-intercepts of $f(x) = x^3 - 5x^2 - x + 5$.

Solution   Find solutions for $f(x) = 0$ by factoring.

$$x^3 - 5x^2 - x + 5 = 0 \qquad \text{Factor by grouping.}$$
$$x^2(x - 5) - (x - 5) = 0 \qquad \text{Factor out the common factor.}$$
$$(x^2 - 1)(x - 5) = 0 \qquad \text{Factor the difference of squares.}$$
$$(x + 1)(x - 1)(x - 5) = 0 \qquad \text{Set each factor equal to zero.}$$
$$x + 1 = 0 \quad \text{or} \quad x - 1 = 0 \quad \text{or} \quad x - 5 = 0$$
$$x = -1 \qquad x = 1 \qquad x = 5$$

There are three x-intercepts: $(-1, 0)$, $(1, 0)$, and $(5, 0)$. See **Figure 4**.

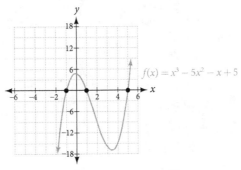

Figure 4

Example 4   **Finding the y- and x-Intercepts of a Polynomial in Factored Form**

Find the y- and x-intercepts of $g(x) = (x - 2)^2(2x + 3)$.

Solution   The y-intercept can be found by evaluating $g(0)$.

$$g(0) = (0 - 2)^2(2(0) + 3)$$
$$= 12$$

So the y-intercept is (0, 12).

The x-intercepts can be found by solving $g(x) = 0$.

$$(x - 2)^2(2x + 3) = 0$$

$$(x - 2)^2 = 0 \qquad (2x + 3) = 0$$
$$x - 2 = 0 \quad \text{or} \quad x = -\frac{3}{2}$$
$$x = 2$$

So the x-intercepts are $(2, 0)$ and $\left(-\frac{3}{2}, 0\right)$.

*Analysis   We can always check that our answers are reasonable by using a graphing calculator to graph the polynomial as shown in **Figure 5**.*

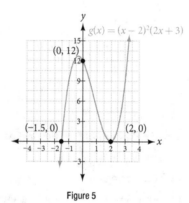

Figure 5

Example 5   **Finding the x-Intercepts of a Polynomial Function Using a Graph**

Find the x-intercepts of $h(x) = x^3 + 4x^2 + x - 6$.

Solution   This polynomial is not in factored form, has no common factors, and does not appear to be factorable using techniques previously discussed. Fortunately, we can use technology to find the intercepts. Keep in mind that some values make graphing difficult by hand. In these cases, we can take advantage of graphing utilities.

Looking at the graph of this function, as shown in **Figure 6**, it appears that there are x-intercepts at $x = -3, -2,$ and 1.

Figure 6

We can check whether these are correct by substituting these values for x and verifying that

$$h(-3) = h(-2) = h(1) = 0.$$

Since $h(x) = x^3 + 4x^2 + x - 6$, we have:

$$h(-3) = (-3)^3 + 4(-3)^2 + (-3) - 6 = -27 + 36 - 3 - 6 = 0$$
$$h(-2) = (-2)^3 + 4(-2)^2 + (-2) - 6 = -8 + 16 - 2 - 6 = 0$$
$$h(1) = (1)^3 + 4(1)^2 + (1) - 6 = 1 + 4 + 1 - 6 = 0$$

Each x-intercept corresponds to a zero of the polynomial function and each zero yields a factor, so we can now write the polynomial in factored form.

$$h(x) = x^3 + 4x^2 + x - 6$$
$$= (x + 3)(x + 2)(x - 1)$$

Try It #1

Find the y- and x-intercepts of the function $f(x) = x^4 - 19x^2 + 30x$.

Identifying Zeros and Their Multiplicities

Graphs behave differently at various x-intercepts. Sometimes, the graph will cross over the horizontal axis at an intercept. Other times, the graph will touch the horizontal axis and "bounce" off.

Suppose, for example, we graph the function shown.

$$f(x) = (x + 3)(x - 2)^2(x + 1)^3.$$

Notice in **Figure 7** that the behavior of the function at each of the x-intercepts is different.

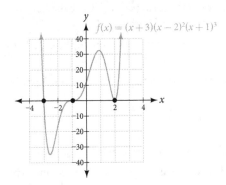

Figure 7 Identifying the behavior of the graph at an x-intercept by examining the multiplicity of the zero.

The x-intercept $x = -3$ is the solution of equation $(x + 3) = 0$. The graph passes directly through the x-intercept at $x = -3$. The factor is linear (has a degree of 1), so the behavior near the intercept is like that of a line—it passes directly through the intercept. We call this a single zero because the zero corresponds to a single factor of the function.

The x-intercept $x = 2$ is the repeated solution of equation $(x - 2)^2 = 0$. The graph touches the axis at the intercept and changes direction. The factor is quadratic (degree 2), so the behavior near the intercept is like that of a quadratic—it bounces off of the horizontal axis at the intercept.

$$(x - 2)^2 = (x - 2)(x - 2)$$

The factor is repeated, that is, the factor $(x - 2)$ appears twice. The number of times a given factor appears in the factored form of the equation of a polynomial is called the **multiplicity**. The zero associated with this factor, $x = 2$, has multiplicity 2 because the factor $(x - 2)$ occurs twice.

The x-intercept $x = -1$ is the repeated solution of factor $(x + 1)^3 = 0$. The graph passes through the axis at the intercept, but flattens out a bit first. This factor is cubic (degree 3), so the behavior near the intercept is like that of a cubic—with the same S-shape near the intercept as the toolkit function $f(x) = x^3$. We call this a triple zero, or a zero with multiplicity 3.

For zeros with even multiplicities, the graphs *touch* or are tangent to the x-axis. For zeros with odd multiplicities, the graphs *cross* or intersect the x-axis. See **Figure 8** for examples of graphs of polynomial functions with multiplicity 1, 2, and 3.

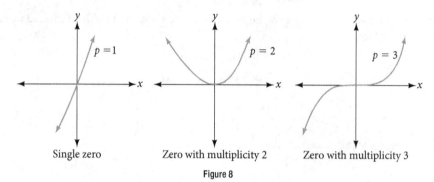

Single zero   Zero with multiplicity 2   Zero with multiplicity 3

Figure 8

For higher even powers, such as 4, 6, and 8, the graph will still touch and bounce off of the horizontal axis but, for each increasing even power, the graph will appear flatter as it approaches and leaves the x-axis.

For higher odd powers, such as 5, 7, and 9, the graph will still cross through the horizontal axis, but for each increasing odd power, the graph will appear flatter as it approaches and leaves the x-axis.

---

graphical behavior of polynomials at x-intercepts

If a polynomial contains a factor of the form $(x - h)^p$, the behavior near the x-intercept h is determined by the power p. We say that $x = h$ is a zero of **multiplicity** p.

The graph of a polynomial function will touch the x-axis at zeros with even multiplicities. The graph will cross the x-axis at zeros with odd multiplicities.

The sum of the multiplicities is the degree of the polynomial function.

How To...

Given a graph of a polynomial function of degree n, identify the zeros and their multiplicities.

1. If the graph crosses the x-axis and appears almost linear at the intercept, it is a single zero.
2. If the graph touches the x-axis and bounces off of the axis, it is a zero with even multiplicity.
3. If the graph crosses the x-axis at a zero, it is a zero with odd multiplicity.
4. The sum of the multiplicities is n.

Example 6   **Identifying Zeros and Their Multiplicities**

Use the graph of the function of degree 6 in **Figure 9** to identify the zeros of the function and their possible multiplicities.

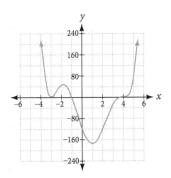

Figure 9

Solution The polynomial function is of degree 6. The sum of the multiplicities must be 6.

Starting from the left, the first zero occurs at $x = -3$. The graph touches the x-axis, so the multiplicity of the zero must be even. The zero of -3 most likely has multiplicity 2.

The next zero occurs at $x = -1$. The graph looks almost linear at this point. This is a single zero of multiplicity 1.

The last zero occurs at $x = 4$. The graph crosses the x-axis, so the multiplicity of the zero must be odd. We know that the multiplicity is likely 3 and that the sum of the multiplicities is likely 6.

Try It #2

Use the graph of the function of degree 5 in **Figure 10** to identify the zeros of the function and their multiplicities.

Figure 10

Determining End Behavior

As we have already learned, the behavior of a graph of a polynomial function of the form

$$f(x) = a_n x^n + a_{n-1} x^{n-1} + \ldots + a_1 x + a_0$$

will either ultimately rise or fall as x increases without bound and will either rise or fall as x decreases without bound. This is because for very large inputs, say 100 or 1,000, the leading term dominates the size of the output. The same is true for very small inputs, say -100 or $-1,000$.

Recall that we call this behavior the *end behavior* of a function. As we pointed out when discussing quadratic equations, when the leading term of a polynomial function, $a_n x^n$, is an even power function, as x increases or decreases without bound, $f(x)$ increases without bound. When the leading term is an odd power function, as x decreases without bound, $f(x)$ also decreases without bound; as x increases without bound, $f(x)$ also increases without bound. If the leading term is negative, it will change the direction of the end behavior. **Figure 11** summarizes all four cases.

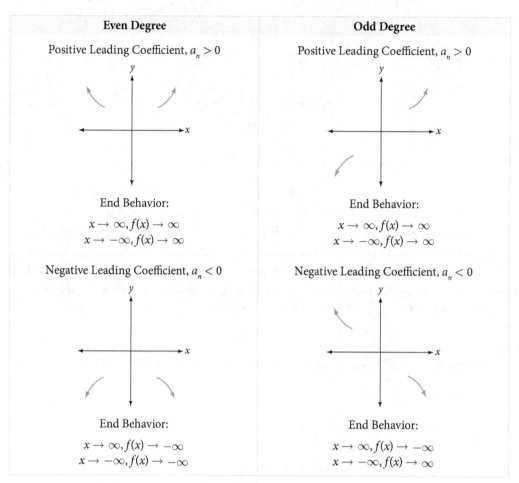

Figure 11

Understanding the Relationship Between Degree and Turning Points

In addition to the end behavior, recall that we can analyze a polynomial function's local behavior. It may have a turning point where the graph changes from increasing to decreasing (rising to falling) or decreasing to increasing (falling to rising). Look at the graph of the polynomial function $f(x) = x^4 - x^3 - 4x^2 + 4x$ in **Figure 12**. The graph has three turning points.

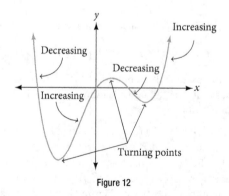

Figure 12

This function f is a 4th degree polynomial function and has 3 turning points. The maximum number of turning points of a polynomial function is always one less than the degree of the function.

---

interpreting turning points

A turning point is a point of the graph where the graph changes from increasing to decreasing (rising to falling) or decreasing to increasing (falling to rising).

A polynomial of degree n will have at most $n - 1$ turning points.

Example 7   **Finding the Maximum Number of Turning Points Using the Degree of a Polynomial Function**

Find the maximum number of turning points of each polynomial function.

  a. $f(x) = -x^3 + 4x^5 - 3x^2 + 1$ **b.** $f(x) = -(x - 1)^2(1 + 2x^2)$

Solution

  a. $f(x) = -x^3 + 4x^5 - 3x^2 + 1$

   First, rewrite the polynomial function in descending order: $f(x) = 4x^5 - x^3 - 3x^2 + 1$
   Identify the degree of the polynomial function. This polynomial function is of degree 5.
   The maximum number of turning points is $5 - 1 = 4$.

  b. $f(x) = -(x - 1)^2(1 + 2x^2)$

   First, identify the leading term of the polynomial function if the function were expanded.

$$f(x) = -(x - 1)^2(1 + 2x^2)$$
$$a_n = -(x^2)(2x^2) - 2x^4$$

   Then, identify the degree of the polynomial function. This polynomial function is of degree 4.

   The maximum number of turning points is $4 - 1 = 3$.

Graphing Polynomial Functions

We can use what we have learned about multiplicities, end behavior, and turning points to sketch graphs of polynomial functions. Let us put this all together and look at the steps required to graph polynomial functions.

How To...

Given a polynomial function, sketch the graph.

1. Find the intercepts.
2. Check for symmetry. If the function is an even function, its graph is symmetrical about the y-axis, that is, $f(-x) = f(x)$. If a function is an odd function, its graph is symmetrical about the origin, that is, $f(-x) = -f(x)$.
3. Use the multiplicities of the zeros to determine the behavior of the polynomial at the x-intercepts.
4. Determine the end behavior by examining the leading term.
5. Use the end behavior and the behavior at the intercepts to sketch a graph.
6. Ensure that the number of turning points does not exceed one less than the degree of the polynomial.
7. Optionally, use technology to check the graph.

Example 8   **Sketching the Graph of a Polynomial Function**

Sketch a graph of $f(x) = -2(x + 3)^2(x - 5)$.

Solution   This graph has two x-intercepts. At $x = -3$, the factor is squared, indicating a multiplicity of 2. The graph will bounce at this x-intercept. At $x = 5$, the function has a multiplicity of one, indicating the graph will cross through the axis at this intercept.

The y-intercept is found by evaluating $f(0)$.

$$f(0) = -2(0 + 3)^2(0 - 5)$$
$$= -2 \cdot 9 \cdot (-5)$$
$$= 90$$

The y-intercept is $(0, 90)$.

Additionally, we can see the leading term, if this polynomial were multiplied out, would be $-2x^3$, so the end behavior is that of a vertically reflected cubic, with the outputs decreasing as the inputs approach infinity, and the outputs increasing as the inputs approach negative infinity. See **Figure 13**.

Figure 13

To sketch this, we consider that:

- As $x \to -\infty$ the function $f(x) \to \infty$, so we know the graph starts in the second quadrant and is decreasing toward the x-axis.

- Since $f(-x) = -2(-x+3)^2(-x-5)$ is not equal to $f(x)$, the graph does not display symmetry.

- At $(-3, 0)$, the graph bounces off of the x-axis, so the function must start increasing. At $(0, 90)$, the graph crosses the y-axis at the y-intercept. See **Figure 14**.

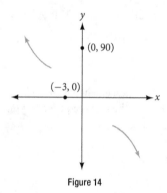

Figure 14

Somewhere after this point, the graph must turn back down or start decreasing toward the horizontal axis because the graph passes through the next intercept at $(5, 0)$. See **Figure 15**.

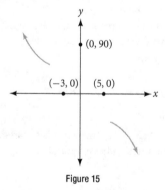

Figure 15

As $x \to \infty$ the function $f(x) \to -\infty$, so we know the graph continues to decrease, and we can stop drawing the graph in the fourth quadrant.

Using technology, we can create the graph for the polynomial function, shown in **Figure 16**, and verify that the resulting graph looks like our sketch in **Figure 15**.

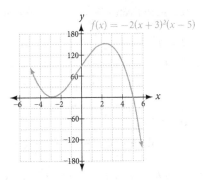

Figure 16 The complete graph of the polynomial function $f(x) = -2(x + 3)^2(x - 5)$

Try It #3

Sketch a graph of $f(x) = \frac{1}{4}x(x - 1)^4(x + 3)^3$.

Using the Intermediate Value Theorem

In some situations, we may know two points on a graph but not the zeros. If those two points are on opposite sides of the x-axis, we can confirm that there is a zero between them. Consider a polynomial function f whose graph is smooth and continuous. The **Intermediate Value Theorem** states that for two numbers a and b in the domain of f, if $a < b$ and $f(a) \neq f(b)$, then the function f takes on every value between $f(a)$ and $f(b)$. (While the theorem is intuitive, the proof is actually quite complicated and require higher mathematics.) We can apply this theorem to a special case that is useful in graphing polynomial functions. If a point on the graph of a continuous function f at $x = a$ lies above the x-axis and another point at $x = b$ lies below the x-axis, there must exist a third point between $x = a$ and $x = b$ where the graph crosses the x-axis. Call this point $(c, f(c))$. This means that we are assured there is a solution c where $f(c) = 0$.

In other words, the Intermediate Value Theorem tells us that when a polynomial function changes from a negative value to a positive value, the function must cross the x-axis. **Figure 17** shows that there is a zero between a and b.

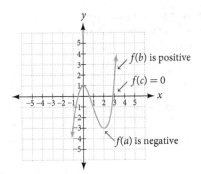

Figure 17 Using the Intermediate Value Theorem to show there exists a zero

---

Intermediate Value Theorem

Let f be a polynomial function. The **Intermediate Value Theorem** states that if $f(a)$ and $f(b)$ have opposite signs, then there exists at least one value c between a and b for which $f(c) = 0$.

Example 9   **Using the Intermediate Value Theorem**

Show that the function $f(x) = x^3 - 5x^2 + 3x + 6$ has at least two real zeros between $x = 1$ and $x = 4$.

Solution   As a start, evaluate $f(x)$ at the integer values $x = 1, 2, 3,$ and 4. See **Table 2**.

x	1	2	3	4
$f(x)$	5	0	-3	2

Table 2

We see that one zero occurs at $x = 2$. Also, since $f(3)$ is negative and $f(4)$ is positive, by the Intermediate Value Theorem, there must be at least one real zero between 3 and 4.

We have shown that there are at least two real zeros between $x = 1$ and $x = 4$.

Analysis   *We can also see on the graph of the function in **Figure 18** that there are two real zeros between $x = 1$ and $x = 4$.*

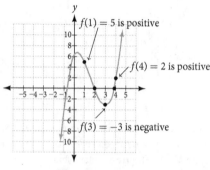

Figure 18

Try It #4

Show that the function $f(x) = 7x^5 - 9x^4 - x^2$ has at least one real zero between $x = 1$ and $x = 2$.

Writing Formulas for Polynomial Functions

Now that we know how to find zeros of polynomial functions, we can use them to write formulas based on graphs. Because a polynomial function written in factored form will have an x-intercept where each factor is equal to zero, we can form a function that will pass through a set of x-intercepts by introducing a corresponding set of factors.

> **factored form of polynomials**
>
> If a polynomial of lowest degree p has horizontal intercepts at $x = x_1, x_2, ..., x_n$, then the polynomial can be written in the factored form: $f(x) = a(x - x_1)^{p_1}(x - x_2)^{p_2} ... (x - x_n)^{p_n}$ where the powers p_i on each factor can be determined by the behavior of the graph at the corresponding intercept, and the stretch factor a can be determined given a value of the function other than the x-intercept.

How To...

Given a graph of a polynomial function, write a formula for the function.

1. Identify the x-intercepts of the graph to find the factors of the polynomial.
2. Examine the behavior of the graph at the x-intercepts to determine the multiplicity of each factor.
3. Find the polynomial of least degree containing all the factors found in the previous step.
4. Use any other point on the graph (the y-intercept may be easiest) to determine the stretch factor.

Example 10   **Writing a Formula for a Polynomial Function from the Graph**

Write a formula for the polynomial function shown in **Figure 19**.

Figure 19

Solution   This graph has three x-intercepts: $x = -3, 2,$ and 5. The y-intercept is located at $(0, -2)$. At $x = -3$ and $x = 5$, the graph passes through the axis linearly, suggesting the corresponding factors of the polynomial will be linear. At $x = 2$, the graph bounces at the intercept, suggesting the corresponding factor of the polynomial will be second degree (quadratic). Together, this gives us

$$f(x) = a(x + 3)(x - 2)^2(x - 5)$$

To determine the stretch factor, we utilize another point on the graph. We will use the y-intercept $(0, -2)$, to solve for a.

$$f(0) = a(0 + 3)(0 - 2)^2(0 - 5)$$
$$-2 = a(0 + 3)(0 - 2)^2(0 - 5)$$
$$-2 = -60a$$
$$a = \frac{1}{30}$$

The graphed polynomial appears to represent the function $f(x) = \frac{1}{30}(x + 3)(x - 2)^2\, (x - 5)$.

Try It #5

Given the graph shown in **Figure 20**, write a formula for the function shown.

Figure 20

Using Local and Global Extrema

With quadratics, we were able to algebraically find the maximum or minimum value of the function by finding the vertex. For general polynomials, finding these turning points is not possible without more advanced techniques from calculus. Even then, finding where extrema occur can still be algebraically challenging. For now, we will estimate the locations of turning points using technology to generate a graph.

Each turning point represents a local minimum or maximum. Sometimes, a turning point is the highest or lowest point on the entire graph. In these cases, we say that the turning point is a **global maximum** or a **global minimum**. These are also referred to as the absolute maximum and absolute minimum values of the function.

local and global extrema

A local maximum or local minimum at $x = a$ (sometimes called the relative maximum or minimum, respectively) is the output at the highest or lowest point on the graph in an open interval around $x = a$. If a function has a local maximum at a, then $f(a) \geq f(x)$ for all x in an open interval around $x = a$. If a function has a local minimum at a, then $f(a) \leq f(x)$ for all x in an open interval around $x = a$.

A **global maximum** or **global minimum** is the output at the highest or lowest point of the function. If a function has a global maximum at a, then $f(a) \geq f(x)$ for all x. If a function has a global minimum at a, then $f(a) \leq f(x)$ for all x.

We can see the difference between local and global extrema in **Figure 21**.

Figure 21

Q & A...

Do all polynomial functions have a global minimum or maximum?

No. Only polynomial functions of even degree have a global minimum or maximum. For example, $f(x) = x$ has neither a global maximum nor a global minimum.

Example 11   **Using Local Extrema to Solve Applications**

An open-top box is to be constructed by cutting out squares from each corner of a 14 cm by 20 cm sheet of plastic then folding up the sides. Find the size of squares that should be cut out to maximize the volume enclosed by the box.

Solution   We will start this problem by drawing a picture like that in **Figure 22**, labeling the width of the cut-out squares with a variable, w.

Figure 22

Notice that after a square is cut out from each end, it leaves a $(14 - 2w)$ cm by $(20 - 2w)$ cm rectangle for the base of the box, and the box will be w cm tall. This gives the volume

$$V(w) = (20 - 2w)(14 - 2w)w$$
$$= 280w - 68w^2 + 4w^3$$

Notice, since the factors are w, $20 - 2w$ and $14 - 2w$, the three zeros are 10, 7, and 0, respectively. Because a height of 0 cm is not reasonable, we consider the only the zeros 10 and 7. The shortest side is 14 and we are cutting off two squares, so values w may take on are greater than zero or less than 7. This means we will restrict the domain of this function to $0 < w < 7$. Using technology to sketch the graph of $V(w)$ on this reasonable domain, we get a graph like that in **Figure 23**. We can use this graph to estimate the maximum value for the volume, restricted to values for w that are reasonable for this problem—values from 0 to 7.

Figure 23

From this graph, we turn our focus to only the portion on the reasonable domain, $[0, 7]$. We can estimate the maximum value to be around 340 cubic cm, which occurs when the squares are about 2.75 cm on each side. To improve this estimate, we could use advanced features of our technology, if available, or simply change our window to zoom in on our graph to produce **Figure 24**.

Figure 24

From this zoomed-in view, we can refine our estimate for the maximum volume to about 339 cubic cm, when the squares measure approximately 2.7 cm on each side.

Try It #6

Use technology to find the maximum and minimum values on the interval $[-1, 4]$ of the function $f(x) = -0.2(x - 2)^3(x + 1)^2(x - 4)$.

Access the following online resource for additional instruction and practice with graphing polynomial functions.

- Intermediate Value Theorem (http://openstaxcollege.org/l/ivt)

5.3   SECTION EXERCISES

VERBAL

1. What is the difference between an x-intercept and a zero of a polynomial function f?

2. If a polynomial function of degree n has n distinct zeros, what do you know about the graph of the function?

3. Explain how the Intermediate Value Theorem can assist us in finding a zero of a function.

4. Explain how the factored form of the polynomial helps us in graphing it.

5. If the graph of a polynomial just touches the x-axis and then changes direction, what can we conclude about the factored form of the polynomial?

ALGEBRAIC

For the following exercises, find the x- or t-intercepts of the polynomial functions.

6. $C(t) = 2(t - 4)(t + 1)(t - 6)$

7. $C(t) = 3(t + 2)(t - 3)(t + 5)$

8. $C(t) = 4t(t - 2)^2(t + 1)$

9. $C(t) = 2t(t - 3)(t + 1)^2$

10. $C(t) = 2t^4 - 8t^3 + 6t^2$

11. $C(t) = 4t^4 + 12t^3 - 40t^2$

12. $f(x) = x^4 - x^2$

13. $f(x) = x^3 + x^2 - 20x$

14. $f(x) = x^3 + 6x^2 - 7x$

15. $f(x) = x^3 + x^2 - 4x - 4$

16. $f(x) = x^3 + 2x^2 - 9x - 18$

17. $f(x) = 2x^3 - x^2 - 8x + 4$

18. $f(x) = x^6 - 7x^3 - 8$

19. $f(x) = 2x^4 + 6x^2 - 8$

20. $f(x) = x^3 - 3x^2 - x + 3$

21. $f(x) = x^6 - 2x^4 - 3x^2$

22. $f(x) = x^6 - 3x^4 - 4x^2$

23. $f(x) = x^5 - 5x^3 + 4x$

For the following exercises, use the Intermediate Value Theorem to confirm that the given polynomial has at least one zero within the given interval.

24. $f(x) = x^3 - 9x$, between $x = -4$ and $x = -2$.

25. $f(x) = x^3 - 9x$, between $x = 2$ and $x = 4$.

26. $f(x) = x^5 - 2x$, between $x = 1$ and $x = 2$.

27. $f(x) = -x^4 + 4$, between $x = 1$ and $x = 3$.

28. $f(x) = -2x^3 - x$, between $x = -1$ and $x = 1$.

29. $f(x) = x^3 - 100x + 2$, between $x = 0.01$ and $x = 0.1$

For the following exercises, find the zeros and give the multiplicity of each.

30. $f(x) = (x + 2)^3(x - 3)^2$

31. $f(x) = x^2(2x + 3)^5(x - 4)^2$

32. $f(x) = x^3 (x - 1)^3(x + 2)$

33. $f(x) = x^2(x^2 + 4x + 4)$

34. $f(x) = (2x + 1)^3(9x^2 - 6x + 1)$

35. $f(x) = (3x + 2)^5(x^2 - 10x + 25)$

36. $f(x) = x(4x^2 - 12x + 9)(x^2 + 8x + 16)$

37. $f(x) = x^6 - x^5 - 2x^4$

38. $f(x) = 3x^4 + 6x^3 + 3x^2$

39. $f(x) = 4x^5 - 12x^4 + 9x^3$

40. $f(x) = 2x^4(x^3 - 4x^2 + 4x)$

41. $f(x) = 4x^4(9x^4 - 12x^3 + 4x^2)$

GRAPHICAL

For the following exercises, graph the polynomial functions. Note x- and y-intercepts, multiplicity, and end behavior.

42. $f(x) = (x + 3)^2(x - 2)$

43. $g(x) = (x + 4)(x - 1)^2$

44. $h(x) = (x - 1)^3(x + 3)^2$

45. $k(x) = (x - 3)^3(x - 2)^2$

46. $m(x) = -2x(x - 1)(x + 3)$

47. $n(x) = -3x(x + 2)(x - 4)$

For the following exercises, use the graphs to write the formula for a polynomial function of least degree.

48.

49.

50.

51.

52.
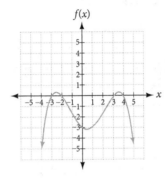

For the following exercises, use the graph to identify zeros and multiplicity.

53.

54.

55.

56.
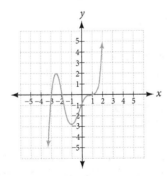

For the following exercises, use the given information about the polynomial graph to write the equation.

57. Degree 3. Zeros at $x = -2$, $x = 1$, and $x = 3$. y-intercept at $(0, -4)$.

58. Degree 3. Zeros at $x = -5$, $x = -2$, and $x = 1$. y-intercept at $(0, 6)$

59. Degree 5. Roots of multiplicity 2 at $x = 3$ and $x = 1$, and a root of multiplicity 1 at $x = -3$. y-intercept at $(0, 9)$

60. Degree 4. Root of multiplicity 2 at $x = 4$, and roots of multiplicity 1 at $x = 1$ and $x = -2$. y-intercept at $(0, -3)$.

61. Degree 5. Double zero at $x = 1$, and triple zero at $x = 3$. Passes through the point $(2, 15)$.

62. Degree 3. Zeros at $x = 4$, $x = 3$, and $x = 2$. y-intercept at $(0, -24)$.

63. Degree 3. Zeros at $x = -3$, $x = -2$ and $x = 1$. y-intercept at $(0, 12)$.

64. Degree 5. Roots of multiplicity 2 at $x = -3$ and $x = 2$ and a root of multiplicity 1 at $x = -2$. y-intercept at $(0, 4)$.

65. Degree 4. Roots of multiplicity 2 at $x = \frac{1}{2}$ and roots of multiplicity 1 at $x = 6$ and $x = -2$. y-intercept at $(0, 18)$.

66. Double zero at $x = -3$ and triple zero at $x = 0$. Passes through the point $(1, 32)$.

TECHNOLOGY

For the following exercises, use a calculator to approximate local minima and maxima or the global minimum and maximum.

67. $f(x) = x^3 - x - 1$

68. $f(x) = 2x^3 - 3x - 1$

69. $f(x) = x^4 + x$

70. $f(x) = -x^4 + 3x - 2$

71. $f(x) = x^4 - x^3 + 1$

EXTENSIONS

For the following exercises, use the graphs to write a polynomial function of least degree.

72.

73.

74.

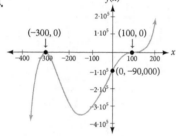

REAL-WORLD APPLICATIONS

For the following exercises, write the polynomial function that models the given situation.

75. A rectangle has a length of 10 units and a width of 8 units. Squares of x by x units are cut out of each corner, and then the sides are folded up to create an open box. Express the volume of the box as a polynomial function in terms of x.

76. Consider the same rectangle of the preceding problem. Squares of $2x$ by $2x$ units are cut out of each corner. Express the volume of the box as a polynomial in terms of x.

77. A square has sides of 12 units. Squares $x + 1$ by $x + 1$ units are cut out of each corner, and then the sides are folded up to create an open box. Express the volume of the box as a function in terms of x.

78. A cylinder has a radius of $x + 2$ units and a height of 3 units greater. Express the volume of the cylinder as a polynomial function.

79. A right circular cone has a radius of $3x + 6$ and a height 3 units less. Express the volume of the cone as a polynomial function. The volume of a cone is $V = \frac{1}{3}\pi r^2 h$ for radius r and height h.

LEARNING OBJECTIVES

In this section, you will:

- Use long division to divide polynomials.
- Use synthetic division to divide polynomials.

5.4   DIVIDING POLYNOMIALS

Figure 1 Lincoln Memorial, Washington, D.C. (credit: Ron Cogswell, Flickr)

The exterior of the Lincoln Memorial in Washington, D.C., is a large rectangular solid with length 61.5 meters (m), width 40 m, and height 30 m.[15] We can easily find the volume using elementary geometry.

$$V = l \cdot w \cdot h$$
$$= 61.5 \cdot 40 \cdot 30$$
$$= 73{,}800$$

So the volume is 73,800 cubic meters (m³). Suppose we knew the volume, length, and width. We could divide to find the height.

$$h = \frac{V}{l \cdot w}$$
$$= \frac{73{,}800}{61.5 \cdot 40}$$
$$= 30$$

As we can confirm from the dimensions above, the height is 30 m. We can use similar methods to find any of the missing dimensions. We can also use the same method if any or all of the measurements contain variable expressions. For example, suppose the volume of a rectangular solid is given by the polynomial $3x^4 - 3x^3 - 33x^2 + 54x$. The length of the solid is given by $3x$; the width is given by $x - 2$. To find the height of the solid, we can use polynomial division, which is the focus of this section.

Using Long Division to Divide Polynomials

We are familiar with the long division algorithm for ordinary arithmetic. We begin by dividing into the digits of the dividend that have the greatest place value. We divide, multiply, subtract, include the digit in the next place value position, and repeat. For example, let's divide 178 by 3 using long division.

Long Division

$$\begin{array}{r} 59 \\ 3\overline{)178} \\ -15 \\ \hline 28 \\ -27 \\ \hline 1 \end{array}$$

Step 1: $5 \times 3 = 15$ and $17 - 15 = 2$
Step 2:   Bring down the 8
Step 3: $9 \times 3 = 27$ and $28 - 27 = 1$
Answer: $59\ R\ 1$ or $59\frac{1}{3}$

---

15. National Park Service. "Lincoln Memorial Building Statistics." http://www.nps.gov/linc/historyculture/lincoln-memorial-building-statistics.htm. Accessed 4/3/2014/

Another way to look at the solution is as a sum of parts. This should look familiar, since it is the same method used to check division in elementary arithmetic.

$$\text{dividend} = (\text{divisor} \cdot \text{quotient}) + \text{remainder}$$
$$178 = (3 \cdot 59) + 1$$
$$= 177 + 1$$
$$= 178$$

We call this the **Division Algorithm** and will discuss it more formally after looking at an example.

Division of polynomials that contain more than one term has similarities to long division of whole numbers. We can write a polynomial dividend as the product of the divisor and the quotient added to the remainder. The terms of the polynomial division correspond to the digits (and place values) of the whole number division. This method allows us to divide two polynomials. For example, if we were to divide $2x^3 - 3x^2 + 4x + 5$ by $x + 2$ using the long division algorithm, it would look like this:

$$x + 2\overline{)2x^3 - 3x^2 + 4x + 5}$$ Set up the division problem.

$$\begin{array}{r} 2x^2 \\ x + 2\overline{)2x^3 - 3x^2 + 4x + 5} \end{array}$$ $2x^3$ divided by x is $2x^2$.

$$\begin{array}{r} 2x^2 \\ x + 2\overline{)2x^3 - 3x^2 + 4x + 5} \\ -(2x^3 + 4x^2) \\ \hline -7x^2 + 4x \end{array}$$ Multiply $x + 2$ by $2x^2$.
Subtract.
Bring down the next term.

$$\begin{array}{r} 2x^2 - 7x \\ x + 2\overline{)2x^3 - 3x^2 + 4x + 5} \\ -(2x^3 + 4x^2) \\ \hline -7x^2 + 4x \\ -(-7x^2 + 14x) \\ \hline 18x + 5 \end{array}$$ $-7x^2$ divided by x is $-7x$.
Multiply $x + 2$ by $-7x$.
Subtract. Bring down the next term.

$$\begin{array}{r} 2x^2 - 7x + 18 \\ x + 2\overline{)2x^3 - 3x^2 + 4x + 5} \\ -(2x^3 + 4x^2) \\ \hline -7x^2 + 4x \\ -(-7x^2 + 14x) \\ \hline 18x + 5 \\ -18x + 36 \\ \hline -31 \end{array}$$ $18x$ divided by x is 18.
Multiply $x + 2$ by 18.
Subtract.

We have found

$$\frac{2x^3 - 3x^2 + 4x + 5}{x + 2} = 2x^2 - 7x + 18 - \frac{31}{x + 2}$$

or

$$\frac{2x^3 - 3x^2 + 4x + 5}{x + 2} = (x + 2)(2x^2 - 7x + 18) - 31$$

We can identify the dividend, the divisor, the quotient, and the remainder.

$$\underset{\text{Dividend}}{2x^3 - 3x^2 + 4x + 5} = \underset{\text{Divisor}}{(x + 2)} \ \underset{\text{Quotient}}{(2x^2 - 7x + 18)} + \underset{\text{Remainder}}{(-31)}$$

Writing the result in this manner illustrates the Division Algorithm.

> **_the Division Algorithm_**
>
> The **Division Algorithm** states that, given a polynomial dividend $f(x)$ and a non-zero polynomial divisor $d(x)$ where the degree of $d(x)$ is less than or equal to the degree of $f(x)$, there exist unique polynomials $q(x)$ and $r(x)$ such that
>
> $$f(x) = d(x)q(x) + r(x)$$
>
> $q(x)$ is the quotient and $r(x)$ is the remainder. The remainder is either equal to zero or has degree strictly less than $d(x)$.
>
> If $r(x) = 0$, then $d(x)$ divides evenly into $f(x)$. This means that, in this case, both $d(x)$ and $q(x)$ are factors of $f(x)$.

How To...

Given a polynomial and a binomial, use long division to divide the polynomial by the binomial.

1. Set up the division problem.
2. Determine the first term of the quotient by dividing the leading term of the dividend by the leading term of the divisor.
3. Multiply the answer by the divisor and write it below the like terms of the dividend.
4. Subtract the bottom binomial from the top binomial.
5. Bring down the next term of the dividend.
6. Repeat steps 2–5 until reaching the last term of the dividend.
7. If the remainder is non-zero, express as a fraction using the divisor as the denominator.

Example 1   **Using Long Division to Divide a Second-Degree Polynomial**

Divide $5x^2 + 3x - 2$ by $x + 1$.

Solution

$$x + 1 \overline{)5x^2 + 3x - 2}$$ Set up division problem.

$$\begin{array}{r} 5x \\ x + 1 \overline{)5x^2 + 3x - 2} \end{array}$$ $5x^2$ divided by x is $5x$.

$$\begin{array}{r} 5x \\ x + 1 \overline{)5x^2 + 3x - 2} \\ -(5x^2 + 5x) \\ \hline -2x - 2 \end{array}$$ Multiply $x + 1$ by $5x$.
Subtract. Bring down the next term.

$$\begin{array}{r} 5x - 2 \\ x + 1 \overline{)5x^2 + 3x - 2} \\ -(5x^2 + 5x) \\ \hline -2x - 2 \\ -(-2x - 2) \\ \hline 0 \end{array}$$ $-2x$ divided by x is -2.

Multiply $x + 1$ by -2.
Subtract.

The quotient is $5x - 2$. The remainder is 0. We write the result as

$$\frac{5x^2 + 3x - 2}{x + 1} = 5x - 2$$

or

$$5x^2 + 3x - 2 = (x + 1)(5x - 2)$$

Analysis   This division problem had a remainder of 0. This tells us that the dividend is divided evenly by the divisor, and that the divisor is a factor of the dividend.

Example 2   **Using Long Division to Divide a Third-Degree Polynomial**

Divide $6x^3 + 11x^2 - 31x + 15$ by $3x - 2$.

Solution

$$
\begin{array}{r}
2x^2 + 5x - 7 \\
3x - 2 \overline{)6x^3 + 11x^2 - 31x + 1} \\
\underline{-(6x^3 - 4x^2)} \\
15x^2 - 31x \\
\underline{-(15x^2 + 10x)} \\
-21x + 15 \\
\underline{-(-21x + 14)} \\
1
\end{array}
$$

$6x^3$ divided by $3x$ is $2x^2$.

Multiply $3x - 2$ by $2x^2$.

Subtract. Bring down the next term. $15x^2$ divided by $3x$ is $5x$.

Multiply $3x - 2$ by $5x$.

Subtract. Bring down the next term. $-21x$ divided by $3x$ is -7.

Multiply $3x - 2$ by -7.

Subtract. The remainder is 1.

There is a remainder of 1. We can express the result as:

$$\frac{6x^3 + 11x^2 - 31x + 15}{3x - 2} = 2x^2 + 5x - 7 + \frac{1}{3x - 2}$$

Analysis   *We can check our work by using the Division Algorithm to rewrite the solution. Then multiply.*

$$(3x - 2)(2x^2 + 5x - 7) + 1 = 6x^3 + 11x^2 - 31x + 15$$

Notice, as we write our result,

- *the dividend is $6x^3 + 11x^2 - 31x + 15$*
- *the divisor is $3x - 2$*
- *the quotient is $2x^2 + 5x - 7$*
- *the remainder is 1*

Try It #1

Divide $16x^3 - 12x^2 + 20x - 3$ by $4x + 5$.

Using Synthetic Division to Divide Polynomials

As we've seen, long division of polynomials can involve many steps and be quite cumbersome. **Synthetic division** is a shorthand method of dividing polynomials for the special case of dividing by a linear factor whose leading coefficient is 1.

To illustrate the process, recall the example at the beginning of the section.

Divide $2x^3 - 3x^2 + 4x + 5$ by $x + 2$ using the long division algorithm.

The final form of the process looked like this:

$$
\begin{array}{r}
2x^2 + x + 18 \\
x + 2 \overline{)2x^3 - 3x^2 + 4x + 5} \\
\underline{-(2x^3 + 4x^2)} \\
-7x^2 + 4x \\
\underline{-(-7x^2 - 14x)} \\
18x + 5 \\
\underline{-(18x + 36)} \\
-31
\end{array}
$$

There is a lot of repetition in the table. If we don't write the variables but, instead, line up their coefficients in columns under the division sign and also eliminate the partial products, we already have a simpler version of the entire problem.

$$
\begin{array}{r}
2\overline{)2 \quad -3 \quad 4 \quad 5} \\
\underline{-2 \quad -4} \\
-7 \quad 14 \\
\underline{18 \; -36} \\
-31
\end{array}
$$

Synthetic division carries this simplification even a few more steps. Collapse the table by moving each of the rows up to fill any vacant spots. Also, instead of dividing by 2, as we would in division of whole numbers, then multiplying and subtracting the middle product, we change the sign of the "divisor" to -2, multiply and add. The process starts by bringing down the leading coefficient.

$$
\begin{array}{r|rrrr}
-2 & 2 & -3 & 4 & 5 \\
& & -4 & 14 & -36 \\
\hline
& 2 & -7 & 18 & -31
\end{array}
$$

We then multiply it by the "divisor" and add, repeating this process column by column, until there are no entries left. The bottom row represents the coefficients of the quotient; the last entry of the bottom row is the remainder. In this case, the quotient is $2x^2 - 7x + 18$ and the remainder is -31. The process will be made more clear in **Example 3**.

---

synthetic division

Synthetic division is a shortcut that can be used when the divisor is a binomial in the form $x - k$ where k is a real number. In **synthetic division**, only the coefficients are used in the division process.

How To…

Given two polynomials, use synthetic division to divide.

1. Write k for the divisor.
2. Write the coefficients of the dividend.
3. Bring the lead coefficient down.
4. Multiply the lead coefficient by k. Write the product in the next column.
5. Add the terms of the second column.
6. Multiply the result by k. Write the product in the next column.
7. Repeat steps 5 and 6 for the remaining columns.
8. Use the bottom numbers to write the quotient. The number in the last column is the remainder and has degree 0, the next number from the right has degree 1, the next number from the right has degree 2, and so on.

Example 3   **Using Synthetic Division to Divide a Second-Degree Polynomial**

Use synthetic division to divide $5x^2 - 3x - 36$ by $x - 3$.

Solution Begin by setting up the synthetic division. Write k and the coefficients.

$$
\begin{array}{r|rrr}
3 & 5 & -3 & -36
\end{array}
$$

Bring down the lead coefficient. Multiply the lead coefficient by k.

$$
\begin{array}{r|rrr}
3 & 5 & -3 & -36 \\
& & 15 & \\
\hline
& 5 & &
\end{array}
$$

Continue by adding the numbers in the second column. Multiply the resulting number by k. Write the result in the next column. Then add the numbers in the third column.

$$
\begin{array}{r|rrr}
3 & 5 & -3 & -36 \\
& & 15 & 36 \\
\hline
& 5 & 12 & 0
\end{array}
$$

The result is $5x + 12$. The remainder is 0. So $x - 3$ is a factor of the original polynomial.

Analysis *Just as with long division, we can check our work by multiplying the quotient by the divisor and adding the remainder.*

$$(x - 3)(5x + 12) + 0 = 5x^2 - 3x - 36$$

Example 4   **Using Synthetic Division to Divide a Third-Degree Polynomial**

Use synthetic division to divide $4x^3 + 10x^2 - 6x - 20$ by $x + 2$.

Solution The binomial divisor is $x + 2$ so $k = -2$. Add each column, multiply the result by -2, and repeat until the last column is reached.

$$
\begin{array}{r|rrrr}
-2 & 4 & 10 & -6 & -20 \\
   &   & -8 & -4 & 20 \\
\hline
   & 4 & 2 & -10 & 0
\end{array}
$$

The result is $4x^2 + 2x - 10$. The remainder is 0. Thus, $x + 2$ is a factor of $4x^3 + 10x^2 - 6x - 20$.

Analysis *The graph of the polynomial function $f(x) = 4x^3 + 10x^2 - 6x - 20$ in* **Figure 2** *shows a zero at $x = k = -2$. This confirms that $x + 2$ is a factor of $4x^3 + 10x^2 - 6x - 20$.*

Figure 2

Example 5   **Using Synthetic Division to Divide a Fourth-Degree Polynomial**

Use synthetic division to divide $-9x^4 + 10x^3 + 7x^2 - 6$ by $x - 1$.

Solution Notice there is no x-term. We will use a zero as the coefficient for that term.

$$
\begin{array}{r|rrrrr}
1 & -9 & 10 & 7 & 0 & -6 \\
  &   & -9 & 1 & 8 & 8 \\
\hline
  & -9 & 1 & 8 & 8 & 2
\end{array}
$$

The result is $-9x^3 + x^2 + 8x + 8 + \dfrac{2}{x - 1}$.

Try It #2

Use synthetic division to divide $3x^4 + 18x^3 - 3x + 40$ by $x + 7$.

Using Polynomial Division to Solve Application Problems

Polynomial division can be used to solve a variety of application problems involving expressions for area and volume. We looked at an application at the beginning of this section. Now we will solve that problem in the following example.

Example 6   **Using Polynomial Division in an Application Problem**

The volume of a rectangular solid is given by the polynomial $3x^4 - 3x^3 - 33x^2 + 54x$. The length of the solid is given by $3x$ and the width is given by $x - 2$. Find the height, t, of the solid.

Solution   There are a few ways to approach this problem. We need to divide the expression for the volume of the solid by the expressions for the length and width. Let us create a sketch as in **Figure 3**.

Figure 3

We can now write an equation by substituting the known values into the formula for the volume of a rectangular solid.

$$V = l \cdot w \cdot h$$

$$3x^4 - 3x^3 - 33x^2 + 54x = 3x \cdot (x - 2) \cdot h$$

To solve for h, first divide both sides by $3x$.

$$\frac{3x \cdot (x - 2) \cdot h}{3x} = \frac{3x^4 - 3x^3 - 33x^2 + 54x}{3x}$$

$$(x - 2)h = x^3 - x^2 - 11x + 18$$

Now solve for h using synthetic division.

$$h = \frac{x^3 - x^2 - 11x + 18}{x - 2}$$

$$
\begin{array}{r|rrrr}
2 & 1 & -1 & -11 & 18 \\
  &   & 2 & 2   & -18 \\
\hline
  & 1 & 1 & -9 & 0
\end{array}
$$

The quotient is $x^2 + x - 9$ and the remainder is 0. The height of the solid is $x^2 + x - 9$.

Try It #3

The area of a rectangle is given by $3x^3 + 14x^2 - 23x + 6$. The width of the rectangle is given by $x + 6$. Find an expression for the length of the rectangle.

Access these online resources for additional instruction and practice with polynomial division.

- Dividing a Trinomial by a Binomial Using Long Division (http://openstaxcollege.org/l/dividetribild)
- Dividing a Polynomial by a Binomial Using Long Division (http://openstaxcollege.org/l/dividepolybild)
- Ex 2: Dividing a Polynomial by a Binomial Using Synthetic Division (http://openstaxcollege.org/l/dividepolybisd2)
- Ex 4: Dividing a Polynomial by a Binomial Using Synthetic Division (http://openstaxcollege.org/l/dividepolybisd4)

5.4 SECTION EXERCISES

VERBAL

1. If division of a polynomial by a binomial results in a remainder of zero, what can be conclude?

2. If a polynomial of degree n is divided by a binomial of degree 1, what is the degree of the quotient?

ALGEBRAIC

For the following exercises, use long division to divide. Specify the quotient and the remainder.

3. $(x^2 + 5x - 1) \div (x - 1)$

4. $(2x^2 - 9x - 5) \div (x - 5)$

5. $(3x^2 + 23x + 14) \div (x + 7)$

6. $(4x^2 - 10x + 6) \div (4x + 2)$

7. $(6x^2 - 25x - 25) \div (6x + 5)$

8. $(-x^2 - 1) \div (x + 1)$

9. $(2x^2 - 3x + 2) \div (x + 2)$

10. $(x^3 - 126) \div (x - 5)$

11. $(3x^2 - 5x + 4) \div (3x + 1)$

12. $(x^3 - 3x^2 + 5x - 6) \div (x - 2)$

13. $(2x^3 + 3x^2 - 4x + 15) \div (x + 3)$

For the following exercises, use synthetic division to find the quotient.

14. $(3x^3 - 2x^2 + x - 4) \div (x + 3)$

15. $(2x^3 - 6x^2 - 7x + 6) \div (x - 4)$

16. $(6x^3 - 10x^2 - 7x - 15) \div (x + 1)$

17. $(4x^3 - 12x^2 - 5x - 1) \div (2x + 1)$

18. $(9x^3 - 9x^2 + 18x + 5) \div (3x - 1)$

19. $(3x^3 - 2x^2 + x - 4) \div (x + 3)$

20. $(-6x^3 + x^2 - 4) \div (2x - 3)$

21. $(2x^3 + 7x^2 - 13x - 3) \div (2x - 3)$

22. $(3x^3 - 5x^2 + 2x + 3) \div (x + 2)$

23. $(4x^3 - 5x^2 + 13) \div (x + 4)$

24. $(x^3 - 3x + 2) \div (x + 2)$

25. $(x^3 - 21x^2 + 147x - 343) \div (x - 7)$

26. $(x^3 - 15x^2 + 75x - 125) \div (x - 5)$

27. $(9x^3 - x + 2) \div (3x - 1)$

28. $(6x^3 - x^2 + 5x + 2) \div (3x + 1)$

29. $(x^4 + x^3 - 3x^2 - 2x + 1) \div (x + 1)$

30. $(x^4 - 3x^2 + 1) \div (x - 1)$

31. $(x^4 + 2x^3 - 3x^2 + 2x + 6) \div (x + 3)$

32. $(x^4 - 10x^3 + 37x^2 - 60x + 36) \div (x - 2)$

33. $(x^4 - 8x^3 + 24x^2 - 32x + 16) \div (x - 2)$

34. $(x^4 + 5x^3 - 3x^2 - 13x + 10) \div (x + 5)$

35. $(x^4 - 12x^3 + 54x^2 - 108x + 81) \div (x - 3)$

36. $(4x^4 - 2x^3 - 4x + 2) \div (2x - 1)$

37. $(4x^4 + 2x^3 - 4x^2 + 2x + 2) \div (2x + 1)$

For the following exercises, use synthetic division to determine whether the first expression is a factor of the second. If it is, indicate the factorization.

38. $x - 2, 4x^3 - 3x^2 - 8x + 4$

39. $x - 2, 3x^4 - 6x^3 - 5x + 10$

40. $x + 3, -4x^3 + 5x^2 + 8$

41. $x - 2, 4x^4 - 15x^2 - 4$

42. $x - \dfrac{1}{2}, 2x^4 - x^3 + 2x - 1$

43. $x + \dfrac{1}{3}, 3x^4 + x^3 - 3x + 1$

GRAPHICAL

For the following exercises, use the graph of the third-degree polynomial and one factor to write the factored form of the polynomial suggested by the graph. The leading coefficient is one.

44. Factor is $x^2 - x + 3$

45. Factor is $x^2 + 2x + 4$

46. Factor is $x^2 + 2x + 5$

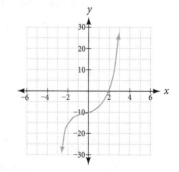

47. Factor is $x^2 + x + 1$

48. Factor is $x^2 + 2x + 2$

For the following exercises, use synthetic division to find the quotient and remainder.

49. $\dfrac{4x^3 - 33}{x - 2}$

50. $\dfrac{2x^3 + 25}{x + 3}$

51. $\dfrac{3x^3 + 2x - 5}{x - 1}$

52. $\dfrac{-4x^3 - x^2 - 12}{x + 4}$

53. $\dfrac{x^4 - 22}{x + 2}$

TECHNOLOGY

For the following exercises, use a calculator with CAS to answer the questions.

54. Consider $\dfrac{x^k - 1}{x - 1}$ with $k = 1, 2, 3$. What do you expect the result to be if $k = 4$?

55. Consider $\dfrac{x^k + 1}{x + 1}$ for $k = 1, 3, 5$. What do you expect the result to be if $k = 7$?

56. Consider $\dfrac{x^4 - k^4}{x - k}$ for $k = 1, 2, 3$. What do you expect the result to be if $k = 4$?

57. Consider $\dfrac{x^k}{x + 1}$ with $k = 1, 2, 3$. What do you expect the result to be if $k = 4$?

58. Consider $\dfrac{x^k}{x - 1}$ with $k = 1, 2, 3$. What do you expect the result to be if $k = 4$?

EXTENSIONS

For the following exercises, use synthetic division to determine the quotient involving a complex number.

59. $\dfrac{x + 1}{x - i}$

60. $\dfrac{x^2 + 1}{x - i}$

61. $\dfrac{x + 1}{x + i}$

62. $\dfrac{x^2 + 1}{x + i}$

63. $\dfrac{x^3 + 1}{x - i}$

REAL-WORLD APPLICATIONS

For the following exercises, use the given length and area of a rectangle to express the width algebraically.

64. Length is $x + 5$, area is $2x^2 + 9x - 5$.

65. Length is $2x + 5$, area is $4x^3 + 10x^2 + 6x + 15$

66. Length is $3x - 4$, area is $6x^4 - 8x^3 + 9x^2 - 9x - 4$

For the following exercises, use the given volume of a box and its length and width to express the height of the box algebraically.

67. Volume is $12x^3 + 20x^2 - 21x - 36$, length is $2x + 3$, width is $3x - 4$.

68. Volume is $18x^3 - 21x^2 - 40x + 48$, length is $3x - 4$, width is $3x - 4$.

69. Volume is $10x^3 + 27x^2 + 2x - 24$, length is $5x - 4$, width is $2x + 3$.

70. Volume is $10x^3 + 30x^2 - 8x - 24$, length is 2, width is $x + 3$.

For the following exercises, use the given volume and radius of a cylinder to express the height of the cylinder algebraically.

71. Volume is $\pi(25x^3 - 65x^2 - 29x - 3)$, radius is $5x + 1$.

72. Volume is $\pi(4x^3 + 12x^2 - 15x - 50)$, radius is $2x + 5$.

73. Volume is $\pi(3x^4 + 24x^3 + 46x^2 - 16x - 32)$, radius is $x + 4$.

LEARNING OBJECTIVES

In this section, you will:

- Evaluate a polynomial using the Remainder Theorem.
- Use the Factor Theorem to solve a polynomial equation.
- Use the Rational Zero Theorem to find rational zeros.
- Find zeros of a polynomial function.
- Use the Linear Factorization Theorem to find polynomials with given zeros.
- Use Decartes' Rule of Signs.
- Solve real-world applications of polynomial equations.

5.5   ZEROS OF POLYNOMIAL FUNCTIONS

A new bakery offers decorated sheet cakes for children's birthday parties and other special occasions. The bakery wants the volume of a small cake to be 351 cubic inches. The cake is in the shape of a rectangular solid. They want the length of the cake to be four inches longer than the width of the cake and the height of the cake to be one-third of the width. What should the dimensions of the cake pan be?

This problem can be solved by writing a cubic function and solving a cubic equation for the volume of the cake. In this section, we will discuss a variety of tools for writing polynomial functions and solving polynomial equations.

Evaluating a Polynomial Using the Remainder Theorem

In the last section, we learned how to divide polynomials. We can now use polynomial division to evaluate polynomials using the **Remainder Theorem**. If the polynomial is divided by $x - k$, the remainder may be found quickly by evaluating the polynomial function at k, that is, $f(k)$ Let's walk through the proof of the theorem.

Recall that the Division Algorithm states that, given a polynomial dividend $f(x)$ and a non-zero polynomial divisor $d(x)$ where the degree of $d(x)$ is less than or equal to the degree of $f(x)$, there exist unique polynomials $q(x)$ and $r(x)$ such that

$$f(x) = d(x)q(x) + r(x)$$

If the divisor, $d(x)$, is $x - k$, this takes the form

$$f(x) = (x - k)q(x) + r$$

Since the divisor $x - k$ is linear, the remainder will be a constant, r. And, if we evaluate this for $x = k$, we have

$$f(k) = (k - k)q(k) + r$$
$$= 0 \cdot q(k) + r$$
$$= r$$

In other words, $f(k)$ is the remainder obtained by dividing $f(x)$ by $x - k$.

---

the Remainder Theorem

If a polynomial $f(x)$ is divided by $x - k$, then the remainder is the value $f(k)$.

How To...

Given a polynomial function f, evaluate $f(x)$ at $x = k$ using the Remainder Theorem.

1. Use synthetic division to divide the polynomial by $x - k$.
2. The remainder is the value $f(k)$.

Example 1   **Using the Remainder Theorem to Evaluate a Polynomial**

Use the Remainder Theorem to evaluate $f(x) = 6x^4 - x^3 - 15x^2 + 2x - 7$ at $x = 2$.

Solution   To find the remainder using the Remainder Theorem, use synthetic division to divide the polynomial by $x - 2$.

$$
\begin{array}{r|rrrrr}
2 & 6 & -1 & -15 & 2 & -7 \\
  &   & 12 & 22 & 14 & 32 \\
\hline
  & 6 & 11 & 7 & 16 & 25
\end{array}
$$

The remainder is 25. Therefore, $f(2) = 25$.

Analysis   We can check our answer by evaluating $f(2)$.

$$f(x) = 6x^4 - x^3 - 15x^2 + 2x - 7$$
$$f(2) = 6(2)^4 - (2)^3 - 15(2)^2 + 2(2) - 7$$
$$= 25$$

Try It #1

Use the Remainder Theorem to evaluate $f(x) = 2x^5 - 3x^4 - 9x^3 + 8x^2 + 2$ at $x = -3$.

Using the Factor Theorem to Solve a Polynomial Equation

The **Factor Theorem** is another theorem that helps us analyze polynomial equations. It tells us how the zeros of a polynomial are related to the factors. Recall that the Division Algorithm.

$$f(x) = (x - k)q(x) + r$$

If k is a zero, then the remainder r is $f(k) = 0$ and $f(x) = (x - k)q(x) + 0$ or $f(x) = (x - k)q(x)$.

Notice, written in this form, $x - k$ is a factor of $f(x)$. We can conclude if k is a zero of $f(x)$, then $x - k$ is a factor of $f(x)$.

Similarly, if $x - k$ is a factor of $f(x)$, then the remainder of the Division Algorithm $f(x) = (x - k)q(x) + r$ is 0. This tells us that k is a zero.

This pair of implications is the Factor Theorem. As we will soon see, a polynomial of degree n in the complex number system will have n zeros. We can use the Factor Theorem to completely factor a polynomial into the product of n factors. Once the polynomial has been completely factored, we can easily determine the zeros of the polynomial.

---

the Factor Theorem

According to the **Factor Theorem**, k is a zero of $f(x)$ if and only if $(x - k)$ is a factor of $f(x)$.

How To...

Given a factor and a third-degree polynomial, use the Factor Theorem to factor the polynomial.

1. Use synthetic division to divide the polynomial by $(x - k)$.
2. Confirm that the remainder is 0.
3. Write the polynomial as the product of $(x - k)$ and the quadratic quotient.
4. If possible, factor the quadratic.
5. Write the polynomial as the product of factors.

Example 2   **Using the Factor Theorem to Solve a Polynomial Equation**

Show that $(x + 2)$ is a factor of $x^3 - 6x^2 - x + 30$. Find the remaining factors. Use the factors to determine the zeros of the polynomial.

Solution   We can use synthetic division to show that $(x + 2)$ is a factor of the polynomial.

$$
\begin{array}{r|rrrr}
-2 & 1 & -6 & -1 & 30 \\
   &   & -2 & 16 & -30 \\
\hline
   & 1 & -8 & 15 & 0
\end{array}
$$

The remainder is zero, so $(x + 2)$ is a factor of the polynomial. We can use the Division Algorithm to write the polynomial as the product of the divisor and the quotient:

$$(x + 2)(x^2 - 8x + 15)$$

We can factor the quadratic factor to write the polynomial as

$$(x + 2)(x - 3)(x - 5)$$

By the Factor Theorem, the zeros of $x^3 - 6x^2 - x + 30$ are −2, 3, and 5.

Try It #2

Use the Factor Theorem to find the zeros of $f(x) = x^3 + 4x^2 - 4x - 16$ given that $(x - 2)$ is a factor of the polynomial.

Using the Rational Zero Theorem to Find Rational Zeros

Another use for the Remainder Theorem is to test whether a rational number is a zero for a given polynomial. But first we need a pool of rational numbers to test. The **Rational Zero Theorem** helps us to narrow down the number of possible rational zeros using the ratio of the factors of the constant term and factors of the leading coefficient of the polynomial.

Consider a quadratic function with two zeros, $x = \frac{2}{5}$ and $x = \frac{3}{4}$. By the Factor Theorem, these zeros have factors associated with them. Let us set each factor equal to 0, and then construct the original quadratic function absent its stretching factor.

$x - \frac{2}{5} = 0$ or $x - \frac{3}{4} = 0$ Set each factor equal to 0.

$5x - 2 = 0$ or $4x - 3 = 0$ Multiply both sides of the equation to eliminate fractions.

$f(x) = (5x - 2)(4x - 3)$ Create the quadratic function, multiplying the factors.

$f(x) = 20x^2 - 23x + 6$ Expand the polynomial.

$f(x) = (5 \cdot 4)x^2 - 23x + (2 \cdot 3)$

Notice that two of the factors of the constant term, 6, are the two numerators from the original rational roots: 2 and 3. Similarly, two of the factors from the leading coefficient, 20, are the two denominators from the original rational roots: 5 and 4.

We can infer that the numerators of the rational roots will always be factors of the constant term and the denominators will be factors of the leading coefficient. This is the essence of the Rational Zero Theorem; it is a means to give us a pool of possible rational zeros.

---

the Rational Zero Theorem

The **Rational Zero Theorem** states that, if the polynomial $f(x) = a_n x^n + a_{n-1} x^{n-1} + \dots + a_1 x + a_0$ has integer coefficients, then every rational zero of $f(x)$ has the form $\frac{p}{q}$ where p is a factor of the constant term a_0 and q is a factor of the leading coefficient a_n.

When the leading coefficient is 1, the possible rational zeros are the factors of the constant term.

How To...

Given a polynomial function $f(x)$, use the Rational Zero Theorem to find rational zeros.

1. Determine all factors of the constant term and all factors of the leading coefficient.

2. Determine all possible values of $\frac{p}{q}$, where p is a factor of the constant term and q is a factor of the leading coefficient. Be sure to include both positive and negative candidates.

3. Determine which possible zeros are actual zeros by evaluating each case of $f\left(\frac{p}{q}\right)$.

Example 3   **Listing All Possible Rational Zeros**

List all possible rational zeros of $f(x) = 2x^4 - 5x^3 + x^2 - 4$.

Solution   The only possible rational zeros of $f(x)$ are the quotients of the factors of the last term, -4, and the factors of the leading coefficient, 2.

The constant term is -4; the factors of -4 are $p = \pm 1, \pm 2, \pm 4$.

The leading coefficient is 2; the factors of 2 are $q = \pm 1, \pm 2$.

If any of the four real zeros are rational zeros, then they will be of one of the following factors of -4 divided by one of the factors of 2.

$$\frac{p}{q} = \pm\frac{1}{1}, \pm\frac{1}{2} \quad \frac{p}{q} = \pm\frac{2}{1}, \pm\frac{2}{2} \quad \frac{p}{q} = \pm\frac{4}{1}, \pm\frac{4}{2}$$

Note that $\frac{2}{2} = 1$ and $\frac{4}{2} = 2$, which have already been listed. So we can shorten our list.

$$\frac{p}{q} = \frac{\text{Factors of the last}}{\text{Factors of the first}} = \pm 1, \pm 2, \pm 4, \pm\frac{1}{2}$$

Example 4   **Using the Rational Zero Theorem to Find Rational Zeros**

Use the Rational Zero Theorem to find the rational zeros of $f(x) = 2x^3 + x^2 - 4x + 1$.

Solution   The Rational Zero Theorem tells us that if $\frac{p}{q}$ is a zero of $f(x)$, then p is a factor of 1 and q is a factor of 2.

$$\frac{p}{q} = \frac{\text{factor of constant term}}{\text{factor of leading coefficient}}$$

$$= \frac{\text{factor of 1}}{\text{factor of 2}}$$

The factors of 1 are ± 1 and the factors of 2 are ± 1 and ± 2. The possible values for $\frac{p}{q}$ are ± 1 and $\pm\frac{1}{2}$. These are the possible rational zeros for the function. We can determine which of the possible zeros are actual zeros by substituting these values for x in $f(x)$.

$$f(-1) = 2(-1)^3 + (-1)^2 - 4(-1) + 1 = 4$$
$$f(1) = 2(1)^3 + (1)^2 - 4(1) + 1 = 0$$
$$f\left(-\frac{1}{2}\right) = 2\left(-\frac{1}{2}\right)^3 + \left(-\frac{1}{2}\right)^2 - 4\left(-\frac{1}{2}\right) + 1 = 3$$
$$f\left(\frac{1}{2}\right) = 2\left(\frac{1}{2}\right)^3 + \left(\frac{1}{2}\right)^2 - 4\left(\frac{1}{2}\right) + 1 = -\frac{1}{2}$$

Of those, -1, $-\frac{1}{2}$, and $\frac{1}{2}$ are not zeros of $f(x)$. 1 is the only rational zero of $f(x)$.

Try It #3

Use the Rational Zero Theorem to find the rational zeros of $f(x) = x^3 - 5x^2 + 2x + 1$.

Finding the Zeros of Polynomial Functions

The Rational Zero Theorem helps us to narrow down the list of possible rational zeros for a polynomial function. Once we have done this, we can use synthetic division repeatedly to determine all of the **zeros** of a polynomial function.

How To…

Given a polynomial function f, use synthetic division to find its zeros.

1. Use the Rational Zero Theorem to list all possible rational zeros of the function.
2. Use synthetic division to evaluate a given possible zero by synthetically dividing the candidate into the polynomial. If the remainder is 0, the candidate is a zero. If the remainder is not zero, discard the candidate.
3. Repeat step two using the quotient found with synthetic division. If possible, continue until the quotient is a quadratic.
4. Find the zeros of the quadratic function. Two possible methods for solving quadratics are factoring and using the quadratic formula.

Example 5   **Finding the Zeros of a Polynomial Function with Repeated Real Zeros**

Find the zeros of $f(x) = 4x^3 - 3x - 1$.

Solution   The Rational Zero Theorem tells us that if $\frac{p}{q}$ is a zero of $f(x)$, then p is a factor of -1 and q is a factor of 4.

$$\frac{p}{q} = \frac{\text{factor of constant term}}{\text{factor of leading coefficient}}$$

$$= \frac{\text{factor of } -1}{\text{factor of } 4}$$

The factors of -1 are ± 1 and the factors of 4 are $\pm 1, \pm 2,$ and ± 4. The possible values for $\frac{p}{q}$ are $\pm 1, \pm \frac{1}{2},$ and $\pm \frac{1}{4}$. These are the possible rational zeros for the function. We will use synthetic division to evaluate each possible zero until we find one that gives a remainder of 0. Let's begin with 1.

$$
\begin{array}{r|rrr}
1 & 4 & 0 & -3 & -1 \\
  &   & 4 & 4 & 1 \\
\hline
  & 4 & 4 & 1 & 0
\end{array}
$$

Dividing by $(x - 1)$ gives a remainder of 0, so 1 is a zero of the function. The polynomial can be written as

$$(x - 1)(4x^2 + 4x + 1).$$

The quadratic is a perfect square. $f(x)$ can be written as

$$(x - 1)(2x + 1)^2.$$

We already know that 1 is a zero. The other zero will have a multiplicity of 2 because the factor is squared. To find the other zero, we can set the factor equal to 0.

$$2x + 1 = 0$$
$$x = -\frac{1}{2}$$

The zeros of the function are 1 and $-\frac{1}{2}$ with multiplicity 2.

Analysis   *Look at the graph of the function f in* **Figure 1.** *Notice, at $x = -0.5$, the graph bounces off the x-axis, indicating the even multiplicity (2, 4, 6…) for the zero -0.5. At $x = 1$, the graph crosses the x-axis, indicating the odd multiplicity (1, 3, 5…) for the zero $x = 1$.*

Figure 1

Using the Fundamental Theorem of Algebra

Now that we can find rational zeros for a polynomial function, we will look at a theorem that discusses the number of complex zeros of a polynomial function. The **Fundamental Theorem of Algebra** tells us that every polynomial function has at least one complex zero. This theorem forms the foundation for solving polynomial equations.

Suppose f is a polynomial function of degree four, and $f(x) = 0$. The Fundamental Theorem of Algebra states that there is at least one complex solution, call it c_1. By the Factor Theorem, we can write $f(x)$ as a product of $x - c_1$ and a polynomial quotient. Since $x - c_1$ is linear, the polynomial quotient will be of degree three. Now we apply the Fundamental Theorem of Algebra to the third-degree polynomial quotient. It will have at least one complex zero, call it c_2. So we can write the polynomial quotient as a product of $x - c_2$ and a new polynomial quotient of degree two. Continue to apply the Fundamental Theorem of Algebra until all of the zeros are found. There will be four of them and each one will yield a factor of $f(x)$.

---

the Fundamental Theorem of Algebra

The Fundamental Theorem of Algebra states that, if $f(x)$ is a polynomial of degree $n > 0$, then $f(x)$ has at least one complex zero.

We can use this theorem to argue that, if $f(x)$ is a polynomial of degree $n > 0$, and a is a non-zero real number, then $f(x)$ has exactly n linear factors

$$f(x) = a(x - c_1)(x - c_2)...(x - c_n)$$

where $c_1, c_2, ..., c_n$ are complex numbers. Therefore, $f(x)$ has n roots if we allow for multiplicities.

Q & A...

Does every polynomial have at least one imaginary zero?

No. Real numbers are a subset of complex numbers, but not the other way around. A complex number is not necessarily imaginary. Real numbers are also complex numbers.

Example 6   **Finding the Zeros of a Polynomial Function with Complex Zeros**

Find the zeros of $f(x) = 3x^3 + 9x^2 + x + 3$.

Solution   The Rational Zero Theorem tells us that if $\frac{p}{q}$ is a zero of $f(x)$, then p is a factor of 3 and q is a factor of 3.

$$\frac{p}{q} = \frac{\text{factor of constant term}}{\text{factor of leading coefficient}}$$

$$= \frac{\text{factor of 3}}{\text{factor of 3}}$$

The factors of 3 are ± 1 and ± 3. The possible values for $\frac{p}{q}$, and therefore the possible rational zeros for the function, are $\pm 3, \pm 1,$ and $\pm \frac{1}{3}$. We will use synthetic division to evaluate each possible zero until we find one that gives a remainder of 0. Let's begin with -3.

$$
\begin{array}{r|rrrr}
-3 & 3 & 9 & 1 & 3 \\
   &   & -9 & 0 & -3 \\
\hline
   & 3 & 0 & 1 & 0 \\
\end{array}
$$

Dividing by $(x + 3)$ gives a remainder of 0, so -3 is a zero of the function. The polynomial can be written as

$$(x + 3)(3x^2 + 1)$$

We can then set the quadratic equal to 0 and solve to find the other zeros of the function.

$$3x^2 + 1 = 0$$

$$x^2 = -\frac{1}{3}$$

$$x = \pm\sqrt{-\frac{1}{3}} = \pm\frac{i\sqrt{3}}{3}$$

The zeros of $f(x)$ are -3 and $\pm\frac{i\sqrt{3}}{3}$.

Analysis   Look at the graph of the function f in **Figure 2**. Notice that, at $x = -3$, the graph crosses the x-axis, indicating an odd multiplicity (1) for the zero $x = -3$. Also note the presence of the two turning points. This means that, since there is a 3rd degree polynomial, we are looking at the maximum number of turning points. So, the end behavior of increasing without bound to the right and decreasing without bound to the left will continue. Thus, all the x-intercepts for the function are shown. So either the multiplicity of $x = -3$ is 1 and there are two complex solutions, which is what we found, or the multiplicity at $x = -3$ is three. Either way, our result is correct.

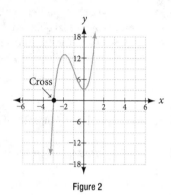

Figure 2

Try It #4

Find the zeros of $f(x) = 2x^3 + 5x^2 - 11x + 4$.

Using the Linear Factorization Theorem to Find Polynomials with Given Zeros

A vital implication of the Fundamental Theorem of Algebra, as we stated above, is that a polynomial function of degree n will have n zeros in the set of complex numbers, if we allow for multiplicities. This means that we can factor the polynomial function into n factors. The **Linear Factorization Theorem** tells us that a polynomial function will have the same number of factors as its degree, and that each factor will be in the form $(x - c)$, where c is a complex number.

Let f be a polynomial function with real coefficients, and suppose $a + bi$, $b \neq 0$, is a zero of $f(x)$. Then, by the Factor Theorem, $x - (a + bi)$ is a factor of $f(x)$. For f to have real coefficients, $x - (a - bi)$ must also be a factor of $f(x)$. This is true because any factor other than $x - (a - bi)$, when multiplied by $x - (a + bi)$, will leave imaginary components in the product. Only multiplication with conjugate pairs will eliminate the imaginary parts and result in real coefficients. In other words, if a polynomial function f with real coefficients has a complex zero $a + bi$, then the complex conjugate $a - bi$ must also be a zero of $f(x)$. This is called the Complex Conjugate Theorem.

> **complex conjugate theorem**
>
> According to the **Linear Factorization Theorem,** a polynomial function will have the same number of factors as its degree, and each factor will be in the form $(x - c)$, where c is a complex number.
>
> If the polynomial function f has real coefficients and a complex zero in the form $a + bi$, then the complex conjugate of the zero, $a - bi$, is also a zero.

How To…

Given the zeros of a polynomial function f and a point $(c, f(c))$ on the graph of f, use the Linear Factorization Theorem to find the polynomial function.

1. Use the zeros to construct the linear factors of the polynomial.
2. Multiply the linear factors to expand the polynomial.
3. Substitute $(c, f(c))$ into the function to determine the leading coefficient.
4. Simplify.

Example 7   **Using the Linear Factorization Theorem to Find a Polynomial with Given Zeros**

Find a fourth degree polynomial with real coefficients that has zeros of $-3, 2, i$, such that $f(-2) = 100$.

Solution   Because $x = i$ is a zero, by the Complex Conjugate Theorem $x = -i$ is also a zero. The polynomial must have factors of $(x + 3)$, $(x - 2)$, $(x - i)$, and $(x + i)$. Since we are looking for a degree 4 polynomial, and now have four zeros, we have all four factors. Let's begin by multiplying these factors.

$$f(x) = a(x + 3)(x - 2)(x - i)(x + i)$$

$$f(x) = a(x^2 + x - 6)(x^2 + 1)$$

$$f(x) = a(x^4 + x^3 - 5x^2 + x - 6)$$

We need to find a to ensure $f(-2) = 100$. Substitute $x = -2$ and $f(2) = 100$ into $f(x)$.

$$100 = a((-2)^4 + (-2)^3 - 5(-2)^2 + (-2) - 6)$$

$$100 = a(-20)$$

$$-5 = a$$

So the polynomial function is

$$f(x) = -5(x^4 + x^3 - 5x^2 + x - 6)$$

or

$$f(x) = -5x^4 - 5x^3 + 25x^2 - 5x + 30$$

Analysis   *We found that both i and $-i$ were zeros, but only one of these zeros needed to be given. If i is a zero of a polynomial with real coefficients, then $-i$ must also be a zero of the polynomial because $-i$ is the complex conjugate of i.*

Q & A...

If $2 + 3i$ were given as a zero of a polynomial with real coefficients, would $2 - 3i$ also need to be a zero?

Yes. When any complex number with an imaginary component is given as a zero of a polynomial with real coefficients, the conjugate must also be a zero of the polynomial.

Try It #5

Find a third degree polynomial with real coefficients that has zeros of 5 and $-2i$ such that $f(1) = 10$.

Using Descartes' Rule of Signs

There is a straightforward way to determine the possible numbers of positive and negative real zeros for any polynomial function. If the polynomial is written in descending order, **Descartes' Rule of Signs** tells us of a relationship between the number of sign changes in $f(x)$ and the number of positive real zeros. For example, the polynomial function below has one sign change.

$$f(x) = x^4 + x^3 + x^2 + x - 1$$

This tells us that the function must have 1 positive real zero.

There is a similar relationship between the number of sign changes in $f(-x)$ and the number of negative real zeros.

$$f(-x) = (-x)^4 + (-x)^3 + (-x)^2 + (-x) - 1$$

$$f(-x) = +x^4 - x^3 + x^2 - x - 1$$

In this case, $f(-x)$ has 3 sign changes. This tells us that $f(x)$ could have 3 or 1 negative real zeros.

Descartes' Rule of Signs

According to **Descartes' Rule of Signs,** if we let $f(x) = a_n x^n + a_{n-1} x^{n-1} + ... + a_1 x + a_0$ be a polynomial function with real coefficients:

- The number of positive real zeros is either equal to the number of sign changes of $f(x)$ or is less than the number of sign changes by an even integer.
- The number of negative real zeros is either equal to the number of sign changes of $f(-x)$ or is less than the number of sign changes by an even integer.

Example 8   **Using Descartes' Rule of Signs**

Use Descartes' Rule of Signs to determine the possible numbers of positive and negative real zeros for $f(x) = -x^4 - 3x^3 + 6x^2 - 4x - 12$.

Solution   Begin by determining the number of sign changes.

$$f(x) = -x^4 - 3x^3 + 6x^2 - 4x - 12$$

Figure 3

There are two sign changes, so there are either 2 or 0 positive real roots. Next, we examine $f(-x)$ to determine the number of negative real roots.

$$f(-x) = -(-x)^4 - 3(-x)^3 + 6(-x)^2 - 4(-x) - 12$$

$$f(-x) = -x^4 + 3x^3 + 6x^2 + 4x - 12$$

$$f(-x) = -x^4 + 3x^3 + 6x^2 + 4x - 12$$

Figure 4

Again, there are two sign changes, so there are either 2 or 0 negative real roots.

There are four possibilities, as we can see in **Table 1.**

Positive Real Zeros	Negative Real Zeros	Complex Zeros	Total Zeros
2	2	0	4
2	0	2	4
0	2	2	4
0	0	4	4

Table 1

Analysis *We can confirm the numbers of positive and negative real roots by examining a graph of the function. See* **Figure 5.** *We can see from the graph that the function has 0 positive real roots and 2 negative real roots.*

Figure 5

Try It #6

Use Descartes' Rule of Signs to determine the maximum possible numbers of positive and negative real zeros for $f(x) = 2x^4 - 10x^3 + 11x^2 - 15x + 12$. Use a graph to verify the numbers of positive and negative real zeros for the function.

Solving Real-World Applications

We have now introduced a variety of tools for solving polynomial equations. Let's use these tools to solve the bakery problem from the beginning of the section.

Example 9   **Solving Polynomial Equations**

A new bakery offers decorated sheet cakes for children's birthday parties and other special occasions. The bakery wants the volume of a small cake to be 351 cubic inches. The cake is in the shape of a rectangular solid. They want the length of the cake to be four inches longer than the width of the cake and the height of the cake to be one-third of the width. What should the dimensions of the cake pan be?

Solution   Begin by writing an equation for the volume of the cake. The volume of a rectangular solid is given by $V = lwh$. We were given that the length must be four inches longer than the width, so we can express the length of the cake as $l = w + 4$. We were given that the height of the cake is one-third of the width, so we can express the height of the cake as $h = \dfrac{1}{3} w$. Let's write the volume of the cake in terms of width of the cake.

$$V = (w + 4)(w)\left(\frac{1}{3}w\right)$$

$$V = \frac{1}{3}w^3 + \frac{4}{3}w^2$$

Substitute the given volume into this equation.

$$351 = \frac{1}{3}w^3 + \frac{4}{3}w^2 \qquad \text{Substitute 351 for } V.$$

$$1053 = w^3 + 4w^2 \qquad \text{Multiply both sides by 3.}$$

$$0 = w^3 + 4w^2 - 1053 \qquad \text{Subtract 1053 from both sides.}$$

Descartes' rule of signs tells us there is one positive solution. The Rational Zero Theorem tells us that the possible rational zeros are ± 3, ± 9, ± 13, ± 27, ± 39, ± 81, ± 117, ± 351, and ± 1053. We can use synthetic division to test these possible zeros. Only positive numbers make sense as dimensions for a cake, so we need not test any negative values. Let's begin by testing values that make the most sense as dimensions for a small sheet cake. Use synthetic division to check $x = 1$.

$$
\begin{array}{r|rrrr}
1 & 1 & 4 & 0 & -1053 \\
  &   & 1 & 5 & 5 \\
\hline
  & 1 & 5 & 5 & -1048 \\
\end{array}
$$

Since 1 is not a solution, we will check $x = 3$.

$$
\begin{array}{r|rrrr}
3 & 1 & 4 & 0 & -1053 \\
  &   & 3 & 21 & 63 \\
\hline
  & 1 & 7 & 21 & -990 \\
\end{array}
$$

Since 3 is not a solution either, we will test $x = 9$.

$$
\begin{array}{r|rrrr}
9 & 1 & 4 & 0 & -1053 \\
  &   & 9 & 117 & 1053 \\
\hline
  & 1 & 13 & 117 & 0 \\
\end{array}
$$

Synthetic division gives a remainder of 0, so 9 is a solution to the equation. We can use the relationships between the width and the other dimensions to determine the length and height of the sheet cake pan.

$$l = w + 4 = 9 + 4 = 13 \text{ and } h = \frac{1}{3}w = \frac{1}{3}(9) = 3$$

The sheet cake pan should have dimensions 13 inches by 9 inches by 3 inches.

Try It #7

A shipping container in the shape of a rectangular solid must have a volume of 84 cubic meters. The client tells the manufacturer that, because of the contents, the length of the container must be one meter longer than the width, and the height must be one meter greater than twice the width. What should the dimensions of the container be?

Access these online resources for additional instruction and practice with zeros of polynomial functions.

- Real Zeros, Factors, and Graphs of Polynomial Functions (http://openstaxcollege.org/l/realzeros)
- Complex Factorization Theorem (http://openstaxcollege.org/l/factortheorem)
- Find the Zeros of a Polynomial Function (http://openstaxcollege.org/l/findthezeros)
- Find the Zeros of a Polynomial Function 2 (http://openstaxcollege.org/l/findthezeros2)
- Find the Zeros of a Polynomial Function 3 (http://openstaxcollege.org/l/findthezeros3)

5.5   SECTION EXERCISES

VERBAL

1. Describe a use for the Remainder Theorem.

2. Explain why the Rational Zero Theorem does not guarantee finding zeros of a polynomial function.

3. What is the difference between rational and real zeros?

4. If Descartes' Rule of Signs reveals a no change of signs or one sign of changes, what specific conclusion can be drawn?

5. If synthetic division reveals a zero, why should we try that value again as a possible solution?

ALGEBRAIC

For the following exercises, use the Remainder Theorem to find the remainder.

6. $(x^4 - 9x^2 + 14) \div (x - 2)$

7. $(3x^3 - 2x^2 + x - 4) \div (x + 3)$

8. $(x^4 + 5x^3 - 4x - 17) \div (x + 1)$

9. $(-3x^2 + 6x + 24) \div (x - 4)$

10. $(5x^5 - 4x^4 + 3x^3 - 2x^2 + x - 1) \div (x + 6)$

11. $(x^4 - 1) \div (x - 4)$

12. $(3x^3 + 4x^2 - 8x + 2) \div (x - 3)$

13. $(4x^3 + 5x^2 - 2x + 7) \div (x + 2)$

For the following exercises, use the Factor Theorem to find all real zeros for the given polynomial function and one factor.

14. $f(x) = 2x^3 - 9x^2 + 13x - 6; x - 1$

15. $f(x) = 2x^3 + x^2 - 5x + 2; x + 2$

16. $f(x) = 3x^3 + x^2 - 20x + 12; x + 3$

17. $f(x) = 2x^3 + 3x^2 + x + 6; x + 2$

18. $f(x) = -5x^3 + 16x^2 - 9; x - 3$

19. $x^3 + 3x^2 + 4x + 12; x + 3$

20. $4x^3 - 7x + 3; x - 1$

21. $2x^3 + 5x^2 - 12x - 30, 2x + 5$

For the following exercises, use the Rational Zero Theorem to find all real zeros.

22. $x^3 - 3x^2 - 10x + 24 = 0$

23. $2x^3 + 7x^2 - 10x - 24 = 0$

24. $x^3 + 2x^2 - 9x - 18 = 0$

25. $x^3 + 5x^2 - 16x - 80 = 0$

26. $x^3 - 3x^2 - 25x + 75 = 0$

27. $2x^3 - 3x^2 - 32x - 15 = 0$

28. $2x^3 + x^2 - 7x - 6 = 0$

29. $2x^3 - 3x^2 - x + 1 = 0$

30. $3x^3 - x^2 - 11x - 6 = 0$

31. $2x^3 - 5x^2 + 9x - 9 = 0$

32. $2x^3 - 3x^2 + 4x + 3 = 0$

33. $x^4 - 2x^3 - 7x^2 + 8x + 12 = 0$

34. $x^4 + 2x^3 - 9x^2 - 2x + 8 = 0$

35. $4x^4 + 4x^3 - 25x^2 - x + 6 = 0$

36. $2x^4 - 3x^3 - 15x^2 + 32x - 12 = 0$

37. $x^4 + 2x^3 - 4x^2 - 10x - 5 = 0$

38. $4x^3 - 3x + 1 = 0$

39. $8x^4 + 26x^3 + 39x^2 + 26x + 6$

For the following exercises, find all complex solutions (real and non-real).

40. $x^3 + x^2 + x + 1 = 0$

41. $x^3 - 8x^2 + 25x - 26 = 0$

42. $x^3 + 13x^2 + 57x + 85 = 0$

43. $3x^3 - 4x^2 + 11x + 10 = 0$

44. $x^4 + 2x^3 + 22x^2 + 50x - 75 = 0$

45. $2x^3 - 3x^2 + 32x + 17 = 0$

GRAPHICAL

Use Descartes' Rule to determine the possible number of positive and negative solutions. Then graph to confirm which of those possibilities is the actual combination.

46. $f(x) = x^3 - 1$

47. $f(x) = x^4 - x^2 - 1$

48. $f(x) = x^3 - 2x^2 - 5x + 6$

49. $f(x) = x^3 - 2x^2 + x - 1$

50. $f(x) = x^4 + 2x^3 - 12x^2 + 14x - 5$

51. $f(x) = 2x^3 + 37x^2 + 200x + 300$

52. $f(x) = x^3 - 2x^2 - 16x + 32$ **53.** $f(x) = 2x^4 - 5x^3 - 5x^2 + 5x + 3$ **54.** $f(x) = 2x^4 - 5x^3 - 14x^2 + 20x + 8$

55. $f(x) = 10x^4 - 21x^2 + 11$

NUMERIC

For the following exercises, list all possible rational zeros for the functions.

56. $f(x) = x^4 + 3x^3 - 4x + 4$ **57.** $f(x) = 2x^3 + 3x^2 - 8x + 5$ **58.** $f(x) = 3x^3 + 5x^2 - 5x + 4$

59. $f(x) = 6x^4 - 10x^2 + 13x + 1$ **60.** $f(x) = 4x^5 - 10x^4 + 8x^3 + x^2 - 8$

TECHNOLOGY

For the following exercises, use your calculator to graph the polynomial function. Based on the graph, find the rational zeros. All real solutions are rational.

61. $f(x) = 6x^3 - 7x^2 + 1$ **62.** $f(x) = 4x^3 - 4x^2 - 13x - 5$

63. $f(x) = 8x^3 - 6x^2 - 23x + 6$ **64.** $f(x) = 12x^4 + 55x^3 + 12x^2 - 117x + 54$

65. $f(x) = 16x^4 - 24x^3 + x^2 - 15x + 25$

EXTENSIONS

For the following exercises, construct a polynomial function of least degree possible using the given information.

66. Real roots: $-1, 1, 3$ and $(2, f(2)) = (2, 4)$ **67.** Real roots: $-1, 1$ (with multiplicity 2 and 1) and $(2, f(2)) = (2, 4)$

68. Real roots: $-2, \frac{1}{2}$ (with multiplicity 2) and $(-3, f(-3)) = (-3, 5)$ **69.** Real roots: $-\frac{1}{2}, 0, \frac{1}{2}$ and $(-2, f(-2)) = (-2, 6)$

70. Real roots: $-4, -1, 1, 4$ and $(-2, f(-2)) = (-2, 10)$

REAL-WORLD APPLICATIONS

For the following exercises, find the dimensions of the box described.

71. The length is twice as long as the width. The height is 2 inches greater than the width. The volume is 192 cubic inches.

72. The length, width, and height are consecutive whole numbers. The volume is 120 cubic inches.

73. The length is one inch more than the width, which is one inch more than the height. The volume is 86.625 cubic inches.

74. The length is three times the height and the height is one inch less than the width. The volume is 108 cubic inches.

75. The length is 3 inches more than the width. The width is 2 inches more than the height. The volume is 120 cubic inches.

For the following exercises, find the dimensions of the right circular cylinder described.

76. The radius is 3 inches more than the height. The volume is 16π cubic meters.

77. The height is one less than one half the radius. The volume is 72π cubic meters.

78. The radius and height differ by one meter. The radius is larger and the volume is 48π cubic meters.

79. The radius and height differ by two meters. The height is greater and the volume is 28.125π cubic meters.

80. The radius is $\frac{1}{3}$ meter greater than the height. The volume is $\frac{98}{9\pi}\pi$ cubic meters.

LEARNING OBJECTIVES

In this section, you will:

- Use arrow notation.
- Solve applied problems involving rational functions.
- Find the domains of rational functions.
- Identify vertical asymptotes.
- Identify horizontal asymptotes.
- Graph rational functions.

5.6   RATIONAL FUNCTIONS

Suppose we know that the cost of making a product is dependent on the number of items, x, produced. This is given by the equation $C(x) = 15{,}000x - 0.1x^2 + 1000$. If we want to know the average cost for producing x items, we would divide the cost function by the number of items, x.

The average cost function, which yields the average cost per item for x items produced, is

$$f(x) = \frac{15{,}000x - 0.1x^2 + 1000}{x}$$

Many other application problems require finding an average value in a similar way, giving us variables in the denominator. Written without a variable in the denominator, this function will contain a negative integer power.

In the last few sections, we have worked with polynomial functions, which are functions with non-negative integers for exponents. In this section, we explore rational functions, which have variables in the denominator.

Using Arrow Notation

We have seen the graphs of the basic reciprocal function and the squared reciprocal function from our study of toolkit functions. Examine these graphs, as shown in **Figure 1**, and notice some of their features.

Graphs of Toolkit Functions

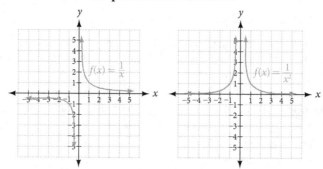

Figure 1

Several things are apparent if we examine the graph of $f(x) = \dfrac{1}{x}$.

1. On the left branch of the graph, the curve approaches the x-axis $(y = 0)$ as $x \to -\infty$.
2. As the graph approaches $x = 0$ from the left, the curve drops, but as we approach zero from the right, the curve rises.
3. Finally, on the right branch of the graph, the curves approaches the x-axis $(y = 0)$ as $x \to \infty$.

To summarize, we use **arrow notation** to show that x or $f(x)$ is approaching a particular value. See **Table 1**.

Symbol	Meaning
$x \to a^-$	x approaches a from the left ($x < a$ but close to a)
$x \to a^+$	x approaches a from the right ($x > a$ but close to a)
$x \to \infty$	x approaches infinity (x increases without bound)
$x \to -\infty$	x approaches negative infinity (x decreases without bound)
$f(x) \to \infty$	The output approaches infinity (the output increases without bound)
$f(x) \to -\infty$	The output approaches negative infinity (the output decreases without bound)
$f(x) \to a$	The output approaches a

Table 1 Arrow Notation

Local Behavior of $f(x) = \dfrac{1}{x}$

Let's begin by looking at the reciprocal function, $f(x) = \dfrac{1}{x}$. We cannot divide by zero, which means the function is undefined at $x = 0$; so zero is not in the domain. As the input values approach zero from the left side (becoming very small, negative values), the function values decrease without bound (in other words, they approach negative infinity). We can see this behavior in **Table 2**.

x	-0.1	-0.01	-0.001	-0.0001
$f(x) = \dfrac{1}{x}$	-10	-100	-1000	$-10{,}000$

Table 2

We write in arrow notation

$$\text{as } x \to 0^-, \ f(x) \to -\infty$$

As the input values approach zero from the right side (becoming very small, positive values), the function values increase without bound (approaching infinity). We can see this behavior in **Table 3**.

x	0.1	0.01	0.001	0.0001
$f(x) = \dfrac{1}{x}$	10	100	1000	$10{,}000$

Table 3

We write in arrow notation

$$\text{As } x \to 0^+, \ f(x) \to \infty.$$

See **Figure 2**.

Figure 2

This behavior creates a **vertical asymptote**, which is a vertical line that the graph approaches but never crosses. In this case, the graph is approaching the vertical line $x = 0$ as the input becomes close to zero. See **Figure 3**.

Figure 3

---

vertical asymptote

A **vertical asymptote** of a graph is a vertical line $x = a$ where the graph tends toward positive or negative infinity as the inputs approach a. We write

$$\text{As } x \to a, f(x) \to \infty, \text{ or as } x \to a, f(x) \to -\infty.$$

End Behavior of $f(x) = \dfrac{1}{x}$

As the values of x approach infinity, the function values approach 0. As the values of x approach negative infinity, the function values approach 0. See **Figure 4**. Symbolically, using arrow notation

$$\text{As } x \to \infty, f(x) \to 0, \text{ and as } x \to -\infty, f(x) \to 0.$$

Figure 4

Based on this overall behavior and the graph, we can see that the function approaches 0 but never actually reaches 0; it seems to level off as the inputs become large. This behavior creates a **horizontal asymptote**, a horizontal line that the graph approaches as the input increases or decreases without bound. In this case, the graph is approaching the horizontal line $y = 0$. See **Figure 5**.

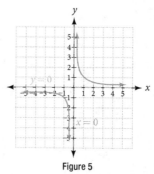

Figure 5

horizontal asymptote

A **horizontal asymptote** of a graph is a horizontal line $y = b$ where the graph approaches the line as the inputs increase or decrease without bound. We write

$$\text{As } x \to \infty \text{ or } x \to -\infty, f(x) \to b.$$

Example 1   **Using Arrow Notation**

Use arrow notation to describe the end behavior and local behavior of the function graphed in **Figure 6**.

Figure 6

Solution   Notice that the graph is showing a vertical asymptote at $x = 2$, which tells us that the function is undefined at $x = 2$.

$$\text{As } x \to 2^-, f(x) \to -\infty, \text{ and as } x \to 2^+, \ f(x) \to \infty.$$

And as the inputs decrease without bound, the graph appears to be leveling off at output values of 4, indicating a horizontal asymptote at $y = 4$. As the inputs increase without bound, the graph levels off at 4.

$$\text{As } x \to \infty, f(x) \to 4 \text{ and as } x \to -\infty, f(x) \to 4.$$

Try It #1

Use arrow notation to describe the end behavior and local behavior for the reciprocal squared function.

Example 2   **Using Transformations to Graph a Rational Function**

Sketch a graph of the reciprocal function shifted two units to the left and up three units. Identify the horizontal and vertical asymptotes of the graph, if any.

Solution   Shifting the graph left 2 and up 3 would result in the function

$$f(x) = \frac{1}{x + 2} + 3$$

or equivalently, by giving the terms a common denominator,

$$f(x) = \frac{3x + 7}{x + 2}$$

The graph of the shifted function is displayed in **Figure 7**.

Figure 7

Notice that this function is undefined at $x = -2$, and the graph also is showing a vertical asymptote at $x = -2$.

$$\text{As } x \to -2^{-}, f(x) \to -\infty, \text{ and as } x \to -2^{+}, f(x) \to \infty.$$

As the inputs increase and decrease without bound, the graph appears to be leveling off at output values of 3, indicating a horizontal asymptote at $y = 3$.

$$\text{As } x \to \pm\infty, f(x) \to 3.$$

Analysis *Notice that horizontal and vertical asymptotes are shifted left 2 and up 3 along with the function.*

Try It #2

Sketch the graph, and find the horizontal and vertical asymptotes of the reciprocal squared function that has been shifted right 3 units and down 4 units.

Solving Applied Problems Involving Rational Functions

In **Example 2**, we shifted a toolkit function in a way that resulted in the function $f(x) = \dfrac{3x + 7}{x + 2}$. This is an example of a rational function. A **rational function** is a function that can be written as the quotient of two polynomial functions. Many real-world problems require us to find the ratio of two polynomial functions. Problems involving rates and concentrations often involve rational functions.

---

rational function

A **rational function** is a function that can be written as the quotient of two polynomial functions $P(x)$ and $Q(x)$.

$$f(x) = \frac{P(x)}{Q(x)} = \frac{a_p x^p + a_{p-1} x^{p-1} + \dots + a_1 x + a_0}{b_q x^q + b_{q-1} x^{q-1} + \dots + b_1 x + b_0}, \quad Q(x) \neq 0$$

Example 3   **Solving an Applied Problem Involving a Rational Function**

A large mixing tank currently contains 100 gallons of water into which 5 pounds of sugar have been mixed. A tap will open pouring 10 gallons per minute of water into the tank at the same time sugar is poured into the tank at a rate of 1 pound per minute. Find the concentration (pounds per gallon) of sugar in the tank after 12 minutes. Is that a greater concentration than at the beginning?

Solution   Let t be the number of minutes since the tap opened. Since the water increases at 10 gallons per minute, and the sugar increases at 1 pound per minute, these are constant rates of change. This tells us the amount of water in the tank is changing linearly, as is the amount of sugar in the tank. We can write an equation independently for each:

$$\text{water: } W(t) = 100 + 10t \text{ in gallons}$$

$$\text{sugar: } S(t) = 5 + 1t \text{ in pounds}$$

The concentration, C, will be the ratio of pounds of sugar to gallons of water

$$C(t) = \frac{5 + t}{100 + 10t}$$

The concentration after 12 minutes is given by evaluating $C(t)$ at $t = 12$.

$$C(12) = \frac{5 + 12}{100 + 10(12)}$$

$$= \frac{17}{220}$$

This means the concentration is 17 pounds of sugar to 220 gallons of water.

At the beginning, the concentration is

$$C(0) = \frac{5 + 0}{100 + 10(0)}$$

$$= \frac{1}{20}$$

Since $\dfrac{17}{220} \approx 0.08 > \dfrac{1}{20} = 0.05$, the concentration is greater after 12 minutes than at the beginning.

Try It #3

There are 1,200 freshmen and 1,500 sophomores at a prep rally at noon. After 12 p.m., 20 freshmen arrive at the rally every five minutes while 15 sophomores leave the rally. Find the ratio of freshmen to sophomores at 1 p.m.

Finding the Domains of Rational Functions

A vertical asymptote represents a value at which a rational function is undefined, so that value is not in the domain of the function. A reciprocal function cannot have values in its domain that cause the denominator to equal zero. In general, to find the domain of a rational function, we need to determine which inputs would cause division by zero.

---

domain of a rational function

The domain of a rational function includes all real numbers except those that cause the denominator to equal zero.

How To...

Given a rational function, find the domain.

1. Set the denominator equal to zero.
2. Solve to find the x-values that cause the denominator to equal zero.
3. The domain is all real numbers except those found in Step 2.

Example 4 **Finding the Domain of a Rational Function**

Find the domain of $f(x) = \dfrac{x + 3}{x^2 - 9}$.

Solution Begin by setting the denominator equal to zero and solving.

$$x^2 - 9 = 0$$
$$x^2 = 9$$
$$x = \pm 3$$

The denominator is equal to zero when $x = \pm 3$. The domain of the function is all real numbers except $x = \pm 3$.

*Analysis A graph of this function, as shown in **Figure 8**, confirms that the function is not defined when $x = \pm 3$.*

Figure 8

There is a vertical asymptote at $x = 3$ and a hole in the graph at $x = -3$. We will discuss these types of holes in greater detail later in this section.

Try It #4

Find the domain of $f(x) = \dfrac{4x}{5(x - 1)(x - 5)}$.

Identifying Vertical Asymptotes of Rational Functions

By looking at the graph of a rational function, we can investigate its local behavior and easily see whether there are asymptotes. We may even be able to approximate their location. Even without the graph, however, we can still determine whether a given rational function has any asymptotes, and calculate their location.

Vertical Asymptotes

The vertical asymptotes of a rational function may be found by examining the factors of the denominator that are not common to the factors in the numerator. Vertical asymptotes occur at the zeros of such factors.

How To...

Given a rational function, identify any vertical asymptotes of its graph.

1. Factor the numerator and denominator.
2. Note any restrictions in the domain of the function.
3. Reduce the expression by canceling common factors in the numerator and the denominator.
4. Note any values that cause the denominator to be zero in this simplified version. These are where the vertical asymptotes occur.
5. Note any restrictions in the domain where asymptotes do not occur. These are removable discontinuities or "holes."

Example 5   **Identifying Vertical Asymptotes**

Find the vertical asymptotes of the graph of $k(x) = \dfrac{5 + 2x^2}{2 - x - x^2}$.

Solution   First, factor the numerator and denominator.

$$k(x) = \frac{5 + 2x^2}{2 - x - x^2}$$

$$= \frac{5 + 2x^2}{(2 + x)(1 - x)}$$

To find the vertical asymptotes, we determine where this function will be undefined by setting the denominator equal to zero:

$$(2 + x)(1 - x) = 0$$

$$x = -2, 1$$

Neither $x = -2$ nor $x = 1$ are zeros of the numerator, so the two values indicate two vertical asymptotes. The graph in **Figure 9** confirms the location of the two vertical asymptotes.

Figure 9

Removable Discontinuities

Occasionally, a graph will contain a hole: a single point where the graph is not defined, indicated by an open circle. We call such a hole a **removable discontinuity**.

For example, the function $f(x) = \dfrac{x^2 - 1}{x^2 - 2x - 3}$ may be re-written by factoring the numerator and the denominator.

$$f(x) = \frac{(x + 1)(x - 1)}{(x + 1)(x - 3)}$$

Notice that $x + 1$ is a common factor to the numerator and the denominator. The zero of this factor, $x = -1$, is the location of the removable discontinuity. Notice also that $x - 3$ is not a factor in both the numerator and denominator. The zero of this factor, $x = 3$, is the vertical asymptote. See **Figure 10**. [Note that removable discontinuities may not be visible when we use a graphing calculator, depending upon the window selected.]

Removable discontinuity at $x = -1$

Vertical asymptote at $x = 3$

Figure 10

---

removable discontinuities of rational functions

A **removable discontinuity** occurs in the graph of a rational function at $x = a$ if a is a zero for a factor in the denominator that is common with a factor in the numerator. We factor the numerator and denominator and check for common factors. If we find any, we set the common factor equal to 0 and solve. This is the location of the removable discontinuity. This is true if the multiplicity of this factor is greater than or equal to that in the denominator. If the multiplicity of this factor is greater in the denominator, then there is still an asymptote at that value.

Example 6   **Identifying Vertical Asymptotes and Removable Discontinuities for a Graph**

Find the vertical asymptotes and removable discontinuities of the graph of $k(x) = \dfrac{x - 2}{x^2 - 4}$.

Solution   Factor the numerator and the denominator.

$$k(x) = \frac{x - 2}{(x - 2)(x + 2)}$$

Notice that there is a common factor in the numerator and the denominator, $x - 2$. The zero for this factor is $x = 2$. This is the location of the removable discontinuity.

Notice that there is a factor in the denominator that is not in the numerator, $x + 2$. The zero for this factor is $x = -2$. The vertical asymptote is $x = -2$. See **Figure 11**.

Figure 11

The graph of this function will have the vertical asymptote at $x = -2$, but at $x = 2$ the graph will have a hole.

Try It #5

Find the vertical asymptotes and removable discontinuities of the graph of $f(x) = \dfrac{x^2 - 25}{x^3 - 6x^2 + 5x}$.

Identifying Horizontal Asymptotes of Rational Functions

While vertical asymptotes describe the behavior of a graph as the *output* gets very large or very small, horizontal asymptotes help describe the behavior of a graph as the *input* gets very large or very small. Recall that a polynomial's end behavior will mirror that of the leading term. Likewise, a rational function's end behavior will mirror that of the ratio of the function that is the ratio of the leading term.

There are three distinct outcomes when checking for horizontal asymptotes:

Case 1: If the degree of the denominator > degree of the numerator, there is a horizontal asymptote at $y = 0$.

$$\text{Example: } f(x) = \frac{4x + 2}{x^2 + 4x - 5}$$

In this case, the end behavior is $f(x) \approx \dfrac{4x}{x^2} = \dfrac{4}{x}$. This tells us that, as the inputs increase or decrease without bound, this function will behave similarly to the function $g(x) = \dfrac{4}{x}$, and the outputs will approach zero, resulting in a horizontal asymptote at $y = 0$. See **Figure 12**. Note that this graph crosses the horizontal asymptote.

Figure 12 Horizontal asymptote $y = 0$ when $f(x) = \frac{p(x)}{q(x)}$, $q(x) \neq 0$ where degree of $p <$ degree of q.

Case 2: If the degree of the denominator < degree of the numerator by one, we get a slant asymptote.

$$\text{Example: } f(x) = \frac{3x^2 - 2x + 1}{x - 1}$$

In this case, the end behavior is $f(x) \approx \dfrac{3x^2}{x} = 3x$. This tells us that as the inputs increase or decrease without bound, this function will behave similarly to the function $g(x) = 3x$. As the inputs grow large, the outputs will grow and not level off, so this graph has no horizontal asymptote. However, the graph of $g(x) = 3x$ looks like a diagonal line, and since f will behave similarly to g, it will approach a line close to $y = 3x$. This line is a slant asymptote.

To find the equation of the slant asymptote, divide $\dfrac{3x^2 - 2x + 1}{x - 1}$. The quotient is $3x + 1$, and the remainder is 2. The slant asymptote is the graph of the line $g(x) = 3x + 1$. See **Figure 13**.

Figure 13 Slant asymptote when $f(x) = \dfrac{p(x)}{q(x)}$, $q(x) \neq 0$ where degree of $p >$ degree of q by 1.

Case 3: If the degree of the denominator = degree of the numerator, there is a horizontal asymptote at $y = \dfrac{a_n}{b_n}$, where a_n and b_n are the leading coefficients of $p(x)$ and $q(x)$ for $f(x) = \dfrac{p(x)}{q(x)}$, $q(x) \neq 0$.

$$\text{Example: } f(x) = \frac{3x^2 + 2}{x^2 + 4x - 5}$$

In this case, the end behavior is $f(x) \approx \dfrac{3x^2}{x^2} = 3$. This tells us that as the inputs grow large, this function will behave like the function $g(x) = 3$, which is a horizontal line. As $x \to \pm\infty$, $f(x) \to 3$, resulting in a horizontal asymptote at $y = 3$. See **Figure 14**. Note that this graph crosses the horizontal asymptote.

Figure 14 Horizontal asymptote when $f(x) = \dfrac{p(x)}{q(x)}$, $q(x) \neq 0$ where degree of $p =$ degree of q.

Notice that, while the graph of a rational function will never cross a vertical asymptote, the graph may or may not cross a horizontal or slant asymptote. Also, although the graph of a rational function may have many vertical asymptotes, the graph will have at most one horizontal (or slant) asymptote.

It should be noted that, if the degree of the numerator is larger than the degree of the denominator by more than one, the end behavior of the graph will mimic the behavior of the reduced end behavior fraction. For instance, if we had the function

$$f(x) = \frac{3x^5 - x^2}{x + 3}$$

with end behavior

$$f(x) \approx \frac{3x^5}{x} = 3x^4,$$

the end behavior of the graph would look similar to that of an even polynomial with a positive leading coefficient.

$$x \to \pm\infty, \ f(x) \to \infty$$

> ***horizontal asymptotes of rational functions***
>
> The horizontal asymptote of a rational function can be determined by looking at the degrees of the numerator and denominator.
>
> - Degree of numerator *is less than* degree of denominator: horizontal asymptote at $y = 0$.
> - Degree of numerator *is greater than degree of denominator by one*: no horizontal asymptote; slant asymptote.
> - Degree of numerator *is equal to* degree of denominator: horizontal asymptote at ratio of leading coefficients.

Example 7   **Identifying Horizontal and Slant Asymptotes**

For the functions listed, identify the horizontal or slant asymptote.

a. $g(x) = \dfrac{6x^3 - 10x}{2x^3 + 5x^2}$ **b.** $h(x) = \dfrac{x^2 - 4x + 1}{x + 2}$ **c.** $k(x) = \dfrac{x^2 + 4x}{x^3 - 8}$

Solution   For these solutions, we will use $f(x) = \dfrac{p(x)}{q(x)}, q(x) \neq 0$.

a. $g(x) = \dfrac{6x^3 - 10x}{2x^3 + 5x^2}$: The degree of p = degree of q = 3, so we can find the horizontal asymptote by taking the ratio of the leading terms. There is a horizontal asymptote at $y = \frac{6}{2}$ or $y = 3$.

b. $h(x) = \dfrac{x^2 - 4x + 1}{x + 2}$: The degree of p = 2 and degree of q = 1. Since $p > q$ by 1, there is a slant asymptote found at $\dfrac{x^2 - 4x + 1}{x + 2}$.

$$2 \begin{array}{|rrr} 1 & -4 & 1 \\ & -2 & 12 \\ \hline 1 & -6 & 13 \end{array}$$

The quotient is $x - 2$ and the remainder is 13. There is a slant asymptote at $y = x - 2$.

c. $k(x) = \dfrac{x^2 + 4x}{x^3 - 8}$: The degree of p = 2 < degree of q = 3, so there is a horizontal asymptote $y = 0$.

Example 8   **Identifying Horizontal Asymptotes**

In the sugar concentration problem earlier, we created the equation $C(t) = \dfrac{5 + t}{100 + 10t}$.

Find the horizontal asymptote and interpret it in context of the problem.

Solution   Both the numerator and denominator are linear (degree 1). Because the degrees are equal, there will be a horizontal asymptote at the ratio of the leading coefficients. In the numerator, the leading term is t, with coefficient 1. In the denominator, the leading term is $10t$, with coefficient 10. The horizontal asymptote will be at the ratio of these values:

$$t \to \infty, C(t) \to \frac{1}{10}$$

This function will have a horizontal asymptote at $y = \dfrac{1}{10}$.

This tells us that as the values of t increase, the values of C will approach $\dfrac{1}{10}$. In context, this means that, as more time goes by, the concentration of sugar in the tank will approach one-tenth of a pound of sugar per gallon of water or $\dfrac{1}{10}$ pounds per gallon.

Example 9   **Identifying Horizontal and Vertical Asymptotes**

Find the horizontal and vertical asymptotes of the function

$$f(x) = \frac{(x - 2)(x + 3)}{(x - 1)(x + 2)(x - 5)}$$

Solution   First, note that this function has no common factors, so there are no potential removable discontinuities.

The function will have vertical asymptotes when the denominator is zero, causing the function to be undefined. The denominator will be zero at $x = 1, -2$, and 5, indicating vertical asymptotes at these values.

The numerator has degree 2, while the denominator has degree 3. Since the degree of the denominator is greater than the degree of the numerator, the denominator will grow faster than the numerator, causing the outputs to tend towards zero as the inputs get large, and so as $x \to \pm\infty$, $f(x) \to 0$. This function will have a horizontal asymptote at $y = 0$. See **Figure 15**.

Figure 15

Functions involving roots are often called radical functions. While it is not possible to find an inverse of most polynomial functions, some basic polynomials do have inverses. Such functions are called **invertible functions**, and we use the notation $f^{-1}(x)$.

Warning: $f^{-1}(x)$ is not the same as the reciprocal of the function $f(x)$. This use of "−1" is reserved to denote inverse functions. To denote the reciprocal of a function $f(x)$, we would need to write $(f(x))^{-1} = \dfrac{1}{f(x)}$.

An important relationship between inverse functions is that they "undo" each other. If f^{-1} is the inverse of a function f, then f is the inverse of the function f^{-1}. In other words, whatever the function f does to x, f^{-1} undoes it—and viceversa. More formally, we write

$$f^{-1}(f(x)) = x, \text{ for all } x \text{ in the domain of } f$$

and

$$f(f^{-1}(x)) = x, \text{ for all } x \text{ in the domain of } f^{-1}$$

Note that the inverse switches the domain and range of th original function.

> ***verifying two functions are inverses of one another***
>
> Two functions, f and g, are inverses of one another if for all x in the domain of f and g.
>
> $$g(f(x)) = f(g(x)) = x$$

How To…

Given a polynomial function, find the inverse of the function by restricting the domain in such a way that the new function is one-to-one.

1. Replace $f(x)$ with y.
2. Interchange x and y.
3. Solve for y, and rename the function $f^{-1}(x)$.

Example 1   **Verifying Inverse Functions**

Show that $f(x) = \dfrac{1}{x+1}$ and $f^{-1}(x) = \dfrac{1}{x} - 1$ are inverses, for $x \neq 0, -1$.

Solution   We must show that $f^{-1}(f(x)) = x$ and $f(f^{-1}(x)) = x$.

$$f^{-1}(f(x)) = f^{-1}\left(\frac{1}{x+1}\right)$$

$$= \frac{1}{\dfrac{1}{x+1}} - 1$$

$$= (x+1) - 1$$

$$= x$$

$$f(f^{-1}(x)) = f\left(\frac{1}{x} - 1\right)$$

$$= \frac{1}{\left(\dfrac{1}{x} - 1\right) + 1}$$

$$= \frac{1}{\dfrac{1}{x}}$$

$$= x$$

Therefore, $f(x) = \dfrac{1}{x+1}$ and $f^{-1}(x) = \dfrac{1}{x} - 1$ are inverses.

Try It #1

Show that $f(x) = \dfrac{x+5}{3}$ and $f^{-1}(x) = 3x - 5$ are inverses.

Example 2   **Finding the Inverse of a Cubic Function**

Find the inverse of the function $f(x) = 5x^3 + 1$.

Solution   This is a transformation of the basic cubic toolkit function, and based on our knowledge of that function, we know it is one-to-one. Solving for the inverse by solving for x.

$$y = 5x^3 + 1$$
$$x = 5y^3 + 1$$
$$x - 1 = 5y^3$$
$$\frac{x-1}{5} = y^3$$
$$f^{-1}(x) = \sqrt[3]{\frac{x-1}{5}}$$

Analysis   Look at the graph of f and f^{-1}. Notice that one graph is the reflection of the other about the line $y = x$. This is always the case when graphing a function and its inverse function.

*Also, since the method involved interchanging x and y, notice corresponding points. If (a, b) is on the graph of f, then (b, a) is on the graph of f^{-1}. Since $(0, 1)$ is on the graph of f, then $(1, 0)$ is on the graph of f^{-1}. Similarly, since $(1, 6)$ is on the graph of f, then $(6, 1)$ is on the graph of f^{-1}. See **Figure 4**.*

Figure 4

Try It #2

Find the inverse function of $f(x) = \sqrt[3]{x + 4}$

Restricting the Domain to Find the Inverse of a Polynomial Function

So far, we have been able to find the inverse functions of cubic functions without having to restrict their domains. However, as we know, not all cubic polynomials are one-to-one. Some functions that are not one-to-one may have their domain restricted so that they are one-to-one, but only over that domain. The function over the restricted domain would then have an inverse function. Since quadratic functions are not one-to-one, we must restrict their domain in order to find their inverses.

> **restricting the domain**
>
> If a function is not one-to-one, it cannot have an inverse. If we restrict the domain of the function so that it becomes one-to-one, thus creating a new function, this new function will have an inverse.

Given a polynomial function, restrict the domain of a function that is not one-to-one and then find the inverse.

1. Restrict the domain by determining a domain on which the original function is one-to-one.
2. Replace $f(x)$ with y.
3. Interchange x and y.
4. Solve for y, and rename the function or pair of functions $f^{-1}(x)$.
5. Revise the formula for $f^{-1}(x)$ by ensuring that the outputs of the inverse function correspond to the restricted domain of the original function.

Example 3   **Restricting the Domain to Find the Inverse of a Polynomial Function**

Find the inverse function of f:

    a. $f(x) = (x-4)^2, x \geq 4$ **b.** $f(x) = (x-4)^2, x \leq 4$

Solution   The original function $f(x) = (x-4)^2$ is not one-to-one, but the function is restricted to a domain of $x \geq 4$ or $x \leq 4$ on which it is one-to-one. See **Figure 5**.

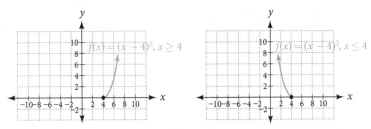

Figure 5

To find the inverse, start by replacing $f(x)$ with the simple variable y.

$$y = (x-4)^2 \quad \text{Interchange } x \text{ and } y.$$
$$x = (y-4)^2 \quad \text{Take the square root.}$$
$$\pm\sqrt{x} = y - 4 \quad \text{Add 4 to both sides.}$$
$$4 \pm \sqrt{x} = y$$

This is not a function as written. We need to examine the restrictions on the domain of the original function to determine the inverse. Since we reversed the roles of x and y for the original $f(x)$, we looked at the domain: the values x could assume. When we reversed the roles of x and y, this gave us the values y could assume. For this function, $x \geq 4$, so for the inverse, we should have $y \geq 4$, which is what our inverse function gives.

    a. The domain of the original function was restricted to $x \geq 4$, so the outputs of the inverse need to be the same, $f(x) \geq 4$, and we must use the $+$ case:

$$f^{-1}(x) = 4 + \sqrt{x}$$

    b. The domain of the original function was restricted to $x \leq 4$, so the outputs of the inverse need to be the same, $f(x) \leq 4$, and we must use the $-$ case:

$$f^{-1}(x) = 4 - \sqrt{x}$$

Analysis   *On the graphs in **Figure 6**, we see the original function graphed on the same set of axes as its inverse function. Notice that together the graphs show symmetry about the line $y = x$. The coordinate pair $(4, 0)$ is on the graph of f and the coordinate pair $(0, 4)$ is on the graph of f^{-1}. For any coordinate pair, if (a, b) is on the graph of f, then (b, a) is on the graph of f^{-1}. Finally, observe that the graph of f intersects the graph of f^{-1} on the line $y = x$. Points of intersection for the graphs of f and f^{-1} will always lie on the line $y = x$.*

Figure 6

Example 4   **Finding the Inverse of a Quadratic Function When the Restriction Is Not Specified**

Restrict the domain and then find the inverse of

$$f(x) = (x - 2)^2 - 3.$$

Solution   We can see this is a parabola with vertex at $(2, -3)$ that opens upward. Because the graph will be decreasing on one side of the vertex and increasing on the other side, we can restrict this function to a domain on which it will be one-to-one by limiting the domain to $x \geq 2$.

To find the inverse, we will use the vertex form of the quadratic. We start by replacing $f(x)$ with a simple variable, y, then solve for x.

$$y = (x - 2)^2 - 3 \qquad \text{Interchange } x \text{ and } y.$$
$$x = (y - 2)^2 - 3 \qquad \text{Add 3 to both sides.}$$
$$x + 3 = (y - 2)^2 \qquad \text{Take the square root.}$$
$$\pm\sqrt{x + 3} = y - 2 \qquad \text{Add 2 to both sides.}$$
$$2 \pm \sqrt{x + 3} = y \qquad \text{Rename the function.}$$
$$f^{-1}(x) = 2 \pm \sqrt{x + 3}$$

Now we need to determine which case to use. Because we restricted our original function to a domain of $x \geq 2$, the outputs of the inverse should be the same, telling us to utilize the $+$ case

$$f^{-1}(x) = 2 + \sqrt{x + 3}$$

If the quadratic had not been given in vertex form, rewriting it into vertex form would be the first step. This way we may easily observe the coordinates of the vertex to help us restrict the domain.

Analysis   Notice that we arbitrarily decided to restrict the domain on $x \geq 2$. We could just have easily opted to restrict the domain on $x \leq 2$, in which case $f^{-1}(x) = 2 - \sqrt{x + 3}$. Observe the original function graphed on the same set of axes as its inverse function in **Figure 7**. *Notice that both graphs show symmetry about the line $y = x$. The coordinate pair $(2, -3)$ is on the graph of f and the coordinate pair $(-3, 2)$ is on the graph of f^{-1}. Observe from the graph of both functions on the same set of axes that*

$$\text{domain of } f = \text{range of } f^{-1} = [2, \infty)$$

and

$$\text{domain of } f^{-1} = \text{range of } f = [-3, \infty)$$

Finally, observe that the graph of f intersects the graph of f^{-1} along the line $y = x$.

Figure 7

Try It #3

Find the inverse of the function $f(x) = x^2 + 1$, on the domain $x \geq 0$.

Solving Applications of Radical Functions

Notice that the functions from previous examples were all polynomials, and their inverses were radical functions. If we want to find the inverse of a radical function, we will need to restrict the domain of the answer because the range of the original function is limited.

How To...

Given a radical function, find the inverse.

1. Determine the range of the original function.
2. Replace $f(x)$ with y, then solve for x.
3. If necessary, restrict the domain of the inverse function to the range of the original function.

Example 5   **Finding the Inverse of a Radical Function**

Restrict the domain of the function $f(x) = \sqrt{x-4}$ and then find the inverse.

Solution   Note that the original function has range $f(x) \geq 0$. Replace $f(x)$ with y, then solve for x.

$$y = \sqrt{x-4} \qquad \text{Replace } f(x) \text{ with } y.$$
$$x = \sqrt{y-4} \qquad \text{Interchange } x \text{ and } y.$$
$$x = \sqrt{y-4} \qquad \text{Square each side.}$$
$$x^2 = y - 4 \qquad \text{Add 4.}$$
$$x^2 + 4 = y \qquad \text{Rename the function } f^{-1}(x).$$
$$f^{-1}(x) = x^2 + 4$$

Recall that the domain of this function must be limited to the range of the original function.

$$f^{-1}(x) = x^2 + 4, x \geq 0$$

Analysis   Notice in **Figure 8** that the inverse is a reflection of the original function over the line $y = x$. Because the original function has only positive outputs, the inverse function has only positive inputs.

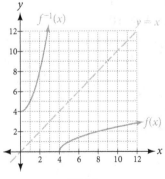

Figure 8

Try It #4

Restrict the domain and then find the inverse of the function $f(x) = \sqrt{2x+3}$.

442   CHAPTER 5 **POLYNOMIAL AND RATIONAL FUNCTIONS**

Solving Applications of Radical Functions

Radical functions are common in physical models, as we saw in the section opener. We now have enough tools to be able to solve the problem posed at the start of the section.

Example 6   **Solving an Application with a Cubic Function**

A mound of gravel is in the shape of a cone with the height equal to twice the radius. The volume of the cone in terms of the radius is given by

$$V = \frac{2}{3}\pi r^3$$

Find the inverse of the function $V = \frac{2}{3}\pi r^3$ that determines the volume V of a cone and is a function of the radius r. Then use the inverse function to calculate the radius of such a mound of gravel measuring 100 cubic feet. Use $\pi = 3.14$.

Solution   Start with the given function for V. Notice that the meaningful domain for the function is $r > 0$ since negative radii would not make sense in this context nor would a radius of 0. Also note the range of the function (hence, the domain of the inverse function) is $V \geq 0$. Solve for r in terms of V, using the method outlined previously. Note that in real-world applications, we do not swap the variables when finding inverses. Instead, we change which variable is considered to be the independent variable.

$$V = \frac{2}{3}\pi r^3$$

$$r^3 = \frac{3V}{2\pi} \qquad \text{Solve for } r^3.$$

$$r = \sqrt[3]{\frac{3V}{2\pi}} \qquad \text{Solve for } r.$$

This is the result stated in the section opener. Now evaluate this for $V = 100$ and $\pi = 3.14$.

$$r = \sqrt[3]{\frac{3V}{2\pi}}$$

$$= \sqrt[3]{\frac{3 \cdot 100}{2 \cdot 3.14}}$$

$$\approx \sqrt[3]{47.7707}$$

$$\approx 3.63$$

Therefore, the radius is about 3.63 ft.

Determining the Domain of a Radical Function Composed with Other Functions

When radical functions are composed with other functions, determining domain can become more complicated.

Example 7   **Finding the Domain of a Radical Function Composed with a Rational Function**

Find the domain of the function $f(x) = \sqrt{\frac{(x+2)(x-3)}{(x-1)}}$.

Solution   Because a square root is only defined when the quantity under the radical is non-negative, we need to determine where $\frac{(x+2)(x-3)}{(x-1)} \geq 0$. The output of a rational function can change signs (change from positive to negative or vice versa) at x-intercepts and at vertical asymptotes. For this equation, the graph could change signs at $x = -2$, 1, and 3.

To determine the intervals on which the rational expression is positive, we could test some values in the expression or sketch a graph. While both approaches work equally well, for this example we will use a graph as shown in **Figure 9**.

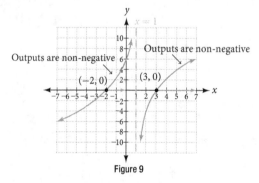

Figure 9

This function has two x-intercepts, both of which exhibit linear behavior near the x-intercepts. There is one vertical asymptote, corresponding to a linear factor; this behavior is similar to the basic reciprocal toolkit function, and there is no horizontal asymptote because the degree of the numerator is larger than the degree of the denominator. There is a y-intercept at $(0, 6)$.

From the y-intercept and x-intercept at $x = -2$, we can sketch the left side of the graph. From the behavior at the asymptote, we can sketch the right side of the graph.

From the graph, we can now tell on which intervals the outputs will be non-negative, so that we can be sure that the original function $f(x)$ will be defined. $f(x)$ has domain $-2 \leq x < 1$ or $x \geq 3$, or in interval notation, $[-2, 1) \cup [3, \infty)$.

Finding Inverses of Rational Functions

As with finding inverses of quadratic functions, it is sometimes desirable to find the inverse of a rational function, particularly of rational functions that are the ratio of linear functions, such as in concentration applications.

Example 8   **Finding the Inverse of a Rational Function**

The function $C = \dfrac{20 + 0.4n}{100 + n}$ represents the concentration C of an acid solution after n mL of 40% solution has been added to 100 mL of a 20% solution. First, find the inverse of the function; that is, find an expression for n in terms of C. Then use your result to determine how much of the 40% solution should be added so that the final mixture is a 35% solution.

Solution   We first want the inverse of the function in order to determine how many mL we need for a given concentration. We will solve for n in terms of C.

$$C = \frac{20 + 0.4n}{100 + n}$$

$$C(100 + n) = 20 + 0.4n$$

$$100C + Cn = 20 + 0.4n$$

$$100C - 20 = 0.4n - Cn$$

$$100C - 20 = (0.4 - C)n$$

$$n = \frac{100C - 20}{0.4 - C}$$

Now evaluate this function at 35%, which is $C = 0.35$.

$$n = \frac{100(0.35) - 20}{0.4 - 0.35}$$

$$= \frac{15}{0.05}$$

$$= 300$$

We can conclude that 300 mL of the 40% solution should be added.

Try It #5

Find the inverse of the function $f(x) = \dfrac{x + 3}{x - 2}$.

Access these online resources for additional instruction and practice with inverses and radical functions.

- Graphing the Basic Square Root Function (http://openstaxcollege.org/l/graphsquareroot)
- Find the Inverse of a Square Root Function (http://openstaxcollege.org/l/inversesquare)
- Find the Inverse of a Rational Function (http://openstaxcollege.org/l/inverserational)
- Find the Inverse of a Rational Function and an Inverse Function Value (http://openstaxcollege.org/l/rationalinverse)
- Inverse Functions (http://openstaxcollege.org/l/inversefunction)

5.7 SECTION EXERCISES

VERBAL

1. Explain why we cannot find inverse functions for all polynomial functions.

2. Why must we restrict the domain of a quadratic function when finding its inverse?

3. When finding the inverse of a radical function, what restriction will we need to make?

4. The inverse of a quadratic function will always take what form?

ALGEBRAIC

For the following exercises, find the inverse of the function on the given domain.

5. $f(x) = (x-4)^2$, $[4, \infty)$

6. $f(x) = (x+2)^2$, $[-2, \infty)$

7. $f(x) = (x+1)^2 - 3$, $[-1, \infty)$

8. $f(x) = 3x^2 + 5$, $(-\infty, 0]$

9. $f(x) = 12 - x^2$, $[0, \infty)$

10. $f(x) = 9 - x^2$, $[0, \infty)$

11. $f(x) = 2x^2 + 4$, $[0, \infty)$

For the following exercises, find the inverse of the functions.

12. $f(x) = x^3 + 5$

13. $f(x) = 3x^3 + 1$

14. $f(x) = 4 - x^3$

15. $f(x) = 4 - 2x^3$

For the following exercises, find the inverse of the functions.

16. $f(x) = \sqrt{2x+1}$

17. $f(x) = \sqrt{3-4x}$

18. $f(x) = 9 + \sqrt{4x-4}$

19. $f(x) = \sqrt{6x-8} + 5$

20. $f(x) = 9 + 2\sqrt[3]{x}$

21. $f(x) = 3 - \sqrt[3]{x}$

22. $f(x) = \dfrac{2}{x+8}$

23. $f(x) = \dfrac{3}{x-4}$

24. $f(x) = \dfrac{x+3}{x+7}$

25. $f(x) = \dfrac{x-2}{x+7}$

26. $f(x) = \dfrac{3x+4}{5-4x}$

27. $f(x) = \dfrac{5x+1}{2-5x}$

28. $f(x) = x^2 + 2x$, $[-1, \infty)$

29. $f(x) = x^2 + 4x + 1$, $[-2, \infty)$

30. $f(x) = x^2 - 6x + 3$, $[3, \infty)$

GRAPHICAL

For the following exercises, find the inverse of the function and graph both the function and its inverse.

31. $f(x) = x^2 + 2$, $x \geq 0$

32. $f(x) = 4 - x^2$, $x \geq 0$

33. $f(x) = (x+3)^2$, $x \geq -3$

34. $f(x) = (x-4)^2$, $x \geq 4$

35. $f(x) = x^3 + 3$

36. $f(x) = 1 - x^3$

37. $f(x) = x^2 + 4x$, $x \geq -2$

38. $f(x) = x^2 - 6x + 1$, $x \geq 3$

39. $f(x) = \dfrac{2}{x}$

40. $f(x) = \dfrac{1}{x^2}$, $x \geq 0$

For the following exercises, use a graph to help determine the domain of the functions.

41. $f(x) = \sqrt{\dfrac{(x+1)(x-1)}{x}}$

42. $f(x) = \sqrt{\dfrac{(x+2)(x-3)}{x-1}}$

43. $f(x) = \sqrt{\dfrac{x(x+3)}{x-4}}$

44. $f(x) = \sqrt{\dfrac{x^2 - x - 20}{x-2}}$

45. $f(x) = \sqrt{\dfrac{9-x^2}{x+4}}$

TECHNOLOGY

For the following exercises, use a calculator to graph the function. Then, using the graph, give three points on the graph of the inverse with y-coordinates given.

46. $f(x) = x^3 - x - 2, y = 1, 2, 3$

47. $f(x) = x^3 + x - 2, y = 0, 1, 2$

48. $f(x) = x^3 + 3x - 4, y = 0, 1, 2$

49. $f(x) = x^3 + 8x - 4, y = -1, 0, 1$

50. $f(x) = x^4 + 5x + 1, y = -1, 0, 1$

EXTENSIONS

For the following exercises, find the inverse of the functions with a, b, c positive real numbers.

51. $f(x) = ax^3 + b$

52. $f(x) = x^2 + bx$

53. $f(x) = \sqrt{ax^2 + b}$

54. $f(x) = \sqrt[3]{ax + b}$

55. $f(x) = \dfrac{ax + b}{x + c}$

REAL-WORLD APPLICATIONS

For the following exercises, determine the function described and then use it to answer the question.

56. An object dropped from a height of 200 meters has a height, $h(t)$, in meters after t seconds have lapsed, such that $h(t) = 200 - 4.9t^2$. Express t as a function of height, h, and find the time to reach a height of 50 meters.

57. An object dropped from a height of 600 feet has a height, $h(t)$, in feet after t seconds have elapsed, such that $h(t) = 600 - 16t^2$. Express t as a function of height h, and find the time to reach a height of 400 feet.

58. The volume, V, of a sphere in terms of its radius, r, is given by $V(r) = \dfrac{4}{3}\pi r^3$. Express r as a function of V, and find the radius of a sphere with volume of 200 cubic feet.

59. The surface area, A, of a sphere in terms of its radius, r, is given by $A(r) = 4\pi r^2$. Express r as a function of V, and find the radius of a sphere with a surface area of 1000 square inches.

60. A container holds 100 ml of a solution that is 25 ml acid. If n ml of a solution that is 60% acid is added, the function $C(n) = \dfrac{25 + 0.6n}{100 + n}$ gives the concentration, C, as a function of the number of ml added, n. Express n as a function of C and determine the number of ml that need to be added to have a solution that is 50% acid.

61. The period T, in seconds, of a simple pendulum as a function of its length l, in feet, is given by $T(l) = 2\pi \sqrt{\dfrac{l}{32.2}}$. Express l as a function of T and determine the length of a pendulum with period of 2 seconds.

62. The volume of a cylinder, V, in terms of radius, r, and height, h, is given by $V = \pi r^2 h$. If a cylinder has a height of 6 meters, express the radius as a function of V and find the radius of a cylinder with volume of 300 cubic meters.

63. The surface area, A, of a cylinder in terms of its radius, r, and height, h, is given by $A = 2\pi r^2 + 2\pi rh$. If the height of the cylinder is 4 feet, express the radius as a function of V and find the radius if the surface area is 200 square feet.

64. The volume of a right circular cone, V, in terms of its radius, r, and its height, h, is given by $V = \dfrac{1}{3}\pi r^2 h$. Express r in terms of h if the height of the cone is 12 feet and find the radius of a cone with volume of 50 cubic inches.

65. Consider a cone with height of 30 feet. Express the radius, r, in terms of the volume, V, and find the radius of a cone with volume of 1000 cubic feet.

LEARNING OBJECTIVES

In this section, you will:

- Solve direct variation problems.
- Solve inverse variation problems.
- Solve problems involving joint variation.

5.8   MODELING USING VARIATION

A used-car company has just offered their best candidate, Nicole, a position in sales. The position offers 16% commission on her sales. Her earnings depend on the amount of her sales. For instance, if she sells a vehicle for $4,600, she will earn $736. She wants to evaluate the offer, but she is not sure how. In this section, we will look at relationships, such as this one, between earnings, sales, and commission rate.

Solving Direct Variation Problems

In the example above, Nicole's earnings can be found by multiplying her sales by her commission. The formula $e = 0.16s$ tells us her earnings, e, come from the product of 0.16, her commission, and the sale price of the vehicle. If we create a table, we observe that as the sales price increases, the earnings increase as well, which should be intuitive. See **Table 1**.

s, sales price	$e = 0.16s$	Interpretation
$4,600	$e = 0.16(4,600) = 736$	A sale of a $4,600 vehicle results in $736 earnings.
$9,200	$e = 0.16(9,200) = 1,472$	A sale of a $9,200 vehicle results in $1472 earnings.
$18,400	$e = 0.16(18,400) = 2,944$	A sale of a $18,400 vehicle results in $2944 earnings.

Table 1

Notice that earnings are a multiple of sales. As sales increase, earnings increase in a predictable way. Double the sales of the vehicle from $4,600 to $9,200, and we double the earnings from $736 to $1,472. As the input increases, the output increases as a multiple of the input. A relationship in which one quantity is a constant multiplied by another quantity is called **direct variation**. Each variable in this type of relationship **varies directly** with the other.

Figure 1 represents the data for Nicole's potential earnings. We say that earnings vary directly with the sales price of the car. The formula $y = kx^n$ is used for direct variation. The value k is a nonzero constant greater than zero and is called the **constant of variation**. In this case, $k = 0.16$ and $n = 1$. We saw functions like this one when we discussed power functions.

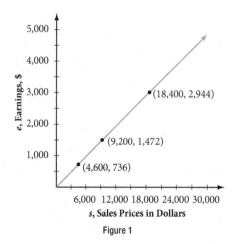

Figure 1

---

direct variation

If x and y are related by an equation of the form

$$y = kx^n$$

then we say that the relationship is **direct variation** and y **varies directly** with, or is proportional to, the nth power of x. In direct variation relationships, there is a nonzero constant ratio $k = \dfrac{y}{x^n}$, where k is called the **constant of variation**, which help defines the relationship between the variables.

How To…

Given a description of a direct variation problem, solve for an unknown.

1. Identify the input, x, and the output, y.
2. Determine the constant of variation. You may need to divide y by the specified power of x to determine the constant of variation.
3. Use the constant of variation to write an equation for the relationship.
4. Substitute known values into the equation to find the unknown.

Example 1   **Solving a Direct Variation Problem**

The quantity y varies directly with the cube of x. If $y = 25$ when $x = 2$, find y when x is 6.

Solution   The general formula for direct variation with a cube is $y = kx^3$. The constant can be found by dividing y by the cube of x.

$$k = \frac{y}{x^3}$$

$$= \frac{25}{2^3}$$

$$= \frac{25}{8}$$

Now use the constant to write an equation that represents this relationship.

$$y = \frac{25}{8}x^3$$

Substitute $x = 6$ and solve for y.

$$y = \frac{25}{8}(6)^3$$

$$= 675$$

Analysis   *The graph of this equation is a simple cubic, as shown in* **Figure 2**.

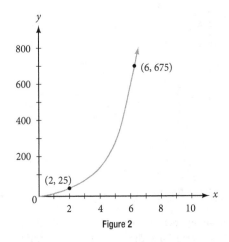

Figure 2

Q & A...

Do the graphs of all direct variation equations look like Example 1?

No. Direct variation equations are power functions—they may be linear, quadratic, cubic, quartic, radical, etc. But all of the graphs pass through (0,0).

Try It #1

The quantity y varies directly with the square of x. If $y = 24$ when $x = 3$, find y when x is 4.

Solving Inverse Variation Problems

Water temperature in an ocean varies inversely to the water's depth. Between the depths of 250 feet and 500 feet, the formula $T = \dfrac{14,000}{d}$ gives us the temperature in degrees Fahrenheit at a depth in feet below Earth's surface. Consider the Atlantic Ocean, which covers 22% of Earth's surface. At a certain location, at the depth of 500 feet, the temperature may be 28°F. If we create **Table 2**, we observe that, as the depth increases, the water temperature decreases.

d, depth	$T = \dfrac{14,000}{d}$	Interpretation
500 ft	$\dfrac{14,000}{500} = 28$	At a depth of 500 ft, the water temperature is 28° F.
1,000 ft	$\dfrac{14,000}{1,000} = 14$	At a depth of 1,000 ft, the water temperature is 14° F.
2,000 ft	$\dfrac{14,000}{2,000} = 7$	At a depth of 2,000 ft, the water temperature is 7° F.

Table 2

We notice in the relationship between these variables that, as one quantity increases, the other decreases. The two quantities are said to be **inversely proportional** and each term **varies inversely** with the other. Inversely proportional relationships are also called **inverse variations**.

For our example, **Figure 3** depicts the inverse variation. We say the water temperature varies inversely with the depth of the water because, as the depth increases, the temperature decreases. The formula $y = \dfrac{k}{x}$ for inverse variation in this case uses $k = 14,000$.

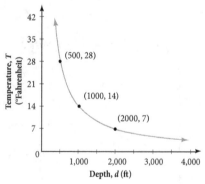

Figure 3

---

inverse variation

If x and y are related by an equation of the form

$$y = \frac{k}{x^n}$$

where k is a nonzero constant, then we say that y **varies inversely** with the nth power of x. In **inversely proportional** relationships, or **inverse variations**, there is a constant multiple $k = x^n y$.

Example 2 Writing a Formula for an Inversely Proportional Relationship

A tourist plans to drive 100 miles. Find a formula for the time the trip will take as a function of the speed the tourist drives.

Solution Recall that multiplying speed by time gives distance. If we let t represent the drive time in hours, and v represent the velocity (speed or rate) at which the tourist drives, then $vt =$ distance. Because the distance is fixed at 100 miles, $vt = 100$ so $t = \dfrac{100}{v}$. Because time is a function of velocity, we can write $t(v)$.

$$t(v) = \frac{100}{v}$$
$$= 100v^{-1}$$

We can see that the constant of variation is 100 and, although we can write the relationship using the negative exponent, it is more common to see it written as a fraction. We say that the time varies inversely with velocity.

How To...

Given a description of an indirect variation problem, solve for an unknown.

1. Identify the input, x, and the output, y.
2. Determine the constant of variation. You may need to multiply y by the specified power of x to determine the constant of variation.
3. Use the constant of variation to write an equation for the relationship.
4. Substitute known values into the equation to find the unknown.

Example 3 Solving an Inverse Variation Problem

A quantity y varies inversely with the cube of x. If $y = 25$ when $x = 2$, find y when x is 6.

Solution The general formula for inverse variation with a cube is $y = \dfrac{k}{x^3}$. The constant can be found by multiplying y by the cube of x.

$$k = x^3 y$$
$$= 2^3 \cdot 25$$
$$= 200$$

Now we use the constant to write an equation that represents this relationship.

$$y = \frac{k}{x^3}, k = 200$$
$$y = \frac{200}{x^3}$$

Substitute $x = 6$ and solve for y.

$$y = \frac{200}{6^3}$$
$$= \frac{25}{27}$$

Analysis The graph of this equation is a rational function, as shown in **Figure 4**.

Figure 4

Try It #2

A quantity y varies inversely with the square of x. If $y = 8$ when $x = 3$, find y when x is 4.

Solving Problems Involving Joint Variation

Many situations are more complicated than a basic direct variation or inverse variation model. One variable often depends on multiple other variables. When a variable is dependent on the product or quotient of two or more variables, this is called **joint variation**. For example, the cost of busing students for each school trip varies with the number of students attending and the distance from the school. The variable c, cost, varies jointly with the number of students, n, and the distance, d.

---

joint variation

Joint variation occurs when a variable varies directly or inversely with multiple variables.

For instance, if x varies directly with both y and z, we have $x = kyz$. If x varies directly with y and inversely with z, we have $x = \dfrac{ky}{z}$. Notice that we only use one constant in a joint variation equation.

Example 4   **Solving Problems Involving Joint Variation**

A quantity x varies directly with the square of y and inversely with the cube root of z. If $x = 6$ when $y = 2$ and $z = 8$, find x when $y = 1$ and $z = 27$.

Solution   Begin by writing an equation to show the relationship between the variables.

$$x = \frac{ky^2}{\sqrt[3]{z}}$$

Substitute $x = 6$, $y = 2$, and $z = 8$ to find the value of the constant k.

$$6 = \frac{k2^2}{\sqrt[3]{8}}$$

$$6 = \frac{4k}{2}$$

$$3 = k$$

Now we can substitute the value of the constant into the equation for the relationship.

$$x = \frac{3y^2}{\sqrt[3]{z}}$$

To find x when $y = 1$ and $z = 27$, we will substitute values for y and z into our equation.

$$x = \frac{3(1)^2}{\sqrt[3]{27}}$$

$$= 1$$

Try It #3

A quantity x varies directly with the square of y and inversely with z. If $x = 40$ when $y = 4$ and $z = 2$, find x when $y = 10$ and $z = 25$.

Access these online resources for additional instruction and practice with direct and inverse variation.

- Direct Variation (http://openstaxcollege.org/l/directvariation)
- Inverse Variation (http://openstaxcollege.org/l/inversevariatio)
- Direct and Inverse Variation (http://openstaxcollege.org/l/directinverse)

5.8 SECTION EXERCISES

VERBAL

1. What is true of the appearance of graphs that reflect a direct variation between two variables?

2. If two variables vary inversely, what will an equation representing their relationship look like?

3. Is there a limit to the number of variables that can jointly vary? Explain.

ALGEBRAIC

For the following exercises, write an equation describing the relationship of the given variables.

4. y varies directly as x and when $x = 6$, $y = 12$.

5. y varies directly as the square of x and when $x = 4$, $y = 80$.

6. y varies directly as the square root of x and when $x = 36$, $y = 24$.

7. y varies directly as the cube of x and when $x = 36$, $y = 24$.

8. y varies directly as the cube root of x and when $x = 27$, $y = 15$.

9. y varies directly as the fourth power of x and when $x = 1$, $y = 6$.

10. y varies inversely as x and when $x = 4$, $y = 2$.

11. y varies inversely as the square of x and when $x = 3$, $y = 2$.

12. y varies inversely as the cube of x and when $x = 2$, $y = 5$.

13. y varies inversely as the fourth power of x and when $x = 3$, $y = 1$.

14. y varies inversely as the square root of x and when $x = 25$, $y = 3$.

15. y varies inversely as the cube root of x and when $x = 64$, $y = 5$.

16. y varies jointly with x and z and when $x = 2$ and $z = 3$, $y = 36$.

17. y varies jointly as x, z, and w and when $x = 1$, $z = 2$, $w = 5$, then $y = 100$.

18. y varies jointly as the square of x and the square of z and when $x = 3$ and $z = 4$, then $y = 72$.

19. y varies jointly as x and the square root of z and when $x = 2$ and $z = 25$, then $y = 100$.

20. y varies jointly as the square of x the cube of z and the square root of w. When $x = 1$, $z = 2$, and $w = 36$, then $y = 48$.

21. y varies jointly as x and z and inversely as w. When $x = 3$, $z = 5$, and $w = 6$, then $y = 10$.

22. y varies jointly as the square of x and the square root of z and inversely as the cube of w. When $x = 3$, $z = 4$, and $w = 3$, then $y = 6$.

23. y varies jointly as x and z and inversely as the square root of w and the square of t. When $x = 3$, $z = 1$, $w = 25$, and $t = 2$, then $y = 6$.

NUMERIC

For the following exercises, use the given information to find the unknown value.

24. y varies directly as x. When $x = 3$, then $y = 12$. Find y when $x = 20$.

25. y varies directly as the square of x. When $x = 2$, then $y = 16$. Find y when $x = 8$.

26. y varies directly as the cube of x. When $x = 3$, then $y = 5$. Find y when $x = 4$.

27. y varies directly as the square root of x. When $x = 16$, then $y = 4$. Find y when $x = 36$.

28. y varies directly as the cube root of x. When $x = 125$, then $y = 15$. Find y when $x = 1{,}000$.

29. y varies inversely with x. When $x = 3$, then $y = 2$. Find y when $x = 1$.

30. y varies inversely with the square of x. When $x = 4$, then $y = 3$. Find y when $x = 2$.

31. y varies inversely with the cube of x. When $x = 3$, then $y = 1$. Find y when $x = 1$.

32. y varies inversely with the square root of x. When $x = 64$, then $y = 12$. Find y when $x = 36$.

33. y varies inversely with the cube root of x. When $x = 27$, then $y = 5$. Find y when $x = 125$.

34. y varies jointly as x and z. When $x = 4$ and $z = 2$, then $y = 16$. Find y when $x = 3$ and $z = 3$.

35. y varies jointly as x, z, and w. When $x = 2$, $z = 1$, and $w = 12$, then $y = 72$. Find y when $x = 1$, $z = 2$, and $w = 3$.

36. y varies jointly as x and the square of z. When $x = 2$ and $z = 4$, then $y = 144$. Find y when $x = 4$ and $z = 5$.

37. y varies jointly as the square of x and the square root of z. When $x = 2$ and $z = 9$, then $y = 24$. Find y when $x = 3$ and $z = 25$.

38. y varies jointly as x and z and inversely as w. When $x = 5$, $z = 2$, and $w = 20$, then $y = 4$. Find y when $x = 3$ and $z = 8$, and $w = 48$.

39. y varies jointly as the square of x and the cube of z and inversely as the square root of w. When $x = 2$, $z = 2$, and $w = 64$, then $y = 12$. Find y when $x = 1$, $z = 3$, and $w = 4$.

40. y varies jointly as the square of x and of z and inversely as the square root of w and of t. When $x = 2$, $z = 3$, $w = 16$, and $t = 3$, then $y = 1$. Find y when $x = 3$, $z = 2$, $w = 36$, and $t = 5$.

TECHNOLOGY

For the following exercises, use a calculator to graph the equation implied by the given variation.

41. y varies directly with the square of x and when $x = 2$, $y = 3$.

42. y varies directly as the cube of x and when $x = 2$, $y = 4$.

43. y varies directly as the square root of x and when $x = 36$, $y = 2$.

44. y varies inversely with x and when $x = 6$, $y = 2$.

45. y varies inversely as the square of x and when $x = 1$, $y = 4$.

EXTENSIONS

For the following exercises, use Kepler's Law, which states that the square of the time, T, required for a planet to orbit the Sun varies directly with the cube of the mean distance, a, that the planet is from the Sun.

46. Using the Earth's time of 1 year and mean distance of 93 million miles, find the equation relating T and a.

47. Use the result from the previous exercise to determine the time required for Mars to orbit the Sun if its mean distance is 142 million miles.

48. Using Earth's distance of 150 million kilometers, find the equation relating T and a.

49. Use the result from the previous exercise to determine the time required for Venus to orbit the Sun if its mean distance is 108 million kilometers.

50. Using Earth's distance of 1 astronomical unit (A.U.), determine the time for Saturn to orbit the Sun if its mean distance is 9.54 A.U.

REAL-WORLD APPLICATIONS

For the following exercises, use the given information to answer the questions.

51. The distance s that an object falls varies directly with the square of the time, t, of the fall. If an object falls 16 feet in one second, how long for it to fall 144 feet?

52. The velocity v of a falling object varies directly to the time, t, of the fall. If after 2 seconds, the velocity of the object is 64 feet per second, what is the velocity after 5 seconds?

53. The rate of vibration of a string under constant tension varies inversely with the length of the string. If a string is 24 inches long and vibrates 128 times per second, what is the length of a string that vibrates 64 times per second?

54. The volume of a gas held at constant temperature varies indirectly as the pressure of the gas. If the volume of a gas is 1200 cubic centimeters when the pressure is 200 millimeters of mercury, what is the volume when the pressure is 300 millimeters of mercury?

55. The weight of an object above the surface of the Earth varies inversely with the square of the distance from the center of the Earth. If a body weighs 50 pounds when it is 3960 miles from Earth's center, what would it weigh it were 3970 miles from Earth's center?

56. The intensity of light measured in foot-candles varies inversely with the square of the distance from the light source. Suppose the intensity of a light bulb is 0.08 foot-candles at a distance of 3 meters. Find the intensity level at 8 meters.

57. The current in a circuit varies inversely with its resistance measured in ohms. When the current in a circuit is 40 amperes, the resistance is 10 ohms. Find the current if the resistance is 12 ohms.

58. The force exerted by the wind on a plane surface varies jointly with the square of the velocity of the wind and with the area of the plane surface. If the area of the surface is 40 square feet surface and the wind velocity is 20 miles per hour, the resulting force is 15 pounds. Find the force on a surface of 65 square feet with a velocity of 30 miles per hour.

59. The horsepower (hp) that a shaft can safely transmit varies jointly with its speed (in revolutions per minute (rpm)) and the cube of the diameter. If the shaft of a certain material 3 inches in diameter can transmit 45 hp at 100 rpm, what must the diameter be in order to transmit 60 hp at 150 rpm?

60. The kinetic energy K of a moving object varies jointly with its mass m and the square of its velocity v. If an object weighing 40 kilograms with a velocity of 15 meters per second has a kinetic energy of 1000 joules, find the kinetic energy if the velocity is increased to 20 meters per second.

CHAPTER 5   REVIEW

Key Terms

arrow notation a way to represent the local and end behavior of a function by using arrows to indicate that an input or output approaches a value

axis of symmetry a vertical line drawn through the vertex of a parabola, that opens up or down, around which the parabola is symmetric; it is defined by $x = -\dfrac{b}{2a}$.

coefficient a nonzero real number multiplied by a variable raised to an exponent

constant of variation the non-zero value k that helps define the relationship between variables in direct or inverse variation

continuous function a function whose graph can be drawn without lifting the pen from the paper because there are no breaks in the graph

degree the highest power of the variable that occurs in a polynomial

Descartes' Rule of Signs a rule that determines the maximum possible numbers of positive and negative real zeros based on the number of sign changes of $f(x)$ and $f(-x)$

direct variation the relationship between two variables that are a constant multiple of each other; as one quantity increases, so does the other

Division Algorithm given a polynomial dividend $f(x)$ and a non-zero polynomial divisor $d(x)$ where the degree of $d(x)$ is less than or equal to the degree of $f(x)$, there exist unique polynomials $q(x)$ and $r(x)$ such that $f(x) = d(x)\,q(x) + r(x)$ where $q(x)$ is the quotient and $r(x)$ is the remainder. The remainder is either equal to zero or has degree strictly less than $d(x)$.

end behavior the behavior of the graph of a function as the input decreases without bound and increases without bound

Factor Theorem k is a zero of polynomial function $f(x)$ if and only if $(x - k)$ is a factor of $f(x)$

Fundamental Theorem of Algebra a polynomial function with degree greater than 0 has at least one complex zero

general form of a quadratic function the function that describes a parabola, written in the form $f(x) = ax^2 + bx + c$, where a, b, and c are real numbers and $a \neq 0$.

global maximum highest turning point on a graph; $f(a)$ where $f(a) \geq f(x)$ for all x.

global minimum lowest turning point on a graph; $f(a)$ where $f(a) \leq f(x)$ for all x.

horizontal asymptote a horizontal line $y = b$ where the graph approaches the line as the inputs increase or decrease without bound.

Intermediate Value Theorem for two numbers a and b in the domain of f, if $a < b$ and $f(a) \neq f(b)$, then the function f takes on every value between $f(a)$ and $f(b)$; specifically, when a polynomial function changes from a negative value to a positive value, the function must cross the x-axis

inverse variation the relationship between two variables in which the product of the variables is a constant

inversely proportional a relationship where one quantity is a constant divided by the other quantity; as one quantity increases, the other decreases

invertible function any function that has an inverse function

imaginary number a number in the form bi where $i = \sqrt{-1}$

joint variation a relationship where a variable varies directly or inversely with multiple variables

leading coefficient the coefficient of the leading term

leading term the term containing the highest power of the variable

Linear Factorization Theorem allowing for multiplicities, a polynomial function will have the same number of factors as its degree, and each factor will be in the form $(x - c)$, where c is a complex number

multiplicity the number of times a given factor appears in the factored form of the equation of a polynomial; if a polynomial contains a factor of the form $(x - h)^p$, $x = h$ is a zero of multiplicity p.

polynomial function a function that consists of either zero or the sum of a finite number of non-zero terms, each of which is a product of a number, called the coefficient of the term, and a variable raised to a non-negative integer power.

power function a function that can be represented in the form $f(x) = kx^p$ where k is a constant, the base is a variable, and the exponent, p, is a constant

rational function a function that can be written as the ratio of two polynomials

Rational Zero Theorem the possible rational zeros of a polynomial function have the form $\frac{p}{q}$ where p is a factor of the constant term and q is a factor of the leading coefficient.

Remainder Theorem if a polynomial $f(x)$ is divided by $x - k$, then the remainder is equal to the value $f(k)$

removable discontinuity a single point at which a function is undefined that, if filled in, would make the function continuous; it appears as a hole on the graph of a function

roots in a given function, the values of x at which $y = 0$, also called zeros

smooth curve a graph with no sharp corners

standard form of a quadratic function the function that describes a parabola, written in the form $f(x) = a(x - h)^2 + k$, where (h, k) is the vertex.

synthetic division a shortcut method that can be used to divide a polynomial by a binomial of the form $x - k$

term of a polynomial function any $a_i x^i$ of a polynomial function in the form $f(x) = a_n x^n + ... + a_2 x^2 + a_1 x + a_0$

turning point the location at which the graph of a function changes direction

varies directly a relationship where one quantity is a constant multiplied by the other quantity

varies inversely a relationship where one quantity is a constant divided by the other quantity

vertex the point at which a parabola changes direction, corresponding to the minimum or maximum value of the quadratic function

vertex form of a quadratic function another name for the standard form of a quadratic function

vertical asymptote a vertical line $x = a$ where the graph tends toward positive or negative infinity as the inputs approach a

zeros in a given function, the values of x at which $y = 0$, also called roots

Key Equations

general form of a quadratic function	$f(x) = ax^2 + bx + c$
standard form of a quadratic function	$f(x) = a(x - h)^2 + k$
general form of a polynomial function	$f(x) = a_n x^n + ... + a_2 x^2 + a_1 x + a_0$
Division Algorithm	$f(x) = d(x)q(x) + r(x)$ where $q(x) \neq 0$
Rational Function	$f(x) = \dfrac{P(x)}{Q(x)} = \dfrac{a_p x^p + a_{p-1} x^{p-1} + ... + a_1 x + a_0}{b_q x^q + b_{q-1} x^{q-1} + ... + b_1 x + b_0}, Q(x) \neq 0$
Direct variation	$y = kx^n$, k is a nonzero constant.
Inverse variation	$y = \dfrac{k}{x^n}$, k is a nonzero constant.

Key Concepts

5.1 Quadratic Functions

- A polynomial function of degree two is called a quadratic function.
- The graph of a quadratic function is a parabola. A parabola is a U-shaped curve that can open either up or down.
- The axis of symmetry is the vertical line passing through the vertex. The zeros, or x-intercepts, are the points at which the parabola crosses the x-axis. The y-intercept is the point at which the parabola crosses the y-axis. See **Example 1**, **Example 7**, and **Example 8**.
- Quadratic functions are often written in general form. Standard or vertex form is useful to easily identify the vertex of a parabola. Either form can be written from a graph. See **Example 2**.
- The vertex can be found from an equation representing a quadratic function. See **Example 3**.
- The domain of a quadratic function is all real numbers. The range varies with the function. See **Example 4**.
- A quadratic function's minimum or maximum value is given by the y-value of the vertex.
- The minimum or maximum value of a quadratic function can be used to determine the range of the function and to solve many kinds of real-world problems, including problems involving area and revenue. See **Example 5** and **Example 6**.
- The vertex and the intercepts can be identified and interpreted to solve real-world problems. See **Example 9**.

5.2 Power Functions and Polynomial Functions

- A power function is a variable base raised to a number power. See **Example 1**.
- The behavior of a graph as the input decreases beyond bound and increases beyond bound is called the end behavior.
- The end behavior depends on whether the power is even or odd. See **Example 2** and **Example 3**.
- A polynomial function is the sum of terms, each of which consists of a transformed power function with positive whole number power. See **Example 4**.
- The degree of a polynomial function is the highest power of the variable that occurs in a polynomial. The term containing the highest power of the variable is called the leading term. The coefficient of the leading term is called the leading coefficient. See **Example 5**.
- The end behavior of a polynomial function is the same as the end behavior of the power function represented by the leading term of the function. See **Example 6** and **Example 7**.
- A polynomial of degree n will have at most n x-intercepts and at most $n - 1$ turning points. See **Example 8**, **Example 9**, **Example 10**, **Example 11**, and **Example 12**.

5.3 Graphs of Polynomial Functions

- Polynomial functions of degree 2 or more are smooth, continuous functions. See **Example 1**.
- To find the zeros of a polynomial function, if it can be factored, factor the function and set each factor equal to zero. See **Example 2**, **Example 3**, and **Example 4**.
- Another way to find the x-intercepts of a polynomial function is to graph the function and identify the points at which the graph crosses the x-axis. See **Example 5**.
- The multiplicity of a zero determines how the graph behaves at the x-intercepts. See **Example 6**.
- The graph of a polynomial will cross the horizontal axis at a zero with odd multiplicity.
- The graph of a polynomial will touch the horizontal axis at a zero with even multiplicity.
- The end behavior of a polynomial function depends on the leading term.
- The graph of a polynomial function changes direction at its turning points.
- A polynomial function of degree n has at most $n - 1$ turning points. See **Example 7**.
- To graph polynomial functions, find the zeros and their multiplicities, determine the end behavior, and ensure that the final graph has at most $n - 1$ turning points. See **Example 8** and **Example 10**.
- Graphing a polynomial function helps to estimate local and global extremas. See **Example 11**.

- The Intermediate Value Theorem tells us that if $f(a)$ and $f(b)$ have opposite signs, then there exists at least one value c between a and b for which $f(c) = 0$. See **Example 9**.

5.4 Dividing Polynomials

- Polynomial long division can be used to divide a polynomial by any polynomial with equal or lower degree. See **Example 1** and **Example 2**.

- The Division Algorithm tells us that a polynomial dividend can be written as the product of the divisor and the quotient added to the remainder.

- Synthetic division is a shortcut that can be used to divide a polynomial by a binomial in the form $x - k$. See **Example 3**, **Example 4**, and **Example 5**.

- Polynomial division can be used to solve application problems, including area and volume. See **Example 6**.

5.5 Zeros of Polynomial Functions

- To find $f(k)$, determine the remainder of the polynomial $f(x)$ when it is divided by $x - k$. this is known as the Remainder Theorem. See **Example 1**.

- According to the Factor Theorem, k is a zero of $f(x)$ if and only if $(x - k)$ is a factor of $f(x)$. See **Example 2**.

- According to the Rational Zero Theorem, each rational zero of a polynomial function with integer coefficients will be equal to a factor of the constant term divided by a factor of the leading coefficient. See **Example 3** and **Example 4**.

- When the leading coefficient is 1, the possible rational zeros are the factors of the constant term.

- Synthetic division can be used to find the zeros of a polynomial function. See **Example 5**.

- According to the Fundamental Theorem, every polynomial function has at least one complex zero. See **Example 6**.

- Every polynomial function with degree greater than 0 has at least one complex zero.

- Allowing for multiplicities, a polynomial function will have the same number of factors as its degree. Each factor will be in the form $(x - c)$, where c is a complex number. See **Example 7**.

- The number of positive real zeros of a polynomial function is either the number of sign changes of the function or less than the number of sign changes by an even integer.

- The number of negative real zeros of a polynomial function is either the number of sign changes of $f(-x)$ or less than the number of sign changes by an even integer. See **Example 8**.

- Polynomial equations model many real-world scenarios. Solving the equations is easiest done by synthetic division. See **Example 9**.

5.6 Rational Functions

- We can use arrow notation to describe local behavior and end behavior of the toolkit functions $f(x) = \frac{1}{x}$ and $f(x) = \frac{1}{x^2}$. See **Example 1**.

- A function that levels off at a horizontal value has a horizontal asymptote. A function can have more than one vertical asymptote. See **Example 2**.

- Application problems involving rates and concentrations often involve rational functions. See **Example 3**.

- The domain of a rational function includes all real numbers except those that cause the denominator to equal zero. See **Example 4**.

- The vertical asymptotes of a rational function will occur where the denominator of the function is equal to zero and the numerator is not zero. See **Example 5**.

- A removable discontinuity might occur in the graph of a rational function if an input causes both numerator and denominator to be zero. See **Example 6**.

- A rational function's end behavior will mirror that of the ratio of the leading terms of the numerator and denominator functions. See **Example 7**, **Example 8**, **Example 9**, and **Example 10**.

- Graph rational functions by finding the intercepts, behavior at the intercepts and asymptotes, and end behavior. See **Example 11**.

- If a rational function has x-intercepts at $x = x_1, x_2, ..., x_n$, vertical asymptotes at $x = v_1, v_2, ..., v_m$, and no $x_i =$ any v_j, then the function can be written in the form

$$f(x) = a\frac{(x - x_1)^{p_1}(x - x_2)^{p_2}...(x - x_n)^{p_n}}{(x - v_1)^{q_1}(x - v_2)^{q_2}...(x - v_m)^{q_n}}$$

See **Example 12**.

5.7 Inverses and Radical Functions

- The inverse of a quadratic function is a square root function.

- If f^{-1} is the inverse of a function f, then f is the inverse of the function f^{-1}. See **Example 1**.

- While it is not possible to find an inverse of most polynomial functions, some basic polynomials are invertible. See **Example 2**.

- To find the inverse of certain functions, we must restrict the function to a domain on which it will be one-to-one. See **Example 3** and **Example 4**.

- When finding the inverse of a radical function, we need a restriction on the domain of the answer. See **Example 5** and **Example 7**.

- Inverse and radical and functions can be used to solve application problems. See **Example 6** and **Example 8**.

5.8 Modeling Using Variation

- A relationship where one quantity is a constant multiplied by another quantity is called direct variation. See **Example 1**.

- Two variables that are directly proportional to one another will have a constant ratio.

- A relationship where one quantity is a constant divided by another quantity is called inverse variation. See **Example 2**.

- Two variables that are inversely proportional to one another will have a constant multiple. See **Example 3**.

- In many problems, a variable varies directly or inversely with multiple variables. We call this type of relationship joint variation. See **Example 4**.

CHAPTER 5 REVIEW EXERCISES

QUADRATIC FUNCTIONS

For the following exercises, write the quadratic function in standard form. Then, give the vertex and axes intercepts. Finally, graph the function.

1. $f(x) = x^2 - 4x - 5$

2. $f(x) = -2x^2 - 4x$

For the following problems, find the equation of the quadratic function using the given information.

3. The vertex is $(-2, 3)$ and a point on the graph is $(3, 6)$.

4. The vertex is $(-3, 6.5)$ and a point on the graph is $(2, 6)$.

For the following exercises, complete the task.

5. A rectangular plot of land is to be enclosed by fencing. One side is along a river and so needs no fence. If the total fencing available is 600 meters, find the dimensions of the plot to have maximum area.

6. An object projected from the ground at a 45 degree angle with initial velocity of 120 feet per second has height, h, in terms of horizontal distance traveled, x, given by $h(x) = \dfrac{-32}{(120)^2} x^2 + x$. Find the maximum height the object attains.

POWER FUNCTIONS AND POLYNOMIAL FUNCTIONS

For the following exercises, determine if the function is a polynomial function and, if so, give the degree and leading coefficient.

7. $f(x) = 4x^5 - 3x^3 + 2x - 1$

8. $f(x) = 5^{x+1} - x^2$

9. $f(x) = x^2(3 - 6x + x^2)$

For the following exercises, determine end behavior of the polynomial function.

10. $f(x) = 2x^4 + 3x^3 - 5x^2 + 7$

11. $f(x) = 4x^3 - 6x^2 + 2$

12. $f(x) = 2x^2(1 + 3x - x^2)$

GRAPHS OF POLYNOMIAL FUNCTIONS

For the following exercises, find all zeros of the polynomial function, noting multiplicities.

13. $f(x) = (x + 3)^2(2x - 1)(x + 1)^3$

14. $f(x) = x^5 + 4x^4 + 4x^3$

15. $f(x) = x^3 - 4x^2 + x - 4$

For the following exercises, based on the given graph, determine the zeros of the function and note multiplicity.

16.

17.

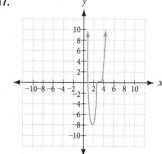

18. Use the Intermediate Value Theorem to show that at least one zero lies between 2 and 3 for the function
$f(x) = x^3 - 5x + 1$

DIVIDING POLYNOMIALS

For the following exercises, use long division to find the quotient and remainder.

19. $\dfrac{x^3 - 2x^2 + 4x + 4}{x - 2}$

20. $\dfrac{3x^4 - 4x^2 + 4x + 8}{x + 1}$

For the following exercises, use synthetic division to find the quotient. If the divisor is a factor, then write the factored form.

21. $\dfrac{x^3 - 2x^2 + 5x - 1}{x + 3}$

22. $\dfrac{x^3 + 4x + 10}{x - 3}$

23. $\dfrac{2x^3 + 6x^2 - 11x - 12}{x + 4}$

24. $\dfrac{3x^4 + 3x^3 + 2x + 2}{x + 1}$

ZEROS OF POLYNOMIAL FUNCTIONS

For the following exercises, use the Rational Zero Theorem to help you solve the polynomial equation.

25. $2x^3 - 3x^2 - 18x - 8 = 0$

26. $3x^3 + 11x^2 + 8x - 4 = 0$

27. $2x^4 - 17x^3 + 46x^2 - 43x + 12 = 0$

28. $4x^4 + 8x^3 + 19x^2 + 32x + 12 = 0$

For the following exercises, use Descartes' Rule of Signs to find the possible number of positive and negative solutions.

29. $x^3 - 3x^2 - 2x + 4 = 0$

30. $2x^4 - x^3 + 4x^2 - 5x + 1 = 0$

RATIONAL FUNCTIONS

For the following exercises, find the intercepts and the vertical and horizontal asymptotes, and then use them to sketch a graph of the function.

31. $f(x) = \dfrac{x + 2}{x - 5}$

32. $f(x) = \dfrac{x^2 + 1}{x^2 - 4}$

33. $f(x) = \dfrac{3x^2 - 27}{x^2 + x - 2}$

34. $f(x) = \dfrac{x + 2}{x^2 - 9}$

For the following exercises, find the slant asymptote.

35. $f(x) = \dfrac{x^2 - 1}{x + 2}$

36. $f(x) = \dfrac{2x^3 - x^2 + 4}{x^2 + 1}$

INVERSES AND RADICAL FUNCTIONS

For the following exercises, find the inverse of the function with the domain given.

37. $f(x) = (x - 2)^2, x \geq 2$

38. $f(x) = (x + 4)^2 - 3, x \geq -4$

39. $f(x) = x^2 + 6x - 2, x \geq -3$

40. $f(x) = 2x^3 - 3$

41. $f(x) = \sqrt{4x + 5} - 3$

42. $f(x) = \dfrac{x - 3}{2x + 1}$

MODELING USING VARIATION

For the following exercises, find the unknown value.

43. y varies directly as the square of x. If when $x = 3$, $y = 36$, find y if $x = 4$.

44. y varies inversely as the square root of x. If when $x = 25$, $y = 2$, find y if $x = 4$.

45. y varies jointly as the cube of x and as z. If when $x = 1$ and $z = 2$, $y = 6$, find y if $x = 2$ and $z = 3$.

46. y varies jointly as x and the square of z and inversely as the cube of w. If when $x = 3$, $z = 4$, and $w = 2$, $y = 48$, find y if $x = 4$, $z = 5$, and $w = 3$.

For the following exercises, solve the application problem.

47. The weight of an object above the surface of the earth varies inversely with the distance from the center of the earth. If a person weighs 150 pounds when he is on the surface of the earth (3,960 miles from center), find the weight of the person if he is 20 miles above the surface.

48. The volume V of an ideal gas varies directly with the temperature T and inversely with the pressure P. A cylinder contains oxygen at a temperature of 310 degrees K and a pressure of 18 atmospheres in a volume of 120 liters. Find the pressure if the volume is decreased to 100 liters and the temperature is increased to 320 degrees K.

CHAPTER 5 PRACTICE TEST

Give the degree and leading coefficient of the following polynomial function.

1. $f(x) = x^3(3 - 6x^2 - 2x^2)$

Determine the end behavior of the polynomial function.

2. $f(x) = 8x^3 - 3x^2 + 2x - 4$ **3.** $f(x) = -2x^2(4 - 3x - 5x^2)$

Write the quadratic function in standard form. Determine the vertex and axes intercepts and graph the function.

4. $f(x) = x^2 + 2x - 8$

Given information about the graph of a quadratic function, find its equation.

5. Vertex (2, 0) and point on graph (4, 12).

Solve the following application problem.

6. A rectangular field is to be enclosed by fencing. In addition to the enclosing fence, another fence is to divide the field into two parts, running parallel to two sides. If 1,200 feet of fencing is available, find the maximum area that can be enclosed.

Find all zeros of the following polynomial functions, noting multiplicities.

7. $f(x) = (x - 3)^3(3x - 1)(x - 1)^2$ **8.** $f(x) = 2x^6 - 6x^5 + 18x^4$

Based on the graph, determine the zeros of the function and multiplicities.

9.

Use long division to find the quotient.

10. $\dfrac{2x^3 + 3x - 4}{x + 2}$

Use synthetic division to find the quotient. If the divisor is a factor, write the factored form.

11. $\dfrac{x^4 + 3x^2 - 4}{x - 2}$ **12.** $\dfrac{2x^3 + 5x^2 - 7x - 12}{x + 3}$

Use the Rational Zero Theorem to help you find the zeros of the polynomial functions.

13. $f(x) = 2x^3 + 5x^2 - 6x - 9$ **14.** $f(x) = 4x^4 + 8x^3 + 21x^2 + 17x + 4$

15. $f(x) = 4x^4 + 16x^3 + 13x^2 - 15x - 18$ **16.** $f(x) = x^5 + 6x^4 + 13x^3 + 14x^2 + 12x + 8$

Given the following information about a polynomial function, find the function.

17. It has a double zero at $x = 3$ and zeroes at $x = 1$ and $x = -2$. It's y-intercept is $(0, 12)$.

18. It has a zero of multiplicity 3 at $x = \frac{1}{2}$ and another zero at $x = -3$. It contains the point $(1, 8)$.

Use Descartes' Rule of Signs to determine the possible number of positive and negative solutions.

19. $8x^3 - 21x^2 + 6 = 0$

For the following rational functions, find the intercepts and horizontal and vertical asymptotes, and sketch a graph.

20. $f(x) = \dfrac{x + 4}{x^2 - 2x - 3}$ **21.** $f(x) = \dfrac{x^2 + 2x - 3}{x^2 - 4}$

Find the slant asymptote of the rational function.

22. $f(x) = \dfrac{x^2 + 3x - 3}{x - 1}$

Find the inverse of the function.

23. $f(x) = \sqrt{x - 2} + 4$ **24.** $f(x) = 3x^3 - 4$

25. $f(x) = \dfrac{2x + 3}{3x - 1}$

Find the unknown value.

26. y varies inversely as the square of x and when $x = 3$, $y = 2$. Find y if $x = 1$.

27. y varies jointly with x and the cube root of z. If when $x = 2$ and $z = 27$, $y = 12$, find y if $x = 5$ and $z = 8$.

Solve the following application problem.

28. The distance a body falls varies directly as the square of the time it falls. If an object falls 64 feet in 2 seconds, how long will it take to fall 256 feet?

Exponential and Logarithmic Functions

Figure 1 Electron micrograph of *E. Coli* bacteria (credit: "Mattosaurus," Wikimedia Commons)

CHAPTER OUTLINE

6.1 Exponential Functions

6.2 Graphs of Exponential Functions

6.3 Logarithmic Functions

6.4 Graphs of Logarithmic Functions

6.5 Logarithmic Properties

6.6 Exponential and Logarithmic Equations

6.7 Exponential and Logarithmic Models

6.8 Fitting Exponential Models to Data

Introduction

Focus in on a square centimeter of your skin. Look closer. Closer still. If you could look closely enough, you would see hundreds of thousands of microscopic organisms. They are bacteria, and they are not only on your skin, but in your mouth, nose, and even your intestines. In fact, the bacterial cells in your body at any given moment outnumber your own cells. But that is no reason to feel bad about yourself. While some bacteria can cause illness, many are healthy and even essential to the body.

Bacteria commonly reproduce through a process called binary fission, during which one bacterial cell splits into two. When conditions are right, bacteria can reproduce very quickly. Unlike humans and other complex organisms, the time required to form a new generation of bacteria is often a matter of minutes or hours, as opposed to days or years.[16]

For simplicity's sake, suppose we begin with a culture of one bacterial cell that can divide every hour. **Table 1** shows the number of bacterial cells at the end of each subsequent hour. We see that the single bacterial cell leads to over one thousand bacterial cells in just ten hours! And if we were to extrapolate the table to twenty-four hours, we would have over 16 million!

Hour	0	1	2	3	4	5	6	7	8	9	10
Bacteria	1	2	4	8	16	32	64	128	256	512	1024

Table 1

In this chapter, we will explore exponential functions, which can be used for, among other things, modeling growth patterns such as those found in bacteria. We will also investigate logarithmic functions, which are closely related to exponential functions. Both types of functions have numerous real-world applications when it comes to modeling and interpreting data.

16 Todar, PhD, Kenneth. Todar's Online Textbook of Bacteriology. http://textbookofbacteriology.net/growth_3.html.

463

LEARNING OBJECTIVES

In this section, you will:

- Evaluate exponential functions.
- Find the equation of an exponential function.
- Use compound interest formulas.
- Evaluate exponential functions with base *e*.

6.1 EXPONENTIAL FUNCTIONS

India is the second most populous country in the world with a population of about 1.25 billion people in 2013. The population is growing at a rate of about 1.2% each year[17]. If this rate continues, the population of India will exceed China's population by the year 2031. When populations grow rapidly, we often say that the growth is "exponential," meaning that something is growing very rapidly. To a mathematician, however, the term *exponential growth* has a very specific meaning. In this section, we will take a look at *exponential functions*, which model this kind of rapid growth.

Identifying Exponential Functions

When exploring linear growth, we observed a constant rate of change—a constant number by which the output increased for each unit increase in input. For example, in the equation $f(x) = 3x + 4$, the slope tells us the output increases by 3 each time the input increases by 1. The scenario in the India population example is different because we have a *percent* change per unit time (rather than a constant change) in the number of people.

Defining an Exponential Function

A study found that the percent of the population who are vegans in the United States doubled from 2009 to 2011. In 2011, 2.5% of the population was vegan, adhering to a diet that does not include any animal products—no meat, poultry, fish, dairy, or eggs. If this rate continues, vegans will make up 10% of the U.S. population in 2015, 40% in 2019, and 80% in 2050.

What exactly does it mean to *grow exponentially*? What does the word *double* have in common with *percent increase*? People toss these words around errantly. Are these words used correctly? The words certainly appear frequently in the media.

- **Percent change** refers to a *change* based on a *percent* of the original amount.
- **Exponential growth** refers to an *increase* based on a constant multiplicative rate of change over equal increments of time, that is, a *percent* increase of the original amount over time.
- **Exponential decay** refers to a *decrease* based on a constant multiplicative rate of change over equal increments of time, that is, a *percent* decrease of the original amount over time.

For us to gain a clear understanding of exponential growth, let us contrast exponential growth with linear growth. We will construct two functions. The first function is exponential. We will start with an input of 0, and increase each input by 1. We will double the corresponding consecutive outputs. The second function is linear. We will start with an input of 0, and increase each input by 1. We will add 2 to the corresponding consecutive outputs. See **Table 1.**

x	$f(x) = 2^x$	$g(x) = 2x$
0	1	0
1	2	2
2	4	4
3	8	6
4	16	8
5	32	10
6	64	12

Table 1

17 http://www.worldometers.info/world-population/. Accessed February 24, 2014.

From **Table 2** we can infer that for these two functions, exponential growth dwarfs linear growth.

- **Exponential growth** refers to the original value from the range increases by the *same percentage* over equal increments found in the domain.

- **Linear growth** refers to the original value from the range increases by the *same amount* over equal increments found in the domain.

Apparently, the difference between "the same percentage" and "the same amount" is quite significant. For exponential growth, over equal increments, the constant multiplicative rate of change resulted in doubling the output whenever the input increased by one. For linear growth, the constant additive rate of change over equal increments resulted in adding 2 to the output whenever the input was increased by one.

The general form of the exponential function is $f(x) = ab^x$, where a is any nonzero number, b is a positive real number not equal to 1.

- If $b > 1$, the function grows at a rate proportional to its size.

- If $0 < b < 1$, the function decays at a rate proportional to its size.

Let's look at the function $f(x) = 2^x$ from our example. We will create a table (**Table 2**) to determine the corresponding outputs over an interval in the domain from -3 to 3.

x	-3	-2	-1	0	1	2	3
$f(x) = 2^x$	$2^{-3} = \dfrac{1}{8}$	$2^{-2} = \dfrac{1}{4}$	$2^{-1} = \dfrac{1}{2}$	$2^0 = 1$	$2^1 = 2$	$2^2 = 4$	$2^3 = 8$

Table 2

Let us examine the graph of f by plotting the ordered pairs we observe on the table in **Figure 1**, and then make a few observations.

Figure 1

Let's define the behavior of the graph of the exponential function $f(x) = 2^x$ and highlight some its key characteristics.

- the domain is $(-\infty, \infty)$,

- the range is $(0, \infty)$,

- as $x \to \infty, f(x) \to \infty$,

- as $x \to -\infty, f(x) \to 0$,

- $f(x)$ is always increasing,

- the graph of $f(x)$ will never touch the x-axis because base two raised to any exponent never has the result of zero.

- $y = 0$ is the horizontal asymptote.

- the y-intercept is 1.

---

exponential function

For any real number x, an exponential function is a function with the form

$$f(x) = ab^x$$

where

- a is the a non-zero real number called the initial value and
- b is any positive real number such that $b \neq 1$.
- The domain of f is all real numbers.
- The range of f is all positive real numbers if $a > 0$.
- The range of f is all negative real numbers if $a < 0$.
- The y-intercept is $(0, a)$, and the horizontal asymptote is $y = 0$.

Example 1   **Identifying Exponential Functions**

Which of the following equations are *not* exponential functions?

$$f(x) = 4^{3(x-2)} \qquad g(x) = x^3 \qquad h(x) = \left(\frac{1}{3}\right)^x \qquad j(x) = (-2)^x$$

Solution   By definition, an exponential function has a constant as a base and an independent variable as an exponent. Thus, $g(x) = x^3$ does not represent an exponential function because the base is an independent variable. In fact, $g(x) = x^3$ is a power function.

Recall that the base b of an exponential function is always a positive constant, and $b \neq 1$. Thus, $j(x) = (-2)^x$ does not represent an exponential function because the base, -2, is less than 0.

Try It #1

Which of the following equations represent exponential functions?

- $f(x) = 2x^2 - 3x + 1$
- $g(x) = 0.875^x$
- $h(x) = 1.75x + 2$
- $j(x) = 1095.6^{-2x}$

Evaluating Exponential Functions

Recall that the base of an exponential function must be a positive real number other than 1. Why do we limit the base b to positive values? To ensure that the outputs will be real numbers. Observe what happens if the base is not positive:

- Let $b = -9$ and $x = \frac{1}{2}$. Then $f(x) = f\left(\frac{1}{2}\right) = (-9)^{\frac{1}{2}} = \sqrt{-9}$, which is not a real number.

Why do we limit the base to positive values other than 1? Because base 1 results in the constant function. Observe what happens if the base is 1:

- Let $b = 1$. Then $f(x) = 1^x = 1$ for any value of x.

To evaluate an exponential function with the form $f(x) = b^x$, we simply substitute x with the given value, and calculate the resulting power. For example:

Let $f(x) = 2^x$. What is $f(3)$?

$$f(x) = 2^x$$
$$f(3) = 2^3 \qquad \text{Substitute } x = 3.$$
$$= 8 \qquad \text{Evaluate the power.}$$

To evaluate an exponential function with a form other than the basic form, it is important to follow the order of operations.

For example:

Let $f(x) = 30(2)^x$. What is $f(3)$?

$$f(x) = 30(2)^x$$
$$f(3) = 30(2)^3 \qquad \text{Substitute } x = 3.$$
$$= 30(8) \qquad \text{Simplify the power first.}$$
$$= 240 \qquad \text{Multiply.}$$

Note that if the order of operations were not followed, the result would be incorrect:

$$f(3) = 30(2)^3 \neq 60^3 = 216{,}000$$

Example 2   **Evaluating Exponential Functions**

Let $f(x) = 5(3)^{x+1}$. Evaluate $f(2)$ without using a calculator.

Solution   Follow the order of operations. Be sure to pay attention to the parentheses.

$$f(x) = 5(3)^{x+1}$$
$$f(2) = 5(3)^{2+1} \qquad \text{Substitute } x = 2.$$
$$= 5(3)^3 \qquad \text{Add the exponents.}$$
$$= 5(27) \qquad \text{Simplify the power.}$$
$$= 135 \qquad \text{Multiply.}$$

Try It #2

Let $f(x) = 8(1.2)^{x-5}$. Evaluate $f(3)$ using a calculator. Round to four decimal places.

Defining Exponential Growth

Because the output of exponential functions increases very rapidly, the term "exponential growth" is often used in everyday language to describe anything that grows or increases rapidly. However, exponential growth can be defined more precisely in a mathematical sense. If the growth rate is proportional to the amount present, the function models exponential growth.

---

exponential growth

A function that models **exponential growth** grows by a rate proportional to the amount present. For any real number x and any positive real numbers a and b such that $b \neq 1$, an exponential growth function has the form

$$f(x) = ab^x$$

where

- a is the initial or starting value of the function.
- b is the growth factor or growth multiplier per unit x.

In more general terms, we have an *exponential function*, in which a constant base is raised to a variable exponent. To differentiate between linear and exponential functions, let's consider two companies, A and B. Company A has 100 stores and expands by opening 50 new stores a year, so its growth can be represented by the function $A(x) = 100 + 50x$. Company B has 100 stores and expands by increasing the number of stores by 50% each year, so its growth can be represented by the function $B(x) = 100(1 + 0.5)^x$.

A few years of growth for these companies are illustrated in **Table 3**.

Year, x	Stores, Company A	Stores, Company B
0	$100 + 50(0) = 100$	$100(1 + 0.5)^0 = 100$
1	$100 + 50(1) = 150$	$100(1 + 0.5)^1 = 150$
2	$100 + 50(2) = 200$	$100(1 + 0.5)^2 = 225$
3	$100 + 50(3) = 250$	$100(1 + 0.5)^3 = 337.5$
x	$A(x) = 100 + 50x$	$B(x) = 100(1 + 0.5)^x$

Table 3

The graphs comparing the number of stores for each company over a five-year period are shown in **Figure 2**. We can see that, with exponential growth, the number of stores increases much more rapidly than with linear growth.

Figure 2 The graph shows the numbers of stores Companies A and B opened over a five-year period.

Notice that the domain for both functions is $[0, \infty)$, and the range for both functions is $[100, \infty)$. After year 1, Company B always has more stores than Company A.

Now we will turn our attention to the function representing the number of stores for Company B, $B(x) = 100(1 + 0.5)^x$. In this exponential function, 100 represents the initial number of stores, 0.50 represents the growth rate, and $1 + 0.5 = 1.5$ represents the growth factor. Generalizing further, we can write this function as $B(x) = 100(1.5)^x$, where 100 is the initial value, 1.5 is called the *base*, and x is called the *exponent*.

Example 3   **Evaluating a Real-World Exponential Model**

At the beginning of this section, we learned that the population of India was about 1.25 billion in the year 2013, with an annual growth rate of about 1.2%. This situation is represented by the growth function $P(t) = 1.25(1.012)^t$, where t is the number of years since 2013. To the nearest thousandth, what will the population of India be in 2031?

Solution   To estimate the population in 2031, we evaluate the models for $t = 18$, because 2031 is 18 years after 2013. Rounding to the nearest thousandth,

$$P(18) = 1.25(1.012)^{18} \approx 1.549$$

There will be about 1.549 billion people in India in the year 2031.

Try It #3

The population of China was about 1.39 billion in the year 2013, with an annual growth rate of about 0.6%. This situation is represented by the growth function $P(t) = 1.39(1.006)^t$, where t is the number of years since 2013. To the nearest thousandth, what will the population of China be for the year 2031? How does this compare to the population prediction we made for India in **Example 3**?

Finding Equations of Exponential Functions

In the previous examples, we were given an exponential function, which we then evaluated for a given input. Sometimes we are given information about an exponential function without knowing the function explicitly. We must use the information to first write the form of the function, then determine the constants a and b, and evaluate the function.

How To...

Given two data points, write an exponential model.

1. If one of the data points has the form $(0, a)$, then a is the initial value. Using a, substitute the second point into the equation $f(x) = a(b)^x$, and solve for b.
2. If neither of the data points have the form $(0, a)$, substitute both points into two equations with the form $f(x) = a(b)^x$. Solve the resulting system of two equations in two unknowns to find a and b.
3. Using the a and b found in the steps above, write the exponential function in the form $f(x) = a(b)^x$.

Example 4   **Writing an Exponential Model When the Initial Value Is Known**

In 2006, 80 deer were introduced into a wildlife refuge. By 2012, the population had grown to 180 deer. The population was growing exponentially. Write an algebraic function $N(t)$ representing the population (N) of deer over time t.

Solution   We let our independent variable t be the number of years after 2006. Thus, the information given in the problem can be written as input-output pairs: (0, 80) and (6, 180). Notice that by choosing our input variable to be measured as years after 2006, we have given ourselves the initial value for the function, $a = 80$. We can now substitute the second point into the equation $N(t) = 80b^t$ to find b:

$$N(t) = 80b^t$$

$$180 = 80b^6 \qquad \text{Substitute using point (6, 180).}$$

$$\frac{9}{4} = b^6 \qquad \text{Divide and write in lowest terms.}$$

$$b = \left(\frac{9}{4}\right)^{\frac{1}{6}} \qquad \text{Isolate } b \text{ using properties of exponents.}$$

$$b \approx 1.1447 \qquad \text{Round to 4 decimal places.}$$

NOTE: *Unless otherwise stated, do not round any intermediate calculations. Then round the final answer to four places for the remainder of this section.*

The exponential model for the population of deer is $N(t) = 80(1.1447)^t$. (Note that this exponential function models short-term growth. As the inputs gets large, the output will get increasingly larger, so much so that the model may not be useful in the long term.)

We can graph our model to observe the population growth of deer in the refuge over time. Notice that the graph in **Figure 3** passes through the initial points given in the problem, (0, 80) and (6, 180). We can also see that the domain for the function is $[0, \infty)$, and the range for the function is $[80, \infty)$.

Figure 3 Graph showing the population of deer over time, $N(t) = 80(1.1447)^t$, t years after 2006.

Try It #4

A wolf population is growing exponentially. In 2011, 129 wolves were counted. By 2013, the population had reached 236 wolves. What two points can be used to derive an exponential equation modeling this situation? Write the equation representing the population N of wolves over time t.

Example 5   **Writing an Exponential Model When the Initial Value is Not Known**

Find an exponential function that passes through the points $(-2, 6)$ and $(2, 1)$.

Solution   Because we don't have the initial value, we substitute both points into an equation of the form $f(x) = ab^x$, and then solve the system for a and b.

- Substituting $(-2, 6)$ gives $6 = ab^{-2}$
- Substituting $(2, 1)$ gives $1 = ab^2$

Use the first equation to solve for a in terms of b:

$$6 = ab^{-2}$$

$$\frac{6}{b^{-2}} = a \qquad \qquad \text{Divide.}$$

$$a = 6b^2 \qquad \qquad \text{Use properties of exponents to rewrite the denominator.}$$

Substitute a in the second equation, and solve for b:

$$1 = ab^2$$

$$1 = 6b^2b^2 = 6b^4 \qquad \text{Substitute } a.$$

$$b = \left(\frac{1}{6}\right)^{\frac{1}{4}} \qquad \text{Use properties of exponents to isolate } b.$$

$$b \approx 0.6389 \qquad \text{Round 4 decimal places.}$$

Use the value of b in the first equation to solve for the value of a:

$$a = 6b^2 \approx 6(0.6389)^2 \approx 2.4492$$

Thus, the equation is $f(x) = 2.4492(0.6389)^x$.

We can graph our model to check our work. Notice that the graph in **Figure 4** passes through the initial points given in the problem, $(-2, 6)$ and $(2, 1)$. The graph is an example of an exponential decay function.

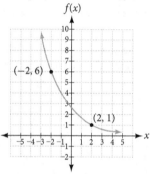

Figure 4 The graph of $f(x) = 2.4492(0.6389)^x$ models exponential decay.

Try It #5

Given the two points $(1, 3)$ and $(2, 4.5)$, find the equation of the exponential function that passes through these two points.

Q & A...

Do two points always determine a unique exponential function?

Yes, provided the two points are either both above the x-axis or both below the x-axis and have different x-coordinates. But keep in mind that we also need to know that the graph is, in fact, an exponential function. Not every graph that looks exponential really is exponential. We need to know the graph is based on a model that shows the same percent growth with each unit increase in x, which in many real world cases involves time.

How To...

Given the graph of an exponential function, write its equation.

1. First, identify two points on the graph. Choose the y-intercept as one of the two points whenever possible. Try to choose points that are as far apart as possible to reduce round-off error.
2. If one of the data points is the y-intercept $(0, a)$, then a is the initial value. Using a, substitute the second point into the equation $f(x) = a(b)^x$, and solve for b.
3. If neither of the data points have the form $(0, a)$, substitute both points into two equations with the form $f(x) = a(b)^x$. Solve the resulting system of two equations in two unknowns to find a and b.
4. Write the exponential function, $f(x) = a(b)^x$.

Example 6   Writing an Exponential Function Given Its Graph

Find an equation for the exponential function graphed in **Figure 5**.

Figure 5

Solution   We can choose the y-intercept of the graph, $(0, 3)$, as our first point. This gives us the initial value, $a = 3$. Next, choose a point on the curve some distance away from $(0, 3)$ that has integer coordinates. One such point is $(2, 12)$.

$y = ab^x$	Write the general form of an exponential equation.
$y = 3b^x$	Substitute the initial value 3 for a.
$12 = 3b^2$	Substitute in 12 for y and 2 for x.
$4 = b^2$	Divide by 3.
$b = \pm 2$	Take the square root.

Because we restrict ourselves to positive values of b, we will use $b = 2$. Substitute a and b into the standard form to yield the equation $f(x) = 3(2)^x$.

Try It #6

Find an equation for the exponential function graphed in **Figure 6**.

Figure 6

How To...

Given two points on the curve of an exponential function, use a graphing calculator to find the equation.

1. Press [**STAT**].
2. Clear any existing entries in columns **L1** or **L2**.
3. In **L1**, enter the x-coordinates given.
4. In **L2**, enter the corresponding y-coordinates.
5. Press [**STAT**] again. Cursor right to **CALC**, scroll down to **ExpReg** (**Exponential Regression**), and press [**ENTER**].
6. The screen displays the values of a and b in the exponential equation $y = a \cdot b^x$

Example 7   Using a Graphing Calculator to Find an Exponential Function

Use a graphing calculator to find the exponential equation that includes the points $(2, 24.8)$ and $(5, 198.4)$.

Solution   Follow the guidelines above. First press [**STAT**], [**EDIT**], [**1: Edit...**], and clear the lists **L1** and **L2**. Next, in the **L1** column, enter the x-coordinates, 2 and 5. Do the same in the **L2** column for the y-coordinates, 24.8 and 198.4. Now press [**STAT**], [**CALC**], [**0: ExpReg**] and press [**ENTER**]. The values $a = 6.2$ and $b = 2$ will be displayed. The exponential equation is $y = 6.2 \cdot 2^x$.

Try It #7

Use a graphing calculator to find the exponential equation that includes the points $(3, 75.98)$ and $(6, 481.07)$.

Applying the Compound-Interest Formula

Savings instruments in which earnings are continually reinvested, such as mutual funds and retirement accounts, use **compound interest**. The term *compounding* refers to interest earned not only on the original value, but on the accumulated value of the account.

The **annual percentage rate (APR)** of an account, also called the **nominal rate**, is the yearly interest rate earned by an investment account. The term *nominal* is used when the compounding occurs a number of times other than once per year. In fact, when interest is compounded more than once a year, the effective interest rate ends up being *greater* than the nominal rate! This is a powerful tool for investing.

We can calculate the compound interest using the compound interest formula, which is an exponential function of the variables time t, principal P, APR r, and number of compounding periods in a year n:

$$A(t) = P\left(1 + \frac{r}{n}\right)^{nt}$$

For example, observe **Table 4**, which shows the result of investing $1,000 at 10% for one year. Notice how the value of the account increases as the compounding frequency increases.

Frequency	Value after 1 year
Annually	$1100
Semiannually	$1102.50
Quarterly	$1103.81
Monthly	$1104.71
Daily	$1105.16

Table 4

---

the compound interest formula

Compound interest can be calculated using the formula

$$A(t) = P\left(1 + \frac{r}{n}\right)^{nt}$$

where

- $A(t)$ is the account value,
- t is measured in years,
- P is the starting amount of the account, often called the principal, or more generally present value,
- r is the annual percentage rate (APR) expressed as a decimal, and
- n is the number of compounding periods in one year.

Example 8   **Calculating Compound Interest**

If we invest $3,000 in an investment account paying 3% interest compounded quarterly, how much will the account be worth in 10 years?

Solution   Because we are starting with $3,000, $P = 3000$. Our interest rate is 3%, so $r = 0.03$. Because we are compounding quarterly, we are compounding 4 times per year, so $n = 4$. We want to know the value of the account in 10 years, so we are looking for $A(10)$, the value when $t = 10$.

$$A(t) = P\left(1 + \frac{r}{n}\right)^{nt} \qquad \text{Use the compound interest formula.}$$

$$A(10) = 3000\left(1 + \frac{0.03}{4}\right)^{4 \cdot 10} \qquad \text{Substitute using given values.}$$

$$\approx \$4{,}045.05 \qquad \text{Round to two decimal places.}$$

The account will be worth about $4,045.05 in 10 years.

Try It #8

An initial investment of $100,000 at 12% interest is compounded weekly (use 52 weeks in a year). What will the investment be worth in 30 years?

Example 9   **Using the Compound Interest Formula to Solve for the Principal**

A 529 Plan is a college-savings plan that allows relatives to invest money to pay for a child's future college tuition; the account grows tax-free. Lily wants to set up a 529 account for her new granddaughter and wants the account to grow to $40,000 over 18 years. She believes the account will earn 6% compounded semi-annually (twice a year). To the nearest dollar, how much will Lily need to invest in the account now?

Solution   The nominal interest rate is 6%, so $r = 0.06$. Interest is compounded twice a year, so $n = 2$.

We want to find the initial investment, P, needed so that the value of the account will be worth $40,000 in 18 years. Substitute the given values into the compound interest formula, and solve for P.

$$A(t) = P\left(1 + \frac{r}{n}\right)^{nt} \qquad \text{Use the compound interest formula.}$$

$$40{,}000 = P\left(1 + \frac{0.06}{2}\right)^{2(18)} \qquad \text{Substitute using given values } A, r, n, \text{ and } t.$$

$$40{,}000 = P(1.03)^{36} \qquad \text{Simplify.}$$

$$\frac{40{,}000}{(1.03)^{36}} = P \qquad \text{Isolate } P.$$

$$P \approx \$13{,}801 \qquad \text{Divide and round to the nearest dollar.}$$

Lily will need to invest $13,801 to have $40,000 in 18 years.

Try It #9

Refer to **Example 9**. To the nearest dollar, how much would Lily need to invest if the account is compounded quarterly?

Evaluating Functions with Base e

As we saw earlier, the amount earned on an account increases as the compounding frequency increases. **Table 5** shows that the increase from annual to semi-annual compounding is larger than the increase from monthly to daily compounding. This might lead us to ask whether this pattern will continue.

Examine the value of $1 invested at 100% interest for 1 year, compounded at various frequencies, listed in **Table 5**.

Frequency	$A(t) = \left(1 + \frac{1}{n}\right)^{n}$	Value
Annually	$\left(1 + \frac{1}{1}\right)^{1}$	$2
Semiannually	$\left(1 + \frac{1}{2}\right)^{2}$	$2.25
Quarterly	$\left(1 + \frac{1}{4}\right)^{4}$	$2.441406
Monthly	$\left(1 + \frac{1}{12}\right)^{12}$	$2.613035
Daily	$\left(1 + \frac{1}{365}\right)^{365}$	$2.714567
Hourly	$\left(1 + \frac{1}{8766}\right)^{8766}$	$2.718127
Once per minute	$\left(1 + \frac{1}{525960}\right)^{525960}$	$2.718279
Once per second	$\left(1 + \frac{1}{31557600}\right)^{31557600}$	$2.718282

Table 5

These values appear to be approaching a limit as n increases without bound. In fact, as n gets larger and larger, the expression $\left(1 + \dfrac{1}{n}\right)^n$ approaches a number used so frequently in mathematics that it has its own name: the letter e. This value is an irrational number, which means that its decimal expansion goes on forever without repeating. Its approximation to six decimal places is shown below.

---

the number e

The letter e represents the irrational number

$$\left(1 + \frac{1}{n}\right)^n, \text{ as } n \text{ increases without bound}$$

The letter e is used as a base for many real-world exponential models. To work with base e, we use the approximation, $e \approx 2.718282$. The constant was named by the Swiss mathematician Leonhard Euler (1707–1783) who first investigated and discovered many of its properties.

Example 10   **Using a Calculator to Find Powers of e**

Calculate $e^{3.14}$. Round to five decimal places.

Solution   On a calculator, press the button labeled [e^x]. The window shows [e^(]. Type 3.14 and then close parenthesis, [)]. Press [**ENTER**]. Rounding to 5 decimal places, $e^{3.14} \approx 23.10387$. Caution: Many scientific calculators have an "**Exp**" button, which is used to enter numbers in scientific notation. It is not used to find powers of e.

Try It #10

Use a calculator to find $e^{-0.5}$. Round to five decimal places.

Investigating Continuous Growth

So far we have worked with rational bases for exponential functions. For most real-world phenomena, however, e is used as the base for exponential functions. Exponential models that use e as the base are called *continuous growth or decay models*. We see these models in finance, computer science, and most of the sciences, such as physics, toxicology, and fluid dynamics.

---

the continuous growth/decay formula

For all real numbers t, and all positive numbers a and r, continuous growth or decay is represented by the formula

$$A(t) = ae^{rt}$$

where

- a is the initial value,
- r is the continuous growth rate per unit time,
- and t is the elapsed time.

If $r > 0$, then the formula represents continuous growth. If $r < 0$, then the formula represents continuous decay.

For business applications, the continuous growth formula is called the continuous compounding formula and takes the form

$$A(t) = Pe^{rt}$$

where

- P is the principal or the initial invested,
- r is the growth or interest rate per unit time,
- and t is the period or term of the investment.

How To…

Given the initial value, rate of growth or decay, and time t, solve a continuous growth or decay function.

1. Use the information in the problem to determine a, the initial value of the function.
2. Use the information in the problem to determine the growth rate r.
   a. If the problem refers to continuous growth, then $r > 0$.
   b. If the problem refers to continuous decay, then $r < 0$.
3. Use the information in the problem to determine the time t.
4. Substitute the given information into the continuous growth formula and solve for $A(t)$.

Example 11 **Calculating Continuous Growth**

A person invested $1,000 in an account earning a nominal 10% per year compounded continuously. How much was in the account at the end of one year?

Solution   Since the account is growing in value, this is a continuous compounding problem with growth rate $r = 0.10$. The initial investment was $1,000, so $P = 1000$. We use the continuous compounding formula to find the value after $t = 1$ year:

$$A(t) = Pe^{rt} \qquad \text{Use the continuous compounding formula.}$$
$$= 1000(e)^{0.1} \qquad \text{Substitute known values for } P, r, \text{ and } t.$$
$$\approx 1105.17 \qquad \text{Use a calculator to approximate.}$$

The account is worth $1,105.17 after one year.

Try It #11

A person invests $100,000 at a nominal 12% interest per year compounded continuously. What will be the value of the investment in 30 years?

Example 12 **Calculating Continuous Decay**

Radon-222 decays at a continuous rate of 17.3% per day. How much will 100 mg of Radon-222 decay to in 3 days?

Solution   Since the substance is decaying, the rate, 17.3%, is negative. So, $r = -0.173$. The initial amount of radon-222 was 100 mg, so $a = 100$. We use the continuous decay formula to find the value after $t = 3$ days:

$$A(t) = ae^{rt} \qquad \text{Use the continuous growth formula.}$$
$$= 100e^{-0.173(3)} \qquad \text{Substitute known values for } a, r, \text{ and } t.$$
$$\approx 59.5115 \qquad \text{Use a calculator to approximate.}$$

So 59.5115 mg of radon-222 will remain.

Try It #12

Using the data in **Example 12**, how much radon-222 will remain after one year?

Access these online resources for additional instruction and practice with exponential functions.

- Exponential Growth Function (http://openstaxcollege.org/l/expgrowth)
- Compound Interest (http://openstaxcollege.org/l/compoundint)

6.1   SECTION EXERCISES

VERBAL

1. Explain why the values of an increasing exponential function will eventually overtake the values of an increasing linear function.

2. Given a formula for an exponential function, is it possible to determine whether the function grows or decays exponentially just by looking at the formula? Explain.

3. The Oxford Dictionary defines the word *nominal* as a value that is "stated or expressed but not necessarily corresponding exactly to the real value."[18] Develop a reasonable argument for why the term *nominal rate* is used to describe the annual percentage rate of an investment account that compounds interest.

ALGEBRAIC

For the following exercises, identify whether the statement represents an exponential function. Explain.

4. The average annual population increase of a pack of wolves is 25.

5. A population of bacteria decreases by a factor of $\frac{1}{8}$ every 24 hours.

6. The value of a coin collection has increased by 3.25% annually over the last 20 years.

7. For each training session, a personal trainer charges his clients $5 less than the previous training session.

8. The height of a projectile at time t is represented by the function $h(t) = -4.9t^2 + 18t + 40$.

For the following exercises, consider this scenario: For each year t, the population of a forest of trees is represented by the function $A(t) = 115(1.025)^t$. In a neighboring forest, the population of the same type of tree is represented by the function $B(t) = 82(1.029)^t$. (Round answers to the nearest whole number.)

9. Which forest's population is growing at a faster rate?

10. Which forest had a greater number of trees initially? By how many?

11. Assuming the population growth models continue to represent the growth of the forests, which forest will have a greater number of trees after 20 years? By how many?

12. Assuming the population growth models continue to represent the growth of the forests, which forest will have a greater number of trees after 100 years? By how many?

13. Discuss the above results from the previous four exercises. Assuming the population growth models continue to represent the growth of the forests, which forest will have the greater number of trees in the long run? Why? What are some factors that might influence the long-term validity of the exponential growth model?

For the following exercises, determine whether the equation represents exponential growth, exponential decay, or neither. Explain.

14. $y = 300(1 - t)^5$

15. $y = 220(1.06)^x$

16. $y = 16.5(1.025)^{\frac{1}{x}}$

17. $y = 11{,}701(0.97)^t$

For the following exercises, find the formula for an exponential function that passes through the two points given.

18. $(0, 6)$ and $(3, 750)$

19. $(0, 2000)$ and $(2, 20)$

20. $\left(-1, \frac{3}{2}\right)$ and $(3, 24)$

21. $(-2, 6)$ and $(3, 1)$

22. $(3, 1)$ and $(5, 4)$

18. Oxford Dictionary. http://oxforddictionaries.com/us/definition/american_english/nomina.

For the following exercises, determine whether the table could represent a function that is linear, exponential, or neither. If it appears to be exponential, find a function that passes through the points.

23.

x	1	2	3	4
f(x)	70	40	10	−20

24.

x	1	2	3	4
h(x)	70	49	34.3	24.01

25.

x	1	2	3	4
m(x)	80	61	42.9	25.61

26.

x	1	2	3	4
f(x)	10	20	40	80

27.

x	1	2	3	4
g(x)	−3.25	2	7.25	12.5

For the following exercises, use the compound interest formula, $A(t) = P\left(1 + \frac{r}{n}\right)^{nt}$.

28. After a certain number of years, the value of an investment account is represented by the equation $10,250\left(1 + \frac{0.04}{12}\right)^{120}$. What is the value of the account?

29. What was the initial deposit made to the account in the previous exercise?

30. How many years had the account from the previous exercise been accumulating interest?

31. An account is opened with an initial deposit of $6,500 and earns 3.6% interest compounded semi-annually. What will the account be worth in 20 years?

32. How much more would the account in the previous exercise have been worth if the interest were compounding weekly?

33. Solve the compound interest formula for the principal, P.

34. Use the formula found in Exercise #31 to calculate the initial deposit of an account that is worth $14,472.74 after earning 5.5% interest compounded monthly for 5 years. (Round to the nearest dollar.)

35. How much more would the account in Exercises #31 and #34 be worth if it were earning interest for 5 more years?

36. Use properties of rational exponents to solve the compound interest formula for the interest rate, r.

37. Use the formula found in the previous exercise to calculate the interest rate for an account that was compounded semi-annually, had an initial deposit of $9,000 and was worth $13,373.53 after 10 years.

38. Use the formula found in the previous exercise to calculate the interest rate for an account that was compounded monthly, had an initial deposit of $5,500, and was worth $38,455 after 30 years.

For the following exercises, determine whether the equation represents continuous growth, continuous decay, or neither. Explain.

39. $y = 3742(e)^{0.75t}$

40. $y = 150(e)^{\frac{3.25}{t}}$

41. $y = 2.25(e)^{-2t}$

42. Suppose an investment account is opened with an initial deposit of $12,000 earning 7.2% interest compounded continuously. How much will the account be worth after 30 years?

43. How much less would the account from Exercise 42 be worth after 30 years if it were compounded monthly instead?

NUMERIC

For the following exercises, evaluate each function. Round answers to four decimal places, if necessary.

44. $f(x) = 2(5)^x$, for $f(-3)$

45. $f(x) = -4^{2x+3}$, for $f(-1)$

46. $f(x) = e^x$, for $f(3)$

47. $f(x) = -2e^{x-1}$, for $f(-1)$

48. $f(x) = 2.7(4)^{-x+1} + 1.5$, for $f(-2)$

49. $f(x) = 1.2e^{2x} - 0.3$, for $f(3)$

50. $f(x) = -\frac{3}{2}(3)^{-x} + \frac{3}{2}$, for $f(2)$

TECHNOLOGY

For the following exercises, use a graphing calculator to find the equation of an exponential function given the points on the curve.

51. $(0, 3)$ and $(3, 375)$ **52.** $(3, 222.62)$ and $(10, 77.456)$ **53.** $(20, 29.495)$ and $(150, 730.89)$

54. $(5, 2.909)$ and $(13, 0.005)$ **55.** $(11,310.035)$ and $(25,356.3652)$

EXTENSIONS

56. The *annual percentage yield* (APY) of an investment account is a representation of the actual interest rate earned on a compounding account. It is based on a compounding period of one year. Show that the APY of an account that compounds monthly can be found with the formula $\text{APY} = \left(1 + \frac{r}{12}\right)^{12} - 1$.

57. Repeat the previous exercise to find the formula for the APY of an account that compounds daily. Use the results from this and the previous exercise to develop a function $I(n)$ for the APY of any account that compounds n times per year.

58. Recall that an exponential function is any equation written in the form $f(x) = a \cdot b^x$ such that a and b are positive numbers and $b \neq 1$. Any positive number b can be written as $b = e^n$ for some value of n. Use this fact to rewrite the formula for an exponential function that uses the number e as a base.

59. In an exponential decay function, the base of the exponent is a value between 0 and 1. Thus, for some number $b > 1$, the exponential decay function can be written as $f(x) = a \cdot \left(\frac{1}{b}\right)^x$. Use this formula, along with the fact that $b = e^n$, to show that an exponential decay function takes the form $f(x) = a(e)^{-nx}$ for some positive number n.

60. The formula for the amount A in an investment account with a nominal interest rate r at any time t is given by $A(t) = a(e)^{rt}$, where a is the amount of principal initially deposited into an account that compounds continuously. Prove that the percentage of interest earned to principal at any time t can be calculated with the formula $I(t) = e^{rt} - 1$.

REAL-WORLD APPLICATIONS

61. The fox population in a certain region has an annual growth rate of 9% per year. In the year 2012, there were 23,900 fox counted in the area. What is the fox population predicted to be in the year 2020?

62. A scientist begins with 100 milligrams of a radioactive substance that decays exponentially. After 35 hours, 50 mg of the substance remains. How many milligrams will remain after 54 hours?

63. In the year 1985, a house was valued at $110,000. By the year 2005, the value had appreciated to $145,000. What was the annual growth rate between 1985 and 2005? Assume that the value continued to grow by the same percentage. What was the value of the house in the year 2010?

64. A car was valued at $38,000 in the year 2007. By 2013, the value had depreciated to $11,000 If the car's value continues to drop by the same percentage, what will it be worth by 2017?

65. Jamal wants to save $54,000 for a down payment on a home. How much will he need to invest in an account with 8.2% APR, compounding daily, in order to reach his goal in 5 years?

66. Kyoko has $10,000 that she wants to invest. Her bank has several investment accounts to choose from, all compounding daily. Her goal is to have $15,000 by the time she finishes graduate school in 6 years. To the nearest hundredth of a percent, what should her minimum annual interest rate be in order to reach her goal? (*Hint:* solve the compound interest formula for the interest rate.)

67. Alyssa opened a retirement account with 7.25% APR in the year 2000. Her initial deposit was $13,500. How much will the account be worth in 2025 if interest compounds monthly? How much more would she make if interest compounded continuously?

68. An investment account with an annual interest rate of 7% was opened with an initial deposit of $4,000 Compare the values of the account after 9 years when the interest is compounded annually, quarterly, monthly, and continuously.

In this section, you will:

- Graph exponential functions.
- Graph exponential functions using transformations.

6.2 GRAPHS OF EXPONENTIAL FUNCTIONS

As we discussed in the previous section, exponential functions are used for many real-world applications such as finance, forensics, computer science, and most of the life sciences. Working with an equation that describes a real-world situation gives us a method for making predictions. Most of the time, however, the equation itself is not enough. We learn a lot about things by seeing their pictorial representations, and that is exactly why graphing exponential equations is a powerful tool. It gives us another layer of insight for predicting future events.

Graphing Exponential Functions

Before we begin graphing, it is helpful to review the behavior of exponential growth. Recall the table of values for a function of the form $f(x) = b^x$ whose base is greater than one. We'll use the function $f(x) = 2^x$. Observe how the output values in **Table 1** change as the input increases by 1.

x	-3	-2	-1	0	1	2	3
$f(x) = 2x$	$\frac{1}{8}$	$\frac{1}{4}$	$\frac{1}{2}$	1	2	4	8

Table 1

Each output value is the product of the previous output and the base, 2. We call the base 2 the *constant ratio*. In fact, for any exponential function with the form $f(x) = ab^x$, b is the constant ratio of the function. This means that as the input increases by 1, the output value will be the product of the base and the previous output, regardless of the value of a.

Notice from the table that

- the output values are positive for all values of x;
- as x increases, the output values increase without bound; and
- as x decreases, the output values grow smaller, approaching zero.

Figure 1 shows the exponential growth function $f(x) = 2^x$.

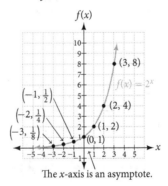

The x-axis is an asymptote.

Figure 1 Notice that the graph gets close to the x-axis, but never touches it.

The domain of $f(x) = 2^x$ is all real numbers, the range is $(0, \infty)$, and the horizontal asymptote is $y = 0$.

To get a sense of the behavior of exponential decay, we can create a table of values for a function of the form $f(x) = b^x$ whose base is between zero and one. We'll use the function $g(x) = \left(\frac{1}{2}\right)^x$. Observe how the output values in **Table 2** change as the input increases by 1.

x	-3	-2	-1	0	1	2	3
$g(x) = \left(\frac{1}{2}\right)^x$	8	4	2	1	$\frac{1}{2}$	$\frac{1}{4}$	$\frac{1}{8}$

Table 2

Again, because the input is increasing by 1, each output value is the product of the previous output and the base, or constant ratio $\frac{1}{2}$.

Notice from the table that

- the output values are positive for all values of x;
- as x increases, the output values grow smaller, approaching zero; and
- as x decreases, the output values grow without bound.

Figure 2 shows the exponential decay function, $g(x) = \left(\frac{1}{2}\right)^x$.

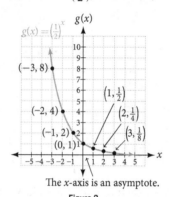

The x-axis is an asymptote.

Figure 2

The domain of $g(x) = \left(\frac{1}{2}\right)^x$ is all real numbers, the range is $(0, \infty)$, and the horizontal asymptote is $y = 0$.

---

characteristics of the graph of the parent function $f(x) = b^x$

An exponential function with the form $f(x) = b^x$, $b > 0$, $b \neq 1$, has these characteristics:

- one-to-one function
- horizontal asymptote: $y = 0$
- domain: $(-\infty, \infty)$
- range: $(0, \infty)$
- x-intercept: none
- y-intercept: $(0, 1)$
- increasing if $b > 1$
- decreasing if $b < 1$

Figure 3 compares the graphs of exponential growth and decay functions.

Figure 3

How To...

Given an exponential function of the form $f(x) = b^x$, graph the function.

1. Create a table of points.
2. Plot at least 3 point from the table, including the y-intercept $(0, 1)$.
3. Draw a smooth curve through the points.
4. State the domain, $(-\infty, \infty)$, the range, $(0, \infty)$, and the horizontal asymptote, $y = 0$.

<u>Example 1</u>   **Sketching the Graph of an Exponential Function of the Form $f(x) = b^x$**

Sketch a graph of $f(x) = 0.25^x$. State the domain, range, and asymptote.

Solution   Before graphing, identify the behavior and create a table of points for the graph.

- Since $b = 0.25$ is between zero and one, we know the function is decreasing. The left tail of the graph will increase without bound, and the right tail will approach the asymptote $y = 0$.

- Create a table of points as in **Table 3**.

x	-3	-2	-1	0	1	2	3
$f(x) = 0.25^x$	64	16	4	1	0.25	0.0625	0.015625

Table 3

- Plot the y-intercept, $(0, 1)$, along with two other points. We can use $(-1, 4)$ and $(1, 0.25)$.

Draw a smooth curve connecting the points as in **Figure 4**.

Figure 4

The domain is $(-\infty, \infty)$; the range is $(0, \infty)$; the horizontal asymptote is $y = 0$.

Try It #1

Sketch the graph of $f(x) = 4^x$. State the domain, range, and asymptote.

Graphing Transformations of Exponential Functions

Transformations of exponential graphs behave similarly to those of other functions. Just as with other parent functions, we can apply the four types of transformations—shifts, reflections, stretches, and compressions—to the parent function $f(x) = b^x$ without loss of shape. For instance, just as the quadratic function maintains its parabolic shape when shifted, reflected, stretched, or compressed, the exponential function also maintains its general shape regardless of the transformations applied.

Graphing a Vertical Shift

The first transformation occurs when we add a constant d to the parent function $f(x) = b^x$, giving us a vertical shift d units in the same direction as the sign. For example, if we begin by graphing a parent function, $f(x) = 2^x$, we can then graph two vertical shifts alongside it, using $d = 3$: the upward shift, $g(x) = 2^x + 3$ and the downward shift, $h(x) = 2^x - 3$. Both vertical shifts are shown in **Figure 5**.

Figure 5

Observe the results of shifting $f(x) = 2^x$ vertically:

- The domain, $(-\infty, \infty)$ remains unchanged.

- When the function is shifted up 3 units to $g(x) = 2^x + 3$:

  ◦ The y-intercept shifts up 3 units to $(0, 4)$.

  ◦ The asymptote shifts up 3 units to $y = 3$.

  ◦ The range becomes $(3, \infty)$.

- When the function is shifted down 3 units to $h(x) = 2^x - 3$:

  ◦ The y-intercept shifts down 3 units to $(0, -2)$.

  ◦ The asymptote also shifts down 3 units to $y = -3$.

  ◦ The range becomes $(-3, \infty)$.

Graphing a Horizontal Shift

The next transformation occurs when we add a constant c to the input of the parent function $f(x) = b^x$, giving us a horizontal shift c units in the *opposite* direction of the sign. For example, if we begin by graphing the parent function $f(x) = 2^x$, we can then graph two horizontal shifts alongside it, using $c = 3$: the shift left, $g(x) = 2^{x+3}$, and the shift right, $h(x) = 2^{x-3}$. Both horizontal shifts are shown in **Figure 6**.

Figure 6

Observe the results of shifting $f(x) = 2^x$ horizontally:

- The domain, $(-\infty, \infty)$, remains unchanged.

- The asymptote, $y = 0$, remains unchanged.

- The y-intercept shifts such that:

  ◦ When the function is shifted left 3 units to $g(x) = 2^{x+3}$, the y-intercept becomes $(0, 8)$. This is because $2^{x+3} = (8)2^x$, so the initial value of the function is 8.

  ◦ When the function is shifted right 3 units to $h(x) = 2^{x-3}$, the y-intercept becomes $\left(0, \dfrac{1}{8}\right)$. Again, see that $2^{x-3} = \left(\dfrac{1}{8}\right)2^x$, so the initial value of the function is $\dfrac{1}{8}$.

> ### *shifts of the parent function $f(x) = b^x$*
>
> For any constants c and d, the function $f(x) = b^{x+c} + d$ shifts the parent function $f(x) = b^x$
>
> - vertically d units, in the *same* direction of the sign of d.
>
> - horizontally c units, in the *opposite* direction of the sign of c.
>
> - The y-intercept becomes $(0, b^c + d)$.
>
> - The horizontal asymptote becomes $y = d$.
>
> - The range becomes (d, ∞).
>
> - The domain, $(-\infty, \infty)$, remains unchanged.

How To...

Given an exponential function with the form $f(x) = b^{x+c} + d$, graph the translation.

1. Draw the horizontal asymptote $y = d$.
2. Identify the shift as $(-c, d)$. Shift the graph of $f(x) = b^x$ left c units if c is positive, and right c units if c is negative.
3. Shift the graph of $f(x) = b^x$ up d units if d is positive, and down d units if d is negative.
4. State the domain, $(-\infty, \infty)$, the range, (d, ∞), and the horizontal asymptote $y = d$.

Example 2   **Graphing a Shift of an Exponential Function**

Graph $f(x) = 2^{x+1} - 3$. State the domain, range, and asymptote.

Solution   We have an exponential equation of the form $f(x) = b^{x+c} + d$, with $b = 2$, $c = 1$, and $d = -3$.

Draw the horizontal asymptote $y = d$, so draw $y = -3$.

Identify the shift as $(-c, d)$, so the shift is $(-1, -3)$.

Shift the graph of $f(x) = b^x$ left 1 units and down 3 units.

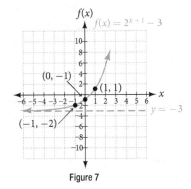

Figure 7

The domain is $(-\infty, \infty)$; the range is $(-3, \infty)$; the horizontal asymptote is $y = -3$.

Try It #2

Graph $f(x) = 2^{x-1} + 3$. State domain, range, and asymptote.

How To...

Given an equation of the form $f(x) = b^{x+c} + d$ for x, use a graphing calculator to approximate the solution.

1. Press [**Y=**]. Enter the given exponential equation in the line headed "$Y_1 =$".
2. Enter the given value for $f(x)$ in the line headed "$Y_2 =$".
3. Press [**WINDOW**]. Adjust the y-axis so that it includes the value entered for "$Y_2 =$".
4. Press [**GRAPH**] to observe the graph of the exponential function along with the line for the specified value of $f(x)$.
5. To find the value of x, we compute the point of intersection. Press [**2ND**] then [**CALC**]. Select "intersect" and press [**ENTER**] three times. The point of intersection gives the value of x for the indicated value of the function.

Example 3   **Approximating the Solution of an Exponential Equation**

Solve $42 = 1.2(5)^x + 2.8$ graphically. Round to the nearest thousandth.

Solution   Press [**Y=**] and enter $1.2(5)^x + 2.8$ next to $Y_1=$. Then enter 42 next to $Y_2=$. For a window, use the values −3 to 3 for x and −5 to 55 for y. Press [**GRAPH**]. The graphs should intersect somewhere near $x = 2$.

For a better approximation, press [**2ND**] then [**CALC**]. Select [**5: intersect**] and press [**ENTER**] three times. The x-coordinate of the point of intersection is displayed as 2.1661943. (Your answer may be different if you use a different window or use a different value for **Guess?**) To the nearest thousandth, $x \approx 2.166$.

Try It #3

Solve $4 = 7.85(1.15)^x - 2.27$ graphically. Round to the nearest thousandth.

Graphing a Stretch or Compression

While horizontal and vertical shifts involve adding constants to the input or to the function itself, a stretch or compression occurs when we multiply the parent function $f(x) = b^x$ by a constant $|a| > 0$. For example, if we begin by graphing the parent function $f(x) = 2^x$, we can then graph the stretch, using $a = 3$, to get $g(x) = 3(2)^x$ as shown on the left in **Figure 8**, and the compression, using $a = \frac{1}{3}$, to get $h(x) = \frac{1}{3}(2)^x$ as shown on the right in **Figure 8**.

Figure 8 (a) $g(x) = 3(2)^x$ stretches the graph of $f(x) = 2^x$ vertically by a factor of 3.
(b) $h(x) = \frac{1}{3}(2)^x$ compresses the graph of $f(x) = 2^x$ vertically by a factor of $\frac{1}{3}$.

---

stretches and compressions of the parent function $f(x) = b^x$

For any factor $a > 0$, the function $f(x) = a(b)^x$

- is stretched vertically by a factor of a if $|a| > 1$.

- is compressed vertically by a factor of a if $|a| < 1$.

- has a y-intercept of $(0, a)$.

- has a horizontal asymptote at $y = 0$, a range of $(0, \infty)$, and a domain of $(-\infty, \infty)$, which are unchanged from the parent function.

Example 4   **Graphing the Stretch of an Exponential Function**

Sketch a graph of $f(x) = 4\left(\frac{1}{2}\right)^x$. State the domain, range, and asymptote.

Solution   Before graphing, identify the behavior and key points on the graph.

- Since $b = \frac{1}{2}$ is between zero and one, the left tail of the graph will increase without bound as x decreases, and the right tail will approach the x-axis as x increases.

- Since $a = 4$, the graph of $f(x) = \left(\frac{1}{2}\right)^x$ will be stretched by a factor of 4.

- Create a table of points as shown in **Table 4**.

x	-3	-2	-1	0	1	2	3
$f(x) = 4\left(\frac{1}{2}\right)^x$	32	16	8	4	2	1	0.5

Table 4

- Plot the y-intercept, $(0, 4)$, along with two other points. We can use $(-1, 8)$ and $(1, 2)$.

Given a common logarithm of the form $y = \log(x)$, evaluate it mentally.

1. Rewrite the argument x as a power of 10: $10^y = x$.
2. Use previous knowledge of powers of 10 to identify y by asking, "To what exponent must 10 be raised in order to get x?"

Example 5   **Finding the Value of a Common Logarithm Mentally**

Evaluate $y = \log(1,000)$ without using a calculator.

Solution   First we rewrite the logarithm in exponential form: $10^y = 1,000$. Next, we ask, "To what exponent must 10 be raised in order to get 1,000?" We know $10^3 = 1,000$ therefore, $\log(1,000) = 3$.

Try It #5

Evaluate $y = \log(1,000,000)$.

Given a common logarithm with the form $y = \log(x)$, evaluate it using a calculator.

1. Press [**LOG**].
2. Enter the value given for x, followed by [ )].
3. Press [**ENTER**].

Example 6   **Finding the Value of a Common Logarithm Using a Calculator**

Evaluate $y = \log(321)$ to four decimal places using a calculator.

Solution

- Press [**LOG**].
- Enter 321, followed by [ )].
- Press [**ENTER**].

Rounding to four decimal places, $\log(321) \approx 2.5065$.

Analysis   Note that $10^2 = 100$ and that $10^3 = 1000$. Since 321 is between 100 and 1000, we know that $\log(321)$ must be between $\log(100)$ and $\log(1000)$. This gives us the following:

$$100 \; < \; 321 \; < \; 1000$$
$$2 \; < \; 2.5065 \; < \; 3$$

Try It #6

Evaluate $y = \log(123)$ to four decimal places using a calculator.

Example 7   **Rewriting and Solving a Real-World Exponential Model**

The amount of energy released from one earthquake was 500 times greater than the amount of energy released from another. The equation $10^x = 500$ represents this situation, where x is the difference in magnitudes on the Richter Scale. To the nearest thousandth, what was the difference in magnitudes?

Solution   We begin by rewriting the exponential equation in logarithmic form.

$$10^x = 500$$

$$\log(500) = x \qquad \text{Use the definition of the common log.}$$

Next we evaluate the logarithm using a calculator:

- Press [**LOG**].
- Enter 500, followed by [ )].
- Press [**ENTER**].
- To the nearest thousandth, $\log(500) \approx 2.699$.

The difference in magnitudes was about 2.699.

Try It #7

The amount of energy released from one earthquake was 8,500 times greater than the amount of energy released from another. The equation $10^x = 8500$ represents this situation, where x is the difference in magnitudes on the Richter Scale. To the nearest thousandth, what was the difference in magnitudes?

Using Natural Logarithms

The most frequently used base for logarithms is e. Base e logarithms are important in calculus and some scientific applications; they are called **natural logarithms**. The base e logarithm, $\log_e(x)$, has its own notation, $\ln(x)$.

Most values of $\ln(x)$ can be found only using a calculator. The major exception is that, because the logarithm of 1 is always 0 in any base, $\ln 1 = 0$. For other natural logarithms, we can use the $\ln$ key that can be found on most scientific calculators. We can also find the natural logarithm of any power of e using the inverse property of logarithms.

---

definition of the natural logarithm

A **natural logarithm** is a logarithm with base e. We write $\log_e(x)$ simply as $\ln(x)$. The natural logarithm of a positive number x satisfies the following definition.

For $x > 0$,

$$y = \ln(x) \text{ is equivalent to } e^y = x$$

We read $\ln(x)$ as, "the logarithm with base e of x" or "the natural logarithm of x."

The logarithm y is the exponent to which e must be raised to get x.

Since the functions $y = e$ and $y = \ln(x)$ are inverse functions, $\ln(e^x) = x$ for all x and $e = x$ for $x > 0$.

How To...

Given a natural logarithm with the form $y = \ln(x)$, evaluate it using a calculator.

1. Press [**LN**].
2. Enter the value given for x, followed by [ )].
3. Press [**ENTER**].

Example 8   **Evaluating a Natural Logarithm Using a Calculator**

Evaluate $y = \ln(500)$ to four decimal places using a calculator.

Solution

- Press [**LN**].
- Enter 500, followed by [ )].
- Press [**ENTER**].

Rounding to four decimal places, $\ln(500) \approx 6.2146$

Try It #8

Evaluate $\ln(-500)$.

Access this online resource for additional instruction and practice with logarithms.

- Introduction to Logarithms (http://openstaxcollege.org/l/intrologarithms)

6.3 SECTION EXERCISES

VERBAL

1. What is a base b logarithm? Discuss the meaning by interpreting each part of the equivalent equations $b^y = x$ and $\log_b(x) = y$ for $b > 0$, $b \neq 1$.

2. How is the logarithmic function $f(x) = \log_b(x)$ related to the exponential function $g(x) = b^x$? What is the result of composing these two functions?

3. How can the logarithmic equation $\log_b x = y$ be solved for x using the properties of exponents?

4. Discuss the meaning of the common logarithm. What is its relationship to a logarithm with base b, and how does the notation differ?

5. Discuss the meaning of the natural logarithm. What is its relationship to a logarithm with base b, and how does the notation differ?

ALGEBRAIC

For the following exercises, rewrite each equation in exponential form.

6. $\log_4(q) = m$

7. $\log_a(b) = c$

8. $\log_{16}(y) = x$

9. $\log_x(64) = y$

10. $\log_y(x) = -11$

11. $\log_{15}(a) = b$

12. $\log_y(137) = x$

13. $\log_{13}(142) = a$

14. $\log(v) = t$

15. $\ln(w) = n$

For the following exercises, rewrite each equation in logarithmic form.

16. $4^x = y$

17. $c^d = k$

18. $m^{-7} = n$

19. $19^x = y$

20. $x^{-\frac{10}{13}} = y$

21. $n^4 = 103$

22. $\left(\dfrac{7}{5}\right)^m = n$

23. $y^x = \dfrac{39}{100}$

24. $10^a = b$

25. $e^k = h$

For the following exercises, solve for x by converting the logarithmic equation to exponential form.

26. $\log_3(x) = 2$

27. $\log_2(x) = -3$

28. $\log_5(x) = 2$

29. $\log_3(x) = 3$

30. $\log_2(x) = 6$

31. $\log_9(x) = \dfrac{1}{2}$

32. $\log_{18}(x) = 2$

33. $\log_6(x) = -3$

34. $\log(x) = 3$

35. $\ln(x) = 2$

For the following exercises, use the definition of common and natural logarithms to simplify.

36. $\log(100^8)$

37. $10^{\log(32)}$

38. $2\log(0.0001)$

39. $e^{\ln(1.06)}$

40. $\ln(e^{-5.03})$

41. $e^{\ln(10.125)} + 4$

NUMERIC

For the following exercises, evaluate the base b logarithmic expression without using a calculator.

42. $\log_3\left(\dfrac{1}{27}\right)$

43. $\log_6\left(\sqrt{6}\right)$

44. $\log_2\left(\dfrac{1}{8}\right) + 4$

45. $6\log_8(4)$

For the following exercises, evaluate the common logarithmic expression without using a calculator.

46. $\log(10,000)$

47. $\log(0.001)$

48. $\log(1) + 7$

49. $2\log(100^{-3})$

For the following exercises, evaluate the natural logarithmic expression without using a calculator.

50. $\ln\left(e^{\frac{1}{3}}\right)$ **51.** $\ln(1)$ **52.** $\ln(e^{-0.225}) - 3$ **53.** $25\ln\left(e^{\frac{2}{5}}\right)$

TECHNOLOGY

For the following exercises, evaluate each expression using a calculator. Round to the nearest thousandth.

54. $\log(0.04)$ **55.** $\ln(15)$ **56.** $\ln\left(\dfrac{4}{5}\right)$ **57.** $\log\left(\sqrt{2}\right)$ **58.** $\ln\left(\sqrt{2}\right)$

EXTENSIONS

59. Is $x = 0$ in the domain of the function $f(x) = \log(x)$? If so, what is the value of the function when $x = 0$? Verify the result.

60. Is $f(x) = 0$ in the range of the function $f(x) = \log(x)$? If so, for what value of x? Verify the result.

61. Is there a number x such that $\ln x = 2$? If so, what is that number? Verify the result.

62. Is the following true: $\dfrac{\log_3(27)}{\log_4\left(\dfrac{1}{64}\right)} = -1$? Verify the result.

63. Is the following true: $\dfrac{\ln(e^{1.725})}{\ln(1)} = 1.725$? Verify the result.

REAL-WORLD APPLICATIONS

64. The exposure index EI for a 35 millimeter camera is a measurement of the amount of light that hits the film. It is determined by the equation $EI = \log_2\left(\dfrac{f^2}{t}\right)$, where f is the "f-stop" setting on the camera, and t is the exposure time in seconds. Suppose the f-stop setting is 8 and the desired exposure time is 2 seconds. What will the resulting exposure index be?

65. Refer to the previous exercise. Suppose the light meter on a camera indicates an EI of -2, and the desired exposure time is 16 seconds. What should the f-stop setting be?

66. The intensity levels I of two earthquakes measured on a seismograph can be compared by the formula

$$\log\frac{I_1}{I_2} = M_1 - M_2$$

where M is the magnitude given by the Richter Scale. In August 2009, an earthquake of magnitude 6.1 hit Honshu, Japan. In March 2011, that same region experienced yet another, more devastating earthquake, this time with a magnitude of 9.0.[23] How many times greater was the intensity of the 2011 earthquake? Round to the nearest whole number.

23   http://earthquake.usgs.gov/earthquakes/world/historical.php. Accessed 3/4/2014.

LEARNING OBJECTIVES

In this section, you will:

- Identify the domain of a logarithmic function.
- Graph logarithmic functions.

6.4   GRAPHS OF LOGARITHMIC FUNCTIONS

In **Graphs of Exponential Functions**, we saw how creating a graphical representation of an exponential model gives us another layer of insight for predicting future events. How do logarithmic graphs give us insight into situations? Because every logarithmic function is the inverse function of an exponential function, we can think of every output on a logarithmic graph as the input for the corresponding inverse exponential equation. In other words, logarithms give the *cause* for an *effect*.

To illustrate, suppose we invest \$2,500 in an account that offers an annual interest rate of 5%, compounded continuously. We already know that the balance in our account for any year t can be found with the equation $A = 2500e^{0.05t}$.

But what if we wanted to know the year for any balance? We would need to create a corresponding new function by interchanging the input and the output; thus we would need to create a logarithmic model for this situation. By graphing the model, we can see the output (year) for any input (account balance). For instance, what if we wanted to know how many years it would take for our initial investment to double? **Figure 1** shows this point on the logarithmic graph.

Logarithmic Model Showing Years as a Function of the Balance in the Account

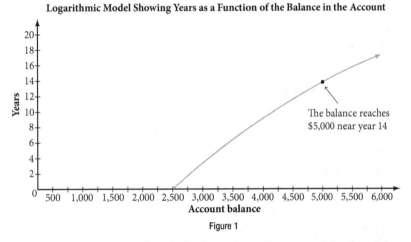

Figure 1

In this section we will discuss the values for which a logarithmic function is defined, and then turn our attention to graphing the family of logarithmic functions.

Finding the Domain of a Logarithmic Function

Before working with graphs, we will take a look at the domain (the set of input values) for which the logarithmic function is defined.

Recall that the exponential function is defined as $y = b^x$ for any real number x and constant $b > 0$, $b \neq 1$, where

- The domain of y is $(-\infty, \infty)$.
- The range of y is $(0, \infty)$.

In the last section we learned that the logarithmic function $y = \log_b(x)$ is the inverse of the exponential function $y = b^x$. So, as inverse functions:

- The domain of $y = \log_b(x)$ is the range of $y = b^x$: $(0, \infty)$.
- The range of $y = \log_b(x)$ is the domain of $y = b^x$: $(-\infty, \infty)$.

Transformations of the parent function $y = \log_b(x)$ behave similarly to those of other functions. Just as with other parent functions, we can apply the four types of transformations—shifts, stretches, compressions, and reflections—to the parent function without loss of shape.

In **Graphs of Exponential Functions** we saw that certain transformations can change the *range* of $y = b^x$. Similarly, applying transformations to the parent function $y = \log_b(x)$ can change the *domain*. When finding the domain of a logarithmic function, therefore, it is important to remember that the domain consists *only of positive real numbers*. That is, the argument of the logarithmic function must be greater than zero.

For example, consider $f(x) = \log_4(2x - 3)$. This function is defined for any values of x such that the argument, in this case $2x - 3$, is greater than zero. To find the domain, we set up an inequality and solve for x :

$$2x - 3 > 0 \qquad \text{Show the argument greater than zero.}$$
$$2x > 3 \qquad \text{Add 3.}$$
$$x > 1.5 \qquad \text{Divide by 2.}$$

In interval notation, the domain of $f(x) = \log_4(2x - 3)$ is $(1.5, \infty)$.

How To...

Given a logarithmic function, identify the domain.

1. Set up an inequality showing the argument greater than zero.
2. Solve for x.
3. Write the domain in interval notation.

Example 1   **Identifying the Domain of a Logarithmic Shift**

What is the domain of $f(x) = \log_2(x + 3)$?

Solution   The logarithmic function is defined only when the input is positive, so this function is defined when $x + 3 > 0$.

Solving this inequality,
$$x + 3 > 0 \qquad \text{The input must be positive.}$$
$$x > -3 \qquad \text{Subtract 3.}$$

The domain of $f(x) = \log_2(x + 3)$ is $(-3, \infty)$.

Try It #1

What is the domain of $f(x) = \log_5(x - 2) + 1$?

Example 2   **Identifying the Domain of a Logarithmic Shift and Reflection**

What is the domain of $f(x) = \log(5 - 2x)$?

Solution   The logarithmic function is defined only when the input is positive, so this function is defined when $5 - 2x > 0$.

Solving this inequality,
$$5 - 2x > 0 \qquad \text{The input must be positive.}$$
$$-2x > -5 \qquad \text{Subtract 5.}$$
$$x < \frac{5}{2} \qquad \text{Divide by } -2 \text{ and switch the inequality.}$$

The domain of $f(x) = \log(5 - 2x)$ is $\left(-\infty, \frac{5}{2}\right)$.

Try It #2

What is the domain of $f(x) = \log(x - 5) + 2$?

Graphing Logarithmic Functions

Now that we have a feel for the set of values for which a logarithmic function is defined, we move on to graphing logarithmic functions. The family of logarithmic functions includes the parent function $y = \log_b(x)$ along with all its transformations: shifts, stretches, compressions, and reflections.

We begin with the parent function $y = \log_b(x)$. Because every logarithmic function of this form is the inverse of an exponential function with the form $y = b^x$, their graphs will be reflections of each other across the line $y = x$. To illustrate this, we can observe the relationship between the input and output values of $y = 2^x$ and its equivalent $x = \log_2(y)$ in **Table 1**.

x	−3	−2	−1	0	1	2	3
$2^x = y$	$\frac{1}{8}$	$\frac{1}{4}$	$\frac{1}{2}$	1	2	4	8
$\log_2(y) = x$	−3	−2	−1	0	1	2	3

Table 1

Using the inputs and outputs from **Table 1**, we can build another table to observe the relationship between points on the graphs of the inverse functions $f(x) = 2^x$ and $g(x) = \log_2(x)$. See **Table 2**.

$f(x) = 2^x$	$\left(-3, \frac{1}{8}\right)$	$\left(-2, \frac{1}{4}\right)$	$\left(-1, \frac{1}{2}\right)$	(0, 1)	(1, 2)	(2, 4)	(3, 8)
$g(x) = \log_2(x)$	$\left(\frac{1}{8}, -3\right)$	$\left(\frac{1}{4}, -2\right)$	$\left(\frac{1}{2}, -1\right)$	(1, 0)	(2, 1)	(4, 2)	(8, 3)

Table 2

As we'd expect, the x- and y-coordinates are reversed for the inverse functions. **Figure 2** shows the graph of f and g.

Figure 2 Notice that the graphs of $f(x) = 2^x$ and $g(x) = \log_2(x)$ are reflections about the line $y = x$.

Observe the following from the graph:

- $f(x) = 2^x$ has a y-intercept at $(0, 1)$ and $g(x) = \log_2(x)$ has an x-intercept at $(1, 0)$.
- The domain of $f(x) = 2^x$, $(-\infty, \infty)$, is the same as the range of $g(x) = \log_2(x)$.
- The range of $f(x) = 2^x$, $(0, \infty)$, is the same as the domain of $g(x) = \log_2(x)$.

characteristics of the graph of the parent function, f(x) = log_b(x)

For any real number x and constant $b > 0$, $b \neq 1$, we can see the following characteristics in the graph of $f(x) = \log_b(x)$:

- one-to-one function
- vertical asymptote: $x = 0$
- domain: $(0, \infty)$
- range: $(-\infty, \infty)$
- x-intercept: $(1, 0)$ and key point $(b, 1)$
- y-intercept: none
- increasing if $b > 1$
- decreasing if $0 < b < 1$

See **Figure 3**.

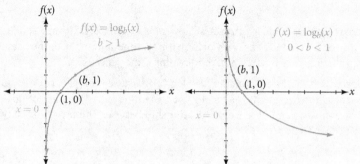

Figure 3

Figure 4 shows how changing the base b in $f(x) = \log_b(x)$ can affect the graphs. Observe that the graphs compress vertically as the value of the base increases. (*Note*: recall that the function $\ln(x)$ has base $e \approx 2.718$.)

Figure 4 The graphs of three logarithmic functions with different bases, all greater than 1.

How To...

Given a logarithmic function with the form $f(x) = \log_b(x)$, graph the function.

1. Draw and label the vertical asymptote, $x = 0$.
2. Plot the x-intercept, $(1, 0)$.
3. Plot the key point $(b, 1)$.
4. Draw a smooth curve through the points.
5. State the domain, $(0, \infty)$, the range, $(-\infty, \infty)$, and the vertical asymptote, $x = 0$.

Example 3   **Graphing a Logarithmic Function with the Form f(x) = log_b(x).**

Graph $f(x) = \log_5(x)$. State the domain, range, and asymptote.

Solution   Before graphing, identify the behavior and key points for the graph.

- Since $b = 5$ is greater than one, we know the function is increasing. The left tail of the graph will approach the vertical asymptote $x = 0$, and the right tail will increase slowly without bound.

- The x-intercept is $(1, 0)$.

- The key point $(5, 1)$ is on the graph.

- We draw and label the asymptote, plot and label the points, and draw a smooth curve through the points (see **Figure 5**).

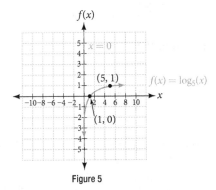

Figure 5

The domain is $(0, \infty)$, the range is $(-\infty, \infty)$, and the vertical asymptote is $x = 0$.

Try It #3

Graph $f(x) = \log_{\frac{1}{5}}(x)$. State the domain, range, and asymptote.

Graphing Transformations of Logarithmic Functions

As we mentioned in the beginning of the section, transformations of logarithmic graphs behave similarly to those of other parent functions. We can shift, stretch, compress, and reflect the parent function $y = \log_b(x)$ without loss of shape.

Graphing a Horizontal Shift of $f(x) = \log_b(x)$

When a constant c is added to the input of the parent function $f(x) = \log_b(x)$, the result is a horizontal shift c units in the *opposite* direction of the sign on c. To visualize horizontal shifts, we can observe the general graph of the parent function $f(x) = \log_b(x)$ and for $c > 0$ alongside the shift left, $g(x) = \log_b(x + c)$, and the shift right, $h(x) = \log_b(x - c)$. See **Figure 6**.

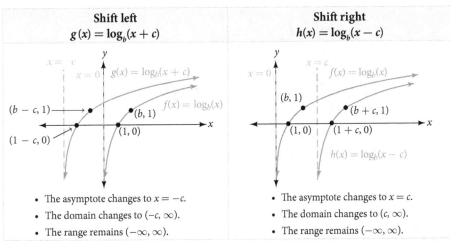

Figure 6

---

horizontal shifts of the parent function $y = \log_b(x)$

For any constant c, the function $f(x) = \log_b(x + c)$

- shifts the parent function $y = \log_b(x)$ left c units if $c > 0$.
- shifts the parent function $y = \log_b(x)$ right c units if $c < 0$.
- has the vertical asymptote $x = -c$.
- has domain $(-c, \infty)$.
- has range $(-\infty, \infty)$.

How To…

Given a logarithmic function with the form $f(x) = \log_b(x + c)$, graph the translation.

1. Identify the horizontal shift:
   a. If $c > 0$, shift the graph of $f(x) = \log_b(x)$ left c units.
   b. If $c < 0$, shift the graph of $f(x) = \log_b(x)$ right c units.
2. Draw the vertical asymptote $x = -c$.
3. Identify three key points from the parent function. Find new coordinates for the shifted functions by subtracting c from the x coordinate.
4. Label the three points.
5. The domain is $(-c, \infty)$, the range is $(-\infty, \infty)$, and the vertical asymptote is $x = -c$.

Example 4   **Graphing a Horizontal Shift of the Parent Function y = log_b(x)**

Sketch the horizontal shift $f(x) = \log_3(x - 2)$ alongside its parent function. Include the key points and asymptotes on the graph. State the domain, range, and asymptote.

Solution   Since the function is $f(x) = \log_3(x - 2)$, we notice $x + (-2) = x - 2$.

Thus $c = -2$, so $c < 0$. This means we will shift the function $f(x) = \log_3(x)$ right 2 units.

The vertical asymptote is $x = -(-2)$ or $x = 2$.

Consider the three key points from the parent function, $\left(\frac{1}{3}, -1\right)$, $(1, 0)$, and $(3, 1)$.

The new coordinates are found by adding 2 to the x coordinates.

Label the points $\left(\frac{7}{3}, -1\right)$, $(3, 0)$, and $(5, 1)$.

The domain is $(2, \infty)$, the range is $(-\infty, \infty)$, and the vertical asymptote is $x = 2$.

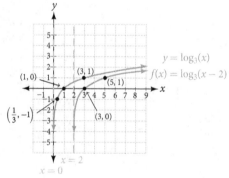

Figure 7

....

Try It #4

Sketch a graph of $f(x) = \log_3(x + 4)$ alongside its parent function. Include the key points and asymptotes on the graph. State the domain, range, and asymptote.

....

Graphing a Vertical Shift of $y = \log_b(x)$

When a constant d is added to the parent function $f(x) = \log_b(x)$, the result is a vertical shift d units in the direction of the sign on d. To visualize vertical shifts, we can observe the general graph of the parent function $f(x) = \log_b(x)$ alongside the shift up, $g(x) = \log_b(x) + d$ and the shift down, $h(x) = \log_b(x) - d$. See **Figure 8**.

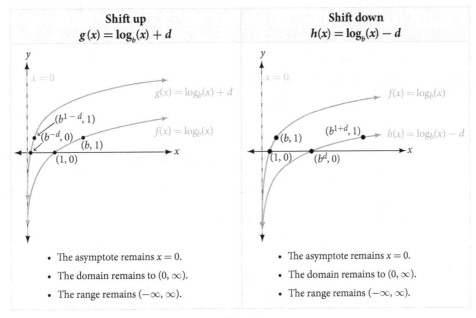

Shift up $g(x) = \log_b(x) + d$	**Shift down** $h(x) = \log_b(x) - d$
• The asymptote remains $x = 0$.	• The asymptote remains $x = 0$.
• The domain remains to $(0, \infty)$.	• The domain remains to $(0, \infty)$.
• The range remains $(-\infty, \infty)$.	• The range remains $(-\infty, \infty)$.

Figure 8

---

vertical shifts of the parent function $y = \log_b(x)$

For any constant d, the function $f(x) = \log_b(x) + d$

- shifts the parent function $y = \log_b(x)$ up d units if $d > 0$.
- shifts the parent function $y = \log_b(x)$ down d units if $d < 0$.
- has the vertical asymptote $x = 0$.
- has domain $(0, \infty)$.
- has range $(-\infty, \infty)$.

How To...

Given a logarithmic function with the form $f(x) = \log_b(x) + d$, graph the translation.

1. Identify the vertical shift:

    a. If $d > 0$, shift the graph of $f(x) = \log_b(x)$ up d units.
    b. If $d < 0$, shift the graph of $f(x) = \log_b(x)$ down d units.

2. Draw the vertical asymptote $x = 0$.
3. Identify three key points from the parent function. Find new coordinates for the shifted functions by adding d to the y coordinate.
4. Label the three points.
5. The domain is $(0, \infty)$, the range is $(-\infty, \infty)$, and the vertical asymptote is $x = 0$.

Example 5   **Graphing a Vertical Shift of the Parent Function** $y = \log_b(x)$

Sketch a graph of $f(x) = \log_3(x) - 2$ alongside its parent function. Include the key points and asymptote on the graph. State the domain, range, and asymptote.

Solution   Since the function is $f(x) = \log_3(x) - 2$, we will notice $d = -2$. Thus $d < 0$.

This means we will shift the function $f(x) = \log_3(x)$ down 2 units.

The vertical asymptote is $x = 0$.

Consider the three key points from the parent function, $\left(\frac{1}{3}, -1\right)$, $(1, 0)$, and $(3, 1)$.

The new coordinates are found by subtracting 2 from the y coordinates.

Label the points $\left(\frac{1}{3}, -3\right)$, $(1, -2)$, and $(3, -1)$.

The domain is $(0, \infty)$, the range is $(-\infty, \infty)$, and the vertical asymptote is $x = 0$.

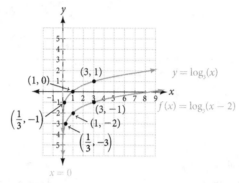

Figure 9

The domain is $(0, \infty)$, the range is $(-\infty, \infty)$, and the vertical asymptote is $x = 0$.

Try It #5

Sketch a graph of $f(x) = \log_2(x) + 2$ alongside its parent function. Include the key points and asymptote on the graph. State the domain, range, and asymptote.

Graphing Stretches and Compressions of $y = \log_b(x)$

When the parent function $f(x) = \log_b(x)$ is multiplied by a constant $a > 0$, the result is a vertical stretch or compression of the original graph. To visualize stretches and compressions, we set $a > 1$ and observe the general graph of the parent function $f(x) = \log_b(x)$ alongside the vertical stretch, $g(x) = a\log_b(x)$ and the vertical compression, $h(x) = \frac{1}{a}\log_b(x)$. See **Figure 10**.

Vertical Stretch	Vertical Compression
$g(x) = a\log_b(x), a > 1$	$h(x) = \frac{1}{a}\log_b(x), a > 1$

- The asymptote remains $x = 0$.
- The x-intercept remains $(1, 0)$.
- The domain remains $(0, \infty)$.
- The range remains $(-\infty, \infty)$.

- The asymptote remains $x = 0$.
- The x-intercept remains $(1, 0)$.
- The domain remains $(0, \infty)$.
- The range remains $(-\infty, \infty)$.

Figure 10

vertical stretches and compressions of the parent function $y = log_b(x)$

For any constant $a > 1$, the function

$$f(x) = a log_b(x)$$

- stretches the parent function $y = log_b(x)$ vertically by a factor of a if $a > 1$.
- compresses the parent function $y = log_b(x)$ vertically by a factor of a if $0 < a < 1$.
- has the vertical asymptote $x = 0$.
- has the x-intercept $(1, 0)$.
- has domain $(0, \infty)$.
- has range $(-\infty, \infty)$.

How To…

Given a logarithmic function with the form $f(x) = a log_b(x)$, $a > 0$, graph the translation.

1. Identify the vertical stretch or compressions:

   a. If $|a| > 1$, the graph of $f(x) = log_b(x)$ is stretched by a factor of a units.

   b. If $|a| < 1$, the graph of $f(x) = log_b(x)$ is compressed by a factor of a units.

2. Draw the vertical asymptote $x = 0$.
3. Identify three key points from the parent function. Find new coordinates for the shifted functions by multiplying the y coordinates by a.
4. Label the three points.
5. The domain is $(0, \infty)$, the range is $(-\infty, \infty)$, and the vertical asymptote is $x = 0$.

Example 6   **Graphing a Stretch or Compression of the Parent Function $y = log_b(x)$**

Sketch a graph of $f(x) = 2 log_4(x)$ alongside its parent function. Include the key points and asymptote on the graph. State the domain, range, and asymptote.

Solution   Since the function is $f(x) = 2 log_4(x)$, we will notice $a = 2$.

This means we will stretch the function $f(x) = log_4(x)$ by a factor of 2.

The vertical asymptote is $x = 0$.

Consider the three key points from the parent function, $\left(\frac{1}{4}, -1 \right)$, $(1, 0)$, and $(4, 1)$.

The new coordinates are found by multiplying the y coordinates by 2.

Label the points $\left(\frac{1}{4}, -2 \right)$, $(1, 0)$, and $(4, 2)$.

The domain is $(0, \infty)$, the range is $(-\infty, \infty)$, and the vertical asymptote is $x = 0$. See **Figure 11**.

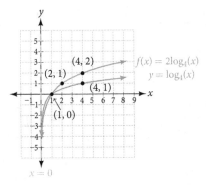

Figure 11

Try It #6

Sketch a graph of $f(x) = \frac{1}{2}\log_4(x)$ alongside its parent function. Include the key points and asymptote on the graph. State the domain, range, and asymptote.

Example 7   Combining a Shift and a Stretch

Sketch a graph of $f(x) = 5\log(x + 2)$. State the domain, range, and asymptote.

Solution   Remember: what happens inside parentheses happens first. First, we move the graph left 2 units, then stretch the function vertically by a factor of 5, as in **Figure 12**. The vertical asymptote will be shifted to $x = -2$. The x-intercept will be $(-1, 0)$. The domain will be $(-2, \infty)$. Two points will help give the shape of the graph: $(-1, 0)$ and $(8, 5)$. We chose $x = 8$ as the x-coordinate of one point to graph because when $x = 8$, $x + 2 = 10$, the base of the common logarithm.

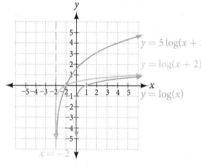

Figure 12

The domain is $(-2, \infty)$, the range is $(-\infty, \infty)$, and the vertical asymptote is $x = -2$.

Try It #7

Sketch a graph of the function $f(x) = 3\log(x - 2) + 1$. State the domain, range, and asymptote.

Graphing Reflections of $f(x) = \log b(x)$

When the parent function $f(x) = \log_b(x)$ is multiplied by -1, the result is a reflection about the x-axis. When the *input* is multiplied by -1, the result is a reflection about the y-axis. To visualize reflections, we restrict $b > 1$, and observe the general graph of the parent function $f(x) = \log_b(x)$ alongside the reflection about the x-axis, $g(x) = -\log_b(x)$ and the reflection about the y-axis, $h(x) = \log_b(-x)$.

Reflection about the x-axis $g(x) = \log_b(x), b > 1$	Reflection about the y-axis $h(x) = \log_b(-x), b > 1$
• The reflected function is decreasing as x moves from zero to infinity.	• The reflected function is decreasing as x moves from infinity to zero.
• The asymptote remains $x = 0$.	• The asymptote remains $x = 0$.
• The x-intercept remains $(1, 0)$.	• The x-intercept remains $(-1, 0)$.
• The key point changes to $(b^{-1}, 1)$.	• The key point changes to $(-b, 1)$.
• The domain remains $(0, \infty)$.	• The domain changes to $(-\infty, 0)$.
• The range remains $(-\infty, \infty)$.	• The range remains $(-\infty, \infty)$.

Figure 13

> **reflections of the parent function $y = \log_b(x)$**
>
> The function $f(x) = -\log_b(x)$
> - reflects the parent function $y = \log_b(x)$ about the x-axis.
> - has domain, $(0, \infty)$, range, $(-\infty, \infty)$, and vertical asymptote, $x = 0$, which are unchanged from the parent function.
>
> The function $f(x) = \log_b(-x)$
> - reflects the parent function $y = \log_b(x)$ about the y-axis.
> - has domain $(-\infty, 0)$.
> - has range, $(-\infty, \infty)$, and vertical asymptote, $x = 0$, which are unchanged from the parent function.

How To...

Given a logarithmic function with the parent function $f(x) = \log_b(x)$, graph a translation.

If $f(x) = -\log_b(x)$	**If $f(x) = \log_b(-x)$**
1. Draw the vertical asymptote, $x = 0$.	1. Draw the vertical asymptote, $x = 0$.
2. Plot the x-intercept, $(1, 0)$.	2. Plot the x-intercept, $(1, 0)$.
3. Reflect the graph of the parent function $f(x) = \log_b(x)$ about the x-axis.	3. Reflect the graph of the parent function $f(x) = \log_b(x)$ about the y-axis.
4. Draw a smooth curve through the points.	4. Draw a smooth curve through the points.
5. State the domain, $(0, \infty)$, the range, $(-\infty, \infty)$, and the vertical asymptote $x = 0$.	5. State the domain, $(-\infty, 0)$, the range, $(-\infty, \infty)$, and the vertical asymptote $x = 0$.

Table 3

Example 8   **Graphing a Reflection of a Logarithmic Function**

Sketch a graph of $f(x) = \log(-x)$ alongside its parent function. Include the key points and asymptote on the graph. State the domain, range, and asymptote.

Solution   Before graphing $f(x) = \log(-x)$, identify the behavior and key points for the graph.

- Since $b = 10$ is greater than one, we know that the parent function is increasing. Since the *input* value is multiplied by -1, f is a reflection of the parent graph about the y-axis. Thus, $f(x) = \log(-x)$ will be decreasing as x moves from negative infinity to zero, and the right tail of the graph will approach the vertical asymptote $x = 0$.

- The x-intercept is $(-1, 0)$.

- We draw and label the asymptote, plot and label the points, and draw a smooth curve through the points.

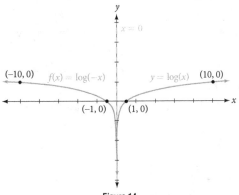

Figure 14

The domain is $(-\infty, 0)$, the range is $(-\infty, \infty)$, and the vertical asymptote is $x = 0$.

Try It #8

Graph $f(x) = -\log(-x)$. State the domain, range, and asymptote.

How To…

Given a logarithmic equation, use a graphing calculator to approximate solutions.

1. Press [**Y=**]. Enter the given logarithm equation or equations as $Y_1=$ and, if needed, $Y_2=$.
2. Press [**GRAPH**] to observe the graphs of the curves and use [**WINDOW**] to find an appropriate view of the graphs, including their point(s) of intersection.
3. To find the value of x, we compute the point of intersection. Press [**2ND**] then [**CALC**]. Select "intersect" and press [**ENTER**] three times. The point of intersection gives the value of x, for the point(s) of intersection.

Example 9   **Approximating the Solution of a Logarithmic Equation**

Solve $4\ln(x) + 1 = -2\ln(x - 1)$ graphically. Round to the nearest thousandth.

Solution   Press [**Y=**] and enter $4\ln(x) + 1$ next to $Y_1=$. Then enter $-2\ln(x - 1)$ next to $Y_2=$. For a window, use the values 0 to 5 for x and −10 to 10 for y. Press [**GRAPH**]. The graphs should intersect somewhere a little to right of $x = 1$.

For a better approximation, press [**2ND**] then [**CALC**]. Select [**5: intersect**] and press [**ENTER**] three times. The x-coordinate of the point of intersection is displayed as 1.3385297. (Your answer may be different if you use a different window or use a different value for **Guess?**) So, to the nearest thousandth, $x \approx 1.339$.

Try It #9

Solve $5\log(x + 2) = 4 - \log(x)$ graphically. Round to the nearest thousandth.

Summarizing Translations of the Logarithmic Function

Now that we have worked with each type of translation for the logarithmic function, we can summarize each in **Table 4** to arrive at the general equation for translating exponential functions.

Translations of the Parent Function $y = \log_b(x)$	
Translation	**Form**
Shift • Horizontally c units to the left • Vertically d units up	$y = \log_b(x + c) + d$
Stretch and Compress • Stretch if $\lvert a \rvert > 1$ • Compression if $\lvert a \rvert < 1$	$y = a\log_b(x)$
Reflect about the x-axis	$y = -\log_b(x)$
Reflect about the y-axis	$y = \log_b(-x)$
General equation for all translations	$y = a\log_b(x + c) + d$

Table 4

translations of logarithmic functions

All translations of the parent logarithmic function, $y = \log_b(x)$, have the form

$$f(x) = a\log_b(x + c) + d$$

where the parent function, $y = \log_b(x), b > 1$, is

• shifted vertically up d units.
• shifted horizontally to the left c units.
• stretched vertically by a factor of $\lvert a \rvert$ if $\lvert a \rvert > 0$.
• compressed vertically by a factor of $\lvert a \rvert$ if $0 < \lvert a \rvert < 1$.
• reflected about the x-axis when $a < 0$.

For $f(x) = \log(-x)$, the graph of the parent function is reflected about the y-axis.

Example 10 **Finding the Vertical Asymptote of a Logarithm Graph**

What is the vertical asymptote of $f(x) = -2\log_3(x + 4) + 5$?

Solution The vertical asymptote is at $x = -4$.

Analysis The coefficient, the base, and the upward translation do not affect the asymptote. The shift of the curve 4 units to the left shifts the vertical asymptote to $x = -4$.

Try It #10

What is the vertical asymptote of $f(x) = 3 + \ln(x - 1)$?

Example 11 **Finding the Equation from a Graph**

Find a possible equation for the common logarithmic function graphed in **Figure 15**.

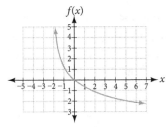

Figure 15

Solution This graph has a vertical asymptote at $x = -2$ and has been vertically reflected. We do not know yet the vertical shift or the vertical stretch. We know so far that the equation will have form:

$$f(x) = -a\log(x + 2) + k$$

It appears the graph passes through the points $(-1, 1)$ and $(2, -1)$. Substituting $(-1, 1)$,

$$\begin{aligned} 1 &= -a\log(-1 + 2) + k && \text{Substitute } (-1, 1). \\ 1 &= -a\log(1) + k && \text{Arithmetic.} \\ 1 &= k && \log(1) = 0. \end{aligned}$$

Next, substituting in $(2, -1)$,

$$\begin{aligned} -1 &= -a\log(2 + 2) + 1 && \text{Plug in } (2, -1). \\ -2 &= -a\log(4) && \text{Arithmetic.} \\ a &= \frac{2}{\log(4)} && \text{Solve for } a. \end{aligned}$$

This gives us the equation $f(x) = -\dfrac{2}{\log(4)}\log(x + 2) + 1$.

*Analysis You can verify this answer by comparing the function values in **Table 5** with the points on the graph in **Figure 15**.*

x	-1	0	1	2	3
$f(x)$	1	0	-0.58496	-1	-1.3219
x	4	5	6	7	8
$f(x)$	-1.5850	-1.8074	-2	-2.1699	-2.3219

Table 5

Try It #11

Give the equation of the natural logarithm graphed in **Figure 16**.

Figure 16

Q & A...

Is it possible to tell the domain and range and describe the end behavior of a function just by looking at the graph?

Yes, if we know the function is a general logarithmic function. For example, look at the graph in **Figure 16**. The graph approaches $x = -3$ (or thereabouts) more and more closely, so $x = -3$ is, or is very close to, the vertical asymptote. It approaches from the right, so the domain is all points to the right, $\{x \mid x > -3\}$. The range, as with all general logarithmic functions, is all real numbers. And we can see the end behavior because the graph goes down as it goes left and up as it goes right. The end behavior is that as $x \rightarrow -3^+$, $f(x) \rightarrow -\infty$ and as $x \rightarrow \infty$, $f(x) \rightarrow \infty$.

Access these online resources for additional instruction and practice with graphing logarithms.

- Graph an Exponential Function and Logarithmic Function (http://openstaxcollege.org/l/graphexplog)
- Match Graphs with Exponential and Logarithmic Functions (http://openstaxcollege.org/l/matchexplog)
- Find the Domain of Logarithmic Functions (http://openstaxcollege.org/l/domainlog)

6.4 SECTION EXERCISES

VERBAL

1. The inverse of every logarithmic function is an exponential function and vice-versa. What does this tell us about the relationship between the coordinates of the points on the graphs of each?

2. What type(s) of translation(s), if any, affect the range of a logarithmic function?

3. What type(s) of translation(s), if any, affect the domain of a logarithmic function?

4. Consider the general logarithmic function $f(x) = \log_b(x)$. Why can't x be zero?

5. Does the graph of a general logarithmic function have a horizontal asymptote? Explain.

ALGEBRAIC

For the following exercises, state the domain and range of the function.

6. $f(x) = \log_3(x + 4)$

7. $h(x) = \ln\left(\dfrac{1}{2} - x\right)$

8. $g(x) = \log_5(2x + 9) - 2$

9. $h(x) = \ln(4x + 17) - 5$

10. $f(x) = \log_2(12 - 3x) - 3$

For the following exercises, state the domain and the vertical asymptote of the function.

11. $f(x) = \log_b(x - 5)$

12. $g(x) = \ln(3 - x)$

13. $f(x) = \log(3x + 1)$

14. $f(x) = 3\log(-x) + 2$

15. $g(x) = -\ln(3x + 9) - 7$

For the following exercises, state the domain, vertical asymptote, and end behavior of the function.

16. $f(x) = \ln(2 - x)$

17. $f(x) = \log\left(x - \dfrac{3}{7}\right)$

18. $h(x) = -\log(3x - 4) + 3$

19. $g(x) = \ln(2x + 6) - 5$

20. $f(x) = \log_3(15 - 5x) + 6$

For the following exercises, state the domain, range, and x- and y-intercepts, if they exist. If they do not exist, write DNE.

21. $h(x) = \log_4(x - 1) + 1$

22. $f(x) = \log(5x + 10) + 3$

23. $g(x) = \ln(-x) - 2$

24. $f(x) = \log_2(x + 2) - 5$

25. $h(x) = 3\ln(x) - 9$

GRAPHICAL

For the following exercises, match each function in **Figure 17** with the letter corresponding to its graph.

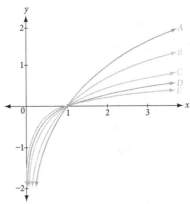

Figure 17

26.

27. $d(x) = \log(x)$

28. $f(x) = \ln(x)$

29. $g(x) = \log_2(x)$

30. $h(x) = \log_5(x)$

31. $j(x) = \log_{25}(x)$

For the following exercises, match each function in **Figure 18** with the letter corresponding to its graph.

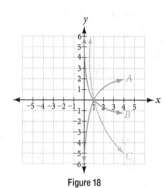

Figure 18

32.

33. $f(x) = \log_{\frac{1}{3}}(x)$

34. $g(x) = \log_2(x)$

35. $h(x) = \log_{\frac{3}{4}}(x)$

For the following exercises, sketch the graphs of each pair of functions on the same axis.

36. $f(x) = \log(x)$ and $g(x) = 10^x$

37. $f(x) = \log(x)$ and $g(x) = \log_{\frac{1}{2}}(x)$

38. $f(x) = \log_4(x)$ and $g(x) = \ln(x)$

39. $f(x) = e^x$ and $g(x) = \ln(x)$

For the following exercises, match each function in **Figure 19** with the letter corresponding to its graph.

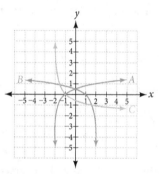

Figure 19

40. $f(x) = \log_4(-x + 2)$

41. $g(x) = -\log_4(x + 2)$

42. $h(x) = \log_4(x + 2)$

For the following exercises, sketch the graph of the indicated function.

43. $f(x) = \log_2(x + 2)$

44. $f(x) = 2\log(x)$

45. $f(x) = \ln(-x)$

46. $g(x) = \log(4x + 16) + 4$

47. $g(x) = \log(6 - 3x) + 1$

48. $h(x) = -\frac{1}{2}\ln(x + 1) - 3$

For the following exercises, write a logarithmic equation corresponding to the graph shown.

49. Use $y = \log_2(x)$ as the parent function.

50. Use $f(x) = \log_3(x)$ as the parent function.

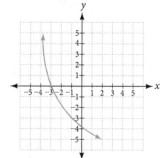

51. Use $f(x) = \log_4(x)$ as the parent function.

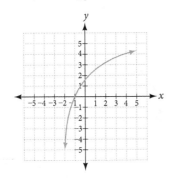

52. Use $f(x) = \log_5(x)$ as the parent function.

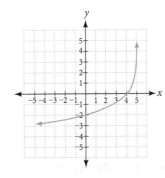

TECHNOLOGY

For the following exercises, use a graphing calculator to find approximate solutions to each equation.

53. $\log(x - 1) + 2 = \ln(x - 1) + 2$ **54.** $\log(2x - 3) + 2 = -\log(2x - 3) + 5$ **55.** $\ln(x - 2) = -\ln(x + 1)$

56. $2\ln(5x + 1) = \frac{1}{2}\ln(-5x) + 1$ **57.** $\frac{1}{3}\log(1 - x) = \log(x + 1) + \frac{1}{3}$

EXTENSIONS

58. Let b be any positive real number such that $b \neq 1$. What must $\log_b 1$ be equal to? Verify the result.

59. Explore and discuss the graphs of $f(x) = \log_{\frac{1}{2}}(x)$ and $g(x) = -\log_2(x)$. Make a conjecture based on the result.

60. Prove the conjecture made in the previous exercise.

61. What is the domain of the function $f(x) = \ln\left(\frac{x + 2}{x - 4}\right)$? Discuss the result.

62. Use properties of exponents to find the x-intercepts of the function $f(x) = \log(x^2 + 4x + 4)$ algebraically. Show the steps for solving, and then verify the result by graphing the function.

LEARNING OBJECTIVES

In this section, you will:

- Use the product rule for logarithms.
- Use the quotient rule for logarithms.
- Use the power rule for logarithms.
- Expand logarithmic expressions.
- Condense logarithmic expressions.
- Use the change-of-base formula for logarithms.

6.5   LOGARITHMIC PROPERTIES

Figure 1 The pH of hydrochloric acid is tested with litmus paper. (credit: David Berardan)

In chemistry, pH is used as a measure of the acidity or alkalinity of a substance. The pH scale runs from 0 to 14. Substances with a pH less than 7 are considered acidic, and substances with a pH greater than 7 are said to be alkaline. Our bodies, for instance, must maintain a pH close to 7.35 in order for enzymes to work properly. To get a feel for what is acidic and what is alkaline, consider the following pH levels of some common substances:

- Battery acid: 0.8
- Stomach acid: 2.7
- Orange juice: 3.3
- Pure water: 7 (at 25° C)

- Human blood: 7.35
- Fresh coconut: 7.8
- Sodium hydroxide (lye): 14

To determine whether a solution is acidic or alkaline, we find its pH, which is a measure of the number of active positive hydrogen ions in the solution. The pH is defined by the following formula, where a is the concentration of hydrogen ion in the solution

$$\text{pH} = -\log([H^+])$$
$$= \log\left(\frac{1}{([H^+])}\right)$$

The equivalence of $-\log([H^+])$ and $\log\left(\frac{1}{[H^+]}\right)$ is one of the logarithm properties we will examine in this section.

Using the Product Rule for Logarithms

Recall that the logarithmic and exponential functions "undo" each other. This means that logarithms have similar properties to exponents. Some important properties of logarithms are given here. First, the following properties are easy to prove.

$$\log_b(1) = 0$$
$$\log_b(b) = 1$$

For example, $\log_5 1 = 0$ since $5^0 = 1$. And $\log_5 5 = 1$ since $5^1 = 5$.

Next, we have the inverse property.

$$\log_b(b^x) = x$$

$$b^{\log_b(x)} = x, x > 0$$

For example, to evaluate log(100), we can rewrite the logarithm as $\log_{10}(10^2)$, and then apply the inverse property $\log_b(b^x) = x$ to get $\log_{10}(10^2) = 2$.

To evaluate $e^{\ln(7)}$, we can rewrite the logarithm as $e^{\log_e(7)}$, and then apply the inverse property $b^{\log_b(x)} = x$ to get $e^{\log_e(7)} = 7$.

Finally, we have the one-to-one property.

$$\log_b M = \log_b N \text{ if and only if } M = N$$

We can use the one-to-one property to solve the equation $\log_3(3x) = \log_3(2x + 5)$ for x. Since the bases are the same, we can apply the one-to-one property by setting the arguments equal and solving for x :

$$3x = 2x + 5 \qquad \text{Set the arguments equal.}$$

$$x = 5 \qquad \text{Subtract } 2x.$$

But what about the equation $\log_3(3x) + \log_3(2x + 5) = 2$? The one-to-one property does not help us in this instance. Before we can solve an equation like this, we need a method for combining terms on the left side of the equation.

Recall that we use the *product rule of exponents* to combine the product of exponents by adding: $x^a x^b = x^{a+b}$. We have a similar property for logarithms, called the **product rule for logarithms**, which says that the logarithm of a product is equal to a sum of logarithms. Because logs are exponents, and we multiply like bases, we can add the exponents. We will use the inverse property to derive the product rule below.

Given any real number x and positive real numbers M, N, and b, where $b \neq 1$, we will show

$$\log_b(MN) = \log_b(M) + \log_b(N).$$

Let $m = \log_b(M)$ and $n = \log_b(N)$. In exponential form, these equations are $b^m = M$ and $b^n = N$. It follows that

$$\log_b(MN) = \log_b(b^m b^n) \qquad \text{Substitute for } M \text{ and } N.$$

$$= \log_b(b^{m+n}) \qquad \text{Apply the product rule for exponents.}$$

$$= m + n \qquad \text{Apply the inverse property of logs.}$$

$$= \log_b(M) + \log_b(N) \qquad \text{Substitute for } m \text{ and } n.$$

Note that repeated applications of the product rule for logarithms allow us to simplify the logarithm of the product of any number of factors. For example, consider $\log_b(wxyz)$. Using the product rule for logarithms, we can rewrite this logarithm of a product as the sum of logarithms of its factors:

$$\log_b(wxyz) = \log_b(w) + \log_b(x) + \log_b(y) + \log_b(z)$$

---

the product rule for logarithms

The **product rule for logarithms** can be used to simplify a logarithm of a product by rewriting it as a sum of individual logarithms.

$$\log_b(MN) = \log_b(M) + \log_b(N) \text{ for } b > 0$$

How To...

Given the logarithm of a product, use the product rule of logarithms to write an equivalent sum of logarithms.

1. Factor the argument completely, expressing each whole number factor as a product of primes.

2. Write the equivalent expression by summing the logarithms of each factor.

Example 1   **Using the Product Rule for Logarithms**

Expand $\log_3(30x(3x + 4))$.

Solution   We begin by factoring the argument completely, expressing 30 as a product of primes.

$$\log_3(30x(3x + 4)) = \log_3(2 \cdot 3 \cdot 5 \cdot x \cdot (3x + 4))$$

Next we write the equivalent equation by summing the logarithms of each factor.

$$\log_3(30x(3x + 4)) = \log_3(2) + \log_3(3) + \log_3(5) + \log_3(x) + \log_3(3x + 4)$$

Try It #1

Expand $\log_b(8k)$.

Using the Quotient Rule for Logarithms

For quotients, we have a similar rule for logarithms. Recall that we use the *quotient rule of exponents* to combine the quotient of exponents by subtracting: $\frac{x^a}{x^b} = x^{a-b}$. The **quotient rule for logarithms** says that the logarithm of a quotient is equal to a difference of logarithms. Just as with the product rule, we can use the inverse property to derive the quotient rule.

Given any real number x and positive real numbers M, N, and b, where $b \neq 1$, we will show

$$\log_b\left(\frac{M}{N}\right) = \log_b(M) - \log_b(N).$$

Let $m = \log_b(M)$ and $n = \log_b(N)$. In exponential form, these equations are $b^m = M$ and $b^n = N$. It follows that

$$\log_b\left(\frac{M}{N}\right) = \log_b\left(\frac{b^m}{b^n}\right) \qquad \text{Substitute for } M \text{ and } N.$$

$$= \log_b(b^{m-n}) \qquad \text{Apply the quotient rule for exponents.}$$

$$= m - n \qquad \text{Apply the inverse property of logs.}$$

$$= \log_b(M) - \log_b(N) \qquad \text{Substitute for } m \text{ and } n.$$

For example, to expand $\log\left(\frac{2x^2 + 6x}{3x + 9}\right)$, we must first express the quotient in lowest terms. Factoring and canceling we get,

$$\log\left(\frac{2x^2 + 6x}{3x + 9}\right) = \log\left(\frac{2x(x + 3)}{3(x + 3)}\right) \qquad \text{Factor the numerator and denominator.}$$

$$= \log\left(\frac{2x}{3}\right) \qquad \text{Cancel the common factors.}$$

Next we apply the quotient rule by subtracting the logarithm of the denominator from the logarithm of the numerator. Then we apply the product rule.

$$\log\left(\frac{2x}{3}\right) = \log(2x) - \log(3)$$

$$= \log(2) + \log(x) - \log(3)$$

---

the quotient rule for logarithms

The **quotient rule for logarithms** can be used to simplify a logarithm or a quotient by rewriting it as the difference of individual logarithms.

$$\log_b\left(\frac{M}{N}\right) = \log_b(M) - \log_b(N)$$

How To…

Given the logarithm of a quotient, use the quotient rule of logarithms to write an equivalent difference of logarithms.

1. Express the argument in lowest terms by factoring the numerator and denominator and canceling common terms.
2. Write the equivalent expression by subtracting the logarithm of the denominator from the logarithm of the numerator.
3. Check to see that each term is fully expanded. If not, apply the product rule for logarithms to expand completely.

Example 2   **Using the Quotient Rule for Logarithms**

Expand $\log_2\left(\dfrac{15x(x-1)}{(3x+4)(2-x)}\right)$.

Solution   First we note that the quotient is factored and in lowest terms, so we apply the quotient rule.

$$\log_2\left(\frac{15x(x-1)}{(3x+4)(2-x)}\right) = \log_2(15x(x-1)) - \log_2((3x+4)(2-x))$$

Notice that the resulting terms are logarithms of products. To expand completely, we apply the product rule, noting that the prime factors of the factor 15 are 3 and 5.

$$\log_2(15x(x-1)) - \log_2((3x+4)(2-x)) = [\log_2(3) + \log_2(5) + \log_2(x) + \log_2(x-1)] - [\log_2(3x+4) + \log_2(2-x)]$$

$$= \log_2(3) + \log_2(5) + \log_2(x) + \log_2(x-1) - \log_2(3x+4) - \log_2(2-x)$$

Analysis   There are exceptions to consider in this and later examples. First, because denominators must never be zero, this expression is not defined for $x = -\dfrac{4}{3}$ and $x = 2$. Also, since the argument of a logarithm must be positive, we note as we observe the expanded logarithm, that $x > 0$, $x > 1$, $x > -\dfrac{4}{3}$, and $x < 2$. Combining these conditions is beyond the scope of this section, and we will not consider them here or in subsequent exercises.

Try It #2

Expand $\log_3\left(\dfrac{7x^2+21x}{7x(x-1)(x-2)}\right)$.

Using the Power Rule for Logarithms

We've explored the product rule and the quotient rule, but how can we take the logarithm of a power, such as x^2? One method is as follows:

$$\log_b(x^2) = \log_b(x \cdot x)$$
$$= \log_b(x) + \log_b(x)$$
$$= 2\log_b(x)$$

Notice that we used the product rule for logarithms to find a solution for the example above. By doing so, we have derived the **power rule for logarithms**, which says that the log of a power is equal to the exponent times the log of the base. Keep in mind that, although the input to a logarithm may not be written as a power, we may be able to change it to a power. For example,

$$100 = 10^2 \qquad \sqrt{3} = 3^{\frac{1}{2}} \qquad \frac{1}{e} = e^{-1}$$

---

the power rule for logarithms

The **power rule for logarithms** can be used to simplify the logarithm of a power by rewriting it as the product of the exponent times the logarithm of the base.

$$\log_b(M^n) = n\log_b(M)$$

How To…

Given the logarithm of a power, use the power rule of logarithms to write an equivalent product of a factor and a logarithm.

1. Express the argument as a power, if needed.
2. Write the equivalent expression by multiplying the exponent times the logarithm of the base.

Example 3   **Expanding a Logarithm with Powers**

Expand $\log_2(x^5)$.

Solution   The argument is already written as a power, so we identify the exponent, 5, and the base, x, and rewrite the equivalent expression by multiplying the exponent times the logarithm of the base.

$$\log_2(x^5) = 5\log_2(x)$$

Try It #3

Expand $\ln(x^2)$.

Example 4   **Rewriting an Expression as a Power before Using the Power Rule**

Expand $\log_3(25)$ using the power rule for logs.

Solution   Expressing the argument as a power, we get $\log_3(25) = \log_3(5^2)$.

Next we identify the exponent, 2, and the base, 5, and rewrite the equivalent expression by multiplying the exponent times the logarithm of the base.

$$\log_3(5^2) = 2\log_3(5)$$

Try It #4

Expand $\ln\left(\dfrac{1}{x^2}\right)$.

Example 5   **Using the Power Rule in Reverse**

Rewrite $4\ln(x)$ using the power rule for logs to a single logarithm with a leading coefficient of 1.

Solution   Because the logarithm of a power is the product of the exponent times the logarithm of the base, it follows that the product of a number and a logarithm can be written as a power. For the expression $4\ln(x)$, we identify the factor, 4, as the exponent and the argument, x, as the base, and rewrite the product as a logarithm of a power: $4\ln(x) = \ln(x^4)$.

Try It #5

Rewrite $2\log_3(4)$ using the power rule for logs to a single logarithm with a leading coefficient of 1.

Expanding Logarithmic Expressions

Taken together, the product rule, quotient rule, and power rule are often called "laws of logs." Sometimes we apply more than one rule in order to simplify an expression. For example:

$$\log_b\left(\frac{6x}{y}\right) = \log_b(6x) - \log_b(y)$$
$$= \log_b(6) + \log_b(x) - \log_b(y)$$

We can use the power rule to expand logarithmic expressions involving negative and fractional exponents. Here is an alternate proof of the quotient rule for logarithms using the fact that a reciprocal is a negative power:

$$\log_b\left(\frac{A}{C}\right) = \log_b(AC^{-1})$$
$$= \log_b(A) + \log_b(C^{-1})$$
$$= \log_b(A) + (-1)\log_b(C)$$
$$= \log_b(A) - \log_b(C)$$

We can also apply the product rule to express a sum or difference of logarithms as the logarithm of a product. With practice, we can look at a logarithmic expression and expand it mentally, writing the final answer. Remember, however, that we can only do this with products, quotients, powers, and roots—never with addition or subtraction inside the argument of the logarithm.

Example 6 **Expanding Logarithms Using Product, Quotient, and Power Rules**

Rewrite $\ln\left(\dfrac{x^4 y}{7}\right)$ as a sum or difference of logs.

Solution   First, because we have a quotient of two expressions, we can use the quotient rule:

$$\ln\left(\frac{x^4 y}{7}\right) = \ln(x^4 y) - \ln(7)$$

Then seeing the product in the first term, we use the product rule:

$$\ln(x^4 y) - \ln(7) = \ln(x^4) + \ln(y) - \ln(7)$$

Finally, we use the power rule on the first term:

$$\ln(x^4) + \ln(y) - \ln(7) = 4\ln(x) + \ln(y) - \ln(7)$$

Try It #6

Expand $\log\left(\dfrac{x^2 y^3}{z^4}\right)$.

Example 7   **Using the Power Rule for Logarithms to Simplify the Logarithm of a Radical Expression**

Expand $\log(\sqrt{x})$.

Solution

$$\log(\sqrt{x}) = \log(x)^{\frac{1}{2}}$$

$$= \frac{1}{2}\log(x)$$

Try It #7

Expand $\ln(\sqrt[3]{x^2})$.

Q & A...

Can we expand $\ln(x^2 + y^2)$?

No. There is no way to expand the logarithm of a sum or difference inside the argument of the logarithm.

Example 8   **Expanding Complex Logarithmic Expressions**

Expand $\log_6\left(\dfrac{64x^3(4x+1)}{(2x-1)}\right)$.

Solution   We can expand by applying the Product and Quotient Rules.

$$\log_6\left(\frac{64x^3(4x+1)}{(2x-1)}\right) = \log_6(64) + \log_6(x^3) + \log_6(4x+1) - \log_6(2x-1) \qquad \text{Apply the Quotient Rule.}$$

$$= \log_6(2^6) + \log_6(x^3) + \log_6(4x+1) - \log_6(2x-1) \qquad \text{Simplify by writing 64 as } 2^6.$$

$$= 6\log_6(2) + 3\log_6(x) + \log_6(4x+1) - \log_6(2x-1) \qquad \text{Apply the Power Rule.}$$

Try It #8

Expand $\ln\left(\dfrac{\sqrt{(x-1)(2x+1)^2}}{x^2-9}\right)$.

Condensing Logarithmic Expressions

We can use the rules of logarithms we just learned to condense sums, differences, and products with the same base as a single logarithm. It is important to remember that the logarithms must have the same base to be combined. We will learn later how to change the base of any logarithm before condensing.

How To…

Given a sum, difference, or product of logarithms with the same base, write an equivalent expression as a single logarithm.

1. Apply the power property first. Identify terms that are products of factors and a logarithm, and rewrite each as the logarithm of a power.
2. Next apply the product property. Rewrite sums of logarithms as the logarithm of a product.
3. Apply the quotient property last. Rewrite differences of logarithms as the logarithm of a quotient.

Example 9   **Using the Product and Quotient Rules to Combine Logarithms**

Write $\log_3(5) + \log_3(8) - \log_3(2)$ as a single logarithm.

Solution   Using the product and quotient rules

$$\log_3(5) + \log_3(8) = \log_3(5 \cdot 8) = \log_3(40)$$

This reduces our original expression to

$$\log_3(40) - \log_3(2)$$

Then, using the quotient rule

$$\log_3(40) - \log_3(2) = \log_3\left(\frac{40}{2}\right) = \log_3(20)$$

Try It #9

Condense $\log(3) - \log(4) + \log(5) - \log(6)$.

Example 10   **Condensing Complex Logarithmic Expressions**

Condense $\log_2(x^2) + \frac{1}{2}\log_2(x - 1) - 3\log_2((x + 3)^2)$.

Solution   We apply the power rule first:

$$\log_2(x^2) + \frac{1}{2}\log_2(x - 1) - 3\log_2((x + 3)^2) = \log_2(x^2) + \log_2\left(\sqrt{x - 1}\right) - \log_2((x + 3)^6)$$

Next we apply the product rule to the sum:

$$\log_2(x^2) + \log_2\left(\sqrt{x - 1}\right) - \log_2((x + 3)^6) = \log_2\left(x^2\sqrt{x - 1}\right) - \log_2((x + 3)^6)$$

Finally, we apply the quotient rule to the difference:

$$\log_2\left(x^2\sqrt{x - 1}\right) - \log_2((x + 3)^6) = \log_2\left(\frac{x^2\sqrt{x - 1}}{(x + 3)^6}\right)$$

Try It #10

Rewrite $\log(5) + 0.5\log(x) - \log(7x - 1) + 3\log(x - 1)$ as a single logarithm.

Example 11   **Rewriting as a Single Logarithm**

Rewrite $2\log(x) - 4\log(x + 5) + \frac{1}{x}\log(3x + 5)$ as a single logarithm.

Solution   We apply the power rule first:

$$2\log(x) - 4\log(x + 5) + \frac{1}{x}\log(3x + 5) = \log(x^2) - \log((x + 5)^4) + \log\left((3x + 5)^{x^{-1}}\right)$$

Next we apply the product rule to the sum:

$$\log(x^2) - \log((x + 5)^4) + \log\left((3x + 5)^{x^{-1}}\right) = \log(x^2) - \log\left((x + 5)^4(3x + 5)^{x^{-1}}\right)$$

Finally, we apply the quotient rule to the difference:

$$\log(x^2) - \log\left((x + 5)^4(3x + 5)^{x^{-1}}\right) = \log\left(\frac{x^2}{(x + 5)^4(3x + 5)^{x^{-1}}}\right)$$

Try It #11

Condense $4(3\log(x) + \log(x + 5) - \log(2x + 3))$.

Example 12 **Applying of the Laws of Logs**

Recall that, in chemistry, $\text{pH} = -\log[H^+]$. If the concentration of hydrogen ions in a liquid is doubled, what is the effect on pH?

Solution Suppose C is the original concentration of hydrogen ions, and P is the original pH of the liquid. Then $P = -\log(C)$. If the concentration is doubled, the new concentration is $2C$. Then the pH of the new liquid is

$$\text{pH} = -\log(2C)$$

Using the product rule of logs

$$\text{pH} = -\log(2C) = -(\log(2) + \log(C)) = -\log(2) - \log(C)$$

Since $P = -\log(C)$, the new pH is

$$\text{pH} = P - \log(2) \approx P - 0.301$$

When the concentration of hydrogen ions is doubled, the pH decreases by about 0.301.

Try It #12

How does the pH change when the concentration of positive hydrogen ions is decreased by half?

Using the Change-of-Base Formula for Logarithms

Most calculators can evaluate only common and natural logs. In order to evaluate logarithms with a base other than 10 or e, we use the **change-of-base formula** to rewrite the logarithm as the quotient of logarithms of any other base; when using a calculator, we would change them to common or natural logs.

To derive the change-of-base formula, we use the one-to-one property and **power rule for logarithms**.

Given any positive real numbers M, b, and n, where $n \neq 1$ and $b \neq 1$, we show

$$\log_b(M) = \frac{\log_n(M)}{\log_n(b)}$$

Let $y = \log_b(M)$. By taking the log base n of both sides of the equation, we arrive at an exponential form, namely $b^y = M$. It follows that

$$\log_n(b^y) = \log_n(M) \qquad \text{Apply the one-to-one property.}$$

$$y\log_n(b) = \log_n(M) \qquad \text{Apply the power rule for logarithms.}$$

$$y = \frac{\log_n(M)}{\log_n(b)} \qquad \text{Isolate } y.$$

$$\log_b(M) = \frac{\log_n(M)}{\log_n(b)} \qquad \text{Substitute for } y.$$

For example, to evaluate $\log_5(36)$ using a calculator, we must first rewrite the expression as a quotient of common or natural logs. We will use the common log.

$$\log_5(36) = \frac{\log(36)}{\log(5)} \qquad \text{Apply the change of base formula using base 10.}$$

$$\approx 2.2266 \qquad \text{Use a calculator to evaluate to 4 decimal places.}$$

---

the change-of-base formula

The **change-of-base formula** can be used to evaluate a logarithm with any base.
For any positive real numbers M, b, and n, where $n \neq 1$ and $b \neq 1$,

$$\log_b(M) = \frac{\log_n(M)}{\log_n(b)}.$$

It follows that the change-of-base formula can be used to rewrite a logarithm with any base as the quotient of common or natural logs.

$$\log_b(M) = \frac{\ln(M)}{\ln(b)} \quad \text{and} \quad \log_b(M) = \frac{\log_n(M)}{\log_n(b)}$$

How To...

Given a logarithm with the form $\log_b(M)$, use the change-of-base formula to rewrite it as a quotient of logs with any positive base n, where $n \neq 1$.

1. Determine the new base n, remembering that the common log, $\log(x)$, has base 10, and the natural log, $\ln(x)$, has base e.
2. Rewrite the log as a quotient using the change-of-base formula
   a. The numerator of the quotient will be a logarithm with base n and argument M.
   b. The denominator of the quotient will be a logarithm with base n and argument b.

Example 13   **Changing Logarithmic Expressions to Expressions Involving Only Natural Logs**

Change $\log_5(3)$ to a quotient of natural logarithms.

Solution   Because we will be expressing $\log_5(3)$ as a quotient of natural logarithms, the new base, $n = e$.
We rewrite the log as a quotient using the change-of-base formula. The numerator of the quotient will be the natural log with argument 3. The denominator of the quotient will be the natural log with argument 5.

$$\log_b(M) = \frac{\ln(M)}{\ln(b)} \qquad \log_5(3) = \frac{\ln(3)}{\ln(5)}$$

Try It #13

Change $\log_{0.5}(8)$ to a quotient of natural logarithms.

Q & A...

Can we change common logarithms to natural logarithms?

Yes. Remember that $\log(9)$ means $\log_{10}(9)$. So, $\log(9) = \frac{\ln(9)}{\ln(10)}$.

Example 14   **Using the Change-of-Base Formula with a Calculator**

Evaluate $\log_2(10)$ using the change-of-base formula with a calculator.

Solution   According to the change-of-base formula, we can rewrite the log base 2 as a logarithm of any other base. Since our calculators can evaluate the natural log, we might choose to use the natural logarithm, which is the log base e.

$$\log_2(10) = \frac{\ln(10)}{\ln(2)} \qquad \text{Apply the change of base formula using base } e.$$
$$\approx 3.3219 \qquad \text{Use a calculator to evaluate to 4 decimal places.}$$

Try It #14

Evaluate $\log_5(100)$ using the change-of-base formula.

Access this online resource for additional instruction and practice with laws of logarithms.

- The Properties of Logarithms (http://openstaxcollege.org/l/proplog)
- Expand Logarithmic Expressions (http://openstaxcollege.org/l/expandlog)
- Evaluate a Natural Logarithmic Expression (http://openstaxcollege.org/l/evaluatelog)

6.5 SECTION EXERCISES

VERBAL

1. How does the power rule for logarithms help when solving logarithms with the form $\log_b(\sqrt[n]{x})$?

2. What does the change-of-base formula do? Why is it useful when using a calculator?

ALGEBRAIC

For the following exercises, expand each logarithm as much as possible. Rewrite each expression as a sum, difference, or product of logs.

3. $\log_b(7x \cdot 2y)$

4. $\ln(3ab \cdot 5c)$

5. $\log_b\left(\dfrac{13}{17}\right)$

6. $\log_4\left(\dfrac{\frac{x}{z}}{w}\right)$

7. $\ln\left(\dfrac{1}{4^k}\right)$

8. $\log_2(y^x)$

For the following exercises, condense to a single logarithm if possible.

9. $\ln(7) + \ln(x) + \ln(y)$

10. $\log_3(2) + \log_3(a) + \log_3(11) + \log_3(b)$

11. $\log_b(28) - \log_b(7)$

12. $\ln(a) - \ln(d) - \ln(c)$

13. $-\log_b\left(\dfrac{1}{7}\right)$

14. $\dfrac{1}{3}\ln(8)$

For the following exercises, use the properties of logarithms to expand each logarithm as much as possible. Rewrite each expression as a sum, difference, or product of logs.

15. $\log\left(\dfrac{x^{15}y^{13}}{z^{19}}\right)$

16. $\ln\left(\dfrac{a^{-2}}{b^{-4}c^5}\right)$

17. $\log(\sqrt{x^3y^{-4}})$

18. $\ln\left(y\sqrt{\dfrac{y}{1-y}}\right)$

19. $\log(x^2 y^3 \sqrt[3]{x^2 y^5})$

For the following exercises, condense each expression to a single logarithm using the properties of logarithms.

20. $\log(2x^4) + \log(3x^5)$

21. $\ln(6x^9) - \ln(3x^2)$

22. $2\log(x) + 3\log(x+1)$

23. $\log(x) - \dfrac{1}{2}\log(y) + 3\log(z)$

24. $4\log_7(c) + \dfrac{\log_7(a)}{3} + \dfrac{\log_7(b)}{3}$

For the following exercises, rewrite each expression as an equivalent ratio of logs using the indicated base.

25. $\log_7(15)$ to base e

26. $\log_{14}(55.875)$ to base 10

For the following exercises, suppose $\log_5(6) = a$ and $\log_5(11) = b$. Use the change-of-base formula along with properties of logarithms to rewrite each expression in terms of a and b. Show the steps for solving.

27. $\log_{11}(5)$

28. $\log_6(55)$

29. $\log_{11}\left(\dfrac{6}{11}\right)$

NUMERIC

For the following exercises, use properties of logarithms to evaluate without using a calculator.

30. $\log_3\left(\dfrac{1}{9}\right) - 3\log_3(3)$

31. $6\log_8(2) + \dfrac{\log_8(64)}{3\log_8(4)}$

32. $2\log_9(3) - 4\log_9(3) + \log_9\left(\dfrac{1}{729}\right)$

For the following exercises, use the change-of-base formula to evaluate each expression as a quotient of natural logs. Use a calculator to approximate each to five decimal places.

33. $\log_3(22)$

34. $\log_8(65)$

35. $\log_6(5.38)$

36. $\log_4\left(\dfrac{15}{2}\right)$

37. $\log_{\frac{1}{2}}(4.7)$

EXTENSIONS

38. Use the product rule for logarithms to find all x values such that $\log_{12}(2x + 6) + \log_{12}(x + 2) = 2$. Show the steps for solving.

39. Use the quotient rule for logarithms to find all x values such that $\log_6(x + 2) - \log_6(x - 3) = 1$. Show the steps for solving.

40. Can the power property of logarithms be derived from the power property of exponents using the equation $b^x = m$? If not, explain why. If so, show the derivation.

41. Prove that $\log_b(n) = \dfrac{1}{\log_n(b)}$ for any positive integers $b > 1$ and $n > 1$.

42. Does $\log_{81}(2401) = \log_3(7)$? Verify the claim algebraically.

LEARNING OBJECTIVES

In this section, you will:

- Use like bases to solve exponential equations.
- Use logarithms to solve exponential equations.
- Use the definition of a logarithm to solve logarithmic equations.
- Use the one-to-one property of logarithms to solve logarithmic equations.
- Solve applied problems involving exponential and logarithmic equations.

6.6   EXPONENTIAL AND LOGARITHMIC EQUATIONS

Figure 1 Wild rabbits in Australia. The rabbit population grew so quickly in Australia that the event became known as the "rabbit plague." (credit: Richard Taylor, Flickr)

In 1859, an Australian landowner named Thomas Austin released 24 rabbits into the wild for hunting. Because Australia had few predators and ample food, the rabbit population exploded. In fewer than ten years, the rabbit population numbered in the millions.

Uncontrolled population growth, as in the wild rabbits in Australia, can be modeled with exponential functions. Equations resulting from those exponential functions can be solved to analyze and make predictions about exponential growth. In this section, we will learn techniques for solving exponential functions.

Using Like Bases to Solve Exponential Equations

The first technique involves two functions with like bases. Recall that the one-to-one property of exponential functions tells us that, for any real numbers b, S, and T, where $b > 0$, $b \neq 1$, $b^S = b^T$ if and only if $S = T$.

In other words, when an exponential equation has the same base on each side, the exponents must be equal. This also applies when the exponents are algebraic expressions. Therefore, we can solve many exponential equations by using the rules of exponents to rewrite each side as a power with the same base. Then, we use the fact that exponential functions are one-to-one to set the exponents equal to one another, and solve for the unknown.

For example, consider the equation $3^{4x-7} = \dfrac{3^{2x}}{3}$. To solve for x, we use the division property of exponents to rewrite the right side so that both sides have the common base, 3. Then we apply the one-to-one property of exponents by setting the exponents equal to one another and solving for x:

$$3^{4x-7} = \frac{3^{2x}}{3}$$

$$3^{4x-7} = \frac{3^{2x}}{3^1} \qquad \text{Rewrite 3 as } 3^1.$$

$$3^{4x-7} = 3^{2x-1} \qquad \text{Use the division property of exponents.}$$

$$4x - 7 = 2x - 1 \qquad \text{Apply the one-to-one property of exponents.}$$

$$2x = 6 \qquad\qquad \text{Subtract } 2x \text{ and add 7 to both sides.}$$

$$x = 3 \qquad\qquad\; \text{Divide by 3.}$$

> **using the one-to-one property of exponential functions to solve exponential equations**
>
> For any algebraic expressions S and T, and any positive real number $b \neq 1$,
>
> $$b^S = b^T \text{ if and only if } S = T$$

How To...

Given an exponential equation with the form $b^S = b^T$, where S and T are algebraic expressions with an unknown, solve for the unknown.

1. Use the rules of exponents to simplify, if necessary, so that the resulting equation has the form $b^S = b^T$.
2. Use the one-to-one property to set the exponents equal.
3. Solve the resulting equation, $S = T$, for the unknown.

Example 1   **Solving an Exponential Equation with a Common Base**

Solve $2^{x-1} = 2^{2x-4}$.

Solution

$$2^{x-1} = 2^{2x-4} \qquad \text{The common base is 2.}$$
$$x - 1 = 2x - 4 \qquad \text{By the one-to-one property the exponents must be equal.}$$
$$x = 3 \qquad \text{Solve for } x.$$

Try It #1

Solve $5^{2x} = 5^{3x+2}$.

Rewriting Equations So All Powers Have the Same Base

Sometimes the common base for an exponential equation is not explicitly shown. In these cases, we simply rewrite the terms in the equation as powers with a common base, and solve using the one-to-one property.

For example, consider the equation $256 = 4^{x-5}$. We can rewrite both sides of this equation as a power of 2. Then we apply the rules of exponents, along with the one-to-one property, to solve for x:

$$256 = 4^{x-5}$$
$$2^8 = (2^2)^{x-5} \qquad \text{Rewrite each side as a power with base 2.}$$
$$2^8 = 2^{2x-10} \qquad \text{Use the one-to-one property of exponents.}$$
$$8 = 2x - 10 \qquad \text{Apply the one-to-one property of exponents.}$$
$$18 = 2x \qquad \text{Add 10 to both sides.}$$
$$x = 9 \qquad \text{Divide by 2.}$$

How To...

Given an exponential equation with unlike bases, use the one-to-one property to solve it.

1. Rewrite each side in the equation as a power with a common base.
2. Use the rules of exponents to simplify, if necessary, so that the resulting equation has the form $b^S = b^T$.
3. Use the one-to-one property to set the exponents equal.
4. Solve the resulting equation, $S = T$, for the unknown.

Example 2   **Solving Equations by Rewriting Them to Have a Common Base**

Solve $8^{x+2} = 16^{x+1}$.

Solution

$$8^{x+2} = 16^{x+1}$$
$$(2^3)^{x+2} = (2^4)^{x+1} \qquad \text{Write 8 and 16 as powers of 2.}$$
$$2^{3x+6} = 2^{4x+4} \qquad \text{To take a power of a power, multiply exponents .}$$
$$3x + 6 = 4x + 4 \qquad \text{Use the one-to-one property to set the exponents equal.}$$
$$x = 2 \qquad \text{Solve for } x.$$

Try It #2

Solve $5^{2x} = 25^{3x+2}$.

Example 3   **Solving Equations by Rewriting Roots with Fractional Exponents to Have a Common Base**

Solve $2^{5x} = \sqrt{2}$.

Solution

$$2^{5x} = 2^{\frac{1}{2}}$$ Write the square root of 2 as a power of 2.

$$5x = \frac{1}{2}$$ Use the one-to-one property.

$$x = \frac{1}{10}$$ Solve for x.

Try It #3

Solve $5^x = \sqrt{5}$.

Q & A...

Do all exponential equations have a solution? If not, how can we tell if there is a solution during the problem-solving process?

No. Recall that the range of an exponential function is always positive. While solving the equation, we may obtain an expression that is undefined.

Example 4   **Solving an Equation with Positive and Negative Powers**

Solve $3^{x+1} = -2$.

Solution   This equation has no solution. There is no real value of x that will make the equation a true statement because any power of a positive number is positive.

Analysis   **Figure 2** *shows that the two graphs do not cross so the left side is never equal to the right side. Thus the equation has no solution.*

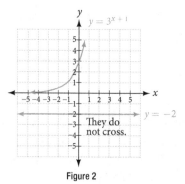

Figure 2

Try It #4

Solve $2^x = -100$.

Solving Exponential Equations Using Logarithms

Sometimes the terms of an exponential equation cannot be rewritten with a common base. In these cases, we solve by taking the logarithm of each side. Recall, since $\log(a) = \log(b)$ is equivalent to $a = b$, we may apply logarithms with the same base on both sides of an exponential equation.

Given an exponential equation in which a common base cannot be found, solve for the unknown.

1. Apply the logarithm of both sides of the equation.

    a. If one of the terms in the equation has base 10, use the common logarithm.

    b. If none of the terms in the equation has base 10, use the natural logarithm.

2. Use the rules of logarithms to solve for the unknown.

Example 5   **Solving an Equation Containing Powers of Different Bases**

Solve $5^{x+2} = 4^x$.

Solution

$5^{x+2} = 4^x$	There is no easy way to get the powers to have the same base .
$\ln(5^{x+2}) = \ln(4^x)$	Take ln of both sides.
$(x+2)\ln(5) = x\ln(4)$	Use laws of logs.
$x\ln(5) + 2\ln(5) = x\ln(4)$	Use the distributive law.
$x\ln(5) - x\ln(4) = -2\ln(5)$	Get terms containing x on one side, terms without x on the other.
$x(\ln(5) - \ln(4)) = -2\ln(5)$	On the left hand side, factor out an x.
$x\ln\left(\dfrac{5}{4}\right) = \ln\left(\dfrac{1}{25}\right)$	Use the laws of logs.
$x = \dfrac{\ln\left(\dfrac{1}{25}\right)}{\ln\left(\dfrac{5}{4}\right)}$	Divide by the coefficient of x.

Try It #5

Solve $2^x = 3^{x+1}$.

Q & A...

Is there any way to solve $2^x = 3^x$?

Yes. The solution is 0.

Equations Containing e

One common type of exponential equations are those with base e. This constant occurs again and again in nature, in mathematics, in science, in engineering, and in finance. When we have an equation with a base e on either side, we can use the natural logarithm to solve it.

How To...

Given an equation of the form $y = Ae^{kt}$, solve for t.

1. Divide both sides of the equation by A.

2. Apply the natural logarithm of both sides of the equation.

3. Divide both sides of the equation by k.

Example 6   **Solve an Equation of the Form $y = Ae^{kt}$**

Solve $100 = 20e^{2t}$.

Solution

$100 = 20e^{2t}$	
$5 = e^{2t}$	Divide by the coefficient of the power .
$\ln(5) = 2t$	Take ln of both sides. Use the fact that $\ln(x)$ and e^x are inverse functions.
$t = \dfrac{\ln(5)}{2}$	Divide by the coefficient of t.

Analysis Using laws of logs, we can also write this answer in the form $t = \ln\sqrt{5}$. If we want a decimal approximation of the answer, we use a calculator.

Try It #6

Solve $3e^{0.5t} = 11$.

Q & A...

Does every equation of the form $y = Ae^{kt}$ have a solution?

No. There is a solution when $k \neq 0$, and when y and A are either both 0 or neither 0, and they have the same sign. An example of an equation with this form that has no solution is $2 = -3e^t$.

Example 7   **Solving an Equation That Can Be Simplified to the Form $y = Ae^{kt}$**

Solve $4e^{2x} + 5 = 12$.

Solution

$$4e^{2x} + 5 = 12$$

$$4e^{2x} = 7 \qquad \text{Combine like terms.}$$

$$e^{2x} = \frac{7}{4} \qquad \text{Divide by the coefficient of the power.}$$

$$2x = \ln\left(\frac{7}{4}\right) \qquad \text{Take ln of both sides.}$$

$$x = \frac{1}{2}\ln\left(\frac{7}{4}\right) \qquad \text{Solve for } x.$$

Try It #7

Solve $3 + e^{2t} = 7e^{2t}$.

Extraneous Solutions

Sometimes the methods used to solve an equation introduce an **extraneous solution**, which is a solution that is correct algebraically but does not satisfy the conditions of the original equation. One such situation arises in solving when the logarithm is taken on both sides of the equation. In such cases, remember that the argument of the logarithm must be positive. If the number we are evaluating in a logarithm function is negative, there is no output.

Example 8   **Solving Exponential Functions in Quadratic Form**

Solve $e^{2x} - e^x = 56$.

Solution

$$e^{2x} - e^x = 56$$

$$e^{2x} - e^x - 56 = 0 \qquad \text{Get one side of the equation equal to zero.}$$

$$(e^x + 7)(e^x - 8) = 0 \qquad \text{Factor by the FOIL method.}$$

$$e^x + 7 = 0 \text{ or } e^x - 8 = 0 \qquad \text{If a product is zero, then one factor must be zero.}$$

$$e^x = -7 \text{ or } e^x = 8 \qquad \text{Isolate the exponentials.}$$

$$e^x = 8 \qquad \text{Reject the equation in which the power equals a negative number.}$$

$$x = \ln(8) \qquad \text{Solve the equation in which the power equals a positive number.}$$

Analysis When we plan to use factoring to solve a problem, we always get zero on one side of the equation, because zero has the unique property that when a product is zero, one or both of the factors must be zero. We reject the equation $e^x = -7$ because a positive number never equals a negative number. The solution $\ln(-7)$ is not a real number, and in the real number system this solution is rejected as an extraneous solution.

Try It #8

Solve $e^{2x} = e^x + 2$.

Q & A...

Does every logarithmic equation have a solution?

No. Keep in mind that we can only apply the logarithm to a positive number. Always check for extraneous solutions.

Using the Definition of a Logarithm to Solve Logarithmic Equations

We have already seen that every logarithmic equation $\log_b(x) = y$ is equivalent to the exponential equation $b^y = x$. We can use this fact, along with the rules of logarithms, to solve logarithmic equations where the argument is an algebraic expression.

For example, consider the equation $\log_2(2) + \log_2(3x - 5) = 3$. To solve this equation, we can use rules of logarithms to rewrite the left side in compact form and then apply the definition of logs to solve for x:

$$\log_2(2) + \log_2(3x - 5) = 3$$

$$\log_2(2(3x - 5)) = 3 \qquad \text{Apply the product rule of logarithms.}$$

$$\log_2(6x - 10) = 3 \qquad \text{Distribute.}$$

$$2^3 = 6x - 10 \qquad \text{Apply the definition of a logarithm.}$$

$$8 = 6x - 10 \qquad \text{Calculate } 2^3.$$

$$18 = 6x \qquad \text{Add 10 to both sides.}$$

$$x = 3 \qquad \text{Divide by 6.}$$

---

using the definition of a logarithm to solve logarithmic equations

For any algebraic expression S and real numbers b and c, where $b > 0$, $b \neq 1$,

$$\log_b(S) = c \text{ if and only if } b^c = S$$

Example 9   Using Algebra to Solve a Logarithmic Equation

Solve $2\ln(x) + 3 = 7$.

Solution

$$2\ln(x) + 3 = 7$$

$$2\ln(x) = 4 \qquad \text{Subtract 3.}$$

$$\ln(x) = 2 \qquad \text{Divide by 2.}$$

$$x = e^2 \qquad \text{Rewrite in exponential form.}$$

Try It #9

Solve $6 + \ln(x) = 10$.

Example 10   Using Algebra Before and After Using the Definition of the Natural Logarithm

Solve $2\ln(6x) = 7$.

Solution

$$2\ln(6x) = 7$$

$$\ln(6x) = \frac{7}{2} \qquad \text{Divide by 2.}$$

$$6x = e^{\frac{7}{2}} \qquad \text{Use the definition of ln.}$$

$$x = \frac{1}{6}e^{\frac{7}{2}} \qquad \text{Divide by 6.}$$

Try It #10

Solve $2\ln(x + 1) = 10$.

Example 11   **Using a Graph to Understand the Solution to a Logarithmic Equation**

Solve $\ln(x) = 3$.

Solution

$$\ln(x) = 3$$
$$x = e^3 \qquad\qquad \text{Use the definition of the natural logarithm.}$$

Figure 3 represents the graph of the equation. On the graph, the x-coordinate of the point at which the two graphs intersect is close to 20. In other words $e^3 \approx 20$. A calculator gives a better approximation: $e^3 \approx 20.0855$.

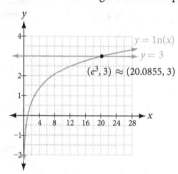

Figure 3 The graphs of $y = \ln(x)$ and $y = 3$ cross at the point $(e^3, 3)$, which is approximately (20.0855, 3).

Try It #11

Use a graphing calculator to estimate the approximate solution to the logarithmic equation $2^x = 1000$ to 2 decimal places.

Using the One-to-One Property of Logarithms to Solve Logarithmic Equations

As with exponential equations, we can use the one-to-one property to solve logarithmic equations. The one-to-one property of logarithmic functions tells us that, for any real numbers $x > 0$, $S > 0$, $T > 0$ and any positive real number b, where $b \neq 1$,

$$\log_b(S) = \log_b(T) \text{ if and only if } S = T.$$

For example,

$$\text{If } \log_2(x - 1) = \log_2(8), \text{ then } x - 1 = 8.$$

So, if $x - 1 = 8$, then we can solve for x, and we get $x = 9$. To check, we can substitute $x = 9$ into the original equation: $\log_2(9 - 1) = \log_2(8) = 3$. In other words, when a logarithmic equation has the same base on each side, the arguments must be equal. This also applies when the arguments are algebraic expressions. Therefore, when given an equation with logs of the same base on each side, we can use rules of logarithms to rewrite each side as a single logarithm. Then we use the fact that logarithmic functions are one-to-one to set the arguments equal to one another and solve for the unknown.

For example, consider the equation $\log(3x - 2) - \log(2) = \log(x + 4)$. To solve this equation, we can use the rules of logarithms to rewrite the left side as a single logarithm, and then apply the one-to-one property to solve for x:

$$\log(3x - 2) - \log(2) = \log(x + 4)$$
$$\log\left(\frac{3x - 2}{2}\right) = \log(x + 4) \qquad \text{Apply the quotient rule of logarithms.}$$
$$\frac{3x - 2}{2} = x + 4 \qquad \text{Apply the one to one property of a logarithm.}$$
$$3x - 2 = 2x + 8 \qquad \text{Multiply both sides of the equation by 2.}$$
$$x = 10 \qquad \text{Subtract } 2x \text{ and add 2.}$$

To check the result, substitute $x = 10$ into $\log(3x - 2) - \log(2) = \log(x + 4)$.

$$\log(3(10) - 2) - \log(2) = \log((10) + 4)$$

$$\log(28) - \log(2) = \log(14)$$

$$\log\left(\frac{28}{2}\right) = \log(14) \qquad \text{The solution checks.}$$

---

using the one-to-one property of logarithms to solve logarithmic equations

For any algebraic expressions S and T and any positive real number b, where $b \neq 1$,

$$\log_b(S) = \log_b(T) \text{ if and only if } S = T$$

Note, when solving an equation involving logarithms, always check to see if the answer is correct or if it is an extraneous solution.

How To...

Given an equation containing logarithms, solve it using the one-to-one property.

1. Use the rules of logarithms to combine like terms, if necessary, so that the resulting equation has the form $\log_b S = \log_b T$.
2. Use the one-to-one property to set the arguments equal.
3. Solve the resulting equation, $S = T$, for the unknown.

Example 12   **Solving an Equation Using the One-to-One Property of Logarithms**

Solve $\ln(x^2) = \ln(2x + 3)$.

Solution

$$\ln(x^2) = \ln(2x + 3)$$

$$x^2 = 2x + 3 \qquad \text{Use the one-to-one property of the logarithm.}$$

$$x^2 - 2x - 3 = 0 \qquad \text{Get zero on one side before factoring.}$$

$$(x - 3)(x + 1) = 0 \qquad \text{Factor using FOIL.}$$

$$x - 3 = 0 \text{ or } x + 1 = 0 \qquad \text{If a product is zero, one of the factors must be zero.}$$

$$x = 3 \text{ or } x = -1 \qquad \text{Solve for } x.$$

Analysis *There are two solutions: 3 or −1. The solution −1 is negative, but it checks when substituted into the original equation because the argument of the logarithm functions is still positive.*

Try It #12

Solve $\ln(x^2) = \ln(1)$.

Solving Applied Problems Using Exponential and Logarithmic Equations

In previous sections, we learned the properties and rules for both exponential and logarithmic functions. We have seen that any exponential function can be written as a logarithmic function and vice versa. We have used exponents to solve logarithmic equations and logarithms to solve exponential equations. We are now ready to combine our skills to solve equations that model real-world situations, whether the unknown is in an exponent or in the argument of a logarithm.

One such application is in science, in calculating the time it takes for half of the unstable material in a sample of a radioactive substance to decay, called its half-life. **Table 1** lists the half-life for several of the more common radioactive substances.

Substance	Use	Half-life
gallium-67	nuclear medicine	80 hours
cobalt-60	manufacturing	5.3 years
technetium-99m	nuclear medicine	6 hours
americium-241	construction	432 years
carbon-14	archeological dating	5,715 years
uranium-235	atomic power	703,800,000 years

Table 1

We can see how widely the half-lives for these substances vary. Knowing the half-life of a substance allows us to calculate the amount remaining after a specified time. We can use the formula for radioactive decay:

$$A(t) = A_0 e^{\frac{\ln(0.5)}{T}t}$$
$$A(t) = A_0 e^{\ln(0.5)\frac{t}{T}}$$
$$A(t) = A_0 (e^{\ln(0.5)})^{\frac{t}{T}}$$
$$A(t) = A_0 \left(\frac{1}{2}\right)^{\frac{t}{T}}$$

where
- A_0 is the amount initially present
- T is the half-life of the substance
- t is the time period over which the substance is studied
- y is the amount of the substance present after time t

Example 13   **Using the Formula for Radioactive Decay to Find the Quantity of a Substance**

How long will it take for ten percent of a 1,000-gram sample of uranium-235 to decay?

Solution $y = 1000e^{\frac{\ln(0.5)}{703,800,000}t}$

$900 = 1000e^{\frac{\ln(0.5)}{703,800,000}t}$ After 10% decays, 900 grams are left.

$0.9 = e^{\frac{\ln(0.5)}{703,800,000}t}$ Divide by 1000.

$\ln(0.9) = \ln\left(e^{\frac{\ln(0.5)}{703,800,000}t}\right)$ Take ln of both sides.

$\ln(0.9) = \frac{\ln(0.5)}{703,800,000}t$ $\ln(e^M) = M$

$t = 703,800,000 \times \frac{\ln(0.9)}{\ln(0.5)}$ years   Solve for t.

$t \approx 106,979,777$ years

Analysis   Ten percent of 1,000 grams is 100 grams. If 100 grams decay, the amount of uranium-235 remaining is 900 grams.

Try It #13

How long will it take before twenty percent of our 1,000-gram sample of uranium-235 has decayed?

Access these online resources for additional instruction and practice with exponential and logarithmic equations.

- Solving Logarithmic Equations (http://openstaxcollege.org/l/solvelogeq)
- Solving Exponential Equations with Logarithms (http://openstaxcollege.org/l/solveexplog)

6.6   SECTION EXERCISES

VERBAL

1. How can an exponential equation be solved?

2. When does an extraneous solution occur? How can an extraneous solution be recognized?

3. When can the one-to-one property of logarithms be used to solve an equation? When can it not be used?

ALGEBRAIC

For the following exercises, use like bases to solve the exponential equation.

4. $4^{-3v-2} = 4^{-v}$

5. $64 \cdot 4^{3x} = 16$

6. $3^{2x+1} \cdot 3^x = 243$

7. $2^{-3n} \cdot \dfrac{1}{4} = 2^{n+2}$

8. $625 \cdot 5^{3x+3} = 125$

9. $\dfrac{36^{3b}}{36^{2b}} = 216^{2-b}$

10. $\left(\dfrac{1}{64}\right)^{3n} \cdot 8 = 2^6$

For the following exercises, use logarithms to solve.

11. $9^{x-10} = 1$

12. $2e^{6x} = 13$

13. $e^{r+10} - 10 = -42$

14. $2 \cdot 10^{9a} = 29$

15. $-8 \cdot 10^{p+7} - 7 = -24$

16. $7e^{3n-5} + 5 = -89$

17. $e^{-3k} + 6 = 44$

18. $-5e^{9x-8} - 8 = -62$

19. $-6e^{9x+8} + 2 = -74$

20. $2^{x+1} = 5^{2x-1}$

21. $e^{2x} - e^x - 132 = 0$

22. $7e^{8x+8} - 5 = -95$

23. $10e^{8x+3} + 2 = 8$

24. $4e^{3x+3} - 7 = 53$

25. $8e^{-5x-2} - 4 = -90$

26. $3^{2x+1} = 7^{x-2}$

27. $e^{2x} - e^x - 6 = 0$

28. $3e^{3-3x} + 6 = -31$

For the following exercises, use the definition of a logarithm to rewrite the equation as an exponential equation.

29. $\log\left(\dfrac{1}{100}\right) = -2$

30. $\log_{324}(18) = \dfrac{1}{2}$

For the following exercises, use the definition of a logarithm to solve the equation.

31. $5\log_7(n) = 10$

32. $-8\log_9(x) = 16$

33. $4 + \log_2(9k) = 2$

34. $2\log(8n+4) + 6 = 10$

35. $10 - 4\ln(9 - 8x) = 6$

For the following exercises, use the one-to-one property of logarithms to solve.

36. $\ln(10 - 3x) = \ln(-4x)$

37. $\log_{13}(5n - 2) = \log_{13}(8 - 5n)$

38. $\log(x + 3) - \log(x) = \log(74)$

39. $\ln(-3x) = \ln(x^2 - 6x)$

40. $\log_4(6 - m) = \log_4 3(m)$

41. $\ln(x - 2) - \ln(x) = \ln(54)$

42. $\log_9(2n^2 - 14n) = \log_9(-45 + n^2)$

43. $\ln(x^2 - 10) + \ln(9) = \ln(10)$

For the following exercises, solve each equation for x.

44. $\log(x + 12) = \log(x) + \log(12)$

45. $\ln(x) + \ln(x - 3) = \ln(7x)$

46. $\log_2(7x + 6) = 3$

47. $\ln(7) + \ln(2 - 4x^2) = \ln(14)$

48. $\log_8(x + 6) - \log_8(x) = \log_8(58)$

49. $\ln(3) - \ln(3 - 3x) = \ln(4)$

50. $\log_3(3x) - \log_3(6) = \log_3(77)$

GRAPHICAL

For the following exercises, solve the equation for x, if there is a solution. Then graph both sides of the equation, and observe the point of intersection (if it exists) to verify the solution.

51. $\log_9(x) - 5 = -4$

52. $\log_3(x) + 3 = 2$

53. $\ln(3x) = 2$

54. $\ln(x - 5) = 1$

55. $\log(4) + \log(-5x) = 2$

56. $-7 + \log_3(4 - x) = -6$

57. $\ln(4x - 10) - 6 = -5$

58. $\log(4 - 2x) = \log(-4x)$

59. $\log_{11}(-2x^2 - 7x) = \log_{11}(x - 2)$

60. $\ln(2x + 9) = \ln(-5x)$

61. $\log_9(3 - x) = \log_9(4x - 8)$

62. $\log(x^2 + 13) = \log(7x + 3)$

63. $\dfrac{3}{\log_2(10)} - \log(x - 9) = \log(44)$

64. $\ln(x) - \ln(x + 3) = \ln(6)$

For the following exercises, solve for the indicated value, and graph the situation showing the solution point.

65. An account with an initial deposit of $6,500 earns 7.25% annual interest, compounded continuously. How much will the account be worth after 20 years?

66. The formula for measuring sound intensity in decibels D is defined by the equation $D = 10 \log\left(\dfrac{I}{I_0}\right)$, where I is the intensity of the sound in watts per square meter and $I_0 = 10^{-12}$ is the lowest level of sound that the average person can hear. How many decibels are emitted from a jet plane with a sound intensity of $8.3 \cdot 10^2$ watts per square meter?

67. The population of a small town is modeled by the equation $P = 1650e^{0.5t}$ where t is measured in years. In approximately how many years will the town's population reach 20,000?

TECHNOLOGY

For the following exercises, solve each equation by rewriting the exponential expression using the indicated logarithm. Then use a calculator to approximate the variable to 3 decimal places.

68. $1000(1.03)^t = 5000$ using the common log.

69. $e^{5x} = 17$ using the natural log

70. $3(1.04)^{3t} = 8$ using the common log

71. $3^{4x-5} = 38$ using the common log

72. $50e^{-0.12t} = 10$ using the natural log

For the following exercises, use a calculator to solve the equation. Unless indicated otherwise, round all answers to the nearest ten-thousandth.

73. $7e^{3x-5} + 7.9 = 47$

74. $\ln(3) + \ln(4.4x + 6.8) = 2$

75. $\log(-0.7x - 9) = 1 + 5\log(5)$

76. Atmospheric pressure P in pounds per square inch is represented by the formula $P = 14.7e^{-0.21x}$, where x is the number of miles above sea level. To the nearest foot, how high is the peak of a mountain with an atmospheric pressure of 8.369 pounds per square inch? (*Hint*: there are 5280 feet in a mile)

77. The magnitude M of an earthquake is represented by the equation $M = \dfrac{2}{3} \log\left(\dfrac{E}{E_0}\right)$ where E is the amount of energy released by the earthquake in joules and $E_0 = 10^{4.4}$ is the assigned minimal measure released by an earthquake. To the nearest hundredth, what would the magnitude be of an earthquake releasing $1.4 \cdot 10^{13}$ joules of energy?

EXTENSIONS

78. Use the definition of a logarithm along with the one-to-one property of logarithms to prove that $b^{\log_b x} = x$.

79. Recall the formula for continually compounding interest, $y = Ae^{kt}$. Use the definition of a logarithm along with properties of logarithms to solve the formula for time t such that t is equal to a single logarithm.

80. Recall the compound interest formula $A = a\left(1 + \dfrac{r}{k}\right)^{kt}$. Use the definition of a logarithm along with properties of logarithms to solve the formula for time t.

81. Newton's Law of Cooling states that the temperature T of an object at any time t can be described by the equation $T = T_s + (T_0 - T_s)e^{-kt}$, where T_s is the temperature of the surrounding environment, T_0 is the initial temperature of the object, and k is the cooling rate. Use the definition of a logarithm along with properties of logarithms to solve the formula for time t such that t is equal to a single logarithm.

LEARNING OBJECTIVES

In this section, you will:

- Model exponential growth and decay.
- Use Newton's Law of Cooling.
- Use logistic-growth models.
- Choose an appropriate model for data.
- Express an exponential model in base e.

6.7 EXPONENTIAL AND LOGARITHMIC MODELS

Figure 1 A nuclear research reactor inside the Neely Nuclear Research Center on the Georgia Institute of Technology campus. (credit: Georgia Tech Research Institute)

We have already explored some basic applications of exponential and logarithmic functions. In this section, we explore some important applications in more depth, including radioactive isotopes and Newton's Law of Cooling.

Modeling Exponential Growth and Decay

In real-world applications, we need to model the behavior of a function. In mathematical modeling, we choose a familiar general function with properties that suggest that it will model the real-world phenomenon we wish to analyze. In the case of rapid growth, we may choose the exponential growth function:

$$y = A_0 e^{kt}$$

where A_0 is equal to the value at time zero, e is Euler's constant, and k is a positive constant that determines the rate (percentage) of growth. We may use the exponential growth function in applications involving **doubling time**, the time it takes for a quantity to double. Such phenomena as wildlife populations, financial investments, biological samples, and natural resources may exhibit growth based on a doubling time. In some applications, however, as we will see when we discuss the logistic equation, the logistic model sometimes fits the data better than the exponential model.

On the other hand, if a quantity is falling rapidly toward zero, without ever reaching zero, then we should probably choose the exponential decay model. Again, we have the form $y = A_0 e^{kt}$ where A_0 is the starting value, and e is Euler's constant. Now k is a negative constant that determines the rate of decay. We may use the exponential decay model when we are calculating **half-life**, or the time it takes for a substance to exponentially decay to half of its original quantity. We use half-life in applications involving radioactive isotopes.

In our choice of a function to serve as a mathematical model, we often use data points gathered by careful observation and measurement to construct points on a graph and hope we can recognize the shape of the graph. Exponential growth and decay graphs have a distinctive shape, as we can see in **Figure 2** and **Figure 3**. It is important to remember that, although parts of each of the two graphs seem to lie on the x-axis, they are really a tiny distance above the x-axis.

Figure 2 A graph showing exponential growth. The equation is $y = 2e^{3x}$.

Figure 3 A graph showing exponential decay. The equation is $y = 3e^{-2x}$.

Exponential growth and decay often involve very large or very small numbers. To describe these numbers, we often use orders of magnitude. The **order of magnitude** is the power of ten, when the number is expressed in scientific notation, with one digit to the left of the decimal. For example, the distance to the nearest star, Proxima Centauri, measured in kilometers, is 40,113,497,200,000 kilometers. Expressed in scientific notation, this is $4.01134972 \times 10^{13}$. So, we could describe this number as having order of magnitude 10^{13}.

characteristics of the exponential function, $y = A_0 e^{kt}$

An exponential function with the form $y = A_0 e^{kt}$ has the following characteristics:

- one-to-one function
- horizontal asymptote: $y = 0$
- domain: $(-\infty, \infty)$
- range: $(0, \infty)$
- x-intercept: none
- y-intercept: $(0, A_0)$
- increasing if $k > 0$ (see **Figure 4**)
- decreasing if $k < 0$ (see **Figure 4**)

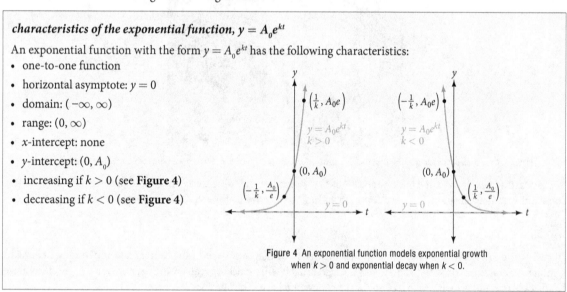

Figure 4 An exponential function models exponential growth when $k > 0$ and exponential decay when $k < 0$.

Example 1   **Graphing Exponential Growth**

A population of bacteria doubles every hour. If the culture started with 10 bacteria, graph the population as a function of time.

Solution   When an amount grows at a fixed percent per unit time, the growth is exponential. To find A_0 we use the fact that A_0 is the amount at time zero, so $A_0 = 10$. To find k, use the fact that after one hour ($t = 1$) the population doubles from 10 to 20. The formula is derived as follows

$$20 = 10e^{k \cdot 1}$$
$$2 = e^k \qquad \text{Divide by 10}$$
$$\ln 2 = k \qquad \text{Take the natural logarithm}$$

so $k = \ln(2)$. Thus the equation we want to graph is $y = 10e^{(\ln 2)t} = 10(e^{\ln 2})^t = 10 \cdot 2^t$. The graph is shown in **Figure 5**.

Figure 5 The graph of $y = 10e^{(\ln 2)t}$.

Analysis The population of bacteria after ten hours is 10,240. We could describe this amount is being of the order of magnitude 10^4. The population of bacteria after twenty hours is 10,485,760 which is of the order of magnitude 10^7, so we could say that the population has increased by three orders of magnitude in ten hours.

Half-Life

We now turn to exponential decay. One of the common terms associated with exponential decay, as stated above, is **half-life**, the length of time it takes an exponentially decaying quantity to decrease to half its original amount. Every radioactive isotope has a half-life, and the process describing the exponential decay of an isotope is called radioactive decay.

To find the half-life of a function describing exponential decay, solve the following equation:

$$\frac{1}{2}A_0 = A_0 e^{kt}$$

We find that the half-life depends only on the constant k and not on the starting quantity A_0.

The formula is derived as follows

$$\frac{1}{2}A_0 = A_0 e^{kt}$$

$$\frac{1}{2} = e^{kt} \qquad \text{Divide by } A_0.$$

$$\ln\left(\frac{1}{2}\right) = kt \qquad \text{Take the natural log.}$$

$$-\ln(2) = kt \qquad \text{Apply laws of logarithms.}$$

$$-\frac{\ln(2)}{k} = t \qquad \text{Divide by } k.$$

Since t, the time, is positive, k must, as expected, be negative. This gives us the half-life formula

$$t = -\frac{\ln(2)}{k}$$

How To...

Given the half-life, find the decay rate.

1. Write $A = A_0 e^{kt}$.

2. Replace A by $\frac{1}{2}A_0$ and replace t by the given half-life.

3. Solve to find k. Express k as an exact value (do not round).

Note: It is also possible to find the decay rate using $k = -\dfrac{\ln(2)}{t}$.

Example 2   **Finding the Function that Describes Radioactive Decay**

The half-life of carbon-14 is 5,730 years. Express the amount of carbon-14 remaining as a function of time, t.

Solution   This formula is derived as follows.

$$A = A_0 e^{kt}$$ The continuous growth formula.

$$0.5A_0 = A_0 e^{k \cdot 5730}$$ Substitute the half-life for t and $0.5A_0$ for $f(t)$.

$$0.5 = e^{5730k}$$ Divide by A_0.

$$\ln(0.5) = 5730k$$ Take the natural log of both sides.

$$k = \frac{\ln(0.5)}{5730}$$ Divide by the coefficient of k.

$$A = A_0 e^{\left(\frac{\ln(0.5)}{5730}\right)t}$$ Substitute for k in the continuous growth formula.

The function that describes this continuous decay is $f(t) = A_0 e^{\left(\frac{\ln(0.5)}{5730}\right)t}$. We observe that the coefficient of t, $\frac{\ln(0.5)}{5730} \approx -1.2097$ is negative, as expected in the case of exponential decay.

Try It #14

The half-life of plutonium-244 is 80,000,000 years. Find function gives the amount of carbon-14 remaining as a function of time, measured in years.

Radiocarbon Dating

The formula for radioactive decay is important in radiocarbon dating, which is used to calculate the approximate date a plant or animal died. Radiocarbon dating was discovered in 1949 by Willard Libby, who won a Nobel Prize for his discovery. It compares the difference between the ratio of two isotopes of carbon in an organic artifact or fossil to the ratio of those two isotopes in the air. It is believed to be accurate to within about 1% error for plants or animals that died within the last 60,000 years.

Carbon-14 is a radioactive isotope of carbon that has a half-life of 5,730 years. It occurs in small quantities in the carbon dioxide in the air we breathe. Most of the carbon on earth is carbon-12, which has an atomic weight of 12 and is not radioactive. Scientists have determined the ratio of carbon-14 to carbon-12 in the air for the last 60,000 years, using tree rings and other organic samples of known dates—although the ratio has changed slightly over the centuries.

As long as a plant or animal is alive, the ratio of the two isotopes of carbon in its body is close to the ratio in the atmosphere. When it dies, the carbon-14 in its body decays and is not replaced. By comparing the ratio of carbon-14 to carbon-12 in a decaying sample to the known ratio in the atmosphere, the date the plant or animal died can be approximated.

Since the half-life of carbon-14 is 5,730 years, the formula for the amount of carbon-14 remaining after t years is

$$A \approx A_0 e^{\left(\frac{\ln(0.5)}{5730}\right)t}$$

where
- A is the amount of carbon-14 remaining
- A_0 is the amount of carbon-14 when the plant or animal began decaying.

This formula is derived as follows:

$$A = A_0 e^{kt}$$ The continuous growth formula.

$$0.5A_0 = A_0 e^{k \cdot 5730}$$ Substitute the half-life for t and $0.5A_0$ for $f(t)$.

$$0.5 = e^{5730k}$$ Divide by A_0.

$$\ln(0.5) = 5730k$$ Take the natural log of both sides.

$$k = \frac{\ln(0.5)}{5730}$$ Divide by the coefficient of k.

$$A = A_0 e^{\left(\frac{\ln(0.5)}{5730}\right)t}$$ Substitute for r in the continuous growth formula.

To find the age of an object, we solve this equation for t:

$$t = \frac{\ln\left(\frac{A}{A_0}\right)}{-0.000121}$$

Out of necessity, we neglect here the many details that a scientist takes into consideration when doing carbon-14 dating, and we only look at the basic formula. The ratio of carbon-14 to carbon-12 in the atmosphere is approximately 0.0000000001%. Let r be the ratio of carbon-14 to carbon-12 in the organic artifact or fossil to be dated, determined by a method called liquid scintillation. From the equation $A \approx A_0 e^{-0.000121t}$ we know the ratio of the percentage of carbon-14 in the object we are dating to the percentage of carbon-14 in the atmosphere is $r = \dfrac{A}{A_0} \approx e^{-0.000121t}$. We solve this equation for t, to get

$$t = \frac{\ln(r)}{-0.000121}$$

How To…

Given the percentage of carbon-14 in an object, determine its age.

1. Express the given percentage of carbon-14 as an equivalent decimal, k.

2. Substitute for k in the equation $t = \dfrac{\ln(r)}{-0.000121}$ and solve for the age, t.

Example 3   **Finding the Age of a Bone**

A bone fragment is found that contains 20% of its original carbon-14. To the nearest year, how old is the bone?

Solution   We substitute 20% = 0.20 for k in the equation and solve for t :

$$t = \frac{\ln(r)}{-0.000121} \qquad \text{Use the general form of the equation.}$$

$$= \frac{\ln(0.20)}{-0.000121} \qquad \text{Substitute for } r.$$

$$\approx 13301 \qquad \text{Round to the nearest year.}$$

The bone fragment is about 13,301 years old.

Analysis   The instruments that measure the percentage of carbon-14 are extremely sensitive and, as we mention above, a scientist will need to do much more work than we did in order to be satisfied. Even so, carbon dating is only accurate to about 1%, so this age should be given as 13,301 years ± 1% or 13,301 years ± 133 years.

Try It #15

Cesium-137 has a half-life of about 30 years. If we begin with 200 mg of cesium-137, will it take more or less than 230 years until only 1 milligram remains?

Calculating Doubling Time

For decaying quantities, we determined how long it took for half of a substance to decay. For growing quantities, we might want to find out how long it takes for a quantity to double. As we mentioned above, the time it takes for a quantity to double is called the **doubling time**.

Given the basic exponential growth equation $A = A_0 e^{kt}$, doubling time can be found by solving for when the original quantity has doubled, that is, by solving $2A_0 = A_0 e^{kt}$.

The formula is derived as follows:

$$2A_0 = A_0 e^{kt}$$

$$2 = e^{kt} \qquad \text{Divide by } A_0.$$

$$\ln(2) = kt \qquad \text{Take the natural logarithm.}$$

$$t = \frac{\ln(2)}{k} \qquad \text{Divide by the coefficient of } t.$$

Thus the doubling time is

$$t = \frac{\ln(2)}{k}$$

Example 4   **Finding a Function That Describes Exponential Growth**

According to Moore's Law, the doubling time for the number of transistors that can be put on a computer chip is approximately two years. Give a function that describes this behavior.

Solution   The formula is derived as follows:

$$t = \frac{\ln(2)}{k}$$ The doubling time formula.

$$2 = \frac{\ln(2)}{k}$$ Use a doubling time of two years.

$$k = \frac{\ln(2)}{2}$$ Multiply by k and divide by 2.

$$A = A_0 e^{\frac{\ln(2)}{2}t}$$ Substitute k into the continuous growth formula.

The function is $A_0 e^{\frac{\ln(2)}{2}t}$.

Try It #16

Recent data suggests that, as of 2013, the rate of growth predicted by Moore's Law no longer holds. Growth has slowed to a doubling time of approximately three years. Find the new function that takes that longer doubling time into account.

Using Newton's Law of Cooling

Exponential decay can also be applied to temperature. When a hot object is left in surrounding air that is at a lower temperature, the object's temperature will decrease exponentially, leveling off as it approaches the surrounding air temperature. On a graph of the temperature function, the leveling off will correspond to a horizontal asymptote at the temperature of the surrounding air. Unless the room temperature is zero, this will correspond to a vertical shift of the generic exponential decay function. This translation leads to **Newton's Law of Cooling**, the scientific formula for temperature as a function of time as an object's temperature is equalized with the ambient temperature

$$T(t) = Ae^{kt} + T_s$$

This formula is derived as follows:

$$T(t) = Ab^{ct} + T_s$$

$$T(t) = Ae^{\ln(b^{ct})} + T_s$$ Laws of logarithms.

$$T(t) = Ae^{ct\ln(b)} + T_s$$ Laws of logarithms.

$$T(t) = Ae^{kt} + T_s$$ Rename the constant $c\ln(b)$, calling it k.

Newton's law of cooling

The temperature of an object, T, in surrounding air with temperature T_s will behave according to the formula

$$T(t) = Ae^{kt} + T_s$$

where
- t is time
- A is the difference between the initial temperature of the object and the surroundings
- k is a constant, the continuous rate of cooling of the object

---

How To...

Given a set of conditions, apply Newton's Law of Cooling.

1. Set T_s equal to the y-coordinate of the horizontal asymptote (usually the ambient temperature).
2. Substitute the given values into the continuous growth formula $T(t) = Ae^{kt} + T_s$ to find the parameters A and k.
3. Substitute in the desired time to find the temperature or the desired temperature to find the time.

Example 5   **Using Newton's Law of Cooling**

A cheesecake is taken out of the oven with an ideal internal temperature of 165°F, and is placed into a 35°F refrigerator. After 10 minutes, the cheesecake has cooled to 150°F. If we must wait until the cheesecake has cooled to 70°F before we eat it, how long will we have to wait?

Solution   Because the surrounding air temperature in the refrigerator is 35 degrees, the cheesecake's temperature will decay exponentially toward 35, following the equation

$$T(t) = Ae^{kt} + 35$$

We know the initial temperature was 165, so $T(0) = 165$.

$$165 = Ae^{k0} + 35 \qquad \text{Substitute } (0, 165).$$

$$A = 130 \qquad \text{Solve for } A.$$

We were given another data point, $T(10) = 150$, which we can use to solve for k.

$$150 = 130e^{k10} + 35 \qquad \text{Substitute } (10, 150).$$

$$115 = 130e^{k10} \qquad \text{Subtract 35.}$$

$$\frac{115}{130} = e^{10k} \qquad \text{Divide by 130.}$$

$$\ln\left(\frac{115}{130}\right) = 10k \qquad \text{Take the natural log of both sides.}$$

$$k = \frac{\ln\left(\dfrac{115}{130}\right)}{10} \approx -0.0123 \quad \text{Divide by the coefficient of } k.$$

This gives us the equation for the cooling of the cheesecake: $T(t) = 130e^{-0.0123t} + 35$.

Now we can solve for the time it will take for the temperature to cool to 70 degrees.

$$70 = 130e^{-0.0123t} + 35 \qquad \text{Substitute in 70 for } T(t).$$

$$35 = 130e^{-0.0123t} \qquad \text{Subtract 35.}$$

$$\frac{35}{130} = e^{-0.0123t} \qquad \text{Divide by 130.}$$

$$\ln\left(\frac{35}{130}\right) = -0.0123t \qquad \text{Take the natural log of both sides}$$

$$t = \frac{\ln\left(\dfrac{35}{130}\right)}{-0.0123} \approx 106.68 \quad \text{Divide by the coefficient of } t.$$

It will take about 107 minutes, or one hour and 47 minutes, for the cheesecake to cool to 70°F.

Try It #17

A pitcher of water at 40 degrees Fahrenheit is placed into a 70 degree room. One hour later, the temperature has risen to 45 degrees. How long will it take for the temperature to rise to 60 degrees?

Using Logistic Growth Models

Exponential growth cannot continue forever. Exponential models, while they may be useful in the short term, tend to fall apart the longer they continue. Consider an aspiring writer who writes a single line on day one and plans to double the number of lines she writes each day for a month. By the end of the month, she must write over 17 billion lines, or one-half-billion pages. It is impractical, if not impossible, for anyone to write that much in such a short period of time. Eventually, an exponential model must begin to approach some limiting value, and then the growth is forced to slow. For this reason, it is often better to use a model with an upper bound instead of an exponential growth model, though the exponential growth model is still useful over a short term, before approaching the limiting value.

The **logistic growth model** is approximately exponential at first, but it has a reduced rate of growth as the output approaches the model's upper bound, called the **carrying capacity**. For constants a, b, and c, the logistic growth of a population over time x is represented by the model

$$f(x) = \frac{c}{1 + ae^{-bx}}$$

The graph in **Figure 6** shows how the growth rate changes over time. The graph increases from left to right, but the growth rate only increases until it reaches its point of maximum growth rate, at which point the rate of increase decreases.

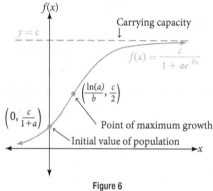

Figure 6

---

logistic growth

The logistic growth model is

$$f(x) = \frac{c}{1 + ae^{-bx}}$$

where

- $\dfrac{c}{1 + a}$ is the initial value
- c is the *carrying capacity, or limiting value*
- b is a constant determined by the rate of growth.

Example 6   **Using the Logistic-Growth Model**

An influenza epidemic spreads through a population rapidly, at a rate that depends on two factors: The more people who have the flu, the more rapidly it spreads, and also the more uninfected people there are, the more rapidly it spreads. These two factors make the logistic model a good one to study the spread of communicable diseases. And, clearly, there is a maximum value for the number of people infected: the entire population.

For example, at time $t = 0$ there is one person in a community of 1,000 people who has the flu. So, in that community, at most 1,000 people can have the flu. Researchers find that for this particular strain of the flu, the logistic growth constant is $b = 0.6030$. Estimate the number of people in this community who will have had this flu after ten days. Predict how many people in this community will have had this flu after a long period of time has passed.

Solution   We substitute the given data into the logistic growth model

$$f(x) = \frac{c}{1 + ae^{-bx}}$$

Because at most 1,000 people, the entire population of the community, can get the flu, we know the limiting value is $c = 1000$. To find a, we use the formula that the number of cases at time $t = 0$ is $\dfrac{c}{1 + a} = 1$, from which it follows that $a = 999$. This model predicts that, after ten days, the number of people who have had the flu is $f(x) = \dfrac{1000}{1 + 999e^{-0.6030x}} \approx 293.8$. Because the actual number must be a whole number (a person has either had the flu or not) we round to 294. In the long term, the number of people who will contract the flu is the limiting value, $c = 1000$.

*Analysis   Remember that, because we are dealing with a virus, we cannot predict with certainty the number of people infected. The model only approximates the number of people infected and will not give us exact or actual values. The graph in **Figure 7** gives a good picture of how this model fits the data.*

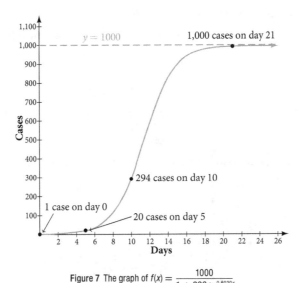

Figure 7 The graph of $f(x) = \dfrac{1000}{1 + 999e^{-0.6030x}}$

Try It #18

Using the model in **Example 6**, estimate the number of cases of flu on day 15.

Choosing an Appropriate Model for Data

Now that we have discussed various mathematical models, we need to learn how to choose the appropriate model for the raw data we have. Many factors influence the choice of a mathematical model, among which are experience, scientific laws, and patterns in the data itself. Not all data can be described by elementary functions. Sometimes, a function is chosen that approximates the data over a given interval. For instance, suppose data were gathered on the number of homes bought in the United States from the years 1960 to 2013. After plotting these data in a scatter plot, we notice that the shape of the data from the years 2000 to 2013 follow a logarithmic curve. We could restrict the interval from 2000 to 2010, apply regression analysis using a logarithmic model, and use it to predict the number of home buyers for the year 2015.

Three kinds of functions that are often useful in mathematical models are linear functions, exponential functions, and logarithmic functions. If the data lies on a straight line, or seems to lie approximately along a straight line, a linear model may be best. If the data is non-linear, we often consider an exponential or logarithmic model, though other models, such as quadratic models, may also be considered.

In choosing between an exponential model and a logarithmic model, we look at the way the data curves. This is called the concavity. If we draw a line between two data points, and all (or most) of the data between those two points lies above that line, we say the curve is concave down. We can think of it as a bowl that bends downward and therefore cannot hold water. If all (or most) of the data between those two points lies below the line, we say the curve is concave up. In this case, we can think of a bowl that bends upward and can therefore hold water. An exponential curve, whether rising or falling, whether representing growth or decay, is always concave up away from its horizontal asymptote. A logarithmic curve is always concave away from its vertical asymptote. In the case of positive data, which is the most common case, an exponential curve is always concave up, and a logarithmic curve always concave down.

A logistic curve changes concavity. It starts out concave up and then changes to concave down beyond a certain point, called a point of inflection.

After using the graph to help us choose a type of function to use as a model, we substitute points, and solve to find the parameters. We reduce round-off error by choosing points as far apart as possible.

Example 7   **Choosing a Mathematical Model**

Does a linear, exponential, logarithmic, or logistic model best fit the values listed in **Table 1**? Find the model, and use a graph to check your choice.

x	1	2	3	4	5	6	7	8	9
y	0	1.386	2.197	2.773	3.219	3.584	3.892	4.159	4.394

Table 1

Solution   First, plot the data on a graph as in **Figure 8**. For the purpose of graphing, round the data to two significant digits.

Figure 8

Clearly, the points do not lie on a straight line, so we reject a linear model. If we draw a line between any two of the points, most or all of the points between those two points lie above the line, so the graph is concave down, suggesting a logarithmic model. We can try $y = a\ln(bx)$. Plugging in the first point, $(1,0)$, gives $0 = a\ln b$.

We reject the case that $a = 0$ (if it were, all outputs would be 0), so we know $\ln(b) = 0$. Thus $b = 1$ and $y = a\ln(x)$. Next we can use the point $(9, 4.394)$ to solve for a:

$$y = a\ln(x)$$

$$4.394 = a\ln(9)$$

$$a = \frac{4.394}{\ln(9)}$$

Because $a = \frac{4.394}{\ln(9)} \approx 2$, an appropriate model for the data is $y = 2\ln(x)$.

To check the accuracy of the model, we graph the function together with the given points as in **Figure 9**.

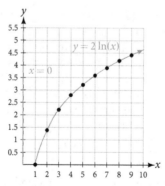

Figure 9 The graph of $y = 2\ln x$.

We can conclude that the model is a good fit to the data.

Compare **Figure 9** to the graph of $y = \ln(x^2)$ shown in **Figure 10**.

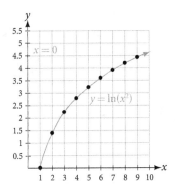

Figure 10 The graph of $y = \ln(x^2)$

The graphs appear to be identical when $x > 0$. A quick check confirms this conclusion: $y = \ln(x^2) = 2\ln(x)$ for $x > 0$.

However, if $x < 0$, the graph of $y = \ln(x^2)$ includes a "extra" branch, as shown in **Figure 11**. This occurs because, while $y = 2\ln(x)$ cannot have negative values in the domain (as such values would force the argument to be negative), the function $y = \ln(x^2)$ can have negative domain values.

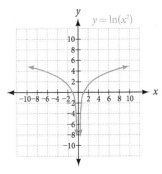

Figure 11

Try It #19

Does a linear, exponential, or logarithmic model best fit the data in **Table 2**? Find the model.

x	1	2	3	4	5	6	7	8	9
y	3.297	5.437	8.963	14.778	24.365	40.172	66.231	109.196	180.034

Table 2

Expressing an Exponential Model in Base e

While powers and logarithms of any base can be used in modeling, the two most common bases are 10 and e. In science and mathematics, the base e is often preferred. We can use laws of exponents and laws of logarithms to change any base to base e.

How To...

Given a model with the form $y = ab^x$, change it to the form $y = A_0 e^{kx}$.

1. Rewrite $y = ab^x$ as $y = ae^{\ln(b^x)}$.

2. Use the power rule of logarithms to rewrite y as $y = ae^{x\ln(b)} = ae^{\ln(b)x}$.

3. Note that $a = A_0$ and $k = \ln(b)$ in the equation $y = A_0 e^{kx}$.

Example 8   **Changing to base e**

Change the function $y = 2.5(3.1)^x$ so that this same function is written in the form $y = A_0 e^{kx}$.

Solution

The formula is derived as follows

$$y = 2.5(3.1)^x$$
$$= 2.5 e^{\ln(3.1^x)} \qquad \text{Insert exponential and its inverse.}$$
$$= 2.5 e^{x \ln 3.1} \qquad \text{Laws of logs.}$$
$$= 2.5 e^{(\ln 3.1)x} \qquad \text{Commutative law of multiplication}$$

Try It #20

Change the function $y = 3(0.5)^x$ to one having e as the base.

Access these online resources for additional instruction and practice with exponential and logarithmic models.

- Logarithm Application – pH (http://openstaxcollege.org/l/logph)
- Exponential Model – Age Using Half-Life (http://openstaxcollege.org/l/expmodelhalf)
- Newton's Law of Cooling (http://openstaxcollege.org/l/newtoncooling)
- Exponential Growth Given Doubling Time (http://openstaxcollege.org/l/expgrowthdbl)
- Exponential Growth – Find Initial Amount Given Doubling Time (http://openstaxcollege.org/l/initialdouble)

6.7 SECTION EXERCISES

VERBAL

1. With what kind of exponential model would *half-life* be associated? What role does half-life play in these models?

3. With what kind of exponential model would *doubling time* be associated? What role does doubling time play in these models?

5. What is an order of magnitude? Why are orders of magnitude useful? Give an example to explain.

2. What is carbon dating? Why does it work? Give an example in which carbon dating would be useful.

4. Define Newton's Law of Cooling. Then name at least three real-world situations where Newton's Law of Cooling would be applied.

NUMERIC

6. The temperature of an object in degrees Fahrenheit after t minutes is represented by the equation $T(t) = 68e^{-0.0174t} + 72$. To the nearest degree, what is the temperature of the object after one and a half hours?

For the following exercises, use the logistic growth model $f(x) = \dfrac{150}{1 + 8e^{-2x}}$.

7. Find and interpret $f(0)$. Round to the nearest tenth.

9. Find the carrying capacity.

11. Determine whether the data from the table could best be represented as a function that is linear, exponential, or logarithmic. Then write a formula for a model that represents the data.

12. Rewrite $f(x) = 1.68(0.65)^x$ as an exponential equation with base e to five significant digits.

8. Find and interpret $f(4)$. Round to the nearest tenth.

10. Graph the model.

x	–2	–1	0	1	2	3	4	5
$f(x)$	0.694	0.833	1	1.2	1.44	1.728	2.074	2.488

TECHNOLOGY

For the following exercises, enter the data from each table into a graphing calculator and graph the resulting scatter plots. Determine whether the data from the table could represent a function that is linear, exponential, or logarithmic.

13.

x	1	2	3	4	5	6	7	8	9	10
$f(x)$	2	4.079	5.296	6.159	6.828	7.375	7.838	8.238	8.592	8.908

14.

x	1	2	3	4	5	6	7	8	9	10
$f(x)$	2.4	2.88	3.456	4.147	4.977	5.972	7.166	8.6	10.32	12.383

15.

x	4	5	6	7	8	9	10	11	12	13
$f(x)$	9.429	9.972	10.415	10.79	11.115	11.401	11.657	11.889	12.101	12.295

16.

x	1.25	2.25	3.56	4.2	5.65	6.75	7.25	8.6	9.25	10.5
$f(x)$	5.75	8.75	12.68	14.6	18.95	22.25	23.75	27.8	29.75	33.5

For the following exercises, use a graphing calculator and this scenario: the population of a fish farm in t years is modeled by the equation $P(t) = \dfrac{1000}{1 + 9e^{-0.6t}}$.

17. Graph the function.

19. To the nearest tenth, what is the doubling time for the fish population?

21. To the nearest tenth, how long will it take for the population to reach 900?

18. What is the initial population of fish?

20. To the nearest whole number, what will the fish population be after 2 years?

22. What is the carrying capacity for the fish population? Justify your answer using the graph of P.

EXTENSIONS

23. A substance has a half-life of 2.045 minutes. If the initial amount of the substance was 132.8 grams, how many half-lives will have passed before the substance decays to 8.3 grams? What is the total time of decay?

24. The formula for an increasing population is given by $P(t) = P_0 e^{rt}$ where P_0 is the initial population and $r > 0$. Derive a general formula for the time t it takes for the population to increase by a factor of M.

25. Recall the formula for calculating the magnitude of an earthquake, $M = \frac{2}{3} \log \left(\frac{S}{S_0} \right)$. Show each step for solving this equation algebraically for the seismic moment S.

26. What is the y-intercept of the logistic growth model $y = \frac{c}{1 + ae^{-rx}}$? Show the steps for calculation. What does this point tell us about the population?

27. Prove that $b^x = e^{x \ln(b)}$ for positive $b \neq 1$.

REAL-WORLD APPLICATIONS

For the following exercises, use this scenario: A doctor prescribes 125 milligrams of a therapeutic drug that decays by about 30% each hour.

28. To the nearest hour, what is the half-life of the drug?

29. Write an exponential model representing the amount of the drug remaining in the patient's system after t hours. Then use the formula to find the amount of the drug that would remain in the patient's system after 3 hours. Round to the nearest milligram.

30. Using the model found in the previous exercise, find $f(10)$ and interpret the result. Round to the nearest hundredth.

For the following exercises, use this scenario: A tumor is injected with 0.5 grams of Iodine-125, which has a decay rate of 1.15% per day.

31. To the nearest day, how long will it take for half of the Iodine-125 to decay?

32. Write an exponential model representing the amount of Iodine-125 remaining in the tumor after t days. Then use the formula to find the amount of Iodine-125 that would remain in the tumor after 60 days. Round to the nearest tenth of a gram.

33. A scientist begins with 250 grams of a radioactive substance. After 250 minutes, the sample has decayed to 32 grams. Rounding to five significant digits, write an exponential equation representing this situation. To the nearest minute, what is the half-life of this substance?

34. The half-life of Radium-226 is 1590 years. What is the annual decay rate? Express the decimal result to four significant digits and the percentage to two significant digits.

35. The half-life of Erbium-165 is 10.4 hours. What is the hourly decay rate? Express the decimal result to four significant digits and the percentage to two significant digits.

36. A wooden artifact from an archeological dig contains 60 percent of the carbon-14 that is present in living trees. To the nearest year, about how many years old is the artifact? (The half-life of carbon-14 is 5730 years.)

37. A research student is working with a culture of bacteria that doubles in size every twenty minutes. The initial population count was 1350 bacteria. Rounding to five significant digits, write an exponential equation representing this situation. To the nearest whole number, what is the population size after 3 hours?

For the following exercises, use this scenario: A biologist recorded a count of 360 bacteria present in a culture after 5 minutes and 1,000 bacteria present after 20 minutes.

38. To the nearest whole number, what was the initial population in the culture?

39. Rounding to six significant digits, write an exponential equation representing this situation. To the nearest minute, how long did it take the population to double?

For the following exercises, use this scenario: A pot of boiling soup with an internal temperature of 100° Fahrenheit was taken off the stove to cool in a 69° F room. After fifteen minutes, the internal temperature of the soup was 95° F.

40. Use Newton's Law of Cooling to write a formula that models this situation.

41. To the nearest minute, how long will it take the soup to cool to 80° F?

42. To the nearest degree, what will the temperature be after 2 and a half hours?

For the following exercises, use this scenario: A turkey is taken out of the oven with an internal temperature of 165° Fahrenheit and is allowed to cool in a 75° F room. After half an hour, the internal temperature of the turkey is 145° F.

43. Write a formula that models this situation.

44. To the nearest degree, what will the temperature be after 50 minutes?

45. To the nearest minute, how long will it take the turkey to cool to 110° F?

For the following exercises, find the value of the number shown on each logarithmic scale. Round all answers to the nearest thousandth.

46.

47.

48. Plot each set of approximate values of intensity of sounds on a logarithmic scale: Whisper: $10^{-10}\dfrac{W}{m^2}$, Vacuum: $10^{-4}\dfrac{W}{m^2}$, Jet: $10^2\dfrac{W}{m^2}$

49. Recall the formula for calculating the magnitude of an earthquake, $M = \dfrac{2}{3}\log\left(\dfrac{S}{S_0}\right)$. One earthquake has magnitude 3.9 on the MMS scale. If a second earthquake has 750 times as much energy as the first, find the magnitude of the second quake. Round to the nearest hundredth.

For the following exercises, use this scenario: The equation $N(t) = \dfrac{500}{1 + 49e^{-0.7t}}$ models the number of people in a town who have heard a rumor after t days.

50. How many people started the rumor?

51. To the nearest whole number, how many people will have heard the rumor after 3 days?

52. As t increases without bound, what value does $N(t)$ approach? Interpret your answer.

For the following exercise, choose the correct answer choice.

53. A doctor and injects a patient with 13 milligrams of radioactive dye that decays exponentially. After 12 minutes, there are 4.75 milligrams of dye remaining in the patient's system. Which is an appropriate model for this situation?

  a. $f(t) = 13(0.0805)^t$ **b.** $f(t) = 13e^{0.9195t}$ **c.** $f(t) = 13e^{(-0.0839t)}$ **d.** $f(t) = \dfrac{4.75}{1 + 13e^{-0.83925t}}$

LEARNING OBJECTIVES

In this section, you will:

- Build an exponential model from data.
- Build a logarithmic model from data.
- Build a logistic model from data.

6.8   FITTING EXPONENTIAL MODELS TO DATA

In previous sections of this chapter, we were either given a function explicitly to graph or evaluate, or we were given a set of points that were guaranteed to lie on the curve. Then we used algebra to find the equation that fit the points exactly. In this section, we use a modeling technique called *regression analysis* to find a curve that models data collected from real-world observations. With regression analysis, we don't expect all the points to lie perfectly on the curve. The idea is to find a model that best fits the data. Then we use the model to make predictions about future events.

Do not be confused by the word *model*. In mathematics, we often use the terms *function, equation*, and *model* interchangeably, even though they each have their own formal definition. The term *model* is typically used to indicate that the equation or function approximates a real-world situation.

We will concentrate on three types of regression models in this section: exponential, logarithmic, and logistic. Having already worked with each of these functions gives us an advantage. Knowing their formal definitions, the behavior of their graphs, and some of their real-world applications gives us the opportunity to deepen our understanding. As each regression model is presented, key features and definitions of its associated function are included for review. Take a moment to rethink each of these functions, reflect on the work we've done so far, and then explore the ways regression is used to model real-world phenomena.

Building an Exponential Model from Data

As we've learned, there are a multitude of situations that can be modeled by exponential functions, such as investment growth, radioactive decay, atmospheric pressure changes, and temperatures of a cooling object. What do these phenomena have in common? For one thing, all the models either increase or decrease as time moves forward. But that's not the whole story. It's the *way* data increase or decrease that helps us determine whether it is best modeled by an exponential equation. Knowing the behavior of exponential functions in general allows us to recognize when to use exponential regression, so let's review exponential growth and decay.

Recall that exponential functions have the form $y = ab^x$ or $y = A_0 e^{kx}$. When performing regression analysis, we use the form most commonly used on graphing utilities, $y = ab^x$. Take a moment to reflect on the characteristics we've already learned about the exponential function $y = ab^x$ (assume $a > 0$):

- b must be greater than zero and not equal to one.
- The initial value of the model is $y = a$.
  - If $b > 1$, the function models exponential growth. As x increases, the outputs of the model increase slowly at first, but then increase more and more rapidly, without bound.
  - If $0 < b < 1$, the function models exponential decay. As x increases, the outputs for the model decrease rapidly at first and then level off to become asymptotic to the x-axis. In other words, the outputs never become equal to or less than zero.

As part of the results, your calculator will display a number known as the *correlation coefficient*, labeled by the variable r, or r^2. (You may have to change the calculator's settings for these to be shown.) The values are an indication of the "goodness of fit" of the regression equation to the data. We more commonly use the value of r^2 instead of r, but the closer either value is to 1, the better the regression equation approximates the data.

> ***exponential regression***
>
> *Exponential regression* is used to model situations in which growth begins slowly and then accelerates rapidly without bound, or where decay begins rapidly and then slows down to get closer and closer to zero. We use the command "**ExpReg**" on a graphing utility to fit an exponential function to a set of data points. This returns an equation of the form, $y = ab^x$
>
> Note that:
>
> - b must be non-negative.
> - when $b > 1$, we have an exponential growth model.
> - when $0 < b < 1$, we have an exponential decay model.

How To...

Given a set of data, perform exponential regression using a graphing utility.

1. Use the **STAT** then **EDIT** menu to enter given data.

   a. Clear any existing data from the lists.
   b. List the input values in the L1 column.
   c. List the output values in the L2 column.

2. Graph and observe a scatter plot of the data using the **STATPLOT** feature.

   a. Use **ZOOM [9]** to adjust axes to fit the data.
   b. Verify the data follow an exponential pattern.

3. Find the equation that models the data.

   a. Select "**ExpReg**" from the **STAT** then **CALC** menu.
   b. Use the values returned for a and b to record the model, $y = ab^x$.

4. Graph the model in the same window as the scatterplot to verify it is a good fit for the data.

Example 1   **Using Exponential Regression to Fit a Model to Data**

In 2007, a university study was published investigating the crash risk of alcohol impaired driving. Data from 2,871 crashes were used to measure the association of a person's blood alcohol level (BAC) with the risk of being in an accident.

Table 1 shows results from the study[24]. The *relative risk* is a measure of how many times more likely a person is to crash. So, for example, a person with a BAC of 0.09 is 3.54 times as likely to crash as a person who has not been drinking alcohol.

BAC	0	0.01	0.03	0.05	0.07	0.09
Relative Risk of Crashing	1	1.03	1.06	1.38	2.09	3.54
BAC	0.11	0.13	0.15	0.17	0.19	0.21
Relative Risk of Crashing	6.41	12.6	22.1	39.05	65.32	99.78

Table 1

a. Let x represent the BAC level, and let y represent the corresponding relative risk. Use exponential regression to fit a model to these data.

b. After 6 drinks, a person weighing 160 pounds will have a BAC of about 0.16. How many times more likely is a person with this weight to crash if they drive after having a 6-pack of beer? Round to the nearest hundredth.

24 Source: *Indiana University Center for Studies of Law in Action,* 2007

Solution

a. Using the **STAT** then **EDIT** menu on a graphing utility, list the **BAC** values in L1 and the relative risk values in L2. Then use the **STATPLOT** feature to verify that the scatterplot follows the exponential pattern shown in **Figure 1**:

Figure 1

Use the "**ExpReg**" command from the **STAT** then **CALC** menu to obtain the exponential model,

$$y = 0.58304829(2.20720213\text{E}10)^x$$

Converting from scientific notation, we have:

$$y = 0.58304829(22{,}072{,}021{,}300)^x$$

Notice that $r^2 \approx 0.97$ which indicates the model is a good fit to the data. To see this, graph the model in the same window as the scatterplot to verify it is a good fit as shown in **Figure 2**:

Figure 2

b. Use the model to estimate the risk associated with a BAC of 0.16. Substitute 0.16 for x in the model and solve for y.

$$y = 0.58304829(22{,}072{,}021{,}300)^x \qquad \text{Use the regression model found in part (a).}$$

$$= 0.58304829(22{,}072{,}021{,}300)^{0.16} \qquad \text{Substitute 0.16 for } x.$$

$$\approx 26.35 \qquad\qquad\qquad\qquad\qquad \text{Round to the nearest hundredth.}$$

If a 160-pound person drives after having 6 drinks, he or she is about 26.35 times more likely to crash than if driving while sober.

Try It #1

Table 2 shows a recent graduate's credit card balance each month after graduation.

Month	1	2	3	4	5	6	7	8
Debt ($)	620.00	761.88	899.80	1039.93	1270.63	1589.04	1851.31	2154.92

Table 2

a. Use exponential regression to fit a model to these data.
b. If spending continues at this rate, what will the graduate's credit card debt be one year after graduating?

Q & A...

Is it reasonable to assume that an exponential regression model will represent a situation indefinitely?

No. Remember that models are formed by real-world data gathered for regression. It is usually reasonable to make estimates within the interval of original observation (interpolation). However, when a model is used to make predictions, it is important to use reasoning skills to determine whether the model makes sense for inputs far beyond the original observation interval (extrapolation).

Building a Logarithmic Model from Data

Just as with exponential functions, there are many real-world applications for logarithmic functions: intensity of sound, pH levels of solutions, yields of chemical reactions, production of goods, and growth of infants. As with exponential models, data modeled by logarithmic functions are either always increasing or always decreasing as time moves forward. Again, it is the *way* they increase or decrease that helps us determine whether a logarithmic model is best.

Recall that logarithmic functions increase or decrease rapidly at first, but then steadily slow as time moves on. By reflecting on the characteristics we've already learned about this function, we can better analyze real world situations that reflect this type of growth or decay. When performing logarithmic regression analysis, we use the form of the logarithmic function most commonly used on graphing utilities, $y = a + b\ln(x)$. For this function

- All input values, x, must be greater than zero.
- The point $(1, a)$ is on the graph of the model.
- If $b > 0$, the model is increasing. Growth increases rapidly at first and then steadily slows over time.
- If $b < 0$, the model is decreasing. Decay occurs rapidly at first and then steadily slows over time.

---

logarithmic regression

Logarithmic regression is used to model situations where growth or decay accelerates rapidly at first and then slows over time. We use the command "LnReg" on a graphing utility to fit a logarithmic function to a set of data points. This returns an equation of the form,

$$y = a + b\ln(x)$$

Note that:

- all input values, x, must be non-negative.
- when $b > 0$, the model is increasing.
- when $b < 0$, the model is decreasing.

How To...

Given a set of data, perform logarithmic regression using a graphing utility.

1. Use the **STAT** then **EDIT** menu to enter given data.
   a. Clear any existing data from the lists.
   b. List the input values in the L1 column.
   c. List the output values in the L2 column.

2. Graph and observe a scatter plot of the data using the **STATPLOT** feature.

   a. Use **ZOOM** [9] to adjust axes to fit the data.

   b. Verify the data follow a logarithmic pattern.

3. Find the equation that models the data.

   a. Select "**LnReg**" from the **STAT** then **CALC** menu.

   b. Use the values returned for *a* and *b* to record the model, $y = a + b\ln(x)$.

4. Graph the model in the same window as the scatterplot to verify it is a good fit for the data.

Example 2   **Using Logarithmic Regression to Fit a Model to Data**

Due to advances in medicine and higher standards of living, life expectancy has been increasing in most developed countries since the beginning of the 20th century.

Table 3 shows the average life expectancies, in years, of Americans from 1900–2010[25].

Year	1900	1910	1920	1930	1940	1950
Life Expectancy (Years)	47.3	50.0	54.1	59.7	62.9	68.2

Year	1960	1970	1980	1990	2000	2010
Life Expectancy (Years)	69.7	70.8	73.7	75.4	76.8	78.7

Table 3

a. Let *x* represent time in decades starting with $x = 1$ for the year 1900, $x = 2$ for the year 1910, and so on. Let *y* represent the corresponding life expectancy. Use logarithmic regression to fit a model to these data.

b. Use the model to predict the average American life expectancy for the year 2030.

Solution

a. Using the **STAT** then **EDIT** menu on a graphing utility, list the years using values 1–12 in L1 and the corresponding life expectancy in L2. Then use the **STATPLOT** feature to verify that the scatterplot follows a logarithmic pattern as shown in **Figure 3**:

Figure 3

Use the "**LnReg**" command from the **STAT** then **CALC** menu to obtain the logarithmic model,

$$y = 42.52722583 + 13.85752327\ln(x)$$

Next, graph the model in the same window as the scatterplot to verify it is a good fit as shown in **Figure 4**:

25 Source: *Center for Disease Control and Prevention*, 2013

Figure 4

   b. To predict the life expectancy of an American in the year 2030, substitute $x = 14$ for the in the model and solve for y:

$$y = 42.52722583 + 13.85752327\ln(x) \quad \text{Use the regression model found in part (a).}$$

$$= 42.52722583 + 13.85752327\ln(14) \quad \text{Substitute 14 for } x.$$

$$\approx 79.1 \qquad\qquad\qquad\qquad\qquad\qquad \text{Round to the nearest tenth}$$

If life expectancy continues to increase at this pace, the average life expectancy of an American will be 79.1 by the year 2030.

Try It #2

Sales of a video game released in the year 2000 took off at first, but then steadily slowed as time moved on. **Table 4** shows the number of games sold, in thousands, from the years 2000–2010.

Year	Number Sold (Thousands)
2000	142
2001	149
2002	154
2003	155
2004	159
2005	161
2006	163
2007	164
2008	164
2009	166
2010	167

Table 4

   a. Let x represent time in years starting with $x = 1$ for the year 2000. Let y represent the number of games sold in thousands. Use logarithmic regression to fit a model to these data.

   b. If games continue to sell at this rate, how many games will sell in 2015? Round to the nearest thousand.

Building a Logistic Model from Data

Like exponential and logarithmic growth, logistic growth increases over time. One of the most notable differences with logistic growth models is that, at a certain point, growth steadily slows and the function approaches an upper bound, or *limiting value*. Because of this, logistic regression is best for modeling phenomena where there are limits in expansion, such as availability of living space or nutrients.

It is worth pointing out that logistic functions actually model resource-limited exponential growth. There are many examples of this type of growth in real-world situations, including population growth and spread of disease, rumors, and even stains in fabric. When performing logistic regression analysis, we use the form most commonly used on graphing utilities:

$$y = \frac{c}{1 + ae^{-bx}}$$

Recall that:

- $\frac{c}{1 + a}$ is the initial value of the model.
- when $b > 0$, the model increases rapidly at first until it reaches its point of maximum growth rate, $\left(\frac{\ln(a)}{b}, \frac{c}{2} \right)$. At that point, growth steadily slows and the function becomes asymptotic to the upper bound $y = c$.
- c is the limiting value, sometimes called the *carrying capacity*, of the model.

---

logistic regression

Logistic regression is used to model situations where growth accelerates rapidly at first and then steadily slows to an upper limit. We use the command "Logistic" on a graphing utility to fit a logistic function to a set of data points. This returns an equation of the form

$$y = \frac{c}{1 + ae^{-bx}}$$

Note that

- The initial value of the model is $\frac{c}{1 + a}$.
- Output values for the model grow closer and closer to $y = c$ as time increases.

How To...

Given a set of data, perform logistic regression using a graphing utility.

1. Use the **STAT** then **EDIT** menu to enter given data.
   a. Clear any existing data from the lists.
   b. List the input values in the L1 column.
   c. List the output values in the L2 column.

2. Graph and observe a scatter plot of the data using the **STATPLOT** feature.
   a. Use **ZOOM** [9] to adjust axes to fit the data.
   b. Verify the data follow a logistic pattern.

3. Find the equation that models the data.
   a. Select "**Logistic**" from the **STAT** then **CALC** menu.
   b. Use the values returned for a, b, and c to record the model, $y = \frac{c}{1 + ae^{-bx}}$.

4. Graph the model in the same window as the scatterplot to verify it is a good fit for the data.

Example 3   **Using Logistic Regression to Fit a Model to Data**

Mobile telephone service has increased rapidly in America since the mid 1990s. Today, almost all residents have cellular service. **Table 5** shows the percentage of Americans with cellular service between the years 1995 and 2012[26].

26 Source: *The World Bankn*, 2013

Year	Americans with Cellular Service (%)	Year	Americans with Cellular Service (%)
1995	12.69	2004	62.852
1996	16.35	2005	68.63
1997	20.29	2006	76.64
1998	25.08	2007	82.47
1999	30.81	2008	85.68
2000	38.75	2009	89.14
2001	45.00	2010	91.86
2002	49.16	2011	95.28
2003	55.15	2012	98.17

Table 5

a. Let x represent time in years starting with $x = 0$ for the year 1995. Let y represent the corresponding percentage of residents with cellular service. Use logistic regression to fit a model to these data.

b. Use the model to calculate the percentage of Americans with cell service in the year 2013. Round to the nearest tenth of a percent.

c. Discuss the value returned for the upper limit c. What does this tell you about the model? What would the limiting value be if the model were exact?

Solution

a. Using the **STAT** then **EDIT** menu on a graphing utility, list the years using values 0–15 in L1 and the corresponding percentage in L2. Then use the **STATPLOT** feature to verify that the scatterplot follows a logistic pattern as shown in **Figure 5**:

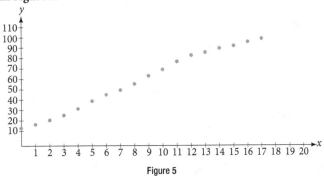

Figure 5

Use the "**Logistic**" command from the **STAT** then **CALC** menu to obtain the logistic model,

$$y = \frac{105.7379526}{1 + 6.88328979e^{-0.2595440013x}}$$

Next, graph the model in the same window as shown in **Figure 6** the scatterplot to verify it is a good fit:

Figure 6

b. To approximate the percentage of Americans with cellular service in the year 2013, substitute $x = 18$ for the in the model and solve for y:

$$y = \frac{105.7379526}{1 + 6.88328979e^{-0.2595440013x}}$$ Use the regression model found in part (a).

$$= \frac{105.7379526}{1 + 6.88328979e^{-0.2595440013(18)}}$$ Substitute 18 for x.

$$\approx 99.3\%$$ Round to the nearest tenth

According to the model, about 99.3% of Americans had cellular service in 2013.

c. The model gives a limiting value of about 105. This means that the maximum possible percentage of Americans with cellular service would be 105%, which is impossible. (How could over 100% of a population have cellular service?) If the model were exact, the limiting value would be $c = 100$ and the model's outputs would get very close to, but never actually reach 100%. After all, there will always be someone out there without cellular service!

Try It #3

Table 6 shows the population, in thousands, of harbor seals in the Wadden Sea over the years 1997 to 2012.

Year	Seal Population (Thousands)	Year	Seal Population (Thousands)
1997	3.493	2005	19.590
1998	5.282	2006	21.955
1999	6.357	2007	22.862
2000	9.201	2008	23.869
2001	11.224	2009	24.243
2002	12.964	2010	24.344
2003	16.226	2011	24.919
2004	18.137	2012	25.108

Table 6

a. Let x represent time in years starting with $x = 0$ for the year 1997. Let y represent the number of seals in thousands. Use logistic regression to fit a model to these data.

b. Use the model to predict the seal population for the year 2020.

c. To the nearest whole number, what is the limiting value of this model?

Access this online resource for additional instruction and practice with exponential function models.

- Exponential Regression on a Calculator (http://openstaxcollege.org/l/pregresscalc)

6.8 SECTION EXERCISES

VERBAL

1. What situations are best modeled by a logistic equation? Give an example, and state a case for why the example is a good fit.

2. What is a carrying capacity? What kind of model has a carrying capacity built into its formula? Why does this make sense?

3. What is regression analysis? Describe the process of performing regression analysis on a graphing utility.

4. What might a scatterplot of data points look like if it were best described by a logarithmic model?

5. What does the *y*-intercept on the graph of a logistic equation correspond to for a population modeled by that equation?

GRAPHICAL

For the following exercises, match the given function of best fit with the appropriate scatterplot in **Figure 7** through **Figure 11**. Answer using the letter beneath the matching graph.

(a)

Figure 7

(b)

Figure 8

(c)

Figure 9

(d)

Figure 10

(e)

Figure 11

6. $y = 10.209e^{-0.294x}$

7. $y = 5.598 - 1.912\ln(x)$

8. $y = 2.104(1.479)^x$

9. $y = 4.607 + 2.733\ln(x)$

10. $y = \dfrac{14.005}{1 + 2.79e^{-0.812x}}$

NUMERIC

11. To the nearest whole number, what is the initial value of a population modeled by the logistic equation $P(t) = \dfrac{175}{1 + 6.995e^{-0.68t}}$? What is the carrying capacity?

12. Rewrite the exponential model $A(t) = 1550(1.085)^x$ as an equivalent model with base e. Express the exponent to four significant digits.

13. A logarithmic model is given by the equation $h(p) = 67.682 - 5.792\ln(p)$. To the nearest hundredth, for what value of p does $h(p) = 62$?

14. A logistic model is given by the equation $P(t) = \dfrac{90}{1 + 5e^{-0.42t}}$. To the nearest hundredth, for what value of t does $P(t) = 45$?

15. What is the y-intercept on the graph of the logistic model given in the previous exercise?

TECHNOLOGY

For the following exercises, use this scenario: The population P of a koi pond over x months is modeled by the function $P(x) = \dfrac{68}{1 + 16e^{-0.28x}}$.

16. Graph the population model to show the population over a span of 3 years.

17. What was the initial population of koi?

18. How many koi will the pond have after one and a half years?

19. How many months will it take before there are 20 koi in the pond?

20. Use the intersect feature to approximate the number of months it will take before the population of the pond reaches half its carrying capacity.

For the following exercises, use this scenario: The population P of an endangered species habitat for wolves is modeled by the function $P(x) = \dfrac{558}{1 + 54.8e^{-0.462x}}$, where x is given in years.

21. Graph the population model to show the population over a span of 10 years.

22. What was the initial population of wolves transported to the habitat?

23. How many wolves will the habitat have after 3 years?

24. How many years will it take before there are 100 wolves in the habitat?

25. Use the intersect feature to approximate the number of years it will take before the population of the habitat reaches half its carrying capacity.

For the following exercises, refer to **Table 7**.

x	1	2	3	4	5	6
$f(x)$	1125	1495	2310	3294	4650	6361

Table 7

26. Use a graphing calculator to create a scatter diagram of the data.

27. Use the regression feature to find an exponential function that best fits the data in the table.

28. Write the exponential function as an exponential equation with base e.

29. Graph the exponential equation on the scatter diagram.

30. Use the intersect feature to find the value of x for which $f(x) = 4000$.

For the following exercises, refer to **Table 8**.

x	1	2	3	4	5	6
$f(x)$	555	383	307	210	158	122

Table 8

31. Use a graphing calculator to create a scatter diagram of the data.

32. Use the regression feature to find an exponential function that best fits the data in the table.

33. Write the exponential function as an exponential equation with base e.

34. Graph the exponential equation on the scatter diagram.

35. Use the intersect feature to find the value of x for which $f(x) = 250$.

For the following exercises, refer to **Table 9**.

x	1	2	3	4	5	6
$f(x)$	5.1	6.3	7.3	7.7	8.1	8.6

Table 9

36. Use a graphing calculator to create a scatter diagram of the data.

37. Use the **LOG**arithm option of the **REG**ression feature to find a logarithmic function of the form $y = a + b\ln(x)$ that best fits the data in the table.

38. Use the logarithmic function to find the value of the function when $x = 10$.

39. Graph the logarithmic equation on the scatter diagram.

40. Use the intersect feature to find the value of x for which $f(x) = 7$.

For the following exercises, refer to **Table 10**.

x	1	2	3	4	5	6	7	8
$f(x)$	7.5	6	5.2	4.3	3.9	3.4	3.1	2.9

Table 10

41. Use a graphing calculator to create a scatter diagram of the data.

42. Use the **LOG**arithm option of the **REG**ression feature to find a logarithmic function of the form $y = a + b\ln(x)$ that best fits the data in the table.

43. Use the logarithmic function to find the value of the function when $x = 10$.

44. Graph the logarithmic equation on the scatter diagram.

45. Use the intersect feature to find the value of x for which $f(x) = 8$.

For the following exercises, refer to **Table 11**.

x	1	2	3	4	5	6	7	8	9	10
$f(x)$	8.7	12.3	15.4	18.5	20.7	22.5	23.3	24	24.6	24.8

Table 11

46. Use a graphing calculator to create a scatter diagram of the data.

47. Use the **LOGISTIC** regression option to find a logistic growth model of the form $y = \dfrac{c}{1 + ae^{-bx}}$ that best fits the data in the table.

48. Graph the logistic equation on the scatter diagram.

49. To the nearest whole number, what is the predicted carrying capacity of the model?

50. Use the intersect feature to find the value of x for which the model reaches half its carrying capacity.

For the following exercises, refer to **Table 12**.

x	0	2	4	5	7	8	10	11	15	17
$f(x)$	12	28.6	52.8	70.3	99.9	112.5	125.8	127.9	135.1	135.9

Table 12

51. Use a graphing calculator to create a scatter diagram of the data.

52. Use the **LOGISTIC** regression option to find a logistic growth model of the form $y = \dfrac{c}{1 + ae^{-bx}}$ that best fits the data in the table.

53. Graph the logistic equation on the scatter diagram.

54. To the nearest whole number, what is the predicted carrying capacity of the model?

55. Use the intersect feature to find the value of x for which the model reaches half its carrying capacity.

EXTENSIONS

56. Recall that the general form of a logistic equation for a population is given by $P(t) = \dfrac{c}{1 + ae^{-bt}}$, such that the initial population at time $t = 0$ is $P(0) = P_0$. Show algebraically that
$$\frac{c - P(t)}{P(t)} = \frac{c - P_0}{P_0}e^{-bt}.$$

57. Use a graphing utility to find an exponential regression formula $f(x)$ and a logarithmic regression formula $g(x)$ for the points $(1.5, 1.5)$ and $(8.5, 8.5)$. Round all numbers to 6 decimal places. Graph the points and both formulas along with the line $y = x$ on the same axis. Make a conjecture about the relationship of the regression formulas.

58. Verify the conjecture made in the previous exercise. Round all numbers to six decimal places when necessary.

59. Find the inverse function $f^{-1}(x)$ for the logistic function $f(x) = \dfrac{c}{1 + ae^{-bx}}$. Show all steps.

60. Use the result from the previous exercise to graph the logistic model $P(t) = \dfrac{20}{1 + 4e^{-0.5t}}$ along with its inverse on the same axis. What are the intercepts and asymptotes of each function?

CHAPTER 6 REVIEW

Key Terms

annual percentage rate (APR) the yearly interest rate earned by an investment account, also called *nominal rate*

carrying capacity in a logistic model, the limiting value of the output

change-of-base formula a formula for converting a logarithm with any base to a quotient of logarithms with any other base.

common logarithm the exponent to which 10 must be raised to get x; $\log_{10}(x)$ is written simply as $\log(x)$.

compound interest interest earned on the total balance, not just the principal

doubling time the time it takes for a quantity to double

exponential growth a model that grows by a rate proportional to the amount present

extraneous solution a solution introduced while solving an equation that does not satisfy the conditions of the original equation

half-life the length of time it takes for a substance to exponentially decay to half of its original quantity

logarithm the exponent to which b must be raised to get x; written $y = \log_b(x)$

logistic growth model a function of the form $f(x) = \dfrac{c}{1 + ae^{-bx}}$ where $\dfrac{c}{1+a}$ is the initial value, c is the carrying capacity, or limiting value, and b is a constant determined by the rate of growth

natural logarithm the exponent to which the number e must be raised to get x; $\log_e(x)$ is written as $\ln(x)$.

Newton's Law of Cooling the scientific formula for temperature as a function of time as an object's temperature is equalized with the ambient temperature

nominal rate the yearly interest rate earned by an investment account, also called *annual percentage rate*

order of magnitude the power of ten, when a number is expressed in scientific notation, with one non-zero digit to the left of the decimal

power rule for logarithms a rule of logarithms that states that the log of a power is equal to the product of the exponent and the log of its base

product rule for logarithms a rule of logarithms that states that the log of a product is equal to a sum of logarithms

quotient rule for logarithms a rule of logarithms that states that the log of a quotient is equal to a difference of logarithms

Key Equations

definition of the exponential function	$f(x) = b^x$, where $b > 0$, $b \neq 1$
definition of exponential growth	$f(x) = ab^x$, where $a > 0$, $b > 0$, $b \neq 1$
compound interest formula	$A = a\left(1 + \dfrac{r}{k}\right)^{kt}$, where $A(t)$ is the account value at time t t is the number of years P is the initial investment, often called the principal r is the annual percentage rate (APR), or nominal rate n is the number of compounding periods in one year
continuous growth formula	$A(t) = ae^{rt}$, where t is the number of unit time periods of growth a is the starting amount (in the continuous compounding formula a is replaced with P, the principal) e is the mathematical constant, $e \approx 2.718282$
General Form for the Translation of the Parent Function $f(x) = b^x$	$f(x) = ab^{x+c} + d$
Definition of the logarithmic function	For $x > 0$, $b > 0$, $b \neq 1$, $y = \log_b(x)$ if and only if $b^y = x$.
Definition of the common logarithm	For $x > 0$, $y = \log(x)$ if and only if $10^y = x$.

Definition of the natural logarithm	For $x > 0$, $y = \ln(x)$ if and only if $e^y = x$.
General Form for the Translation of the Parent Logarithmic Function $f(x) = \log_b(x)$	$f(x) = a\log_b(x + c) + d$
The Product Rule for Logarithms	$\log_b(MN) = \log_b(M) + \log_b(N)$
The Quotient Rule for Logarithms	$\log_b\left(\dfrac{M}{N}\right) = \log_b M - \log_b N$
The Power Rule for Logarithms	$\log_b(M^n) = n\log_b M$
The Change-of-Base Formula	$\log_b M = \dfrac{\log_n M}{\log_n b} \qquad n > 0, n \neq 1, b \neq 1$
One-to-one property for exponential functions	For any algebraic expressions S and T and any positive real number b, where $b^S = b^T$ if and only if $S = T$.
Definition of a logarithm	For any algebraic expression S and positive real numbers b and c, where $b \neq 1$, $\log_b(S) = c$ if and only if $b^c = S$.
One-to-one property for logarithmic functions	For any algebraic expressions S and T and any positive real number b, where $b \neq 1$, $\log_b S = \log_b T$ if and only if $S = T$.
Half-life formula	If $A = A_0 e^{kt}$, $k < 0$, the half-life is $t = -\dfrac{\ln(2)}{k}$. $\quad t = \dfrac{\ln\left(\dfrac{A}{A_0}\right)}{-0.000121}$
Carbon-14 dating	A_0 is the amount of carbon-14 when the plant or animal died, A is the amount of carbon-14 remaining today, t is the age of the fossil in years
Doubling time formula	If $A = A_0 e^{kt}$, $k > 0$, the doubling time is $t = \dfrac{\ln(2)}{k}$
Newton's Law of Cooling	$T(t) = Ae^{kt} + T_s$, where T_s is the ambient temperature, $A = T(0) - T_s$, and k is the continuous rate of cooling.

Key Concepts

6.1 Exponential Functions

- An exponential function is defined as a function with a positive constant other than 1 raised to a variable exponent. See **Example 1**.
- A function is evaluated by solving at a specific value. See **Example 2** and **Example 3**.
- An exponential model can be found when the growth rate and initial value are known. See **Example 4**.
- An exponential model can be found when the two data points from the model are known. See **Example 5**.
- An exponential model can be found using two data points from the graph of the model. See **Example 6**.
- An exponential model can be found using two data points from the graph and a calculator. See **Example 7**.
- The value of an account at any time t can be calculated using the compound interest formula when the principal, annual interest rate, and compounding periods are known. See **Example 8**.
- The initial investment of an account can be found using the compound interest formula when the value of the account, annual interest rate, compounding periods, and life span of the account are known. See **Example 9**.
- The number e is a mathematical constant often used as the base of real world exponential growth and decay models. Its decimal approximation is $e \approx 2.718282$.
- Scientific and graphing calculators have the key $[e^x]$ or $[\exp(x)]$ for calculating powers of e. See **Example 10**.
- Continuous growth or decay models are exponential models that use e as the base. Continuous growth and decay models can be found when the initial value and growth or decay rate are known. See **Example 11** and **Example 12**.

6.2 Graphs of Exponential Functions

- The graph of the function $f(x) = b^x$ has a y-intercept at $(0, 1)$, domain $(-\infty, \infty)$, range $(0, \infty)$, and horizontal asymptote $y = 0$. See **Example 1**.
- If $b > 1$, the function is increasing. The left tail of the graph will approach the asymptote $y = 0$, and the right tail will increase without bound.
- If $0 < b < 1$, the function is decreasing. The left tail of the graph will increase without bound, and the right tail will approach the asymptote $y = 0$.
- The equation $f(x) = b^x + d$ represents a vertical shift of the parent function $f(x) = b^x$.
- The equation $f(x) = b^{x+c}$ represents a horizontal shift of the parent function $f(x) = b^x$. See **Example 2**.
- Approximate solutions of the equation $f(x) = b^{x+c} + d$ can be found using a graphing calculator. See **Example 3**.
- The equation $f(x) = ab^x$, where $a > 0$, represents a vertical stretch if $|a| > 1$ or compression if $0 < |a| < 1$ of the parent function $f(x) = b^x$. See **Example 4**.
- When the parent function $f(x) = b^x$ is multiplied by -1, the result, $f(x) = -b^x$, is a reflection about the x-axis. When the input is multiplied by -1, the result, $f(x) = b^{-x}$, is a reflection about the y-axis. See **Example 5**.
- All translations of the exponential function can be summarized by the general equation $f(x) = ab^{x+c} + d$. See **Table 3**.
- Using the general equation $f(x) = ab^{x+c} + d$, we can write the equation of a function given its description. See **Example 6**.

6.3 Logarithmic Functions

- The inverse of an exponential function is a logarithmic function, and the inverse of a logarithmic function is an exponential function.
- Logarithmic equations can be written in an equivalent exponential form, using the definition of a logarithm. See **Example 1**.
- Exponential equations can be written in their equivalent logarithmic form using the definition of a logarithm See **Example 2**.
- Logarithmic functions with base b can be evaluated mentally using previous knowledge of powers of b. See **Example 3** and **Example 4**.
- Common logarithms can be evaluated mentally using previous knowledge of powers of 10. See **Example 5**.
- When common logarithms cannot be evaluated mentally, a calculator can be used. See **Example 6**.
- Real-world exponential problems with base 10 can be rewritten as a common logarithm and then evaluated using a calculator. See **Example 7**.
- Natural logarithms can be evaluated using a calculator **Example 8**.

6.4 Graphs of Logarithmic Functions

- To find the domain of a logarithmic function, set up an inequality showing the argument greater than zero, and solve for x. See **Example 1** and **Example 2**.
- The graph of the parent function $f(x) = \log_b(x)$ has an x-intercept at $(1, 0)$, domain $(0, \infty)$, range $(-\infty, \infty)$, vertical asymptote $x = 0$, and
  - if $b > 1$, the function is increasing.
  - if $0 < b < 1$, the function is decreasing.

  See **Example 3**.
- The equation $f(x) = \log_b(x + c)$ shifts the parent function $y = \log_b(x)$ horizontally
  - left c units if $c > 0$.
  - right c units if $c < 0$.

  See **Example 4**.
- The equation $f(x) = \log_b(x) + d$ shifts the parent function $y = \log_b(x)$ vertically
  - up d units if $d > 0$.
  - down d units if $d < 0$.

  See **Example 5**.

- For any constant $a > 0$, the equation $f(x) = a\log_b(x)$
  - stretches the parent function $y = \log_b(x)$ vertically by a factor of a if $|a| > 1$.
  - compresses the parent function $y = \log_b(x)$ vertically by a factor of a if $|a| < 1$.

  See **Example 6** and **Example 7**.

- When the parent function $y = \log_b(x)$ is multiplied by -1, the result is a reflection about the x-axis. When the input is multiplied by -1, the result is a reflection about the y-axis.
  - The equation $f(x) = -\log_b(x)$ represents a reflection of the parent function about the x-axis.
- The equation $f(x) = \log_b(-x)$ represents a reflection of the parent function about the y-axis.

  See **Example 8**.

  - A graphing calculator may be used to approximate solutions to some logarithmic equations See **Example 9**.
- All translations of the logarithmic function can be summarized by the general equation $f(x) = a\log_b(x + c) + d$. See **Table 4**.
- Given an equation with the general form $f(x) = a\log_b(x + c) + d$, we can identify the vertical asymptote $x = -c$ for the transformation. See **Example 10**.
- Using the general equation $f(x) = a\log_b(x + c) + d$, we can write the equation of a logarithmic function given its graph. See **Example 11**.

6.5 Logarithmic Properties

- We can use the product rule of logarithms to rewrite the log of a product as a sum of logarithms. See **Example 1**.
- We can use the quotient rule of logarithms to rewrite the log of a quotient as a difference of logarithms. See **Example 2**.
- We can use the power rule for logarithms to rewrite the log of a power as the product of the exponent and the log of its base. See **Example 3**, **Example 4**, and **Example 5**.
- We can use the product rule, the quotient rule, and the power rule together to combine or expand a logarithm with a complex input. See **Example 6**, **Example 7**, and **Example 8**.
- The rules of logarithms can also be used to condense sums, differences, and products with the same base as a single logarithm. See **Example 9**, **Example 10**, **Example 11**, and **Example 12**.
- We can convert a logarithm with any base to a quotient of logarithms with any other base using the change-of-base formula. See **Example 13**.
- The change-of-base formula is often used to rewrite a logarithm with a base other than 10 and e as the quotient of natural or common logs. That way a calculator can be used to evaluate. See **Example 14**.

6.6 Exponential and Logarithmic Equations

- We can solve many exponential equations by using the rules of exponents to rewrite each side as a power with the same base. Then we use the fact that exponential functions are one-to-one to set the exponents equal to one another and solve for the unknown.
- When we are given an exponential equation where the bases are explicitly shown as being equal, set the exponents equal to one another and solve for the unknown. See **Example 1**.
- When we are given an exponential equation where the bases are *not* explicitly shown as being equal, rewrite each side of the equation as powers of the same base, then set the exponents equal to one another and solve for the unknown. See **Example 2**, **Example 3**, and **Example 4**.
- When an exponential equation cannot be rewritten with a common base, solve by taking the logarithm of each side. See **Example 5**.
- We can solve exponential equations with base e, by applying the natural logarithm of both sides because exponential and logarithmic functions are inverses of each other. See **Example 6** and **Example 7**.
- After solving an exponential equation, check each solution in the original equation to find and eliminate any extraneous solutions. See **Example 8**.

- When given an equation of the form $\log_b(S) = c$, where S is an algebraic expression, we can use the definition of a logarithm to rewrite the equation as the equivalent exponential equation $b^c = S$, and solve for the unknown. See **Example 9** and **Example 10**.

- We can also use graphing to solve equations with the form $\log_b(S) = c$. We graph both equations $y = \log_b(S)$ and $y = c$ on the same coordinate plane and identify the solution as the x-value of the intersecting point. See **Example 11**.

- When given an equation of the form $\log_b S = \log_b T$, where S and T are algebraic expressions, we can use the one-to-one property of logarithms to solve the equation $S = T$ for the unknown. See **Example 12**.

- Combining the skills learned in this and previous sections, we can solve equations that model real world situations, whether the unknown is in an exponent or in the argument of a logarithm. See **Example 13**.

6.7 Exponential and Logarithmic Models

- The basic exponential function is $f(x) = ab^x$. If $b > 1$, we have exponential growth; if $0 < b < 1$, we have exponential decay.

- We can also write this formula in terms of continuous growth as $A = A_0 e^{kx}$, where A_0 is the starting value. If A_0 is positive, then we have exponential growth when $k > 0$ and exponential decay when $k < 0$. See **Example 1**.

- In general, we solve problems involving exponential growth or decay in two steps. First, we set up a model and use the model to find the parameters. Then we use the formula with these parameters to predict growth and decay. See **Example 2**.

- We can find the age, t, of an organic artifact by measuring the amount, k, of carbon-14 remaining in the artifact and using the formula $t = \dfrac{\ln(k)}{-0.000121}$ to solve for t. See **Example 3**.

- Given a substance's doubling time or half-life we can find a function that represents its exponential growth or decay. See **Example 4**.

- We can use Newton's Law of Cooling to find how long it will take for a cooling object to reach a desired temperature, or to find what temperature an object will be after a given time. See **Example 5**.

- We can use logistic growth functions to model real-world situations where the rate of growth changes over time, such as population growth, spread of disease, and spread of rumors. See **Example 6**.

- We can use real-world data gathered over time to observe trends. Knowledge of linear, exponential, logarithmic, and logistic graphs help us to develop models that best fit our data. See **Example 7**.

- Any exponential function with the form $y = ab^x$ can be rewritten as an equivalent exponential function with the form $y = A_0 e^{kx}$ where $k = \ln b$. See **Example 8**.

6.8 Fitting Exponential Models to Data

- Exponential regression is used to model situations where growth begins slowly and then accelerates rapidly without bound, or where decay begins rapidly and then slows down to get closer and closer to zero.

- We use the command "ExpReg" on a graphing utility to fit function of the form $y = ab^x$ to a set of data points. See **Example 1**.

- Logarithmic regression is used to model situations where growth or decay accelerates rapidly at first and then slows over time.

- We use the command "LnReg" on a graphing utility to fit a function of the form $y = a + b\ln(x)$ to a set of data points. See **Example 2**.

- Logistic regression is used to model situations where growth accelerates rapidly at first and then steadily slows as the function approaches an upper limit.

- We use the command "Logistic" on a graphing utility to fit a function of the form $y = \dfrac{c}{1 + ae^{-bx}}$ to a set of data points. See **Example 3**.

CHAPTER 6 REVIEW EXERCISES

EXPONENTIAL FUNCTIONS

1. Determine whether the function $y = 156(0.825)^t$ represents exponential growth, exponential decay, or neither. Explain

2. The population of a herd of deer is represented by the function $A(t) = 205(1.13)^t$, where t is given in years. To the nearest whole number, what will the herd population be after 6 years?

3. Find an exponential equation that passes through the points $(2, 2.25)$ and $(5, 60.75)$.

4. Determine whether **Table 1** could represent a function that is linear, exponential, or neither. If it appears to be exponential, find a function that passes through the points.

x	1	2	3	4
$f(x)$	3	0.9	0.27	0.081

Table 1

5. A retirement account is opened with an initial deposit of $8,500 and earns 8.12% interest compounded monthly. What will the account be worth in 20 years?

6. Hsu-Mei wants to save $5,000 for a down payment on a car. To the nearest dollar, how much will she need to invest in an account now with 7.5% APR, compounded daily, in order to reach her goal in 3 years?

7. Does the equation $y = 2.294e^{-0.654t}$ represent continuous growth, continuous decay, or neither? Explain.

8. Suppose an investment account is opened with an initial deposit of $10,500 earning 6.25% interest, compounded continuously. How much will the account be worth after 25 years?

GRAPHS OF EXPONENTIAL FUNCTIONS

9. Graph the function $f(x) = 3.5(2)^x$. State the domain and range and give the y-intercept.

10. Graph the function $f(x) = 4\left(\frac{1}{8}\right)^x$ and its reflection about the y-axis on the same axes, and give the y-intercept.

11. The graph of $f(x) = 6.5^x$ is reflected about the y-axis and stretched vertically by a factor of 7. What is the equation of the new function, $g(x)$? State its y-intercept, domain, and range.

12. The graph below shows transformations of the graph of $f(x) = 2^x$. What is the equation for the transformation?

Figure 1

LOGARITHMIC FUNCTIONS

13. Rewrite $\log_{17}(4913) = x$ as an equivalent exponential equation.

14. Rewrite $\ln(s) = t$ as an equivalent exponential equation.

15. Rewrite $a^{-\frac{2}{5}} = b$ as an equivalent logarithmic equation.

16. Rewrite $e^{-3.5} = h$ as an equivalent logarithmic equation.

17. Solve for $x \log_{64}(x) = \left(\frac{1}{3}\right)$ to exponential form.

18. Evaluate $\log_5\left(\frac{1}{125}\right)$ without using a calculator.

19. Evaluate $\log(0.000001)$ without using a calculator.

20. Evaluate $\log(4.005)$ using a calculator. Round to the nearest thousandth.

21. Evaluate $\ln(e^{-0.8648})$ without using a calculator.

22. Evaluate $\ln(\sqrt[3]{18})$ using a calculator. Round to the nearest thousandth.

GRAPHS OF LOGARITHMIC FUNCTIONS

23. Graph the function $g(x) = \log(7x + 21) - 4$.

24. Graph the function $h(x) = 2\ln(9 - 3x) + 1$.

25. State the domain, vertical asymptote, and end behavior of the function $g(x) = \ln(4x + 20) - 17$.

LOGARITHMIC PROPERTIES

26. Rewrite $\ln(7r \cdot 11st)$ in expanded form.

27. Rewrite $\log_8(x) + \log_8(5) + \log_8(y) + \log_8(13)$ in compact form.

28. Rewrite $\log_m\left(\dfrac{67}{83}\right)$ in expanded form.

29. Rewrite $\ln(z) - \ln(x) - \ln(y)$ in compact form.

30. Rewrite $\ln\left(\dfrac{1}{x^5}\right)$ as a product.

31. Rewrite $-\log_y\left(\dfrac{1}{12}\right)$ as a single logarithm.

32. Use properties of logarithms to expand $\log\left(\dfrac{r^2 s^{11}}{t^{14}}\right)$.

33. Use properties of logarithms to expand $\ln\left(2b\sqrt{\dfrac{b+1}{b-1}}\right)$.

34. Condense the expression $5\ln(b) + \ln(c) + \dfrac{\ln(4-a)}{2}$ to a single logarithm.

35. Condense the expression $3\log_7 v + 6\log_7 w - \dfrac{\log_7 u}{3}$ to a single logarithm.

36. Rewrite $\log_3(12.75)$ to base e.

37. Rewrite $5^{12x-17} = 125$ as a logarithm. Then apply the change of base formula to solve for x using the common log. Round to the nearest thousandth.

EXPONENTIAL AND LOGARITHMIC EQUATIONS

38. Solve $216^{3x} \cdot 216^x = 36^{3x+2}$ by rewriting each side with a common base.

39. Solve $\dfrac{125}{\left(\dfrac{1}{625}\right)^{-x-3}} = 5^3$ by rewriting each side with a common base.

40. Use logarithms to find the exact solution for $7 \cdot 17^{-9x} - 7 = 49$. If there is no solution, write *no solution*.

41. Use logarithms to find the exact solution for $3e^{6n-2} + 1 = -60$. If there is no solution, write *no solution*.

42. Find the exact solution for $5e^{3x} - 4 = 6$. If there is no solution, write *no solution*.

43. Find the exact solution for $2e^{5x-2} - 9 = -56$. If there is no solution, write *no solution*.

44. Find the exact solution for $5^{2x-3} = 7^{x+1}$. If there is no solution, write *no solution*.

45. Find the exact solution for $e^{2x} - e^x - 110 = 0$. If there is no solution, write *no solution*.

46. Use the definition of a logarithm to solve. $-5\log_7(10n) = 5$.

47. Use the definition of a logarithm to find the exact solution for $9 + 6\ln(a + 3) = 33$.

48. Use the one-to-one property of logarithms to find an exact solution for $\log_8(7) + \log_8(-4x) = \log_8(5)$. If there is no solution, write *no solution*.

49. Use the one-to-one property of logarithms to find an exact solution for $\ln(5) + \ln(5x^2 - 5) = \ln(56)$. If there is no solution, write *no solution*.

50. The formula for measuring sound intensity in decibels D is defined by the equation $D = 10\log\left(\dfrac{I}{I_0}\right)$, where I is the intensity of the sound in watts per square meter and $I_0 = 10^{-12}$ is the lowest level of sound that the average person can hear. How many decibels are emitted from a large orchestra with a sound intensity of $6.3 \cdot 10^{-3}$ watts per square meter?

51. The population of a city is modeled by the equation $P(t) = 256,114e^{0.25t}$ where t is measured in years. If the city continues to grow at this rate, how many years will it take for the population to reach one million?

52. Find the inverse function f^{-1} for the exponential function $f(x) = 2 \cdot e^{x+1} - 5$.

53. Find the inverse function f^{-1} for the logarithmic function $f(x) = 0.25 \cdot \log_2(x^3 + 1)$.

EXPONENTIAL AND LOGARITHMIC MODELS

For the following exercises, use this scenario: A doctor prescribes 300 milligrams of a therapeutic drug that decays by about 17% each hour.

54. To the nearest minute, what is the half-life of the drug?

55. Write an exponential model representing the amount of the drug remaining in the patient's system after t hours. Then use the formula to find the amount of the drug that would remain in the patient's system after 24 hours. Round to the nearest hundredth of a gram.

For the following exercises, use this scenario: A soup with an internal temperature of 350° Fahrenheit was taken off the stove to cool in a 71°F room. After fifteen minutes, the internal temperature of the soup was 175°F.

56. Use Newton's Law of Cooling to write a formula that models this situation.

57. How many minutes will it take the soup to cool to 85°F?

For the following exercises, use this scenario: The equation $N(t) = \dfrac{1200}{1 + 199e^{-0.625t}}$ models the number of people in a school who have heard a rumor after t days.

58. How many people started the rumor?

59. To the nearest tenth, how many days will it be before the rumor spreads to half the carrying capacity?

60. What is the carrying capacity?

For the following exercises, enter the data from each table into a graphing calculator and graph the resulting scatter plots. Determine whether the data from the table would likely represent a function that is linear, exponential, or logarithmic.

61.

x	1	2	3	4	5	6	7	8	9	10
$f(x)$	3.05	4.42	6.4	9.28	13.46	19.52	28.3	41.04	59.5	86.28

62.

x	0.5	1	3	5	7	10	12	13	15	17	20
$f(x)$	18.05	17	15.33	14.55	14.04	13.5	13.22	13.1	12.88	12.69	12.45

63. Find a formula for an exponential equation that goes through the points $(-2, 100)$ and $(0, 4)$. Then express the formula as an equivalent equation with base e.

FITTING EXPONENTIAL MODELS TO DATA

64. What is the carrying capacity for a population modeled by the logistic equation $P(t) = \dfrac{250,000}{1 + 499e^{-0.45t}}$? What is the initial population for the model?

65. The population of a culture of bacteria is modeled by the logistic equation $P(t) = \dfrac{14,250}{1 + 29e^{-0.62t}}$, where t is in days. To the nearest tenth, how many days will it take the culture to reach 75% of its carrying capacity?

For the following exercises, use a graphing utility to create a scatter diagram of the data given in the table. Observe the shape of the scatter diagram to determine whether the data is best described by an exponential, logarithmic, or logistic model. Then use the appropriate regression feature to find an equation that models the data. When necessary, round values to five decimal places.

66.

x	1	2	3	4	5	6	7	8	9	10
$f(x)$	409.4	260.7	170.4	110.6	74	44.7	32.4	19.5	12.7	8.1

67.

x	0.15	0.25	0.5	0.75	1	1.5	2	2.25	2.75	3	3.5
$f(x)$	36.21	28.88	24.39	18.28	16.5	12.99	9.91	8.57	7.23	5.99	4.81

68.

x	0	2	4	5	7	8	10	11	15	17
$f(x)$	9	22.6	44.2	62.1	96.9	113.4	133.4	137.6	148.4	149.3

CHAPTER 6 PRACTICE TEST

1. The population of a pod of bottlenose dolphins is modeled by the function $A(t) = 8(1.17)^t$, where t is given in years. To the nearest whole number, what will the pod population be after 3 years?

2. Find an exponential equation that passes through the points (0, 4) and (2, 9).

3. Drew wants to save $2,500 to go to the next World Cup. To the nearest dollar, how much will he need to invest in an account now with 6.25% APR, compounding daily, in order to reach his goal in 4 years?

4. An investment account was opened with an initial deposit of $9,600 and earns 7.4% interest, compounded continuously. How much will the account be worth after 15 years?

5. Graph the function $f(x) = 5(0.5)^{-x}$ and its reflection across the y-axis on the same axes, and give the y-intercept.

6. The graph below shows transformations of the graph of $f(x) = \left(\frac{1}{2}\right)^x$. What is the equation for the transformation?

7. Rewrite $\log_{8.5}(614.125) = a$ as an equivalent exponential equation.

8. Rewrite $e^{\frac{1}{2}} = m$ as an equivalent logarithmic equation.

9. Solve for x by converting the logarithmic equation $\log_{\frac{1}{7}}(x) = 2$ to exponential form.

10. Evaluate $\log(10,000,000)$ without using a calculator.

11. Evaluate $\ln(0.716)$ using a calculator. Round to the nearest thousandth.

12. Graph the function $g(x) = \log(12 - 6x) + 3$.

13. State the domain, vertical asymptote, and end behavior of the function $f(x) = \log_5(39 - 13x) + 7$.

14. Rewrite $\log(17a \cdot 2b)$ as a sum.

15. Rewrite $\log_t(96) - \log_t(8)$ in compact form.

16. Rewrite $\log_8\left(a^{\frac{1}{b}}\right)$ as a product.

17. Use properties of logarithm to expand $\ln\left(y^3 z^2 \cdot \sqrt[3]{x - 4}\right)$.

18. Condense the expression
$4\ln(c) + \ln(d) + \dfrac{\ln(a)}{3} + \dfrac{\ln(b + 3)}{3}$ to a single logarithm.

19. Rewrite $16^{3x - 5} = 1000$ as a logarithm. Then apply the change of base formula to solve for x using the natural log. Round to the nearest thousandth.

20. Solve $\left(\frac{1}{81}\right)^x \cdot \dfrac{1}{243} = \left(\frac{1}{9}\right)^{-3x - 1}$ by rewriting each side with a common base.

21. Use logarithms to find the exact solution for $-9e^{10a - 8} - 5 = -41$. If there is no solution, write *no solution*.

22. Find the exact solution for $10e^{4x + 2} + 5 = 56$. If there is no solution, write *no solution*.

23. Find the exact solution for $-5e^{-4x - 1} - 4 = 64$. If there is no solution, write *no solution*.

24. Find the exact solution for $2^{x - 3} = 6^{2x - 1}$. If there is no solution, write *no solution*.

25. Find the exact solution for $e^{2x} - e^x - 72 = 0$. If there is no solution, write *no solution*.

26. Use the definition of a logarithm to find the exact solution for $4\log(2n) - 7 = -11$.

27. Use the one-to-one property of logarithms to find an exact solution for $\log(4x^2 - 10) + \log(3) = \log(51)$ If there is no solution, write *no solution*.

28. The formula for measuring sound intensity in decibels D is defined by the equation

$$D = 10\log\left(\frac{I}{I_0}\right)$$

where I is the intensity of the sound in watts per square meter and $I_0 = 10^{-12}$ is the lowest level of sound that the average person can hear. How many decibels are emitted from a rock concert with a sound intensity of $4.7 \cdot 10^{-1}$ watts per square meter?

29. A radiation safety officer is working with 112 grams of a radioactive substance. After 17 days, the sample has decayed to 80 grams. Rounding to five significant digits, write an exponential equation representing this situation. To the nearest day, what is the half-life of this substance?

30. Write the formula found in the previous exercise as an equivalent equation with base e. Express the exponent to five significant digits.

31. A bottle of soda with a temperature of 71° Fahrenheit was taken off a shelf and placed in a refrigerator with an internal temperature of 35° F. After ten minutes, the internal temperature of the soda was 63° F. Use Newton's Law of Cooling to write a formula that models this situation. To the nearest degree, what will the temperature of the soda be after one hour?

32. The population of a wildlife habitat is modeled by the equation $P(t) = \dfrac{360}{1 + 6.2e^{-0.35t}}$, where t is given in years. How many animals were originally transported to the habitat? How many years will it take before the habitat reaches half its capacity?

33. Enter the data from **Table 2** into a graphing calculator and graph the resulting scatter plot. Determine whether the data from the table would likely represent a function that is linear, exponential, or logarithmic.

x	1	2	3	4	5	6	7	8	9	10
$f(x)$	3	8.55	11.79	14.09	15.88	17.33	18.57	19.64	20.58	21.42

Table 2

34. The population of a lake of fish is modeled by the logistic equation $P(t) = \dfrac{16,120}{1 + 25e^{-0.75t}}$, where t is time in years. To the nearest hundredth, how many years will it take the lake to reach 80% of its carrying capacity?

For the following exercises, use a graphing utility to create a scatter diagram of the data given in the table. Observe the shape of the scatter diagram to determine whether the data is best described by an exponential, logarithmic, or logistic model. Then use the appropriate regression feature to find an equation that models the data. When necessary, round values to five decimal places.

35.

x	1	2	3	4	5	6	7	8	9	10
$f(x)$	20	21.6	29.2	36.4	46.6	55.7	72.6	87.1	107.2	138.1

36.

x	3	4	5	6	7	8	9	10	11	12	13
$f(x)$	13.98	17.84	20.01	22.7	24.1	26.15	27.37	28.38	29.97	31.07	31.43

37.

x	0	0.5	1	1.5	2	3	4	5	6	7	8
$f(x)$	2.2	2.9	3.9	4.8	6.4	9.3	12.3	15	16.2	17.3	17.9

7

Systems of Equations and Inequalities

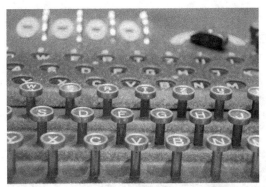

Figure 1 Enigma machines like this one, once owned by Italian dictator Benito Mussolini, were used by government and military officials for enciphering and deciphering top-secret communications during World War II. (credit: Dave Addey, Flickr)

CHAPTER OUTLINE

Introduction

By 1943, it was obvious to the Nazi regime that defeat was imminent unless it could build a weapon with unlimited destructive power, one that had never been seen before in the history of the world. In September, Adolf Hitler ordered German scientists to begin building an atomic bomb. Rumors and whispers began to spread from across the ocean. Refugees and diplomats told of the experiments happening in Norway. However, Franklin D. Roosevelt wasn't sold, and even doubted British Prime Minister Winston Churchill's warning. Roosevelt wanted undeniable proof. Fortunately, he soon received the proof he wanted when a group of mathematicians cracked the "Enigma" code, proving beyond a doubt that Hitler was building an atomic bomb. The next day, Roosevelt gave the order that the United States begin work on the same.

The Enigma is perhaps the most famous cryptographic device ever known. It stands as an example of the pivotal role cryptography has played in society. Now, technology has moved cryptanalysis to the digital world.

Many ciphers are designed using invertible matrices as the method of message transference, as finding the inverse of a matrix is generally part of the process of decoding. In addition to knowing the matrix and its inverse, the receiver must also know the key that, when used with the matrix inverse, will allow the message to be read.

In this chapter, we will investigate matrices and their inverses, and various ways to use matrices to solve systems of equations. First, however, we will study systems of equations on their own: linear and nonlinear, and then partial fractions. We will not be breaking any secret codes here, but we will lay the foundation for future courses.

575

LEARNING OBJECTIVES

In this section, you will:
- Solve systems of equations by graphing.
- Solve systems of equations by substitution.
- Solve systems of equations by addition.
- Identify inconsistent systems of equations containing two variables.
- Express the solution of a system of dependent equations containing two variables.

7.1   SYSTEMS OF LINEAR EQUATIONS: TWO VARIABLES

Figure 1 (credit: Thomas Sørenes)

A skateboard manufacturer introduces a new line of boards. The manufacturer tracks its costs, which is the amount it spends to produce the boards, and its revenue, which is the amount it earns through sales of its boards. How can the company determine if it is making a profit with its new line? How many skateboards must be produced and sold before a profit is possible? In this section, we will consider linear equations with two variables to answer these and similar questions.

Introduction to Systems of Equations

In order to investigate situations such as that of the skateboard manufacturer, we need to recognize that we are dealing with more than one variable and likely more than one equation. A **system of linear equations** consists of two or more linear equations made up of two or more variables such that all equations in the system are considered simultaneously. To find the unique solution to a system of linear equations, we must find a numerical value for each variable in the system that will satisfy all equations in the system at the same time. Some linear systems may not have a solution and others may have an infinite number of solutions. In order for a linear system to have a unique solution, there must be at least as many equations as there are variables. Even so, this does not guarantee a unique solution.

In this section, we will look at systems of linear equations in two variables, which consist of two equations that contain two different variables. For example, consider the following system of linear equations in two variables.

$$2x + y = 15$$
$$3x - y = 5$$

The *solution* to a system of linear equations in two variables is any ordered pair that satisfies each equation independently. In this example, the ordered pair (4, 7) is the solution to the system of linear equations. We can verify the solution by substituting the values into each equation to see if the ordered pair satisfies both equations. Shortly we will investigate methods of finding such a solution if it exists.

$$2(4) + (7) = 15 \quad \text{True}$$
$$3(4) - (7) = 5 \quad \text{True}$$

In addition to considering the number of equations and variables, we can categorize systems of linear equations by the number of solutions. A **consistent system** of equations has at least one solution. A consistent system is considered to be an **independent system** if it has a single solution, such as the example we just explored. The two lines have different slopes and intersect at one point in the plane. A consistent system is considered to be a **dependent system** if the equations have the same slope and the same y-intercepts. In other words, the lines coincide so the equations represent the same line. Every point on the line represents a coordinate pair that satisfies the system. Thus, there are an infinite number of solutions.

Another type of system of linear equations is an **inconsistent system**, which is one in which the equations represent two parallel lines. The lines have the same slope and different y-intercepts. There are no points common to both lines; hence, there is no solution to the system.

> *types of linear systems*
> There are three types of systems of linear equations in two variables, and three types of solutions.
>
> • An **independent system** has exactly one solution pair (x, y). The point where the two lines intersect is the only solution.
> • An **inconsistent system** has no solution. Notice that the two lines are parallel and will never intersect.
> • A **dependent system** has infinitely many solutions. The lines are coincident. They are the same line, so every coordinate pair on the line is a solution to both equations.
>
> **Figure 2** compares graphical representations of each type of system.
>
>
>
> Independent System   Inconsistent System   Dependent System
>
> **Figure 2**

How To…

Given a system of linear equations and an ordered pair, determine whether the ordered pair is a solution.

1. Substitute the ordered pair into each equation in the system.
2. Determine whether true statements result from the substitution in both equations; if so, the ordered pair is a solution.

Example 1   **Determining Whether an Ordered Pair Is a Solution to a System of Equations**

Determine whether the ordered pair $(5, 1)$ is a solution to the given system of equations.

$$x + 3y = 8$$
$$2x - 9 = y$$

Solution   Substitute the ordered pair $(5, 1)$ into both equations.

$$(5) + 3(1) = 8$$
$$8 = 8 \qquad \text{True}$$
$$2(5) - 9 = (1)$$
$$1 = 1 \qquad \text{True}$$

The ordered pair $(5, 1)$ satisfies both equations, so it is the solution to the system.

Analysis *We can see the solution clearly by plotting the graph of each equation. Since the solution is an ordered pair that satisfies both equations, it is a point on both of the lines and thus the point of intersection of the two lines. See **Figure 3**.*

Figure 3

Try It #1

Determine whether the ordered pair (8, 5) is a solution to the following system.

$$5x - 4y = 20$$
$$2x + 1 = 3y$$

Solving Systems of Equations by Graphing

There are multiple methods of solving systems of linear equations. For a system of linear equations in two variables, we can determine both the type of system and the solution by graphing the system of equations on the same set of axes.

Example 2   **Solving a System of Equations in Two Variables by Graphing**

Solve the following system of equations by graphing. Identify the type of system.

$$2x + y = -8$$
$$x - y = -1$$

Solution   Solve the first equation for *y*.

$$2x + y = -8$$
$$y = -2x - 8$$

Solve the second equation for *y*.

$$x - y = -1$$
$$y = x + 1$$

Graph both equations on the same set of axes as in **Figure 4**.

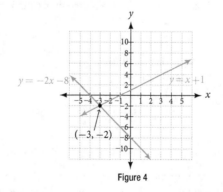

Figure 4

The lines appear to intersect at the point $(-3, -2)$. We can check to make sure that this is the solution to the system by substituting the ordered pair into both equations.

$$2(-3) + (-2) = -8$$
$$-8 = -8 \text{ True}$$
$$(-3) - (-2) = -1$$
$$-1 = -1 \text{ True}$$

The solution to the system is the ordered pair $(-3, -2)$, so the system is independent.

Try It #2

Solve the following system of equations by graphing.

$$2x - 5y = -25$$
$$-4x + 5y = 35$$

Q & A...

Can graphing be used if the system is inconsistent or dependent?

Yes, in both cases we can still graph the system to determine the type of system and solution. If the two lines are parallel, the system has no solution and is inconsistent. If the two lines are identical, the system has infinite solutions and is a dependent system.

Solving Systems of Equations by Substitution

Solving a linear system in two variables by graphing works well when the solution consists of integer values, but if our solution contains decimals or fractions, it is not the most precise method. We will consider two more methods of solving a system of linear equations that are more precise than graphing. One such method is solving a system of equations by the **substitution method**, in which we solve one of the equations for one variable and then substitute the result into the second equation to solve for the second variable. Recall that we can solve for only one variable at a time, which is the reason the substitution method is both valuable and practical.

How To...

Given a system of two equations in two variables, solve using the substitution method.

1. Solve one of the two equations for one of the variables in terms of the other.
2. Substitute the expression for this variable into the second equation, then solve for the remaining variable.
3. Substitute that solution into either of the original equations to find the value of the first variable. If possible, write the solution as an ordered pair.
4. Check the solution in both equations.

Example 3   **Solving a System of Equations in Two Variables by Substitution**

Solve the following system of equations by substitution.

$$-x + y = -5$$
$$2x - 5y = 1$$

Solution   First, we will solve the first equation for y.

$$-x + y = -5$$
$$y = x - 5$$

Now we can substitute the expression $x - 5$ for y in the second equation.

$$2x - 5y = 1$$
$$2x - 5(x - 5) = 1$$
$$2x - 5x + 25 = 1$$
$$-3x = -24$$
$$x = 8$$

Now, we substitute $x = 8$ into the first equation and solve for y.

$$-(8) + y = -5$$
$$y = 3$$

Our solution is (8, 3).

Check the solution by substituting (8, 3) into both equations.

$$-x + y = -5$$
$$-(8) + (3) = -5 \quad \text{True}$$
$$2x - 5y = 1$$
$$2(8) - 5(3) = 1 \quad \text{True}$$

Try It #3

Solve the following system of equations by substitution.

$$x = y + 3$$
$$4 = 3x - 2y$$

Q & A...

Can the substitution method be used to solve any linear system in two variables?

Yes, but the method works best if one of the equations contains a coefficient of 1 or -1 so that we do not have to deal with fractions.

Solving Systems of Equations in Two Variables by the Addition Method

A third method of solving systems of linear equations is the **addition method**. In this method, we add two terms with the same variable, but opposite coefficients, so that the sum is zero. Of course, not all systems are set up with the two terms of one variable having opposite coefficients. Often we must adjust one or both of the equations by multiplication so that one variable will be eliminated by addition.

How To...

Given a system of equations, solve using the addition method.

1. Write both equations with x- and y-variables on the left side of the equal sign and constants on the right.
2. Write one equation above the other, lining up corresponding variables. If one of the variables in the top equation has the opposite coefficient of the same variable in the bottom equation, add the equations together, eliminating one variable. If not, use multiplication by a nonzero number so that one of the variables in the top equation has the opposite coefficient of the same variable in the bottom equation, then add the equations to eliminate the variable.
3. Solve the resulting equation for the remaining variable.
4. Substitute that value into one of the original equations and solve for the second variable.
5. Check the solution by substituting the values into the other equation.

Example 4   **Solving a System by the Addition Method**

Solve the given system of equations by addition.

$$x + 2y = -1$$
$$-x + y = 3$$

Solution   Both equations are already set equal to a constant. Notice that the coefficient of x in the second equation, -1, is the opposite of the coefficient of x in the first equation, 1. We can add the two equations to eliminate x without needing to multiply by a constant.

$$x + 2y = -1$$
$$\underline{-x + y = 3}$$
$$3y = 2$$

Now that we have eliminated x, we can solve the resulting equation for y.

$$3y = 2$$
$$y = \frac{2}{3}$$

Then, we substitute this value for y into one of the original equations and solve for x.

$$-x + y = 3$$
$$-x + \frac{2}{3} = 3$$
$$-x = 3 - \frac{2}{3}$$
$$-x = \frac{7}{3}$$
$$x = -\frac{7}{3}$$

The solution to this system is $\left(-\frac{7}{3}, \frac{2}{3}\right)$.

Check the solution in the first equation.

$$x + 2y = -1$$
$$\left(-\frac{7}{3}\right) + 2\left(\frac{2}{3}\right) = -1$$
$$-\frac{7}{3} + \frac{4}{3} = -1$$
$$-\frac{3}{3} = -1$$
$$-1 = -1 \qquad \text{True}$$

Analysis   We gain an important perspective on systems of equations by looking at the graphical representation. See **Figure 5** *to find that the equations intersect at the solution. We do not need to ask whether there may be a second solution because observing the graph confirms that the system has exactly one solution.*

Figure 5

Example 5   **Using the Addition Method When Multiplication of One Equation Is Required**

Solve the given system of equations by the addition method.

$$3x + 5y = -11$$
$$x - 2y = 11$$

Solution   Adding these equations as presented will not eliminate a variable. However, we see that the first equation has $3x$ in it and the second equation has x. So if we multiply the second equation by -3, the x-terms will add to zero.

$$x - 2y = 11$$
$$-3(x - 2y) = -3(11) \qquad \text{Multiply both sides by } -3.$$
$$-3x + 6y = -33 \qquad \text{Use the distributive property.}$$

Now, let's add them.

$$3x + 5y = -11$$
$$\underline{-3x + 6y = -33}$$
$$11y = -44$$
$$y = -4$$

For the last step, we substitute $y = -4$ into one of the original equations and solve for x.

$$3x + 5y = -11$$
$$3x + 5(-4) = -11$$
$$3x - 20 = -11$$
$$3x = 9$$
$$x = 3$$

Our solution is the ordered pair $(3, -4)$. See **Figure 6**. Check the solution in the original second equation.

$$x - 2y = 11$$
$$(3) - 2(-4) = 3 + 8$$
$$11 = 11 \qquad \text{True}$$

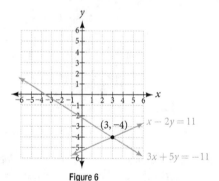

Figure 6

Try It #4

Solve the system of equations by addition.

$$2x - 7y = 2$$
$$3x + y = -20$$

Example 6   **Using the Addition Method When Multiplication of Both Equations Is Required**

Solve the given system of equations in two variables by addition.

$$2x + 3y = -16$$
$$5x - 10y = 30$$

Solution   One equation has $2x$ and the other has $5x$. The least common multiple is $10x$ so we will have to multiply both equations by a constant in order to eliminate one variable. Let's eliminate x by multiplying the first equation by -5 and the second equation by 2.

$$-5(2x + 3y) = -5(-16)$$
$$-10x - 15y = 80$$
$$2(5x - 10y) = 2(30)$$
$$10x - 20y = 60$$

Then, we add the two equations together.

$$-10x - 15y = 80$$
$$\underline{10x - 20y = 60}$$
$$-35y = 140$$
$$y = -4$$

Substitute $y = -4$ into the original first equation.

$$2x + 3(-4) = -16$$
$$2x - 12 = -16$$
$$2x = -4$$
$$x = -2$$

The solution is $(-2, -4)$. Check it in the other equation.

$$5x - 10y = 30$$
$$5(-2) - 10(-4) = 30$$
$$-10 + 40 = 30$$
$$30 = 30$$

See **Figure 7**.

Figure 7

Example 7   **Using the Addition Method in Systems of Equations Containing Fractions**

Solve the given system of equations in two variables by addition.

$$\frac{x}{3} + \frac{y}{6} = 3$$

$$\frac{x}{2} - \frac{y}{4} = 1$$

Solution   First clear each equation of fractions by multiplying both sides of the equation by the least common denominator.

$$6\left(\frac{x}{3} + \frac{y}{6}\right) = 6(3)$$
$$2x + y = 18$$
$$4\left(\frac{x}{2} - \frac{y}{4}\right) = 4(1)$$
$$2x - y = 4$$

Now multiply the second equation by -1 so that we can eliminate the x-variable.

$$-1(2x - y) = -1(4)$$
$$-2x + y = -4$$

Add the two equations to eliminate the x-variable and solve the resulting equation.

$$\begin{aligned} 2x + y &= 18 \\ -2x + y &= -4 \\ \hline 2y &= 14 \\ y &= 7 \end{aligned}$$

Substitute $y = 7$ into the first equation.

$$2x + (7) = 18$$
$$2x = 11$$
$$x = \frac{11}{2}$$
$$= 5.5$$

The solution is $\left(\dfrac{11}{2}, 7\right)$. Check it in the other equation.

$$\frac{x}{2} - \frac{y}{4} = 1$$
$$\frac{\frac{11}{2}}{2} - \frac{7}{4} = 1$$
$$\frac{11}{4} - \frac{7}{4} = 1$$
$$\frac{4}{4} = 1$$

Try It #5

Solve the system of equations by addition.

$$2x + 3y = 8$$
$$3x + 5y = 10$$

Identifying Inconsistent Systems of Equations Containing Two Variables

Now that we have several methods for solving systems of equations, we can use the methods to identify inconsistent systems. Recall that an inconsistent system consists of parallel lines that have the same slope but different y-intercepts. They will never intersect. When searching for a solution to an inconsistent system, we will come up with a false statement, such as $12 = 0$.

Example 8   **Solving an Inconsistent System of Equations**

Solve the following system of equations.

$$x = 9 - 2y$$
$$x + 2y = 13$$

Solution   We can approach this problem in two ways. Because one equation is already solved for x, the most obvious step is to use substitution.

$$x + 2y = 13$$
$$(9 - 2y) + 2y = 13$$
$$9 + 0y = 13$$
$$9 = 13$$

Clearly, this statement is a contradiction because $9 \neq 13$. Therefore, the system has no solution.

The second approach would be to first manipulate the equations so that they are both in slope-intercept form. We manipulate the first equation as follows.

$$x = 9 - 2y$$
$$2y = -x + 9$$
$$y = -\frac{1}{2}x + \frac{9}{2}$$

We then convert the second equation expressed to slope-intercept form.

$$x + 2y = 13$$
$$2y = -x + 13$$
$$y = -\frac{1}{2}x + \frac{13}{2}$$

Comparing the equations, we see that they have the same slope but different y-intercepts. Therefore, the lines are parallel and do not intersect.

$$y = -\frac{1}{2}x + \frac{9}{2}$$
$$y = -\frac{1}{2}x + \frac{13}{2}$$

Analysis *Writing the equations in slope-intercept form confirms that the system is inconsistent because all lines will intersect eventually unless they are parallel. Parallel lines will never intersect; thus, the two lines have no points in common. The graphs of the equations in this example are shown in* **Figure 8**.

Figure 8

Try It #6

Solve the following system of equations in two variables.

$$2y - 2x = 2$$
$$2y - 2x = 6$$

Expressing the Solution of a System of Dependent Equations Containing Two Variables

Recall that a dependent system of equations in two variables is a system in which the two equations represent the same line. Dependent systems have an infinite number of solutions because all of the points on one line are also on the other line. After using substitution or addition, the resulting equation will be an identity, such as $0 = 0$.

Example 9   **Finding a Solution to a Dependent System of Linear Equations**

Find a solution to the system of equations using the addition method.

$$x + 3y = 2$$
$$3x + 9y = 6$$

Solution   With the addition method, we want to eliminate one of the variables by adding the equations. In this case, let's focus on eliminating x. If we multiply both sides of the first equation by -3, then we will be able to eliminate the x-variable.

$$x + 3y = 2$$
$$(-3)(x + 3y) = (-3)(2)$$
$$-3x - 9y = -6$$

Now add the equations.

$$-3x - 9y = -6$$
$$+\ \ 3x + 9y = 6$$
$$\overline{0 = 0}$$

We can see that there will be an infinite number of solutions that satisfy both equations.

Analysis   If we rewrote both equations in the slope-intercept form, we might know what the solution would look like before adding. Let's look at what happens when we convert the system to slope-intercept form.

$$x + 3y = 2$$
$$3y = -x + 2$$
$$y = -\frac{1}{3}x + \frac{2}{3}$$
$$3x + 9y = 6$$
$$9y = -3x + 6$$
$$y = -\frac{3}{9}x + \frac{6}{9}$$
$$y = -\frac{1}{3}x + \frac{2}{3}$$

See **Figure 9**. Notice the results are the same. The general solution to the system is $\left(x, -\frac{1}{3}x + \frac{2}{3}\right)$.

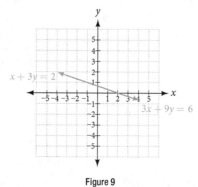

Figure 9

Try It #7

Solve the following system of equations in two variables.

$$y - 2x = 5$$
$$-3y + 6x = -15$$

Using Systems of Equations to Investigate Profits

Using what we have learned about systems of equations, we can return to the skateboard manufacturing problem at the beginning of the section. The skateboard manufacturer's **revenue function** is the function used to calculate the amount of money that comes into the business. It can be represented by the equation $R = xp$, where $x =$ quantity and $p =$ price. The revenue function is shown in orange in **Figure 10**.

The **cost function** is the function used to calculate the costs of doing business. It includes fixed costs, such as rent and salaries, and variable costs, such as utilities. The cost function is shown in blue in **Figure 10**. The x-axis represents quantity in hundreds of units. The y-axis represents either cost or revenue in hundreds of dollars.

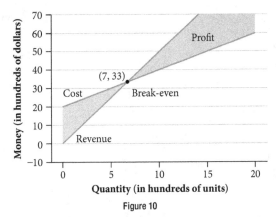

Figure 10

The point at which the two lines intersect is called the **break-even point**. We can see from the graph that if 700 units are produced, the cost is $3,300 and the revenue is also $3,300. In other words, the company breaks even if they produce and sell 700 units. They neither make money nor lose money.

The shaded region to the right of the break-even point represents quantities for which the company makes a profit. The shaded region to the left represents quantities for which the company suffers a loss. The **profit function** is the revenue function minus the cost function, written as $P(x) = R(x) - C(x)$. Clearly, knowing the quantity for which the cost equals the revenue is of great importance to businesses.

Example 10   **Finding the Break-Even Point and the Profit Function Using Substitution**

Given the cost function $C(x) = 0.85x + 35,000$ and the revenue function $R(x) = 1.55x$, find the break-even point and the profit function.

Solution   Write the system of equations using y to replace function notation.

$$y = 0.85x + 35,000$$

$$y = 1.55x$$

Substitute the expression $0.85x + 35,000$ from the first equation into the second equation and solve for x.

$$0.85x + 35,000 = 1.55x$$

$$35,000 = 0.7x$$

$$50,000 = x$$

Then, we substitute $x = 50,000$ into either the cost function or the revenue function.

$$1.55(50,000) = 77,500$$

The break-even point is (50,000, 77,500).

The profit function is found using the formula $P(x) = R(x) - C(x)$.

$$P(x) = 1.55x - (0.85x + 35,000)$$

$$= 0.7x - 35,000$$

The profit function is $P(x) = 0.7x - 35,000$.

Analysis   The cost to produce 50,000 units is $77,500, and the revenue from the sales of 50,000 units is also $77,500. To make a profit, the business must produce and sell more than 50,000 units. See Figure 11.

We see from the graph in Figure 12 that the profit function has a negative value until $x = 50,000$, when the graph crosses the x-axis. Then, the graph emerges into positive y-values and continues on this path as the profit function is a straight line. This illustrates that the break-even point for businesses occurs when the profit function is 0. The area to the left of the break-even point represents operating at a loss.

Figure 11   Figure 12

Example 11 Writing and Solving a System of Equations in Two Variables

The cost of a ticket to the circus is $25.00 for children and $50.00 for adults. On a certain day, attendance at the circus is 2,000 and the total gate revenue is $70,000. How many children and how many adults bought tickets?

Solution   Let c = the number of children and a = the number of adults in attendance.

The total number of people is 2,000. We can use this to write an equation for the number of people at the circus that day.

$$c + a = 2,000$$

The revenue from all children can be found by multiplying $25.00 by the number of children, $25c$. The revenue from all adults can be found by multiplying $50.00 by the number of adults, $50a$. The total revenue is $70,000.

We can use this to write an equation for the revenue.

$$25c + 50a = 70,000$$

We now have a system of linear equations in two variables.

$$c + a = 2,000$$
$$25c + 50a = 70,000$$

In the first equation, the coefficient of both variables is 1. We can quickly solve the first equation for either c or a. We will solve for a.

$$c + a = 2,000$$
$$a = 2,000 - c$$

Substitute the expression $2,000 - c$ in the second equation for a and solve for c.

$$25c + 50(2,000 - c) = 70,000$$
$$25c + 100,000 - 50c = 70,000$$
$$-25c = -30,000$$
$$c = 1,200$$

Substitute $c = 1,200$ into the first equation to solve for a.

$$1,200 + a = 2,000$$
$$a = 800$$

We find that 1,200 children and 800 adults bought tickets to the circus that day.

Try It #8

Meal tickets at the circus cost $4.00 for children and $12.00 for adults. If 1,650 meal tickets were bought for a total of $14,200, how many children and how many adults bought meal tickets?

Access these online resources for additional instruction and practice with systems of linear equations.

- Solving Systems of Equations Using Substitution (http://openstaxcollege.org/l/syssubst)
- Solving Systems of Equations Using Elimination (http://openstaxcollege.org/l/syselim)
- Applications of Systems of Equations (http://openstaxcollege.org/l/sysapp)

7.1 SECTION EXERCISES

VERBAL

1. Can a system of linear equations have exactly two solutions? Explain why or why not.

2. If you are performing a break-even analysis for a business and their cost and revenue equations are dependent, explain what this means for the company's profit margins.

3. If you are solving a break-even analysis and get a negative break-even point, explain what this signifies for the company?

4. If you are solving a break-even analysis and there is no break-even point, explain what this means for the company. How should they ensure there is a break-even point?

5. Given a system of equations, explain at least two different methods of solving that system.

ALGEBRAIC

For the following exercises, determine whether the given ordered pair is a solution to the system of equations.

6. $5x - y = 4$
$x + 6y = 2$ and $(4, 0)$

7. $-3x - 5y = 13$
$-x + 4y = 10$ and $(-6, 1)$

8. $3x + 7y = 1$
$2x + 4y = 0$ and $(2, 3)$

9. $-2x + 5y = 7$
$2x + 9y = 7$ and $(-1, 1)$

10. $x + 8y = 43$
$3x - 2y = -1$ and $(3, 5)$

For the following exercises, solve each system by substitution.

11. $x + 3y = 5$
$2x + 3y = 4$

12. $3x - 2y = 18$
$5x + 10y = -10$

13. $4x + 2y = -10$
$3x + 9y = 0$

14. $2x + 4y = -3.8$
$9x - 5y = 1.3$

15. $-2x + 3y = 1.2$
$-3x - 6y = 1.8$

16. $x - 0.2y = 1$
$-10x + 2y = 5$

17. $3\ x + 5y = 9$
$30x + 50y = -90$

18. $-3x + y = 2$
$12x - 4y = -8$

19. $\dfrac{1}{2}x + \dfrac{1}{3}y = 16$
$\dfrac{1}{6}x + \dfrac{1}{4}y = 9$

20. $-\dfrac{1}{4}x + \dfrac{3}{2}y = 11$
$-\dfrac{1}{8}x + \dfrac{1}{3}y = 3$

For the following exercises, solve each system by addition.

21. $-2x + 5y = -42$
$7x + 2y = 30$

22. $6x - 5y = -34$
$2x + 6y = 4$

23. $5x - y = -2.6$
$-4x - 6y = 1.4$

24. $7x - 2y = 3$
$4x + 5y = 3.25$

25. $-x + 2y = -1$
$5x - 10y = 6$

26. $7x + 6y = 2$
$-28x - 24y = -8$

27. $\dfrac{5}{6}x + \dfrac{1}{4}y = 0$
$\dfrac{1}{8}x - \dfrac{1}{2}y = -\dfrac{43}{120}$

28. $\dfrac{1}{3}x + \dfrac{1}{9}y = \dfrac{2}{9}$
$-\dfrac{1}{2}x + \dfrac{4}{5}y = -\dfrac{1}{3}$

29. $-0.2x + 0.4y = 0.6$
$x - 2y = -3$

30. $-0.1x + 0.2y = 0.6$
$5x - 10y = 1$

For the following exercises, solve each system by any method.

31. $5x + 9y = 16$
$x + 2y = 4$

32. $6x - 8y = -0.6$
$3x + 2y = 0.9$

33. $5x - 2y = 2.25$
$7x - 4y = 3$

34. $x - \dfrac{5}{12}y = -\dfrac{55}{12}$
$-6x + \dfrac{5}{2}y = \dfrac{55}{2}$

35. $7x - 4y = \dfrac{7}{6}$

$\quad\; 2x + 4y = \dfrac{1}{3}$

36. $3x + 6y = 11$

$\quad\; 2x + 4y = 9$

37. $\dfrac{7}{3}x - \dfrac{1}{6}y = 2$

$\quad\; -\dfrac{21}{6}x + \dfrac{3}{12}y = -3$

38. $\dfrac{1}{2}x + \dfrac{1}{3}y = \dfrac{1}{3}$

$\quad\; \dfrac{3}{2}x + \dfrac{1}{4}y = -\dfrac{1}{8}$

39. $2.2x + 1.3y = -0.1$

$\quad\; 4.2x + 4.2y = 2.1$

40. $0.1x + 0.2y = 2$

$\quad\; 0.35x - 0.3y = 0$

GRAPHICAL

For the following exercises, graph the system of equations and state whether the system is consistent, inconsistent, or dependent and whether the system has one solution, no solution, or infinite solutions.

41. $3x - y = 0.6$

$\quad\; x - 2y = 1.3$

42. $-x + 2y = 4$

$\quad\; 2x - 4y = 1$

43. $x + 2y = 7$

$\quad\; 2x + 6y = 12$

44. $3x - 5y = 7$

$\quad\; x - 2y = 3$

45. $3x - 2y = 5$

$\quad\; -9x + 6y = -15$

TECHNOLOGY

For the following exercises, use the intersect function on a graphing device to solve each system. Round all answers to the nearest hundredth.

46. $0.1x + 0.2y = 0.3$

$\quad\; -0.3x + 0.5y = 1$

47. $-0.01x + 0.12y = 0.62$

$\quad\; 0.15x + 0.20y = 0.52$

48. $0.5x + 0.3y = 4$

$\quad\; 0.25x - 0.9y = 0.46$

49. $0.15x + 0.27y = 0.39$

$\quad\; -0.34x + 0.56y = 1.8$

50. $-0.71x + 0.92y = 0.13$

$\quad\; 0.83x + 0.05y = 2.1$

EXTENSIONS

For the following exercises, solve each system in terms of A, B, C, D, E, and F where A – F are nonzero numbers. Note that $A \neq B$ and $AE \neq BD$.

51. $x + y = A$

$\quad\; x - y = B$

52. $x + Ay = 1$

$\quad\; x + By = 1$

53. $Ax + y = 0$

$\quad\; Bx + y = 1$

54. $Ax + By = C$

$\quad\; x + y = 1$

55. $Ax + By = C$

$\quad\; Dx + Ey = F$

REAL-WORLD APPLICATIONS

For the following exercises, solve for the desired quantity.

56. A stuffed animal business has a total cost of production $C = 12x + 30$ and a revenue function $R = 20x$. Find the break-even point.

57. A fast-food restaurant has a cost of production $C(x) = 11x + 120$ and a revenue function $R(x) = 5x$. When does the company start to turn a profit?

58. A cell phone factory has a cost of production $C(x) = 150x + 10{,}000$ and a revenue function $R(x) = 200x$. What is the break-even point?

59. A musician charges $C(x) = 64x + 20{,}000$, where x is the total number of attendees at the concert. The venue charges \$80 per ticket. After how many people buy tickets does the venue break even, and what is the value of the total tickets sold at that point?

60. A guitar factory has a cost of production $C(x) = 75x + 50{,}000$. If the company needs to break even after 150 units sold, at what price should they sell each guitar? Round up to the nearest dollar, and write the revenue function.

For the following exercises, use a system of linear equations with two variables and two equations to solve.

61. Find two numbers whose sum is 28 and difference is 13.

62. A number is 9 more than another number. Twice the sum of the two numbers is 10. Find the two numbers.

63. The startup cost for a restaurant is $120,000, and each meal costs $10 for the restaurant to make. If each meal is then sold for $15, after how many meals does the restaurant break even?

64. A moving company charges a flat rate of $150, and an additional $5 for each box. If a taxi service would charge $20 for each box, how many boxes would you need for it to be cheaper to use the moving company, and what would be the total cost?

65. A total of 1,595 first- and second-year college students gathered at a pep rally. The number of freshmen exceeded the number of sophomores by 15. How many freshmen and sophomores were in attendance?

66. 276 students enrolled in a freshman-level chemistry class. By the end of the semester, 5 times the number of students passed as failed. Find the number of students who passed, and the number of students who failed.

67. There were 130 faculty at a conference. If there were 18 more women than men attending, how many of each gender attended the conference?

68. A jeep and BMW enter a highway running east-west at the same exit heading in opposite directions. The jeep entered the highway 30 minutes before the BMW did, and traveled 7 mph slower than the BMW. After 2 hours from the time the BMW entered the highway, the cars were 306.5 miles apart. Find the speed of each car, assuming they were driven on cruise control.

69. If a scientist mixed 10% saline solution with 60% saline solution to get 25 gallons of 40% saline solution, how many gallons of 10% and 60% solutions were mixed?

70. An investor earned triple the profits of what she earned last year. If she made $500,000.48 total for both years, how much did she earn in profits each year?

71. An investor who dabbles in real estate invested 1.1 million dollars into two land investments. On the first investment, Swan Peak, her return was a 110% increase on the money she invested. On the second investment, Riverside Community, she earned 50% over what she invested. If she earned $1 million in profits, how much did she invest in each of the land deals?

72. If an investor invests a total of $25,000 into two bonds, one that pays 3% simple interest, and the other that pays $2\frac{7}{8}$ % interest, and the investor earns $737.50 annual interest, how much was invested in each account?

73. If an investor invests $23,000 into two bonds, one that pays 4% in simple interest, and the other paying 2% simple interest, and the investor earns $710.00 annual interest, how much was invested in each account?

74. CDs cost $5.96 more than DVDs at All Bets Are Off Electronics. How much would 6 CDs and 2 DVDs cost if 5 CDs and 2 DVDs cost $127.73?

75. A store clerk sold 60 pairs of sneakers. The high-tops sold for $98.99 and the low-tops sold for $129.99. If the receipts for the two types of sales totaled $6,404.40, how many of each type of sneaker were sold?

76. A concert manager counted 350 ticket receipts the day after a concert. The price for a student ticket was $12.50, and the price for an adult ticket was $16.00. The register confirms that $5,075 was taken in. How many student tickets and adult tickets were sold?

77. Admission into an amusement park for 4 children and 2 adults is $116.90. For 6 children and 3 adults, the admission is $175.35. Assuming a different price for children and adults, what is the price of the child's ticket and the price of the adult ticket?

LEARNING OBJECTIVES

In this section, you will:

- Solve systems of three equations in three variables.
- Identify inconsistent systems of equations containing three variables.
- Express the solution of a system of dependent equations containing three variables.

7.2   SYSTEMS OF LINEAR EQUATIONS: THREE VARIABLES

Figure 1   (credit: "Elembis," Wikimedia Commons)

John received an inheritance of $12,000 that he divided into three parts and invested in three ways: in a money-market fund paying 3% annual interest; in municipal bonds paying 4% annual interest; and in mutual funds paying 7% annual interest. John invested $4,000 more in municipal funds than in municipal bonds. He earned $670 in interest the first year. How much did John invest in each type of fund?

Understanding the correct approach to setting up problems such as this one makes finding a solution a matter of following a pattern. We will solve this and similar problems involving three equations and three variables in this section. Doing so uses similar techniques as those used to solve systems of two equations in two variables. However, finding solutions to systems of three equations requires a bit more organization and a touch of visual gymnastics.

Solving Systems of Three Equations in Three Variables

In order to solve systems of equations in three variables, known as three-by-three systems, the primary tool we will be using is called Gaussian elimination, named after the prolific German mathematician Karl Friedrich Gauss. While there is no definitive order in which operations are to be performed, there are specific guidelines as to what type of moves can be made. We may number the equations to keep track of the steps we apply. The goal is to eliminate one variable at a time to achieve upper triangular form, the ideal form for a three-by-three system because it allows for straightforward back-substitution to find a solution (x, y, z), which we call an ordered triple. A system in upper triangular form looks like the following:

$$Ax + By + Cz = D$$
$$Ey + Fz = G$$
$$Hz = K$$

The third equation can be solved for z, and then we back-substitute to find y and x. To write the system in upper triangular form, we can perform the following operations:

1. Interchange the order of any two equations.
2. Multiply both sides of an equation by a nonzero constant.
3. Add a nonzero multiple of one equation to another equation.

The **solution set** to a three-by-three system is an ordered triple $\{(x, y, z)\}$. Graphically, the ordered triple defines the point that is the intersection of three planes in space. You can visualize such an intersection by imagining any corner in a rectangular room. A corner is defined by three planes: two adjoining walls and the floor (or ceiling). Any point where two walls and the floor meet represents the intersection of three planes.

---

number of possible solutions

Figure 2 and **Figure 3** illustrate possible solution scenarios for three-by-three systems.

- Systems that have a single solution are those which, after elimination, result in a **solution set** consisting of an ordered triple $\{(x, y, z)\}$. Graphically, the ordered triple defines a point that is the intersection of three planes in space.

- Systems that have an infinite number of solutions are those which, after elimination, result in an expression that is always true, such as $0 = 0$. Graphically, an infinite number of solutions represents a line or coincident plane that serves as the intersection of three planes in space.

- Systems that have no solution are those that, after elimination, result in a statement that is a contradiction, such as $3 = 0$. Graphically, a system with no solution is represented by three planes with no point in common.

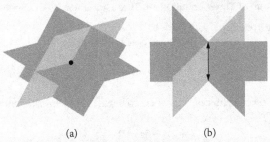

(a)   (b)

Figure 2 (a)Three planes intersect at a single point, representing a three-by-three system with a single solution.
(b) Three planes intersect in a line, representing a three-by-three system with infinite solutions.

(a)   (b)   (c)

Figure 3 All three figures represent three-by-three systems with no solution. (a) The three planes intersect with each other, but not at a common point.
(b) Two of the planes are parallel and intersect with the third plane, but not with each other.
(c) All three planes are parallel, so there is no point of intersection.

Example 1   **Determining Whether an Ordered Triple Is a Solution to a System**

Determine whether the ordered triple $(3, -2, 1)$ is a solution to the system.

$$x + y + z = 2$$
$$6x - 4y + 5z = 31$$
$$5x + 2y + 2z = 13$$

Solution   We will check each equation by substituting in the values of the ordered triple for x, y, and z.

$$x + y + z = 2$$
$$(3) + (-2) + (1) = 2$$
$$\text{True}$$

$$6x - 4y + 5z = 31$$
$$6(3) - 4(-2) + 5(1) = 31$$
$$18 + 8 + 5 = 31$$
$$\text{True}$$

$$5x + 2y + 2z = 13$$
$$5(3) + 2(-2) + 2(1) = 13$$
$$15 - 4 + 2 = 13$$
$$\text{True}$$

The ordered triple $(3, -2, 1)$ is indeed a solution to the system.

How To...

Given a linear system of three equations, solve for three unknowns.

1. Pick any pair of equations and solve for one variable.
2. Pick another pair of equations and solve for the same variable.
3. You have created a system of two equations in two unknowns. Solve the resulting two-by-two system.
4. Back-substitute known variables into any one of the original equations and solve for the missing variable.

Example 2   Solving a System of Three Equations in Three Variables by Elimination

Find a solution to the following system:

$$x - 2y + 3z = 9 \qquad (1)$$
$$-x + 3y - z = -6 \qquad (2)$$
$$2x - 5y + 5z = 17 \qquad (3)$$

Solution   There will always be several choices as to where to begin, but the most obvious first step here is to eliminate x by adding equations (1) and (2).

$$x - 2y + 3z = 9 \qquad (1)$$
$$\underline{-x + 3y - z = -6 \qquad (2)}$$
$$y + 2z = 3 \qquad (3)$$

The second step is multiplying equation (1) by -2 and adding the result to equation (3). These two steps will eliminate the variable x.

$$-2x + 4y - 6z = -18 \qquad \text{(1) multiplied by } -2$$
$$\underline{2x - 5y + 5z = 17 \qquad (3)}$$
$$-y - z = -1 \qquad (5)$$

In equations (4) and (5), we have created a new two-by-two system. We can solve for z by adding the two equations.

$$y + 2z = 3 \qquad (4)$$
$$\underline{-y - z = -1 \qquad (5)}$$
$$z = 2 \qquad (6)$$

Choosing one equation from each new system, we obtain the upper triangular form:

$$x - 2y + 3z = 9 \qquad (1)$$
$$y + 2z = 3 \qquad (4)$$
$$z = 2 \qquad (6)$$

Next, we back-substitute $z = 2$ into equation (4) and solve for y.

$$y + 2(2) = 3$$
$$y + 4 = 3$$
$$y = -1$$

Finally, we can back-substitute $z = 2$ and $y = -1$ into equation (1). This will yield the solution for x.

$$x - 2(-1) + 3(2) = 9$$
$$x + 2 + 6 = 9$$
$$x = 1$$

The solution is the ordered triple (1, −1, 2). See **Figure 4**.

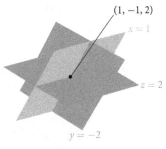

Figure 4

Example 3   **Solving a Real-World Problem Using a System of Three Equations in Three Variables**

In the problem posed at the beginning of the section, John invested his inheritance of $12,000 in three different funds: part in a money-market fund paying 3% interest annually; part in municipal bonds paying 4% annually; and the rest in mutual funds paying 7% annually. John invested $4,000 more in mutual funds than he invested in municipal bonds. The total interest earned in one year was $670. How much did he invest in each type of fund?

Solution   To solve this problem, we use all of the information given and set up three equations. First, we assign a variable to each of the three investment amounts:

$$x = \text{amount invested in money-market fund}$$

$$y = \text{amount invested in municipal bonds}$$

$$z = \text{amount invested in mutual funds}$$

The first equation indicates that the sum of the three principal amounts is $12,000.

$$x + y + z = 12{,}000$$

We form the second equation according to the information that John invested $4,000 more in mutual funds than he invested in municipal bonds.

$$z = y + 4{,}000$$

The third equation shows that the total amount of interest earned from each fund equals $670.

$$0.03x + 0.04y + 0.07z = 670$$

Then, we write the three equations as a system.

$$x + y + z = 12{,}000$$

$$-y + z = 4{,}000$$

$$0.03x + 0.04y + 0.07z = 670$$

To make the calculations simpler, we can multiply the third equation by 100. Thus,

$$x + y + z = 12{,}000 \qquad (1)$$

$$-y + z = 4{,}000 \qquad (2)$$

$$3x + 4y + 7z = 67{,}000 \qquad (3)$$

Step 1. Interchange equation (2) and equation (3) so that the two equations with three variables will line up.

$$x + y + z = 12{,}000$$

$$3x + 4y + 7z = 67{,}000$$

$$-y + z = 4{,}000$$

Step 2. Multiply equation (1) by −3 and add to equation (2). Write the result as row 2.

$$x + y + z = 12{,}000$$

$$y + 4z = 31{,}000$$

$$-y + z = 4{,}000$$

Step 3. Add equation (2) to equation (3) and write the result as equation (3).

$$x + y + z = 12{,}000$$
$$y + 4z = 31{,}000$$
$$5z = 35{,}000$$

Step 4. Solve for z in equation (3). Back-substitute that value in equation (2) and solve for y. Then, back-substitute the values for z and y into equation (1) and solve for x.

$$5z = 35{,}000$$
$$z = 7{,}000$$
$$y + 4(7{,}000) = 31{,}000$$
$$y = 3{,}000$$
$$x + 3{,}000 + 7{,}000 = 12{,}000$$
$$x = 2{,}000$$

John invested \$2,000 in a money-market fund, \$3,000 in municipal bonds, and \$7,000 in mutual funds.

Try It #1

Solve the system of equations in three variables.

$$2x + y - 2z = -1$$
$$3x - 3y - z = 5$$
$$x - 2y + 3z = 6$$

Identifying Inconsistent Systems of Equations Containing Three Variables

Just as with systems of equations in two variables, we may come across an inconsistent system of equations in three variables, which means that it does not have a solution that satisfies all three equations. The equations could represent three parallel planes, two parallel planes and one intersecting plane, or three planes that intersect the other two but not at the same location. The process of elimination will result in a false statement, such as $3 = 7$ or some other contradiction.

Example 4   Solving an Inconsistent System of Three Equations in Three Variables

Solve the following system.

$$x - 3y + z = 4 \qquad (1)$$
$$-x + 2y - 5z = 3 \qquad (2)$$
$$5x - 13y + 13z = 8 \qquad (3)$$

Solution   Looking at the coefficients of x, we can see that we can eliminate x by adding equation (1) to equation (2).

$$
\begin{array}{ll}
x - 3y + z = 4 & (1) \\
-x + 2y - 5z = 3 & (2) \\
\hline
-y - 4z = 7 & (4)
\end{array}
$$

Next, we multiply equation (1) by -5 and add it to equation (3).

$$
\begin{array}{ll}
-5x + 15y - 5z = -20 & \text{(1) multiplied by } -5 \\
5x - 13y + 13z = 8 & (3) \\
\hline
2y + 8z = -12 & (5)
\end{array}
$$

Then, we multiply equation (4) by 2 and add it to equation (5).

$$
\begin{array}{ll}
-2y - 8z = 14 & \text{(4) multiplied by } 2 \\
2y + 8z = -12 & (5) \\
\hline
0 = 2 &
\end{array}
$$

The final equation $0 = 2$ is a contradiction, so we conclude that the system of equations in inconsistent and, therefore, has no solution.

Analysis In this system, each plane intersects the other two, but not at the same location. Therefore, the system is inconsistent.

Try It #2

Solve the system of three equations in three variables.

$$x + y + z = 2$$
$$y - 3z = 1$$
$$2x + y + 5z = 0$$

Expressing the Solution of a System of Dependent Equations Containing Three Variables

We know from working with systems of equations in two variables that a dependent system of equations has an infinite number of solutions. The same is true for dependent systems of equations in three variables. An infinite number of solutions can result from several situations. The three planes could be the same, so that a solution to one equation will be the solution to the other two equations. All three equations could be different but they intersect on a line, which has infinite solutions. Or two of the equations could be the same and intersect the third on a line.

Example 5   Finding the Solution to a Dependent System of Equations

Find the solution to the given system of three equations in three variables.

$$2x + y - 3z = 0 \qquad (1)$$
$$4x + 2y - 6z = 0 \qquad (2)$$
$$x - y + z = 0 \qquad (3)$$

Solution   First, we can multiply equation (1) by -2 and add it to equation (2).

$$-4x - 2y + 6z = 0 \qquad \text{equation (1) multiplied by } -2$$
$$\underline{4x + 2y - 6z = 0 \qquad\qquad (2)}$$
$$0 = 0$$

We do not need to proceed any further. The result we get is an identity, $0 = 0$, which tells us that this system has an infinite number of solutions. There are other ways to begin to solve this system, such as multiplying equation (3) by -2, and adding it to equation (1). We then perform the same steps as above and find the same result, $0 = 0$.

When a system is dependent, we can find general expressions for the solutions. Adding equations (1) and (3), we have

$$2x + y - 3z = 0$$
$$x - y + z = 0$$
$$3x - 2z = 0$$

We then solve the resulting equation for z.

$$3x - 2z = 0$$
$$z = \frac{3}{2}x$$

We back-substitute the expression for z into one of the equations and solve for y.

$$2x + y - 3\left(\frac{3}{2}x\right) = 0$$

$$2x + y - \frac{9}{2}x = 0$$

$$y = \frac{9}{2}x - 2x$$

$$y = \frac{5}{2}x$$

So the general solution is $\left(x, \frac{5}{2}x, \frac{3}{2}x\right)$. In this solution, x can be any real number. The values of y and z are dependent on the value selected for x.

Analysis   As shown in **Figure 5**, two of the planes are the same and they intersect the third plane on a line. The solution set is infinite, as all points along the intersection line will satisfy all three equations.

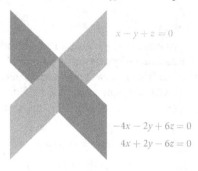

$$x - y + z = 0$$

$$-4x - 2y + 6z = 0$$

$$4x + 2y - 6z = 0$$

Figure 5

Q & A...

Does the generic solution to a dependent system always have to be written in terms of x?

No, you can write the generic solution in terms of any of the variables, but it is common to write it in terms of x and if needed x and y.

Try It #3

Solve the following system.

$$x + y + z = 7$$

$$3x - 2y - z = 4$$

$$x + 6y + 5z = 24$$

Access these online resources for additional instruction and practice with systems of equations in three variables.

- Ex 1: System of Three Equations with Three Unknowns Using Elimination (http://openstaxcollege.org/l/systhree)
- Ex. 2: System of Three Equations with Three Unknowns Using Elimination (http://openstaxcollege.org/l/systhelim)

7.2   SECTION EXERCISES

VERBAL

1. Can a linear system of three equations have exactly two solutions? Explain why or why not

2. If a given ordered triple solves the system of equations, is that solution unique? If so, explain why. If not, give an example where it is not unique.

3. If a given ordered triple does not solve the system of equations, is there no solution? If so, explain why. If not, give an example.

4. Using the method of addition, is there only one way to solve the system?

5. Can you explain whether there can be only one method to solve a linear system of equations? If yes, give an example of such a system of equations. If not, explain why not.

ALGEBRAIC

For the following exercises, determine whether the ordered triple given is the solution to the system of equations.

6. $2x - 6y + 6z = -12$
$x + 4y + 5z = -1$ and $(0, 1, -1)$
$-x + 2y + 3z = -1$

7. $6x - y + 3z = 6$
$3x + 5y + 2z = 0$ and $(3, -3, -5)$
$x + y = 0$

8. $6x - 7y + z = 2$
$-x - y + 3z = 4$ and $(4, 2, -6)$
$2x + y - z = 1$

9. $x - y = 0$
$x - z = 5$ and $(4, 4, -1)$
$x - y + z = -1$

10. $-x - y + 2z = 3$
$5x + 8y - 3z = 4$ and $(4, 1, -7)$
$-x + 3y - 5z = -5$

For the following exercises, solve each system by substitution.

11. $3x - 4y + 2z = -15$
$2x + 4y + z = 16$
$2x + 3y + 5z = 20$

12. $5x - 2y + 3z = 20$
$2x - 4y - 3z = -9$
$x + 6y - 8z = 21$

13. $5x + 2y + 4z = 9$
$-3x + 2y + z = 10$
$4x - 3y + 5z = -3$

14. $4x - 3y + 5z = 31$
$-x + 2y + 4z = 20$
$x + 5y - 2z = -29$

15. $5x - 2y + 3z = 4$
$-4x + 6y - 7z = -1$
$3x + 2y - z = 4$

16. $4x + 6y + 9z = 0$
$-5x + 2y - 6z = 3$
$7x - 4y + 3z = -3$

For the following exercises, solve each system by Gaussian elimination.

17. $2x - y + 3z = 17$
$-5x + 4y - 2z = -46$
$2y + 5z = -7$

18. $5x - 6y + 3z = 50$
$-x + 4y = 10$
$2x - z = 10$

19. $2x + 3y - 6z = 1$
$-4x - 6y + 12z = -2$
$x + 2y + 5z = 10$

20. $4x + 6y - 2z = 8$
$6x + 9y - 3z = 12$
$-2x - 3y + z = -4$

21. $2x + 3y - 4z = 5$
$-3x + 2y + z = 11$
$-x + 5y + 3z = 4$

22. $10x + 2y - 14z = 8$
$-x - 2y - 4z = -1$
$-12x - 6y + 6z = -12$

23. $x + y + z = 14$
$2y + 3z = -14$
$-16y - 24z = -112$

24. $5x - 3y + 4z = -1$
$-4x + 2y - 3z = 0$
$-x + 5y + 7z = -11$

25. $x + y + z = 0$
$2x - y + 3z = 0$
$x - z = 0$

26. $3x + 2y - 5z = 6$
$5x - 4y + 3z = -12$
$4x + 5y - 2z = 15$

27. $x + y + z = 0$
$2x - y + 3z = 0$
$x - z = 1$

28. $3x - \frac{1}{2}y - z = -\frac{1}{2}$
$4x + z = 3$
$-x + \frac{3}{2}y = \frac{5}{2}$

29. $6x - 5y + 6z = 38$
$\frac{1}{5}x - \frac{1}{2}y + \frac{3}{5}z = 1$
$-4x - \frac{3}{2}y - z = -74$

30. $\frac{1}{2}x - \frac{1}{5}y + \frac{2}{5}z = -\frac{13}{10}$
$\frac{1}{4}x - \frac{2}{5}y - \frac{1}{5}z = -\frac{7}{20}$
$-\frac{1}{2}x - \frac{3}{4}y - \frac{1}{2}z = -\frac{5}{4}$

31. $-\frac{1}{3}x - \frac{1}{2}y - \frac{1}{4}z = \frac{3}{4}$
$-\frac{1}{2}x - \frac{1}{4}y - \frac{1}{2}z = 2$
$-\frac{1}{4}x - \frac{3}{4}y - \frac{1}{2}z = -\frac{1}{2}$

32. $\frac{1}{2}x - \frac{1}{4}y + \frac{3}{4}z = 0$
$\frac{1}{4}x - \frac{1}{10}y + \frac{2}{5}z = -2$
$\frac{1}{8}x + \frac{1}{5}y - \frac{1}{8}z = 2$

33. $\frac{4}{5}x - \frac{7}{8}y + \frac{1}{2}z = 1$
$-\frac{4}{5}x - \frac{3}{4}y + \frac{1}{3}z = -8$
$-\frac{2}{5}x - \frac{7}{8}y + \frac{1}{2}z = -5$

34. $-\frac{1}{3}x - \frac{1}{8}y + \frac{1}{6}z = -\frac{4}{3}$
$-\frac{2}{3}x - \frac{7}{8}y + \frac{1}{3}z = -\frac{23}{3}$
$-\frac{1}{3}x - \frac{5}{8}y + \frac{5}{6}z = 0$

35. $-\frac{1}{4}x - \frac{5}{4}y + \frac{5}{2}z = -5$
$-\frac{1}{2}x - \frac{5}{3}y + \frac{5}{4}z = \frac{55}{12}$
$-\frac{1}{3}x - \frac{1}{3}y + \frac{1}{3}z = \frac{5}{3}$

36. $\frac{1}{40}x + \frac{1}{60}y + \frac{1}{80}z = \frac{1}{100}$
$-\frac{1}{2}x - \frac{1}{3}y - \frac{1}{4}z = -\frac{1}{5}$
$\frac{3}{8}x + \frac{3}{12}y + \frac{3}{16}z = \frac{3}{20}$

37. $0.1x - 0.2y + 0.3z = 2$
$0.5x - 0.1y + 0.4z = 8$
$0.7x - 0.2y + 0.3z = 8$

38. $0.2x + 0.1y - 0.3z = 0.2$
$0.8x + 0.4y - 1.2z = 0.1$
$1.6x + 0.8y - 2.4z = 0.2$

39. $1.1x + 0.7y - 3.1z = -1.79$
$2.1x + 0.5y - 1.6z = -0.13$
$0.5x + 0.4y - 0.5z = -0.07$

40. $0.5x - 0.5y + 0.5z = 10$
$0.2x - 0.2y + 0.2z = 4$
$0.1x - 0.1y + 0.1z = 2$

41. $0.1x + 0.2y + 0.3z = 0.37$
$0.1x - 0.2y - 0.3z = -0.27$
$0.5x - 0.1y - 0.3z = -0.03$

42. $0.5x - 0.5y - 0.3z = 0.13$
$0.4x - 0.1y - 0.3z = 0.11$
$0.2x - 0.8y - 0.9z = -0.32$

43. $0.5x + 0.2y - 0.3z = 1$
$0.4x - 0.6y + 0.7z = 0.8$
$0.3x - 0.1y - 0.9z = 0.6$

44. $0.3x + 0.3y + 0.5z = 0.6$
$0.4x + 0.4y + 0.4z = 1.8$
$0.4x + 0.2y + 0.1z = 1.6$

45. $0.8x + 0.8y + 0.8z = 2.4$
$0.3x - 0.5y + 0.2z = 0$
$0.1x + 0.2y + 0.3z = 0.6$

EXTENSIONS

For the following exercises, solve the system for x, y, and z.

46. $x + y + z = 3$
$\frac{x-1}{2} + \frac{y-3}{2} + \frac{z+1}{2} = 0$
$\frac{x-2}{3} + \frac{y+4}{3} + \frac{z-3}{3} = \frac{2}{3}$

47. $5x - 3y - \frac{z+1}{2} = \frac{1}{2}$
$6x + \frac{y-9}{2} + 2z = -3$
$\frac{x+8}{2} - 4y + z = 4$

48. $\frac{x+4}{7} - \frac{y-1}{6} + \frac{z+2}{3} = 1$
$\frac{x-2}{4} + \frac{y+1}{8} - \frac{z+8}{12} = 0$
$\frac{x+6}{3} - \frac{y+2}{3} + \frac{z+4}{2} = 3$

49. $\frac{x-3}{6} + \frac{y+2}{2} - \frac{z-3}{3} = 2$
$\frac{x+2}{4} + \frac{y-5}{2} + \frac{z+4}{2} = 1$
$\frac{x+6}{2} - \frac{y-3}{2} + z + 1 = 9$

50. $\frac{x-1}{3} + \frac{y+3}{4} + \frac{z+2}{6} = 1$
$4x + 3y - 2z = 11$
$0.02x + 0.015y - 0.01z = 0.065$

REAL-WORLD APPLICATIONS

51. Three even numbers sum up to 108. The smaller is half the larger and the middle number is $\frac{3}{4}$ the larger. What are the three numbers?

52. Three numbers sum up to 147. The smallest number is half the middle number, which is half the largest number. What are the three numbers?

53. At a family reunion, there were only blood relatives, consisting of children, parents, and grandparents, in attendance. There were 400 people total. There were twice as many parents as grandparents, and 50 more children than parents. How many children, parents, and grandparents were in attendance?

54. An animal shelter has a total of 350 animals comprised of cats, dogs, and rabbits. If the number of rabbits is 5 less than one-half the number of cats, and there are 20 more cats than dogs, how many of each animal are at the shelter?

55. Your roommate, Sarah, offered to buy groceries for you and your other roommate. The total bill was $82. She forgot to save the individual receipts but remembered that your groceries were $0.05 cheaper than half of her groceries, and that your other roommate's groceries were $2.10 more than your groceries. How much was each of your share of the groceries?

56. Your roommate, John, offered to buy household supplies for you and your other roommate. You live near the border of three states, each of which has a different sales tax. The total amount of money spent was $100.75. Your supplies were bought with 5% tax, John's with 8% tax, and your third roommate's with 9% sales tax. The total amount of money spent without taxes is $93.50. If your supplies before tax were $1 more than half of what your third roommate's supplies were before tax, how much did each of you spend? Give your answer both with and without taxes.

57. Three coworkers work for the same employer. Their jobs are warehouse manager, office manager, and truck driver. The sum of the annual salaries of the warehouse manager and office manager is $82,000. The office manager makes $4,000 more than the truck driver annually. The annual salaries of the warehouse manager and the truck driver total $78,000. What is the annual salary of each of the co-workers?

58. At a carnival, $2,914.25 in receipts were taken at the end of the day. The cost of a child's ticket was $20.50, an adult ticket was $29.75, and a senior citizen ticket was $15.25. There were twice as many senior citizens as adults in attendance, and 20 more children than senior citizens. How many children, adult, and senior citizen tickets were sold?

59. A local band sells out for their concert. They sell all 1,175 tickets for a total purse of $28,112.50. The tickets were priced at $20 for student tickets, $22.50 for children, and $29 for adult tickets. If the band sold twice as many adult as children tickets, how many of each type was sold?

60. In a bag, a child has 325 coins worth $19.50. There were three types of coins: pennies, nickels, and dimes. If the bag contained the same number of nickels as dimes, how many of each type of coin was in the bag?

61. Last year, at Haven's Pond Car Dealership, for a particular model of BMW, Jeep, and Toyota, one could purchase all three cars for a total of $140,000. This year, due to inflation, the same cars would cost $151,830. The cost of the BMW increased by 8%, the Jeep by 5%, and the Toyota by 12%. If the price of last year's Jeep was $7,000 less than the price of last year's BMW, what was the price of each of the three cars last year?

62. A recent college graduate took advantage of his business education and invested in three investments immediately after graduating. He invested $80,500 into three accounts, one that paid 4% simple interest, one that paid $3\frac{1}{8}$% simple interest, and one that paid $2\frac{1}{2}$% simple interest. He earned $2,670 interest at the end of one year. If the amount of the money invested in the second account was four times the amount invested in the third account, how much was invested in each account?

63. You inherit one million dollars. You invest it all in three accounts for one year. The first account pays 3% compounded annually, the second account pays 4% compounded annually, and the third account pays 2% compounded annually. After one year, you earn $34,000 in interest. If you invest four times the money into the account that pays 3% compared to 2%, how much did you invest in each account?

64. You inherit one hundred thousand dollars. You invest it all in three accounts for one year. The first account pays 4% compounded annually, the second account pays 3% compounded annually, and the third account pays 2% compounded annually. After one year, you earn $3,650 in interest. If you invest five times the money in the account that pays 4% compared to 3%, how much did you invest in each account?

65. The top three countries in oil consumption in a certain year are as follows: the United States, Japan, and China. In millions of barrels per day, the three top countries consumed 39.8% of the world's consumed oil. The United States consumed 0.7% more than four times China's consumption. The United States consumed 5% more than triple Japan's consumption. What percent of the world oil consumption did the United States, Japan, and China consume?[27]

66. The top three countries in oil production in the same year are Saudi Arabia, the United States, and Russia. In millions of barrels per day, the top three countries produced 31.4% of the world's produced oil. Saudi Arabia and the United States combined for 22.1% of the world's production, and Saudi Arabia produced 2% more oil than Russia. What percent of the world oil production did Saudi Arabia, the United States, and Russia produce?[28]

67. The top three sources of oil imports for the United States in the same year were Saudi Arabia, Mexico, and Canada. The three top countries accounted for 47% of oil imports. The United States imported 1.8% more from Saudi Arabia than they did from Mexico, and 1.7% more from Saudi Arabia than they did from Canada. What percent of the United States oil imports were from these three countries?[29]

68. The top three oil producers in the United States in a certain year are the Gulf of Mexico, Texas, and Alaska. The three regions were responsible for 64% of the United States oil production. The Gulf of Mexico and Texas combined for 47% of oil production. Texas produced 3% more than Alaska. What percent of United States oil production came from these regions?[30]

69. At one time, in the United States, 398 species of animals were on the endangered species list. The top groups were mammals, birds, and fish, which comprised 55% of the endangered species. Birds accounted for 0.7% more than fish, and fish accounted for 1.5% more than mammals. What percent of the endangered species came from mammals, birds, and fish?

70. Meat consumption in the United States can be broken into three categories: red meat, poultry, and fish. If fish makes up 4% less than one-quarter of poultry consumption, and red meat consumption is 18.2% higher than poultry consumption, what are the percentages of meat consumption?[31]

27 "Oil reserves, production and consumption in 2001," accessed April 6, 2014, http://scaruffi.com/politics/oil.html.
28 "Oil reserves, production and consumption in 2001," accessed April 6, 2014, http://scaruffi.com/politics/oil.html.
29 "Oil reserves, production and consumption in 2001," accessed April 6, 2014, http://scaruffi.com/politics/oil.html.
30 "USA: The coming global oil crisis," accessed April 6, 2014, http://www.oilcrisis.com/us/.
31 "The United States Meat Industry at a Glance," accessed April 6, 2014, http://www.meatami.com/ht/d/sp/i/47465/pid/ 47465.

LEARNING OBJECTIVES

In this section, you will:
- Solve a system of nonlinear equations using substitution.
- Solve a system of nonlinear equations using elimination.
- Graph a nonlinear inequality.
- Graph a system of nonlinear inequalities.

7.3   SYSTEMS OF NONLINEAR EQUATIONS AND INEQUALITIES: TWO VARIABLES

Halley's Comet (**Figure 1**) orbits the sun about once every 75 years. Its path can be considered to be a very elongated ellipse. Other comets follow similar paths in space. These orbital paths can be studied using systems of equations. These systems, however, are different from the ones we considered in the previous section because the equations are not linear.

Figure 1 Halley's Comet (credit: "NASA Blueshift"/Flickr)

In this section, we will consider the intersection of a parabola and a line, a circle and a line, and a circle and an ellipse. The methods for solving systems of nonlinear equations are similar to those for linear equations.

Solving a System of Nonlinear Equations Using Substitution

A **system of nonlinear equations** is a system of two or more equations in two or more variables containing at least one equation that is not linear. Recall that a linear equation can take the form $Ax + By + C = 0$. Any equation that cannot be written in this form in nonlinear. The substitution method we used for linear systems is the same method we will use for nonlinear systems. We solve one equation for one variable and then substitute the result into the second equation to solve for another variable, and so on. There is, however, a variation in the possible outcomes.

Intersection of a Parabola and a Line

There are three possible types of solutions for a system of nonlinear equations involving a parabola and a line.

> **possible types of solutions for points of intersection of a parabola and a line**
>
> **Figure 2** illustrates possible solution sets for a system of equations involving a parabola and a line.
> - No solution. The line will never intersect the parabola.
> - One solution. The line is tangent to the parabola and intersects the parabola at exactly one point.
> - Two solutions. The line crosses on the inside of the parabola and intersects the parabola at two points.

Figure 2

How To...

Given a system of equations containing a line and a parabola, find the solution.

1. Solve the linear equation for one of the variables.
2. Substitute the expression obtained in step one into the parabola equation.
3. Solve for the remaining variable.
4. Check your solutions in both equations.

Example 1   **Solving a System of Nonlinear Equations Representing a Parabola and a Line**

Solve the system of equations.

$$x - y = -1$$
$$y = x^2 + 1$$

Solution   Solve the first equation for x and then substitute the resulting expression into the second equation.

$$x - y = -1$$
$$x = y - 1 \qquad \text{Solve for } x.$$
$$y = x^2 + 1$$
$$y = (y - 1)^2 + 1 \qquad \text{Substitute expression for } x.$$

Expand the equation and set it equal to zero.

$$y = (y - 1)^2$$
$$= (y^2 - 2y + 1) + 1$$
$$= y^2 - 2y + 2$$
$$0 = y^2 - 3y + 2$$
$$= (y - 2)(y - 1)$$

Solving for y gives $y = 2$ and $y = 1$. Next, substitute each value for y into the first equation to solve for x. Always substitute the value into the linear equation to check for extraneous solutions.

$$x - y = -1$$
$$x - (2) = -1$$
$$x = 1$$
$$x - (1) = -1$$
$$x = 0$$

The solutions are $(1, 2)$ and $(0, 1)$, which can be verified by substituting these (x, y) values into both of the original equations. See **Figure 3**.

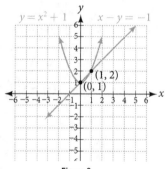

Figure 3

The graph is shown in **Figure 9**. We can see that the solution set consists of all points inside the parabola, but not on the graph itself.

Figure 9

Graphing a System of Nonlinear Inequalities

Now that we have learned to graph nonlinear inequalities, we can learn how to graph systems of nonlinear inequalities. A **system of nonlinear inequalities** is a system of two or more inequalities in two or more variables containing at least one inequality that is not linear. Graphing a system of nonlinear inequalities is similar to graphing a system of linear inequalities. The difference is that our graph may result in more shaded regions that represent a solution than we find in a system of linear inequalities. The solution to a nonlinear system of inequalities is the region of the graph where the shaded regions of the graph of each inequality overlap, or where the regions intersect, called the **feasible region**.

How To…

Given a system of nonlinear inequalities, sketch a graph.

1. Find the intersection points by solving the corresponding system of nonlinear equations.
2. Graph the nonlinear equations.
3. Find the shaded regions of each inequality.
4. Identify the feasible region as the intersection of the shaded regions of each inequality or the set of points common to each inequality.

Example 5   **Graphing a System of Inequalities**

Graph the given system of inequalities.

$$x^2 - y \le 0$$
$$2x^2 + y \le \ 12$$

Solution   These two equations are clearly parabolas. We can find the points of intersection by the elimination process: Add both equations and the variable y will be eliminated. Then we solve for x.

$$x^2 - y = 0$$
$$\underline{2x^2 + y = 12}$$
$$3x^2 = 12$$
$$x^2 = 4$$
$$x = \pm 2$$

Substitute the x-values into one of the equations and solve for y.

$$x^2 - y = 0$$
$$(2)^2 - y = 0$$
$$4 - y = 0$$
$$y = 4$$
$$(-2)^2 - y = 0$$
$$4 - y = 0$$
$$y = 4$$

The two points of intersection are $(2, 4)$ and $(-2, 4)$. Notice that the equations can be rewritten as follows.

$$x^2 - y \leq 0$$
$$x^2 \leq y$$
$$y \geq x^2$$
$$2x^2 + y \leq 12$$
$$y \leq -2x^2 + 12$$

Graph each inequality. See **Figure 10**. The feasible region is the region between the two equations bounded by 2 $x^2 + y \leq 12$ on the top and $x^2 - y \leq 0$ on the bottom.

Figure 10

___

Try It #4

Graph the given system of inequalities.

$$y \geq x^2 - 1$$
$$x - y \geq -1$$

___

Access these online resources for additional instruction and practice with nonlinear equations.

- Solve a System of Nonlinear Equations Using Substitution (http://openstaxcollege.org/l/nonlinsub)
- Solve a System of Nonlinear Equations Using Elimination (http://openstaxcollege.org/l/nonlinelim)

7.3 SECTION EXERCISES

VERBAL

1. Explain whether a system of two nonlinear equations can have exactly two solutions. What about exactly three? If not, explain why not. If so, give an example of such a system, in graph form, and explain why your choice gives two or three answers.

2. When graphing an inequality, explain why we only need to test one point to determine whether an entire region is the solution?

3. When you graph a system of inequalities, will there always be a feasible region? If so, explain why. If not, give an example of a graph of inequalities that does not have a feasible region. Why does it not have a feasible region?

4. If you graph a revenue and cost function, explain how to determine in what regions there is profit.

5. If you perform your break-even analysis and there is more than one solution, explain how you would determine which x-values are profit and which are not.

ALGEBRAIC

For the following exercises, solve the system of nonlinear equations using substitution.

6. $x + y = 4$
$x^2 + y^2 = 9$

7. $y = x - 3$
$x^2 + y^2 = 9$

8. $y = x$
$x^2 + y^2 = 9$

9. $y = -x$
$x^2 + y^2 = 9$

10. $x = 2$
$x^2 - y^2 = 9$

For the following exercises, solve the system of nonlinear equations using elimination.

11. $4x^2 - 9y^2 = 36$
$4x^2 + 9y^2 = 36$

12. $x^2 + y^2 = 25$
$x^2 - y^2 = 1$

13. $2x^2 + 4y^2 = 4$
$2x^2 - 4y^2 = 25x - 10$

14. $y^2 - x^2 = 9$
$3x^2 + 2y^2 = 8$

15. $x^2 + y^2 + \frac{1}{16} = 2500$
$y = 2x^2$

For the following exercises, use any method to solve the system of nonlinear equations.

16. $-2x^2 + y = -5$
$6x - y = 9$

17. $-x^2 + y = 2$
$-x + y = 2$

18. $x^2 + y^2 = 1$
$y = 20x^2 - 1$

19. $x^2 + y^2 = 1$
$y = -x^2$

20. $2x^3 - x^2 = y$
$y = \frac{1}{2} - x$

21. $9x^2 + 25y^2 = 225$
$(x - 6)^2 + y^2 = 1$

22. $x^4 - x^2 = y$
$x^2 + y = 0$

23. $2x^3 - x^2 = y$
$x^2 + y = 0$

For the following exercises, use any method to solve the nonlinear system.

24. $x^2 + y^2 = 9$
$y = 3 - x^2$

25. $x^2 - y^2 = 9$
$x = 3$

26. $x^2 - y^2 = 9$
$y = 3$

27. $x^2 - y^2 = 9$
$x - y = 0$

28. $-x^2 + y = 2$
$-4x + y = -1$

29. $-x^2 + y = 2$
$2y = -x$

30. $x^2 + y^2 = 25$
$x^2 - y^2 = 36$

31. $x^2 + y^2 = 1$
$y^2 = x^2$

32. $16x^2 - 9y^2 + 144 = 0$
$y^2 + x^2 = 16$

33.　$\begin{aligned} 3x^2 - y^2 &= 12 \\ (x-1)^2 + y^2 &= 1 \end{aligned}$　　**34.**　$\begin{aligned} 3x^2 - y^2 &= 12 \\ (x-1)^2 + y^2 &= 4 \end{aligned}$　　**35.** $\begin{aligned} 3x^2 - y^2 &= 12 \\ x^2 + y^2 &= 16 \end{aligned}$

36. $\begin{aligned} x^2 - y^2 - 6x - 4y - 11 &= 0 \\ -x^2 + y^2 &= 5 \end{aligned}$　　**37.** $\begin{aligned} x^2 + y^2 - 6y &= 7 \\ x^2 + y &= 1 \end{aligned}$　　**38.** $\begin{aligned} x^2 + y^2 &= 6 \\ xy &= 1 \end{aligned}$

GRAPHICAL

For the following exercises, graph the inequality.

39. $x^2 + y < 9$　　　　**40.** $x^2 + y^2 < 4$

For the following exercises, graph the system of inequalities. Label all points of intersection.

41. $\begin{aligned} x^2 + y &< 1 \\ y &> 2x \end{aligned}$　　**42.** $\begin{aligned} x^2 + y &< -5 \\ y &> 5x + 10 \end{aligned}$　　**43.** $\begin{aligned} x^2 + y^2 &< 25 \\ 3x^2 - y^2 &> 12 \end{aligned}$

44. $\begin{aligned} x^2 - y^2 &> -4 \\ x^2 + y^2 &< 12 \end{aligned}$　　**45.** $\begin{aligned} x^2 + 3y^2 &> 16 \\ 3x^2 - y^2 &< 1 \end{aligned}$

EXTENSIONS

For the following exercises, graph the inequality.

46. $\begin{aligned} y &\geq e^x \\ y &\leq \ln(x) + 5 \end{aligned}$　　**47.** $\begin{aligned} y &\leq -\log(x) \\ y &\leq e^x \end{aligned}$

For the following exercises, find the solutions to the nonlinear equations with two variables.

48.　$\begin{aligned} \frac{4}{x^2} + \frac{1}{y^2} &= 24 \\ \frac{5}{x^2} - \frac{2}{y^2} + 4 &= 0 \end{aligned}$　　**49.** $\begin{aligned} \frac{6}{x^2} - \frac{1}{y^2} &= 8 \\ \frac{1}{x^2} - \frac{6}{y^2} &= \frac{1}{8} \end{aligned}$　　**50.** $\begin{aligned} x^2 - xy + y^2 - 2 &= 0 \\ x + 3y &= 4 \end{aligned}$

51. $\begin{aligned} x^2 - xy - 2y^2 - 6 &= 0 \\ x^2 + y^2 &= 1 \end{aligned}$　　**52.** $\begin{aligned} x^2 + 4xy - 2y^2 - 6 &= 0 \\ x &= y + 2 \end{aligned}$

TECHNOLOGY

For the following exercises, solve the system of inequalities. Use a calculator to graph the system to confirm the answer.

53. $\begin{aligned} xy &< 1 \\ y &> \sqrt{x} \end{aligned}$　　**54.** $\begin{aligned} x^2 + y &< 3 \\ y &> 2x \end{aligned}$

REAL-WORLD APPLICATIONS

For the following exercises, construct a system of nonlinear equations to describe the given behavior, then solve for the requested solutions.

55. Two numbers add up to 300. One number is twice the square of the other number. What are the numbers?

56. The squares of two numbers add to 360. The second number is half the value of the first number squared. What are the numbers?

57. A laptop company has discovered their cost and revenue functions for each day: $C(x) = 3x^2 - 10x + 200$ and $R(x) = -2x^2 + 100x + 50$. If they want to make a profit, what is the range of laptops per day that they should produce? Round to the nearest number which would generate profit.

58. A cell phone company has the following cost and revenue functions: $C(x) = 8x^2 - 600x + 21{,}500$ and $R(x) = -3x^2 + 480x$. What is the range of cell phones they should produce each day so there is profit? round to the nearest number that generates profit.

Decomposing $\dfrac{P(x)}{Q(x)}$ Where $Q(x)$ Has a Repeated Irreducible Quadratic Factor

Now that we can decompose a simplified rational expression with an irreducible quadratic factor, we will learn how to do partial fraction decomposition when the simplified rational expression has repeated irreducible quadratic factors. The decomposition will consist of partial fractions with linear numerators over each irreducible quadratic factor represented in increasing powers.

---

decomposition of $\dfrac{P(x)}{Q(x)}$: when $Q(x)$ has a repeated irreducible quadratic factor

The partial fraction decomposition of $\dfrac{P(x)}{Q(x)}$, when $Q(x)$ has a repeated irreducible quadratic factor and the degree of $P(x)$ is less than the degree of $Q(x)$, is

$$\frac{P(x)}{(ax^2+bx+c)^n} = \frac{A_1 x + B_1}{(ax^2+bx+c)} + \frac{A_2 x + B_2}{(ax^2+bx+c)^2} + \frac{A_3 x + B_3}{(ax^2+bx+c)^3} + \cdots + \frac{A_n x + B_n}{(ax^2+bx+c)^n}$$

Write the denominators in increasing powers.

How To…

Given a rational expression that has a repeated irreducible factor, decompose it.

1. Use variables like A, B, or C for the constant numerators over linear factors, and linear expressions such as $A_1 x + B_1$, $A_2 x + B_2$, etc., for the numerators of each quadratic factor in the denominator written in increasing powers, such as

$$\frac{P(x)}{Q(x)} = \frac{A}{ax+b} + \frac{A_1 x + B_1}{(ax^2+bx+c)} + \frac{A_2 x + B_2}{(ax^2+bx+c)^2} + \cdots + \frac{A_n x + B_n}{(ax^2+bx+c)^n}$$

2. Multiply both sides of the equation by the common denominator to eliminate fractions.
3. Expand the right side of the equation and collect like terms.
4. Set coefficients of like terms from the left side of the equation equal to those on the right side to create a system of equations to solve for the numerators.

Example 4   **Decomposing a Rational Function with a Repeated Irreducible Quadratic Factor in the Denominator**

Decompose the given expression that has a repeated irreducible factor in the denominator.

$$\frac{x^4 + x^3 + x^2 - x + 1}{x(x^2+1)^2}$$

Solution   The factors of the denominator are x, (x^2+1), and $(x^2+1)^2$. Recall that, when a factor in the denominator is a quadratic that includes at least two terms, the numerator must be of the linear form $Ax + B$. So, let's begin the decomposition.

$$\frac{x^4 + x^3 + x^2 - x + 1}{x(x^2+1)^2} = \frac{A}{x} + \frac{Bx+C}{(x^2+1)} + \frac{Dx+E}{(x^2+1)^2}$$

We eliminate the denominators by multiplying each term by $x(x^2+1)^2$. Thus,

$$x^4 + x^3 + x^2 - x + 1 = A(x^2+1)^2 + (Bx+C)(x)(x^2+1) + (Dx+E)(x)$$

Expand the right side.

$$x^4 + x^3 + x^2 - x + 1 = A(x^4 + 2x^2 + 1) + Bx^4 + Bx^2 + Cx^3 + Cx + Dx^2 + Ex$$

$$= Ax^4 + 2Ax^2 + A + Bx^4 + Bx^2 + Cx^3 + Cx + Dx^2 + Ex$$

Now we will collect like terms.

$$x^4 + x^3 + x^2 - x + 1 = (A+B)x^4 + (C)x^3 + (2A+B+D)x^2 + (C+E)x + A$$

Set up the system of equations matching corresponding coefficients on each side of the equal sign.

$$A + B = 1$$
$$C = 1$$
$$2A + B + D = 1$$
$$C + E = -1$$
$$A = 1$$

We can use substitution from this point. Substitute $A = 1$ into the first equation.

$$1 + B = 1$$
$$B = 0$$

Substitute $A = 1$ and $B = 0$ into the third equation.

$$2(1) + 0 + D = 1$$
$$D = -1$$

Substitute $C = 1$ into the fourth equation.

$$1 + E = -1$$
$$E = -2$$

Now we have solved for all of the unknowns on the right side of the equal sign. We have $A = 1, B = 0, C = 1, D = -1$, and $E = -2$. We can write the decomposition as follows:

$$\frac{x^4 + x^3 + x^2 - x + 1}{x(x^2 + 1)^2} = \frac{1}{x} + \frac{1}{(x^2 + 1)} - \frac{x + 2}{(x^2 + 1)^2}$$

Try It #4

Find the partial fraction decomposition of the expression with a repeated irreducible quadratic factor.

$$\frac{x^3 - 4x^2 + 9x - 5}{(x^2 - 2x + 3)^2}$$

Access these online resources for additional instruction and practice with partial fractions.

- Partial Fraction Decomposition (http://openstaxcollege.org/l/partdecomp)
- Partial Fraction Decomposition With Repeated Linear Factors (http://openstaxcollege.org/l/partdecomprlf)
- Partial Fraction Decomposition With Linear and Quadratic Factors (http://openstaxcollege.org/l/partdecomlqu)

7.4 SECTION EXERCISES

VERBAL

1. Can any quotient of polynomials be decomposed into at least two partial fractions? If so, explain why, and if not, give an example of such a fraction

2. Can you explain why a partial fraction decomposition is unique? (Hint: Think about it as a system of equations.)

3. Can you explain how to verify a partial fraction decomposition graphically?

4. You are unsure if you correctly decomposed the partial fraction correctly. Explain how you could double check your answer.

5. Once you have a system of equations generated by the partial fraction decomposition, can you explain another method to solve it? For example if you had $\dfrac{7x + 13}{3x^2 + 8x + 15} = \dfrac{A}{x + 1} + \dfrac{B}{3x + 5}$, we eventually simplify to $7x + 13 = A(3x + 5) + B(x + 1)$. Explain how you could intelligently choose an x-value that will eliminate either A or B and solve for A and B.

ALGEBRAIC

For the following exercises, find the decomposition of the partial fraction for the nonrepeating linear factors.

6. $\dfrac{5x + 16}{x^2 + 10x + 24}$

7. $\dfrac{3x - 79}{x^2 - 5x - 24}$

8. $\dfrac{-x - 24}{x^2 - 2x - 24}$

9. $\dfrac{10x + 47}{x^2 + 7x + 10}$

10. $\dfrac{x}{6x^2 + 25x + 25}$

11. $\dfrac{32x - 11}{20x^2 - 13x + 2}$

12. $\dfrac{x + 1}{x^2 + 7x + 10}$

13. $\dfrac{5x}{x^2 - 9}$

14. $\dfrac{10x}{x^2 - 25}$

15. $\dfrac{6x}{x^2 - 4}$

16. $\dfrac{2x - 3}{x^2 - 6x + 5}$

17. $\dfrac{4x - 1}{x^2 - x - 6}$

18. $\dfrac{4x + 3}{x^2 + 8x + 15}$

19. $\dfrac{3x - 1}{x^2 - 5x + 6}$

For the following exercises, find the decomposition of the partial fraction for the repeating linear factors.

20. $\dfrac{-5x - 19}{(x + 4)^2}$

21. $\dfrac{x}{(x - 2)^2}$

22. $\dfrac{7x + 14}{(x + 3)^2}$

23. $\dfrac{-24x - 27}{(4x + 5)^2}$

24. $\dfrac{-24x - 27}{(6x - 7)^2}$

25. $\dfrac{5 - x}{(x - 7)^2}$

26. $\dfrac{5x + 14}{2x^2 + 12x + 18}$

27. $\dfrac{5x^2 + 20x + 8}{2x(x + 1)^2}$

28. $\dfrac{4x^2 + 55x + 25}{5x(3x + 5)^2}$

29. $\dfrac{54x^3 + 127x^2 + 80x + 16}{2x^2(3x + 2)^2}$

30. $\dfrac{x^3 - 5x^2 + 12x + 144}{x^2(x^2 + 12x + 36)}$

For the following exercises, find the decomposition of the partial fraction for the irreducible non repeating quadratic factor.

31. $\dfrac{4x^2 + 6x + 11}{(x+2)(x^2 + x + 3)}$

32. $\dfrac{4x^2 + 9x + 23}{(x-1)(x^2 + 6x + 11)}$

33. $\dfrac{-2x^2 + 10x + 4}{(x-1)(x^2 + 3x + 8)}$

34. $\dfrac{x^2 + 3x + 1}{(x+1)(x^2 + 5x - 2)}$

35. $\dfrac{4x^2 + 17x - 1}{(x+3)(x^2 + 6x + 1)}$

36. $\dfrac{4x^2}{(x+5)(x^2 + 7x - 5)}$

37. $\dfrac{4x^2 + 5x + 3}{x^3 - 1}$

38. $\dfrac{-5x^2 + 18x - 4}{x^3 + 8}$

39. $\dfrac{3x^2 - 7x + 33}{x^3 + 27}$

40. $\dfrac{x^2 + 2x + 40}{x^3 - 125}$

41. $\dfrac{4x^2 + 4x + 12}{8x^3 - 27}$

42. $\dfrac{-50x^2 + 5x - 3}{125x^3 - 1}$

43. $\dfrac{-2x^3 - 30x^2 + 36x + 216}{x^4 + 216x}$

For the following exercises, find the decomposition of the partial fraction for the irreducible repeating quadratic factor.

44. $\dfrac{3x^3 + 2x^2 + 14x + 15}{(x^2 + 4)^2}$

45. $\dfrac{x^3 + 6x^2 + 5x + 9}{(x^2 + 1)^2}$

46. $\dfrac{x^3 - x^2 + x - 1}{(x^2 - 3)^2}$

47. $\dfrac{x^2 + 5x + 5}{(x+2)^2}$

48. $\dfrac{x^3 + 2x^2 + 4x}{(x^2 + 2x + 9)^2}$

49. $\dfrac{x^2 + 25}{(x^2 + 3x + 25)^2}$

50. $\dfrac{2x^3 + 11x + 7x + 70}{(2x^2 + x + 14)^2}$

51. $\dfrac{5x + 2}{x(x^2 + 4)^2}$

52. $\dfrac{x^4 + x^3 + 8x^2 + 6x + 36}{x(x^2 + 6)^2}$

53. $\dfrac{2x - 9}{(x^2 - x)^2}$

54. $\dfrac{5x^3 - 2x + 1}{(x^2 + 2x)^2}$

EXTENSIONS

For the following exercises, find the partial fraction expansion.

55. $\dfrac{x^2 + 4}{(x+1)^3}$

56. $\dfrac{x^3 - 4x^2 + 5x + 4}{(x-2)^3}$

For the following exercises, perform the operation and then find the partial fraction decomposition.

57. $\dfrac{7}{x+8} + \dfrac{5}{x-2} - \dfrac{x-1}{x^2 - 6x - 16}$

58. $\dfrac{1}{x-4} - \dfrac{3}{x+6} - \dfrac{2x+7}{x^2 + 2x - 24}$

59. $\dfrac{2x}{x^2 - 16} - \dfrac{1-2x}{x^2 + 6x + 8} - \dfrac{x-5}{x^2 - 4x}$

LEARNING OBJECTIVES

In this section, you will:

- Find the sum and difference of two matrices.
- Find scalar multiples of a matrix.
- Find the product of two matrices.

7.5 MATRICES AND MATRIX OPERATIONS

Figure 1 (credit: "SD Dirk," Flickr)

Two club soccer teams, the Wildcats and the Mud Cats, are hoping to obtain new equipment for an upcoming season. **Table 1** shows the needs of both teams.

	Wildcats	**Mud Cats**
Goals	6	10
Balls	30	24
Jerseys	14	20

Table 1

A goal costs $300; a ball costs $10; and a jersey costs $30. How can we find the total cost for the equipment needed for each team? In this section, we discover a method in which the data in the soccer equipment table can be displayed and used for calculating other information. Then, we will be able to calculate the cost of the equipment.

Finding the Sum and Difference of Two Matrices

To solve a problem like the one described for the soccer teams, we can use a matrix, which is a rectangular array of numbers. A row in a matrix is a set of numbers that are aligned horizontally. A column in a matrix is a set of numbers that are aligned vertically. Each number is an entry, sometimes called an element, of the matrix. Matrices (plural) are enclosed in [ ] or ( ), and are usually named with capital letters. For example, three matrices named A, B, and C are shown below.

$$A = \begin{bmatrix} 1 & 2 \\ 3 & 4 \end{bmatrix}, B = \begin{bmatrix} 1 & 2 & 7 \\ 0 & -5 & 6 \\ 7 & 8 & 2 \end{bmatrix}, C = \begin{bmatrix} -1 & 3 \\ 0 & 2 \\ 3 & 1 \end{bmatrix}$$

Describing Matrices

A matrix is often referred to by its size or dimensions: $m \times n$ indicating m rows and n columns. Matrix entries are defined first by row and then by column. For example, to locate the entry in matrix A identified as a_{ij}, we look for the entry in row i, column j. In matrix A, shown below, the entry in row 2, column 3 is a_{23}.

$$A = \begin{bmatrix} a_{11} & a_{12} & a_{13} \\ a_{21} & a_{22} & a_{23} \\ a_{31} & a_{32} & a_{33} \end{bmatrix}$$

A square matrix is a matrix with dimensions $n \times n$, meaning that it has the same number of rows as columns. The 3×3 matrix above is an example of a square matrix.

A row matrix is a matrix consisting of one row with dimensions $1 \times n$.

$$\begin{bmatrix} a_{11} & a_{12} & a_{13} \end{bmatrix}$$

A column matrix is a matrix consisting of one column with dimensions $m \times 1$.

$$\begin{bmatrix} a_{11} \\ a_{12} \\ a_{13} \end{bmatrix}$$

A matrix may be used to represent a system of equations. In these cases, the numbers represent the coefficients of the variables in the system. Matrices often make solving systems of equations easier because they are not encumbered with variables. We will investigate this idea further in the next section, but first we will look at basic matrix operations.

---

matrices

A **matrix** is a rectangular array of numbers that is usually named by a capital letter: A, B, C, and so on. Each entry in a matrix is referred to as a_{ij}, such that i represents the row and j represents the column. Matrices are often referred to by their dimensions: $m \times n$ indicating m rows and n columns.

Example 1 **Finding the Dimensions of the Given Matrix and Locating Entries**

Given matrix A:

  a. What are the dimensions of matrix A?

  b. What are the entries at a_{31} and a_{22} ?

$$A = \begin{bmatrix} 2 & 1 & 0 \\ 2 & 4 & 7 \\ 3 & 1 & -2 \end{bmatrix}$$

Solution

  a. The dimensions are 3×3 because there are three rows and three columns.

  b. Entry a_{31} is the number at row 3, column 1, which is 3. The entry a_{22} is the number at row 2, column 2, which is 4. Remember, the row comes first, then the column.

Adding and Subtracting Matrices

We use matrices to list data or to represent systems. Because the entries are numbers, we can perform operations on matrices. We add or subtract matrices by adding or subtracting corresponding entries.

In order to do this, the entries must correspond. Therefore, *addition and subtraction of matrices is only possible when the matrices have the same dimensions.* We can add or subtract a 3×3 matrix and another 3×3 matrix, but we cannot add or subtract a 2×3 matrix and a 3×3 matrix because some entries in one matrix will not have a corresponding entry in the other matrix.

---

adding and subtracting matrices

Given matrices A and B of like dimensions, addition and subtraction of A and B will produce matrix C or matrix D of the same dimension.

$$A + B = C \text{ such that } a_{ij} + b_{ij} = c_{ij}$$
$$A - B = D \text{ such that } a_{ij} - b_{ij} = d_{ij}$$

Matrix addition is commutative.

$$A + B = B + A$$

It is also associative.

$$(A + B) + C = A + (B + C)$$

Example 2   **Finding the Sum of Matrices**

Find the sum of A and B, given

$$A = \begin{bmatrix} a & b \\ c & d \end{bmatrix} \text{ and } B = \begin{bmatrix} e & f \\ g & h \end{bmatrix}$$

Solution   Add corresponding entries.

$$A + B = \begin{bmatrix} a & b \\ c & d \end{bmatrix} + \begin{bmatrix} e & f \\ g & h \end{bmatrix}$$

$$= \begin{bmatrix} a + e & b + f \\ c + g & d + h \end{bmatrix}$$

Example 3   **Adding Matrix *A* and Matrix *B***

Find the sum of A and B.

$$A = \begin{bmatrix} 4 & 1 \\ 3 & 2 \end{bmatrix} \text{ and } B = \begin{bmatrix} 5 & 9 \\ 0 & 7 \end{bmatrix}$$

Solution   Add corresponding entries. Add the entry in row 1, column 1, a_{11}, of matrix A to the entry in row 1, column 1, b_{11}, of B. Continue the pattern until all entries have been added.

$$A + B = \begin{bmatrix} 4 & 1 \\ 3 & 2 \end{bmatrix} + \begin{bmatrix} 5 & 9 \\ 0 & 7 \end{bmatrix}$$

$$= \begin{bmatrix} 4 + 5 & 1 + 9 \\ 3 + 0 & 2 + 7 \end{bmatrix}$$

$$= \begin{bmatrix} 9 & 10 \\ 3 & 9 \end{bmatrix}$$

Example 4   **Finding the Difference of Two Matrices**

Find the difference of A and B.

$$A = \begin{bmatrix} -2 & 3 \\ 0 & 1 \end{bmatrix} \text{ and } B = \begin{bmatrix} 8 & 1 \\ 5 & 4 \end{bmatrix}$$

Solution   We subtract the corresponding entries of each matrix.

$$A - B = \begin{bmatrix} -2 & 3 \\ 0 & 1 \end{bmatrix} - \begin{bmatrix} 8 & 1 \\ 5 & 4 \end{bmatrix}$$

$$= \begin{bmatrix} -2 - 8 & 3 - 1 \\ 0 - 5 & 1 - 4 \end{bmatrix}$$

$$= \begin{bmatrix} -10 & 2 \\ -5 & -3 \end{bmatrix}$$

Example 5   **Finding the Sum and Difference of Two 3 x 3 Matrices**

Given A and B:

    a. Find the sum.

    b. Find the difference.

$$A = \begin{bmatrix} 2 & -10 & -2 \\ 14 & 12 & 10 \\ 4 & -2 & 2 \end{bmatrix} \text{ and } B = \begin{bmatrix} 6 & 10 & -2 \\ 0 & -12 & -4 \\ -5 & 2 & -2 \end{bmatrix}$$

Solution

    a. Add the corresponding entries.

$$A + B = \begin{bmatrix} 2 & -10 & -2 \\ 14 & 12 & 10 \\ 4 & -2 & 2 \end{bmatrix} + \begin{bmatrix} 6 & 10 & -2 \\ 0 & -12 & -4 \\ -5 & 2 & -2 \end{bmatrix}$$

$$= \begin{bmatrix} 2+6 & -10+10 & -2-2 \\ 14+0 & 12-12 & 10-4 \\ 4-5 & -2+2 & 2-2 \end{bmatrix}$$

$$= \begin{bmatrix} 8 & 0 & -4 \\ 14 & 0 & 6 \\ -1 & 0 & 0 \end{bmatrix}$$

    b. Subtract the corresponding entries.

$$A - B = \begin{bmatrix} 2 & -10 & -2 \\ 14 & 12 & 10 \\ 4 & -2 & 2 \end{bmatrix} - \begin{bmatrix} 6 & 10 & -2 \\ 0 & -12 & -4 \\ -5 & 2 & -2 \end{bmatrix}$$

$$= \begin{bmatrix} 2-6 & -10-10 & -2+2 \\ 14-0 & 12+12 & 10+4 \\ 4+5 & -2-2 & 2+2 \end{bmatrix}$$

$$= \begin{bmatrix} -4 & -20 & 0 \\ 14 & 24 & 14 \\ 9 & -4 & 4 \end{bmatrix}$$

Try It #1

Add matrix A and matrix B.

$$A = \begin{bmatrix} 2 & 6 \\ 1 & 0 \\ 1 & -3 \end{bmatrix} \text{ and } B = \begin{bmatrix} 3 & -2 \\ 1 & 5 \\ -4 & 3 \end{bmatrix}$$

Finding Scalar Multiples of a Matrix

Besides adding and subtracting whole matrices, there are many situations in which we need to multiply a matrix by a constant called a scalar. Recall that a scalar is a real number quantity that has magnitude, but not direction. For example, time, temperature, and distance are scalar quantities. The process of scalar multiplication involves multiplying each entry in a matrix by a scalar. A **scalar multiple** is any entry of a matrix that results from scalar multiplication.

Consider a real-world scenario in which a university needs to add to its inventory of computers, computer tables, and chairs in two of the campus labs due to increased enrollment. They estimate that 15% more equipment is needed in both labs. The school's current inventory is displayed in **Table 2**.

	Lab A	Lab B
Computers	15	27
Computer Tables	16	34
Chairs	16	34

Table 2

Converting the data to a matrix, we have

$$C_{2013} = \begin{bmatrix} 15 & 27 \\ 16 & 34 \\ 16 & 34 \end{bmatrix}$$

To calculate how much computer equipment will be needed, we multiply all entries in matrix C by 0.15.

$$(0.15)C_{2013} = \begin{bmatrix} (0.15)15 & (0.15)27 \\ (0.15)16 & (0.15)34 \\ (0.15)16 & (0.15)34 \end{bmatrix} = \begin{bmatrix} 2.25 & 4.05 \\ 2.4 & 5.1 \\ 2.4 & 5.1 \end{bmatrix}$$

We must round up to the next integer, so the amount of new equipment needed is

$$\begin{bmatrix} 3 & 5 \\ 3 & 6 \\ 3 & 6 \end{bmatrix}$$

Adding the two matrices as shown below, we see the new inventory amounts.

$$\begin{bmatrix} 15 & 27 \\ 16 & 34 \\ 16 & 34 \end{bmatrix} + \begin{bmatrix} 3 & 5 \\ 3 & 6 \\ 3 & 6 \end{bmatrix} = \begin{bmatrix} 18 & 32 \\ 19 & 40 \\ 19 & 40 \end{bmatrix}$$

This means

$$C_{2014} = \begin{bmatrix} 18 & 32 \\ 19 & 40 \\ 19 & 40 \end{bmatrix}$$

Thus, Lab A will have 18 computers, 19 computer tables, and 19 chairs; Lab B will have 32 computers, 40 computer tables, and 40 chairs.

---

scalar multiplication

Scalar multiplication involves finding the product of a constant by each entry in the matrix. Given

$$A = \begin{bmatrix} a_{11} & a_{12} \\ a_{21} & a_{22} \end{bmatrix}$$

the scalar multiple cA is

$$cA = c\begin{bmatrix} a_{11} & a_{12} \\ a_{21} & a_{22} \end{bmatrix}$$

$$= \begin{bmatrix} ca_{11} & ca_{12} \\ ca_{21} & ca_{22} \end{bmatrix}$$

Scalar multiplication is distributive. For the matrices A, B, and C with scalars a and b,

$$a(A + B) = aA + aB$$
$$(a + b)A = aA + bA$$

Example 6   **Multiplying the Matrix by a Scalar**

Multiply matrix A by the scalar 3.

$$A = \begin{bmatrix} 8 & 1 \\ 5 & 4 \end{bmatrix}$$

Solution   Multiply each entry in A by the scalar 3.

$$3A = 3\begin{bmatrix} 8 & 1 \\ 5 & 4 \end{bmatrix}$$

$$= \begin{bmatrix} 3 \cdot 8 & 3 \cdot 1 \\ 3 \cdot 5 & 3 \cdot 4 \end{bmatrix}$$

$$= \begin{bmatrix} 24 & 3 \\ 15 & 12 \end{bmatrix}$$

Try It #2

Given matrix B, find −2B where

$$A = \begin{bmatrix} 4 & 1 \\ 3 & 2 \end{bmatrix}$$

Example 7 **Finding the Sum of Scalar Multiples**

Find the sum $3A + 2B$.

$$A = \begin{bmatrix} 1 & -2 & 0 \\ 0 & -1 & 2 \\ 4 & 3 & -6 \end{bmatrix} \text{ and } B = \begin{bmatrix} -1 & 2 & 1 \\ 0 & -3 & 2 \\ 0 & 1 & -4 \end{bmatrix}$$

Solution First, find $3A$, then $2B$.

$$3A = \begin{bmatrix} 3\cdot1 & 3(-2) & 3\cdot0 \\ 3\cdot0 & 3(-1) & 3\cdot2 \\ 3\cdot4 & 3\cdot3 & 3(-6) \end{bmatrix}$$

$$= \begin{bmatrix} 3 & -6 & 0 \\ 0 & -3 & 6 \\ 12 & 9 & -18 \end{bmatrix}$$

$$2B = \begin{bmatrix} 2(-1) & 2\cdot2 & 2\cdot1 \\ 2\cdot0 & 2(-3) & 2\cdot2 \\ 2\cdot0 & 2\cdot1 & 2(-4) \end{bmatrix}$$

$$= \begin{bmatrix} -2 & 4 & 2 \\ 0 & -6 & 4 \\ 0 & 2 & -8 \end{bmatrix}$$

Now, add $3A + 2B$.

$$3A + 2B = \begin{bmatrix} 3 & -6 & 0 \\ 0 & -3 & 6 \\ 12 & 9 & -18 \end{bmatrix} + \begin{bmatrix} -2 & 4 & 2 \\ 0 & -6 & 4 \\ 0 & 2 & -8 \end{bmatrix}$$

$$= \begin{bmatrix} 3-2 & -6+4 & 0+2 \\ 0+0 & -3-6 & 6+4 \\ 12+0 & 9+2 & -18-8 \end{bmatrix}$$

$$= \begin{bmatrix} 1 & -2 & 2 \\ 0 & -9 & 10 \\ 12 & 11 & -26 \end{bmatrix}$$

Finding the Product of Two Matrices

In addition to multiplying a matrix by a scalar, we can multiply two matrices. Finding the product of two matrices is only possible when the inner dimensions are the same, meaning that the number of columns of the first matrix is equal to the number of rows of the second matrix. If A is an $m \times r$ matrix and B is an $r \times n$ matrix, then the product matrix AB is an $m \times n$ matrix. For example, the product AB is possible because the number of columns in A is the same as the number of rows in B. If the inner dimensions do not match, the product is not defined.

$$\underset{2\times3}{A} \cdot \underset{3\times3}{B}$$
same

We multiply entries of A with entries of B according to a specific pattern as outlined below. The process of matrix multiplication becomes clearer when working a problem with real numbers.

To obtain the entries in row i of AB, we multiply the entries in row i of A by column j in B and add. For example, given matrices A and B, where the dimensions of A are 2×3 and the dimensions of B are 3×3, the product of AB will be a 2×3 matrix.

$$A = \begin{bmatrix} a_{11} & a_{12} & a_{13} \\ a_{21} & a_{22} & a_{23} \end{bmatrix} \text{ and } B = \begin{bmatrix} b_{11} & b_{12} & b_{13} \\ b_{21} & b_{22} & b_{23} \\ b_{31} & b_{32} & b_{33} \end{bmatrix}$$

Multiply and add as follows to obtain the first entry of the product matrix AB.

1. To obtain the entry in row 1, column 1 of AB, multiply the first row in A by the first column in B, and add.

$$[a_{11} \quad a_{12} \quad a_{13}]\begin{bmatrix} b_{11} \\ b_{21} \\ b_{31} \end{bmatrix} = a_{11} \cdot b_{11} + a_{12} \cdot b_{21} + a_{13} \cdot b_{31}$$

2. To obtain the entry in row 1, column 2 of AB, multiply the first row of A by the second column in B, and add.

$$[a_{11} \quad a_{12} \quad a_{13}]\begin{bmatrix} b_{12} \\ b_{22} \\ b_{32} \end{bmatrix} = a_{11} \cdot b_{12} + a_{12} \cdot b_{22} + a_{13} \cdot b_{32}$$

3. To obtain the entry in row 1, column 3 of AB, multiply the first row of A by the third column in B, and add.

$$[a_{11} \quad a_{12} \quad a_{13}]\begin{bmatrix} b_{13} \\ b_{23} \\ b_{33} \end{bmatrix} = a_{11} \cdot b_{13} + a_{12} \cdot b_{23} + a_{13} \cdot b_{33}$$

We proceed the same way to obtain the second row of AB. In other words, row 2 of A times column 1 of B; row 2 of A times column 2 of B; row 2 of A times column 3 of B. When complete, the product matrix will be

$$AB = \begin{bmatrix} a_{11} \cdot b_{11} + a_{12} \cdot b_{21} + a_{13} \cdot b_{31} & a_{11} \cdot b_{12} + a_{12} \cdot b_{22} + a_{13} \cdot b_{32} & a_{11} \cdot b_{13} + a_{12} \cdot b_{23} + a_{13} \cdot b_{33} \\ a_{21} \cdot b_{11} + a_{22} \cdot b_{21} + a_{23} \cdot b_{31} & a_{21} \cdot b_{12} + a_{22} \cdot b_{22} + a_{23} \cdot b_{32} & a_{21} \cdot b_{13} + a_{22} \cdot b_{23} + a_{23} \cdot b_{33} \end{bmatrix}$$

---

properties of matrix multiplication

For the matrices A, B, and C the following properties hold.

- Matrix multiplication is associative: $(AB)C = A(BC)$.

$$C(A + B) = CA + CB,$$

- Matrix multiplication is distributive:

$$(A + B)C = AC + BC.$$

Note that matrix multiplication is not commutative.

Example 8   **Multiplying Two Matrices**

Multiply matrix A and matrix B.

$$A = \begin{bmatrix} 1 & 2 \\ 3 & 4 \end{bmatrix} \quad \text{and } B = \begin{bmatrix} 5 & 6 \\ 7 & 8 \end{bmatrix}$$

Solution   First, we check the dimensions of the matrices. Matrix A has dimensions 2×2 and matrix B has dimensions 2×2. The inner dimensions are the same so we can perform the multiplication. The product will have the dimensions 2×2.

We perform the operations outlined previously.

$$AB = \begin{bmatrix} 1 & 2 \\ 3 & 4 \end{bmatrix}\begin{bmatrix} 5 & 6 \\ 7 & 8 \end{bmatrix}$$

$$= \begin{bmatrix} 1(5) + 2(7) & 1(6) + 2(8) \\ 3(5) + 4(7) & 3(6) + 4(8) \end{bmatrix}$$

$$= \begin{bmatrix} 19 & 22 \\ 43 & 50 \end{bmatrix}$$

Example 9   **Multiplying Two Matrices**

Given A and B:

a. Find AB.   **b.** Find BA.

$$A = \begin{bmatrix} -1 & 2 & 3 \\ 4 & 0 & 5 \end{bmatrix} \quad \text{and } B = \begin{bmatrix} 5 & -1 \\ -4 & 0 \\ 2 & 3 \end{bmatrix}$$

Solution

a. As the dimensions of A are 2×3 and the dimensions of B are 3×2, these matrices can be multiplied together because the number of columns in A matches the number of rows in B. The resulting product will be a 2×2 matrix, the number of rows in A by the number of columns in B.

$$AB = \begin{bmatrix} -1 & 2 & 3 \\ 4 & 0 & 5 \end{bmatrix} \begin{bmatrix} 5 & -1 \\ -4 & 0 \\ 2 & 3 \end{bmatrix}$$

$$= \begin{bmatrix} -1(5) + 2(-4) + 3(2) & -1(-1) + 2(0) + 3(3) \\ 4(5) + 0(-4) + 5(2) & 4(-1) + 0(0) + 5(3) \end{bmatrix}$$

$$= \begin{bmatrix} -7 & 10 \\ 30 & 11 \end{bmatrix}$$

b. The dimensions of B are 3×2 and the dimensions of A are 2×3. The inner dimensions match so the product is defined and will be a 3×3 matrix.

$$BA = \begin{bmatrix} 5 & -1 \\ -4 & 0 \\ 2 & 3 \end{bmatrix} \begin{bmatrix} -1 & 2 & 3 \\ 4 & 0 & 5 \end{bmatrix}$$

$$= \begin{bmatrix} 5(-1) + -1(4) & 5(2) + -1(0) & 5(3) + -1(5) \\ -4(-1) + 0(4) & -4(2) + 0(0) & -4(3) + 0(5) \\ 2(-1) + 3(4) & 2(2) + 3(0) & 2(3) + 3(5) \end{bmatrix}$$

$$= \begin{bmatrix} -9 & 10 & 10 \\ 4 & -8 & -12 \\ 10 & 4 & 21 \end{bmatrix}$$

Analysis Notice that the products AB and BA are not equal.

$$AB = \begin{bmatrix} -7 & 10 \\ 30 & 11 \end{bmatrix} \neq \begin{bmatrix} -9 & 10 & 10 \\ 4 & -8 & -12 \\ 10 & 4 & 21 \end{bmatrix} = BA$$

This illustrates the fact that matrix multiplication is not commutative.

Q & A...

Is it possible for AB to be defined but not BA?

Yes, consider a matrix A with dimension 3×4 and matrix B with dimension 4×2. For the product AB the inner dimensions are 4 and the product is defined, but for the product BA the inner dimensions are 2 and 3 so the product is undefined.

Example 10   **Using Matrices in Real-World Problems**

Let's return to the problem presented at the opening of this section. We have **Table 3**, representing the equipment needs of two soccer teams.

	Wildcats	Mud Cats
Goals	6	10
Balls	30	24
Jerseys	14	20

Table 3

We are also given the prices of the equipment, as shown in **Table 4.**

Goals	$300
Balls	$10
Jerseys	$30

Table 4

We will convert the data to matrices. Thus, the equipment need matrix is written as

$$E = \begin{bmatrix} 6 & 10 \\ 30 & 24 \\ 14 & 20 \end{bmatrix}$$

The cost matrix is written as

$$C = \begin{bmatrix} 300 & 10 & 30 \end{bmatrix}$$

We perform matrix multiplication to obtain costs for the equipment.

$$CE = \begin{bmatrix} 300 & 10 & 30 \end{bmatrix} \begin{bmatrix} 6 & 10 \\ 30 & 24 \\ 14 & 20 \end{bmatrix}$$

$$= [300(6) + 10(30) + 30(14) \quad 300(10) + 10(24) + 30(20)]$$

$$= [2{,}520 \ 3{,}840]$$

The total cost for equipment for the Wildcats is $2,520, and the total cost for equipment for the Mud Cats is $3,840.

How To...

Given a matrix operation, evaluate using a calculator.

1. Save each matrix as a matrix variable [A], [B], [C], ...
2. Enter the operation into the calculator, calling up each matrix variable as needed.
3. If the operation is defined, the calculator will present the solution matrix; if the operation is undefined, it will display an error message.

Example 11 **Using a Calculator to Perform Matrix Operations**

Find $AB - C$ given

$$A = \begin{bmatrix} -15 & 25 & 32 \\ 41 & -7 & -28 \\ 10 & 34 & -2 \end{bmatrix}, B = \begin{bmatrix} 45 & 21 & -37 \\ -24 & 52 & 19 \\ 6 & -48 & -31 \end{bmatrix}, \text{ and } C = \begin{bmatrix} -100 & -89 & -98 \\ 25 & -56 & 74 \\ -67 & 42 & -75 \end{bmatrix}.$$

Solution On the matrix page of the calculator, we enter matrix A above as the matrix variable [A], matrix B above as the matrix variable [B], and matrix C above as the matrix variable [C].

On the home screen of the calculator, we type in the problem and call up each matrix variable as needed.

$$[A] \times [B] - [C]$$

The calculator gives us the following matrix.

$$\begin{bmatrix} -983 & -462 & 136 \\ 1{,}820 & 1{,}897 & -856 \\ -311 & 2{,}032 & 413 \end{bmatrix}$$

Access these online resources for additional instruction and practice with matrices and matrix operations.

- Dimensions of a Matrix (http://openstaxcollege.org/l/matrixdimen)
- Matrix Addition and Subtraction (http://openstaxcollege.org/l/matrixaddsub)
- Matrix Operations (http://openstaxcollege.org/l/matrixoper)
- Matrix Multiplication (http://openstaxcollege.org/l/matrixmult)

7.5   SECTION EXERCISES

VERBAL

1. Can we add any two matrices together? If so, explain why; if not, explain why not and give an example of two matrices that cannot be added together.

2. Can we multiply any column matrix by any row matrix? Explain why or why not.

3. Can both the products AB and BA be defined? If so, explain how; if not, explain why.

4. Can any two matrices of the same size be multiplied? If so, explain why, and if not, explain why not and give an example of two matrices of the same size that cannot be multiplied together.

5. Does matrix multiplication commute? That is, does $AB = BA$? If so, prove why it does. If not, explain why it does not.

ALGEBRAIC

For the following exercises, use the matrices below and perform the matrix addition or subtraction. Indicate if the operation is undefined.

$$A = \begin{bmatrix} 1 & 3 \\ 0 & 7 \end{bmatrix}, B = \begin{bmatrix} 2 & 14 \\ 22 & 6 \end{bmatrix}, C = \begin{bmatrix} 1 & 5 \\ 8 & 92 \\ 12 & 6 \end{bmatrix}, D = \begin{bmatrix} 10 & 14 \\ 7 & 2 \\ 5 & 61 \end{bmatrix}, E = \begin{bmatrix} 6 & 12 \\ 14 & 5 \end{bmatrix}, F = \begin{bmatrix} 0 & 9 \\ 78 & 17 \\ 15 & 4 \end{bmatrix}$$

6. $A + B$ 7. $C + D$ 8. $A + C$ 9. $B - E$ 10. $C + F$ 11. $D - B$

For the following exercises, use the matrices below to perform scalar multiplication.

$$A = \begin{bmatrix} 4 & 6 \\ 13 & 12 \end{bmatrix}, B = \begin{bmatrix} 3 & 9 \\ 21 & 12 \\ 0 & 64 \end{bmatrix}, C = \begin{bmatrix} 16 & 3 & 7 & 18 \\ 90 & 5 & 3 & 29 \end{bmatrix}, D = \begin{bmatrix} 18 & 12 & 13 \\ 8 & 14 & 6 \\ 7 & 4 & 21 \end{bmatrix}$$

12. $5A$ 13. $3B$ 14. $-2B$ 15. $-4C$ 16. $\frac{1}{2}C$ 17. $100D$

For the following exercises, use the matrices below to perform matrix multiplication.

$$A = \begin{bmatrix} -1 & 5 \\ 3 & 2 \end{bmatrix}, B = \begin{bmatrix} 3 & 6 & 4 \\ -8 & 0 & 12 \end{bmatrix}, C = \begin{bmatrix} 4 & 10 \\ -2 & 6 \\ 5 & 9 \end{bmatrix}, D = \begin{bmatrix} 2 & -3 & 12 \\ 9 & 3 & 1 \\ 0 & 8 & -10 \end{bmatrix}$$

18. AB 19. BC 20. CA 21. BD 22. DC 23. CB

For the following exercises, use the matrices below to perform the indicated operation if possible. If not possible, explain why the operation cannot be performed.

$$A = \begin{bmatrix} 2 & -5 \\ 6 & 7 \end{bmatrix}, B = \begin{bmatrix} -9 & 6 \\ -4 & 2 \end{bmatrix}, C = \begin{bmatrix} 0 & 9 \\ 7 & 1 \end{bmatrix}, D = \begin{bmatrix} -8 & 7 & -5 \\ 4 & 3 & 2 \\ 0 & 9 & 2 \end{bmatrix}, E = \begin{bmatrix} 4 & 5 & 3 \\ 7 & -6 & -5 \\ 1 & 0 & 9 \end{bmatrix}$$

24. $A + B - C$ 25. $4A + 5D$ 26. $2C + B$ 27. $3D + 4E$ 28. $C - 0.5D$ 29. $100D - 10E$

For the following exercises, use the matrices below to perform the indicated operation if possible. If not possible, explain why the operation cannot be performed. (Hint: $A^2 = A \cdot A$)

$$A = \begin{bmatrix} -10 & 20 \\ 5 & 25 \end{bmatrix}, B = \begin{bmatrix} 40 & 10 \\ -20 & 30 \end{bmatrix}, C = \begin{bmatrix} -1 & 0 \\ 0 & -1 \\ 1 & 0 \end{bmatrix}$$

30. AB **31.** BA **32.** CA **33.** BC **34.** A^2 **35.** B^2

36. C^2 **37.** B^2A^2 **38.** A^2B^2 **39.** $(AB)^2$ **40.** $(BA)^2$

For the following exercises, use the matrices below to perform the indicated operation if possible. If not possible, explain why the operation cannot be performed. (Hint: $A^2 = A \cdot A$)

$$A = \begin{bmatrix} 1 & 0 \\ 2 & 3 \end{bmatrix}, B = \begin{bmatrix} -2 & 3 & 4 \\ -1 & 1 & -5 \end{bmatrix}, C = \begin{bmatrix} 0.5 & 0.1 \\ 1 & 0.2 \\ -0.5 & 0.3 \end{bmatrix}, D = \begin{bmatrix} 1 & 0 & -1 \\ -6 & 7 & 5 \\ 4 & 2 & 1 \end{bmatrix}$$

41. AB **42.** BA **43.** BD **44.** DC **45.** D^2 **46.** A^2

47. D^3 **48.** $(AB)C$ **49.** $A(BC)$

TECHNOLOGY

For the following exercises, use the matrices below to perform the indicated operation if possible. If not possible, explain why the operation cannot be performed. Use a calculator to verify your solution.

$$A = \begin{bmatrix} -2 & 0 & 9 \\ 1 & 8 & -3 \\ 0.5 & 4 & 5 \end{bmatrix}, B = \begin{bmatrix} 0.5 & 3 & 0 \\ -4 & 1 & 6 \\ 8 & 7 & 2 \end{bmatrix}, C = \begin{bmatrix} 1 & 0 & 1 \\ 0 & 1 & 0 \\ 1 & 0 & 1 \end{bmatrix}$$

50. AB **51.** BA **52.** CA **53.** BC **54.** ABC

EXTENSIONS

For the following exercises, use the matrix below to perform the indicated operation on the given matrix.

$$B = \begin{bmatrix} 1 & 0 & 0 \\ 0 & 0 & 1 \\ 0 & 1 & 0 \end{bmatrix}$$

55. B^2 **56.** B^3 **57.** B^4 **58.** B^5

59. Using the above questions, find a formula for B^n. Test the formula for B^{201} and B^{202}, using a calculator.

LEARNING OBJECTIVES

In this section, you will:

- Write the augmented matrix of a system of equations.
- Write the system of equations from an augmented matrix.
- Perform row operations on a matrix.
- Solve a system of linear equations using matrices.

7.6   SOLVING SYSTEMS WITH GAUSSIAN ELIMINATION

Figure 1 German mathematician Carl Friedrich Gauss (1777–1855).

Carl Friedrich Gauss lived during the late 18th century and early 19th century, but he is still considered one of the most prolific mathematicians in history. His contributions to the science of mathematics and physics span fields such as algebra, number theory, analysis, differential geometry, astronomy, and optics, among others. His discoveries regarding matrix theory changed the way mathematicians have worked for the last two centuries.

We first encountered Gaussian elimination in **Systems of Linear Equations: Two Variables**. In this section, we will revisit this technique for solving systems, this time using matrices.

Writing the Augmented Matrix of a System of Equations

A matrix can serve as a device for representing and solving a system of equations. To express a system in matrix form, we extract the coefficients of the variables and the constants, and these become the entries of the matrix. We use a vertical line to separate the coefficient entries from the constants, essentially replacing the equal signs. When a system is written in this form, we call it an **augmented matrix**.

For example, consider the following 2×2 system of equations.

$$3x + 4y = 7$$
$$4x - 2y = 5$$

We can write this system as an augmented matrix:

$$\left[\begin{array}{cc|c} 3 & 4 & 7 \\ 4 & -2 & 5 \end{array}\right]$$

We can also write a matrix containing just the coefficients. This is called the **coefficient matrix**.

$$\left[\begin{array}{cc} 3 & 4 \\ 4 & -2 \end{array}\right]$$

A three-by-three system of equations such as

$$3x - y - z = 0$$
$$x + y = 5$$
$$2x - 3z = 2$$

has a coefficient matrix

$$\begin{bmatrix} 3 & -1 & -1 \\ 1 & 1 & 0 \\ 2 & 0 & -3 \end{bmatrix}$$

and is represented by the augmented matrix

$$\left[\begin{array}{ccc|c} 3 & -1 & -1 & 0 \\ 1 & 1 & 0 & 5 \\ 2 & 0 & -3 & 2 \end{array}\right]$$

Notice that the matrix is written so that the variables line up in their own columns: x-terms go in the first column, y-terms in the second column, and z-terms in the third column. It is very important that each equation is written in standard form $ax + by + cz = d$ so that the variables line up. When there is a missing variable term in an equation, the coefficient is 0.

How To...

Given a system of equations, write an augmented matrix.

1. Write the coefficients of the x-terms as the numbers down the first column.
2. Write the coefficients of the y-terms as the numbers down the second column.
3. If there are z-terms, write the coefficients as the numbers down the third column.
4. Draw a vertical line and write the constants to the right of the line.

Example 1   **Writing the Augmented Matrix for a System of Equations**

Write the augmented matrix for the given system of equations.

$$x + 2y - z = 3$$
$$2x - y + 2z = 6$$
$$x - 3y + 3z = 4$$

Solution   The augmented matrix displays the coefficients of the variables, and an additional column for the constants.

$$\left[\begin{array}{ccc|c} 1 & 2 & -1 & 3 \\ 2 & -1 & 2 & 6 \\ 1 & -3 & 3 & 4 \end{array}\right]$$

Try It #1

Write the augmented matrix of the given system of equations.

$$4x - 3y = 11$$
$$3x + 2y = 4$$

Writing a System of Equations from an Augmented Matrix

We can use augmented matrices to help us solve systems of equations because they simplify operations when the systems are not encumbered by the variables. However, it is important to understand how to move back and forth between formats in order to make finding solutions smoother and more intuitive. Here, we will use the information in an augmented matrix to write the system of equations in standard form.

Example 2   **Writing a System of Equations from an Augmented Matrix Form**

Find the system of equations from the augmented matrix.

$$\left[\begin{array}{rrr|r} 1 & -3 & -5 & -2 \\ 2 & -5 & -4 & 5 \\ -3 & 5 & 4 & 6 \end{array}\right]$$

Solution   When the columns represent the variables x, y, and z,

$$\left[\begin{array}{rrr|r} 1 & -3 & -5 & -2 \\ 2 & -5 & -4 & 5 \\ -3 & 5 & 4 & 6 \end{array}\right] \rightarrow \begin{array}{r} x - 3y - 5z = -2 \\ 2x - 5y - 4z = 5 \\ -3x + 5y + 4z = 6 \end{array}$$

Try It #2

Write the system of equations from the augmented matrix.

$$\left[\begin{array}{rrr|r} 1 & -1 & 1 & 5 \\ 2 & -1 & 3 & 1 \\ 0 & 1 & 1 & -9 \end{array}\right]$$

Performing Row Operations on a Matrix

Now that we can write systems of equations in augmented matrix form, we will examine the various **row operations** that can be performed on a matrix, such as addition, multiplication by a constant, and interchanging rows.

Performing row operations on a matrix is the method we use for solving a system of equations. In order to solve the system of equations, we want to convert the matrix to **row-echelon form**, in which there are ones down the **main diagonal** from the upper left corner to the lower right corner, and zeros in every position below the main diagonal as shown.

Row-echelon form

$$\begin{bmatrix} 1 & a & b \\ 0 & 1 & d \\ 0 & 0 & 1 \end{bmatrix}$$

We use row operations corresponding to equation operations to obtain a new matrix that is **row-equivalent** in a simpler form. Here are the guidelines to obtaining row-echelon form.

1. In any nonzero row, the first nonzero number is a 1. It is called a *leading* 1.

2. Any all-zero rows are placed at the bottom on the matrix.

3. Any leading 1 is below and to the right of a previous leading 1.

4. Any column containing a leading 1 has zeros in all other positions in the column.

To solve a system of equations we can perform the following row operations to convert the coefficient matrix to row-echelon form and do back-substitution to find the solution.

1. Interchange rows. (Notation: $R_i \leftrightarrow R_j$)

2. Multiply a row by a constant. (Notation: cR_i)

3. Add the product of a row multiplied by a constant to another row. (Notation: $R_i + cR_j$)

Each of the row operations corresponds to the operations we have already learned to solve systems of equations in three variables. With these operations, there are some key moves that will quickly achieve the goal of writing a matrix in row-echelon form. To obtain a matrix in row-echelon form for finding solutions, we use Gaussian elimination, a method that uses row operations to obtain a 1 as the first entry so that row 1 can be used to convert the remaining rows.

Now we substitute the expression for z into the second equation to solve for y in terms of x.

$$y - 2z = 0$$
$$z = \frac{1-x}{3}$$
$$y - 2\left(\frac{1-x}{3}\right) = 0$$
$$y = \frac{2-2x}{3}$$

The generic solution is $\left(x, \dfrac{2-2x}{3}, \dfrac{1-x}{3}\right)$.

Try It #5

Solve the system using matrices.

$$x + 4y - z = 4$$
$$2x + 5y + 8z = 15$$
$$x + 3y - 3z = 1$$

Q & A...

Can any system of linear equations be solved by Gaussian elimination?

Yes, a system of linear equations of any size can be solved by Gaussian elimination.

How To...

Given a system of equations, solve with matrices using a calculator.

1. Save the augmented matrix as a matrix variable $[A]$, $[B]$, $[C]$, ...
2. Use the **ref(** function in the calculator, calling up each matrix variable as needed.

Example 9   **Solving Systems of Equations with Matrices Using a Calculator**

Solve the system of equations.

$$5x + 3y + 9z = -1$$
$$-2x + 3y - z = -2$$
$$-x - 4y + 5z = 1$$

Solution   Write the augmented matrix for the system of equations.

$$\begin{bmatrix} 5 & 3 & 9 & | & -1 \\ -2 & 3 & -1 & | & -2 \\ -1 & -4 & 5 & | & -1 \end{bmatrix}$$

On the matrix page of the calculator, enter the augmented matrix above as the matrix variable $[A]$.

$$[A] = \begin{bmatrix} 5 & 3 & 9 & -1 \\ -2 & 3 & -1 & -2 \\ -1 & -4 & 5 & 1 \end{bmatrix}$$

Use the **ref(** function in the calculator, calling up the matrix variable $[A]$.

$$\text{ref}([A])$$

Evaluate.

$$\begin{bmatrix} 1 & \frac{3}{5} & \frac{9}{5} & \frac{1}{5} \\ 0 & 1 & \frac{13}{21} & -\frac{4}{7} \\ 0 & 0 & 1 & -\frac{24}{187} \end{bmatrix} \rightarrow \begin{aligned} x + \tfrac{3}{5}y + \tfrac{9}{5}z &= -\tfrac{1}{5} \\ y + \tfrac{13}{21}z &= -\tfrac{4}{7} \\ z &= -\tfrac{24}{187} \end{aligned}$$

Using back-substitution, the solution is $\left(\dfrac{61}{187}, -\dfrac{92}{187}, -\dfrac{24}{187}\right)$.

Example 10   **Applying 2 × 2 Matrices to Finance**

Carolyn invests a total of $12,000 in two municipal bonds, one paying 10.5% interest and the other paying 12% interest. The annual interest earned on the two investments last year was $1,335. How much was invested at each rate?

Solution   We have a system of two equations in two variables. Let $x =$ the amount invested at 10.5% interest, and $y =$ the amount invested at 12% interest.

$$x + y = 12,000$$
$$0.105x + 0.12y = 1,335$$

As a matrix, we have

$$\begin{bmatrix} 1 & 1 & | & 12,000 \\ 0.105 & 0.12 & | & 1,335 \end{bmatrix}$$

Multiply row 1 by -0.105 and add the result to row 2.

$$\begin{bmatrix} 1 & 1 & | & 12,000 \\ 0 & 0.015 & | & 75 \end{bmatrix}$$

Then,

$$0.015y = 75$$
$$y = 5,000$$

So $12,000 - 5,000 = 7,000$.

Thus, $5,000 was invested at 12% interest and $7,000 at 10.5% interest.

Example 11   **Applying 3 × 3 Matrices to Finance**

Ava invests a total of $10,000 in three accounts, one paying 5% interest, another paying 8% interest, and the third paying 9% interest. The annual interest earned on the three investments last year was $770. The amount invested at 9% was twice the amount invested at 5%. How much was invested at each rate?

Solution   We have a system of three equations in three variables. Let x be the amount invested at 5% interest, let y be the amount invested at 8% interest, and let z be the amount invested at 9% interest. Thus,

$$x + y + z = 10,000$$
$$0.05x + 0.08y + 0.09z = 770$$
$$2x - z = 0$$

As a matrix, we have

$$\begin{bmatrix} 1 & 1 & 1 & | & 10,000 \\ 0.05 & 0.08 & 0.09 & | & 770 \\ 2 & 0 & -1 & | & 0 \end{bmatrix}$$

Now, we perform Gaussian elimination to achieve row-echelon form.

$$-0.05R_1 + R_2 = R_2 \rightarrow \begin{bmatrix} 1 & 1 & 1 & | & 10,000 \\ 0 & 0.03 & 0.04 & | & 270 \\ 2 & 0 & -1 & | & 0 \end{bmatrix}$$

$$-2R_1 + R_3 = R_3 \rightarrow \begin{bmatrix} 1 & 1 & 1 & | & 10,000 \\ 0 & 0.03 & 0.04 & | & 270 \\ 0 & -2 & -3 & | & -20,000 \end{bmatrix}$$

$$\frac{1}{0.03}R_2 = R_2 \rightarrow \begin{bmatrix} 0 & 1 & 1 & | & 10,000 \\ 0 & 1 & \frac{4}{3} & | & 9,000 \\ 0 & -2 & -3 & | & -20,000 \end{bmatrix}$$

$$2R_2 + R_3 = R_3 \rightarrow \begin{bmatrix} 1 & 1 & 1 & | & 10,000 \\ 0 & 1 & \frac{4}{3} & | & 9,000 \\ 0 & 0 & -\frac{1}{3} & | & -2,000 \end{bmatrix}$$

The third row tells us $-\frac{1}{3}z = -2,000$; thus $z = 6,000$.

The second row tells us $y + \frac{4}{3}z = 9{,}000$. Substituting $z = 6{,}000$, we get

$$y + \frac{4}{3}(6{,}000) = 9{,}000$$
$$y + 8{,}000 = 9{,}000$$
$$y = 1{,}000$$

The first row tells us $x + y + z = 10{,}000$. Substituting $y = 1{,}000$ and $z = 6{,}000$, we get

$$x + 1{,}000 + 6{,}000 = 10{,}000$$
$$x = 3{,}000$$

The answer is $3,000 invested at 5% interest, $1,000 invested at 8%, and $6,000 invested at 9% interest.

Try It #6

A small shoe company took out a loan of $1,500,000 to expand their inventory. Part of the money was borrowed at 7%, part was borrowed at 8%, and part was borrowed at 10%. The amount borrowed at 10% was four times the amount borrowed at 7%, and the annual interest on all three loans was $130,500. Use matrices to find the amount borrowed at each rate.

Access these online resources for additional instruction and practice with solving systems of linear equations using Gaussian elimination.

- Solve a System of Two Equations Using an Augmented Matrix (http://openstaxcollege.org/l/system2augmat)
- Solve a System of Three Equations Using an Augmented Matrix (http://openstaxcollege.org/l/system3augmat)
- Augmented Matrices on the Calculator (http://openstaxcollege.org/l/augmatcalc)

7.6   SECTION EXERCISES

VERBAL

1. Can any system of linear equations be written as an augmented matrix? Explain why or why not. Explain how to write that augmented matrix.

2. Can any matrix be written as a system of linear equations? Explain why or why not. Explain how to write that system of equations.

3. Is there only one correct method of using row operations on a matrix? Try to explain two different row operations possible to solve the augmented matrix
$$\begin{bmatrix} 9 & 3 & | & 0 \\ 1 & -2 & | & 6 \end{bmatrix}.$$

4. Can a matrix whose entry is 0 on the diagonal be solved? Explain why or why not. What would you do to remedy the situation?

5. Can a matrix that has 0 entries for an entire row have one solution? Explain why or why not.

ALGEBRAIC

For the following exercises, write the augmented matrix for the linear system.

6. $8x - 37y = 8$
$2x + 12y = 3$

7. $16y = 4$
$9x - y = 2$

8. $3x + 2y + 10z = 3$
$-6x + 2y + 5z = 13$
$4x + z = 18$

9. $x + 5y + 8z = 19$
$12x + 3y = 4$
$3x + 4y + 9z = -7$

10. $6x + 12y + 16z = 4$
$19x - 5y + 3z = -9$
$x + 2y = -8$

For the following exercises, write the linear system from the augmented matrix.

11. $\begin{bmatrix} -2 & 5 & | & 5 \\ 6 & -18 & | & 26 \end{bmatrix}$

12. $\begin{bmatrix} 3 & 4 & | & 10 \\ 10 & 17 & | & 439 \end{bmatrix}$

13. $\begin{bmatrix} 3 & 2 & 0 & | & 3 \\ -1 & -9 & 4 & | & -1 \\ 8 & 5 & 7 & | & 8 \end{bmatrix}$

14. $\begin{bmatrix} 8 & 29 & 1 & | & 43 \\ -1 & 7 & 5 & | & 38 \\ 0 & 0 & 3 & | & 10 \end{bmatrix}$

15. $\begin{bmatrix} 4 & 5 & -2 & | & 12 \\ 0 & 1 & 58 & | & 2 \\ 8 & 7 & -3 & | & -5 \end{bmatrix}$

For the following exercises, solve the system by Gaussian elimination.

16. $\begin{bmatrix} 1 & 0 & | & 3 \\ 0 & 0 & | & 0 \end{bmatrix}$

17. $\begin{bmatrix} 1 & 0 & | & 1 \\ 1 & 0 & | & 2 \end{bmatrix}$

18. $\begin{bmatrix} 1 & 2 & | & 3 \\ 4 & 5 & | & 6 \end{bmatrix}$

19. $\begin{bmatrix} -1 & 2 & | & -3 \\ 4 & -5 & | & 6 \end{bmatrix}$

20. $\begin{bmatrix} -2 & 0 & | & 1 \\ 0 & 2 & | & -1 \end{bmatrix}$

21. $2x - 3y = -9$
$5x + 4y = 58$

22. $6x + 2y = -4$
$3x + 4y = -17$

23. $2x + 3y = 12$
$4x + y = 14$

24. $-4x - 3y = -2$
$3x - 5y = -13$

25. $-5x + 8y = 3$
$10x + 6y = 5$

26. $3x + 4y = 12$
$-6x - 8y = -24$

27. $-60x + 45y = 12$
$20x - 15y = -4$

28. $11x + 10y = 43$
$15x + 20y = 65$

29. $2x - y = 2$
$3x + 2y = 17$

30. $-1.06x - 2.25y = 5.51$
$-5.03x - 1.08y = 5.40$

31. $\dfrac{3}{4}x - \dfrac{3}{5}y = 4$
$\dfrac{1}{4}x + \dfrac{2}{3}y = 1$

32. $\dfrac{1}{4}x - \dfrac{2}{3}y = -1$
$\dfrac{1}{2}x + \dfrac{1}{3}y = 3$

33. $\begin{bmatrix} 1 & 0 & 0 & | & 31 \\ 0 & 1 & 1 & | & 45 \\ 0 & 0 & 1 & | & 87 \end{bmatrix}$

34. $\begin{bmatrix} 1 & 0 & 1 & | & 50 \\ 1 & 1 & 0 & | & 20 \\ 0 & 1 & 1 & | & -90 \end{bmatrix}$

35. $\begin{bmatrix} 1 & 2 & 3 & | & 4 \\ 0 & 5 & 6 & | & 7 \\ 0 & 0 & 8 & | & 9 \end{bmatrix}$

Example 3   **Finding the Multiplicative Inverse Using Matrix Multiplication**

Use matrix multiplication to find the inverse of the given matrix.

$$A = \begin{bmatrix} 1 & -2 \\ 2 & -3 \end{bmatrix}$$

Solution   For this method, we multiply A by a matrix containing unknown constants and set it equal to the identity.

$$\begin{bmatrix} 1 & -2 \\ 2 & -3 \end{bmatrix} \begin{bmatrix} a & b \\ c & d \end{bmatrix} = \begin{bmatrix} 1 & 0 \\ 0 & 1 \end{bmatrix}$$

Find the product of the two matrices on the left side of the equal sign.

$$\begin{bmatrix} 1 & -2 \\ 2 & -3 \end{bmatrix} \begin{bmatrix} a & b \\ c & d \end{bmatrix} = \begin{bmatrix} 1a - 2c & 1b - 2d \\ 2a - 3c & 2b - 3d \end{bmatrix}$$

Next, set up a system of equations with the entry in row 1, column 1 of the new matrix equal to the first entry of the identity, 1. Set the entry in row 2, column 1 of the new matrix equal to the corresponding entry of the identity, which is 0.

$$1a - 2c = 1 \qquad R_1$$
$$2a - 3c = 0 \qquad R_2$$

Using row operations, multiply and add as follows: $(-2)R_1 + R_2 \rightarrow R_2$. Add the equations, and solve for c.

$$1a - 2c = 1$$
$$0 + 1c = -2$$
$$c = -2$$

Back-substitute to solve for a.

$$a - 2(-2) = 1$$
$$a + 4 = 1$$
$$a = -3$$

Write another system of equations setting the entry in row 1, column 2 of the new matrix equal to the corresponding entry of the identity, 0. Set the entry in row 2, column 2 equal to the corresponding entry of the identity.

$$1b - 2d = 0 \qquad R_1$$
$$2b - 3d = 1 \qquad R_2$$

Using row operations, multiply and add as follows: $(-2)R_1 + R_2 = R_2$. Add the two equations and solve for d.

$$1b - 2d = 0$$
$$\underline{0 + 1d = 1}$$
$$d = 1$$

Once more, back-substitute and solve for b.

$$b - 2(1) = 0$$
$$b - 2 = 0$$
$$b = 2$$
$$A^{-1} = \begin{bmatrix} -3 & 2 \\ -2 & 1 \end{bmatrix}$$

Finding the Multiplicative Inverse by Augmenting with the Identity

Another way to find the multiplicative inverse is by augmenting with the identity. When matrix A is transformed into I, the augmented matrix I transforms into A^{-1}.

For example, given

$$A = \begin{bmatrix} 2 & 1 \\ 5 & 3 \end{bmatrix}$$

augment A with the identity

$$\begin{bmatrix} 2 & 1 & | & 1 & 0 \\ 5 & 3 & | & 0 & 1 \end{bmatrix}$$

Perform row operations with the goal of turning A into the identity.

1. Switch row 1 and row 2.

$$\begin{bmatrix} 5 & 3 & | & 0 & 1 \\ 2 & 1 & | & 1 & 0 \end{bmatrix}$$

2. Multiply row 2 by -2 and add to row 1.

$$\begin{bmatrix} 1 & 1 & | & -2 & 1 \\ 2 & 1 & | & 1 & 0 \end{bmatrix}$$

3. Multiply row 1 by -2 and add to row 2.

$$\begin{bmatrix} 1 & 1 & | & -2 & 1 \\ 0 & -1 & | & 5 & -2 \end{bmatrix}$$

4. Add row 2 to row 1.

$$\begin{bmatrix} 1 & 0 & | & 3 & -1 \\ 0 & -1 & | & 5 & -2 \end{bmatrix}$$

5. Multiply row 2 by -1.

$$\begin{bmatrix} 1 & 0 & | & 3 & -1 \\ 0 & 1 & | & -5 & 2 \end{bmatrix}$$

The matrix we have found is A^{-1}.

$$A^{-1} = \begin{bmatrix} 3 & -1 \\ -5 & 2 \end{bmatrix}$$

Finding the Multiplicative Inverse of 2 × 2 Matrices Using a Formula

When we need to find the multiplicative inverse of a 2×2 matrix, we can use a special formula instead of using matrix multiplication or augmenting with the identity.

If A is a 2×2 matrix, such as

$$A = \begin{bmatrix} a & b \\ c & d \end{bmatrix}$$

the multiplicative inverse of A is given by the formula

$$A^{-1} = \frac{1}{ad - bc}\begin{bmatrix} d & -b \\ -c & a \end{bmatrix}$$

where $ad - bc \neq 0$. If $ad - bc = 0$, then A has no inverse.

Example 4   Using the Formula to Find the Multiplicative Inverse of Matrix A

Use the formula to find the multiplicative inverse of

$$A = \begin{bmatrix} 1 & -2 \\ 2 & -3 \end{bmatrix}$$

Solution   Using the formula, we have

$$A^{-1} = \frac{1}{(1)(-3) - (-2)(2)}\begin{bmatrix} -3 & 2 \\ -2 & 1 \end{bmatrix}$$

$$= \frac{1}{-3 + 4}\begin{bmatrix} -3 & 2 \\ -2 & 1 \end{bmatrix}$$

$$= \begin{bmatrix} -3 & 2 \\ -2 & 1 \end{bmatrix}$$

Analysis   We can check that our formula works by using one of the other methods to calculate the inverse. Let's augment A with the identity.

$$\begin{bmatrix} 1 & -2 & | & 1 & 0 \\ 2 & -3 & | & 0 & 1 \end{bmatrix}$$

Perform row operations with the goal of turning A into the identity.

1. *Multiply row 1 by −2 and add to row 2.*

$$\left[\begin{array}{cc|cc} 1 & -2 & 1 & 0 \\ 0 & 1 & -2 & 1 \end{array}\right]$$

2. *Multiply row 1 by 2 and add to row 1.*

$$\left[\begin{array}{cc|cc} 1 & 0 & -3 & 2 \\ 0 & 1 & -2 & 1 \end{array}\right]$$

So, we have verified our original solution.

$$A^{-1} = \begin{bmatrix} -3 & 2 \\ -2 & 1 \end{bmatrix}$$

Try It #2

Use the formula to find the inverse of matrix A. Verify your answer by augmenting with the identity matrix.

$$A^{-1} = \begin{bmatrix} 1 & -1 \\ 2 & 3 \end{bmatrix}$$

Example 5   **Finding the Inverse of the Matrix, If It Exists**

Find the inverse, if it exists, of the given matrix.

$$A = \begin{bmatrix} 3 & 6 \\ 1 & 2 \end{bmatrix}$$

Solution   We will use the method of augmenting with the identity.

$$\left[\begin{array}{cc|cc} 3 & 6 & 1 & 0 \\ 1 & 3 & 0 & 1 \end{array}\right]$$

1. Switch row 1 and row 2.

$$\left[\begin{array}{cc|cc} 1 & 3 & 0 & 1 \\ 3 & 6 & 1 & 0 \end{array}\right]$$

2. Multiply row 1 by −3 and add it to row 2.

$$\left[\begin{array}{cc|cc} 1 & 2 & 1 & 0 \\ 0 & 0 & -3 & 1 \end{array}\right]$$

3. There is nothing further we can do. The zeros in row 2 indicate that this matrix has no inverse.

Finding the Multiplicative Inverse of 3 × 3 Matrices

Unfortunately, we do not have a formula similar to the one for a 2 × 2 matrix to find the inverse of a 3 × 3 matrix. Instead, we will augment the original matrix with the identity matrix and use row operations to obtain the inverse.

Given a 3 × 3 matrix

$$A = \begin{bmatrix} 2 & 3 & 1 \\ 3 & 3 & 1 \\ 2 & 4 & 1 \end{bmatrix}$$

augment A with the identity matrix

$$A \,|\, I = \left[\begin{array}{ccc|ccc} 2 & 3 & 1 & 1 & 0 & 0 \\ 3 & 3 & 1 & 0 & 1 & 0 \\ 2 & 4 & 1 & 0 & 0 & 1 \end{array}\right]$$

To begin, we write the augmented matrix with the identity on the right and A on the left. Performing elementary row operations so that the identity matrix appears on the left, we will obtain the inverse matrix on the right. We will find the inverse of this matrix in the next example.

How To...

Given a 3 × 3 matrix, find the inverse

1. Write the original matrix augmented with the identity matrix on the right.
2. Use elementary row operations so that the identity appears on the left.
3. What is obtained on the right is the inverse of the original matrix.
4. Use matrix multiplication to show that $AA^{-1} = I$ and $A^{-1} A = I$.

Example 6   **Finding the Inverse of a 3 × 3 Matrix**

Given the 3×3 matrix A, find the inverse.

$$A = \begin{bmatrix} 2 & 3 & 1 \\ 3 & 3 & 1 \\ 2 & 4 & 1 \end{bmatrix}$$

Solution   Augment A with the identity matrix, and then begin row operations until the identity matrix replaces A. The matrix on the right will be the inverse of A.

$$\left[\begin{array}{ccc|ccc} 2 & 3 & 1 & 1 & 0 & 0 \\ 3 & 3 & 1 & 0 & 1 & 0 \\ 2 & 4 & 1 & 0 & 0 & 1 \end{array}\right] \xrightarrow{\text{Interchange } R_2 \text{ and } R_1} \left[\begin{array}{ccc|ccc} 3 & 3 & 1 & 0 & 1 & 0 \\ 2 & 3 & 1 & 1 & 0 & 0 \\ 2 & 4 & 1 & 0 & 0 & 1 \end{array}\right]$$

$$-R_2 + R_1 = R_1 \rightarrow \left[\begin{array}{ccc|ccc} 1 & 0 & 0 & -1 & 1 & 0 \\ 2 & 3 & 1 & 1 & 0 & 0 \\ 2 & 4 & 1 & 0 & 0 & 1 \end{array}\right]$$

$$-R_2 + R_3 = R_3 \rightarrow \left[\begin{array}{ccc|ccc} 1 & 0 & 0 & -1 & 1 & 0 \\ 2 & 3 & 1 & 1 & 0 & 0 \\ 0 & 1 & 0 & -1 & 0 & 1 \end{array}\right]$$

$$R_3 \leftrightarrow R_2 \rightarrow \left[\begin{array}{ccc|ccc} 1 & 0 & 0 & -1 & 1 & 0 \\ 0 & 1 & 0 & -1 & 0 & 1 \\ 2 & 3 & 1 & 1 & 0 & 0 \end{array}\right]$$

$$-2R_1 + R_3 = R_3 \rightarrow \left[\begin{array}{ccc|ccc} 1 & 0 & 0 & -1 & 1 & 0 \\ 0 & 1 & 0 & -1 & 0 & 1 \\ 0 & 3 & 1 & 3 & -2 & 0 \end{array}\right]$$

$$-3R_2 + R_3 = R_3 \rightarrow \left[\begin{array}{ccc|ccc} 1 & 0 & 0 & -1 & 1 & 0 \\ 0 & 1 & 0 & -1 & 0 & 1 \\ 0 & 0 & 1 & 6 & -2 & -3 \end{array}\right]$$

Thus,

$$A^{-1} = B = \begin{bmatrix} -1 & 1 & 0 \\ -1 & 0 & 1 \\ 6 & -2 & -3 \end{bmatrix}$$

Analysis   To prove that $B = A^{-1}$, let's multiply the two matrices together to see if the product equals the identity, if $AA^{-1} = I$ and $A^{-1}A = I$.

$$AA^{-1} = \begin{bmatrix} 2 & 3 & 1 \\ 3 & 3 & 1 \\ 2 & 4 & 1 \end{bmatrix} \begin{bmatrix} -1 & 1 & 0 \\ -1 & 0 & 1 \\ 6 & -2 & -3 \end{bmatrix}$$

$$= \begin{bmatrix} 2(-1) + 3(-1) + 1(6) & 2(1) + 3(0) + 1(-2) & 2(0) + 3(1) + 1(-3) \\ 3(-1) + 3(-1) + 1(6) & 3(1) + 3(0) + 1(-2) & 3(0) + 3(1) + 1(-3) \\ 2(-1) + 4(-1) + 1(6) & 2(1) + 4(0) + 1(-2) & 2(0) + 4(1) + 1(-3) \end{bmatrix}$$

$$= \begin{bmatrix} 1 & 0 & 0 \\ 0 & 1 & 0 \\ 0 & 0 & 1 \end{bmatrix}$$

$$A^{-1}A = \begin{bmatrix} -1 & 1 & 0 \\ -1 & 0 & 1 \\ 6 & -2 & -3 \end{bmatrix} \begin{bmatrix} 2 & 3 & 1 \\ 3 & 3 & 1 \\ 2 & 4 & 1 \end{bmatrix}$$

$$= \begin{bmatrix} -1(2) + 1(3) + 0(2) & -1(3) + 1(3) + 0(4) & -1(1) + 1(1) + 0(1) \\ -1(2) + 0(3) + 1(2) & -1(3) + 0(3) + 1(4) & -1(1) + 0(1) + 1(1) \\ 6(2) + -2(3) + -3(2) & 6(3) + -2(3) + -3(4) & 6(1) + -2(1) + -3(1) \end{bmatrix}$$

$$= \begin{bmatrix} 1 & 0 & 0 \\ 0 & 1 & 0 \\ 0 & 0 & 1 \end{bmatrix}$$

Try It #3

Find the inverse of the 3×3 matrix.

$$A = \begin{bmatrix} 2 & -17 & 11 \\ -1 & 11 & -7 \\ 0 & 3 & -2 \end{bmatrix}$$

Solving a System of Linear Equations Using the Inverse of a Matrix

Solving a system of linear equations using the inverse of a matrix requires the definition of two new matrices: X is the matrix representing the variables of the system, and B is the matrix representing the constants. Using matrix multiplication, we may define a system of equations with the same number of equations as variables as

$$AX = B$$

To solve a system of linear equations using an inverse matrix, let A be the coefficient matrix, let X be the variable matrix, and let B be the constant matrix. Thus, we want to solve a system $AX = B$. For example, look at the following system of equations.

$$a_1 x + b_1 y = c_1$$
$$a_2 x + b_2 y = c_2$$

From this system, the coefficient matrix is

$$A = \begin{bmatrix} a_1 & b_1 \\ a_2 & b_2 \end{bmatrix}$$

The variable matrix is

$$X = \begin{bmatrix} x \\ y \end{bmatrix}$$

And the constant matrix is

$$B = \begin{bmatrix} c_1 \\ c_2 \end{bmatrix}$$

Then $AX = B$ looks like

$$\begin{bmatrix} a_1 & b_1 \\ a_2 & b_2 \end{bmatrix} \begin{bmatrix} x \\ y \end{bmatrix} = \begin{bmatrix} c_1 \\ c_2 \end{bmatrix}$$

Recall the discussion earlier in this section regarding multiplying a real number by its inverse, $(2^{-1})\, 2 = \left(\frac{1}{2}\right)2 = 1$. To solve a single linear equation $ax = b$ for x, we would simply multiply both sides of the equation by the multiplicative inverse (reciprocal) of a. Thus,

$$ax = b$$
$$\left(\frac{1}{a}\right)ax = \left(\frac{1}{a}\right)b$$
$$(a^{-1})ax = (a^{-1})b$$
$$[(a^{-1})a]x = (a^{-1})b$$
$$1x = (a^{-1})b$$
$$x = (a^{-1})b$$

The only difference between a solving a linear equation and a system of equations written in matrix form is that finding the inverse of a matrix is more complicated, and matrix multiplication is a longer process. However, the goal is the same—to isolate the variable.

We will investigate this idea in detail, but it is helpful to begin with a 2×2 system and then move on to a 3×3 system.

---

solving a system of equations using the inverse of a matrix

Given a system of equations, write the coefficient matrix A, the variable matrix X, and the constant matrix B. Then

$$AX = B$$

Multiply both sides by the inverse of A to obtain the solution.

$$(A^{-1})AX = (A^{-1})B$$

$$[(A^{-1})A]X = (A^{-1})B$$

$$IX = (A^{-1})B$$

$$X = (A^{-1})B$$

Q & A...

If the coefficient matrix does not have an inverse, does that mean the system has no solution?

No, if the coefficient matrix is not invertible, the system could be inconsistent and have no solution, or be dependent and have infinitely many solutions.

Example 7   **Solving a 2 × 2 System Using the Inverse of a Matrix**

Solve the given system of equations using the inverse of a matrix.

$$3x + 8y = 5$$

$$4x + 11y = 7$$

Solution   Write the system in terms of a coefficient matrix, a variable matrix, and a constant matrix.

$$A = \begin{bmatrix} 3 & 8 \\ 4 & 11 \end{bmatrix}, X = \begin{bmatrix} x \\ y \end{bmatrix}, B = \begin{bmatrix} 5 \\ 7 \end{bmatrix}$$

Then

$$\begin{bmatrix} 3 & 8 \\ 4 & 11 \end{bmatrix} \begin{bmatrix} x \\ y \end{bmatrix} = \begin{bmatrix} 5 \\ 7 \end{bmatrix}$$

First, we need to calculate A^{-1}. Using the formula to calculate the inverse of a 2 by 2 matrix, we have:

$$A^{-1} = \frac{1}{ad - bc}\begin{bmatrix} d & -b \\ -c & a \end{bmatrix}$$

$$= \frac{1}{3(11) - 8(4)}\begin{bmatrix} 11 & -8 \\ -4 & 3 \end{bmatrix}$$

$$= \frac{1}{1}\begin{bmatrix} 11 & -8 \\ -4 & 3 \end{bmatrix}$$

So,

$$A^{-1} = \begin{bmatrix} 11 & -8 \\ -4 & 3 \end{bmatrix}$$

Now we are ready to solve. Multiply both sides of the equation by A^{-1}.

$$(A^{-1})AX = (A^{-1})B$$

$$\begin{bmatrix} 11 & -8 \\ -4 & 3 \end{bmatrix}\begin{bmatrix} 3 & 8 \\ 4 & 11 \end{bmatrix}\begin{bmatrix} x \\ y \end{bmatrix} = \begin{bmatrix} 11 & -8 \\ -4 & 3 \end{bmatrix}\begin{bmatrix} 5 \\ 7 \end{bmatrix}$$

$$\begin{bmatrix} 1 & 0 \\ 0 & 1 \end{bmatrix}\begin{bmatrix} x \\ y \end{bmatrix} = \begin{bmatrix} 11(5) + (-8)7 \\ -4(5) + 3(7) \end{bmatrix}$$

$$\begin{bmatrix} x \\ y \end{bmatrix} = \begin{bmatrix} -1 \\ 1 \end{bmatrix}$$

The solution is $(-1, 1)$.

Q & A...

Can we solve for X by finding the product BA^{-1}?

No, recall that matrix multiplication is not commutative, so $A^{-1}B \neq BA^{-1}$. Consider our steps for solving the matrix equation.

$$(A^{-1})AX = (A^{-1})B$$
$$[(A^{-1})A]X = (A^{-1})B$$
$$IX = (A^{-1})B$$
$$X = (A^{-1})B$$

Notice in the first step we multiplied both sides of the equation by A^{-1}, but the A^{-1} was to the left of A on the left side and to the left of B on the right side. Because matrix multiplication is not commutative, order matters.

Example 8   **Solving a 3 × 3 System Using the Inverse of a Matrix**

Solve the following system using the inverse of a matrix.

$$5x + 15y + 56z = 35$$
$$-4x - 11y - 41z = -26$$
$$-x - 3y - 11z = -7$$

Solution   Write the equation $AX = B$.

$$\begin{bmatrix} 5 & 15 & 56 \\ -4 & -11 & -41 \\ -1 & -3 & -11 \end{bmatrix} \begin{bmatrix} x \\ y \\ z \end{bmatrix} = \begin{bmatrix} 35 \\ -26 \\ -7 \end{bmatrix}$$

First, we will find the inverse of A by augmenting with the identity.

$$\left[\begin{array}{ccc|ccc} 5 & 15 & 56 & 1 & 0 & 0 \\ -4 & -11 & -41 & 0 & 1 & 0 \\ -1 & -3 & -11 & 0 & 0 & 1 \end{array}\right]$$

Multiply row 1 by $\frac{1}{5}$.

$$\left[\begin{array}{ccc|ccc} 1 & 3 & \frac{56}{5} & \frac{1}{5} & 0 & 0 \\ -4 & -11 & -41 & 0 & 1 & 0 \\ -1 & -3 & -11 & 0 & 0 & 1 \end{array}\right]$$

Multiply row 1 by 4 and add to row 2.

$$\left[\begin{array}{ccc|ccc} 1 & 3 & \frac{56}{5} & \frac{1}{5} & 0 & 0 \\ 0 & 1 & \frac{19}{5} & \frac{4}{5} & 1 & 0 \\ -1 & -3 & -11 & 0 & 0 & 1 \end{array}\right]$$

Add row 1 to row 3.

$$\left[\begin{array}{ccc|ccc} 1 & 3 & \frac{56}{5} & \frac{1}{5} & 0 & 0 \\ 0 & 1 & \frac{19}{5} & \frac{4}{5} & 1 & 0 \\ 0 & 0 & \frac{1}{5} & \frac{1}{5} & 0 & 1 \end{array}\right]$$

Multiply row 2 by -3 and add to row 1.

$$\left[\begin{array}{ccc|ccc} 1 & 0 & -\frac{1}{5} & -\frac{11}{5} & -3 & 0 \\ 0 & 1 & \frac{19}{5} & \frac{4}{5} & 1 & 0 \\ 0 & 0 & \frac{1}{5} & \frac{1}{5} & 0 & 1 \end{array}\right]$$

Multiply row 3 by 5.

$$\left[\begin{array}{ccc|ccc} 1 & 0 & -\frac{1}{5} & -\frac{11}{5} & -3 & 0 \\ 0 & 1 & \frac{19}{5} & \frac{4}{5} & 1 & 0 \\ 0 & 0 & 1 & 1 & 0 & 5 \end{array}\right]$$

Multiply row 3 by $\frac{1}{5}$ and add to row 1.

$$\left[\begin{array}{ccc|ccc} 1 & 0 & 0 & -2 & -3 & 1 \\ 0 & 1 & \frac{19}{5} & \frac{4}{5} & 1 & 0 \\ 0 & 0 & 1 & 1 & 0 & 5 \end{array}\right]$$

Multiply row 3 by $-\frac{19}{5}$ and add to row 2.

$$\left[\begin{array}{ccc|ccc} 1 & 0 & 0 & -2 & -3 & 1 \\ 0 & 1 & 0 & -3 & 1 & -19 \\ 0 & 0 & 1 & 1 & 0 & 5 \end{array}\right]$$

So,

$$A^{-1} = \left[\begin{array}{ccc} -2 & -3 & 1 \\ -3 & 1 & -19 \\ 1 & 0 & 5 \end{array}\right]$$

Multiply both sides of the equation by A^{-1}. We want $A^{-1}AX = A^{-1}B$:

$$\left[\begin{array}{ccc} -2 & -3 & 1 \\ -3 & 1 & -19 \\ 1 & 0 & 5 \end{array}\right]\left[\begin{array}{ccc} 5 & 15 & 56 \\ -4 & -11 & -41 \\ -1 & -3 & -11 \end{array}\right]\left[\begin{array}{c} x \\ y \\ z \end{array}\right] = \left[\begin{array}{ccc} -2 & -3 & 1 \\ -3 & 1 & -19 \\ 1 & 0 & 5 \end{array}\right]\left[\begin{array}{c} 35 \\ -26 \\ -7 \end{array}\right]$$

Thus,

$$A^{-1}B = \left[\begin{array}{c} -70+78-7 \\ -105-26+133 \\ 35+0-35 \end{array}\right] = \left[\begin{array}{c} 1 \\ 2 \\ 0 \end{array}\right]$$

The solution is (1, 2, 0).

Try It #4

Solve the system using the inverse of the coefficient matrix.

$$2x - 17y + 11z = 0$$
$$-x + 11y - 7z = 8$$
$$3y - 2z = -2$$

How To…

Given a system of equations, solve with matrix inverses using a calculator.

1. Save the coefficient matrix and the constant matrix as matrix variables [A] and [B].
2. Enter the multiplication into the calculator, calling up each matrix variable as needed.
3. If the coefficient matrix is invertible, the calculator will present the solution matrix; if the coefficient matrix is not invertible, the calculator will present an error message.

Example 9   **Using a Calculator to Solve a System of Equations with Matrix Inverses**

Solve the system of equations with matrix inverses using a calculator

$$2x + 3y + z = 32$$
$$3x + 3y + z = -27$$
$$2x + 4y + z = -2$$

Solution On the matrix page of the calculator, enter the coefficient matrix as the matrix variable [A], and enter the constant matrix as the matrix variable [B].

$$[A] = \left[\begin{array}{ccc} 2 & 3 & 1 \\ 3 & 3 & 1 \\ 2 & 4 & 1 \end{array}\right], \quad [B] = \left[\begin{array}{c} 32 \\ -27 \\ -2 \end{array}\right]$$

On the home screen of the calculator, type in the multiplication to solve for X, calling up each matrix variable as needed.

$$[A]^{-1} \times [B]$$

Evaluate the expression.

$$\begin{bmatrix} -59 \\ -34 \\ 252 \end{bmatrix}$$

Access these online resources for additional instruction and practice with solving systems with inverses.

- The Identity Matrix (http://openstaxcollege.org/l/identmatrix)
- Determining Inverse Matrices (http://openstaxcollege.org/l/inversematrix)
- Using a Matrix Equation to Solve a System of Equations (http://openstaxcollege.org/l/matrixsystem)

7.7   SECTION EXERCISES

VERBAL

1. In a previous section, we showed that matrix multiplication is not commutative, that is, $AB \neq BA$ in most cases. Can you explain why matrix multiplication is commutative for matrix inverses, that is, $A^{-1}A = AA^{-1}$?

2. Does every 2×2 matrix have an inverse? Explain why or why not. Explain what condition is necessary for an inverse to exist.

3. Can you explain whether a 2×2 matrix with an entire row of zeros can have an inverse?

4. Can a matrix with an entire column of zeros have an inverse? Explain why or why not.

5. Can a matrix with zeros on the diagonal have an inverse? If so, find an example. If not, prove why not. For simplicity, assume a 2×2 matrix.

ALGEBRAIC

In the following exercises, show that matrix A is the inverse of matrix B.

6. $A = \begin{bmatrix} 1 & 0 \\ -1 & 1 \end{bmatrix}, B = \begin{bmatrix} 1 & 0 \\ 1 & 1 \end{bmatrix}$

7. $A = \begin{bmatrix} 1 & 2 \\ 3 & 4 \end{bmatrix}, B = \begin{bmatrix} -2 & 1 \\ \frac{3}{2} & -\frac{1}{2} \end{bmatrix}$

8. $A = \begin{bmatrix} 4 & 5 \\ 7 & 0 \end{bmatrix}, B = \begin{bmatrix} 0 & \frac{1}{7} \\ \frac{1}{5} & -\frac{4}{35} \end{bmatrix}$

9. $A = \begin{bmatrix} -2 & \frac{1}{2} \\ 3 & -1 \end{bmatrix}, B = \begin{bmatrix} -2 & -1 \\ -6 & -4 \end{bmatrix}$

10. $A = \begin{bmatrix} 1 & 0 & 1 \\ 0 & 1 & -1 \\ 0 & 1 & 1 \end{bmatrix}, B = \frac{1}{2}\begin{bmatrix} 2 & 1 & -1 \\ 0 & 1 & 1 \\ 0 & -1 & 1 \end{bmatrix}$

11. $A = \begin{bmatrix} 1 & 2 & 3 \\ 4 & 0 & 2 \\ 1 & 6 & 9 \end{bmatrix}, B = \frac{1}{4}\begin{bmatrix} 6 & 0 & -2 \\ 17 & -3 & -5 \\ -12 & 2 & 4 \end{bmatrix}$

12. $A = \begin{bmatrix} 3 & 8 & 2 \\ 1 & 1 & 1 \\ 5 & 6 & 12 \end{bmatrix}, B = \frac{1}{36}\begin{bmatrix} -6 & 84 & -6 \\ 7 & -26 & 1 \\ -1 & -22 & 5 \end{bmatrix}$

For the following exercises, find the multiplicative inverse of each matrix, if it exists.

13. $\begin{bmatrix} 3 & -2 \\ 1 & 9 \end{bmatrix}$

14. $\begin{bmatrix} -2 & 2 \\ 3 & 1 \end{bmatrix}$

15. $\begin{bmatrix} -3 & 7 \\ 9 & 2 \end{bmatrix}$

16. $\begin{bmatrix} -4 & -3 \\ -5 & 8 \end{bmatrix}$

17. $\begin{bmatrix} 1 & 1 \\ 2 & 2 \end{bmatrix}$

18. $\begin{bmatrix} 0 & 1 \\ 1 & 0 \end{bmatrix}$

19. $\begin{bmatrix} 0.5 & 1.5 \\ 1 & -0.5 \end{bmatrix}$

20. $\begin{bmatrix} 1 & 0 & 6 \\ -2 & 1 & 7 \\ 3 & 0 & 2 \end{bmatrix}$

21. $\begin{bmatrix} 0 & 1 & -3 \\ 4 & 1 & 0 \\ 1 & 0 & 5 \end{bmatrix}$

22. $\begin{bmatrix} 1 & 2 & -1 \\ -3 & 4 & 1 \\ -2 & -4 & -5 \end{bmatrix}$

23. $\begin{bmatrix} 1 & 9 & -3 \\ 2 & 5 & 6 \\ 4 & -2 & 7 \end{bmatrix}$

24. $\begin{bmatrix} 1 & -2 & 3 \\ -4 & 8 & -12 \\ 1 & 4 & 2 \end{bmatrix}$

25. $\begin{bmatrix} \frac{1}{2} & \frac{1}{2} & \frac{1}{2} \\ \frac{1}{3} & \frac{1}{4} & \frac{1}{5} \\ \frac{1}{6} & \frac{1}{7} & \frac{1}{8} \end{bmatrix}$

26. $\begin{bmatrix} 1 & 2 & 3 \\ 4 & 5 & 6 \\ 7 & 8 & 9 \end{bmatrix}$

Consider a 3×3 system of equations.

$$a_1x + b_1y + c_1z = d_1$$
$$a_2x + b_2y + c_2z = d_2$$
$$a_3x + b_3y + c_3z = d_3$$

$$x = \frac{D_x}{D}, y = \frac{D_y}{D}, z = \frac{D_z}{D}, D \neq 0$$

where

$$D = \begin{vmatrix} a_1 & b_1 & c_1 \\ a_2 & b_2 & c_2 \\ a_3 & b_3 & c_3 \end{vmatrix}, D_x = \begin{vmatrix} d_1 & b_1 & c_1 \\ d_2 & b_2 & c_2 \\ d_3 & b_3 & c_3 \end{vmatrix}, D_y = \begin{vmatrix} a_1 & d_1 & c_1 \\ a_2 & d_2 & c_2 \\ a_3 & d_3 & c_3 \end{vmatrix}, D_z = \begin{vmatrix} a_1 & b_1 & d_1 \\ a_2 & b_2 & d_2 \\ a_3 & b_3 & d_3 \end{vmatrix}$$

If we are writing the determinant D_x, we replace the x column with the constant column. If we are writing the determinant D_y, we replace the y column with the constant column. If we are writing the determinant D_z, we replace the z column with the constant column. Always check the answer.

Example 4 **Solving a 3 × 3 System Using Cramer's Rule**

Find the solution to the given 3×3 system using Cramer's Rule.

$$x + y - z = 6$$
$$3x - 2y + z = -5$$
$$x + 3y - 2z = 14$$

Solution Use Cramer's Rule.

$$D = \begin{vmatrix} 1 & 1 & -1 \\ 3 & -2 & 1 \\ 1 & 3 & -2 \end{vmatrix}, D_x = \begin{vmatrix} 6 & 1 & -1 \\ -5 & -2 & 1 \\ 14 & 3 & -2 \end{vmatrix}, D_y = \begin{vmatrix} 1 & 6 & -1 \\ 3 & -5 & 1 \\ 1 & 14 & -2 \end{vmatrix}, D_z = \begin{vmatrix} 1 & 1 & 6 \\ 3 & -2 & -5 \\ 1 & 3 & 14 \end{vmatrix}$$

Then,

$$x = \frac{D_x}{D} = \frac{-3}{-3} = 1$$

$$y = \frac{D_y}{D} = \frac{-9}{-3} = 3$$

$$z = \frac{D_z}{D} = \frac{6}{-3} = -2$$

The solution is $(1, 3, -2)$.

Try It #3

Use Cramer's Rule to solve the 3×3 matrix.

$$x - 3y + 7z = 13$$
$$x + y + z = 1$$
$$x - 2y + 3z = 4$$

Example 5 **Using Cramer's Rule to Solve an Inconsistent System**

Solve the system of equations using Cramer's Rule.

$$3x - 2y = 4 \qquad (1)$$
$$6x - 4y = 0 \qquad (2)$$

Solution We begin by finding the determinants D, D_x, and D_y.

$$D = \begin{vmatrix} 3 & -2 \\ 6 & -4 \end{vmatrix} = 3(-4) - 6(-2) = 0$$

We know that a determinant of zero means that either the system has no solution or it has an infinite number of solutions. To see which one, we use the process of elimination. Our goal is to eliminate one of the variables.

1. Multiply equation (1) by -2.

2. Add the result to equation (2).

$$-6x + 4y = -8$$
$$\underline{6x - 4y = 0}$$
$$0 = -8$$

We obtain the equation $0 = -8$, which is false. Therefore, the system has no solution. Graphing the system reveals two parallel lines. See **Figure 1**.

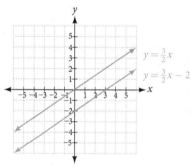

Figure 1

Example 6   Use Cramer's Rule to Solve a Dependent System

Solve the system with an infinite number of solutions.

$$x - 2y + 3z = 0 \qquad (1)$$
$$3x + y - 2z = 0 \qquad (2)$$
$$2x - 4y + 6z = 0 \qquad (3)$$

Solution   Let's find the determinant first. Set up a matrix augmented by the first two columns.

$$\begin{vmatrix} 1 & -2 & 3 \\ 3 & 1 & -2 \\ 2 & -4 & 6 \end{vmatrix} \begin{matrix} 1 & -2 \\ 3 & 1 \\ 2 & -4 \end{matrix}$$

Then,

$$1(1)(6) + (-2)(-2)(2) + 3(3)(-4) - 2(1)(3) - (-4)(-2)(1) - 6(3)(-2) = 0$$

As the determinant equals zero, there is either no solution or an infinite number of solutions. We have to perform elimination to find out.

1. Multiply equation (1) by -2 and add the result to equation (3):

$$-2x + 4y - 6x = 0$$
$$\underline{2x - 4y + 6z = 0}$$
$$0 = 0$$

2. Obtaining an answer of $0 = 0$, a statement that is always true, means that the system has an infinite number of solutions. Graphing the system, we can see that two of the planes are the same and they both intersect the third plane on a line. See **Figure 2**.

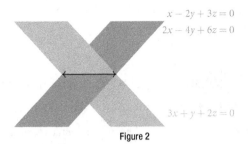

$x - 2y + 3z = 0$

$2x - 4y + 6z = 0$

$3x + y + 2z = 0$

Figure 2

Understanding Properties of Determinants

There are many properties of determinants. Listed here are some properties that may be helpful in calculating the determinant of a matrix.

---

properties of determinants

1. If the matrix is in upper triangular form, the determinant equals the product of entries down the main diagonal.

2. When two rows are interchanged, the determinant changes sign.

3. If either two rows or two columns are identical, the determinant equals zero.

4. If a matrix contains either a row of zeros or a column of zeros, the determinant equals zero.

5. The determinant of an inverse matrix A^{-1} is the reciprocal of the determinant of the matrix A.

6. If any row or column is multiplied by a constant, the determinant is multiplied by the same factor.

Example 7 **Illustrating Properties of Determinants**

Illustrate each of the properties of determinants.

Solution Property 1 states that if the matrix is in upper triangular form, the determinant is the product of the entries down the main diagonal.

$$A = \begin{bmatrix} 1 & 2 & 3 \\ 0 & 2 & 1 \\ 0 & 0 & -1 \end{bmatrix}$$

Augment A with the first two columns.

$$A = \begin{bmatrix} 1 & 2 & 3 & | & 1 & 2 \\ 0 & 2 & 1 & | & 0 & 2 \\ 0 & 0 & -1 & | & 0 & 0 \end{bmatrix}$$

Then

$$\det(A) = 1(2)(-1) + 2(1)(0) + 3(0)(0) - 0(2)(3) - 0(1)(1) + 1(0)(2)$$

$$= -2$$

Property 2 states that interchanging rows changes the sign. Given

$$A = \begin{bmatrix} -1 & 5 \\ 4 & -3 \end{bmatrix}, \det(A) = (-1)(-3) - (4)(5) = 3 - 20 = -17$$

$$B = \begin{bmatrix} 4 & -3 \\ -1 & 5 \end{bmatrix}, \det(B) = (4)(5) - (-1)(-3) = 20 - 3 = 17$$

Property 3 states that if two rows or two columns are identical, the determinant equals zero.

$$A = \begin{vmatrix} 1 & 2 & 2 \\ 2 & 2 & 2 \\ -1 & 2 & 2 \end{vmatrix} \begin{matrix} 1 & 2 \\ 2 & 2 \\ -1 & 2 \end{matrix}$$

$$\det(A) = 1(2)(2) + 2(2)(-1) + 2(2)(2) + 1(2)(2) - 2(2)(1) - 2(2)(2)$$

$$= 4 - 4 + 8 + 4 - 4 - 8 = 0$$

Property 4 states that if a row or column equals zero, the determinant equals zero. Thus,

$$A = \begin{bmatrix} 1 & 2 \\ 0 & 0 \end{bmatrix}, \det(A) = 1(0) - 2(0) = 0$$

Property 5 states that the determinant of an inverse matrix A^{-1} is the reciprocal of the determinant A. Thus,

$$A = \begin{bmatrix} 1 & 2 \\ 3 & 4 \end{bmatrix}, \det(A) = 1(4) - 3(2) = -2$$

$$A^{-1} = \begin{bmatrix} -2 & 1 \\ \frac{3}{2} & -\frac{1}{2} \end{bmatrix}, \det(A^{-1}) = -2\left(-\frac{1}{2}\right) - \left(\frac{3}{2}\right)(1) = -\frac{1}{2}$$

Property 6 states that if any row or column of a matrix is multiplied by a constant, the determinant is multiplied by the same factor. Thus,

$$A = \begin{bmatrix} 1 & 2 \\ 3 & 4 \end{bmatrix}, \det(A) = 1(4) - 2(3) = -2$$

$$B = \begin{bmatrix} 2(1) & 2(2) \\ 3 & 4 \end{bmatrix}, \det(B) = 2(4) - 3(4) = -4$$

Example 8   **Using Cramer's Rule and Determinant Properties to Solve a System**

Find the solution to the given 3×3 system.

$$2x + 4y + 4z = 2 \qquad (1)$$
$$3x + 7y + 7z = -5 \qquad (2)$$
$$x + 2y + 2z = 4 \qquad (3)$$

Solution   Using Cramer's Rule, we have

$$D = \begin{vmatrix} 2 & 4 & 4 \\ 3 & 7 & 7 \\ 1 & 2 & 2 \end{vmatrix}$$

Notice that the second and third columns are identical. According to Property 3, the determinant will be zero, so there is either no solution or an infinite number of solutions. We have to perform elimination to find out.

1. Multiply equation (3) by –2 and add the result to equation (1).

$$-2x - 4y - 4x = -8$$
$$\underline{2x + 4y + 4z = 2}$$
$$0 = -6$$

Obtaining a statement that is a contradiction means that the system has no solution.

Access these online resources for additional instruction and practice with Cramer's Rule.

- Solve a System of Two Equations Using Cramer's Rule (http://openstaxcollege.org/l/system2cramer)
- Solve a Systems of Three Equations using Cramer's Rule (http://openstaxcollege.org/l/system3cramer)

7.8   SECTION EXERCISES

VERBAL

1. Explain why we can always evaluate the determinant of a square matrix.

2. Examining Cramer's Rule, explain why there is no unique solution to the system when the determinant of your matrix is 0. For simplicity, use a 2×2 matrix.

3. Explain what it means in terms of an inverse for a matrix to have a 0 determinant.

4. The determinant of 2×2 matrix A is 3. If you switch the rows and multiply the first row by 6 and the second row by 2, explain how to find the determinant and provide the answer.

ALGEBRAIC

For the following exercises, find the determinant.

5. $\begin{vmatrix} 1 & 2 \\ 3 & 4 \end{vmatrix}$

6. $\begin{vmatrix} -1 & 2 \\ 3 & -4 \end{vmatrix}$

7. $\begin{vmatrix} 2 & -5 \\ -1 & 6 \end{vmatrix}$

8. $\begin{vmatrix} -8 & 4 \\ -1 & 5 \end{vmatrix}$

9. $\begin{vmatrix} 1 & 0 \\ 3 & -4 \end{vmatrix}$

10. $\begin{vmatrix} 10 & 20 \\ 0 & -10 \end{vmatrix}$

11. $\begin{vmatrix} 10 & 0.2 \\ 5 & 0.1 \end{vmatrix}$

12. $\begin{vmatrix} 6 & -3 \\ 8 & 4 \end{vmatrix}$

13. $\begin{vmatrix} -2 & -3 \\ 3.1 & 4{,}000 \end{vmatrix}$

14. $\begin{vmatrix} -1.1 & 0.6 \\ 7.2 & -0.5 \end{vmatrix}$

15. $\begin{vmatrix} -1 & 0 & 0 \\ 0 & 1 & 0 \\ 0 & 0 & -3 \end{vmatrix}$

16. $\begin{vmatrix} -1 & 4 & 0 \\ 0 & 2 & 3 \\ 0 & 0 & -3 \end{vmatrix}$

17. $\begin{vmatrix} 1 & 0 & 1 \\ 0 & 1 & 0 \\ 1 & 0 & 0 \end{vmatrix}$

18. $\begin{vmatrix} 2 & -3 & 1 \\ 3 & -4 & 1 \\ -5 & 6 & 1 \end{vmatrix}$

19. $\begin{vmatrix} -2 & 1 & 4 \\ -4 & 2 & -8 \\ 2 & -8 & -3 \end{vmatrix}$

20. $\begin{vmatrix} 6 & -1 & 2 \\ -4 & -3 & 5 \\ 1 & 9 & -1 \end{vmatrix}$

21. $\begin{vmatrix} 5 & 1 & -1 \\ 2 & 3 & 1 \\ 3 & -6 & -3 \end{vmatrix}$

22. $\begin{vmatrix} 1.1 & 2 & -1 \\ -4 & 0 & 0 \\ 4.1 & -0.4 & 2.5 \end{vmatrix}$

23. $\begin{vmatrix} 2 & -1.6 & 3.1 \\ 1.1 & 3 & -8 \\ -9.3 & 0 & 2 \end{vmatrix}$

24. $\begin{vmatrix} -\frac{1}{2} & \frac{1}{3} & \frac{1}{4} \\ \frac{1}{5} & -\frac{1}{6} & \frac{1}{7} \\ 0 & 0 & \frac{1}{8} \end{vmatrix}$

For the following exercises, solve the system of linear equations using Cramer's Rule.

25. $2x - 3y = -1$
$4x + 5y = 9$

26. $5x - 4y = 2$
$-4x + 7y = 6$

27. $6x - 3y = 2$
$-8x + 9y = -1$

28. $2x + 6y = 12$
$5x - 2y = 13$

29. $4x + 3y = 23$
$2x - y = -1$

30. $10x - 6y = 2$
$-5x + 8y = -1$

31. $4x - 3y = -3$
$2x + 6y = -4$

32. $4x - 5y = 7$
$-3x + 9y = 0$

33. $4x + 10y = 180$
$-3x - 5y = -105$

34. $8x - 2y = -3$
$-4x + 6y = 4$

For the following exercises, solve the system of linear equations using Cramer's Rule.

35. $x + 2y - 4z = -1$
$7x + 3y + 5z = 26$
$-2x - 6y + 7z = -6$

36. $-5x + 2y - 4z = -47$
$4x - 3y - z = -94$
$3x - 3y + 2z = 94$

37. $4x + 5y - z = -7$
$-2x - 9y + 2z = 8$
$5y + 7z = 21$

38. $4x - 3y + 4z = 10$
$5x - 2z = -2$
$3x + 2y - 5z = -9$

39. $4x - 2y + 3z = 6$
$-6x + y = -2$
$2x + 7y + 8z = 24$

40. $5x + 2y - z = 1$
$-7x - 8y + 3z = 1.5$
$6x - 12y + z = 7$

41. $13x - 17y + 16z = 73$
$-11x + 15y + 17z = 61$
$46x + 10y - 30z = -18$

42. $-4x - 3y - 8z = -7$
$2x - 9y + 5z = 0.5$
$5x - 6y - 5z = -2$

43. $4x - 6y + 8z = 10$
$-2x + 3y - 4z = -5$
$x + y + z = 1$

44. $4x - 6y + 8z = 10$
$-2x + 3y - 4z = -5$
$12x + 18y - 24z = -30$

TECHNOLOGY

For the following exercises, use the determinant function on a graphing utility.

45. $\begin{vmatrix} 1 & 0 & 8 & 9 \\ 0 & 2 & 1 & 0 \\ 1 & 0 & 3 & 0 \\ 0 & 2 & 4 & 3 \end{vmatrix}$

46. $\begin{vmatrix} 1 & 0 & 2 & 1 \\ 0 & -9 & 1 & 3 \\ 3 & 0 & -2 & -1 \\ 0 & 1 & 1 & -2 \end{vmatrix}$

47. $\begin{vmatrix} \frac{1}{2} & 1 & 7 & 4 \\ 0 & \frac{1}{2} & 100 & 5 \\ 0 & 0 & 2 & 2{,}000 \\ 0 & 0 & 0 & 2 \end{vmatrix}$

48. $\begin{vmatrix} 1 & 0 & 0 & 0 \\ 2 & 3 & 0 & 0 \\ 4 & 5 & 6 & 0 \\ 7 & 8 & 9 & 0 \end{vmatrix}$

REAL-WORLD APPLICATIONS

For the following exercises, create a system of linear equations to describe the behavior. Then, calculate the determinant. Will there be a unique solution? If so, find the unique solution.

49. Two numbers add up to 56. One number is 20 less than the other.

50. Two numbers add up to 104. If you add two times the first number plus two times the second number, your total is 208

51. Three numbers add up to 106. The first number is 3 less than the second number. The third number is 4 more than the first number.

52. Three numbers add to 216. The sum of the first two numbers is 112. The third number is 8 less than the first two numbers combined.

For the following exercises, create a system of linear equations to describe the behavior. Then, solve the system for all solutions using Cramer's Rule.

53. You invest $10,000 into two accounts, which receive 8% interest and 5% interest. At the end of a year, you had $10,710 in your combined accounts. How much was invested in each account?

54. You invest $80,000 into two accounts, $22,000 in one account, and $58,000 in the other account. At the end of one year, assuming simple interest, you have earned $2,470 in interest. The second account receives half a percent less than twice the interest on the first account. What are the interest rates for your accounts?

55. A movie theater needs to know how many adult tickets and children tickets were sold out of the 1,200 total tickets. If children's tickets are $5.95, adult tickets are $11.15, and the total amount of revenue was $12,756, how many children's tickets and adult tickets were sold?

56. A concert venue sells single tickets for $40 each and couple's tickets for $65. If the total revenue was $18,090 and the 321 tickets were sold, how many single tickets and how many couple's tickets were sold?

57. You decide to paint your kitchen green. You create the color of paint by mixing yellow and blue paints. You cannot remember how many gallons of each color went into your mix, but you know there were 10 gal total. Additionally, you kept your receipt, and know the total amount spent was $29.50. If each gallon of yellow costs $2.59, and each gallon of blue costs $3.19, how many gallons of each color go into your green mix?

58. You sold two types of scarves at a farmers' market and would like to know which one was more popular. The total number of scarves sold was 56, the yellow scarf cost $10, and the purple scarf cost $11. If you had total revenue of $583, how many yellow scarves and how many purple scarves were sold?

59. Your garden produced two types of tomatoes, one green and one red. The red weigh 10 oz, and the green weigh 4 oz. You have 30 tomatoes, and a total weight of 13 lb, 14 oz. How many of each type of tomato do you have?

60. At a market, the three most popular vegetables make up 53% of vegetable sales. Corn has 4% higher sales than broccoli, which has 5% more sales than onions. What percentage does each vegetable have in the market share?

61. At the same market, the three most popular fruits make up 37% of the total fruit sold. Strawberries sell twice as much as oranges, and kiwis sell one more percentage point than oranges. For each fruit, find the percentage of total fruit sold.

62. Three bands performed at a concert venue. The first band charged $15 per ticket, the second band charged $45 per ticket, and the final band charged $22 per ticket. There were 510 tickets sold, for a total of $12,700. If the first band had 40 more audience members than the second band, how many tickets were sold for each band?

63. A movie theatre sold tickets to three movies. The tickets to the first movie were $5, the tickets to the second movie were $11, and the third movie was $12. 100 tickets were sold to the first movie. The total number of tickets sold was 642, for a total revenue of $6,774. How many tickets for each movie were sold?

64. Men aged 20–29, 30–39, and 40–49 made up 78% of the population at a prison last year. This year, the same age groups made up 82.08% of the population. The 20–29 age group increased by 20%, the 30–39 age group increased by 2%, and the 40–49 age group decreased to $\frac{3}{4}$ of their previous population. Originally, the 30–39 age group had 2% more prisoners than the 20–29 age group. Determine the prison population percentage for each age group last year.

65. At a women's prison down the road, the total number of inmates aged 20–49 totaled 5,525. This year, the 20–29 age group increased by 10%, the 30–39 age group decreased by 20%, and the 40–49 age group doubled. There are now 6,040 prisoners. Originally, there were 500 more in the 30–39 age group than the 20–29 age group. Determine the prison population for each age group last year.

For the following exercises, use this scenario: A health-conscious company decides to make a trail mix out of almonds, dried cranberries, and chocolate-covered cashews. The nutritional information for these items is shown in **Table 1**.

	Fat (g)	Protein (g)	Carbohydrates (g)
Almonds (10)	6	2	3
Cranberries (10)	0.02	0	8
Cashews (10)	7	3.5	5.5

Table 1

66. For the special "low-carb" trail mix, there are 1,000 pieces of mix. The total number of carbohydrates is 425 g, and the total amount of fat is 570.2 g. If there are 200 more pieces of cashews than cranberries, how many of each item is in the trail mix?

67. For the "hiking" mix, there are 1,000 pieces in the mix, containing 390.8 g of fat, and 165 g of protein. If there is the same amount of almonds as cashews, how many of each item is in the trail mix?

68. For the "energy-booster" mix, there are 1,000 pieces in the mix, containing 145 g of protein and 625 g of carbohydrates. If the number of almonds and cashews summed together is equivalent to the amount of cranberries, how many of each item is in the trail mix?

CHAPTER 7 REVIEW

Key Terms

addition method an algebraic technique used to solve systems of linear equations in which the equations are added in a way that eliminates one variable, allowing the resulting equation to be solved for the remaining variable; substitution is then used to solve for the first variable

augmented matrix a coefficient matrix adjoined with the constant column separated by a vertical line within the matrix brackets

break-even point the point at which a cost function intersects a revenue function; where profit is zero

coefficient matrix a matrix that contains only the coefficients from a system of equations

column a set of numbers aligned vertically in a matrix

consistent system a system for which there is a single solution to all equations in the system and it is an independent system, or if there are an infinite number of solutions and it is a dependent system

cost function the function used to calculate the costs of doing business; it usually has two parts, fixed costs and variable costs

Cramer's Rule a method for solving systems of equations that have the same number of equations as variables using determinants

dependent system a system of linear equations in which the two equations represent the same line; there are an infinite number of solutions to a dependent system

determinant a number calculated using the entries of a square matrix that determines such information as whether there is a solution to a system of equations

entry an element, coefficient, or constant in a matrix

feasible region the solution to a system of nonlinear inequalities that is the region of the graph where the shaded regions of each inequality intersect

Gaussian elimination using elementary row operations to obtain a matrix in row-echelon form

identity matrix a square matrix containing ones down the main diagonal and zeros everywhere else; it acts as a 1 in matrix algebra

inconsistent system a system of linear equations with no common solution because they represent parallel lines, which have no point or line in common

independent system a system of linear equations with exactly one solution pair (x, y)

main diagonal entries from the upper left corner diagonally to the lower right corner of a square matrix

matrix a rectangular array of numbers

multiplicative inverse of a matrix a matrix that, when multiplied by the original, equals the identity matrix

nonlinear inequality an inequality containing a nonlinear expression

partial fraction decomposition the process of returning a simplified rational expression to its original form, a sum or difference of simpler rational expressions

partial fractions the individual fractions that make up the sum or difference of a rational expression before combining them into a simplified rational expression

profit function the profit function is written as $P(x) = R(x) - C(x)$, revenue minus cost

revenue function the function that is used to calculate revenue, simply written as $R = xp$, where $x =$ quantity and $p =$ price

row a set of numbers aligned horizontally in a matrix

row operations adding one row to another row, multiplying a row by a constant, interchanging rows, and so on, with the goal of achieving row-echelon form

row-echelon form after performing row operations, the matrix form that contains ones down the main diagonal and zeros at every space below the diagonal

row-equivalent two matrices A and B are row-equivalent if one can be obtained from the other by performing basic row operations

scalar multiple an entry of a matrix that has been multiplied by a scalar

solution set the set of all ordered pairs or triples that satisfy all equations in a system of equations

substitution method an algebraic technique used to solve systems of linear equations in which one of the two equations is solved for one variable and then substituted into the second equation to solve for the second variable

system of linear equations a set of two or more equations in two or more variables that must be considered simultaneously.

system of nonlinear equations a system of equations containing at least one equation that is of degree larger than one

system of nonlinear inequalities a system of two or more inequalities in two or more variables containing at least one inequality that is not linear

Key Equations

Identity matrix for a 2 × 2 matrix

$$I_2 = \begin{bmatrix} 1 & 0 \\ 0 & 1 \end{bmatrix}$$

Identity matrix for a 3 × 3 matrix

$$I_3 = \begin{bmatrix} 1 & 0 & 0 \\ 0 & 1 & 0 \\ 0 & 0 & 1 \end{bmatrix}$$

Multiplicative inverse of a 2 × 2 matrix

$$A^{-1} = \frac{1}{ad - bc} \begin{bmatrix} d & -b \\ -c & a \end{bmatrix}, \text{ where } ad - bc \neq 0$$

Key Concepts

7.1 Systems of Linear Equations: Two Variables

- A system of linear equations consists of two or more equations made up of two or more variables such that all equations in the system are considered simultaneously.
- The solution to a system of linear equations in two variables is any ordered pair that satisfies each equation independently. See **Example 1**.
- Systems of equations are classified as independent with one solution, dependent with an infinite number of solutions, or inconsistent with no solution.
- One method of solving a system of linear equations in two variables is by graphing. In this method, we graph the equations on the same set of axes. See **Example 2**.
- Another method of solving a system of linear equations is by substitution. In this method, we solve for one variable in one equation and substitute the result into the second equation. See **Example 3**.
- A third method of solving a system of linear equations is by addition, in which we can eliminate a variable by adding opposite coefficients of corresponding variables. See **Example 4**.
- It is often necessary to multiply one or both equations by a constant to facilitate elimination of a variable when adding the two equations together. See **Example 5**, **Example 6**, and **Example 7**.
- Either method of solving a system of equations results in a false statement for inconsistent systems because they are made up of parallel lines that never intersect. See **Example 8**.
- The solution to a system of dependent equations will always be true because both equations describe the same line. See **Example 9**.
- Systems of equations can be used to solve real-world problems that involve more than one variable, such as those relating to revenue, cost, and profit. See **Example 10** and **Example 11**.

7.2 Systems of Linear Equations: Three Variables

- A solution set is an ordered triple $\{(x, y, z)\}$ that represents the intersection of three planes in space. See **Example 1**.
- A system of three equations in three variables can be solved by using a series of steps that forces a variable to be eliminated. The steps include interchanging the order of equations, multiplying both sides of an equation by a nonzero constant, and adding a nonzero multiple of one equation to another equation. See **Example 2**.
- Systems of three equations in three variables are useful for solving many different types of real-world problems. See **Example 3**.
- A system of equations in three variables is inconsistent if no solution exists. After performing elimination operations, the result is a contradiction. See **Example 4**.
- Systems of equations in three variables that are inconsistent could result from three parallel planes, two parallel planes and one intersecting plane, or three planes that intersect the other two but not at the same location.

- A system of equations in three variables is dependent if it has an infinite number of solutions. After performing elimination operations, the result is an identity. See **Example 5**.

- Systems of equations in three variables that are dependent could result from three identical planes, three planes intersecting at a line, or two identical planes that intersect the third on a line.

7.3 Systems of Nonlinear Equations and Inequalities: Two Variables

- There are three possible types of solutions to a system of equations representing a line and a parabola: (1) no solution, the line does not intersect the parabola; (2) one solution, the line is tangent to the parabola; and (3) two solutions, the line intersects the parabola in two points. See **Example 1**.

- There are three possible types of solutions to a system of equations representing a circle and a line: (1) no solution, the line does not intersect the circle; (2) one solution, the line is tangent to the parabola; (3) two solutions, the line intersects the circle in two points. See **Example 2**.

- There are five possible types of solutions to the system of nonlinear equations representing an ellipse and a circle: (1) no solution, the circle and the ellipse do not intersect; (2) one solution, the circle and the ellipse are tangent to each other; (3) two solutions, the circle and the ellipse intersect in two points; (4) three solutions, the circle and ellipse intersect in three places; (5) four solutions, the circle and the ellipse intersect in four points. See **Example 3**.

- An inequality is graphed in much the same way as an equation, except for $>$ or $<$, we draw a dashed line and shade the region containing the solution set. See **Example 4**.

- Inequalities are solved the same way as equalities, but solutions to systems of inequalities must satisfy both inequalities. See **Example 5**.

7.4 Partial Fractions

- Decompose $\dfrac{P(x)}{Q(x)}$ by writing the partial fractions as $\dfrac{A}{a_1 x + b_1} + \dfrac{B}{a_2 x + b_2}$. Solve by clearing the fractions, expanding the right side, collecting like terms, and setting corresponding coefficients equal to each other, then setting up and solving a system of equations. See **Example 1**.

- The decomposition of $\dfrac{P(x)}{Q(x)}$ with repeated linear factors must account for the factors of the denominator in increasing powers. See **Example 2**.

- The decomposition of $\dfrac{P(x)}{Q(x)}$ with a nonrepeated irreducible quadratic factor needs a linear numerator over the quadratic factor, as in $\dfrac{A}{x} + \dfrac{Bx + C}{(ax^2 + bx + c)}$. See **Example 3**.

- In the decomposition of $\dfrac{P(x)}{Q(x)}$, where $Q(x)$ has a repeated irreducible quadratic factor, when the irreducible quadratic factors are repeated, powers of the denominator factors must be represented in increasing powers as $\dfrac{Ax + B}{(ax^2 + bx + c)} + \dfrac{A_2 x + B_2}{(ax^2 + bx + c)^2} + \dots + \dfrac{A_n x + B_n}{(ax^2 + bx + c)^n}$. See **Example 4**.

7.5 Matrices and Matrix Operations

- A matrix is a rectangular array of numbers. Entries are arranged in rows and columns.

- The dimensions of a matrix refer to the number of rows and the number of columns. A 3×2 matrix has three rows and two columns. See **Example 1**.

- We add and subtract matrices of equal dimensions by adding and subtracting corresponding entries of each matrix. See **Example 2**, **Example 3**, **Example 4**, and **Example 5**.

- Scalar multiplication involves multiplying each entry in a matrix by a constant. See **Example 6**.

- Scalar multiplication is often required before addition or subtraction can occur. See **Example 7**.

- Multiplying matrices is possible when inner dimensions are the same—the number of columns in the first matrix must match the number of rows in the second.

- The product of two matrices, A and B, is obtained by multiplying each entry in row 1 of A by each entry in column 1 of B; then multiply each entry of row 1 of A by each entry in columns 2 of B, and so on. See **Example 8** and **Example 9**.

- Many real-world problems can often be solved using matrices. See **Example 10**.
- We can use a calculator to perform matrix operations after saving each matrix as a matrix variable. See **Example 11**.

7.6 Solving Systems with Gaussian Elimination

- An augmented matrix is one that contains the coefficients and constants of a system of equations. See **Example 1**.
- A matrix augmented with the constant column can be represented as the original system of equations. See **Example 2**.
- Row operations include multiplying a row by a constant, adding one row to another row, and interchanging rows.
- We can use Gaussian elimination to solve a system of equations. See **Example 3**, **Example 4**, and **Example 5**.
- Row operations are performed on matrices to obtain row-echelon form. See **Example 6**.
- To solve a system of equations, write it in augmented matrix form. Perform row operations to obtain row-echelon form. Back-substitute to find the solutions. See **Example 7** and **Example 8**.
- A calculator can be used to solve systems of equations using matrices. See **Example 9**.
- Many real-world problems can be solved using augmented matrices. See **Example 10** and **Example 11**.

7.7 Solving Systems with Inverses

- An identity matrix has the property $AI = IA = A$. See **Example 1**.
- An invertible matrix has the property $AA^{-1} = A^{-1}A = I$. See **Example 2**.
- Use matrix multiplication and the identity to find the inverse of a 2×2 matrix. See **Example 3**.
- The multiplicative inverse can be found using a formula. See **Example 4**.
- Another method of finding the inverse is by augmenting with the identity. See **Example 5**.
- We can augment a 3×3 matrix with the identity on the right and use row operations to turn the original matrix into the identity, and the matrix on the right becomes the inverse. See **Example 6**.
- Write the system of equations as $AX = B$, and multiply both sides by the inverse of A: $A^{-1}AX = A^{-1}B$. See **Example 7** and **Example 8**.
- We can also use a calculator to solve a system of equations with matrix inverses. See **Example 9**.

7.8 Solving Systems with Cramer's Rule

- The determinant for $\begin{bmatrix} a & b \\ c & d \end{bmatrix}$ is $ad - bc$. See **Example 1**.
- Cramer's Rule replaces a variable column with the constant column. Solutions are $x = \dfrac{D_x}{D}, y = \dfrac{D_y}{D}$. See **Example 2**.
- To find the determinant of a 3×3 matrix, augment with the first two columns. Add the three diagonal entries (upper left to lower right) and subtract the three diagonal entries (lower left to upper right). See **Example 3**.
- To solve a system of three equations in three variables using Cramer's Rule, replace a variable column with the constant column for each desired solution: $x = \dfrac{D_x}{D}, y = \dfrac{D_y}{D}, z = \dfrac{D_z}{D}$. See **Example 4**.
- Cramer's Rule is also useful for finding the solution of a system of equations with no solution or infinite solutions. See **Example 5** and **Example 6**.
- Certain properties of determinants are useful for solving problems. For example:
  - If the matrix is in upper triangular form, the determinant equals the product of entries down the main diagonal.
  - When two rows are interchanged, the determinant changes sign.
  - If either two rows or two columns are identical, the determinant equals zero.
  - If a matrix contains either a row of zeros or a column of zeros, the determinant equals zero.
  - The determinant of an inverse matrix A^{-1} is the reciprocal of the determinant of the matrix A.
  - If any row or column is multiplied by a constant, the determinant is multiplied by the same factor. See **Example 7** and **Example 8**.

CHAPTER 7 REVIEW EXERCISES

SYSTEMS OF LINEAR EQUATIONS: TWO VARIABLES

For the following exercises, determine whether the ordered pair is a solution to the system of equations.

1. $3x - y = 4$
    $x + 4y = -3$ and $(-1, 1)$

2. $6x - 2y = 24$
    $-3x + 3y = 18$ and $(9, 15)$

For the following exercises, use substitution to solve the system of equations.

3. $10x + 5y = -5$
    $3x - 2y = -12$

4. $\frac{4}{7}x + \frac{1}{5}y = \frac{43}{70}$
    $\frac{5}{6}x - \frac{1}{3}y = -\frac{2}{3}$

5. $5x + 6y = 14$
    $4x + 8y = 8$

For the following exercises, use addition to solve the system of equations.

6. $3x + 2y = -7$
    $2x + 4y = 6$

7. $3x + 4y = 2$
    $9x + 12y = 3$

8. $8x + 4y = 2$
    $6x - 5y = 0.7$

For the following exercises, write a system of equations to solve each problem. Solve the system of equations.

9. A factory has a cost of production $C(x) = 150x + 15,000$ and a revenue function $R(x) = 200x$. What is the break-even point?

10. A performer charges $C(x) = 50x + 10,000$, where x is the total number of attendees at a show. The venue charges $75 per ticket. After how many people buy tickets does the venue break even, and what is the value of the total tickets sold at that point?

SYSTEMS OF LINEAR EQUATIONS: THREE VARIABLES

For the following exercises, solve the system of three equations using substitution or addition.

11. $0.5x - 0.5y = 10$
    $-0.2y + 0.2x = 4$
    $0.1x + 0.1z = 2$

12. $5x + 3y - z = 5$
    $3x - 2y + 4z = 13$
    $4x + 3y + 5z = 22$

13. $x + y + z = 1$
    $2x + 2y + 2z = 1$
    $3x + 3y = 2$

14. $2x - 3y + z = -1$
    $x + y + z = -4$
    $4x + 2y - 3z = 33$

15. $3x + 2y - z = -10$
    $x - y + 2z = 7$
    $-x + 3y + z = -2$

16. $3x + 4z = -11$
    $x - 2y = 5$
    $4y - z = -10$

17. $2x - 3y + z = 0$
    $2x + 4y - 3z = 0$
    $6x - 2y - z = 0$

18. $6x - 4y - 2z = 2$
    $3x + 2y - 5z = 4$
    $6y - 7z = 5$

For the following exercises, write a system of equations to solve each problem. Solve the system of equations.

19. Three odd numbers sum up to 61. The smaller is one-third the larger and the middle number is 16 less than the larger. What are the three numbers?

20. A local theatre sells out for their show. They sell all 500 tickets for a total purse of $8,070.00. The tickets were priced at $15 for students, $12 for children, and $18 for adults. If the band sold three times as many adult tickets as children's tickets, how many of each type was sold?

SYSTEMS OF NONLINEAR EQUATIONS AND INEQUALITIES: TWO VARIABLES

For the following exercises, solve the system of nonlinear equations.

21. $y = x^2 - 7$
$y = 5x - 13$

22. $y = x^2 - 4$
$y = 5x + 10$

23. $x^2 + y^2 = 16$
$y = x - 8$

24. $x^2 + y^2 = 25$
$y = x^2 + 5$

25. $x^2 + y^2 = 4$
$y - x^2 = 3$

For the following exercises, graph the inequality.

26. $y > x^2 - 1$

27. $\frac{1}{4}x^2 + y^2 < 4$

For the following exercises, graph the system of inequalities.

28. $x^2 + y^2 + 2x < 3$
$y > -x^2 - 3$

29. $x^2 - 2x + y^2 - 4x < 4$
$y < -x + 4$

30. $x^2 + y^2 < 1$
$y^2 < x$

PARTIAL FRACTIONS

For the following exercises, decompose into partial fractions.

31. $\dfrac{-2x + 6}{x^2 + 3x + 2}$

32. $\dfrac{10x + 2}{4x^2 + 4x + 1}$

33. $\dfrac{7x + 20}{x^2 + 10x + 25}$

34. $\dfrac{x - 18}{x^2 - 12x + 36}$

35. $\dfrac{-x^2 + 36x + 70}{x^3 - 125}$

36. $\dfrac{-5x^2 + 6x - 2}{x^3 + 27}$

37. $\dfrac{x^3 - 4x^2 + 3x + 11}{(x^2 - 2)^2}$

38. $\dfrac{4x^4 - 2x^3 + 22x^2 - 6x + 48}{x(x^2 + 4)^2}$

MATRICES AND MATRIX OPERATIONS

For the following exercises, perform the requested operations on the given matrices.

$$A = \begin{bmatrix} 4 & -2 \\ 1 & 3 \end{bmatrix}, B = \begin{bmatrix} 6 & 7 & -3 \\ 11 & -2 & 4 \end{bmatrix}, C = \begin{bmatrix} 6 & 7 \\ 11 & -2 \\ 14 & 0 \end{bmatrix}, D = \begin{bmatrix} 1 & -4 & 9 \\ 10 & 5 & -7 \\ 2 & 8 & 5 \end{bmatrix}, E = \begin{bmatrix} 7 & -14 & 3 \\ 2 & -1 & 3 \\ 0 & 1 & 9 \end{bmatrix}$$

39. $-4A$

40. $10D - 6E$

41. $B + C$

42. AB

43. BA

44. BC

45. CB

46. DE

47. ED

48. EC

49. CE

50. A^3

SOLVING SYSTEMS WITH GAUSSIAN ELIMINATION

For the following exercises, write the system of linear equations from the augmented matrix. Indicate whether there will be a unique solution.

51. $\begin{bmatrix} 1 & 0 & -3 & | & 7 \\ 0 & 1 & 2 & | & -5 \\ 0 & 0 & 0 & | & 0 \end{bmatrix}$

52. $\begin{bmatrix} 1 & 0 & 5 & | & -9 \\ 0 & 1 & -2 & | & 4 \\ 0 & 0 & 0 & | & 3 \end{bmatrix}$

For the following exercises, write the augmented matrix from the system of linear equations.

53. $-2x + 2y + z = 7$
$2x - 8y + 5z = 0$
$19x - 10y + 22z = 3$

54. $4x + 2y - 3z = 14$
$-12x + 3y + z = 100$
$9x - 6y + 2z = 31$

55. $x + 3z = 12$
$-x + 4y = 0$
$y + 2z = -7$

For the following exercises, solve the system of linear equations using Gaussian elimination.

56. $3x - 4y = -7$
$-6x + 8y = 14$

57. $3x - 4y = 1$
$-6x + 8y = 6$

58. $-1.1x - 2.3y = 6.2$
$-5.2x - 4.1y = 4.3$

59. $2x + 3y + 2z = 1$
$-4x - 6y - 4z = -2$
$10x + 15y + 10z = 0$

60. $-x + 2y - 4z = 8$
$3y + 8z = -4$
$-7x + y + 2z = 1$

SOLVING SYSTEMS WITH INVERSES

For the following exercises, find the inverse of the matrix.

61. $\begin{bmatrix} -0.2 & 1.4 \\ 1.2 & -0.4 \end{bmatrix}$

62. $\begin{bmatrix} \dfrac{1}{2} & -\dfrac{1}{2} \\ -\dfrac{1}{4} & \dfrac{3}{4} \end{bmatrix}$

63. $\begin{bmatrix} 12 & 9 & -6 \\ -1 & 3 & 2 \\ -4 & -3 & 2 \end{bmatrix}$

64. $\begin{bmatrix} 2 & 1 & 3 \\ 1 & 2 & 3 \\ 3 & 2 & 1 \end{bmatrix}$

For the following exercises, find the solutions by computing the inverse of the matrix.

65. $0.3x - 0.1y = -10$
$-0.1x + 0.3y = 14$

66. $0.4x - 0.2y = -0.6$
$-0.1x + 0.05y = 0.3$

67. $4x + 3y - 3z = -4.3$
$5x - 4y - z = -6.1$
$x + z = -0.7$

68. $-2x - 3y + 2z = 3$
$-x + 2y + 4z = -5$
$-2y + 5z = -3$

For the following exercises, write a system of equations to solve each problem. Solve the system of equations.

69. Students were asked to bring their favorite fruit to class. 90% of the fruits consisted of banana, apple, and oranges. If oranges were half as popular as bananas and apples were 5% more popular than bananas, what are the percentages of each individual fruit?

70. A sorority held a bake sale to raise money and sold brownies and chocolate chip cookies. They priced the brownies at $2 and the chocolate chip cookies at $1. They raised $250 and sold 175 items. How many brownies and how many cookies were sold?

SOLVING SYSTEMS WITH CRAMER'S RULE

For the following exercises, find the determinant.

71. $\begin{bmatrix} 100 & 0 \\ 0 & 0 \end{bmatrix}$

72. $\begin{bmatrix} 0.2 & -0.6 \\ 0.7 & -1.1 \end{bmatrix}$

73. $\begin{bmatrix} -1 & 4 & 3 \\ 0 & 2 & 3 \\ 0 & 0 & -3 \end{bmatrix}$

74. $\begin{bmatrix} \sqrt{2} & 0 & 0 \\ 0 & \sqrt{2} & 0 \\ 0 & 0 & \sqrt{2} \end{bmatrix}$

For the following exercises, use Cramer's Rule to solve the linear systems of equations.

75. $4x - 2y = 23$
$-5x - 10y = -35$

76. $0.2x - 0.1y = 0$
$-0.3x + 0.3y = 2.5$

77. $-0.5x + 0.1y = 0.3$
$-0.25x + 0.05y = 0.15$

78. $x + 6y + 3z = 4$
$2x + y + 2z = 3$
$3x - 2y + z = 0$

79. $4x - 3y + 5z = -\dfrac{5}{2}$
$7x - 9y - 3z = \dfrac{3}{2}$
$x - 5y - 5z = \dfrac{5}{2}$

80. $\dfrac{3}{10}x - \dfrac{1}{5}y - \dfrac{3}{10}z = -\dfrac{1}{50}$
$\dfrac{1}{10}x - \dfrac{1}{10}y - \dfrac{1}{2}z = -\dfrac{9}{50}$
$\dfrac{2}{5}x - \dfrac{1}{2}y - \dfrac{3}{5}z = -\dfrac{1}{5}$

CHAPTER 7 PRACTICE TEST

Is the following ordered pair a solution to the system of equations?

1. $-5x - y = 12$
  $x + 4y = 9$ with $(-3, 3)$

For the following exercises, solve the systems of linear and nonlinear equations using substitution or elimination. Indicate if no solution exists.

2. $\frac{1}{2}x - \frac{1}{3}y = 4$ **3.** $-\frac{1}{2}x - 4y = 4$ **4.** $5x - y = 1$ **5.** $4x - 6y - 2z = \frac{1}{10}$
  $\frac{3}{2}x - y = 0$ $2x + 16y = 2$ $-10x + 2y = -2$ $x - 7y + 5z = -\frac{1}{4}$

                                                                                             $3x + 6y - 9z = \frac{6}{5}$

6. $x + z = 20$ **7.** $5x - 4y - 3z = 0$ **8.** $y = x^2 + 2x - 3$ **9.** $y^2 + x^2 = 25$
   $x + y + z = 20$ $2x + y + 2z = 0$ $y = x - 1$ $y^2 - 2x^2 = 1$
   $x + 2y + z = 10$ $x - 6y - 7z = 0$

For the following exercises, graph the following inequalities.

10. $y < x^2 + 9$ **11.** $x^2 + y^2 > 4$
                                                $y < x^2 + 1$

For the following exercises, write the partial fraction decomposition.

12. $\dfrac{-8x - 30}{x^2 + 10x + 25}$ **13.** $\dfrac{13x + 2}{(3x + 1)^2}$ **14.** $\dfrac{x^4 - x^3 + 2x - 1}{x(x^2 + 1)^2}$

For the following exercises, perform the given matrix operations.

15. $5\begin{bmatrix} 4 & 9 \\ -2 & 3 \end{bmatrix} + \frac{1}{2}\begin{bmatrix} -6 & 12 \\ 4 & -8 \end{bmatrix}$ **16.** $\begin{bmatrix} 1 & 4 & -7 \\ -2 & 9 & 5 \\ 12 & 0 & -4 \end{bmatrix}\begin{bmatrix} 3 & -4 \\ 1 & 3 \\ 5 & 10 \end{bmatrix}$ **17.** $\begin{bmatrix} \frac{1}{2} & \frac{1}{3} \\ \frac{1}{4} & \frac{1}{5} \end{bmatrix}^{-1}$

18. $\det\begin{vmatrix} 0 & 0 \\ 400 & 4{,}000 \end{vmatrix}$ **19.** $\det\begin{vmatrix} \frac{1}{2} & -\frac{1}{2} & 0 \\ -\frac{1}{2} & 0 & \frac{1}{2} \\ 0 & \frac{1}{2} & 0 \end{vmatrix}$

20. If $\det(A) = -6$, what would be the determinant if you switched rows 1 and 3, multiplied the second row by 12, and took the inverse?

21. Rewrite the system of linear equations as an augmented matrix.
  $14x - 2y + 13z = 140$
  $-2x + 3y - 6z = -1$
  $x - 5y + 12z = 11$

22. Rewrite the augmented matrix as a system of linear equations.
$$\begin{bmatrix} 1 & 0 & 3 & | & 12 \\ -2 & 4 & 9 & | & -5 \\ -6 & 1 & 2 & | & 8 \end{bmatrix}$$

CHAPTER 7 SYSTEMS OF EQUATIONS AND INEQUALITIES

For the following exercises, use Gaussian elimination to solve the systems of equations.

23. $x - 6y = 4$
$2x - 12y = 0$

24. $2x + y + z = -3$
$x - 2y + 3z = 6$
$x - y - z = 6$

For the following exercises, use the inverse of a matrix to solve the systems of equations.

25. $4x - 5y = -50$
$-x + 2y = 80$

26. $\frac{1}{100}x - \frac{3}{100}y + \frac{1}{20}z = -49$

$\frac{3}{100}x - \frac{7}{100}y - \frac{1}{100}z = 13$

$\frac{9}{100}x - \frac{9}{100}y - \frac{9}{100}z = 99$

For the following exercises, use Cramer's Rule to solve the systems of equations.

27. $200x - 300y = 2$
$400x + 715y = 4$

28. $0.1x + 0.1y - 0.1z = -1.2$
$0.1x - 0.2y + 0.4z = -1.2$
$0.5x - 0.3y + 0.8z = -5.9$

For the following exercises, solve using a system of linear equations.

29. A factory producing cell phones has the following cost and revenue functions:
$C(x) = x^2 + 75x + 2{,}688$ and $R(x) = x^2 + 160x$. What is the range of cell phones they should produce each day so there is profit? Round to the nearest number that generates profit.

30. A small fair charges $1.50 for students, $1 for children, and $2 for adults. In one day, three times as many children as adults attended. A total of 800 tickets were sold for a total revenue of $1,050. How many of each type of ticket was sold?

8

Analytic Geometry

Figure 1 (a) Greek philosopher Aristotle (384–322 BCE) (b) German mathematician and astronomer Johannes Kepler (1571–1630)

CHAPTER OUTLINE

Introduction

The Greek mathematician Menaechmus (c. 380–c. 320 BCE) is generally credited with discovering the shapes formed by the intersection of a plane and a right circular cone. Depending on how he tilted the plane when it intersected the cone, he formed different shapes at the intersection—beautiful shapes with near-perfect symmetry.

It was also said that Aristotle may have had an intuitive understanding of these shapes, as he observed the orbit of the planet to be circular. He presumed that the planets moved in circular orbits around Earth, and for nearly 2000 years this was the commonly held belief.

It was not until the Renaissance movement that Johannes Kepler noticed that the orbits of the planet were not circular in nature. His published law of planetary motion in the 1600s changed our view of the solar system forever. He claimed that the sun was at one end of the orbits, and the planets revolved around the sun in an oval-shaped path.

In this chapter, we will investigate the two-dimensional figures that are formed when a right circular cone is intersected by a plane. We will begin by studying each of three figures created in this manner. We will develop defining equations for each figure and then learn how to use these equations to solve a variety of problems.

681

LEARNING OBJECTIVES

In this section, you will:

- Write equations of ellipses in standard form.
- Graph ellipses centered at the origin.
- Graph ellipses not centered at the origin.
- Solve applied problems involving ellipses.

8.1   THE ELLIPSE

Figure 1   The National Statuary Hall in Washington, D.C. (credit: Greg Palmer, Flickr)

Can you imagine standing at one end of a large room and still being able to hear a whisper from a person standing at the other end? The National Statuary Hall in Washington, D.C., shown in **Figure 1**, is such a room.[32] It is an oval-shaped room called a *whispering chamber* because the shape makes it possible for sound to travel along the walls. In this section, we will investigate the shape of this room and its real-world applications, including how far apart two people in Statuary Hall can stand and still hear each other whisper.

Writing Equations of Ellipses in Standard Form

A conic section, or **conic**, is a shape resulting from intersecting a right circular cone with a plane. The angle at which the plane intersects the cone determines the shape, as shown in **Figure 2**.

Ellipse   Hyperbola   Parabola
Figure 2

Conic sections can also be described by a set of points in the coordinate plane. Later in this chapter, we will see that the graph of any quadratic equation in two variables is a conic section. The signs of the equations and the coefficients of the variable terms determine the shape. This section focuses on the four variations of the standard form of the

32 Architect of the Capitol. http://www.aoc.gov. Accessed April 15, 2014.

equation for the ellipse. An **ellipse** is the set of all points (x, y) in a plane such that the sum of their distances from two fixed points is a constant. Each fixed point is called a **focus** (plural: **foci**).

We can draw an ellipse using a piece of cardboard, two thumbtacks, a pencil, and string. Place the thumbtacks in the cardboard to form the foci of the ellipse. Cut a piece of string longer than the distance between the two thumbtacks (the length of the string represents the constant in the definition). Tack each end of the string to the cardboard, and trace a curve with a pencil held taut against the string. The result is an ellipse. See **Figure 3**.

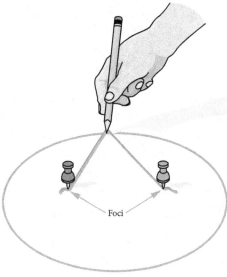

Figure 3

Every ellipse has two axes of symmetry. The longer axis is called the **major axis**, and the shorter axis is called the **minor axis**. Each endpoint of the major axis is the **vertex** of the ellipse (plural: **vertices**), and each endpoint of the minor axis is a co-vertex of the ellipse. The **center of an ellipse** is the midpoint of both the major and minor axes. The axes are perpendicular at the center. The foci always lie on the major axis, and the sum of the distances from the foci to any point on the ellipse (the constant sum) is greater than the distance between the foci. See **Figure 4**.

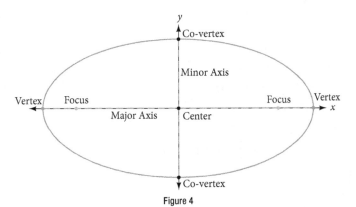

Figure 4

In this section, we restrict ellipses to those that are positioned vertically or horizontally in the coordinate plane. That is, the axes will either lie on or be parallel to the x- and y-axes. Later in the chapter, we will see ellipses that are rotated in the coordinate plane.

To work with horizontal and vertical ellipses in the coordinate plane, we consider two cases: those that are centered at the origin and those that are centered at a point other than the origin. First we will learn to derive the equations of ellipses, and then we will learn how to write the equations of ellipses in standard form. Later we will use what we learn to draw the graphs.

To derive the equation of an ellipse centered at the origin, we begin with the foci $(-c, 0)$ and $(c, 0)$. The ellipse is the set of all points (x, y) such that the sum of the distances from (x, y) to the foci is constant, as shown in **Figure 5**.

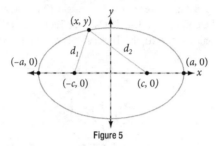

Figure 5

If $(a, 0)$ is a vertex of the ellipse, the distance from $(-c, 0)$ to $(a, 0)$ is $a - (-c) = a + c$. The distance from $(c, 0)$ to $(a, 0)$ is $a - c$. The sum of the distances from the foci to the vertex is

$$(a + c) + (a - c) = 2a$$

If (x, y) is a point on the ellipse, then we can define the following variables:

$$d_1 = \text{the distance from } (-c, 0) \text{ to } (x, y)$$
$$d_2 = \text{the distance from } (c, 0) \text{ to } (x, y)$$

By the definition of an ellipse, $d_1 + d_2$ is constant for any point (x, y) on the ellipse. We know that the sum of these distances is $2a$ for the vertex $(a, 0)$. It follows that $d_1 + d_2 = 2a$ for any point on the ellipse. We will begin the derivation by applying the distance formula. The rest of the derivation is algebraic.

$$d_1 + d_2 = \sqrt{(x - (-c))^2 + (y - 0)^2} + \sqrt{(x - c)^2 + (y - 0)^2} = 2a \quad \text{Distance formula}$$

$$\sqrt{(x + c)^2 + y^2} + \sqrt{(x - c)^2 + y^2} = 2a \qquad\qquad\qquad\qquad \text{Simplify expressions.}$$

$$\sqrt{(x + c)^2 + y^2} = 2a - \sqrt{(x - c)^2 + y^2} \qquad\qquad\qquad \text{Move radical to opposite side.}$$

$$(x + c)^2 + y^2 = \left[2a - \sqrt{(x - c)^2 + y^2}\right]^2 \qquad\qquad\qquad \text{Square both sides.}$$

$$x^2 + 2cx + c^2 + y^2 = 4a^2 - 4a\sqrt{(x - c)^2 + y^2} + (x - c)^2 + y^2 \qquad \text{Expand the squares.}$$

$$x^2 + 2cx + c^2 + y^2 = 4a^2 - 4a\sqrt{(x - c)^2 + y^2} + x^2 - 2cx + c^2 + y^2 \qquad \text{Expand remaining squares.}$$

$$2cx = 4a^2 - 4a\sqrt{(x - c)^2 + y^2} - 2cx \qquad\qquad\qquad \text{Combine like terms.}$$

$$4cx - 4a^2 = -4a\sqrt{(x - c)^2 + y^2} \qquad\qquad\qquad\qquad \text{Isolate the radical.}$$

$$cx - a^2 = -a\sqrt{(x - c)^2 + y^2} \qquad\qquad\qquad\qquad \text{Divide by 4.}$$

$$\left[cx - a^2\right]^2 = a^2\left[\sqrt{(x - c)^2 + y^2}\right]^2 \qquad\qquad\qquad \text{Square both sides.}$$

$$c^2 x^2 - 2a^2 cx + a^4 = a^2(x^2 - 2cx + c^2 + y^2) \qquad\qquad \text{Expand the squares.}$$

$$c^2 x^2 - 2a^2 cx + a^4 = a^2 x^2 - 2a^2 cx + a^2 c^2 + a^2 y^2 \qquad\qquad \text{Distribute } a^2.$$

$$a^2 x^2 - c^2 x^2 + a^2 y^2 = a^4 - a^2 c^2 \qquad\qquad\qquad\qquad \text{Rewrite.}$$

$$x^2(a^2 - c^2) + a^2 y^2 = a^2(a^2 - c^2) \qquad\qquad\qquad\qquad \text{Factor common terms.}$$

$$x^2 b^2 + a^2 y^2 = a^2 b^2 \qquad\qquad\qquad\qquad\qquad\qquad \text{Set } b^2 = a^2 - c^2.$$

$$\frac{x^2 b^2}{a^2 b^2} + \frac{a^2 y^2}{a^2 b^2} = \frac{a^2 b^2}{a^2 b^2} \qquad\qquad\qquad\qquad\qquad \text{Divide both sides by } a^2 b^2.$$

$$\frac{x^2}{a^2} + \frac{y^2}{b^2} = 1 \qquad\qquad\qquad\qquad\qquad\qquad\qquad \text{Simplify.}$$

Thus, the standard equation of an ellipse is $\dfrac{x^2}{a^2} + \dfrac{y^2}{b^2} = 1$. This equation defines an ellipse centered at the origin. If $a > b$, the ellipse is stretched further in the horizontal direction, and if $b > a$, the ellipse is stretched further in the vertical direction.

Writing Equations of Ellipses Centered at the Origin in Standard Form

Standard forms of equations tell us about key features of graphs. Take a moment to recall some of the standard forms of equations we've worked with in the past: linear, quadratic, cubic, exponential, logarithmic, and so on. By learning to interpret standard forms of equations, we are bridging the relationship between algebraic and geometric representations of mathematical phenomena.

The key features of the ellipse are its center, vertices, co-vertices, foci, and lengths and positions of the major and minor axes. Just as with other equations, we can identify all of these features just by looking at the standard form of the equation. There are four variations of the standard form of the ellipse. These variations are categorized first by the location of the center (the origin or not the origin), and then by the position (horizontal or vertical). Each is presented along with a description of how the parts of the equation relate to the graph. Interpreting these parts allows us to form a mental picture of the ellipse.

---

standard forms of the equation of an ellipse with center $(0, 0)$

The standard form of the equation of an ellipse with center $(0, 0)$ and major axis on the x-axis is

$$\frac{x^2}{a^2} + \frac{y^2}{b^2} = 1$$

where

- $a > b$
- the length of the major axis is $2a$
- the coordinates of the vertices are $(\pm a, 0)$
- the length of the minor axis is $2b$
- the coordinates of the co-vertices are $(0, \pm b)$
- the coordinates of the foci are $(\pm c, 0)$, where $c^2 = a^2 - b^2$. See **Figure 6a.**

The standard form of the equation of an ellipse with center $(0, 0)$ and major axis on the y-axis is

$$\frac{x^2}{b^2} + \frac{y^2}{a^2} = 1$$

where

- $a > b$
- the length of the major axis is $2a$
- the coordinates of the vertices are $(0, \pm a)$
- the length of the minor axis is $2b$
- the coordinates of the co-vertices are $(\pm b, 0)$
- the coordinates of the foci are $(0, \pm c)$, where $c^2 = a^2 - b^2$. See **Figure 6b.**

Note that the vertices, co-vertices, and foci are related by the equation $c^2 = a^2 - b^2$. When we are given the coordinates of the foci and vertices of an ellipse, we can use this relationship to find the equation of the ellipse in standard form.

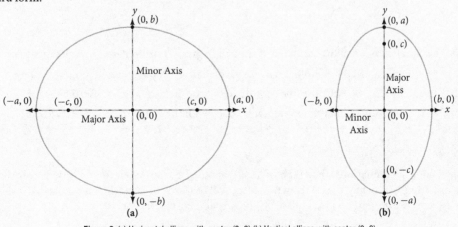

Figure 6 (a) Horizontal ellipse with center $(0, 0)$ (b) Vertical ellipse with center $(0, 0)$

How To...

Given the vertices and foci of an ellipse centered at the origin, write its equation in standard form.

1. Determine whether the major axis lies on the *x*- or *y*-axis.

    a. If the given coordinates of the vertices and foci have the form ($\pm a$, 0) and ($\pm c$, 0) respectively, then the major axis is the *x*-axis. Use the standard form $\frac{x^2}{a^2} + \frac{y^2}{b^2} = 1$.

    b. If the given coordinates of the vertices and foci have the form (0, $\pm a$) and ($\pm c$, 0), respectively, then the major axis is the *y*-axis. Use the standard form $\frac{x^2}{b^2} + \frac{y^2}{a^2} = 1$.

2. Use the equation $c^2 = a^2 - b^2$, along with the given coordinates of the vertices and foci, to solve for b^2.

3. Substitute the values for a^2 and b^2 into the standard form of the equation determined in Step 1.

Example 1   **Writing the Equation of an Ellipse Centered at the Origin in Standard Form**

What is the standard form equation of the ellipse that has vertices (± 8, 0) and foci (± 5, 0)?

Solution   The foci are on the *x*-axis, so the major axis is the *x*-axis. Thus, the equation will have the form

$$\frac{x^2}{a^2} + \frac{y^2}{b^2} = 1$$

The vertices are (± 8, 0), so $a = 8$ and $a^2 = 64$.

The foci are (± 5, 0), so $c = 5$ and $c^2 = 25$.

We know that the vertices and foci are related by the equation $c^2 = a^2 - b^2$. Solving for b^2, we have:

$$c^2 = a^2 - b^2$$
$$25 = 64 - b^2 \qquad \text{Substitute for } c^2 \text{ and } a^2.$$
$$b^2 = 39 \qquad \text{Solve for } b^2.$$

Now we need only substitute $a^2 = 64$ and $b^2 = 39$ into the standard form of the equation. The equation of the ellipse is $\frac{x^2}{64} + \frac{y^2}{39} = 1$.

Try It #1

What is the standard form equation of the ellipse that has vertices (0, ± 4) and foci (0, $\pm \sqrt{15}$)?

Q & A...

Can we write the equation of an ellipse centered at the origin given coordinates of just one focus and vertex?

Yes. Ellipses are symmetrical, so the coordinates of the vertices of an ellipse centered around the origin will always have the form ($\pm a$, 0) or (0, $\pm a$). Similarly, the coordinates of the foci will always have the form ($\pm c$, 0) or (0, $\pm c$). Knowing this, we can use *a* and *c* from the given points, along with the equation $c^2 = a^2 - b^2$, to find b^2.

Writing Equations of Ellipses Not Centered at the Origin

Like the graphs of other equations, the graph of an ellipse can be translated. If an ellipse is translated *h* units horizontally and *k* units vertically, the center of the ellipse will be (*h*, *k*). This translation results in the standard form of the equation we saw previously, with *x* replaced by ($x - h$) and *y* replaced by ($y - k$).

standard forms of the equation of an ellipse with center (h, k)

The standard form of the equation of an ellipse with center (h, k) and major axis parallel to the x-axis is

$$\frac{(x - h)^2}{a^2} + \frac{(y - k)^2}{b^2} = 1$$

where

- $a > b$
- the length of the major axis is $2a$
- the coordinates of the vertices are $(h \pm a, k)$
- the length of the minor axis is $2b$
- the coordinates of the co-vertices are $(h, k \pm b)$
- the coordinates of the foci are $(h \pm c, k)$, where $c^2 = a^2 - b^2$. See **Figure 7a**.

The standard form of the equation of an ellipse with center (h, k) and major axis parallel to the y-axis is

$$\frac{(x - h)^2}{b^2} + \frac{(y - k)^2}{a^2} = 1$$

where

- $a > b$
- the length of the major axis is $2a$
- the coordinates of the vertices are $(h, k \pm a)$
- the length of the minor axis is $2b$
- the coordinates of the co-vertices are $(h \pm b, k)$
- the coordinates of the foci are $(h, k \pm c)$, where $c^2 = a^2 - b^2$. See **Figure 7b**.

Just as with ellipses centered at the origin, ellipses that are centered at a point (h, k) have vertices, co-vertices, and foci that are related by the equation $c^2 = a^2 - b^2$. We can use this relationship along with the midpoint and distance formulas to find the equation of the ellipse in standard form when the vertices and foci are given.

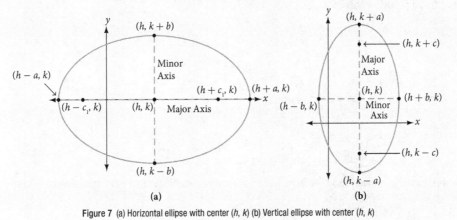

Figure 7 (a) Horizontal ellipse with center (h, k) (b) Vertical ellipse with center (h, k)

How To…

Given the vertices and foci of an ellipse not centered at the origin, write its equation in standard form.

1. Determine whether the major axis is parallel to the x- or y-axis.
   a. If the y-coordinates of the given vertices and foci are the same, then the major axis is parallel to the x-axis. Use the standard form $\frac{(x - h)^2}{a^2} + \frac{(y - k)^2}{b^2} = 1$.
   b. If the x-coordinates of the given vertices and foci are the same, then the major axis is parallel to the y-axis. Use the standard form $\frac{(x - h)^2}{b^2} + \frac{(y - k)^2}{a^2} = 1$.

2. Identify the center of the ellipse (h, k) using the midpoint formula and the given coordinates for the vertices.

3. Find a^2 by solving for the length of the major axis, $2a$, which is the distance between the given vertices.

4. Find c^2 using h and k, found in Step 2, along with the given coordinates for the foci.

5. Solve for b^2 using the equation $c^2 = a^2 - b^2$.

6. Substitute the values for h, k, a^2, and b^2 into the standard form of the equation determined in Step 1.

Example 2   **Writing the Equation of an Ellipse Centered at a Point Other Than the Origin**

What is the standard form equation of the ellipse that has vertices $(-2, -8)$ and $(-2, 2)$ and foci $(-2, -7)$ and $(-2, 1)$?

Solution   The x-coordinates of the vertices and foci are the same, so the major axis is parallel to the y-axis. Thus, the equation of the ellipse will have the form

$$\frac{(x - h)^2}{b^2} + \frac{(y - k)^2}{a^2} = 1$$

First, we identify the center, (h, k). The center is halfway between the vertices, $(-2, -8)$ and $(-2, 2)$.

Applying the midpoint formula, we have:

$$(h, k) = \left(\frac{-2 + (-2)}{2}, \frac{-8 + 2}{2} \right)$$
$$= (-2, -3)$$

Next, we find a^2. The length of the major axis, $2a$, is bounded by the vertices. We solve for a by finding the distance between the y-coordinates of the vertices.

$$2a = 2 - (-8)$$
$$2a = 10$$
$$a = 5$$

So $a^2 = 25$.

Now we find c^2. The foci are given by $(h, k \pm c)$. So, $(h, k - c) = (-2, -7)$ and $(h, k + c) = (-2, 1)$. We substitute $k = -3$ using either of these points to solve for c.

$$k + c = 1$$
$$-3 + c = 1$$
$$c = 4$$

So $c^2 = 16$.

Next, we solve for b^2 using the equation $c^2 = a^2 - b^2$.

$$c^2 = a^2 - b^2$$
$$16 = 25 - b^2$$
$$b^2 = 9$$

Finally, we substitute the values found for h, k, a^2, and b^2 into the standard form equation for an ellipse:

$$\frac{(x + 2)^2}{9} + \frac{(y + 3)^2}{25} = 1$$

Try It #2

What is the standard form equation of the ellipse that has vertices $(-3, 3)$ and $(5, 3)$ and foci $(1 - 2\sqrt{3}, 3)$ and $(1 + 2\sqrt{3}, 3)$?

Graphing Ellipses Centered at the Origin

Just as we can write the equation for an ellipse given its graph, we can graph an ellipse given its equation. To graph ellipses centered at the origin, we use the standard form $\frac{x^2}{a^2} + \frac{y^2}{b^2} = 1$, $a > b$ for horizontal ellipses and $\frac{x^2}{b^2} + \frac{y^2}{a^2} = 1$, $a > b$ for vertical ellipses.

Example 6   **Solving Applied Problems Involving Hyperbolas**

The design layout of a cooling tower is shown in **Figure 11**. The tower stands 179.6 meters tall. The diameter of the top is 72 meters. At their closest, the sides of the tower are 60 meters apart.

Figure 11 Project design for a natural draft cooling tower

Find the equation of the hyperbola that models the sides of the cooling tower. Assume that the center of the hyperbola—indicated by the intersection of dashed perpendicular lines in the figure—is the origin of the coordinate plane. Round final values to four decimal places.

Solution   We are assuming the center of the tower is at the origin, so we can use the standard form of a horizontal hyperbola centered at the origin: $\dfrac{x^2}{a^2} - \dfrac{y^2}{b^2} = 1$, where the branches of the hyperbola form the sides of the cooling tower. We must find the values of a^2 and b^2 to complete the model.

First, we find a^2. Recall that the length of the transverse axis of a hyperbola is $2a$. This length is represented by the distance where the sides are closest, which is given as 65.3 meters. So, $2a = 60$. Therefore, $a = 30$ and $a^2 = 900$.

To solve for b^2, we need to substitute for x and y in our equation using a known point. To do this, we can use the dimensions of the tower to find some point (x, y) that lies on the hyperbola. We will use the top right corner of the tower to represent that point. Since the y-axis bisects the tower, our x-value can be represented by the radius of the top, or 36 meters. The y-value is represented by the distance from the origin to the top, which is given as 79.6 meters. Therefore,

$$\frac{x^2}{a^2} - \frac{y^2}{b^2} = 1 \qquad\qquad \text{Standard form of horizontal hyperbola.}$$

$$b^2 = \frac{y^2}{\dfrac{x^2}{a^2} - 1} \qquad\qquad \text{Isolate } b^2$$

$$= \frac{(79.6)^2}{\dfrac{(36)^2}{900} - 1} \qquad\qquad \text{Substitute for } a^2, x, \text{ and } y$$

$$\approx 14400.3636 \qquad\qquad \text{Round to four decimal places}$$

The sides of the tower can be modeled by the hyperbolic equation

$$\frac{x^2}{900} - \frac{y^2}{14400.3636} = 1, \text{ or } \frac{x^2}{30^2} - \frac{y^2}{120.0015^2} = 1$$

Try It #6

A design for a cooling tower project is shown in **Figure 12**. Find the equation of the hyperbola that models the sides of the cooling tower. Assume that the center of the hyperbola—indicated by the intersection of dashed perpendicular lines in the figure—is the origin of the coordinate plane. Round final values to four decimal places.

Figure 12

Access these online resources for additional instruction and practice with hyperbolas.

- Conic Sections: The Hyperbola Part 1 of 2 (http://openstaxcollege.org/l/hyperbola1)
- Conic Sections: The Hyperbola Part 2 of 2 (http://openstaxcollege.org/l/hyperbola2)
- Graph a Hyperbola with Center at Origin (http://openstaxcollege.org/l/hyperbolaorigin)
- Graph a Hyperbola with Center not at Origin (http://openstaxcollege.org/l/hbnotorigin)

8.2 SECTION EXERCISES

VERBAL

1. Define a hyperbola in terms of its foci.

2. What can we conclude about a hyperbola if its asymptotes intersect at the origin?

3. What must be true of the foci of a hyperbola?

4. If the transverse axis of a hyperbola is vertical, what do we know about the graph?

5. Where must the center of hyperbola be relative to its foci?

ALGEBRAIC

For the following exercises, determine whether the following equations represent hyperbolas. If so, write in standard form.

6. $3y^2 + 2x = 6$

7. $\dfrac{x^2}{36} - \dfrac{y^2}{9} = 1$

8. $5y^2 + 4x^2 = 6x$

9. $25x^2 - 16y^2 = 400$

10. $-9x^2 + 18x + y^2 + 4y - 14 = 0$

For the following exercises, write the equation for the hyperbola in standard form if it is not already, and identify the vertices and foci, and write equations of asymptotes.

11. $\dfrac{x^2}{25} - \dfrac{y^2}{36} = 1$

12. $\dfrac{x^2}{100} - \dfrac{y^2}{9} = 1$

13. $\dfrac{y^2}{4} - \dfrac{x^2}{81} = 1$

14. $9y^2 - 4x^2 = 1$

15. $\dfrac{(x-1)^2}{9} - \dfrac{(y-2)^2}{16} = 1$

16. $\dfrac{(y-6)^2}{36} - \dfrac{(x+1)^2}{16} = 1$

17. $\dfrac{(x-2)^2}{49} - \dfrac{(y+7)^2}{49} = 1$

18. $4x^2 - 8x - 9y^2 - 72y + 112 = 0$

19. $-9x^2 - 54x + 9y^2 - 54y + 81 = 0$

20. $4x^2 - 24x - 36y^2 - 360y + 864 = 0$

21. $-4x^2 + 24x + 16y^2 - 128y + 156 = 0$

22. $-4x^2 + 40x + 25y^2 - 100y + 100 = 0$

23. $x^2 + 2x - 100y^2 - 1000y + 2401 = 0$

24. $-9x^2 + 72x + 16y^2 + 16y + 4 = 0$

25. $4x^2 + 24x - 25y^2 + 200y - 464 = 0$

For the following exercises, find the equations of the asymptotes for each hyperbola.

26. $\dfrac{y^2}{3^2} - \dfrac{x^2}{3^2} = 1$

27. $\dfrac{(x-3)^2}{5^2} - \dfrac{(y+4)^2}{2^2} = 1$

28. $\dfrac{(y-3)^2}{3^2} - \dfrac{(x+5)^2}{6^2} = 1$

29. $9x^2 - 18x - 16y^2 + 32y - 151 = 0$

30. $16y^2 + 96y - 4x^2 + 16x + 112 = 0$

GRAPHICAL

For the following exercises, sketch a graph of the hyperbola, labeling vertices and foci.

31. $\dfrac{x^2}{49} - \dfrac{y^2}{16} = 1$

32. $\dfrac{x^2}{64} - \dfrac{y^2}{4} = 1$

33. $\dfrac{y^2}{9} - \dfrac{x^2}{25} = 1$

34. $81x^2 - 9y^2 = 1$

35. $\dfrac{(y+5)^2}{9} - \dfrac{(x-4)^2}{25} = 1$

36. $\dfrac{(x-2)^2}{8} - \dfrac{(y+3)^2}{27} = 1$

37. $\dfrac{(y-3)^2}{9} - \dfrac{(x-3)^2}{9} = 1$

38. $-4x^2 - 8x + 16y^2 - 32y - 52 = 0$

39. $x^2 - 8x - 25y^2 - 100y - 109 = 0$

40. $-x^2 + 8x + 4y^2 - 40y + 88 = 0$

41. $64x^2 + 128x - 9y^2 - 72y - 656 = 0$

42. $16x^2 + 64x - 4y^2 - 8y - 4 = 0$

43. $-100x^2 + 1000x + y^2 - 10y - 2575 = 0$

44. $4x^2 + 16x - 4y^2 + 16y + 16 = 0$

For the following exercises, given information about the graph of the hyperbola, find its equation.

45. Vertices at $(3, 0)$ and $(-3, 0)$ and one focus at $(5, 0)$.

46. Vertices at $(0, 6)$ and $(0, -6)$ and one focus at $(0, -8)$.

47. Vertices at $(1, 1)$ and $(11, 1)$ and one focus at $(12, 1)$.

48. Center: $(0, 0)$; vertex: $(0, -13)$; one focus: $(0, \sqrt{313})$.

49. Center: $(4, 2)$; vertex: $(9, 2)$; one focus: $(4 + \sqrt{26}, 2)$.

50. Center: $(3, 5)$; vertex: $(3, 11)$; one focus: $(3, 5 + 2\sqrt{10})$.

For the following exercises, given the graph of the hyperbola, find its equation.

51.

52.

53.

54.

55.

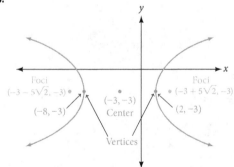

EXTENSIONS

For the following exercises, express the equation for the hyperbola as two functions, with y as a function of x. Express as simply as possible. Use a graphing calculator to sketch the graph of the two functions on the same axes.

56. $\dfrac{x^2}{4} - \dfrac{y^2}{9} = 1$

57. $\dfrac{y^2}{9} - \dfrac{x^2}{1} = 1$

58. $\dfrac{(x-2)^2}{16} - \dfrac{(y+3)^2}{25} = 1$

59. $-4x^2 - 16x + y^2 - 2y - 19 = 0$ **60.** $4x^2 - 24x - y^2 - 4y + 16 = 0$

REAL-WORLD APPLICATIONS

For the following exercises, a hedge is to be constructed in the shape of a hyperbola near a fountain at the center of the yard. Find the equation of the hyperbola and sketch the graph.

61. The hedge will follow the asymptotes $y = x$ and $y = -x$, and its closest distance to the center fountain is 5 yards.

62. The hedge will follow the asymptotes $y = 2x$ and $y = -2x$, and its closest distance to the center fountain is 6 yards.

63. The hedge will follow the asymptotes $y = \dfrac{1}{2}x$ and $y = -\dfrac{1}{2}x$, and its closest distance to the center fountain is 10 yards.

64. The hedge will follow the asymptotes $y = \dfrac{2}{3}x$ and $y = -\dfrac{2}{3}x$, and its closest distance to the center fountain is 12 yards.

65. The hedge will follow the asymptotes $y = \dfrac{3}{4}x$ and $y = -\dfrac{3}{4}x$, and its closest distance to the center fountain is 20 yards.

For the following exercises, assume an object enters our solar system and we want to graph its path on a coordinate system with the sun at the origin and the x-axis as the axis of symmetry for the object's path. Give the equation of the flight path of each object using the given information.

66. The object enters along a path approximated by the line $y = x - 2$ and passes within 1 au (astronomical unit) of the sun at its closest approach, so that the sun is one focus of the hyperbola. It then departs the solar system along a path approximated by the line $y = -x + 2$.

67. The object enters along a path approximated by the line $y = 2x - 2$ and passes within 0.5 au of the sun at its closest approach, so the sun is one focus of the hyperbola. It then departs the solar system along a path approximated by the line $y = -2x + 2$.

68. The object enters along a path approximated by the line $y = 0.5x + 2$ and passes within 1 au of the sun at its closest approach, so the sun is one focus of the hyperbola. It then departs the solar system along a path approximated by the line $y = -0.5x - 2$.

69. The object enters along a path approximated by the line $y = \dfrac{1}{3}x - 1$ and passes within 1 au of the sun at its closest approach, so the sun is one focus of the hyperbola. It then departs the solar system along a path approximated by the line $y = -\dfrac{1}{3}x + 1$.

70. The object enters along a path approximated by the line $y = 3x - 9$ and passes within 1 au of the sun at its closest approach, so the sun is one focus of the hyperbola. It then departs the solar system along a path approximated by the line $y = -3x + 9$.

LEARNING OBJECTIVES

In this section, you will:

- Graph parabolas with vertices at the origin.
- Write equations of parabolas in standard form.
- Graph parabolas with vertices not at the origin.
- Solve applied problems involving parabolas.

8.3   THE PARABOLA

Figure 1 The Olympic torch concludes its journey around the world when it is used to light the
Olympic cauldron during the opening ceremony. (credit: Ken Hackman, U.S. Air Force)

Did you know that the Olympic torch is lit several months before the start of the games? The ceremonial method for lighting the flame is the same as in ancient times. The ceremony takes place at the Temple of Hera in Olympia, Greece, and is rooted in Greek mythology, paying tribute to Prometheus, who stole fire from Zeus to give to all humans. One of eleven acting priestesses places the torch at the focus of a parabolic mirror (see **Figure 1**), which focuses light rays from the sun to ignite the flame.

Parabolic mirrors (or reflectors) are able to capture energy and focus it to a single point. The advantages of this property are evidenced by the vast list of parabolic objects we use every day: satellite dishes, suspension bridges, telescopes, microphones, spotlights, and car headlights, to name a few. Parabolic reflectors are also used in alternative energy devices, such as solar cookers and water heaters, because they are inexpensive to manufacture and need little maintenance. In this section we will explore the parabola and its uses, including low-cost, energy-efficient solar designs.

Graphing Parabolas with Vertices at the Origin

In **The Ellipse**, we saw that an ellipse is formed when a plane cuts through a right circular cone. If the plane is parallel to the edge of the cone, an unbounded curve is formed. This curve is a **parabola**. See **Figure 2**.

Figure 2 Parabola

Like the ellipse and hyperbola, the parabola can also be defined by a set of points in the coordinate plane. A parabola is the set of all points (x, y) in a plane that are the same distance from a fixed line, called the **directrix**, and a fixed point (the **focus**) not on the directrix.

In **Quadratic Functions**, we learned about a parabola's vertex and axis of symmetry. Now we extend the discussion to include other key features of the parabola. See **Figure 3**. Notice that the axis of symmetry passes through the focus and vertex and is perpendicular to the directrix. The vertex is the midpoint between the directrix and the focus.

The line segment that passes through the focus and is parallel to the directrix is called the **latus rectum**. The endpoints of the latus rectum lie on the curve. By definition, the distance d from the focus to any point P on the parabola is equal to the distance from P to the directrix.

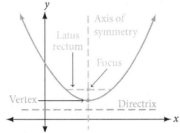

Figure 3 Key features of the parabola

To work with parabolas in the coordinate plane, we consider two cases: those with a vertex at the origin and those with a vertex at a point other than the origin. We begin with the former.

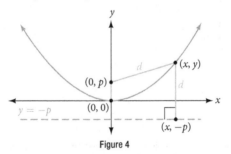

Figure 4

Let (x, y) be a point on the parabola with vertex $(0, 0)$, focus $(0, p)$, and directrix $y = -p$ as shown in **Figure 4**. The distance d from point (x, y) to point $(x, -p)$ on the directrix is the difference of the y-values: $d = y + p$. The distance from the focus $(0, p)$ to the point (x, y) is also equal to d and can be expressed using the distance formula.

$$d = \sqrt{(x - 0)^2 + (y - p)^2}$$
$$= \sqrt{x^2 + (y - p)^2}$$

Set the two expressions for d equal to each other and solve for y to derive the equation of the parabola. We do this because the distance from (x, y) to $(0, p)$ equals the distance from (x, y) to $(x, -p)$.

$$\sqrt{x^2 + (y - p)^2} = y + p$$

We then square both sides of the equation, expand the squared terms, and simplify by combining like terms.

$$x^2 + (y - p)^2 = (y + p)^2$$
$$x^2 + y^2 - 2py + p^2 = y^2 + 2py + p^2$$
$$x^2 - 2py = 2py$$
$$x^2 = 4py$$

The equations of parabolas with vertex $(0, 0)$ are $y^2 = 4px$ when the x-axis is the axis of symmetry and $x^2 = 4py$ when the y-axis is the axis of symmetry. These standard forms are given below, along with their general graphs and key features.

standard forms of parabolas with vertex (0, 0)

Table 1 and **Figure 5** summarize the standard features of parabolas with a vertex at the origin.

Axis of Symmetry	Equation	Focus	Directrix	Endpoints of Latus Rectum
x-axis	$y^2 = 4px$	$(p, 0)$	$x = -p$	$(p, \pm 2p)$
y-axis	$x^2 = 4py$	$(0, p)$	$y = -p$	$(\pm 2p, p)$

Table 1

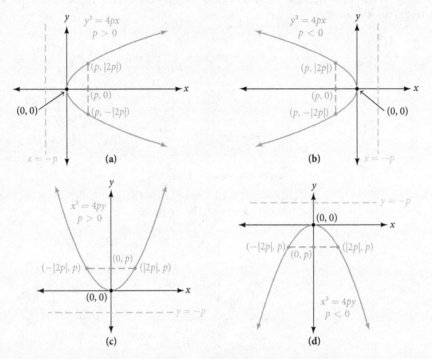

Figure 5 (a) When $p > 0$ and the axis of symmetry is the x-axis, the parabola opens right. (b) When $p < 0$ and the axis of symmetry is the x-axis, the parabola opens left. (c) When $p < 0$ and the axis of symmetry is the y-axis, the parabola opens up. (d) When $p < 0$ and the axis of symmetry is the y-axis, the parabola opens down.

The key features of a parabola are its vertex, axis of symmetry, focus, directrix, and latus rectum. See **Figure 5**. When given a standard equation for a parabola centered at the origin, we can easily identify the key features to graph the parabola.

A line is said to be tangent to a curve if it intersects the curve at exactly one point. If we sketch lines tangent to the parabola at the endpoints of the latus rectum, these lines intersect on the axis of symmetry, as shown in **Figure 6**.

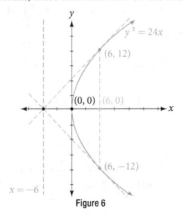

Figure 6

How To...

Given a standard form equation for a parabola centered at (0, 0), sketch the graph.

1. Determine which of the standard forms applies to the given equation: $y^2 = 4px$ or $x^2 = 4py$.
2. Use the standard form identified in Step 1 to determine the axis of symmetry, focus, equation of the directrix, and endpoints of the latus rectum.

   a. If the equation is in the form $y^2 = 4px$, then
   - the axis of symmetry is the *x*-axis, $y = 0$
   - set $4p$ equal to the coefficient of x in the given equation to solve for p. If $p > 0$, the parabola opens right. If $p < 0$, the parabola opens left.
   - use p to find the coordinates of the focus, $(p, 0)$
   - use p to find the equation of the directrix, $x = -p$
   - use p to find the endpoints of the latus rectum, $(p, \pm 2p)$. Alternately, substitute $x = p$ into the original equation.

   b. If the equation is in the form $x^2 = 4py$, then
   - the axis of symmetry is the *y*-axis, $x = 0$
   - set $4p$ equal to the coefficient of y in the given equation to solve for p. If $p > 0$, the parabola opens up. If $p < 0$, the parabola opens down.
   - use p to find the coordinates of the focus, $(0, p)$
   - use p to find equation of the directrix, $y = -p$
   - use p to find the endpoints of the latus rectum, $(\pm 2p, p)$

3. Plot the focus, directrix, and latus rectum, and draw a smooth curve to form the parabola.

Example 1   **Graphing a Parabola with Vertex (0, 0) and the *x*-axis as the Axis of Symmetry**

Graph $y^2 = 24x$. Identify and label the focus, directrix, and endpoints of the latus rectum.

Solution   The standard form that applies to the given equation is $y^2 = 4px$. Thus, the axis of symmetry is the *x*-axis. It follows that:

- $24 = 4p$, so $p = 6$. Since $p > 0$, the parabola opens right
- the coordinates of the focus are $(p, 0) = (6, 0)$
- the equation of the directrix is $x = -p = -6$
- the endpoints of the latus rectum have the same *x*-coordinate at the focus. To find the endpoints, substitute $x = 6$ into the original equation: $(6, \pm 12)$

Next we plot the focus, directrix, and latus rectum, and draw a smooth curve to form the parabola. **Figure 7**

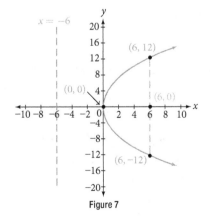

Figure 7

Try It #1

Graph $y^2 = -16x$. Identify and label the focus, directrix, and endpoints of the latus rectum.

Example 2   **Graphing a Parabola with Vertex (0, 0) and the y-axis as the Axis of Symmetry**

Graph $x^2 = -6y$. Identify and label the focus, directrix, and endpoints of the latus rectum.

Solution   The standard form that applies to the given equation is $x^2 = 4py$. Thus, the axis of symmetry is the y-axis. It follows that:

- $-6 = 4p$, so $p = -\frac{3}{2}$. Since $p < 0$, the parabola opens down.
- the coordinates of the focus are $(0, p) = \left(0, -\frac{3}{2}\right)$
- the equation of the directrix is $y = -p = \frac{3}{2}$
- the endpoints of the latus rectum can be found by substituting $y = \frac{3}{2}$ into the original equation, $\left(\pm 3, -\frac{3}{2}\right)$

Next we plot the focus, directrix, and latus rectum, and draw a smooth curve to form the parabola.

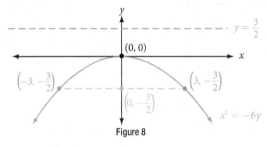

Figure 8

Try It #2

Graph $x^2 = 8y$. Identify and label the focus, directrix, and endpoints of the latus rectum.

Writing Equations of Parabolas in Standard Form

In the previous examples, we used the standard form equation of a parabola to calculate the locations of its key features. We can also use the calculations in reverse to write an equation for a parabola when given its key features.

How To…

Given its focus and directrix, write the equation for a parabola in standard form.

1. Determine whether the axis of symmetry is the x- or y-axis.

   a. If the given coordinates of the focus have the form $(p, 0)$, then the axis of symmetry is the x-axis.
      Use the standard form $y^2 = 4px$.

   b. If the given coordinates of the focus have the form $(0, p)$, then the axis of symmetry is the y-axis.
      Use the standard form $x^2 = 4py$.

2. Multiply $4p$.

3. Substitute the value from Step 2 into the equation determined in Step 1.

Example 3   **Writing the Equation of a Parabola in Standard Form Given its Focus and Directrix**

What is the equation for the parabola with focus $\left(-\frac{1}{2}, 0\right)$ and directrix $x = \frac{1}{2}$?

Solution   The focus has the form $(p, 0)$, so the equation will have the form $y^2 = 4px$.

- Multiplying $4p$, we have $4p = 4(-\frac{1}{2}) = -2$.
- Substituting for $4p$, we have $y^2 = 4px = -2x$.

Therefore, the equation for the parabola is $y^2 = -2x$.

Try It #3

What is the equation for the parabola with focus $\left(0, \dfrac{7}{2}\right)$ and directrix $y = -\dfrac{7}{2}$?

Graphing Parabolas with Vertices Not at the Origin

Like other graphs we've worked with, the graph of a parabola can be translated. If a parabola is translated h units horizontally and k units vertically, the vertex will be (h, k). This translation results in the standard form of the equation we saw previously with x replaced by $(x - h)$ and y replaced by $(y - k)$.

To graph parabolas with a vertex (h, k) other than the origin, we use the standard form $(y - k)^2 = 4p(x - h)$ for parabolas that have an axis of symmetry parallel to the x-axis, and $(x - h)^2 = 4p(y - k)$ for parabolas that have an axis of symmetry parallel to the y-axis. These standard forms are given below, along with their general graphs and key features.

standard forms of parabolas with vertex (h, k)

Table 2 and **Figure 9** summarize the standard features of parabolas with a vertex at a point (h, k).

Axis of Symmetry	Equation	Focus	Directrix	Endpoints of Latus Rectum
$y = k$	$(y - k)^2 = 4p(x - h)$	$(h + p, k)$	$x = h - p$	$(h + p, k \pm 2p)$
$x = h$	$(x - h)^2 = 4p(y - k)$	$(h, k + p)$	$y = k - p$	$(h \pm 2p, k + p)$

Table 2

Figure 9 (a) When $p > 0$, the parabola opens right. (b) When $p < 0$, the parabola opens left.
(c) When $p > 0$, the parabola opens up. (d) When $p < 0$, the parabola opens down.

How To…

Given a standard form equation for a parabola centered at (h, k), sketch the graph.

1. Determine which of the standard forms applies to the given equation: $(y - k)^2 = 4p(x - h)$ or $(x - h)^2 = 4p(y - k)$.

2. Use the standard form identified in Step 1 to determine the vertex, axis of symmetry, focus, equation of the directrix, and endpoints of the latus rectum.

   a. If the equation is in the form $(y - k)^2 = 4p(x - h)$, then:
- use the given equation to identify h and k for the vertex, (h, k)
- use the value of k to determine the axis of symmetry, $y = k$
- set $4p$ equal to the coefficient of $(x - h)$ in the given equation to solve for p. If $p > 0$, the parabola opens right. If $p < 0$, the parabola opens left.
- use h, k, and p to find the coordinates of the focus, $(h + p, k)$
- use h and p to find the equation of the directrix, $x = h - p$
- use h, k, and p to find the endpoints of the latus rectum, $(h + p, k \pm 2p)$

   b. If the equation is in the form $(x - h)^2 = 4p(y - k)$, then:
- use the given equation to identify h and k for the vertex, (h, k)
- use the value of h to determine the axis of symmetry, $x = h$
- set $4p$ equal to the coefficient of $(y - k)$ in the given equation to solve for p. If $p > 0$, the parabola opens up. If $p < 0$, the parabola opens down.
- use h, k, and p to find the coordinates of the focus, $(h, k + p)$
- use k and p to find the equation of the directrix, $y = k - p$
- use h, k, and p to find the endpoints of the latus rectum, $(h \pm 2p, k + p)$

3. Plot the vertex, axis of symmetry, focus, directrix, and latus rectum, and draw a smooth curve to form the parabola.

Example 4   **Graphing a Parabola with Vertex (h, k) and Axis of Symmetry Parallel to the x-axis**

Graph $(y - 1)^2 = -16(x + 3)$. Identify and label the vertex, axis of symmetry, focus, directrix, and endpoints of the latus rectum.

Solution   The standard form that applies to the given equation is $(y - k)^2 = 4p(x - h)$. Thus, the axis of symmetry is parallel to the x-axis. It follows that:

- the vertex is $(h, k) = (-3, 1)$
- the axis of symmetry is $y = k = 1$
- $-16 = 4p$, so $p = -4$. Since $p < 0$, the parabola opens left.
- the coordinates of the focus are $(h + p, k) = (-3 + (-4), 1) = (-7, 1)$
- the equation of the directrix is $x = h - p = -3 - (-4) = 1$
- the endpoints of the latus rectum are $(h + p, k \pm 2p) = (-3 + (-4), 1 \pm 2(-4))$, or $(-7, -7)$ and $(-7, 9)$

Next we plot the vertex, axis of symmetry, focus, directrix, and latus rectum, and draw a smooth curve to form the parabola. See **Figure 10**.

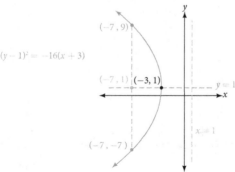

Figure 10

Try It #4

Graph $(y + 1)^2 = 4(x - 8)$. Identify and label the vertex, axis of symmetry, focus, directrix, and endpoints of the latus rectum.

Example 5   **Graphing a Parabola from an Equation Given in General Form**

Graph $x^2 - 8x - 28y - 208 = 0$. Identify and label the vertex, axis of symmetry, focus, directrix, and endpoints of the latus rectum.

Solution   Start by writing the equation of the parabola in standard form. The standard form that applies to the given equation is $(x - h)^2 = 4p\,(y - k)$. Thus, the axis of symmetry is parallel to the y-axis. To express the equation of the parabola in this form, we begin by isolating the terms that contain the variable x in order to complete the square.

$$x^2 - 8x - 28y - 208 = 0$$
$$x^2 - 8x = 28y + 208$$
$$x^2 - 8x + 16 = 28y + 208 + 16$$
$$(x - 4)^2 = 28y + 224$$
$$(x - 4)^2 = 28(y + 8)$$
$$(x - 4)^2 = 4 \cdot 7 \cdot (y + 8)$$

It follows that:

- the vertex is $(h, k) = (4, -8)$
- the axis of symmetry is $x = h = 4$
- since $p = 7$, $p > 0$ and so the parabola opens up
- the coordinates of the focus are $(h, k + p) = (4, -8 + 7) = (4, -1)$
- the equation of the directrix is $y = k - p = -8 - 7 = -15$
- the endpoints of the latus rectum are $(h \pm 2p, k + p) = (4 \pm 2(7), -8 + 7)$, or $(-10, -1)$ and $(18, -1)$

Next we plot the vertex, axis of symmetry, focus, directrix, and latus rectum, and draw a smooth curve to form the parabola. See **Figure 11**.

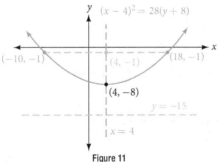

Figure 11

Try It #5

Graph $(x + 2)^2 = -20\,(y - 3)$. Identify and label the vertex, axis of symmetry, focus, directrix, and endpoints of the latus rectum.

Solving Applied Problems Involving Parabolas

As we mentioned at the beginning of the section, parabolas are used to design many objects we use every day, such as telescopes, suspension bridges, microphones, and radar equipment. Parabolic mirrors, such as the one used to light the Olympic torch, have a very unique reflecting property. When rays of light parallel to the parabola's axis of symmetry are directed toward any surface of the mirror, the light is reflected directly to the focus. See **Figure 12**. This is why the Olympic torch is ignited when it is held at the focus of the parabolic mirror.

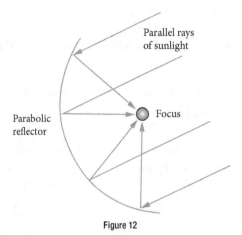

Figure 12

Parabolic mirrors have the ability to focus the sun's energy to a single point, raising the temperature hundreds of degrees in a matter of seconds. Thus, parabolic mirrors are featured in many low-cost, energy efficient solar products, such as solar cookers, solar heaters, and even travel-sized fire starters.

Example 6   **Solving Applied Problems Involving Parabolas**

A cross-section of a design for a travel-sized solar fire starter is shown in **Figure 13**. The sun's rays reflect off the parabolic mirror toward an object attached to the igniter. Because the igniter is located at the focus of the parabola, the reflected rays cause the object to burn in just seconds.

a. Find the equation of the parabola that models the fire starter. Assume that the vertex of the parabolic mirror is the origin of the coordinate plane.

b. Use the equation found in part (a) to find the depth of the fire starter.

Figure 13 Cross-section of a travel-sized solar fire starter

Solution

a. The vertex of the dish is the origin of the coordinate plane, so the parabola will take the standard form $x^2 = 4py$, where $p > 0$. The igniter, which is the focus, is 1.7 inches above the vertex of the dish. Thus we have $p = 1.7$.

$$x^2 = 4py \qquad \text{Standard form of upward-facing parabola with vertex } (0, 0)$$
$$x^2 = 4(1.7)y \qquad \text{Substitute 1.7 for } p.$$
$$x^2 = 6.8y \qquad \text{Multiply.}$$

b. The dish extends $\dfrac{4.5}{2} = 2.25$ inches on either side of the origin. We can substitute 2.25 for x in the equation from part (a) to find the depth of the dish.

$$x^2 = 6.8y \qquad \text{Equation found in part (a).}$$
$$(2.25)^2 = 6.8y \qquad \text{Substitute 2.25 for } x.$$
$$y \approx 0.74 \qquad \text{Solve for } y.$$

The dish is about 0.74 inches deep.

Try It #6

Balcony-sized solar cookers have been designed for families living in India. The top of a dish has a diameter of 1,600 mm. The sun's rays reflect off the parabolic mirror toward the "cooker," which is placed 320 mm from the base.

a. Find an equation that models a cross-section of the solar cooker. Assume that the vertex of the parabolic mirror is the origin of the coordinate plane, and that the parabola opens to the right (i.e., has the *x*-axis as its axis of symmetry).

b. Use the equation found in part (a) to find the depth of the cooker.

Access these online resources for additional instruction and practice with parabolas.

- Conic Sections: The Parabola Part 1 of 2 (http://openstaxcollege.org/l/parabola1)
- Conic Sections: The Parabola Part 2 of 2 (http://openstaxcollege.org/l/parabola2)
- Parabola with Vertical Axis (http://openstaxcollege.org/l/parabolavertcal)
- Parabola with Horizontal Axis (http://openstaxcollege.org/l/parabolahoriz)

8.3 SECTION EXERCISES

VERBAL

1. Define a parabola in terms of its focus and directrix.

2. If the equation of a parabola is written in standard form and p is positive and the directrix is a vertical line, then what can we conclude about its graph?

3. If the equation of a parabola is written in standard form and p is negative and the directrix is a horizontal line, then what can we conclude about its graph?

4. What is the effect on the graph of a parabola if its equation in standard form has increasing values of p?

5. As the graph of a parabola becomes wider, what will happen to the distance between the focus and directrix?

ALGEBRAIC

For the following exercises, determine whether the given equation is a parabola. If so, rewrite the equation in standard form.

6. $y^2 = 4 - x^2$

7. $y = 4x^2$

8. $3x^2 - 6y^2 = 12$

9. $(y - 3)^2 = 8(x - 2)$

10. $y^2 + 12x - 6y - 51 = 0$

For the following exercises, rewrite the given equation in standard form, and then determine the vertex (V), focus (F), and directrix (d) of the parabola.

11. $x = 8y^2$

12. $y = \frac{1}{4}x^2$

13. $y = -4x^2$

14. $x = \frac{1}{8}y^2$

15. $x = 36y^2$

16. $x = \frac{1}{36}y^2$

17. $(x - 1)^2 = 4(y - 1)$

18. $(y - 2)^2 = \frac{4}{5}(x + 4)$

19. $(y - 4)^2 = 2(x + 3)$

20. $(x + 1)^2 = 2(y + 4)$

21. $(x + 4)^2 = 24(y + 1)$

22. $(y + 4)^2 = 16(x + 4)$

23. $y^2 + 12x - 6y + 21 = 0$

24. $x^2 - 4x - 24y + 28 = 0$

25. $5x^2 - 50x - 4y + 113 = 0$

26. $y^2 - 24x + 4y - 68 = 0$

27. $x^2 - 4x + 2y - 6 = 0$

28. $y^2 - 6y + 12x - 3 = 0$

29. $3y^2 - 4x - 6y + 23 = 0$

30. $x^2 + 4x + 8y - 4 = 0$

GRAPHICAL

For the following exercises, graph the parabola, labeling the focus and the directrix.

31. $x = \frac{1}{8}y^2$

32. $y = 36x^2$

33. $y = \frac{1}{36}x^2$

34. $y = -9x^2$

35. $(y - 2)^2 = -\frac{4}{3}(x + 2)$

36. $-5(x + 5)^2 = 4(y + 5)$

37. $-6(y + 5)^2 = 4(x - 4)$

38. $y^2 - 6y - 8x + 1 = 0$

39. $x^2 + 8x + 4y + 20 = 0$

40. $3x^2 + 30x - 4y + 95 = 0$

41. $y^2 - 8x + 10y + 9 = 0$

42. $x^2 + 4x + 2y + 2 = 0$

43. $y^2 + 2y - 12x + 61 = 0$

44. $-2x^2 + 8x - 4y - 24 = 0$

For the following exercises, find the equation of the parabola given information about its graph.

45. Vertex is $(0, 0)$; directrix is $y = 4$, focus is $(0, -4)$.

46. Vertex is $(0, 0)$; directrix is $x = 4$, focus is $(-4, 0)$.

47. Vertex is $(2, 2)$; directrix is $x = 2 - \sqrt{2}$, focus is $(2 + \sqrt{2}, 2)$.

48. Vertex is $(-2, 3)$; directrix is $x = -\frac{7}{2}$, focus is $\left(-\frac{1}{2}, 3\right)$.

49. Vertex is $(\sqrt{2}, -\sqrt{3})$; directrix is $x = 2\sqrt{2}$, focus is $(0, -\sqrt{3})$.

50. Vertex is $(1, 2)$; directrix is $y = \frac{11}{3}$, focus is $\left(1, \frac{1}{3}\right)$.

For the following exercises, determine the equation for the parabola from its graph.

51.

52.

53.

54.

55.

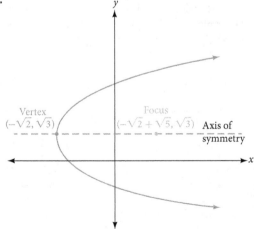

EXTENSIONS

For the following exercises, the vertex and endpoints of the latus rectum of a parabola are given. Find the equation.

56. $V(0, 0)$, Endpoints $(2, 1)$, $(-2, 1)$

57. $V(0, 0)$, Endpoints $(-2, 4)$, $(-2, -4)$

58. $V(1, 2)$, Endpoints $(-5, 5)$, $(7, 5)$

59. $V(-3, -1)$, Endpoints $(0, 5)$, $(0, -7)$

60. $V(4, -3)$, Endpoints $\left(5, -\dfrac{7}{2}\right)$, $\left(3, -\dfrac{7}{2}\right)$

REAL-WORLD APPLICATIONS

61. The mirror in an automobile headlight has a parabolic cross-section with the light bulb at the focus. On a schematic, the equation of the parabola is given as $x^2 = 4y$. At what coordinates should you place the light bulb?

62. If we want to construct the mirror from the previous exercise such that the focus is located at $(0, 0.25)$, what should the equation of the parabola be?

63. A satellite dish is shaped like a paraboloid of revolution. This means that it can be formed by rotating a parabola around its axis of symmetry. The receiver is to be located at the focus. If the dish is 12 feet across at its opening and 4 feet deep at its center, where should the receiver be placed?

64. Consider the satellite dish from the previous exercise. If the dish is 8 feet across at the opening and 2 feet deep, where should we place the receiver?

65. A searchlight is shaped like a paraboloid of revolution. A light source is located 1 foot from the base along the axis of symmetry. If the opening of the searchlight is 3 feet across, find the depth.

66. If the searchlight from the previous exercise has the light source located 6 inches from the base along the axis of symmetry and the opening is 4 feet, find the depth.

67. An arch is in the shape of a parabola. It has a span of 100 feet and a maximum height of 20 feet. Find the equation of the parabola, and determine the height of the arch 40 feet from the center.

68. If the arch from the previous exercise has a span of 160 feet and a maximum height of 40 feet, find the equation of the parabola, and determine the distance from the center at which the height is 20 feet.

69. An object is projected so as to follow a parabolic path given by $y = -x^2 + 96x$, where x is the horizontal distance traveled in feet and y is the height. Determine the maximum height the object reaches.

70. For the object from the previous exercise, assume the path followed is given by $y = -0.5x^2 + 80x$. Determine how far along the horizontal the object traveled to reach maximum height.

LEARNING OBJECTIVES

In this section, you will:

- Identify nondegenerate conic sections given their general form equations.
- Use rotation of axes formulas.
- Write equations of rotated conics in standard form.
- Identify conics without rotating axes.

8.4 ROTATION OF AXIS

As we have seen, conic sections are formed when a plane intersects two right circular cones aligned tip to tip and extending infinitely far in opposite directions, which we also call a cone. The way in which we slice the cone will determine the type of conic section formed at the intersection. A circle is formed by slicing a cone with a plane perpendicular to the axis of symmetry of the cone. An ellipse is formed by slicing a single cone with a slanted plane not perpendicular to the axis of symmetry. A parabola is formed by slicing the plane through the top or bottom of the double-cone, whereas a hyperbola is formed when the plane slices both the top and bottom of the cone. See **Figure 1**.

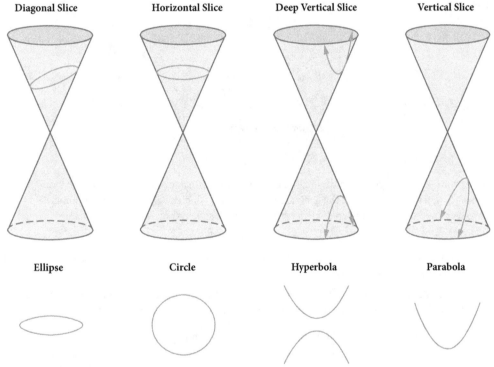

Figure 1 The nondegenerate conic sections

Ellipses, circles, hyperbolas, and parabolas are sometimes called the **nondegenerate conic sections**, in contrast to the **degenerate conic sections**, which are shown in **Figure 2**. A degenerate conic results when a plane intersects the double cone and passes through the apex. Depending on the angle of the plane, three types of degenerate conic sections are possible: a point, a line, or two intersecting lines.

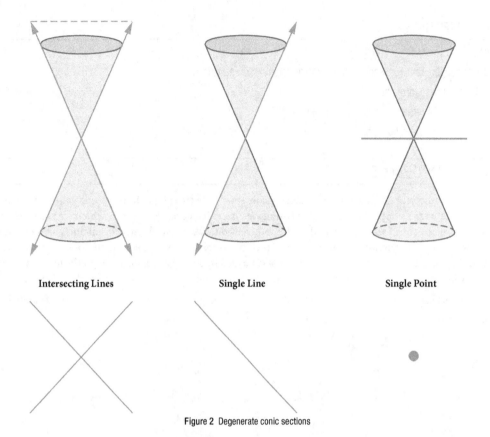

Figure 2 Degenerate conic sections

Identifying Nondegenerate Conics in General Form

In previous sections of this chapter, we have focused on the standard form equations for nondegenerate conic sections. In this section, we will shift our focus to the general form equation, which can be used for any conic. The general form is set equal to zero, and the terms and coefficients are given in a particular order, as shown below.

$$Ax^2 + Bxy + Cy^2 + Dx + Ey + F = 0$$

where A, B, and C are not all zero. We can use the values of the coefficients to identify which type conic is represented by a given equation.

You may notice that the general form equation has an xy term that we have not seen in any of the standard form equations. As we will discuss later, the xy term rotates the conic whenever B is not equal to zero.

Conic Sections	Example
ellipse	$4x^2 + 9y^2 = 1$
circle	$4x^2 + 4y^2 = 1$
hyperbola	$4x^2 - 9y^2 = 1$
parabola	$4x^2 = 9y$ or $4y^2 = 9x$
one line	$4x + 9y = 1$
intersecting lines	$(x - 4)(y + 4) = 0$
parallel lines	$(x - 4)(x - 9) = 0$
a point	$4x^2 + 4y^2 = 0$
no graph	$4x^2 + 4y^2 = -1$

Table 1

general form of conic sections

A **nondegenerate conic section** has the general form

$$Ax^2 + Bxy + Cy^2 + Dx + Ey + F = 0$$

where A, B, and C are not all zero.

Table 2 summarizes the different conic sections where $B = 0$, and A and C are nonzero real numbers. This indicates that the conic has not been rotated.

Conic Sections	Example
ellipse	$Ax^2 + Cy^2 + Dx + Ey + F = 0$, $A \neq C$ and $AC > 0$
circle	$Ax^2 + Cy^2 + Dx + Ey + F = 0$, $A = C$
hyperbola	$Ax^2 - Cy^2 + Dx + Ey + F = 0$ or $-Ax^2 + Cy^2 + Dx + Ey + F = 0$, where A and C are positive
parabola	$Ax^2 + Dx + Ey + F = 0$ or $Cy^2 + Dx + Ey + F = 0$

Table 2

How To...

Given the equation of a conic, identify the type of conic.

1. Rewrite the equation in the general form, $Ax^2 + Bxy + Cy^2 + Dx + Ey + F = 0$.
2. Identify the values of A and C from the general form.
   a. If A and C are nonzero, have the same sign, and are not equal to each other, then the graph is an ellipse.
   b. If A and C are equal and nonzero and have the same sign, then the graph is a circle.
   c. If A and C are nonzero and have opposite signs, then the graph is a hyperbola.
   d. If either A or C is zero, then the graph is a parabola.

Example 1   **Identifying a Conic from Its General Form**

Identify the graph of each of the following nondegenerate conic sections.

   a. $4x^2 - 9y^2 + 36x + 36y - 125 = 0$
   b. $9y^2 + 16x + 36y - 10 = 0$
   c. $3x^2 + 3y^2 - 2x - 6y - 4 = 0$
   d. $-25x^2 - 4y^2 + 100x + 16y + 20 = 0$

Solution

   a. Rewriting the general form, we have

$$Ax^2 + Bxy + Cy^2 + Dx + Ey + F = 0$$
$$4x^2 + 0xy + (-9)y^2 + 36x + 36y + (-125) = 0$$

$A = 4$ and $C = -9$, so we observe that A and C have opposite signs. The graph of this equation is a hyperbola.

   b. Rewriting the general form, we have

$$Ax^2 + Bxy + Cy^2 + Dx + Ey + F = 0$$
$$0x^2 + 0xy + 9y^2 + 16x + 36y + (-10) = 0$$

$A = 0$ and $C = 9$. We can determine that the equation is a parabola, since A is zero.

c. Rewriting the general form, we have

$$Ax^2 + Bxy + Cy^2 + Dx + Ey + F = 0$$

$$3x^2 + 0xy + 3y^2 + (-2)x + (-6)y + (-4) = 0$$

$A = 3$ and $C = 3$. Because $A = C$, the graph of this equation is a circle.

d. Rewriting the general form, we have

$$Ax^2 + Bxy + Cy^2 + Dx + Ey + F = 0$$

$$(-25)x^2 + 0xy + (-4)y^2 + 100x + 16y + 20 = 0$$

$A = -25$ and $C = -4$. Because $AC > 0$ and $A \neq C$, the graph of this equation is an ellipse.

Try It #1

Identify the graph of each of the following nondegenerate conic sections.

a. $16y^2 - x^2 + x - 4y - 9 = 0$ **b.** $16x^2 + 4y^2 + 16x + 49y - 81 = 0$

Finding a New Representation of the Given Equation after Rotating through a Given Angle

Until now, we have looked at equations of conic sections without an xy term, which aligns the graphs with the x- and y- axes. When we add an xy term, we are rotating the conic about the origin. If the x- and y-axes are rotated through an angle, say θ, then every point on the plane may be thought of as having two representations: (x, y) on the Cartesian plane with the original x-axis and y-axis, and (x', y') on the new plane defined by the new, rotated axes, called the x'-axis and y'-axis. See **Figure 3**.

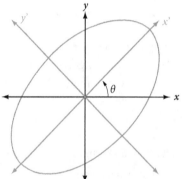

Figure 3 The graph of the rotated ellipse

$$x^2 + y^2 - xy - 15 = 0$$

We will find the relationships between x and y on the Cartesian plane with x' and y' on the new rotated plane. See **Figure 4**.

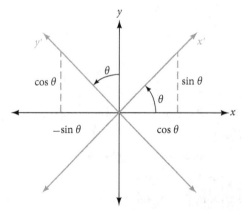

Figure 4 The Cartesian plane with x- and y-axes and the resulting x'– and y'–axes formed by a rotation by an angle θ.

The original coordinate x- and y-axes have unit vectors i and j. The rotated coordinate axes have unit vectors i' and j'. The angle θ is known as the **angle of rotation**. See **Figure 5**. We may write the new unit vectors in terms of the original ones.

$$i' = \cos\theta i + \sin\theta j$$
$$j' = -\sin\theta i + \cos\theta j$$

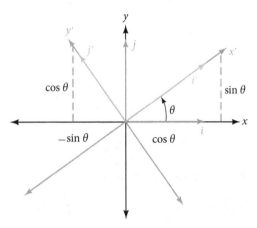

Figure 5 Relationship between the old and new coordinate planes.

Consider a vector u in the new coordinate plane. It may be represented in terms of its coordinate axes.

$u = x' i' + y' j'$	
$u = x'(i\cos\theta + j\sin\theta) + y'(-i\sin\theta + j\cos\theta)$	Substitute.
$u = ix'\cos\theta + jx'\sin\theta - iy'\sin\theta + jy'\cos\theta$	Distribute.
$u = ix'\cos\theta - iy'\sin\theta + jx'\sin\theta + jy'\cos\theta$	Apply commutative property.
$u = (x'\cos\theta - y'\sin\theta)i + (x'\sin\theta + y'\cos\theta)j$	Factor by grouping.

Because $u = x' i' + y' j'$, we have representations of x and y in terms of the new coordinate system.

$$x = x'\cos\theta - y'\sin\theta \qquad \text{and} \qquad y = x'\sin\theta + y'\cos\theta$$

---

equations of rotation

If a point (x, y) on the Cartesian plane is represented on a new coordinate plane where the axes of rotation are formed by rotating an angle θ from the positive x-axis, then the coordinates of the point with respect to the new axes are (x', y'). We can use the following equations of rotation to define the relationship between (x, y) and (x', y'):

$$x = x'\cos\theta - y'\sin\theta \qquad \text{and} \qquad y = x'\sin\theta + y'\cos\theta$$

How To...

Given the equation of a conic, find a new representation after rotating through an angle.

1. Find x and y where $x = x'\cos\theta - y'\sin\theta$ and $y = x'\sin\theta + y'\cos\theta$.
2. Substitute the expression for x and y into in the given equation, then simplify.
3. Write the equations with x' and y' in standard form.

Example 2   **Finding a New Representation of an Equation after Rotating through a Given Angle**

Find a new representation of the equation $2x^2 - xy + 2y^2 - 30 = 0$ after rotating through an angle of $\theta = 45°$.

Solution   Find x and y, where $x = x'\cos\theta - y'\sin\theta$ and $y = x'\sin\theta + y'\cos\theta$.

Because $\theta = 45°$,

$$x = x' \cos(45°) - y' \sin(45°)$$

$$x = x'\left(\frac{1}{\sqrt{2}}\right) - y'\left(\frac{1}{\sqrt{2}}\right)$$

$$x = \frac{x' - y'}{\sqrt{2}}$$

and

$$y = x' \sin(45°) + y' \cos(45°)$$

$$y = x'\left(\frac{1}{\sqrt{2}}\right) + y'\left(\frac{1}{\sqrt{2}}\right)$$

$$y = \frac{x' + y'}{\sqrt{2}}$$

Substitute $x = x' \cos\theta - y' \sin\theta$ and $y = x' \sin\theta + y' \cos\theta$ into $2x^2 - xy + 2y^2 - 30 = 0$.

$$2\left(\frac{x' - y'}{\sqrt{2}}\right)^2 - \left(\frac{x' - y'}{\sqrt{2}}\right)\left(\frac{x' + y'}{\sqrt{2}}\right) + 2\left(\frac{x' + y'}{\sqrt{2}}\right)^2 - 30 = 0$$

Simplify.

$$\cancel{2}\frac{(x' - y')(x' - y')}{\cancel{2}} - \frac{(x' - y')(x' + y')}{2} + \cancel{2}\frac{(x' + y')(x' + y')}{\cancel{2}} - 30 = 0 \qquad \text{FOIL method}$$

$$x'^2 \cancel{- 2x'y'} + y'^2 - \frac{(x'^2 - y'^2)}{2} + x'^2 \cancel{+ 2x'y'} + y'^2 - 30 = 0 \qquad \text{Combine like terms.}$$

$$2x'^2 + 2y'^2 - \frac{(x'^2 - y'^2)}{2} = 30 \qquad \text{Combine like terms.}$$

$$2\left(2x'^2 + 2y'^2 - \frac{(x'^2 - y'^2)}{2}\right) = 2(30) \qquad \text{Multiply both sides by 2.}$$

$$4x'^2 + 4y'^2 - (x'^2 - y'^2) = 60 \qquad \text{Simplify.}$$

$$4x'^2 + 4y'^2 - x'^2 + y'^2 = 60 \qquad \text{Distribute.}$$

$$\frac{3x'^2}{60} + \frac{5y'^2}{60} = \frac{60}{60} \qquad \text{Set equal to 1.}$$

Write the equations with x' and y' in the standard form.

$$\frac{x'^2}{20} + \frac{y'^2}{12} = 1$$

This equation is an ellipse. **Figure 6** shows the graph.

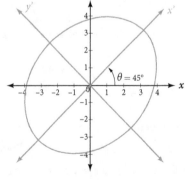

Figure 6

Writing Equations of Rotated Conics in Standard Form

Now that we can find the standard form of a conic when we are given an angle of rotation, we will learn how to transform the equation of a conic given in the form $Ax^2 + Bxy + Cy^2 + Dx + Ey + F = 0$. into standard form by rotating the axes. To do so, we will rewrite the general form as an equation in the x' and y' coordinate system without the $x'y'$ term, by rotating the axes by a measure of θ that satisfies

$$\cot(2\theta) = \frac{A - C}{B}$$

We have learned already that any conic may be represented by the second degree equation

$$Ax^2 + Bxy + Cy^2 + Dx + Ey + F = 0.$$

where A, B, and C are not all zero. However, if $B \neq 0$, then we have an xy term that prevents us from rewriting the equation in standard form. To eliminate it, we can rotate the axes by an acute angle θ where $\cot(2\theta) = \dfrac{A - C}{B}$.

- If $\cot(2\theta) > 0$, then 2θ is in the first quadrant, and θ is between $(0°, 45°)$.
- If $\cot(2\theta) < 0$, then 2θ is in the second quadrant, and θ is between $(45°, 90°)$.
- If $A = C$, then $\theta = 45°$.

How To...

Given an equation for a conic in the $x'y'$ system, rewrite the equation without the $x'y'$ term in terms of x' and y', where the x' and y' axes are rotations of the standard axes by θ degrees.

1. Find $\cot(2\theta)$.
2. Find $\sin\theta$ and $\cos\theta$.
3. Substitute $\sin\theta$ and $\cos\theta$ into $x = x'\cos\theta - y'\sin\theta$ and $y = x'\sin\theta + y'\cos\theta$.
4. Substitute the expression for x and y into in the given equation, and then simplify.
5. Write the equations with x' and y' in the standard form with respect to the rotated axes.

Example 3   **Rewriting an Equation with respect to the x' and y' axes without the $x'y'$ Term**

Rewrite the equation $8x^2 - 12xy + 17y^2 = 20$ in the $x'y'$ system without an $x'y'$ term.

Solution   First, we find $\cot(2\theta)$. See **Figure 7**.

$$8x^2 - 12xy + 17y^2 = 20 \Rightarrow A = 8, B = -12 \text{ and } C = 17$$

$$\cot(2\theta) = \frac{A - C}{B} = \frac{8 - 17}{-12}$$

$$\cot(2\theta) = \frac{-9}{-12} = \frac{3}{4}$$

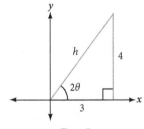

Figure 7

$$\cot(2\theta) = \frac{3}{4} = \frac{\text{adjacent}}{\text{opposite}}$$

So the hypotenuse is

$$3^2 + 4^2 = h^2$$
$$9 + 16 = h^2$$
$$25 = h^2$$
$$h = 5$$

Next, we find $\sin\theta$ and $\cos\theta$.

$$\sin\theta = \sqrt{\frac{1-\cos(2\theta)}{2}} = \sqrt{\frac{1-\frac{3}{5}}{2}} = \sqrt{\frac{\frac{5}{5}-\frac{3}{5}}{2}} = \sqrt{\frac{5-3}{5}\cdot\frac{1}{2}} = \sqrt{\frac{2}{10}} = \sqrt{\frac{1}{5}}$$

$$\sin\theta = \frac{1}{\sqrt{5}}$$

$$\cos\theta = \sqrt{\frac{1+\cos(2\theta)}{2}} = \sqrt{\frac{1+\frac{3}{5}}{2}} = \sqrt{\frac{\frac{5}{5}+\frac{3}{5}}{2}} = \sqrt{\frac{5+3}{5}\cdot\frac{1}{2}} = \sqrt{\frac{8}{10}} = \sqrt{\frac{4}{5}}$$

$$\cos\theta = \frac{2}{\sqrt{5}}$$

Substitute the values of $\sin\theta$ and $\cos\theta$ into $x = x'\cos\theta - y'\sin\theta$ and $y = x'\sin\theta + y'\cos\theta$.

$$x = x'\cos\theta - y'\sin\theta$$

$$x = x'\left(\frac{2}{\sqrt{5}}\right) - y'\left(\frac{1}{\sqrt{5}}\right)$$

$$x = \frac{2x' - y'}{\sqrt{5}}$$

and

$$y = x'\sin\theta + y'\cos\theta$$

$$y = x'\left(\frac{1}{\sqrt{5}}\right) + y'\left(\frac{2}{\sqrt{5}}\right)$$

$$y = \frac{x' + 2y'}{\sqrt{5}}$$

Substitute the expressions for x and y into in the given equation, and then simplify.

$$8\left(\frac{2x' - y'}{\sqrt{5}}\right)^2 - 12\left(\frac{2x' - y'}{\sqrt{5}}\right)\left(\frac{x' + 2y'}{\sqrt{5}}\right) + 17\left(\frac{x' + 2y'}{\sqrt{5}}\right)^2 = 20$$

$$8\left(\frac{(2x' - y')(2x' - y')}{5}\right) - 12\left(\frac{(2x' - y')(x' + 2y')}{5}\right) + 17\left(\frac{(x' + 2y')(x' + 2y')}{5}\right) = 20$$

$$8\left(4x'^2 - 4x'y' + y'^2\right) - 12(2x'^2 + 3x'y' - 2y'^2) + 17(x'^2 + 4x'y' + 4y'^2) = 100$$

$$32x'^2 - 32x'y' + 8y'^2 - 24x'^2 - 36x'y' + 24y'^2 + 17x'^2 + 68x'y' + 68y'^2 = 100$$

$$25x'^2 + 100y'^2 = 100$$

$$\frac{25}{100}x'^2 + \frac{100}{100}y'^2 = \frac{100}{100}$$

Write the equations with x' and y' in the standard form with respect to the new coordinate system.

$$\frac{x'^2}{4} + \frac{y'^2}{1} = 1$$

Figure 8 shows the graph of the ellipse.

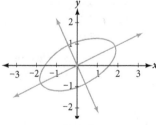

Figure 8

Try It #2
Rewrite the $13x^2 - 6\sqrt{3}xy + 7y^2 = 16$ in the $x'y'$ system without the $x'y'$ term.

Example 4   **Graphing an Equation That Has No $x'y'$ Terms**

Graph the following equation relative to the $x'y'$ system:

$$x^2 + 12xy - 4y^2 = 30$$

Solution   First, we find $\cot(2\theta)$.

$$x^2 + 12xy - 4y^2 = 20 \Rightarrow A = 1, B = 12, \text{ and } C = -4$$

$$\cot(2\theta) = \frac{A - C}{B}$$

$$\cot(2\theta) = \frac{1 - (-4)}{12}$$

$$\cot(2\theta) = \frac{5}{12}$$

Because $\cot(2\theta) = \dfrac{5}{12}$, we can draw a reference triangle as in **Figure 9**.

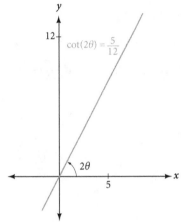

Figure 9

$$\cot(2\theta) = \frac{5}{12} = \frac{\text{adjacent}}{\text{opposite}}$$

Thus, the hypotenuse is

$$5^2 + 12^2 = h^2$$

$$25 + 144 = h^2$$

$$169 = h^2$$

$$h = 13$$

Next, we find $\sin\theta$ and $\cos\theta$. We will use half-angle identities.

$$\sin\theta = \sqrt{\frac{1 - \cos(2\theta)}{2}} = \sqrt{\frac{1 - \frac{5}{13}}{2}} = \sqrt{\frac{\frac{13}{13} - \frac{5}{13}}{2}} = \sqrt{\frac{8}{13} \cdot \frac{1}{2}} = \frac{2}{\sqrt{13}}$$

$$\cos\theta = \sqrt{\frac{1 + \cos(2\theta)}{2}} = \sqrt{\frac{1 + \frac{5}{13}}{2}} = \sqrt{\frac{\frac{13}{13} + \frac{5}{13}}{2}} = \sqrt{\frac{18}{13} \cdot \frac{1}{2}} = \frac{3}{\sqrt{13}}$$

Now we find x and y.

$$x = x'\cos\theta - y'\sin\theta$$

$$x = x'\left(\frac{3}{\sqrt{13}}\right) - y'\left(\frac{2}{\sqrt{13}}\right)$$

$$x = \frac{3x' - 2y'}{\sqrt{13}}$$

and

$$y = x' \sin\theta + y' \cos\theta$$

$$y = x'\left(\frac{2}{\sqrt{13}}\right) + y'\left(\frac{3}{\sqrt{13}}\right)$$

$$y = \frac{2x' + 3y'}{\sqrt{13}}$$

Now we substitute $x = \dfrac{3x' - 2y'}{\sqrt{13}}$ and $y = \dfrac{2x' + 3y'}{\sqrt{13}}$ into $x^2 + 12xy - 4y^2 = 30$.

$$\left(\frac{3x' - 2y'}{\sqrt{13}}\right)^2 + 12\left(\frac{3x' - 2y'}{\sqrt{13}}\right)\left(\frac{2x' + 3y'}{\sqrt{13}}\right) - 4\left(\frac{2x' + 3y'}{\sqrt{13}}\right)^2 = 30$$

$$\left(\frac{1}{13}\right)\left[(3x' - 2y')^2 + 12(3x' - 2y')(2x' + 3y') - 4(2x' + 3y')^2\right] = 30 \qquad \text{Factor.}$$

$$\left(\frac{1}{13}\right)\left[9x'^2 - 12x'y' + 4y'^2 + 12(6x'^2 + 5x'y' - 6y'^2) - 4(4x'^2 + 12x'y' + 9y'^2)\right] = 30 \qquad \text{Multiply.}$$

$$\left(\frac{1}{13}\right)\left[9x'^2 - 12x'y' + 4y'^2 + 72x'^2 + 60x'y' - 72y'^2 - 16x'^2 - 48x'y' - 36y'^2\right] = 30 \qquad \text{Distribute.}$$

$$\left(\frac{1}{13}\right)\left[65x'^2 - 104y'^2\right] = 30 \qquad \text{Combine like terms.}$$

$$65x'^2 - 104y'^2 = 390 \qquad \text{Multiply.}$$

$$\frac{x'^2}{6} - \frac{4y'^2}{15} = 1 \qquad \text{Divide by 390.}$$

Figure 10 shows the graph of the hyperbola $\dfrac{x'^2}{6} - \dfrac{4y'^2}{15} = 1$.

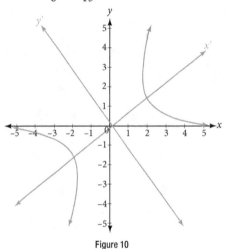

Figure 10

Identifying Conics without Rotating Axes

Now we have come full circle. How do we identify the type of conic described by an equation? What happens when the axes are rotated? Recall, the general form of a conic is

$$Ax^2 + Bxy + Cy^2 + Dx + Ey + F = 0$$

If we apply the rotation formulas to this equation we get the form

$$A'x'^2 + B'x'y' + C'y'^2 + D'x' + E'y' + F' = 0$$

It may be shown that $B^2 - 4AC = B'^2 - 4A'C'$. The expression does not vary after rotation, so we call the expression invariant. The discriminant, $B^2 - 4AC$, is invariant and remains unchanged after rotation. Because the discriminant remains unchanged, observing the discriminant enables us to identify the conic section.

---

using the discriminant to identify a conic

If the equation $Ax^2 + Bxy + Cy^2 + Dx + Ey + F = 0$ is transformed by rotating axes into the equation $A'x'^2 + B'x'y' + C'y'^2 + D'x' + E'y' + F' = 0$, then $B^2 - 4AC = B'^2 - 4A'C'$.

The equation $Ax^2 + Bxy + Cy^2 + Dx + Ey + F = 0$ is an ellipse, a parabola, or a hyperbola, or a degenerate case of one of these.

If the discriminant, $B^2 - 4AC$, is

- < 0, the conic section is an ellipse
- $= 0$, the conic section is a parabola
- > 0, the conic section is a hyperbola

Example 5   **Identifying the Conic without Rotating Axes**

Identify the conic for each of the following without rotating axes.

　a. $5x^2 + 2\sqrt{3}xy + 2y^2 - 5 = 0$

　b. $5x^2 + 2\sqrt{3}xy + 12y^2 - 5 = 0$

Solution

　a. Let's begin by determining A, B, and C.

$$\underset{A}{5x^2} + \underset{B}{2\sqrt{3}xy} + \underset{C}{2y^2} - 5 = 0$$

　　Now, we find the discriminant.

$$B^2 - 4AC = \left(2\sqrt{3}\right)^2 - 4(5)(2)$$
$$= 4(3) - 40$$
$$= 12 - 40$$
$$= -28 < 0$$

　　Therefore, $5x^2 + 2\sqrt{3}xy + 2y^2 - 5 = 0$ represents an ellipse.

　b. Again, let's begin by determining A, B, and C.

$$\underset{A}{5x^2} + \underset{B}{2\sqrt{3}xy} + \underset{C}{12y^2} - 5 = 0$$

　　Now, we find the discriminant.

$$B^2 - 4AC = \left(2\sqrt{3}\right)^2 - 4(5)(12)$$
$$= 4(3) - 240$$
$$= 12 - 240$$
$$= -228 < 0$$

　　Therefore, $5x^2 + 2\sqrt{3}xy + 12y^2 - 5 = 0$ represents an ellipse.

Try It #3

Identify the conic for each of the following without rotating axes.

a. $x^2 - 9xy + 3y^2 - 12 = 0$

b. $10x^2 - 9xy + 4y^2 - 4 = 0$

Access this online resource for additional instruction and practice with conic sections and rotation of axes.

- Introduction to Conic Sections (http://openstaxcollege.org/l/introconic)

8.4   SECTION EXERCISES

VERBAL

1. What effect does the xy term have on the graph of a conic section?

2. If the equation of a conic section is written in the form $Ax^2 + By^2 + Cx + Dy + E = 0$ and $AB = 0$, what can we conclude?

3. If the equation of a conic section is written in the form $Ax^2 + Bxy + Cy^2 + Dx + Ey + F = 0$, and $B^2 - 4AC > 0$, what can we conclude?

4. Given the equation $ax^2 + 4x + 3y^2 - 12 = 0$, what can we conclude if $a > 0$?

5. For the equation $Ax^2 + Bxy + Cy^2 + Dx + Ey + F = 0$, the value of θ that satisfies $\cot(2\theta) = \dfrac{A - C}{B}$ gives us what information?

ALGEBRAIC

For the following exercises, determine which conic section is represented based on the given equation.

6. $9x^2 + 4y^2 + 72x + 36y - 500 = 0$

7. $x^2 - 10x + 4y - 10 = 0$

8. $2x^2 - 2y^2 + 4x - 6y - 2 = 0$

9. $4x^2 - y^2 + 8x - 1 = 0$

10. $4y^2 - 5x + 9y + 1 = 0$

11. $2x^2 + 3y^2 - 8x - 12y + 2 = 0$

12. $4x^2 + 9xy + 4y^2 - 36y - 125 = 0$

13. $3x^2 + 6xy + 3y^2 - 36y - 125 = 0$

14. $-3x^2 + 3\sqrt{3}xy - 4y^2 + 9 = 0$

15. $2x^2 + 4\sqrt{3}xy + 6y^2 - 6x - 3 = 0$

16. $-x^2 + 4\sqrt{2}xy + 2y^2 - 2y + 1 = 0$

17. $8x^2 + 4\sqrt{2}xy + 4y^2 - 10x + 1 = 0$

For the following exercises, find a new representation of the given equation after rotating through the given angle.

18. $3x^2 + xy + 3y^2 - 5 = 0, \theta = 45°$

19. $4x^2 - xy + 4y^2 - 2 = 0, \theta = 45°$

20. $2x^2 + 8xy - 1 = 0, \theta = 30°$

21. $-2x^2 + 8xy + 1 = 0, \theta = 45°$

22. $4x^2 + \sqrt{2}xy + 4y^2 + y + 2 = 0, \theta = 45°$

For the following exercises, determine the angle θ that will eliminate the xy term and write the corresponding equation without the xy term.

23. $x^2 + 3\sqrt{3}xy + 4y^2 + y - 2 = 0$

24. $4x^2 + 2\sqrt{3}xy + 6y^2 + y - 2 = 0$

25. $9x^2 - 3\sqrt{3}xy + 6y^2 + 4y - 3 = 0$

26. $-3x^2 - \sqrt{3}xy - 2y^2 - x = 0$

27. $16x^2 + 24xy + 9y^2 + 6x - 6y + 2 = 0$

28. $x^2 + 4xy + 4y^2 + 3x - 2 = 0$

29. $x^2 + 4xy + y^2 - 2x + 1 = 0$

30. $4x^2 - 2\sqrt{3}xy + 6y^2 - 1 = 0$

GRAPHICAL

For the following exercises, rotate through the given angle based on the given equation. Give the new equation and graph the original and rotated equation.

31. $y = -x^2, \theta = -45°$

32. $x = y^2, \theta = 45°$

33. $\dfrac{x^2}{4} + \dfrac{y^2}{1} = 1, \theta = 45°$

34. $\dfrac{y^2}{16} + \dfrac{x^2}{9} = 1, \theta = 45°$

35. $y^2 - x^2 = 1, \theta = 45°$

36. $y = \dfrac{x^2}{2}, \theta = 30°$

37. $x = (y - 1)^2, \theta = 30°$

38. $\dfrac{x^2}{9} + \dfrac{y^2}{4} = 1, \theta = 30°$

For the following exercises, graph the equation relative to the $x'y'$ system in which the equation has no $x'y'$ term.

39. $xy = 9$

41. $x^2 - 10xy + y^2 - 24 = 0$

43. $6x^2 + 2\sqrt{3}xy + 4y^2 - 21 = 0$

45. $21x^2 + 2\sqrt{3}xy + 19y^2 - 18 = 0$

47. $16x^2 + 24xy + 9y^2 - 60x + 80y = 0$

49. $4x^2 - 4xy + y^2 - 8\sqrt{5}x - 16\sqrt{5}y = 0$

40. $x^2 + 10xy + y^2 - 6 = 0$

42. $4x^2 - 3\sqrt{3}xy + y^2 - 22 = 0$

44. $11x^2 + 10\sqrt{3}xy + y^2 - 64 = 0$

46. $16x^2 + 24xy + 9y^2 - 130x + 90y = 0$

48. $13x^2 - 6\sqrt{3}xy + 7y^2 - 16 = 0$

For the following exercises, determine the angle of rotation in order to eliminate the xy term. Then graph the new set of axes.

50. $6x^2 - 5\sqrt{3}xy + y^2 + 10x - 12y = 0$

52. $6x^2 - 8\sqrt{3}xy + 14y^2 + 10x - 3y = 0$

54. $8x^2 + 3xy + 4y^2 + 2x - 4 = 0$

51. $6x^2 - 5xy + 6y^2 + 20x - y = 0$

53. $4x^2 + 6\sqrt{3}xy + 10y^2 + 20x - 40y = 0$

55. $16x^2 + 24xy + 9y^2 + 20x - 44y = 0$

For the following exercises, determine the value of k based on the given equation.

56. Given $4x^2 + kxy + 16y^2 + 8x + 24y - 48 = 0$, find k for the graph to be a parabola.

58. Given $3x^2 + kxy + 4y^2 - 6x + 20y + 128 = 0$, find k for the graph to be a hyperbola.

60. Given $6x^2 + 12xy + ky^2 + 16x + 10y + 4 = 0$, find k for the graph to be an ellipse.

57. Given $2x^2 + kxy + 12y^2 + 10x - 16y + 28 = 0$, find k for the graph to be an ellipse.

59. Given $kx^2 + 8xy + 8y^2 - 12x + 16y + 18 = 0$, find k for the graph to be a parabola.

In this section, you will:

- Identify a conic in polar form.
- Graph the polar equations of conics.
- Define conics in terms of a focus and a directrix.

8.5   CONIC SECTIONS IN POLAR COORDINATES

Figure 1 Planets orbiting the sun follow elliptical paths. (credit: NASA Blueshift, Flickr)

Most of us are familiar with orbital motion, such as the motion of a planet around the sun or an electron around an atomic nucleus. Within the planetary system, orbits of planets, asteroids, and comets around a larger celestial body are often elliptical. Comets, however, may take on a parabolic or hyperbolic orbit instead. And, in reality, the characteristics of the planets' orbits may vary over time. Each orbit is tied to the location of the celestial body being orbited and the distance and direction of the planet or other object from that body. As a result, we tend to use polar coordinates to represent these orbits.

In an elliptical orbit, the periapsis is the point at which the two objects are closest, and the apoapsis is the point at which they are farthest apart. Generally, the velocity of the orbiting body tends to increase as it approaches the periapsis and decrease as it approaches the apoapsis. Some objects reach an escape velocity, which results in an infinite orbit. These bodies exhibit either a parabolic or a hyperbolic orbit about a body; the orbiting body breaks free of the celestial body's gravitational pull and fires off into space. Each of these orbits can be modeled by a conic section in the polar coordinate system.

Identifying a Conic in Polar Form

Any conic may be determined by three characteristics: a single focus, a fixed line called the directrix, and the ratio of the distances of each to a point on the graph. Consider the parabola $x = 2 + y^2$ shown in **Figure 2**.

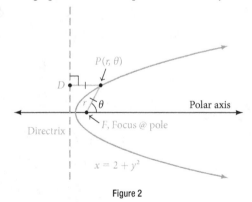

Figure 2

In **The Parabola**, we learned how a parabola is defined by the focus (a fixed point) and the directrix (a fixed line). In this section, we will learn how to define any conic in the polar coordinate system in terms of a fixed point, the focus $P(r, \theta)$ at the pole, and a line, the directrix, which is perpendicular to the polar axis.

If F is a fixed point, the focus, and D is a fixed line, the directrix, then we can let e be a fixed positive number, called the **eccentricity**, which we can define as the ratio of the distances from a point on the graph to the focus and the point on the graph to the directrix. Then the set of all points P such that $e = \dfrac{PF}{PD}$ is a conic. In other words, we can define a conic as the set of all points P with the property that the ratio of the distance from P to F to the distance from P to D is equal to the constant e.

For a conic with eccentricity e,

- if $0 \leq e < 1$, the conic is an ellipse
- if $e = 1$, the conic is a parabola
- if $e > 1$, the conic is an hyperbola

With this definition, we may now define a conic in terms of the directrix, $x = \pm p$, the eccentricity e, and the angle θ. Thus, each conic may be written as a **polar equation**, an equation written in terms of r and θ.

---

the polar equation for a conic

For a conic with a focus at the origin, if the directrix is $x = \pm p$, where p is a positive real number, and the **eccentricity** is a positive real number e, the conic has a **polar equation**

$$r = \frac{ep}{1 \pm e \cos \theta}$$

For a conic with a focus at the origin, if the directrix is $y = \pm p$, where p is a positive real number, and the eccentricity is a positive real number e, the conic has a polar equation

$$r = \frac{ep}{1 \pm e \sin \theta}$$

How To...

Given the polar equation for a conic, identify the type of conic, the directrix, and the eccentricity.

1. Multiply the numerator and denominator by the reciprocal of the constant in the denominator to rewrite the equation in standard form.
2. Identify the eccentricity e as the coefficient of the trigonometric function in the denominator.
3. Compare e with 1 to determine the shape of the conic.
4. Determine the directrix as $x = p$ if cosine is in the denominator and $y = p$ if sine is in the denominator. Set ep equal to the numerator in standard form to solve for x or y.

Example 1   **Identifying a Conic Given the Polar Form**

For each of the following equations, identify the conic with focus at the origin, the directrix, and the eccentricity.

a. $r = \dfrac{6}{3 + 2 \sin \theta}$ **b.** $r = \dfrac{12}{4 + 5 \cos \theta}$ **c.** $r = \dfrac{7}{2 - 2 \sin \theta}$

Solution   For each of the three conics, we will rewrite the equation in standard form. Standard form has a 1 as the constant in the denominator. Therefore, in all three parts, the first step will be to multiply the numerator and denominator by the reciprocal of the constant of the original equation, $\dfrac{1}{c}$, where c is that constant.

a. Multiply the numerator and denominator by $\dfrac{1}{3}$.

$$r = \frac{6}{3 + 2\sin \theta} \cdot \frac{\left(\dfrac{1}{3}\right)}{\left(\dfrac{1}{3}\right)} = \frac{6\left(\dfrac{1}{3}\right)}{3\left(\dfrac{1}{3}\right) + 2\left(\dfrac{1}{3}\right)\sin \theta} = \frac{2}{1 + \dfrac{2}{3}\sin \theta}$$

Because $\sin \theta$ is in the denominator, the directrix is $y = p$. Comparing to standard form, note that $e = \frac{2}{3}$. Therefore, from the numerator,

$$2 = ep$$
$$2 = \frac{2}{3}p$$
$$\left(\frac{3}{2}\right)2 = \left(\frac{3}{2}\right)\frac{2}{3}p$$
$$3 = p$$

Since $e < 1$, the conic is an ellipse. The eccentricity is $e = \frac{2}{3}$ and the directrix is $y = 3$.

b. Multiply the numerator and denominator by $\frac{1}{4}$.

$$r = \frac{12}{4 + 5\cos\theta} \cdot \frac{\left(\frac{1}{4}\right)}{\left(\frac{1}{4}\right)}$$
$$r = \frac{12\left(\frac{1}{4}\right)}{4\left(\frac{1}{4}\right) + 5\left(\frac{1}{4}\right)\cos\theta}$$
$$r = \frac{3}{1 + \frac{5}{4}\cos\theta}$$

Because $\cos\theta$ is in the denominator, the directrix is $x = p$. Comparing to standard form, $e = \frac{5}{4}$. Therefore, from the numerator,

$$3 = ep$$
$$3 = \frac{5}{4}p$$
$$\left(\frac{4}{5}\right)3 = \left(\frac{4}{5}\right)\frac{5}{4}p$$
$$\frac{12}{5} = p$$

Since $e > 1$, the conic is a hyperbola. The eccentricity is $e = \frac{5}{4}$ and the directrix is $x = \frac{12}{5} = 2.4$.

c. Multiply the numerator and denominator by $\frac{1}{2}$.

$$r = \frac{7}{2 - 2\sin\theta} \cdot \frac{\left(\frac{1}{2}\right)}{\left(\frac{1}{2}\right)}$$
$$r = \frac{7\left(\frac{1}{2}\right)}{2\left(\frac{1}{2}\right) - 2\left(\frac{1}{2}\right)\sin\theta}$$
$$r = \frac{\frac{7}{2}}{1 - \sin\theta}$$

Because sine is in the denominator, the directrix is $y = -p$. Comparing to standard form, $e = 1$. Therefore, from the numerator,

$$\frac{7}{2} = ep$$
$$\frac{7}{2} = (1)p$$
$$\frac{7}{2} = p$$

Because $e = 1$, the conic is a parabola. The eccentricity is $e = 1$ and the directrix is $y = -\frac{7}{2} = -3.5$.

Try It #1

Identify the conic with focus at the origin, the directrix, and the eccentricity for $r = \dfrac{2}{3 - \cos \theta}$.

Graphing the Polar Equations of Conics

When graphing in Cartesian coordinates, each conic section has a unique equation. This is not the case when graphing in polar coordinates. We must use the eccentricity of a conic section to determine which type of curve to graph, and then determine its specific characteristics. The first step is to rewrite the conic in standard form as we have done in the previous example. In other words, we need to rewrite the equation so that the denominator begins with 1. This enables us to determine e and, therefore, the shape of the curve. The next step is to substitute values for θ and solve for r to plot a few key points. Setting θ equal to 0, $\dfrac{\pi}{2}$, π, and $\dfrac{3\pi}{2}$ provides the vertices so we can create a rough sketch of the graph.

Example 2   Graphing a Parabola in Polar Form

Graph $r = \dfrac{5}{3 + 3 \cos \theta}$.

Solution   First, we rewrite the conic in standard form by multiplying the numerator and denominator by the reciprocal of 3, which is $\dfrac{1}{3}$.

$$r = \frac{5}{3 + 3 \cos \theta} = \frac{5\left(\dfrac{1}{3}\right)}{3\left(\dfrac{1}{3}\right) + 3\left(\dfrac{1}{3}\right)\cos \theta}$$

$$r = \frac{\dfrac{5}{3}}{1 + \cos \theta}$$

Because $e = 1$, we will graph a parabola with a focus at the origin. The function has a $\cos \theta$, and there is an addition sign in the denominator, so the directrix is $x = p$.

$$\frac{5}{3} = ep$$

$$\frac{5}{3} = (1)p$$

$$\frac{5}{3} = p$$

The directrix is $x = \dfrac{5}{3}$.

Plotting a few key points as in **Table 1** will enable us to see the vertices. See **Figure 3**.

	A	**B**	**C**	**D**
θ	0	$\dfrac{\pi}{2}$	π	$\dfrac{3\pi}{2}$
$r = \dfrac{5}{3 + 3 \cos \theta}$	$\dfrac{5}{6} \approx 0.83$	$\dfrac{5}{3} \approx 1.67$	undefined	$\dfrac{5}{3} \approx 1.67$

Table 1

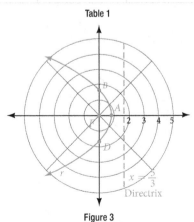

Figure 3

Analysis We can check our result with a graphing utility. See **Figure 4**.

Figure 4

Example 3   **Graphing a Hyperbola in Polar Form**

Graph $r = \dfrac{8}{2 - 3\sin\theta}$.

Solution   First, we rewrite the conic in standard form by multiplying the numerator and denominator by the reciprocal of 2, which is $\dfrac{1}{2}$.

$$r = \frac{8}{2 - 3\sin\theta} = \frac{8\left(\frac{1}{2}\right)}{2\left(\frac{1}{2}\right) - 3\left(\frac{1}{2}\right)\sin\theta}$$

$$r = \frac{4}{1 - \frac{3}{2}\cos\theta}$$

Because $e = \dfrac{3}{2}$, $e > 1$, so we will graph a hyperbola with a focus at the origin. The function has a $\sin\theta$ term and there is a subtraction sign in the denominator, so the directrix is $y = -p$.

$$4 = ep$$

$$4 = \left(\frac{3}{2}\right)p$$

$$4\left(\frac{2}{3}\right) = p$$

$$\frac{8}{3} = p$$

The directrix is $y = -\dfrac{8}{3}$.

Plotting a few key points as in **Table 2** will enable us to see the vertices. See **Figure 5**.

	A	B	C	D
θ	0	$\dfrac{\pi}{2}$	π	$\dfrac{3\pi}{2}$
$r = \dfrac{8}{2 - 3\sin\theta}$	4	-8	4	$\dfrac{8}{5} = 1.6$

Table 2

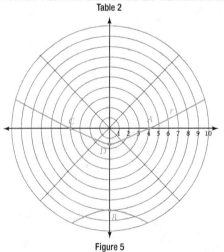

Figure 5

8.3 The Parabola

- A parabola is the set of all points (x, y) in a plane that are the same distance from a fixed line, called the directrix, and a fixed point (the focus) not on the directrix.
- The standard form of a parabola with vertex $(0, 0)$ and the x-axis as its axis of symmetry can be used to graph the parabola. If $p > 0$, the parabola opens right. If $p < 0$, the parabola opens left. See **Example 1**.
- The standard form of a parabola with vertex $(0, 0)$ and the y-axis as its axis of symmetry can be used to graph the parabola. If $p > 0$, the parabola opens up. If $p < 0$, the parabola opens down. See **Example 2**.
- When given the focus and directrix of a parabola, we can write its equation in standard form. See **Example 3**.
- The standard form of a parabola with vertex (h, k) and axis of symmetry parallel to the x-axis can be used to graph the parabola. If $p > 0$, the parabola opens right. If $p < 0$, the parabola opens left. See **Example 4**.
- The standard form of a parabola with vertex (h, k) and axis of symmetry parallel to the y-axis can be used to graph the parabola. If $p > 0$, the parabola opens up. If $p < 0$, the parabola opens down. See **Example 5**.
- Real-world situations can be modeled using the standard equations of parabolas. For instance, given the diameter and focus of a cross-section of a parabolic reflector, we can find an equation that models its sides. See **Example 6**.

8.4 Rotation of Axes

- Four basic shapes can result from the intersection of a plane with a pair of right circular cones connected tail to tail. They include an ellipse, a circle, a hyperbola, and a parabola.
- A nondegenerate conic section has the general form $Ax^2 + Bxy + Cy^2 + Dx + Ey + F = 0$ where A, B and C are not all zero. The values of A, B, and C determine the type of conic. See **Example 1**.
- Equations of conic sections with an xy term have been rotated about the origin. See **Example 2**.
- The general form can be transformed into an equation in the x' and y' coordinate system without the $x'y'$ term. See **Example 3** and **Example 4**.
- An expression is described as invariant if it remains unchanged after rotating. Because the discriminant is invariant, observing it enables us to identify the conic section. See **Example 5**.

8.5 Conic Sections in Polar Coordinates

- Any conic may be determined by a single focus, the corresponding eccentricity, and the directrix. We can also define a conic in terms of a fixed point, the focus $P(r, \theta)$ at the pole, and a line, the directrix, which is perpendicular to the polar axis.
- A conic is the set of all points $e = \dfrac{PF}{PD}$, where eccentricity e is a positive real number. Each conic may be written in terms of its polar equation. See **Example 1**.
- The polar equations of conics can be graphed. See **Example 2**, **Example 3**, and **Example 4**.
- Conics can be defined in terms of a focus, a directrix, and eccentricity. See **Example 5** and **Example 6**.
- We can use the identities $r = \sqrt{x^2 + y^2}$, $x = r\cos\theta$, and $y = r\sin\theta$ to convert the equation for a conic from polar to rectangular form. See **Example 7**.

CHAPTER 8 REVIEW EXERCISES

THE ELLIPSE

For the following exercises, write the equation of the ellipse in standard form. Then identify the center, vertices, and foci.

1. $\dfrac{x^2}{25} + \dfrac{y^2}{64} = 1$

2. $\dfrac{(x-2)^2}{100} + \dfrac{(y+3)^2}{36} = 1$

3. $9x^2 + y^2 + 54x - 4y + 76 = 0$

4. $9x^2 + 36y^2 - 36x + 72y + 36 = 0$

For the following exercises, graph the ellipse, noting center, vertices, and foci.

5. $\dfrac{x^2}{36} + \dfrac{y^2}{9} = 1$

6. $\dfrac{(x-4)^2}{25} + \dfrac{(y+3)^2}{49} = 1$

7. $4x^2 + y^2 + 16x + 4y - 44 = 0$

8. $2x^2 + 3y^2 - 20x + 12y + 38 = 0$

For the following exercises, use the given information to find the equation for the ellipse.

9. Center at $(0, 0)$, focus at $(3, 0)$, vertex at $(-5, 0)$

10. Center at $(2, -2)$, vertex at $(7, -2)$, focus at $(4, -2)$

11. A whispering gallery is to be constructed such that the foci are located 35 feet from the center. If the length of the gallery is to be 100 feet, what should the height of the ceiling be?

THE HYPERBOLA

For the following exercises, write the equation of the hyperbola in standard form. Then give the center, vertices, and foci.

12. $\dfrac{x^2}{81} - \dfrac{y^2}{9} = 1$

13. $\dfrac{(y+1)^2}{16} - \dfrac{(x-4)^2}{36} = 1$

14. $9y^2 - 4x^2 + 54y - 16x + 29 = 0$

15. $3x^2 - y^2 - 12x - 6y - 9 = 0$

For the following exercises, graph the hyperbola, labeling vertices and foci.

16. $\dfrac{x^2}{9} - \dfrac{y^2}{16} = 1$

17. $\dfrac{(y-1)^2}{49} - \dfrac{(x+1)^2}{4} = 1$

18. $x^2 - 4y^2 + 6x + 32y - 91 = 0$

19. $2y^2 - x^2 - 12y - 6 = 0$

For the following exercises, find the equation of the hyperbola.

20. Center at $(0, 0)$, vertex at $(0, 4)$, focus at $(0, -6)$

21. Foci at $(3, 7)$ and $(7, 7)$, vertex at $(6, 7)$

THE PARABOLA

For the following exercises, write the equation of the parabola in standard form. Then give the vertex, focus, and directrix.

22. $y^2 = 12x$

23. $(x+2)^2 = \dfrac{1}{2}(y-1)$

24. $y^2 - 6y - 6x - 3 = 0$

25. $x^2 + 10x - y + 23 = 0$

For the following exercises, graph the parabola, labeling vertex, focus, and directrix.

26. $x^2 + 4y = 0$

27. $(y-1)^2 = \dfrac{1}{2}(x+3)$

28. $x^2 - 8x - 10y + 46 = 0$

29. $2y^2 + 12y + 6x + 15 = 0$

For the following exercises, write the equation of the parabola using the given information.

30. Focus at $(-4, 0)$; directrix is $x = 4$

31. Focus at $\left(2, \dfrac{9}{8}\right)$; directrix is $y = \dfrac{7}{8}$

32. A cable TV receiving dish is the shape of a paraboloid of revolution. Find the location of the receiver, which is placed at the focus, if the dish is 5 feet across at its opening and 1.5 feet deep.

ROTATION OF AXES

For the following exercises, determine which of the conic sections is represented.

33. $16x^2 + 24xy + 9y^2 + 24x - 60y - 60 = 0$

34. $4x^2 + 14xy + 5y^2 + 18x - 6y + 30 = 0$

35. $4x^2 + xy + 2y^2 + 8x - 26y + 9 = 0$

For the following exercises, determine the angle θ that will eliminate the xy term, and write the corresponding equation without the xy term.

36. $x^2 + 4xy - 2y^2 - 6 = 0$

37. $x^2 - xy + y^2 - 6 = 0$

For the following exercises, graph the equation relative to the $x'y'$ system in which the equation has no $x'y'$ term.

38. $9x^2 - 24xy + 16y^2 - 80x - 60y + 100 = 0$

39. $x^2 - xy + y^2 - 2 = 0$

40. $6x^2 + 24xy - y^2 - 12x + 26y + 11 = 0$

CONIC SECTIONS IN POLAR COORDINATES

For the following exercises, given the polar equation of the conic with focus at the origin, identify the eccentricity and directrix.

41. $r = \dfrac{10}{1 - 5\cos\theta}$

42. $r = \dfrac{6}{3 + 2\cos\theta}$

43. $r = \dfrac{1}{4 + 3\sin\theta}$

44. $r = \dfrac{3}{5 - 5\sin\theta}$

For the following exercises, graph the conic given in polar form. If it is a parabola, label the vertex, focus, and directrix. If it is an ellipse or a hyperbola, label the vertices and foci.

45. $r = \dfrac{3}{1 - \sin\theta}$

46. $r = \dfrac{8}{4 + 3\sin\theta}$

47. $r = \dfrac{10}{4 + 5\cos\theta}$

48. $r = \dfrac{9}{3 - 6\cos\theta}$

For the following exercises, given information about the graph of a conic with focus at the origin, find the equation in polar form.

49. Directrix is $x = 3$ and eccentricity $e = 1$

50. Directrix is $y = -2$ and eccentricity $e = 4$

CHAPTER 8 PRACTICE TEST

For the following exercises, write the equation in standard form and state the center, vertices, and foci.

1. $\dfrac{x^2}{9} + \dfrac{y^2}{4} = 1$

2. $9y^2 + 16x^2 - 36y + 32x - 92 = 0$

For the following exercises, sketch the graph, identifying the center, vertices, and foci.

3. $\dfrac{(x-3)^2}{64} + \dfrac{(y-2)^2}{36} = 1$

4. $2x^2 + y^2 + 8x - 6y - 7 = 0$

5. Write the standard form equation of an ellipse with a center at $(1, 2)$, vertex at $(7, 2)$, and focus at $(4, 2)$.

6. A whispering gallery is to be constructed with a length of 150 feet. If the foci are to be located 20 feet away from the wall, how high should the ceiling be?

For the following exercises, write the equation of the hyperbola in standard form, and give the center, vertices, foci, and asymptotes.

7. $\dfrac{x^2}{49} - \dfrac{y^2}{81} = 1$

8. $16y^2 - 9x^2 + 128y + 112 = 0$

For the following exercises, graph the hyperbola, noting its center, vertices, and foci. State the equations of the asymptotes.

9. $\dfrac{(x-3)^2}{25} - \dfrac{(y+3)^2}{1} = 1$

10. $y^2 - x^2 + 4y - 4x - 18 = 0$

11. Write the standard form equation of a hyperbola with foci at $(1, 0)$ and $(1, 6)$, and a vertex at $(1, 2)$.

For the following exercises, write the equation of the parabola in standard form, and give the vertex, focus, and equation of the directrix.

12. $y^2 + 10x = 0$

13. $3x^2 - 12x - y + 11 = 0$

For the following exercises, graph the parabola, labeling the vertex, focus, and directrix.

14. $(x-1)^2 = -4(y+3)$

15. $y^2 + 8x - 8y + 40 = 0$

16. Write the equation of a parabola with a focus at $(2, 3)$ and directrix $y = -1$.

17. A searchlight is shaped like a paraboloid of revolution. If the light source is located 1.5 feet from the base along the axis of symmetry, and the depth of the searchlight is 3 feet, what should the width of the opening be?

For the following exercises, determine which conic section is represented by the given equation, and then determine the angle θ that will eliminate the xy term.

18. $3x^2 - 2xy + 3y^2 = 4$

19. $x^2 + 4xy + 4y^2 + 6x - 8y = 0$

For the following exercises, rewrite in the $x'y'$ system without the $x'y'$ term, and graph the rotated graph.

20. $11x^2 + 10\sqrt{3}xy + y^2 = 4$

21. $16x^2 + 24xy + 9y^2 - 125x = 0$

For the following exercises, identify the conic with focus at the origin, and then give the directrix and eccentricity.

22. $r = \dfrac{3}{2 - \sin\theta}$

23. $r = \dfrac{5}{4 + 6\cos\theta}$

For the following exercises, graph the given conic section. If it is a parabola, label vertex, focus, and directrix. If it is an ellipse or a hyperbola, label vertices and foci.

24. $r = \dfrac{12}{4 - 8\sin\theta}$

25. $r = \dfrac{2}{4 + 4\sin\theta}$

26. Find a polar equation of the conic with focus at the origin, eccentricity of $e = 2$, and directrix: $x = 3$.

Sequences, Probability and Counting Theory

Figure 1 (credit: Robert S. Donovan, Flickr.)

CHAPTER OUTLINE

Introduction

A lottery winner has some big decisions to make regarding what to do with the winnings. Buy a villa in Saint Barthélemy? A luxury convertible? A cruise around the world?

The likelihood of winning the lottery is slim, but we all love to fantasize about what we could buy with the winnings. One of the first things a lottery winner has to decide is whether to take the winnings in the form of a lump sum or as a series of regular payments, called an annuity, over the next 30 years or so.

This decision is often based on many factors, such as tax implications, interest rates, and investment strategies. There are also personal reasons to consider when making the choice, and one can make many arguments for either decision. However, most lottery winners opt for the lump sum.

In this chapter, we will explore the mathematics behind situations such as these. We will take an in-depth look at annuities. We will also look at the branch of mathematics that would allow us to calculate the number of ways to choose lottery numbers and the probability of winning.

755

LEARNING OBJECTIVES

In this section, you will:

- Write the terms of a sequence defined by an explicit formula.
- Write the terms of a sequence defined by a recursive formula.
- Use factorial notation.

9.1 SEQUENCES AND THEIR NOTATIONS

A video game company launches an exciting new advertising campaign. They predict the number of online visits to their website, or hits, will double each day. The model they are using shows 2 hits the first day, 4 hits the second day, 8 hits the third day, and so on. See **Table 1**.

Day	1	2	3	4	5	...
Hits	2	4	8	16	32	...

Table 1

If their model continues, how many hits will there be at the end of the month? To answer this question, we'll first need to know how to determine a list of numbers written in a specific order. In this section, we will explore these kinds of ordered lists.

Writing the Terms of a Sequence Defined by an Explicit Formula

One way to describe an ordered list of numbers is as a **sequence**. A sequence is a function whose domain is a subset of the counting numbers. The sequence established by the number of hits on the website is

$$\{2, 4, 8, 16, 32, \ldots\}.$$

The ellipsis (…) indicates that the sequence continues indefinitely. Each number in the sequence is called a **term**. The first five terms of this sequence are 2, 4, 8, 16, and 32.

Listing all of the terms for a sequence can be cumbersome. For example, finding the number of hits on the website at the end of the month would require listing out as many as 31 terms. A more efficient way to determine a specific term is by writing a formula to define the sequence.

One type of formula is an explicit formula, which defines the terms of a sequence using their position in the sequence. Explicit formulas are helpful if we want to find a specific term of a sequence without finding all of the previous terms. We can use the formula to find the nth term of the sequence, where n is any positive number. In our example, each number in the sequence is double the previous number, so we can use powers of 2 to write a formula for the nth term.

$$\{2,\ 4,\ 8,\ 16,\ 22, \ldots, ?, \ldots\}$$
$$\downarrow\ \downarrow\ \downarrow\ \downarrow\ \downarrow\qquad\ \downarrow$$
$$\{2^1, 2^2, 2^3, 2^4,\ 2^5, \ldots, 2^n, \ldots\}$$

The first term of the sequence is $2^1 = 2$, the second term is $2^2 = 4$, the third term is $2^3 = 8$, and so on. The nth term of the sequence can be found by raising 2 to the nth power. An explicit formula for a sequence is named by a lower case letter $a, b, c\ldots$ with the subscript n. The explicit formula for this sequence is

$$a_n = 2^n.$$

Now that we have a formula for the nth term of the sequence, we can answer the question posed at the beginning of this section. We were asked to find the number of hits at the end of the month, which we will take to be 31 days. To find the number of hits on the last day of the month, we need to find the 31st term of the sequence. We will substitute 31 for n in the formula.

$$a_{31} = 2^{31}$$
$$= 2{,}147{,}483{,}648$$

If the doubling trend continues, the company will get 2,147,483,648 hits on the last day of the month. That is over 2.1 billion hits! The huge number is probably a little unrealistic because it does not take consumer interest and competition into account. It does, however, give the company a starting point from which to consider business decisions.

Another way to represent the sequence is by using a table. The first five terms of the sequence and the nth term of the sequence are shown in **Table 2**.

n	1	2	3	4	5	n
nth term of the sequence, a_n	2	4	8	16	32	2^n

Table 2

Graphing provides a visual representation of the sequence as a set of distinct points. We can see from the graph in **Figure 1** that the number of hits is rising at an exponential rate. This particular sequence forms an exponential function.

Figure 1

Lastly, we can write this particular sequence as

$$\{2, 4, 8, 16, 32, \ldots, 2^n, \ldots\}$$

A sequence that continues indefinitely is called an **infinite sequence**. The domain of an infinite sequence is the set of counting numbers. If we consider only the first 10 terms of the sequence, we could write

$$\{2, 4, 8, 16, 32, \ldots, 2^n, \ldots, 1024\}.$$

This sequence is called a **finite sequence** because it does not continue indefinitely.

---

sequence

A **sequence** is a function whose domain is the set of positive integers. A **finite sequence** is a sequence whose domain consists of only the first n positive integers. The numbers in a sequence are called **terms**. The variable a with a number subscript is used to represent the terms in a sequence and to indicate the position of the term in the sequence.

$$a_1, a_2, a_3, \ldots, a_n, \ldots$$

We call a_1 the first term of the sequence, a_2 the second term of the sequence, a_3 the third term of the sequence, and so on. The term an is called the ***nth term of the sequence***, or the general term of the sequence. An **explicit formula** defines the nth term of a sequence using the position of the term. A sequence that continues indefinitely is an **infinite sequence**.

Q & A...

Does a sequence always have to begin with a_1?

No. In certain problems, it may be useful to define the initial term as a_0 instead of a_1. In these problems, the domain of the function includes 0.

Given an explicit formula, write the first n terms of a sequence.

1. Substitute each value of n into the formula. Begin with $n = 1$ to find the first term, a_1.
2. To find the second term, a_2, use $n = 2$.
3. Continue in the same manner until you have identified all n terms.

Example 1   **Writing the Terms of a Sequence Defined by an Explicit Formula**

Write the first five terms of the sequence defined by the explicit formula $a_n = -3n + 8$.

Solution   Substitute $n = 1$ into the formula. Repeat with values 2 through 5 for n.

$$n = 1 \quad a_1 = -3(1) + 8 = 5$$
$$n = 2 \quad a_2 = -3(2) + 8 = 2$$
$$n = 3 \quad a_3 = -3(3) + 8 = -1$$
$$n = 4 \quad a_4 = -3(4) + 8 = -4$$
$$n = 5 \quad a_5 = -3(5) + 8 = -7$$

The first five terms are {5, 2, −1, −4, −7}.

*Analysis   The sequence values can be listed in a table. A table, such as **Table 3**, is a convenient way to input the function into a graphing utility.*

n	1	2	3	4	5
a_n	5	2	−1	−4	−7

Table 3

A graph can be made from this table of values. From the graph in **Figure 2**, we can see that this sequence represents a linear function, but notice the graph is not continuous because the domain is over the positive integers only.

Figure 2

Try It #1

Write the first five terms of the sequence defined by the explicit formula $t_n = 5n - 4$.

Investigating Alternating Sequences

Sometimes sequences have terms that are alternate. In fact, the terms may actually alternate in sign. The steps to finding terms of the sequence are the same as if the signs did not alternate. However, the resulting terms will not show increase or decrease as n increases. Let's take a look at the following sequence.

$$\{2, -4, 6, -8\}$$

Notice the first term is greater than the second term, the second term is less than the third term, and the third term is greater than the fourth term. This trend continues forever. Do not rearrange the terms in numerical order to interpret the sequence.

How To…

Given an explicit formula with alternating terms, write the first n terms of a sequence.

1. Substitute each value of n into the formula. Begin with $n = 1$ to find the first term, a_1. The sign of the term is given by the $(-1)^n$ in the explicit formula.
2. To find the second term, a_2, use $n = 2$.
3. Continue in the same manner until you have identified all n terms.

Example 2   **Writing the Terms of an Alternating Sequence Defined by an Explicit Formula**

Write the first five terms of the sequence.

$$a_n = \frac{(-1)^n n^2}{n + 1}$$

Solution   Substitute $n = 1$, $n = 2$, and so on in the formula.

$$n = 1 \quad a_1 = \frac{(-1)^1 1^1}{1 + 1} = -\frac{1}{2}$$

$$n = 2 \quad a_2 = \frac{(-1)^2 2^2}{2 + 1} = \frac{4}{3}$$

$$n = 3 \quad a_3 = \frac{(-1)^3 3^2}{3 + 1} = -\frac{9}{4}$$

$$n = 4 \quad a_4 = \frac{(-1)^4 4^2}{4 + 1} = \frac{16}{5}$$

$$n = 5 \quad a_5 = \frac{(-1)^5 5^2}{5 + 1} = -\frac{25}{6}$$

The first five terms are $\left\{ -\dfrac{1}{2}, \dfrac{4}{3}, -\dfrac{9}{4}, \dfrac{16}{5}, -\dfrac{25}{6} \right\}$

Analysis   *The graph of this function, shown in **Figure 3**, looks different from the ones we have seen previously in this section because the terms of the sequence alternate between positive and negative values.*

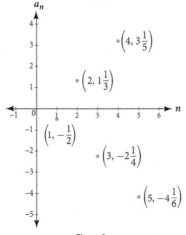

Figure 3

Q & A...

In Example 2, does the (−1) to the power of *n* account for the oscillations of signs?

Yes, the power might be n, $n + 1$, $n − 1$, and so on, but any odd powers will result in a negative term, and any even power will result in a positive term.

Try It #2

Write the first five terms of the sequence. $a_n = \dfrac{4n}{(-2)^n}$

Investigating Piecewise Explicit Formulas

We've learned that sequences are functions whose domain is over the positive integers. This is true for other types of functions, including some piecewise functions. Recall that a piecewise function is a function defined by multiple subsections. A different formula might represent each individual subsection.

How To...

Given an explicit formula for a piecewise function, write the first *n* terms of a sequence

1. Identify the formula to which $n = 1$ applies.
2. To find the first term, a_1, use $n = 1$ in the appropriate formula.
3. Identify the formula to which $n = 2$ applies.
4. To find the second term, a_2, use $n = 2$ in the appropriate formula.
5. Continue in the same manner until you have identified all *n* terms.

Example 3   **Writing the Terms of a Sequence Defined by a Piecewise Explicit Formula**

Write the first six terms of the sequence.

$$a_n = \begin{cases} n^2 & \text{if } n \text{ is not divisible by 3} \\ \dfrac{n}{3} & \text{if } n \text{ is divisible by 3} \end{cases}$$

Solution   Substitute $n = 1$, $n = 2$, and so on in the appropriate formula. Use n^2 when n is not a multiple of 3. Use $\dfrac{n}{3}$ when n is a multiple of 3.

$$a_1 = 1^2 = 1 \qquad \text{1 is not a multiple of 3. Use } n^2.$$
$$a_2 = 2^2 = 4 \qquad \text{2 is not a multiple of 3. Use } n^2.$$
$$a_3 = \frac{3}{3} = 1 \qquad \text{3 is a multiple of 3. Use } \frac{n}{3}.$$
$$a_4 = 4^2 = 16 \qquad \text{4 is not a multiple of 3. Use } n^2.$$
$$a_5 = 5^2 = 25 \qquad \text{5 is not a multiple of 3. Use } n^2.$$
$$a_6 = \frac{6}{3} = 2 \qquad \text{6 is a multiple of 3. Use } \frac{n}{3}.$$

The first six terms are {1, 4, 1, 16, 25, 2}.

Analysis   *Every third point on the graph shown in* **Figure 4** *stands out from the two nearby points. This occurs because the sequence was defined by a piecewise function.*

Figure 4

Try It #3

Write the first six terms of the sequence. $a_n = \begin{cases} 2n^3 & \text{if } n \text{ is odd} \\ \dfrac{5n}{2} & \text{if } n \text{ is even} \end{cases}$

Finding an Explicit Formula

Thus far, we have been given the explicit formula and asked to find a number of terms of the sequence. Sometimes, the explicit formula for the nth term of a sequence is not given. Instead, we are given several terms from the sequence. When this happens, we can work in reverse to find an explicit formula from the first few terms of a sequence. The key to finding an explicit formula is to look for a pattern in the terms. Keep in mind that the pattern may involve alternating terms, formulas for numerators, formulas for denominators, exponents, or bases.

How To...

Given the first few terms of a sequence, find an explicit formula for the sequence.

1. Look for a pattern among the terms.
2. If the terms are fractions, look for a separate pattern among the numerators and denominators.
3. Look for a pattern among the signs of the terms.
4. Write a formula for a_n in terms of n. Test your formula for $n = 1$, $n = 2$, and $n = 3$.

Example 4   **Writing an Explicit Formula for the *n*th Term of a Sequence**

Write an explicit formula for the nth term of each sequence.

a. $\left\{ -\dfrac{2}{11}, \dfrac{3}{13}, -\dfrac{4}{15}, \dfrac{5}{17}, -\dfrac{6}{19}, \ldots \right\}$

b. $\left\{ -\dfrac{2}{25}, -\dfrac{2}{125}, -\dfrac{2}{625}, -\dfrac{2}{3,125}, -\dfrac{2}{15,625}, \ldots \right\}$

c. $\{e^4, e^5, e^6, e^7, e^8, \ldots\}$

Solution   Look for the pattern in each sequence.

a. The terms alternate between positive and negative. We can use $(-1)^n$ to make the terms alternate. The numerator can be represented by $n + 1$. The denominator can be represented by $2n + 9$.

$$a_n = \frac{(-1)^n (n+1)}{2n+9}$$

b. The terms are all negative.

$\left\{ -\dfrac{2}{25}, -\dfrac{2}{125}, -\dfrac{2}{625}, -\dfrac{2}{3,125}, -\dfrac{2}{15,625}, \ldots \right\}$ The numerator is 2

$\left\{ -\dfrac{2}{5^2}, -\dfrac{2}{5^3}, -\dfrac{2}{5^4}, -\dfrac{2}{5^5}, -\dfrac{2}{5^6}, -\dfrac{2}{5^7}, \ldots -\dfrac{2}{5^n} \right\}$ The denominators are increasing powers of 5

So we know that the fraction is negative, the numerator is 2, and the denominator can be represented by 5^{n+1}.

$$a_n = -\frac{2}{5^{n+1}}$$

c. The terms are powers of e. For $n = 1$, the first term is e^4 so the exponent must be $n + 3$.

$$a_n = e^{n+3}$$

Try It #4

Write an explicit formula for the nth term of the sequence.

$$\{9, -81, 729, -6,561, 59,049, \ldots\}$$

Try It #5

Write an explicit formula for the *n*th term of the sequence.

$$\left\{ -\frac{3}{4}, -\frac{9}{8}, -\frac{27}{12}, -\frac{81}{16}, -\frac{243}{20}, \dots \right\}$$

Try It #6

Write an explicit formula for the *n*th term of the sequence.

$$\left\{ \frac{1}{e^2}, \ \frac{1}{e}, \ 1, \ e, \ e^2, \dots \right\}$$

Writing the Terms of a Sequence Defined by a Recursive Formula

Sequences occur naturally in the growth patterns of nautilus shells, pinecones, tree branches, and many other natural structures. We may see the sequence in the leaf or branch arrangement, the number of petals of a flower, or the pattern of the chambers in a nautilus shell. Their growth follows the Fibonacci sequence, a famous sequence in which each term can be found by adding the preceding two terms. The numbers in the sequence are 1, 1, 2, 3, 5, 8, 13, 21, 34,.... Other examples from the natural world that exhibit the Fibonacci sequence are the Calla Lily, which has just one petal, the Black-Eyed Susan with 13 petals, and different varieties of daisies that may have 21 or 34 petals.

Each term of the Fibonacci sequence depends on the terms that come before it. The Fibonacci sequence cannot easily be written using an explicit formula. Instead, we describe the sequence using a **recursive formula**, a formula that defines the terms of a sequence using previous terms.

A recursive formula always has two parts: the value of an initial term (or terms), and an equation defining a_n in terms of preceding terms. For example, suppose we know the following:

$$a_1 = 3$$
$$a_n = 2a_{n-1} - 1, \text{ for } n \geq 2$$

We can find the subsequent terms of the sequence using the first term.

$$a_1 = 3$$
$$a_2 = 2a_1 - 1 = 2(3) - 1 = 5$$
$$a_3 = 2a_2 - 1 = 2(5) - 1 = 9$$
$$a_4 = 2a_3 - 1 = 2(9) - 1 = 17$$

So the first four terms of the sequence are {3, 5, 9, 17} .

The recursive formula for the Fibonacci sequence states the first two terms and defines each successive term as the sum of the preceding two terms.

$$a_1 = 1$$
$$a_2 = 1$$
$$a_n = a_{n-1} + a_{n-2} \text{ for } n \geq 3$$

To find the tenth term of the sequence, for example, we would need to add the eighth and ninth terms. We were told previously that the eighth and ninth terms are 21 and 34, so

$$a_{10} = a_9 + a_8 = 34 + 21 = 55$$

---

recursive formula

A **recursive formula** is a formula that defines each term of a sequence using preceding term(s). Recursive formulas must always state the initial term, or terms, of the sequence.

Must the first two terms always be given in a recursive formula?

No. The Fibonacci sequence defines each term using the two preceding terms, but many recursive formulas define each term using only one preceding term. These sequences need only the first term to be defined.

How To...

Given a recursive formula with only the first term provided, write the first n terms of a sequence.

1. Identify the initial term, a_1, which is given as part of the formula. This is the first term.
2. To find the second term, a_2, substitute the initial term into the formula for a_{n-1}. Solve.
3. To find the third term, a_3, substitute the second term into the formula. Solve.
4. Repeat until you have solved for the nth term.

Example 5   **Writing the Terms of a Sequence Defined by a Recursive Formula**

Write the first five terms of the sequence defined by the recursive formula.

$$a_1 = 9$$
$$a_n = 3a_{n-1} - 20, \text{ for } n \geq 2$$

Solution   The first term is given in the formula. For each subsequent term, we replace a_{n-1} with the value of the preceding term.

$$n = 1 \quad a_1 = 9$$
$$n = 2 \quad a_2 = 3a_1 - 20 = 3(9) - 20 = 27 - 20 = 7$$
$$n = 3 \quad a_3 = 3a_2 - 20 = 3(7) - 20 = 21 - 20 = 1$$
$$n = 4 \quad a_4 = 3a_3 - 20 = 3(1) - 20 = 3 - 20 = -17$$
$$n = 5 \quad a_5 = 3a_4 - 20 = 3(-17) - 20 = -51 - 20 = -71$$

The first five terms are $\{9, 7, 1, -17, -71\}$. See **Figure 5.**

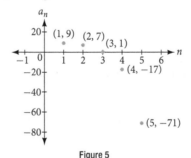

Figure 5

Try It #7

Write the first five terms of the sequence defined by the recursive formula.

$$a_1 = 2$$
$$a_n = 2a_{n-1} + 1, \text{ for } n \geq 2$$

How To...

Given a recursive formula with two initial terms, write the first n terms of a sequence.

1. Identify the initial term, a_1, which is given as part of the formula.
2. Identify the second term, a_2, which is given as part of the formula.
3. To find the third term, substitute the initial term and the second term into the formula. Evaluate.
4. Repeat until you have evaluated the nth term.

Example 6   **Writing the Terms of a Sequence Defined by a Recursive Formula**

Write the first six terms of the sequence defined by the recursive formula.

$$a_1 = 1$$
$$a_2 = 2$$
$$a_n = 3a_{n-1} + 4a_{n-2}, \text{ for } n \geq 3$$

Solution   The first two terms are given. For each subsequent term, we replace a_{n-1} and a_{n-2} with the values of the two preceding terms.

$$n = 3 \qquad a_3 = 3a_2 + 4a_1 = 3(2) + 4(1) = 10$$
$$n = 4 \qquad a_4 = 3a_3 + 4a_2 = 3(10) + 4(2) = 38$$
$$n = 5 \qquad a_5 = 3a_4 + 4a_3 = 3(38) + 4(10) = 154$$
$$n = 6 \qquad a_6 = 3a_5 + 4a_4 = 3(154) + 4(38) = 614$$

The first six terms are {1, 2, 10, 38, 154, 614}. See **Figure 6**.

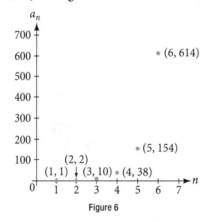

Figure 6

Try It #8

Write the first 8 terms of the sequence defined by the recursive formula.

$$a_1 = 0$$
$$a_2 = 1$$
$$a_3 = 1$$
$$a_n = \frac{a_{n-1}}{a_{n-2}} + a_{n-3}, \text{ for } n \geq 4$$

Using Factorial Notation

The formulas for some sequences include products of consecutive positive integers. **n factorial**, written as $n!$, is the product of the positive integers from 1 to n. For example,

$$4! = 4 \cdot 3 \cdot 2 \cdot 1 = 24$$
$$5! = 5 \cdot 4 \cdot 3 \cdot 2 \cdot 1 = 120$$

An example of formula containing a factorial is $a_n = (n + 1)!$. The sixth term of the sequence can be found by substituting 6 for n.

$$a_6 = (6 + 1)! = 7! = 7 \cdot 6 \cdot 5 \cdot 4 \cdot 3 \cdot 2 \cdot 1 = 5{,}040$$

The factorial of any whole number n is $n(n - 1)!$ We can therefore also think of 5! as $5 \cdot 4!$.

> *factorial*
>
> ***n* factorial** is a mathematical operation that can be defined using a recursive formula. The factorial of *n*, denoted *n*!, is defined for a positive integer *n* as:
>
> $$0! = 1$$
> $$1! = 1$$
> $$n! = n(n-1)(n-2) \cdots (2)(1), \text{ for } n \geq 2$$
>
> The special case 0! is defined as 0! = 1.

Q & A...

Can factorials always be found using a calculator?

No. Factorials get large very quickly—faster than even exponential functions! When the output gets too large for the calculator, it will not be able to calculate the factorial.

Example 7   **Writing the Terms of a Sequence Using Factorials**

Write the first five terms of the sequence defined by the explicit formula $a_n = \dfrac{5n}{(n+2)!}$.

Solution   Substitute $n = 1$, $n = 2$, and so on in the formula.

$$n = 1 \qquad a_1 = \frac{5(1)}{(1+2)!} = \frac{5}{3!} = \frac{5}{3 \cdot 2 \cdot 1} = \frac{5}{6}$$

$$n = 2 \qquad a_2 = \frac{5(2)}{(2+2)!} = \frac{10}{4!} = \frac{10}{4 \cdot 3 \cdot 2 \cdot 1} = \frac{5}{12}$$

$$n = 3 \qquad a_3 = \frac{5(3)}{(3+2)!} = \frac{15}{5!} = \frac{15}{5 \cdot 4 \cdot 3 \cdot 2 \cdot 1} = \frac{1}{8}$$

$$n = 4 \qquad a_4 = \frac{5(4)}{(4+2)!} = \frac{20}{6!} = \frac{20}{6 \cdot 5 \cdot 4 \cdot 3 \cdot 2 \cdot 1} = \frac{1}{36}$$

$$n = 5 \qquad a_5 = \frac{5(5)}{(5+2)!} = \frac{25}{7!} = \frac{25}{7 \cdot 6 \cdot 5 \cdot 4 \cdot 3 \cdot 2 \cdot 1} = \frac{5}{1,008}$$

The first five terms are $\left\{ \dfrac{5}{6}, \dfrac{5}{12}, \dfrac{1}{8}, \dfrac{1}{36}, \dfrac{5}{1,008} \right\}$.

Analysis   **Figure 7** *shows the graph of the sequence. Notice that, since factorials grow very quickly, the presence of the factorial term in the denominator results in the denominator becoming much larger than the numerator as n increases. This means the quotient gets smaller and, as the plot of the terms shows, the terms are decreasing and nearing zero.*

Figure 7

Try It #9

Write the first five terms of the sequence defined by the explicit formula $a_n = \dfrac{(n+1)!}{2n}$.

Access this online resource for additional instruction and practice with sequences.

* Finding Terms in a Sequence (http://openstaxcollege.org/l/findingterms)

9.1   SECTION EXERCISES

VERBAL

1. Discuss the meaning of a sequence. If a finite sequence is defined by a formula, what is its domain? What about an infinite sequence?

2. Describe three ways that a sequence can be defined.

3. Is the ordered set of even numbers an infinite sequence? What about the ordered set of odd numbers? Explain why or why not.

4. What happens to the terms a_n of a sequence when there is a negative factor in the formula that is raised to a power that includes n? What is the term used to describe this phenomenon?

5. What is a factorial, and how is it denoted? Use an example to illustrate how factorial notation can be beneficial.

ALGEBRAIC

For the following exercises, write the first four terms of the sequence.

6. $a_n = 2^n - 2$

7. $a_n = -\dfrac{16}{n+1}$

8. $a_n = -(-5)^{n-1}$

9. $a_n = \dfrac{2^n}{n^3}$

10. $a_n = \dfrac{2n+1}{n^3}$

11. $a_n = 1.25 \cdot (-4)^{n-1}$

12. $a_n = -4 \cdot (-6)^{n-1}$

13. $a_n = \dfrac{n^2}{2n+1}$

14. $a_n = (-10)^n + 1$

15. $a_n = -\left(\dfrac{4 \cdot (-5)^{n-1}}{5} \right)$

For the following exercises, write the first eight terms of the piecewise sequence.

16. $a_n = \begin{cases} (-2)^n - 2 & \text{if } n \text{ is even} \\ (3)^{n-1} & \text{if } n \text{ is odd} \end{cases}$

17. $a_n = \begin{cases} \dfrac{n^2}{2n+1} & \text{if } n \le 5 \\ n^2 - 5 & \text{if } n > 5 \end{cases}$

18. $a_n = \begin{cases} (2n+1)^2 & \text{if } n \text{ is divisible by 4} \\ \dfrac{2}{n} & \text{if } n \text{ is not divisible by 4} \end{cases}$

19. $a_n = \begin{cases} -0.6 \cdot 5^{n-1} & \text{if } n \text{ is prime or 1} \\ 2.5 \cdot (-2)^{n-1} & \text{if } n \text{ is composite} \end{cases}$

20. $a_n = \begin{cases} 4(n^2 - 2) & \text{if } n \le 3 \text{ or } n > 6 \\ \dfrac{n^2 - 2}{4} & \text{if } 3 < n \le 6 \end{cases}$

For the following exercises, write an explicit formula for each sequence.

21. $4, 7, 12, 19, 28, \dots$

22. $-4, 2, -10, 14, -34, \dots$

23. $1, 1, \dfrac{4}{3}, 2, \dfrac{16}{5}, \dots$

24. $0, \dfrac{1-e^1}{1+e^2}, \dfrac{1-e^2}{1+e^3}, \dfrac{1-e^3}{1+e^4}, \dfrac{1-e^4}{1+e^5}, \dots$

25. $1, -\dfrac{1}{2}, \dfrac{1}{4}, -\dfrac{1}{8}, \dfrac{1}{16}, \dots$

For the following exercises, write the first five terms of the sequence.

26. $a_1 = 9, a_n = a_{n-1} + n$

27. $a_1 = 3, a_n = (-3)a_{n-1}$

28. $a_1 = -4, a_n = \dfrac{a_{n-1} + 2n}{a_{n-1} - 1}$

29. $a_1 = -1, a_n = \dfrac{(-3)^{n-1}}{a_{n-1} - 2}$

30. $a_1 = -30, a_n = (2 + a_{n-1})\left(\dfrac{1}{2}\right)^n$

For the following exercises, write the first eight terms of the sequence.

31. $a_1 = \dfrac{1}{24}, a_2 = 1, a_n = (2a_{n-2})(3a_{n-1})$

32. $a_1 = -1, a_2 = 5, a_n = a_{n-2}(3 - a_{n-1})$

33. $a_1 = 2, a_2 = 10, a_n = \dfrac{2(a_{n-1} + 2)}{a_{n-2}}$

For the following exercises, write a recursive formula for each sequence.

34. $-2.5, -5, -10, -20, -40, \ldots$ **35.** $-8, -6, -3, 1, 6, \ldots$ **36.** $2, 4, 12, 48, 240, \ldots$

37. $35, 38, 41, 44, 47, \ldots$ **38.** $15, 3, \dfrac{3}{5}, \dfrac{3}{25}, \dfrac{3}{125}, \ldots$

For the following exercises, evaluate the factorial.

39. $6!$ **40.** $\left(\dfrac{12}{6}\right)!$ **41.** $\dfrac{12!}{6!}$ **42.** $\dfrac{100!}{99!}$

For the following exercises, write the first four terms of the sequence.

43. $a_n = \dfrac{n!}{n^2}$ **44.** $a_n = \dfrac{3 \cdot n!}{4 \cdot n!}$ **45.** $a_n = \dfrac{n!}{n^2 - n - 1}$ **46.** $a_n = \dfrac{100 \cdot n}{n(n-1)!}$

GRAPHICAL

For the following exercises, graph the first five terms of the indicated sequence

47. $a_n = \dfrac{(-1)^n}{n} + n$ **48.** $a_n = \begin{cases} \dfrac{4+n}{2n} & \text{if } n \text{ in even} \\ 3+n & \text{if } n \text{ is odd} \end{cases}$ **49.** $a_1 = 2, a_n = (-a_{n-1} + 1)^2$

50. $a_n = 1, a_n = a_{n-1} + 8$ **51.** $a_n = \dfrac{(n+1)!}{(n-1)!}$

For the following exercises, write an explicit formula for the sequence using the first five points shown on the graph.

52.

53.

54.

For the following exercises, write a recursive formula for the sequence using the first five points shown on the graph.

55.

56.

TECHNOLOGY

Follow these steps to evaluate a sequence defined recursively using a graphing calculator:
- On the home screen, key in the value for the initial term a_1 and press [**ENTER**].
- Enter the recursive formula by keying in all numerical values given in the formula, along with the key strokes [**2ND**] **ANS** for the previous term a_{n-1}. Press [**ENTER**].
- Continue pressing [**ENTER**] to calculate the values for each successive term.

For the following exercises, use the steps above to find the indicated term or terms for the sequence.

57. Find the first five terms of the sequence $a_1 = \dfrac{87}{111}$, $a_n = \dfrac{4}{3}a_{n-1} + \dfrac{12}{37}$. Use the >**Frac** feature to give fractional results.

58. Find the 15th term of the sequence $a_1 = 625$, $a_n = 0.8a_{n-1} + 18$.

59. Find the first five terms of the sequence $a_1 = 2$, $a_n = 2^{[(a_{n-1})-1]} + 1$.

60. Find the first ten terms of the sequence $a_1 = 8$, $a_n = \dfrac{(a_{n-1}+1)!}{a_{n-1}!}$.

61. Find the tenth term of the sequence $a_1 = 2$, $a_n = na_{n-1}$

Follow these steps to evaluate a finite sequence defined by an explicit formula.

Using a TI-84, do the following.
- In the home screen, press [**2ND**] **LIST**.
- Scroll over to **OPS** and choose "**seq(**" from the dropdown list. Press [**ENTER**].
- In the line headed "**Expr:**" type in the explicit formula, using the [**X,T, θ, n**] button for n
- In the line headed "**Variable:**" type in the variable used on the previous step.
- In the line headed "**start:**" key in the value of n that begins the sequence.
- In the line headed "**end:**" key in the value of n that ends the sequence.
- Press [**ENTER**] 3 times to return to the home screen. You will see the sequence syntax on the screen. Press [**ENTER**] to see the list of terms for the finite sequence defined. Use the right arrow key to scroll through the list of terms.

Using a TI-83, do the following.
- In the home screen, press [**2ND**] **LIST**.
- Scroll over to **OPS** and choose "**seq(**" from the dropdown list. Press [**ENTER**].
- Enter the items in the order "**Expr**", "**Variable**", "**start**", "**end**" separated by commas. See the instructions above for the description of each item.
- Press [**ENTER**] to see the list of terms for the finite sequence defined. Use the right arrow key to scroll through the list of terms.

For the following exercises, use the steps above to find the indicated terms for the sequence. Round to the nearest thousandth when necessary.

62. List the first five terms of the sequence.
$$a_n = -\frac{28}{9}n + \frac{5}{3}$$

63. List the first six terms of the sequence.
$$a_n = \frac{n^3 - 3.5n^2 + 4.1n - 1.5}{2.4n}$$

64. List the first five terms of the sequence.
$$a_n = \frac{15n \cdot (-2)^{n-1}}{47}$$

65. List the first four terms of the sequence.
$$a_n = 5.7^n + 0.275(n-1)!$$

66. List the first six terms of the sequence $a_n = \dfrac{n!}{n}$.

EXTENSIONS

67. Consider the sequence defined by $a_n = -6 - 8n$. Is $a_n = -421$ a term in the sequence? Verify the result.

68. What term in the sequence $a_n = \dfrac{n^2 + 4n + 4}{2(n+2)}$ has the value 41? Verify the result.

69. Find a recursive formula for the sequence
$1, 0, -1, -1, 0, 1, 1, 0, -1, -1, 0, 1, 1, ...$
(Hint: find a pattern for an based on the first two terms.)

70. Calculate the first eight terms of the sequences $a_n = \dfrac{(n+2)!}{(n-1)!}$ and $b_n = n^3 + 3n^2 + 2n$, and then make a conjecture about the relationship between these two sequences.

71. Prove the conjecture made in the preceding exercise.

Using Explicit Formulas for Arithmetic Sequences

We can think of an arithmetic sequence as a function on the domain of the natural numbers; it is a linear function because it has a constant rate of change. The common difference is the constant rate of change, or the slope of the function. We can construct the linear function if we know the slope and the vertical intercept.

$$a_n = a_1 + d(n - 1)$$

To find the y-intercept of the function, we can subtract the common difference from the first term of the sequence. Consider the following sequence.

The common difference is -50, so the sequence represents a linear function with a slope of -50. To find the y-intercept, we subtract -50 from 200: $200 - (-50) = 200 + 50 = 250$. You can also find the y-intercept by graphing the function and determining where a line that connects the points would intersect the vertical axis. The graph is shown in **Figure 4**.

Figure 4

Recall the slope-intercept form of a line is $y = mx + b$. When dealing with sequences, we use a_n in place of y and n in place of x. If we know the slope and vertical intercept of the function, we can substitute them for m and b in the slope-intercept form of a line. Substituting -50 for the slope and 250 for the vertical intercept, we get the following equation:

$$a_n = -50n + 250$$

We do not need to find the vertical intercept to write an explicit formula for an arithmetic sequence. Another explicit formula for this sequence is $a_n = 200 - 50(n - 1)$, which simplifies to $a_n = -50n + 250$.

> **explicit formula for an arithmetic sequence**
>
> An explicit formula for the nth term of an arithmetic sequence is given by
>
> $$a_n = a_1 + d(n - 1)$$

How To…

Given the first several terms for an arithmetic sequence, write an explicit formula.

1. Find the common difference, $a_2 - a_1$.
2. Substitute the common difference and the first term into $a_n = a_1 + d(n - 1)$.

Example 5   **Writing the nth Term Explicit Formula for an Arithmetic Sequence**

Write an explicit formula for the arithmetic sequence.

$$\{2, 12, 22, 32, 42, \ldots\}$$

Solution   The common difference can be found by subtracting the first term from the second term.

$$d = a_2 - a_1$$
$$= 12 - 2$$
$$= 10$$

The common difference is 10. Substitute the common difference and the first term of the sequence into the formula and simplify.

$$a_n = 2 + 10(n - 1)$$
$$a_n = 10n - 8$$

Analysis   The graph of this sequence, represented in **Figure 5**, shows a slope of 10 and a vertical intercept of −8.

Figure 5

Try It #6

Write an explicit formula for the following arithmetic sequence.

$$\{50, 47, 44, 41, \ldots\}$$

Finding the Number of Terms in a Finite Arithmetic Sequence

Explicit formulas can be used to determine the number of terms in a finite arithmetic sequence. We need to find the common difference, and then determine how many times the common difference must be added to the first term to obtain the final term of the sequence.

How To...

Given the first three terms and the last term of a finite arithmetic sequence, find the total number of terms.

1. Find the common difference d.
2. Substitute the common difference and the first term into $a_n = a_1 + d(n - 1)$.
3. Substitute the last term for a_n and solve for n.

Example 6   **Finding the Number of Terms in a Finite Arithmetic Sequence**

Find the number of terms in the finite arithmetic sequence.

$$\{8, 1, -6, \ldots, -41\}$$

Solution   The common difference can be found by subtracting the first term from the second term.

$$1 - 8 = -7$$

The common difference is -7. Substitute the common difference and the initial term of the sequence into the nth term formula and simplify.

$$a_n = a_1 + d(n-1)$$
$$a_n = 8 + (-7)(n-1)$$
$$a_n = 15 - 7n$$

Substitute -41 for a_n and solve for n

$$-41 = 15 - 7n$$
$$8 = n$$

There are eight terms in the sequence.

Try It #7

Find the number of terms in the finite arithmetic sequence.

$$\{6, 11, 16, \dots, 56\}$$

Solving Application Problems with Arithmetic Sequences

In many application problems, it often makes sense to use an initial term of a_0 instead of a_1. In these problems, we alter the explicit formula slightly to account for the difference in initial terms. We use the following formula:

$$a_n = a_0 + dn$$

Example 7   **Solving Application Problems with Arithmetic Sequences**

A five-year old child receives an allowance of $1 each week. His parents promise him an annual increase of $2 per week.

   a. Write a formula for the child's weekly allowance in a given year.

   b. What will the child's allowance be when he is 16 years old?

Solution

   a. The situation can be modeled by an arithmetic sequence with an initial term of 1 and a common difference of 2.

   Let A be the amount of the allowance and n be the number of years after age 5. Using the altered explicit formula for an arithmetic sequence we get:

$$A_n = 1 + 2n$$

   b. We can find the number of years since age 5 by subtracting.

$$16 - 5 = 11$$

   We are looking for the child's allowance after 11 years. Substitute 11 into the formula to find the child's allowance at age 16.

$$A_{11} = 1 + 2(11) = 23$$

   The child's allowance at age 16 will be $23 per week.

Try It #8

A woman decides to go for a 10-minute run every day this week and plans to increase the time of her daily run by 4 minutes each week. Write a formula for the time of her run after n weeks. How long will her daily run be 8 weeks from today?

Access this online resource for additional instruction and practice with arithmetic sequences.

   • Arithmetic Sequences (http://openstaxcollege.org/l/arithmeticseq)

9.2   SECTION EXERCISES

VERBAL

1. What is an arithmetic sequence?

2. How is the common difference of an arithmetic sequence found?

3. How do we determine whether a sequence is arithmetic?

4. What are the main differences between using a recursive formula and using an explicit formula to describe an arithmetic sequence?

5. Describe how linear functions and arithmetic sequences are similar. How are they different?

ALGEBRAIC

For the following exercises, find the common difference for the arithmetic sequence provided.

6. $\{5, 11, 17, 23, 29, ...\}$

7. $\left\{0, \frac{1}{2}, 1, \frac{3}{2}, 2, ...\right\}$

For the following exercises, determine whether the sequence is arithmetic. If so find the common difference.

8. $\{11.4, 9.3, 7.2, 5.1, 3, ...\}$

9. $\{4, 16, 64, 256, 1024, ...\}$

For the following exercises, write the first five terms of the arithmetic sequence given the first term and common difference.

10. $a_1 = -25, d = -9$

11. $a_1 = 0, d = \frac{2}{3}$

For the following exercises, write the first five terms of the arithmetic series given two terms.

12. $a_1 = 17, a_7 = -31$

13. $a_{13} = -60, a_{33} = -160$

For the following exercises, find the specified term for the arithmetic sequence given the first term and common difference.

14. First term is 3, common difference is 4, find the 5th term.

15. First term is 4, common difference is 5, find the 4th term.

16. First term is 5, common difference is 6, find the 8th term.

17. First term is 6, common difference is 7, find the 6th term.

18. First term is 7, common difference is 8, find the 7th term.

For the following exercises, find the first term given two terms from an arithmetic sequence.

19. Find the first term or a_1 of an arithmetic sequence if $a_6 = 12$ and $a_{14} = 28$.

20. Find the first term or a_1 of an arithmetic sequence if $a_7 = 21$ and $a_{15} = 42$.

21. Find the first term or a_1 of an arithmetic sequence if $a_8 = 40$ and $a_{23} = 115$.

22. Find the first term or a_1 of an arithmetic sequence if $a_9 = 54$ and $a_{17} = 102$.

23. Find the first term or a_1 of an arithmetic sequence if $a_{11} = 11$ and $a_{21} = 16$.

For the following exercises, find the specified term given two terms from an arithmetic sequence.

24. $a_1 = 33$ and $a_7 = -15$. Find a_4.

25. $a_3 = -17.1$ and $a_{10} = -15.7$. Find a_{21}.

For the following exercises, use the recursive formula to write the first five terms of the arithmetic sequence.

26. $a_1 = 39; a_n = a_{n-1} - 3$

27. $a_1 = -19; a_n = a_{n-1} - 1.4$

Example 2   **Writing the Terms of a Geometric Sequence**

List the first four terms of the geometric sequence with $a_1 = 5$ and $r = -2$.

Solution   Multiply a_1 by -2 to find a_2. Repeat the process, using a_2 to find a_3, and so on.

$$a_1 = 5$$
$$a_2 = -2a_1 = -10$$
$$a_3 = -2a_2 = 20$$
$$a_4 = -2a_3 = -40$$

The first four terms are $\{5, -10, 20, -40\}$.

Try It #3

List the first five terms of the geometric sequence with $a_1 = 18$ and $r = \dfrac{1}{3}$.

Using Recursive Formulas for Geometric Sequences

A recursive formula allows us to find any term of a geometric sequence by using the previous term. Each term is the product of the common ratio and the previous term. For example, suppose the common ratio is 9. Then each term is nine times the previous term. As with any recursive formula, the initial term must be given.

> ***recursive formula for a geometric sequence***
> The recursive formula for a geometric sequence with common ratio r and first term a_1 is
> $$a_n = ra_{n-1}, n \geq 2$$

How To...

Given the first several terms of a geometric sequence, write its recursive formula.

1. State the initial term.
2. Find the common ratio by dividing any term by the preceding term.
3. Substitute the common ratio into the recursive formula for a geometric sequence.

Example 3   **Using Recursive Formulas for Geometric Sequences**

Write a recursive formula for the following geometric sequence.

$$\{6, 9, 13.5, 20.25, ... \}$$

Solution   The first term is given as 6. The common ratio can be found by dividing the second term by the first term.

$$r = \frac{9}{6} = 1.5$$

Substitute the common ratio into the recursive formula for geometric sequences and define a_1.

$$a_n = ra_{n-1}$$
$$a_n = 1.5a_{n-1} \text{ for } n \geq 2$$
$$a_1 = 6$$

Analysis   *The sequence of data points follows an exponential pattern. The common ratio is also the base of an exponential function as shown in* **Figure 2**.

Figure 2

Q & A...

Do we have to divide the second term by the first term to find the common ratio?

No. We can divide any term in the sequence by the previous term. It is, however, most common to divide the second term by the first term because it is often the easiest method of finding the common ratio.

Try It #4

Write a recursive formula for the following geometric sequence.

$$\left\{2, \frac{4}{3}, \frac{8}{9}, \frac{16}{27}, ...\right\}$$

Using Explicit Formulas for Geometric Sequences

Because a geometric sequence is an exponential function whose domain is the set of positive integers, and the common ratio is the base of the function, we can write explicit formulas that allow us to find particular terms.

$$a_n = a_1 r^{n-1}$$

Let's take a look at the sequence {18, 36, 72, 144, 288,...}. This is a geometric sequence with a common ratio of 2 and an exponential function with a base of 2. An explicit formula for this sequence is

$$a_n = 18 \cdot 2^{n-1}$$

The graph of the sequence is shown in **Figure 3**.

Figure 3

---

explicit formula for a geometric sequence

The nth term of a geometric sequence is given by the explicit formula:

$$a_n = a_1 r^{n-1}$$

Example 4   **Writing Terms of Geometric Sequences Using the Explicit Formula**

Given a geometric sequence with $a_1 = 3$ and $a_4 = 24$, find a_2.

Solution   The sequence can be written in terms of the initial term and the common ratio r.

$$3, 3r, 3r^2, 3r^3,...$$

Find the common ratio using the given fourth term.

$$a_n = a_1 r^{n-1}$$

$$a_4 = 3r^3 \qquad \text{Write the fourth term of sequence in terms of } a_1 \text{ and } r$$

$$24 = 3r^3 \qquad \text{Substitute 24 for } a_4$$

$$8 = r^3 \qquad \text{Divide}$$

$$r = 2 \qquad \text{Solve for the common ratio}$$

Find the second term by multiplying the first term by the common ratio.

$$a_2 = 2a_1$$

$$= 2(3)$$

$$= 6$$

Analysis *The common ratio is multiplied by the first term once to find the second term, twice to find the third term, three times to find the fourth term, and so on. The tenth term could be found by multiplying the first term by the common ratio nine times or by multiplying by the common ratio raised to the ninth power.*

Try It #5

Given a geometric sequence with $a_2 = 4$ and $a_3 = 32$, find a_6.

Example 5   **Writing an Explicit Formula for the *n*th Term of a Geometric Sequence**

Write an explicit formula for the *n*th term of the following geometric sequence.

$$\{2, 10, 50, 250,...\}$$

Solution   The first term is 2. The common ratio can be found by dividing the second term by the first term.

$$\frac{10}{2} = 5$$

The common ratio is 5. Substitute the common ratio and the first term of the sequence into the formula.

$$a_n = a_1 r^{(n-1)}$$

$$a_n = 2 \cdot 5^{n-1}$$

The graph of this sequence in **Figure 4** shows an exponential pattern.

Figure 4

Try It #6

Write an explicit formula for the following geometric sequence.

$$\{-1, 3, -9, 27, ...\}$$

Solving Application Problems with Geometric Sequences

In real-world scenarios involving arithmetic sequences, we may need to use an initial term of a_0 instead of a_1. In these problems, we can alter the explicit formula slightly by using the following formula:

$$a_n = a_0 r^n$$

Example 6   Solving Application Problems with Geometric Sequences

In 2013, the number of students in a small school is 284. It is estimated that the student population will increase by 4% each year.

    a. Write a formula for the student population.

    b. Estimate the student population in 2020.

Solution

    a. The situation can be modeled by a geometric sequence with an initial term of 284. The student population will be 104% of the prior year, so the common ratio is 1.04.

        Let P be the student population and n be the number of years after 2013. Using the explicit formula for a geometric sequence we get

$$P_n = 284 \cdot 1.04^n$$

    b. We can find the number of years since 2013 by subtracting.

$$2020 - 2013 = 7$$

        We are looking for the population after 7 years. We can substitute 7 for n to estimate the population in 2020.

$$P_7 = 284 \cdot 1.04^7 \approx 374$$

    The student population will be about 374 in 2020.

Try It #7

A business starts a new website. Initially the number of hits is 293 due to the curiosity factor. The business estimates the number of hits will increase by 2.6% per week.

a. Write a formula for the number of hits.

b. Estimate the number of hits in 5 weeks.

Access these online resources for additional instruction and practice with geometric sequences.

- Geometric Sequences (http://openstaxcollege.org/l/geometricseq)
- Determine the Type of Sequence (http://openstaxcollege.org/l/sequencetype)
- Find the Formula for a Sequence (http://openstaxcollege.org/l/sequenceformula)

Substitute $a_1 = 4{,}374$ and $r = -\dfrac{1}{3}$ into the formula, and simplify to find the sum:

$$S = \frac{a_1}{1 - r}$$

$$S = \frac{4{,}374}{1 - \left(-\dfrac{1}{3}\right)} = 3{,}280.5$$

d. The formula is exponential, so the series is geometric, but $r > 1$. The sum does not exist.

Example 8   **Finding an Equivalent Fraction for a Repeating Decimal**

Find an equivalent fraction for the repeating decimal $0.\overline{3}$

Solution   We notice the repeating decimal $0.\overline{3} = 0.333...$ so we can rewrite the repeating decimal as a sum of terms.

$$0.\overline{3} = 0.3 + 0.03 + 0.003 + ...$$

Looking for a pattern, we rewrite the sum, noticing that we see the first term multiplied to 0.1 in the second term, and the second term multiplied to 0.1 in the third term.

$$0.\overline{3} = 0.3 + (0.1) \underbrace{(0.3)}_{\text{First Term}} + (0.1) \underbrace{(0.1)(0.3)}_{\text{Second Term}}$$

Notice the pattern; we multiply each consecutive term by a common ratio of 0.1 starting with the first term of 0.3. So, substituting into our formula for an infinite geometric sum, we have

$$S = \frac{a_1}{1 - r} = \frac{0.3}{1 - 0.1} = \frac{0.3}{0.9} = \frac{1}{3}.$$

Try It #12

Find the sum, if it exists. $2 + \dfrac{2}{3} + \dfrac{2}{9} + ...$

Try It #13

Find the sum, if it exists. $\displaystyle\sum_{k=1}^{\infty} 0.76k + 1$

Try It #14

Find the sum, if it exists. $\displaystyle\sum_{k=1}^{\infty} \left(-\dfrac{3}{8}\right)^k$

Solving Annuity Problems

At the beginning of the section, we looked at a problem in which a couple invested a set amount of money each month into a college fund for six years. An **annuity** is an investment in which the purchaser makes a sequence of periodic, equal payments. To find the amount of an annuity, we need to find the sum of all the payments and the interest earned. In the example, the couple invests \$50 each month. This is the value of the initial deposit. The account paid 6% annual interest, compounded monthly. To find the interest rate per payment period, we need to divide the 6% annual percentage interest (APR) rate by 12. So the monthly interest rate is 0.5%. We can multiply the amount in the account each month by 100.5% to find the value of the account after interest has been added.

We can find the value of the annuity right after the last deposit by using a geometric series with $a_1 = 50$ and $r = 100.5\% = 1.005$. After the first deposit, the value of the annuity will be \$50. Let us see if we can determine the amount in the college fund and the interest earned.

We can find the value of the annuity after n deposits using the formula for the sum of the first n terms of a geometric series. In 6 years, there are 72 months, so $n = 72$. We can substitute $a_1 = 50$, $r = 1.005$, and $n = 72$ into the formula, and simplify to find the value of the annuity after 6 years.

$$S_{72} = \frac{50(1 - 1.005^{72})}{1 - 1.005} \approx 4{,}320.44$$

After the last deposit, the couple will have a total of \$4,320.44 in the account. Notice, the couple made 72 payments of \$50 each for a total of $72(50) = \$3{,}600$. This means that because of the annuity, the couple earned \$720.44 interest in their college fund.

How To...

Given an initial deposit and an interest rate, find the value of an annuity.

1. Determine a_1, the value of the initial deposit.
2. Determine n, the number of deposits.
3. Determine r.
   a. Divide the annual interest rate by the number of times per year that interest is compounded.
   b. Add 1 to this amount to find r.
4. Substitute values for a_1, r, and n into the formula for the sum of the first n terms of a geometric series,
   $$S_n = \frac{a_1(1 - r^n)}{1 - r}.$$
5. Simplify to find S_n, the value of the annuity after n deposits.

Example 9   **Solving an Annuity Problem**

A deposit of $100 is placed into a college fund at the beginning of every month for 10 years. The fund earns 9% annual interest, compounded monthly, and paid at the end of the month. How much is in the account right after the last deposit?

Solution   The value of the initial deposit is $100, so $a_1 = 100$. A total of 120 monthly deposits are made in the 10 years, so $n = 120$. To find r, divide the annual interest rate by 12 to find the monthly interest rate and add 1 to represent the new monthly deposit.

$$r = 1 + \frac{0.09}{12} = 1.0075$$

Substitute $a_1 = 100$, $r = 1.0075$, and $n = 120$ into the formula for the sum of the first n terms of a geometric series, and simplify to find the value of the annuity.

$$S_{120} = \frac{100(1 - 1.0075^{120})}{1 - 1.0075} \approx 19{,}351.43$$

So the account has $19,351.43 after the last deposit is made.

Try It #15

At the beginning of each month, $200 is deposited into a retirement fund. The fund earns 6% annual interest, compounded monthly, and paid into the account at the end of the month. How much is in the account if deposits are made for 10 years?

Access these online resources for additional instruction and practice with series.

- Arithmetic Series (http://openstaxcollege.org/l/arithmeticser)
- Geometric Series (http://openstaxcollege.org/l/geometricser)
- Summation Notation (http://openstaxcollege.org/l/sumnotation)

9.4 SECTION EXERCISES

VERBAL

1. What is an nth partial sum?

2. What is the difference between an arithmetic sequence and an arithmetic series?

3. What is a geometric series?

4. How is finding the sum of an infinite geometric series different from finding the nth partial sum?

5. What is an annuity?

ALGEBRAIC

For the following exercises, express each description of a sum using summation notation.

6. The sum of terms $m^2 + 3m$ from $m = 1$ to $m = 5$

7. The sum from of $n = 0$ to $n = 4$ of $5n$

8. The sum of $6k - 5$ from $k = -2$ to $k = 1$

9. The sum that results from adding the number 4 five times

For the following exercises, express each arithmetic sum using summation notation.

10. $5 + 10 + 15 + 20 + 25 + 30 + 35 + 40 + 45 + 50$

11. $10 + 18 + 26 + \ldots + 162$

12. $\frac{1}{2} + 1 + \frac{3}{2} + 2 + \ldots + 4$

For the following exercises, use the formula for the sum of the first n terms of each arithmetic sequence.

13. $\frac{3}{2} + 2 + \frac{5}{2} + 3 + \frac{7}{2}$

14. $19 + 25 + 31 + \ldots + 73$

15. $3.2 + 3.4 + 3.6 + \ldots + 5.6$

For the following exercises, express each geometric sum using summation notation.

16. $1 + 3 + 9 + 27 + 81 + 243 + 729 + 2187$

17. $8 + 4 + 2 + \ldots + 0.125$

18. $-\frac{1}{6} + \frac{1}{12} - \frac{1}{24} + \ldots + \frac{1}{768}$

For the following exercises, use the formula for the sum of the first n terms of each geometric sequence, and then state the indicated sum.

19. $9 + 3 + 1 + \frac{1}{3} + \frac{1}{9}$

20. $\sum_{n=1}^{9} 5 \cdot 2^{n-1}$

21. $\sum_{a=1}^{11} 64 \cdot 0.2^{a-1}$

For the following exercises, determine whether the infinite series has a sum. If so, write the formula for the sum. If not, state the reason.

22. $12 + 18 + 24 + 30 + \ldots$

23. $2 + 1.6 + 1.28 + 1.024 + \ldots$

24. $\sum_{m=1}^{\infty} 4^{m-1}$

25. $\sum_{k=1}^{\infty} -\left(-\frac{1}{2}\right)^{k-1}$

GRAPHICAL

For the following exercises, use the following scenario. Javier makes monthly deposits into a savings account. He opened the account with an initial deposit of $50. Each month thereafter he increased the previous deposit amount by $20.

26. Graph the arithmetic sequence showing one year of Javier's deposits.

27. Graph the arithmetic series showing the monthly sums of one year of Javier's deposits.

For the following exercises, use the geometric series $\sum_{k=1}^{\infty} \left(\frac{1}{2}\right)^k$.

28. Graph the first 7 partial sums of the series.

29. What number does S_n seem to be approaching in the graph? Find the sum to explain why this makes sense.

NUMERIC

For the following exercises, find the indicated sum.

30. $\sum_{a=1}^{14} a$

31. $\sum_{n=1}^{6} n(n-2)$

32. $\sum_{k=1}^{17} k^2$

33. $\sum_{k=1}^{7} 2^k$

For the following exercises, use the formula for the sum of the first n terms of an arithmetic series to find the sum.

34. $-1.7 + -0.4 + 0.9 + 2.2 + 3.5 + 4.8$

35. $6 + \frac{15}{2} + 9 + \frac{21}{2} + 12 + \frac{27}{2} + 15$

36. $-1 + 3 + 7 + ... + 31$

37. $\sum_{k=1}^{11} \left(\frac{k}{2} - \frac{1}{2}\right)$

For the following exercises, use the formula for the sum of the first n terms of a geometric series to find the partial sum.

38. S_6 for the series $-2 - 10 - 50 - 250 ...$

39. S_7 for the series $0.4 - 2 + 10 - 50 ...$

40. $\sum_{k=1}^{9} 2^{k-1}$

41. $\sum_{n=1}^{10} -2 \cdot \left(\frac{1}{2}\right)^{n-1}$

For the following exercises, find the sum of the infinite geometric series.

42. $4 + 2 + 1 + \frac{1}{2} ...$

43. $-1 - \frac{1}{4} - \frac{1}{16} - \frac{1}{64} ...$

44. $\sum_{\infty}^{k=1} 3 \cdot \left(\frac{1}{4}\right)^{k-1}$

45. $\sum_{n=1}^{\infty} 4.6 \cdot 0.5^{n-1}$

For the following exercises, determine the value of the annuity for the indicated monthly deposit amount, the number of deposits, and the interest rate.

46. Deposit amount: $50; total deposits: 60; interest rate: 5%, compounded monthly

47. Deposit amount: $150; total deposits: 24; interest rate: 3%, compounded monthly

48. Deposit amount: $450; total deposits: 60; interest rate: 4.5%, compounded quarterly

49. Deposit amount: $100; total deposits: 120; interest rate: 10%, compounded semi-annually

EXTENSIONS

50. The sum of terms $50 - k^2$ from $k = x$ through 7 is 115. What is x?

51. Write an explicit formula for a_k such that $\sum_{k=0}^{6} a_k = 189$. Assume this is an arithmetic series.

52. Find the smallest value of n such that $\sum_{k=1}^{n} (3k - 5) > 100$.

53. How many terms must be added before the series $-1 - 3 - 5 - 7....$ has a sum less than -75?

54. Write $0.\overline{65}$ as an infinite geometric series using summation notation. Then use the formula for finding the sum of an infinite geometric series to convert 0.65 to a fraction.

55. The sum of an infinite geometric series is five times the value of the first term. What is the common ratio of the series?

56. To get the best loan rates available, the Riches want to save enough money to place 20% down on a $160,000 home. They plan to make monthly deposits of $125 in an investment account that offers 8.5% annual interest compounded semi-annually. Will the Riches have enough for a 20% down payment after five years of saving? How much money will they have saved?

57. Karl has two years to save $10,000 to buy a used car when he graduates. To the nearest dollar, what would his monthly deposits need to be if he invests in an account offering a 4.2% annual interest rate that compounds monthly?

REAL-WORLD APPLICATIONS

58. Keisha devised a week-long study plan to prepare for finals. On the first day, she plans to study for 1 hour, and each successive day she will increase her study time by 30 minutes. How many hours will Keisha have studied after one week?

59. A boulder rolled down a mountain, traveling 6 feet in the first second. Each successive second, its distance increased by 8 feet. How far did the boulder travel after 10 seconds?

60. A scientist places 50 cells in a petri dish. Every hour, the population increases by 1.5%. What will the cell count be after 1 day?

61. A pendulum travels a distance of 3 feet on its first swing. On each successive swing, it travels $\frac{3}{4}$ the distance of the previous swing. What is the total distance traveled by the pendulum when it stops swinging?

62. Rachael deposits $1,500 into a retirement fund each year. The fund earns 8.2% annual interest, compounded monthly. If she opened her account when she was 19 years old, how much will she have by the time she is 55? How much of that amount will be interest earned?

LEARNING OBJECTIVES

In this section, you will:

- Solve counting problems using the Addition Principle.
- Solve counting problems using the Multiplication Principle.
- Solve counting problems using permutations involving n distinct objects.
- Solve counting problems using combinations.
- Find the number of subsets of a given set.
- Solve counting problems using permutations involving n non-distinct objects.

9.5   COUNTING PRINCIPLES

A new company sells customizable cases for tablets and smartphones. Each case comes in a variety of colors and can be personalized for an additional fee with images or a monogram. A customer can choose not to personalize or could choose to have one, two, or three images or a monogram. The customer can choose the order of the images and the letters in the monogram. The company is working with an agency to develop a marketing campaign with a focus on the huge number of options they offer. Counting the possibilities is challenging!

We encounter a wide variety of counting problems every day. There is a branch of mathematics devoted to the study of counting problems such as this one. Other applications of counting include secure passwords, horse racing outcomes, and college scheduling choices. We will examine this type of mathematics in this section.

Using the Addition Principle

The company that sells customizable cases offers cases for tablets and smartphones. There are 3 supported tablet models and 5 supported smartphone models. The **Addition Principle** tells us that we can add the number of tablet options to the number of smartphone options to find the total number of options. By the Addition Principle, there are 8 total options, as we can see in **Figure 1**.

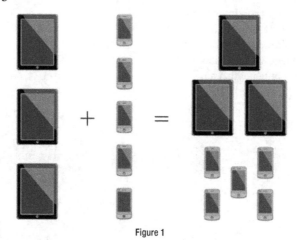

Figure 1

---

the Addition Principle
According to the **Addition Principle**, if one event can occur in m ways and a second event with no common outcomes can occur in n ways, then the first or second event can occur in $m + n$ ways.

Example 1   **Using the Addition Principle**

There are 2 vegetarian entrée options and 5 meat entrée options on a dinner menu. What is the total number of entrée options?

Solution   We can add the number of vegetarian options to the number of meat options to find the total number of entrée options.

Vegetarian + Vegetarian +   Meat   +   Meat   +   Meat   +   Meat   +   Meat
     ↓   ↓   ↓   ↓   ↓   ↓   ↓
 Option 1   +   Option 2   + Option 3   + Option 4   + Option 5   + Option 6   + Option 7

There are 7 total options.

Try It #1

A student is shopping for a new computer. He is deciding among 3 desktop computers and 4 laptop computers. What is the total number of computer options?

Using the Multiplication Principle

The **Multiplication Principle** applies when we are making more than one selection. Suppose we are choosing an appetizer, an entrée, and a dessert. If there are 2 appetizer options, 3 entrée options, and 2 dessert options on a fixed-price dinner menu, there are a total of 12 possible choices of one each as shown in the tree diagram in **Figure 2**.

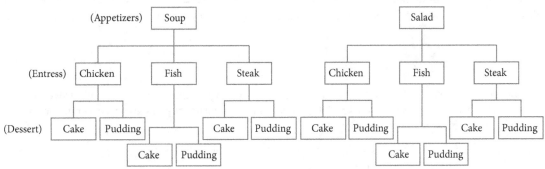

Figure 2

The possible choices are:

1. soup, chicken, cake
2. soup, chicken, pudding
3. soup, fish, cake
4. soup, fish, pudding
5. soup, steak, cake
6. soup, steak, pudding
7. salad, chicken, cake
8. salad, chicken, pudding
9. salad, fish, cake
10. salad, fish, pudding
11. salad, steak, cake
12. salad, steak, pudding

We can also find the total number of possible dinners by multiplying.

We could also conclude that there are 12 possible dinner choices simply by applying the Multiplication Principle.

# of appetizer options × # of entree options × # of dessert options

2   ×   3   ×   2   = 12

> *the Multiplication Principle*
> According to the **Multiplication Principle**, if one event can occur in m ways and a second event can occur in n ways after the first event has occurred, then the two events can occur in $m \times n$ ways. This is also known as the **Fundamental Counting Principle**.

Example 2 **Using the Multiplication Principle**

Diane packed 2 skirts, 4 blouses, and a sweater for her business trip. She will need to choose a skirt and a blouse for each outfit and decide whether to wear the sweater. Use the Multiplication Principle to find the total number of possible outfits.

Solution   To find the total number of outfits, find the product of the number of skirt options, the number of blouse options, and the number of sweater options.

$$\text{\# of skirt options} \times \text{\# of blouse options} \times \text{\# of sweater options}$$
$$2 \quad \times \quad 4 \quad \times \quad 2 \quad = 16$$

There are 16 possible outfits.

Try It #2

A restaurant offers a breakfast special that includes a breakfast sandwich, a side dish, and a beverage. There are 3 types of breakfast sandwiches, 4 side dish options, and 5 beverage choices. Find the total number of possible breakfast specials.

Finding the Number of Permutations of n Distinct Objects

The Multiplication Principle can be used to solve a variety of problem types. One type of problem involves placing objects in order. We arrange letters into words and digits into numbers, line up for photographs, decorate rooms, and more. An ordering of objects is called a **permutation**.

Finding the Number of Permutations of n Distinct Objects Using the Multiplication Principle

To solve permutation problems, it is often helpful to draw line segments for each option. That enables us to determine the number of each option so we can multiply. For instance, suppose we have four paintings, and we want to find the number of ways we can hang three of the paintings in order on the wall. We can draw three lines to represent the three places on the wall.

$$\underline{\hspace{2cm}} \times \underline{\hspace{2cm}} \times \underline{\hspace{2cm}}$$

There are four options for the first place, so we write a 4 on the first line.

$$\underline{\quad 4 \quad} \times \underline{\hspace{2cm}} \times \underline{\hspace{2cm}}$$

After the first place has been filled, there are three options for the second place so we write a 3 on the second line.

$$\underline{\quad 4 \quad} \times \underline{\quad 3 \quad} \times \underline{\hspace{2cm}}$$

After the second place has been filled, there are two options for the third place so we write a 2 on the third line. Finally, we find the product.

$$\underline{\quad 4 \quad} \times \underline{\quad 3 \quad} \times \underline{\quad 2 \quad} = 24$$

There are 24 possible permutations of the paintings.

How To...

Given n distinct options, determine how many permutations there are.

1. Determine how many options there are for the first situation.
2. Determine how many options are left for the second situation.
3. Continue until all of the spots are filled.
4. Multiply the numbers together.

Example 3   **Finding the Number of Permutations Using the Multiplication Principle**

At a swimming competition, nine swimmers compete in a race.

  a. How many ways can they place first, second, and third?

  b. How many ways can they place first, second, and third if a swimmer named Ariel wins first place? (Assume there is only one contestant named Ariel.)

  c. How many ways can all nine swimmers line up for a photo?

Solution

  a. Draw lines for each place.

$$\underline{\text{options for 1st place}} \times \underline{\text{options for 2nd place}} \times \underline{\text{options for 3rd place}}$$

There are 9 options for first place. Once someone has won first place, there are 8 remaining options for second place. Once first and second place have been won, there are 7 remaining options for third place.

$$\underline{\quad 9 \quad} \times \underline{\quad 8 \quad} \times \underline{\quad 7 \quad} = 504$$

Multiply to find that there are 504 ways for the swimmers to place.

  b. Draw lines for describing each place.

$$\underline{\text{options for 1st place}} \times \underline{\text{options for 2nd place}} \times \underline{\text{options for 3rd place}}$$

We know Ariel must win first place, so there is only 1 option for first place. There are 8 remaining options for second place, and then 7 remaining options for third place.

$$\underline{\quad 1 \quad} \times \underline{\quad 8 \quad} \times \underline{\quad 7 \quad} = 56$$

Multiply to find that there are 56 ways for the swimmers to place if Ariel wins first.

  c. Draw lines for describing each place in the photo.

$$\underline{\quad} \times \underline{\quad} \times \underline{\quad} \times \underline{\quad} \times \underline{\quad} \times \underline{\quad} \times \underline{\quad} \times \underline{\quad} \times \underline{\quad}$$

There are 9 choices for the first spot, then 8 for the second, 7 for the third, 6 for the fourth, and so on until only 1 person remains for the last spot.

$$\underline{\ 9\ } \times \underline{\ 8\ } \times \underline{\ 7\ } \times \underline{\ 6\ } \times \underline{\ 5\ } \times \underline{\ 4\ } \times \underline{\ 3\ } \times \underline{\ 2\ } \times \underline{\ 1\ } = 362,880$$

There are 362,880 possible permutations for the swimmers to line up.

Analysis Note that in part c, we found there were 9! ways for 9 people to line up. The number of permutations of n distinct objects can always be found by n!.

Try It #3

A family of five is having portraits taken. Use the Multiplication Principle to find how many ways the family can line up for the portrait.

Try It #4

A family of five is having portraits taken. Use the Multiplication Principle to find how many ways the photographer can line up 3 of the family members.

Try It #5

A family of five is having portraits taken. Use the Multiplication Principle to find how many ways the family can line up for the portrait if the parents are required to stand on each end.

Finding the Number of Permutations of *n* Distinct Objects Using a Formula

For some permutation problems, it is inconvenient to use the Multiplication Principle because there are so many numbers to multiply. Fortunately, we can solve these problems using a formula. Before we learn the formula, let's look at two common notations for permutations. If we have a set of *n* objects and we want to choose *r* objects from the set in order, we write $P(n, r)$. Another way to write this is $_nP_r$, a notation commonly seen on computers and calculators. To calculate $P(n, r)$, we begin by finding $n!$, the number of ways to line up all *n* objects. We then divide by $(n - r)!$ to cancel out the $(n - r)$ items that we do not wish to line up.

Let's see how this works with a simple example. Imagine a club of six people. They need to elect a president, a vice president, and a treasurer. Six people can be elected president, any one of the five remaining people can be elected vice president, and any of the remaining four people could be elected treasurer. The number of ways this may be done is $6 \times 5 \times 4 = 120$. Using factorials, we get the same result.

$$\frac{6!}{3!} = \frac{6 \cdot 5 \cdot 4 \cdot 3!}{3!} = 6 \cdot 5 \cdot 4 = 120$$

There are 120 ways to select 3 officers in order from a club with 6 members. We refer to this as a permutation of 6 taken 3 at a time. The general formula is as follows.

$$P(n, r) = \frac{n!}{(n - r)!}$$

Note that the formula stills works if we are choosing <u>all</u> *n* objects and placing them in order. In that case we would be dividing by $(n - n)!$ or $0!$, which we said earlier is equal to 1. So the number of permutations of *n* objects taken *n* at a time is $\frac{n!}{1}$ or just $n!$.

---

formula for permutations of n distinct objects
Given *n* distinct objects, the number of ways to select *r* objects from the set in order is

$$P(n, r) = \frac{n!}{(n - r)!}$$

How To...

Given a word problem, evaluate the possible permutations.
1. Identify *n* from the given information.
2. Identify *r* from the given information.
3. Replace *n* and *r* in the formula with the given values.
4. Evaluate.

Example 4   **Finding the Number of Permutations Using the Formula**

A professor is creating an exam of 9 questions from a test bank of 12 questions. How many ways can she select and arrange the questions?

Solution   Substitute $n = 12$ and $r = 9$ into the permutation formula and simplify.

$$P(n, r) = \frac{n!}{(n - r)!}$$

$$P(12, 9) = \frac{12!}{(12 - 9)!} = \frac{12!}{3!} = 79,833,600$$

There are 79,833,600 possible permutations of exam questions!

Analysis We can also use a calculator to find permutations. For this problem, we would enter **15**, *press the* [$_nP_r$ *function*], *enter* [**12**], *and then press the equal sign. The* [$_nP_r$ *function*] *may be located under the* [**MATH**] *menu with probability commands.*

Q & A...

Could we have solved Example 4 using the Multiplication Principle?

Yes. We could have multiplied $15 \cdot 14 \cdot 13 \cdot 12 \cdot 11 \cdot 10 \cdot 9 \cdot 8 \cdot 7 \cdot 6 \cdot 5 \cdot 4$ to find the same answer.

Try It #6

A play has a cast of 7 actors preparing to make their curtain call. Use the permutation formula to find how many ways the 7 actors can line up.

Try It #7

A play has a cast of 7 actors preparing to make their curtain call. Use the permutation formula to find how many ways 5 of the 7 actors can be chosen to line up.

Find the Number of Combinations Using the Formula

So far, we have looked at problems asking us to put objects in order. There are many problems in which we want to select a few objects from a group of objects, but we do not care about the order. When we are selecting objects and the order does not matter, we are dealing with **combinations**. A selection of r objects from a set of n objects where the order does not matter can be written as $C(n, r)$. Just as with permutations, $C(n, r)$ can also be written as $_nC_r$. In this case, the general formula is as follows.

$$C(n, r) = \frac{n!}{r!(n - r)!}$$

An earlier problem considered choosing 3 of 4 possible paintings to hang on a wall. We found that there were 24 ways to select 3 of the 4 paintings in order. But what if we did not care about the order? We would expect a smaller number because selecting paintings 1, 2, 3 would be the same as selecting paintings 2, 3, 1. To find the number of ways to select 3 of the 4 paintings, disregarding the order of the paintings, divide the number of permutations by the number of ways to order 3 paintings. There are $3! = 3 \cdot 2 \cdot 1 = 6$ ways to order 3 paintings. There are $\frac{24}{6}$, or 4 ways to select 3 of the 4 paintings.

This number makes sense because every time we are selecting 3 paintings, we are *not* selecting 1 painting. There are 4 paintings we could choose not to select, so there are 4 ways to select 3 of the 4 paintings.

---

formula for combinations of n distinct objects

Given n distinct objects, the number of ways to select r objects from the set is

$$C(n, r) = \frac{n!}{r!(n - r)!}$$

How To...

Given a number of options, determine the possible number of combinations.

1. Identify n from the given information.
2. Identify r from the given information.
3. Replace n and r in the formula with the given values.
4. Evaluate.

Example 5   **Finding the Number of Combinations Using the Formula**

A fast food restaurant offers five side dish options. Your meal comes with two side dishes.

    a. How many ways can you select your side dishes?

    b. How many ways can you select 3 side dishes?

Solution

    a. We want to choose 2 side dishes from 5 options.

$$C(5, 2) = \frac{5!}{2!(5 - 2)!} = 10$$

    b. We want to choose 3 side dishes from 5 options.

$$C(5, 3) = \frac{5!}{3!(5 - 3)!} = 10$$

Analysis   We can also use a graphing calculator to find combinations. Enter **5**, then press $_nC_r$, enter **3**, and then press the equal sign. The $_nC_r$ **function** may be located under the **MATH** menu with probability commands.

Q & A...

Is it a coincidence that parts (a) and (b) in Example 5 have the same answers?

No. When we choose r objects from n objects, we are **not** choosing $(n - r)$ objects. Therefore, $C(n, r) = C(n, n - r)$.

Try It #8

An ice cream shop offers 10 flavors of ice cream. How many ways are there to choose 3 flavors for a banana split?

Finding the Number of Subsets of a Set

We have looked only at combination problems in which we chose exactly r objects. In some problems, we want to consider choosing every possible number of objects. Consider, for example, a pizza restaurant that offers 5 toppings. Any number of toppings can be ordered. How many different pizzas are possible?

To answer this question, we need to consider pizzas with any number of toppings. There is $C(5, 0) = 1$ way to order a pizza with no toppings. There are $C(5, 1) = 5$ ways to order a pizza with exactly one topping. If we continue this process, we get

$$C(5, 0) + C(5, 1) + C(5, 2) + C(5, 3) + C(5, 4) + C(5, 5) = 32$$

There are 32 possible pizzas. This result is equal to 2^5.

We are presented with a sequence of choices. For each of the n objects we have two choices: include it in the subset or not. So for the whole subset we have made n choices, each with two options. So there are a total of $2 \cdot 2 \cdot 2 \cdot ... \cdot 2$ possible resulting subsets, all the way from the empty subset, which we obtain when we say "no" each time, to the original set itself, which we obtain when we say "yes" each time.

> **formula for the number of subsets of a set**
> A set containing n distinct objects has 2^n subsets.

Example 6   Finding the Number of Subsets of a Set

A restaurant offers butter, cheese, chives, and sour cream as toppings for a baked potato. How many different ways are there to order a potato?

Solution   We are looking for the number of subsets of a set with 4 objects. Substitute $n = 4$ into the formula.

$$2^n = 2^4$$
$$= 16$$

There are 16 possible ways to order a potato.

Try It #9

A sundae bar at a wedding has 6 toppings to choose from. Any number of toppings can be chosen. How many different sundaes are possible?

Finding the Number of Permutations of n Non-Distinct Objects

We have studied permutations where all of the objects involved were distinct. What happens if some of the objects are indistinguishable? For example, suppose there is a sheet of 12 stickers. If all of the stickers were distinct, there would be 12! ways to order the stickers. However, 4 of the stickers are identical stars, and 3 are identical moons. Because all of the objects are not distinct, many of the 12! permutations we counted are duplicates. The general formula for this situation is as follows.

$$\frac{n!}{r_1!\,r_2!\,...\,r_k!}$$

In this example, we need to divide by the number of ways to order the 4 stars and the ways to order the 3 moons to find the number of unique permutations of the stickers. There are 4! ways to order the stars and 3! ways to order the moon.

$$\frac{12!}{4!3!} = 3{,}326{,}400$$

There are 3,326,400 ways to order the sheet of stickers.

---

formula for finding the number of permutations of n non-distinct objects

If there are n elements in a set and r_1 are alike, r_2 are alike, r_3 are alike, and so on through r_k, the number of permutations can be found by

$$\frac{n!}{r_1!\,r_2!\,...\,r_k!}$$

Example 7   **Finding the Number of Permutations of *n* Non-Distinct Objects**

Find the number of rearrangements of the letters in the word DISTINCT.

Solution   There are 8 letters. Both I and T are repeated 2 times. Substitute $n = 8$, $r_1 = 2$, and $r_2 = 2$ into the formula.

$$\frac{8!}{2!2!} = 10{,}080$$

There are 10,080 arrangements.

Try It #10

Find the number of rearrangements of the letters in the word CARRIER.

Access these online resources for additional instruction and practice with combinations and permutations.

- Combinations (http://openstaxcollege.org/l/combinations)
- Permutations (http://openstaxcollege.org/l/permutations)

9.5   SECTION EXERCISES

VERBAL

For the following exercises, assume that there are n ways an event A can happen, m ways an event B can happen, and that A and B are non-overlapping.

1. Use the Addition Principle of counting to explain how many ways event A or B can occur.

2. Use the Multiplication Principle of counting to explain how many ways event A and B can occur.

Answer the following questions.

3. When given two separate events, how do we know whether to apply the Addition Principle or the Multiplication Principle when calculating possible outcomes? What conjunctions may help to determine which operations to use?

4. Describe how the permutation of n objects differs from the permutation of choosing r objects from a set of n objects. Include how each is calculated.

5. What is the term for the arrangement that selects r objects from a set of n objects when the order of the r objects is not important? What is the formula for calculating the number of possible outcomes for this type of arrangement?

NUMERIC

For the following exercises, determine whether to use the Addition Principle or the Multiplication Principle. Then perform the calculations.

6. Let the set $A = \{ -5, -3, -1, 2, 3, 4, 5, 6\}$. How many ways are there to choose a negative or an even number from A?

7. Let the set $B = \{ -23, -16, -7, -2, 20, 36, 48, 72\}$. How many ways are there to choose a positive or an odd number from A?

8. How many ways are there to pick a red ace or a club from a standard card playing deck?

9. How many ways are there to pick a paint color from 5 shades of green, 4 shades of blue, or 7 shades of yellow?

10. How many outcomes are possible from tossing a pair of coins?

11. How many outcomes are possible from tossing a coin and rolling a 6-sided die?

12. How many two-letter strings—the first letter from A and the second letter from B—can be formed from the sets $A = \{b, c, d\}$ and $B = \{a, e, i, o, u\}$?

13. How many ways are there to construct a string of 3 digits if numbers can be repeated?

14. How many ways are there to construct a string of 3 digits if numbers cannot be repeated?

For the following exercises, compute the value of the expression.

15. $P(5, 2)$ **16.** $P(8, 4)$ **17.** $P(3, 3)$ **18.** $P(9, 6)$ **19.** $P(11, 5)$

20. $C(8, 5)$ **21.** $C(12, 4)$ **22.** $C(26, 3)$ **23.** $C(7, 6)$ **24.** $C(10, 3)$

For the following exercises, find the number of subsets in each given set.

25. $\{1, 2, 3, 4, 5, 6, 7, 8, 9, 10\}$ **26.** $\{a, b, c, \ldots, z\}$

27. A set containing 5 distinct numbers, 4 distinct letters, and 3 distinct symbols

28. The set of even numbers from 2 to 28

29. The set of two-digit numbers between 1 and 100 containing the digit 0

For the following exercises, find the distinct number of arrangements.

30. The letters in the word "juggernaut"   **31.** The letters in the word "academia"

32. The letters in the word "academia" that begin and end in "a"   **33.** The symbols in the string #,#,#,@,@,$,$,$,%,%,%,%

34. The symbols in the string #,#,#,@,@,$,$,$,%,%,%,% that begin and end with "%"

EXTENSIONS

35. The set, S consists of 900,000,000 whole numbers, each being the same number of digits long. How many digits long is a number from S? (*Hint*: use the fact that a whole number cannot start with the digit 0.)

36. The number of 5-element subsets from a set containing n elements is equal to the number of 6-element subsets from the same set. What is the value of n? (Hint: the order in which the elements for the subsets are chosen is not important.)

37. Can $C(n, r)$ ever equal $P(n, r)$? Explain.

38. Suppose a set A has 2,048 subsets. How many distinct objects are contained in A?

39. How many arrangements can be made from the letters of the word "mountains" if all the vowels must form a string?

REAL-WORLD APPLICATIONS

40. A family consisting of 2 parents and 3 children is to pose for a picture with 2 family members in the front and 3 in the back.

    a. How many arrangements are possible with no restrictions?

    b. How many arrangements are possible if the parents must sit in the front?

    c. How many arrangements are possible if the parents must be next to each other?

41. A cell phone company offers 6 different voice packages and 8 different data packages. Of those, 3 packages include both voice and data. How many ways are there to choose either voice or data, but not both?

42. In horse racing, a "trifecta" occurs when a bettor wins by selecting the first three finishers in the exact order (1st place, 2nd place, and 3rd place). How many different trifectas are possible if there are 14 horses in a race?

43. A wholesale T-shirt company offers sizes small, medium, large, and extra-large in organic or non-organic cotton and colors white, black, gray, blue, and red. How many different T-shirts are there to choose from?

44. Hector wants to place billboard advertisements throughout the county for his new business. How many ways can Hector choose 15 neighborhoods to advertise in if there are 30 neighborhoods in the county?

45. An art store has 4 brands of paint pens in 12 different colors and 3 types of ink. How many paint pens are there to choose from?

46. How many ways can a committee of 3 freshmen and 4 juniors be formed from a group of 8 freshmen and 11 juniors?

47. How many ways can a baseball coach arrange the order of 9 batters if there are 15 players on the team?

48. A conductor needs 5 cellists and 5 violinists to play at a diplomatic event. To do this, he ranks the orchestra's 10 cellists and 16 violinists in order of musical proficiency. What is the ratio of the total cellist rankings possible to the total violinist rankings possible?

49. A motorcycle shop has 10 choppers, 6 bobbers, and 5 café racers—different types of vintage motorcycles. How many ways can the shop choose 3 choppers, 5 bobbers, and 2 café racers for a weekend showcase?

50. A skateboard shop stocks 10 types of board decks, 3 types of trucks, and 4 types of wheels. How many different skateboards can be constructed?

51. Just-For-Kicks Sneaker Company offers an online customizing service. How many ways are there to design a custom pair of Just-For-Kicks sneakers if a customer can choose from a basic shoe up to 11 customizable options?

52. A car wash offers the following optional services to the basic wash: clear coat wax, triple foam polish, undercarriage wash, rust inhibitor, wheel brightener, air freshener, and interior shampoo. How many washes are possible if any number of options can be added to the basic wash?

53. Susan bought 20 plants to arrange along the border of her garden. How many distinct arrangements can she make if the plants are comprised of 6 tulips, 6 roses, and 8 daisies?

54. How many unique ways can a string of Christmas lights be arranged from 9 red, 10 green, 6 white, and 12 gold color bulbs?

LEARNING OBJECTIVES

In this section, you will:

- Apply the Binomial Theorem.

9.6 BINOMIAL THEOREM

A polynomial with two terms is called a binomial. We have already learned to multiply binomials and to raise binomials to powers, but raising a binomial to a high power can be tedious and time-consuming. In this section, we will discuss a shortcut that will allow us to find $(x + y)^n$ without multiplying the binomial by itself n times.

Identifying Binomial Coefficients

In **Counting Principles**, we studied combinations. In the shortcut to finding $(x + y)^n$, we will need to use combinations to find the coefficients that will appear in the expansion of the binomial. In this case, we use the notation $\binom{n}{r}$ instead of $C(n, r)$, but it can be calculated in the same way. So

$$\binom{n}{r} = C(n, r) = \frac{n!}{r!(n - r)!}$$

The combination $\binom{n}{r}$ is called a **binomial coefficient**. An example of a binomial coefficient is $\binom{5}{2} = C(5, 2) = 10$.

---

binomial coefficients

If n and r are integers greater than or equal to 0 with $n \geq r$, then the **binomial coefficient** is

$$\binom{n}{r} = C(n, r) = \frac{n!}{r!(n - r)!}$$

Q & A...

Is a binomial coefficient always a whole number?

Yes. Just as the number of combinations must always be a whole number, a binomial coefficient will always be a whole number.

Example 1   **Finding Binomial Coefficients**

Find each binomial coefficient.

   a. $\binom{5}{3}$ b. $\binom{9}{2}$ c. $\binom{9}{7}$

Solution

Use the formula to calculate each binomial coefficient. You can also use the $_nC_r$ function on your calculator.

$$\binom{n}{r} = C(n, r) = \frac{n!}{r!(n - r)!}$$

   a. $\binom{5}{3} = \frac{5!}{3!(5 - 3)!} = \frac{5 \cdot 4 \cdot 3!}{3!2!} = 10$

   b. $\binom{9}{2} = \frac{9!}{2!(9 - 2)!} = \frac{9 \cdot 8 \cdot 7!}{2!7!} = 36$

   c. $\binom{9}{7} = \frac{9!}{7!(9 - 7)!} = \frac{9 \cdot 8 \cdot 7!}{7!2!} = 36$

Analysis *Notice that we obtained the same result for parts (b) and (c). If you look closely at the solution for these two parts, you will see that you end up with the same two factorials in the denominator, but the order is reversed, just as with combinations.*

$$\binom{n}{r} = \binom{n}{n - r}$$

Try It #1

Find each binomial coefficient.

a. $\begin{pmatrix} 7 \\ 3 \end{pmatrix}$ **b.** $\begin{pmatrix} 11 \\ 4 \end{pmatrix}$

Using the Binomial Theorem

When we expand $(x + y)^n$ by multiplying, the result is called a **binomial expansion**, and it includes binomial coefficients. If we wanted to expand $(x + y)^{52}$, we might multiply $(x + y)$ by itself fifty-two times. This could take hours! If we examine some simple binomial expansions, we can find patterns that will lead us to a shortcut for finding more complicated binomial expansions.

$$(x + y)^2 = x^2 + 2xy + y^2$$
$$(x + y)^3 = x^3 + 3x^2 y + 3xy^2 + y^3$$
$$(x + y)^4 = x^4 + 4x^3 y + 6x^2 y^2 + 4xy^3 + y^4$$

First, let's examine the exponents. With each successive term, the exponent for x decreases and the exponent for y increases. The sum of the two exponents is n for each term.

Next, let's examine the coefficients. Notice that the coefficients increase and then decrease in a symmetrical pattern. The coefficients follow a pattern:

$$\begin{pmatrix} n \\ 0 \end{pmatrix}, \begin{pmatrix} n \\ 1 \end{pmatrix}, \begin{pmatrix} n \\ 2 \end{pmatrix}, ..., \begin{pmatrix} n \\ n \end{pmatrix}.$$

These patterns lead us to the **Binomial Theorem**, which can be used to expand any binomial.

$$(x + y)^n = \sum_{k=0}^{n} \begin{pmatrix} n \\ k \end{pmatrix} x^{n-k} y^k$$

$$= x^n + \begin{pmatrix} n \\ 1 \end{pmatrix} x^{n-1} y + \begin{pmatrix} n \\ 2 \end{pmatrix} x^{n-2} y^2 + ... + \begin{pmatrix} n \\ n-1 \end{pmatrix} xy^{n-1} + y^n$$

Another way to see the coefficients is to examine the expansion of a binomial in general form, $x + y$, to successive powers 1, 2, 3, and 4.

$$(x + y)^1 = x + y$$
$$(x + y)^2 = x^2 + 2xy + y^2$$
$$(x + y)^3 = x^3 + 3x^2 y + 3xy^2 + y^3$$
$$(x + y)^4 = x^4 + 4x^3 y + 6x^2 y^2 + 4xy^3 + y^4$$

Can you guess the next expansion for the binomial $(x + y)^5$?

Pascal's Triangle

Figure 1

See **Figure 1**, which illustrates the following:

- There are $n + 1$ terms in the expansion of $(x + y)^n$.
- The degree (or sum of the exponents) for each term is n.
- The powers on x begin with n and decrease to 0.
- The powers on y begin with 0 and increase to n.
- The coefficients are symmetric.

To determine the expansion on $(x + y)^5$, we see $n = 5$, thus, there will be $5 + 1 = 6$ terms. Each term has a combined degree of 5. In descending order for powers of x, the pattern is as follows:

- Introduce x^5, and then for each successive term reduce the exponent on x by 1 until $x^0 = 1$ is reached.
- Introduce $y^0 = 1$, and then increase the exponent on y by 1 until y^5 is reached.

$$x^5, x^4y, x^3y^2, x^2y^3, xy^4, y^5$$

The next expansion would be

$$(x + y)^5 = x^5 + 5x^4y + 10x^3y^2 + 10x^2y^3 + 5xy^4 + y^5.$$

But where do those coefficients come from? The binomial coefficients are symmetric. We can see these coefficients in an array known as Pascal's Triangle, shown in **Figure 2**.

Figure 2

To generate Pascal's Triangle, we start by writing a 1. In the row below, row 2, we write two 1's. In the 3rd row, flank the ends of the rows with 1's, and add $1 + 1$ to find the middle number, 2. In the nth row, flank the ends of the row with 1's. Each element in the triangle is the sum of the two elements immediately above it.

To see the connection between Pascal's Triangle and binomial coefficients, let us revisit the expansion of the binomials in general form.

$$1 \longrightarrow (x + y)^0 = 1$$
$$1 \quad 1 \longrightarrow (x + y)^1 = x + y$$
$$1 \quad 2 \quad 1 \longrightarrow (x + y)^2 = x^2 + 2xy + y^2$$
$$1 \quad 3 \quad 3 \quad 1 \longrightarrow (x + y)^3 = x^3 + 3x^2y + 3xy^2 + y^3$$
$$1 \quad 4 \quad 6 \quad 4 \quad 1 \longrightarrow (x + y)^4 = x^4 + 4x^3y + 6x^2y^2 + 4xy^3 + y^4$$
$$1 \quad 5 \quad 10 \quad 10 \quad 5 \quad 1 \longrightarrow (x + y)^5 = x^5 + 5x^4y + 10x^3y^2 + 10x^2y^3 + 5xy^4 + y^5$$

---

the Binomial Theorem

The **Binomial Theorem** is a formula that can be used to expand any binomial.

$$(x + y)^n = \sum_{k=0}^{n} \binom{n}{k} x^{n-k} y^k$$

$$= x^n + \binom{n}{1} x^{n-1} y + \binom{n}{2} x^{n-2} y^2 + \dots + \binom{n}{n-1} xy^{n-1} + y^n$$

How To...

Given a binomial, write it in expanded form.

1. Determine the value of n according to the exponent.
2. Evaluate the $k = 0$ through $k = n$ using the Binomial Theorem formula.
3. Simplify.

Example 2   **Expanding a Binomial**

Write in expanded form.

    a. $(x + y)^5$

    b. $(3x - y)^4$

Solution

    a. Substitute $n = 5$ into the formula. Evaluate the $k = 0$ through $k = 5$ terms. Simplify.

$$(x + y)^5 = \binom{5}{0}x^5y^0 + \binom{5}{1}x^4y^1 + \binom{5}{2}x^3y^2 + \binom{5}{3}x^2y^3 + \binom{5}{4}x^1y^4 + \binom{5}{5}x^0y^5$$

$$(x + y)^5 = x^5 + 5x^4y + 10x^3y^2 + 10x^2y^3 + 5xy^4 + y^5$$

    b. Substitute $n = 4$ into the formula. Evaluate the $k = 0$ through $k = 4$ terms. Notice that $3x$ is in the place that was occupied by x and that $-y$ is in the place that was occupied by y. So we substitute them. Simplify.

$$(3x - y)^4 = \binom{4}{0}(3x)^4(-y)^0 + \binom{4}{1}(3x)^3(-y)^1 + \binom{4}{2}(3x)^2(-y)^2 + \binom{4}{3}(3x)^1(-y)^3 + \binom{4}{4}(3x)^0(-y)^4$$

$$(3x - y)^4 = 81x^4 - 108x^3y + 54x^2y^2 - 12xy^3 + y^4$$

Analysis   *Notice the alternating signs in part b. This happens because $(-y)$ raised to odd powers is negative, but $(-y)$ raised to even powers is positive. This will occur whenever the binomial contains a subtraction sign.*

Try It #2

Write in expanded form.

a. $(x - y)^5$ **b.** $(2x + 5y)^3$

Using the Binomial Theorem to Find a Single Term

Expanding a binomial with a high exponent such as $(x + 2y)^{16}$ can be a lengthy process.

Sometimes we are interested only in a certain term of a binomial expansion. We do not need to fully expand a binomial to find a single specific term.

Note the pattern of coefficients in the expansion of $(x + y)^5$.

$$(x + y)^5 = x^5 + \binom{5}{1}x^4y + \binom{5}{2}x^3y^2 + \binom{5}{3}x^2y^3 + \binom{5}{4}xy^4 + y^5$$

The second term is $\binom{5}{1}x^4y$. The third term is $\binom{5}{2}x^3y^2$. We can generalize this result.

$$\binom{n}{r}x^{n-r}y^r$$

---

the $(r + 1)$th term of a binomial expansion

The $(r + 1)$th term of the binomial expansion of $(x + y)^n$ is:

$$\binom{n}{r}x^{n-r}y^r$$

How To...

Given a binomial, write a specific term without fully expanding.

1. Determine the value of n according to the exponent.
2. Determine $(r + 1)$.
3. Determine r.
4. Replace r in the formula for the $(r + 1)$th term of the binomial expansion.

Example 3   **Writing a Given Term of a Binomial Expansion**

Find the tenth term of $(x + 2y)^{16}$ without fully expanding the binomial.

Solution   Because we are looking for the tenth term, $r + 1 = 10$, we will use $r = 9$ in our calculations.

$$\binom{n}{r}x^{n-r}y^r$$

$$\binom{16}{9}x^{16-9}(2y)^9 = 5{,}857{,}280 x^7 y^9$$

Try It #3

Find the sixth term of $(3x - y)^9$ without fully expanding the binomial.

Access these online resources for additional instruction and practice with binomial expansion.

- The Binomial Theorem (http://openstaxcollege.org/l/binomialtheorem)
- Binomial Theorem Example (http://openstaxcollege.org/l/btexample)

9.6 SECTION EXERCISES

VERBAL

1. What is a binomial coefficient, and how it is calculated?

2. What role do binomial coefficients play in a binomial expansion? Are they restricted to any type of number?

3. What is the Binomial Theorem and what is its use?

4. When is it an advantage to use the Binomial Theorem? Explain.

ALGEBRAIC

For the following exercises, evaluate the binomial coefficient.

5. $\binom{6}{2}$ **6.** $\binom{5}{3}$ **7.** $\binom{7}{4}$ **8.** $\binom{9}{7}$ **9.** $\binom{10}{9}$

10. $\binom{25}{11}$ **11.** $\binom{17}{6}$ **12.** $\binom{200}{199}$

For the following exercises, use the Binomial Theorem to expand each binomial.

13. $(4a - b)^3$ **14.** $(5a + 2)^3$ **15.** $(3a + 2b)^3$ **16.** $(2x + 3y)^4$ **17.** $(4x + 2y)^5$

18. $(3x - 2y)^4$ **19.** $(4x - 3y)^5$ **20.** $\left(\frac{1}{x} + 3y\right)^5$ **21.** $(x^{-1} + 2y^{-1})^4$ **22.** $\left(\sqrt{x} - \sqrt{y}\right)^5$

For the following exercises, use the Binomial Theorem to write the first three terms of each binomial.

23. $(a + b)^{17}$ **24.** $(x - 1)^{18}$ **25.** $(a - 2b)^{15}$ **26.** $(x - 2y)^8$ **27.** $(3a + b)^{20}$

28. $(2a + 4b)^7$ **29.** $\left(x^3 - \sqrt{y}\right)^8$

For the following exercises, find the indicated term of each binomial without fully expanding the binomial.

30. The fourth term of $(2x - 3y)^4$

31. The fourth term of $(3x - 2y)^5$

32. The third term of $(6x - 3y)^7$

33. The eighth term of $(7 + 5y)^{14}$

34. The seventh term of $(a + b)^{11}$

35. The fifth term of $(x - y)^7$

36. The tenth term of $(x - 1)^{12}$

37. The ninth term of $(a - 3b^2)^{11}$

38. The fourth term of $\left(x^3 - \frac{1}{2}\right)^{10}$

39. The eighth term of $\left(\frac{y}{2} + \frac{2}{x}\right)^9$

GRAPHICAL

For the following exercises, use the Binomial Theorem to expand the binomial $f(x) = (x + 3)^4$. Then find and graph each indicated sum on one set of axes.

40. Find and graph $f_1(x)$, such that $f_1(x)$ is the first term of the expansion.

41. Find and graph $f_2(x)$, such that $f_2(x)$ is the sum of the first two terms of the expansion.

42. Find and graph $f_3(x)$, such that $f_3(x)$ is the sum of the first three terms of the expansion.

43. Find and graph $f_4(x)$, such that $f_4(x)$ is the sum of the first four terms of the expansion.

44. Find and graph $f_5(x)$, such that $f_5(x)$ is the sum of the first five terms of the expansion.

EXTENSIONS

45. In the expansion of $(5x + 3y)^n$, each term has the form $\binom{n}{k} a^{n-k} b^k$, where k successively takes on the value 0, 1, 2, ..., n. If $\binom{n}{k} = \binom{7}{2}$, what is the corresponding term?

46. In the expansion of $(a + b)^n$, the coefficient of $a^{n-k}b^k$ is the same as the coefficient of which other term?

47. Consider the expansion of $(x + b)^{40}$. What is the exponent of b in the kth term?

48. Find $\binom{n}{k-1} + \binom{n}{k}$ and write the answer as a binomial coefficient in the form $\binom{n}{k}$. Prove it. Hint: Use the fact that, for any integer p, such that $p \geq 1$, $p! = p(p-1)!$.

49. Which expression cannot be expanded using the Binomial Theorem? Explain.

    a. $(x^2 - 2x + 1)$

    b. $\left(\sqrt{a} + 4\sqrt{a} - 5\right)^8$

    c. $(x^3 + 2y^2 - z)^5$

    d. $\left(3x^2 - \sqrt{2y^3}\right)^{12}$

LEARNING OBJECTIVES

In this section, you will:

- Construct probability models.
- Compute probabilities of equally likely outcomes.
- Compute probabilities of the union of two events.
- Use the complement rule to find probabilities.
- Compute probability using counting theory.

9.7 PROBABILITY

Figure 1 An example of a "spaghetti model," which can be used to predict possible paths of a tropical storm.[33]

Residents of the Southeastern United States are all too familiar with charts, known as spaghetti models, such as the one in **Figure 1**. They combine a collection of weather data to predict the most likely path of a hurricane. Each colored line represents one possible path. The group of squiggly lines can begin to resemble strands of spaghetti, hence the name. In this section, we will investigate methods for making these types of predictions.

Constructing Probability Models

Suppose we roll a six-sided number cube. Rolling a number cube is an example of an **experiment**, or an activity with an observable result. The numbers on the cube are possible results, or **outcomes**, of this experiment. The set of all possible outcomes of an experiment is called the **sample space** of the experiment. The sample space for this experiment is {1, 2, 3, 4, 5, 6}. An **event** is any subset of a sample space.

The likelihood of an event is known as **probability**. The probability of an event p is a number that always satisfies $0 \leq p \leq 1$, where 0 indicates an impossible event and 1 indicates a certain event. A **probability model** is a mathematical description of an experiment listing all possible outcomes and their associated probabilities. For instance, if there is a 1% chance of winning a raffle and a 99% chance of losing the raffle, a probability model would look much like **Table 1**.

Outcome	Probability
Winning the raffle	1%
Losing the raffle	99%

Table 1

The sum of the probabilities listed in a probability model must equal 1, or 100%.

33 The figure is for illustrative purposes only and does not model any particular storm.

How To...
Given a probability event where each event is equally likely, construct a probability model.
1. Identify every outcome.
2. Determine the total number of possible outcomes.
3. Compare each outcome to the total number of possible outcomes.

Example 1   **Constructing a Probability Model**

Construct a probability model for rolling a single, fair die, with the event being the number shown on the die.

Solution   Begin by making a list of all possible outcomes for the experiment. The possible outcomes are the numbers that can be rolled: 1, 2, 3, 4, 5, and 6. There are six possible outcomes that make up the sample space.

Assign probabilities to each outcome in the sample space by determining a ratio of the outcome to the number of possible outcomes. There is one of each of the six numbers on the cube, and there is no reason to think that any particular face is more likely to show up than any other one, so the probability of rolling any number is $\frac{1}{6}$.

Outcome	Roll of 1	Roll of 2	Roll of 3	Roll of 4	Roll of 5	Roll of 6
Probability	$\frac{1}{6}$	$\frac{1}{6}$	$\frac{1}{6}$	$\frac{1}{6}$	$\frac{1}{6}$	$\frac{1}{6}$

Table 2

Q & A...
Do probabilities always have to be expressed as fractions?

No. Probabilities can be expressed as fractions, decimals, or percents. Probability must always be a number between 0 and 1, inclusive of 0 and 1.

Try It #1
Construct a probability model for tossing a fair coin.

Computing Probabilities of Equally Likely Outcomes

Let S be a sample space for an experiment. When investigating probability, an event is any subset of S. When the outcomes of an experiment are all equally likely, we can find the probability of an event by dividing the number of outcomes in the event by the total number of outcomes in S. Suppose a number cube is rolled, and we are interested in finding the probability of the event "rolling a number less than or equal to 4." There are 4 possible outcomes in the event and 6 possible outcomes in S, so the probability of the event is $\frac{4}{6} = \frac{2}{3}$.

---

computing the probability of an event with equally likely outcomes

The probability of an event E in an experiment with sample space S with equally likely outcomes is given by

$$P(E) = \frac{\text{number of elements in } E}{\text{number of elements in } S} = \frac{n(E)}{n(S)}$$

E is a subset of S, so it is always true that $0 \leq P(E) \leq 1$.

Example 2   **Computing the Probability of an Event with Equally Likely Outcomes**

A six-sided number cube is rolled. Find the probability of rolling an odd number.

Solution   The event "rolling an odd number" contains three outcomes. There are 6 equally likely outcomes in the sample space. Divide to find the probability of the event.

$$P(E) = \frac{3}{6} = \frac{1}{2}$$

Try It #2

A six-sided number cube is rolled. Find the probability of rolling a number greater than 2.

Computing the Probability of the Union of Two Events

We are often interested in finding the probability that one of multiple events occurs. Suppose we are playing a card game, and we will win if the next card drawn is either a heart or a king. We would be interested in finding the probability of the next card being a heart or a king. The **union of two events** E and F, written $E \cup F$, is the event that occurs if either or both events occur.

$$P(E \cup F) = P(E) + P(F) - P(E \cap F)$$

Suppose the spinner in **Figure 2** is spun. We want to find the probability of spinning orange or spinning a b.

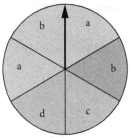

Figure 2

There are a total of 6 sections, and 3 of them are orange. So the probability of spinning orange is $\frac{3}{6} = \frac{1}{2}$. There are a total of 6 sections, and 2 of them have a b. So the probability of spinning a b is $\frac{2}{6} = \frac{1}{3}$. If we added these two probabilities, we would be counting the sector that is both orange and a b twice. To find the probability of spinning an orange or a b, we need to subtract the probability that the sector is both orange and has a b.

$$\frac{1}{2} + \frac{1}{3} - \frac{1}{6} = \frac{2}{3}$$

The probability of spinning orange or a b is $\frac{2}{3}$.

---

probability of the union of two events

The probability of the union of two events E and F (written $E \cup F$) equals the sum of the probability of E and the probability of F minus the probability of E and F occurring together (which is called the intersection of E and F and is written as $E \cap F$).

$$P(E \cup F) = P(E) + P(F) - P(E \cap F)$$

Example 3   **Computing the Probability of the Union of Two Events**

A card is drawn from a standard deck. Find the probability of drawing a heart or a 7.

Solution   A standard deck contains an equal number of hearts, diamonds, clubs, and spades. So the probability of drawing a heart is $\frac{1}{4}$. There are four 7s in a standard deck, and there are a total of 52 cards. So the probability of drawing a 7 is $\frac{1}{13}$.

The only card in the deck that is both a heart and a 7 is the 7 of hearts, so the probability of drawing both a heart and a 7 is $\frac{1}{52}$. Substitute $P(H) = \frac{1}{4}$, $P(7) = \frac{1}{13}$, and $P(H \cap 7) = \frac{1}{52}$ into the formula.

$$P(E \cup F) = P(E) + P(F) - P(E \cap F)$$

$$= \frac{1}{4} + \frac{1}{13} - \frac{1}{52}$$

$$= \frac{4}{13}$$

The probability of drawing a heart or a 7 is $\frac{4}{13}$.

Try It #3

A card is drawn from a standard deck. Find the probability of drawing a red card or an ace.

Computing the Probability of Mutually Exclusive Events

Suppose the spinner in **Figure 2** is spun again, but this time we are interested in the probability of spinning an orange or a d. There are no sectors that are both orange and contain a d, so these two events have no outcomes in common. Events are said to be **mutually exclusive events** when they have no outcomes in common. Because there is no overlap, there is nothing to subtract, so the general formula is

$$P(E \cup F) = P(E) + P(F)$$

Notice that with mutually exclusive events, the intersection of E and F is the empty set. The probability of spinning an orange is $\frac{3}{6} = \frac{1}{2}$ and the probability of spinning a d is $\frac{1}{6}$. We can find the probability of spinning an orange or a d simply by adding the two probabilities.

$$P(E \cup F) = P(E) + P(F)$$
$$= \frac{1}{2} + \frac{1}{6}$$
$$= \frac{2}{3}$$

The probability of spinning an orange or a d is $\frac{2}{3}$.

---

probability of the union of mutually exclusive events

The probability of the union of two mutually exclusive events E and F is given by

$$P(E \cup F) = P(E) + P(F)$$

How To...

Given a set of events, compute the probability of the union of mutually exclusive events.

1. Determine the total number of outcomes for the first event.
2. Find the probability of the first event.
3. Determine the total number of outcomes for the second event.
4. Find the probability of the second event.
5. Add the probabilities.

Example 4   **Computing the Probability of the Union of Mutually Exclusive Events**

A card is drawn from a standard deck. Find the probability of drawing a heart or a spade.

Solution   The events "drawing a heart" and "drawing a spade" are mutually exclusive because they cannot occur at the same time. The probability of drawing a heart is $\frac{1}{4}$, and the probability of drawing a spade is also $\frac{1}{4}$, so the probability of drawing a heart or a spade is

$$\frac{1}{4} + \frac{1}{4} = \frac{1}{2}$$

Try It #4

A card is drawn from a standard deck. Find the probability of drawing an ace or a king.

Using the Complement Rule to Compute Probabilities

We have discussed how to calculate the probability that an event will happen. Sometimes, we are interested in finding the probability that an event will not happen. The **complement of an event** E, denoted E', is the set of outcomes in the sample space that are not in E. For example, suppose we are interested in the probability that a horse will lose a race. If event W is the horse winning the race, then the complement of event W is the horse losing the race.

To find the probability that the horse loses the race, we need to use the fact that the sum of all probabilities in a probability model must be 1.

$$P(E') = 1 - P(E)$$

The probability of the horse winning added to the probability of the horse losing must be equal to 1. Therefore, if the probability of the horse winning the race is $\frac{1}{9}$, the probability of the horse losing the race is simply

$$1 - \frac{1}{9} = \frac{8}{9}$$

---

the complement rule

The probability that the **complement of an event** will occur is given by

$$P(E') = 1 - P(E)$$

Example 5   **Using the Complement Rule to Calculate Probabilities**

Two six-sided number cubes are rolled.

 a. Find the probability that the sum of the numbers rolled is less than or equal to 3.

 b. Find the probability that the sum of the numbers rolled is greater than 3.

Solution   The first step is to identify the sample space, which consists of all the possible outcomes. There are two number cubes, and each number cube has six possible outcomes. Using the Multiplication Principle, we find that there are 6×6, or 36 total possible outcomes. So, for example, 1-1 represents a 1 rolled on each number cube.

1-1	1-2	1-3	1-4	1-5	1-6
2-1	2-2	2-3	2-4	2-5	2-6
3-1	3-2	3-3	3-4	3-5	3-6
4-1	4-2	4-3	4-4	4-5	4-6
5-1	5-2	5-3	5-4	5-5	5-6
6-1	6-2	6-3	6-4	6-5	6-6

Table 3

 a. We need to count the number of ways to roll a sum of 3 or less. These would include the following outcomes: 1-1, 1-2, and 2-1. So there are only three ways to roll a sum of 3 or less. The probability is

$$\frac{3}{36} = \frac{1}{12}$$

 b. Rather than listing all the possibilities, we can use the Complement Rule. Because we have already found the probability of the complement of this event, we can simply subtract that probability from 1 to find the probability that the sum of the numbers rolled is greater than 3.

$$P(E') = 1 - P(E)$$
$$= 1 - \frac{1}{12}$$
$$= \frac{11}{12}$$

Try It #5

Two number cubes are rolled. Use the Complement Rule to find the probability that the sum is less than 10.

Computing Probability Using Counting Theory

Many interesting probability problems involve counting principles, permutations, and combinations. In these problems, we will use permutations and combinations to find the number of elements in events and sample spaces. These problems can be complicated, but they can be made easier by breaking them down into smaller counting problems.

Assume, for example, that a store has 8 cellular phones and that 3 of those are defective. We might want to find the probability that a couple purchasing 2 phones receives 2 phones that are not defective. To solve this problem, we need to calculate all of the ways to select 2 phones that are not defective as well as all of the ways to select 2 phones. There are 5 phones that are not defective, so there are $C(5, 2)$ ways to select 2 phones that are not defective. There are 8 phones, so there are $C(8, 2)$ ways to select 2 phones. The probability of selecting 2 phones that are not defective is:

$$\frac{\text{ways to select 2 phones that are not defective}}{\text{ways to select 2 phones}} = \frac{C(5, 2)}{C(8, 2)}$$

$$= \frac{10}{28}$$

$$= \frac{5}{14}$$

Example 6   Computing Probability Using Counting Theory

A child randomly selects 5 toys from a bin containing 3 bunnies, 5 dogs, and 6 bears.

   a. Find the probability that only bears are chosen.

   b. Find the probability that 2 bears and 3 dogs are chosen.

   c. Find the probability that at least 2 dogs are chosen.

Solution

   a. We need to count the number of ways to choose only bears and the total number of possible ways to select 5 toys. There are 6 bears, so there are $C(6, 5)$ ways to choose 5 bears. There are 14 toys, so there are $C(14, 5)$ ways to choose any 5 toys.

$$\frac{C(6, 5)}{C(14, 5)} = \frac{6}{2{,}002} = \frac{3}{1{,}001}$$

   b. We need to count the number of ways to choose 2 bears and 3 dogs and the total number of possible ways to select 5 toys. There are 6 bears, so there are $C(6, 2)$ ways to choose 2 bears. There are 5 dogs, so there are $C(5, 3)$ ways to choose 3 dogs. Since we are choosing both bears and dogs at the same time, we will use the Multiplication Principle. There are $C(6, 2) \cdot C(5, 3)$ ways to choose 2 bears and 3 dogs. We can use this result to find the probability.

$$\frac{C(6, 2)C(5, 3)}{C(14, 5)} = \frac{15 \cdot 10}{2{,}002} = \frac{75}{1{,}001}$$

   c. It is often easiest to solve "at least" problems using the Complement Rule. We will begin by finding the probability that fewer than 2 dogs are chosen. If less than 2 dogs are chosen, then either no dogs could be chosen, or 1 dog could be chosen.

   When no dogs are chosen, all 5 toys come from the 9 toys that are not dogs. There are $C(9, 5)$ ways to choose toys from the 9 toys that are not dogs. Since there are 14 toys, there are $C(14, 5)$ ways to choose the 5 toys from all of the toys.

$$\frac{C(9, 5)}{C(14,5)} = \frac{63}{1{,}001}$$

   If there is 1 dog chosen, then 4 toys must come from the 9 toys that are not dogs, and 1 must come from the 5 dogs. Since we are choosing both dogs and other toys at the same time, we will use the Multiplication Principle. There are $C(5, 1) \cdot C(9, 4)$ ways to choose 1 dog and 1 other toy.

$$\frac{C(5, 1)C(9, 4)}{C(14, 5)} = \frac{5 \cdot 126}{2{,}002} = \frac{315}{1{,}001}$$

Because these events would not occur together and are therefore mutually exclusive, we add the probabilities to find the probability that fewer than 2 dogs are chosen.

$$\frac{63}{1,001} + \frac{315}{1,001} = \frac{378}{1,001}$$

We then subtract that probability from 1 to find the probability that at least 2 dogs are chosen.

$$1 - \frac{378}{1,001} = \frac{623}{1,001}$$

Try It #6

A child randomly selects 3 gumballs from a container holding 4 purple gumballs, 8 yellow gumballs, and 2 green gumballs.

a. Find the probability that all 3 gumballs selected are purple.

b. Find the probability that no yellow gumballs are selected.

c. Find the probability that at least 1 yellow gumball is selected.

Access these online resources for additional instruction and practice with probability.

- Introduction to Probability (http://openstaxcollege.org/l/introprob)
- Determining Probability (http://openstaxcollege.org/l/determineprob)

9.7   SECTION EXERCISES

VERBAL

1. What term is used to express the likelihood of an event occurring? Are there restrictions on its values? If so, what are they? If not, explain.

2. What is a sample space?

3. What is an experiment?

4. What is the difference between events and outcomes? Give an example of both using the sample space of tossing a coin 50 times.

5. The *union of two sets* is defined as a set of elements that are present in at least one of the sets. How is this similar to the definition used for the *union of two events* from a probability model? How is it different?

NUMERIC

For the following exercises, use the spinner shown in **Figure 3** to find the probabilities indicated.

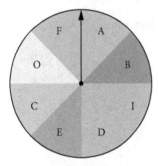

Figure 3

6. Landing on red

7. Landing on a vowel

8. Not landing on blue

9. Landing on purple or a vowel

10. Landing on blue or a vowel

11. Landing on green or blue

12. Landing on yellow or a consonant

13. Not landing on yellow or a consonant

For the following exercises, two coins are tossed.

14. What is the sample space?

15. Find the probability of tossing two heads.

16. Find the probability of tossing exactly one tail.

17. Find the probability of tossing at least one tail.

For the following exercises, four coins are tossed.

18. What is the sample space?

19. Find the probability of tossing exactly two heads.

20. Find the probability of tossing exactly three heads.

21. Find the probability of tossing four heads or four tails.

22. Find the probability of tossing all tails.

23. Find the probability of tossing not all tails.

24. Find the probability of tossing exactly two heads or at least two tails.

25. Find the probability of tossing either two heads or three heads.

For the following exercises, one card is drawn from a standard deck of 52 cards. Find the probability of drawing the following:

26. A club

27. A two

28. Six or seven

29. Red six

30. An ace or a diamond

31. A non-ace

32. A heart or a non-jack

For the following exercises, two dice are rolled, and the results are summed.

33. Construct a table showing the sample space of outcomes and sums.

34. Find the probability of rolling a sum of 3.

35. Find the probability of rolling at least one four or a sum of 8.

36. Find the probability of rolling an odd sum less than 9.

37. Find the probability of rolling a sum greater than or equal to 15.

38. Find the probability of rolling a sum less than 15.

39. Find the probability of rolling a sum less than 6 or greater than 9.

40. Find the probability of rolling a sum between 6 and 9, inclusive.

41. Find the probability of rolling a sum of 5 or 6.

42. Find the probability of rolling any sum other than 5 or 6.

For the following exercises, a coin is tossed, and a card is pulled from a standard deck. Find the probability of the following:

43. A head on the coin or a club

44. A tail on the coin or red ace

45. A head on the coin or a face card

46. No aces

For the following exercises, use this scenario: a bag of M&Ms contains 12 blue, 6 brown, 10 orange, 8 yellow, 8 red, and 4 green M&Ms. Reaching into the bag, a person grabs 5 M&Ms.

47. What is the probability of getting all blue M&Ms?

48. What is the probability of getting 4 blue M&Ms?

49. What is the probability of getting 3 blue M&Ms?

50. What is the probability of getting no brown M&Ms?

EXTENSIONS

Use the following scenario for the exercises that follow: In the game of Keno, a player starts by selecting 20 numbers from the numbers 1 to 80. After the player makes his selections, 20 winning numbers are randomly selected from numbers 1 to 80. A win occurs if the player has correctly selected 3, 4, or 5 of the 20 winning numbers. (Round all answers to the nearest hundredth of a percent.)

51. What is the percent chance that a player selects exactly 3 winning numbers?

52. What is the percent chance that a player selects exactly 4 winning numbers?

53. What is the percent chance that a player selects all 5 winning numbers?

54. What is the percent chance of winning?

55. How much less is a player's chance of selecting 3 winning numbers than the chance of selecting either 4 or 5 winning numbers?

REAL-WORLD APPLICATIONS

Use this data for the exercises that follow: In 2013, there were roughly 317 million citizens in the United States, and about 40 million were elderly (aged 65 and over).[34]

56. If you meet a U.S. citizen, what is the percent chance that the person is elderly? (Round to the nearest tenth of a percent.)

57. If you meet five U.S. citizens, what is the percent chance that exactly one is elderly? (Round to the nearest tenth of a percent.)

58. If you meet five U.S. citizens, what is the percent chance that three are elderly? (Round to the nearest tenth of a percent.)

59. If you meet five U.S. citizens, what is the percent chance that four are elderly? (Round to the nearest thousandth of a percent.)

60. It is predicted that by 2030, one in five U.S. citizens will be elderly. How much greater will the chances of meeting an elderly person be at that time? What policy changes do you foresee if these statistics hold true?

34 United States Census Bureau. http://www.census.gov

CHAPTER 9 REVIEW

Key Terms

Addition Principle if one event can occur in m ways and a second event with no common outcomes can occur in n ways, then the first or second event can occur in $m + n$ ways

annuity an investment in which the purchaser makes a sequence of periodic, equal payments

arithmetic sequence a sequence in which the difference between any two consecutive terms is a constant

arithmetic series the sum of the terms in an arithmetic sequence

binomial coefficient the number of ways to choose r objects from n objects where order does not matter; equivalent to $C(n, r)$, denoted $\binom{n}{r}$

binomial expansion the result of expanding $(x + y)^n$ by multiplying

Binomial Theorem a formula that can be used to expand any binomial

combination a selection of objects in which order does not matter

common difference the difference between any two consecutive terms in an arithmetic sequence

common ratio the ratio between any two consecutive terms in a geometric sequence

complement of an event the set of outcomes in the sample space that are not in the event E

diverge a series is said to diverge if the sum is not a real number

event any subset of a sample space

experiment an activity with an observable result

explicit formula a formula that defines each term of a sequence in terms of its position in the sequence

finite sequence a function whose domain consists of a finite subset of the positive integers $\{1, 2, \dots n\}$ for some positive integer n

Fundamental Counting Principle if one event can occur in m ways and a second event can occur in n ways after the first event has occurred, then the two events can occur in $m \times n$ ways; also known as the Multiplication Principle

geometric sequence a sequence in which the ratio of a term to a previous term is a constant

geometric series the sum of the terms in a geometric sequence

index of summation in summation notation, the variable used in the explicit formula for the terms of a series and written below the sigma with the lower limit of summation

infinite sequence a function whose domain is the set of positive integers

infinite series the sum of the terms in an infinite sequence

lower limit of summation the number used in the explicit formula to find the first term in a series

Multiplication Principle if one event can occur in m ways and a second event can occur in n ways after the first event has occurred, then the two events can occur in $m \times n$ ways; also known as the Fundamental Counting Principle

mutually exclusive events events that have no outcomes in common

n factorial the product of all the positive integers from 1 to n

nth partial sum the sum of the first n terms of a sequence

nth term of a sequence a formula for the general term of a sequence

outcomes the possible results of an experiment

permutation a selection of objects in which order matters

probability a number from 0 to 1 indicating the likelihood of an event

probability model a mathematical description of an experiment listing all possible outcomes and their associated probabilities

recursive formula a formula that defines each term of a sequence using previous term(s)

sample space the set of all possible outcomes of an experiment

sequence a function whose domain is a subset of the positive integers

series the sum of the terms in a sequence

summation notation a notation for series using the Greek letter sigma; it includes an explicit formula and specifies the first and last terms in the series

term a number in a sequence

union of two events the event that occurs if either or both events occur

upper limit of summation the number used in the explicit formula to find the last term in a series

Key Equations

Formula for a factorial	$0! = 1$ $1! = 1$ $n! = n(n-1)(n-2) \cdots (2)(1)$, for $n \geq 2$
recursive formula for nth term of an arithmetic sequence	$a_n = a_{n-1} + d; n \geq 2$
explicit formula for nth term of an arithmetic sequence	$a_n = a_1 + d(n-1)$
recursive formula for nth term of a geometric sequence	$a_n = ra_{n-1}, n \geq 2$
explicit formula for nth term of a geometric sequence	$a_n = a_1 r^{n-1}$
sum of the first n terms of an arithmetic series	$S_n = \dfrac{n(a_1 + a_n)}{2}$
sum of the first n terms of a geometric series	$S_n = \dfrac{a_1(1 - r^n)}{1 - r}, r \neq 1$
sum of an infinite geometric series with $-1 < r < 1$	$S_n = \dfrac{a_1}{1 - r}, r \neq 1$
number of permutations of n distinct objects taken r at a time	$P(n, r) = \dfrac{n!}{(n-r)!}$
number of combinations of n distinct objects taken r at a time	$C(n, r) = \dfrac{n!}{r!(n-r)!}$
number of permutations of n non-distinct objects	$\dfrac{n!}{r_1! r_2! \dots r_k!}$
Binomial Theorem	$(x + y)^n = \displaystyle\sum_{k-0}^{n} \binom{n}{k} x^{n-k}y^k$
$(r + 1)^{\text{th}}$ term of a binomial expansion	$\binom{n}{r} x^{n-r} y^r$
probability of an event with equally likely outcomes	$P(E) = \dfrac{n(E)}{n(S)}$
probability of the union of two events	$P(E \cup F) = P(E) + P(F) - P(E \cap F)$
probability of the union of mutually exclusive events	$P(E \cup F) = P(E) + P(F)$
probability of the complement of an event	$P(E') = 1 - P(E)$

Key Concepts

9.1 Sequences and Their Notations

- A sequence is a list of numbers, called terms, written in a specific order.
- Explicit formulas define each term of a sequence using the position of the term. See **Example 1**, **Example 2**, and **Example 3**.
- An explicit formula for the nth term of a sequence can be written by analyzing the pattern of several terms. See **Example 4**.
- Recursive formulas define each term of a sequence using previous terms.
- Recursive formulas must state the initial term, or terms, of a sequence.
- A set of terms can be written by using a recursive formula. See **Example 5** and **Example 6**.
- A factorial is a mathematical operation that can be defined recursively.
- The factorial of n is the product of all integers from 1 to n See **Example 7**.

9.2 Arithmetic Sequences

- An arithmetic sequence is a sequence where the difference between any two consecutive terms is a constant.
- The constant between two consecutive terms is called the common difference.
- The common difference is the number added to any one term of an arithmetic sequence that generates the subsequent term. See **Example 1**.
- The terms of an arithmetic sequence can be found by beginning with the initial term and adding the common difference repeatedly. See **Example 2** and **Example 3**.
- A recursive formula for an arithmetic sequence with common difference d is given by $a_n = a_{n-1} + d, n \geq 2$. See **Example 4**.
- As with any recursive formula, the initial term of the sequence must be given.
- An explicit formula for an arithmetic sequence with common difference d is given by $a_n = a_1 + d(n-1)$. See **Example 5**.
- An explicit formula can be used to find the number of terms in a sequence. See **Example 6**.
- In application problems, we sometimes alter the explicit formula slightly to $a_n = a_0 + dn$. See **Example 7**.

9.3 Geometric Sequences

- A geometric sequence is a sequence in which the ratio between any two consecutive terms is a constant.
- The constant ratio between two consecutive terms is called the common ratio.
- The common ratio can be found by dividing any term in the sequence by the previous term. See **Example 1**.
- The terms of a geometric sequence can be found by beginning with the first term and multiplying by the common ratio repeatedly. See **Example 2** and **Example 4**.
- A recursive formula for a geometric sequence with common ratio r is given by $a_n = ra_{n-1}$ for $n \geq 2$.
- As with any recursive formula, the initial term of the sequence must be given. See **Example 3**.
- An explicit formula for a geometric sequence with common ratio r is given by $a_n = a_1 r^{n-1}$. See **Example 5**.
- In application problems, we sometimes alter the explicit formula slightly to $a_n = a_0 r^n$. See **Example 6**.

9.4 Series and Their Notations

- The sum of the terms in a sequence is called a series.
- A common notation for series is called summation notation, which uses the Greek letter sigma to represent the sum. See **Example 1**.
- The sum of the terms in an arithmetic sequence is called an arithmetic series.
- The sum of the first n terms of an arithmetic series can be found using a formula. See **Example 2** and **Example 3**.
- The sum of the terms in a geometric sequence is called a geometric series.
- The sum of the first n terms of a geometric series can be found using a formula. See **Example 4** and **Example 5**.
- The sum of an infinite series exists if the series is geometric with $-1 < r < 1$.

- If the sum of an infinite series exists, it can be found using a formula. See **Example 6**, **Example 7**, and **Example 8**.
- An annuity is an account into which the investor makes a series of regularly scheduled payments. The value of an annuity can be found using geometric series. See **Example 9**.

9.5 Counting Principles

- If one event can occur in m ways and a second event with no common outcomes can occur in n ways, then the first or second event can occur in $m + n$ ways. See **Example 1**.
- If one event can occur in m ways and a second event can occur in n ways after the first event has occurred, then the two events can occur in $m \times n$ ways. See **Example 2**.
- A permutation is an ordering of n objects.
- If we have a set of n objects and we want to choose r objects from the set in order, we write $P(n, r)$.
- Permutation problems can be solved using the Multiplication Principle or the formula for $P(n, r)$. See **Example 3** and **Example 4**.
- A selection of objects where the order does not matter is a combination.
- Given n distinct objects, the number of ways to select r objects from the set is $C(n, r)$ and can be found using a formula. See **Example 5**.
- A set containing n distinct objects has 2^n subsets. See **Example 6**.
- For counting problems involving non-distinct objects, we need to divide to avoid counting duplicate permutations. See **Example 7**.

9.6 Binomial Theorem

- $\binom{n}{r}$ is called a binomial coefficient and is equal to $C(n, r)$. See **Example 1**.
- The Binomial Theorem allows us to expand binomials without multiplying. See **Example 2**.
- We can find a given term of a binomial expansion without fully expanding the binomial. See **Example 3**.

9.7 Probability

- Probability is always a number between 0 and 1, where 0 means an event is impossible and 1 means an event is certain.
- The probabilities in a probability model must sum to 1. See **Example 1**.
- When the outcomes of an experiment are all equally likely, we can find the probability of an event by dividing the number of outcomes in the event by the total number of outcomes in the sample space for the experiment. See **Example 2**.
- To find the probability of the union of two events, we add the probabilities of the two events and subtract the probability that both events occur simultaneously. See **Example 3**.
- To find the probability of the union of two mutually exclusive events, we add the probabilities of each of the events. See **Example 4**.
- The probability of the complement of an event is the difference between 1 and the probability that the event occurs. See **Example 5**.
- In some probability problems, we need to use permutations and combinations to find the number of elements in events and sample spaces. See **Example 6**.

CHAPTER 9 REVIEW EXERCISES

SEQUENCES AND THEIR NOTATION

1. Write the first four terms of the sequence defined by the recursive formula $a_1 = 2, a_n = a_{n-1} + n$.

2. Evaluate $\dfrac{6!}{(5-3)!3!}$.

3. Write the first four terms of the sequence defined by the explicit formula $a_n = 10^n + 3$.

4. Write the first four terms of the sequence defined by the explicit formula $a_n = \dfrac{n!}{n(n+1)}$.

ARITHMETIC SEQUENCES

5. Is the sequence $\dfrac{4}{7}, \dfrac{47}{21}, \dfrac{82}{21}, \dfrac{39}{7}, \dots$ arithmetic? If so, find the common difference.

6. Is the sequence 2, 4, 8, 16, ... arithmetic? If so, find the common difference.

7. An arithmetic sequence has the first term $a_1 = 18$ and common difference $d = -8$. What are the first five terms?

8. An arithmetic sequence has terms $a_3 = 11.7$ and $a_8 = -14.6$. What is the first term?

9. Write a recursive formula for the arithmetic sequence $-20, -10, 0, 10, \dots$

10. Write a recursive formula for the arithmetic sequence $0, -\dfrac{1}{2}, -1, -\dfrac{3}{2}, \dots$, and then find the 31st term.

11. Write an explicit formula for the arithmetic sequence $\dfrac{7}{8}, \dfrac{29}{24}, \dfrac{37}{24}, \dfrac{15}{8}, \dots$

12. How many terms are in the finite arithmetic sequence 12, 20, 28, ... , 172?

GEOMETRIC SEQUENCES

13. Find the common ratio for the geometric sequence 2.5, 5, 10, 20, ...

14. Is the sequence 4, 16, 28, 40, ... geometric? If so find the common ratio. If not, explain why.

15. A geometric sequence has terms $a_7 = 16,384$ and $a_9 = 262,144$. What are the first five terms?

16. A geometric sequence has the first term $a_1 = -3$ and common ratio $r = \dfrac{1}{2}$. What is the 8th term?

17. What are the first five terms of the geometric sequence $a_1 = 3, a_n = 4 \cdot a_{n-1}$?

18. Write a recursive formula for the geometric sequence $1, \dfrac{1}{3}, \dfrac{1}{9}, \dfrac{1}{27}, \dots$

19. Write an explicit formula for the geometric sequence $-\dfrac{1}{5}, -\dfrac{1}{15}, -\dfrac{1}{45}, -\dfrac{1}{135}, \dots$

20. How many terms are in the finite geometric sequence $-5, -\dfrac{5}{3}, -\dfrac{5}{9}, \dots, -\dfrac{5}{59,049}$?

SERIES AND THEIR NOTATION

21. Use summation notation to write the sum of terms $\dfrac{1}{2}m + 5$ from $m = 0$ to $m = 5$.

22. Use summation notation to write the sum that results from adding the number 13 twenty times.

23. Use the formula for the sum of the first n terms of an arithmetic series to find the sum of the first eleven terms of the arithmetic series 2.5, 4, 5.5, ...

24. A ladder has 15 tapered rungs, the lengths of which increase by a common difference. The first rung is 5 inches long, and the last rung is 20 inches long. What is the sum of the lengths of the rungs?

25. Use the formula for the sum of the first n terms of a geometric series to find S_9 for the series $12, 6, 3, \frac{3}{2}, \ldots$

26. The fees for the first three years of a hunting club membership are given in **Table 1**. If fees continue to rise at the same rate, how much will the total cost be for the first ten years of membership?

Year	Membership Fees
1	$1500
2	$1950
3	$2535

Table 1

27. Find the sum of the infinite geometric series
$$\sum_{k-1}^{\infty} 45 \cdot \left(-\frac{1}{3}\right)^{k=1}.$$

28. A ball has a bounce-back ratio $\frac{3}{5}$ of the height of the previous bounce. Write a series representing the total distance traveled by the ball, assuming it was initially dropped from a height of 5 feet. What is the total distance? (Hint: the total distance the ball travels on each bounce is the sum of the heights of the rise and the fall.)

29. Alejandro deposits $80 of his monthly earnings into an annuity that earns 6.25% annual interest, compounded monthly. How much money will he have saved after 5 years?

30. The twins Sarah and Scott both opened retirement accounts on their 21st birthday. Sarah deposits $4,800.00 each year, earning 5.5% annual interest, compounded monthly. Scott deposits $3,600.00 each year, earning 8.5% annual interest, compounded monthly. Which twin will earn the most interest by the time they are 55 years old? How much more?

COUNTING PRINCIPLES

31. How many ways are there to choose a number from the set $\{-10, -6, 4, 10, 12, 18, 24, 32\}$ that is divisible by either 4 or 6?

32. In a group of 20 musicians, 12 play piano, 7 play trumpet, and 2 play both piano and trumpet. How many musicians play either piano or trumpet?

33. How many ways are there to construct a 4-digit code if numbers can be repeated?

34. A palette of water color paints has 3 shades of green, 3 shades of blue, 2 shades of red, 2 shades of yellow, and 1 shade of black. How many ways are there to choose one shade of each color?

35. Calculate $P(18, 4)$.

36. In a group of 5 freshman, 10 sophomores, 3 juniors, and 2 seniors, how many ways can a president, vice president, and treasurer be elected?

37. Calculate $C(15, 6)$.

38. A coffee shop has 7 Guatemalan roasts, 4 Cuban roasts, and 10 Costa Rican roasts. How many ways can the shop choose 2 Guatemalan, 2 Cuban, and 3 Costa Rican roasts for a coffee tasting event?

39. How many subsets does the set $\{1, 3, 5, \ldots, 99\}$ have?

40. A day spa charges a basic day rate that includes use of a sauna, pool, and showers. For an extra charge, guests can choose from the following additional services: massage, body scrub, manicure, pedicure, facial, and straight-razor shave. How many ways are there to order additional services at the day spa?

41. How many distinct ways can the word DEADWOOD be arranged?

42. How many distinct rearrangements of the letters of the word DEADWOOD are there if the arrangement must begin and end with the letter D?

BINOMIAL THEOREM

43. Evaluate the binomial coefficient $\begin{pmatrix} 23 \\ 8 \end{pmatrix}$.

44. Use the Binomial Theorem to expand $\left(3x + \dfrac{1}{2}y\right)^6$.

45. Use the Binomial Theorem to write the first three terms of $(2a + b)^{17}$.

46. Find the fourth term of $(3a^2 - 2b)^{11}$ without fully expanding the binomial.

PROBABILITY

For the following exercises, assume two die are rolled.

47. Construct a table showing the sample space.

48. What is the probability that a roll includes a 2?

49. What is the probability of rolling a pair?

50. What is the probability that a roll includes a 2 or results in a pair?

51. What is the probability that a roll doesn't include a 2 or result in a pair?

52. What is the probability of rolling a 5 or a 6?

53. What is the probability that a roll includes neither a 5 nor a 6?

For the following exercises, use the following data: An elementary school survey found that 350 of the 500 students preferred soda to milk. Suppose 8 children from the school are attending a birthday party. (Show calculations and round to the nearest tenth of a percent.)

54. What is the percent chance that all the children attending the party prefer soda?

55. What is the percent chance that at least one of the children attending the party prefers milk?

56. What is the percent chance that exactly 3 of the children attending the party prefer soda?

57. What is the percent chance that exactly 3 of the children attending the party prefer milk?

CHAPTER 9 PRACTICE TEST

1. Write the first four terms of the sequence defined by the recursive formula $a = -14, a_n = \dfrac{2 + a_{n-1}}{2}$.

2. Write the first four terms of the sequence defined by the explicit formula $a_n = \dfrac{n^2 - n - 1}{n!}$.

3. Is the sequence 0.3, 1.2, 2.1, 3, ... arithmetic? If so find the common difference.

4. An arithmetic sequence has the first term $a_1 = -4$ and common difference $d = -\dfrac{4}{3}$. What is the 6th term?

5. Write a recursive formula for the arithmetic sequence $-2, -\dfrac{7}{2}, -5, -\dfrac{13}{2}$, ... and then find the 22nd term.

6. Write an explicit formula for the arithmetic sequence 15.6, 15, 14.4, 13.8, ... and then find the 32nd term.

7. Is the sequence $-2, -1, -\dfrac{1}{2}, -\dfrac{1}{4}$, ... geometric? If so find the common ratio. If not, explain why.

8. What is the 11th term of the geometric sequence $-1.5, -3, -6, -12, ...$?

9. Write a recursive formula for the geometric sequence $1, -\dfrac{1}{2}, \dfrac{1}{4}, -\dfrac{1}{8}, ...$

10. Write an explicit formula for the geometric sequence $4, -\dfrac{4}{3}, \dfrac{4}{9}, -\dfrac{4}{27}, ...$

11. Use summation notation to write the sum of terms $3k^2 - \dfrac{5}{6}k$ from $k = -3$ to $k = 15$.

12. A community baseball stadium has 10 seats in the first row, 13 seats in the second row, 16 seats in the third row, and so on. There are 56 rows in all. What is the seating capacity of the stadium?

13. Use the formula for the sum of the first n terms of a geometric series to find $\displaystyle\sum_{k=1}^{7} -0.2 \cdot (-5)^{k-1}$.

14. Find the sum of the infinite geometric series. $\displaystyle\sum_{k=1}^{\infty} \dfrac{1}{3} \cdot \left(-\dfrac{1}{5}\right)^{k-1}$

15. Rachael deposits $3,600 into a retirement fund each year. The fund earns 7.5% annual interest, compounded monthly. If she opened her account when she was 20 years old, how much will she have by the time she's 55? How much of that amount was interest earned?

16. In a competition of 50 professional ballroom dancers, 22 compete in the fox-trot competition, 18 compete in the tango competition, and 6 compete in both the fox-trot and tango competitions. How many dancers compete in the foxtrot or tango competitions?

17. A buyer of a new sedan can custom order the car by choosing from 5 different exterior colors, 3 different interior colors, 2 sound systems, 3 motor designs, and either manual or automatic transmission. How many choices does the buyer have?

18. To allocate annual bonuses, a manager must choose his top four employees and rank them first to fourth. In how many ways can he create the "Top-Four" list out of the 32 employees?

19. A rock group needs to choose 3 songs to play at the annual Battle of the Bands. How many ways can they choose their set if have 15 songs to pick from?

20. A self-serve frozen yogurt shop has 8 candy toppings and 4 fruit toppings to choose from. How many ways are there to top a frozen yogurt?

21. How many distinct ways can the word EVANESCENCE be arranged if the anagram must end with the letter E?

22. Use the Binomial Theorem to expand $\left(\dfrac{3}{2}x - \dfrac{1}{2}y\right)^5$.

23. Find the seventh term of $\left(x^2 - \dfrac{1}{2}\right)^{13}$ without fully expanding the binomial.

For the following exercises, use the spinner in **Figure 1**.

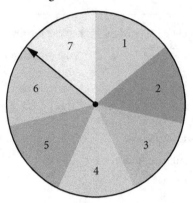

Figure 1

24. Construct a probability model showing each possible outcome and its associated probability. (Use the first letter for colors.)

25. What is the probability of landing on an odd number?

26. What is the probability of landing on blue?

27. What is the probability of landing on blue or an odd number?

28. What is the probability of landing on anything other than blue or an odd number?

29. A bowl of candy holds 16 peppermint, 14 butterscotch, and 10 strawberry flavored candies. Suppose a person grabs a handful of 7 candies. What is the percent chance that exactly 3 are butterscotch? (Show calculations and round to the nearest tenth of a percent.)

Try It Answers

Chapter 1

Section 1.1

1. a. $\frac{11}{1}$ **b.** $\frac{3}{1}$ **c.** $-\frac{4}{1}$ **2. a.** 4 (or 4.0), terminating **b.** $0.\overline{615384}$, repeating **c.** -0.85, terminating **3. a.** rational and repeating **b.** rational and terminating **c.** irrational **d.** rational and repeating **e.** irrational **4. a.** positive, irrational; right **b.** negative, rational; left **c.** positive, rational; right **d.** negative, irrational; left **e.** positive, rational; right

5.

	N	W	I	Q	Q'
a. $-\frac{35}{7}$			$\times$	$\times$	
b. 0		$\times$	$\times$	$\times$	
c. $\sqrt{169}$	$\times$	$\times$	$\times$	$\times$	
d. $\sqrt{24}$					$\times$
e. 4.763763763...				$\times$	

6. a. 10 **b.** 2 **c.** 4.5 **d.** 25 **e.** 26

7. a. 11, commutative property of multiplication, associative property of multiplication, inverse property of multiplication, identity property of multiplication; **b.** 33, distributive property; **c.** 26, distributive property; **d.** $\frac{4}{9}$, commutative property of addition, associative property of addition, inverse property of addition, identity property of addition; **e.** 0, distributive property, inverse property of addition, identity property of addition

8.

	Constants	Variables
a. $2\pi r(r + h)$	$2, \pi$	r, h
b. $2(L + W)$	2	L, W
c. $4y^3 + y$	4	y

9. a. 5 **b.** 11 **c.** 9 **d.** 26
10. a. 4 **b.** 11 **c.** $\frac{121}{3}\pi$ **d.** 1,728 **e.** 3
11. 1,152 cm²

12. a. $-2y - 2z$ or $-2(y + z)$ **b.** $\frac{2}{t} - 1$ **c.** $3pq - 4p + q$ **d.** $7r - 2s + 6$
13. $A = P(1 + rt)$

Section 1.2

1. a. k^{15} **b.** $\left(\frac{2}{y}\right)^5$ **c.** t^{14} **2. a.** s^7 **b.** $(-3)^5$ **c.** $(ef^2)^2$
3. a. $(3y)^{24}$ **b.** t^{35} **c.** $(-g)^{16}$ **4. a.** 1 **b.** $\frac{1}{2}$ **c.** 1 **d.** 1
5. a. $\frac{1}{(-3t)^6}$ **b.** $\frac{1}{f^3}$ **c.** $\frac{2}{5k^3}$ **6. a.** $t^{-5} = \frac{1}{t^5}$ **b.** $\frac{1}{25}$
7. a. $g^{10}h^{15}$ **b.** $125t^3$ **c.** $-27y^{15}$ **d.** $\frac{1}{a^{18}b^{21}}$ **e.** $\frac{r^{12}}{s^8}$
8. a. $\frac{b^{15}}{c^3}$ **b.** $\frac{625}{u^{32}}$ **c.** $-\frac{1}{w^{105}}$ **d.** $\frac{q^{24}}{p^{32}}$ **e.** $\frac{1}{c^{20}d^{12}}$
9. a. $\frac{v^6}{8u^3}$ **b.** $\frac{1}{x^3}$ **c.** $\frac{e^4}{f^4}$ **d.** $\frac{27r}{s}$ **e.** 1 **f.** $\frac{16h^{10}}{49}$ **10. a.** $\$1.52 \times 10^5$
b. 7.158×10^9 **c.** $\$8.55 \times 10^{13}$ **d.** 3.34×10^{-9} **e.** 7.15×10^{-8}
11. a. 703,000 **b.** $-816,000,000,000$ **c.** -0.00000000000039
d. 0.000008 **12. a.** -8.475×10^6 **b.** 8×10^{-8} **c.** 2.976×10^{13}
d. -4.3×10^6 **e.** $\approx 1.24 \times 10^{15}$ **13.** Number of cells: 3×10^{13};
length of a cell: 8×10^{-6} m; total length: 2.4×10^8 m or 240,000,000 m

Section 1.3

1. a. 15 **b.** 3 **c.** 4 **d.** 17 **2.** $5|x||y|\sqrt{2yz}$ Notice the absolute value signs around x and y? That's because their value must be positive.
3. $10|x|$ **4.** $\frac{x\sqrt{2}}{3y^2}$ We do not need the absolute value signs for y^2 because that term will always be nonnegative.
5. $b^4\sqrt{3ab}$ **6.** $13\sqrt{5}$ **7.** 0 **8.** $6\sqrt{6}$ **9.** $14 - 7\sqrt{3}$
10. a. -6 **b.** 6 **c.** $88\sqrt[3]{9}$ **11.** $(\sqrt{9})^5 = 3^5 = 243$ **12.** $x(5y)^{\frac{9}{2}}$
13. $28x^{\frac{23}{15}}$

Section 1.4

1. The degree is 6, the leading term is $-x^6$, and the leading coefficient is -1. **2.** $2x^3 + 7x^2 - 4x - 3$ **3.** $-11x^3 - x^2 + 7x - 9$
4. $3x^4 - 10x^3 - 8x^2 + 21x + 14$ **5.** $3x^2 + 16x - 35$
6. $16x^2 - 8x + 1$ **7.** $4x^2 - 49$ **8.** $6x^2 + 21xy - 29x - 7y + 9$

Section 1.5

1. $(b^2 - a)(x + 6)$ **2.** $(x - 6)(x - 1)$ **3. a.** $(2x + 3)(x + 3)$
b. $(3x - 1)(2x + 1)$ **4.** $(7x - 1)^2$ **5.** $(9y + 10)(9y - 10)$
6. $(6a + b)(36a^2 - 6ab + b^2)$ **7.** $(10x - 1)(100x^2 + 10x + 1)$
8. $(5a - 1)^{-\frac{1}{4}}(17a - 2)$

Section 1.6

1. $\frac{1}{x + 6}$ **2.** $\frac{(x + 5)(x + 6)}{(x + 2)(x + 4)}$ **3.** 1 **4.** $\frac{2(x - 7)}{(x + 5)(x - 3)}$
5. $\frac{x^2 - y^2}{xy^2}$

Chapter 2

Section 2.1

1.

x	$y = \frac{1}{2}x + 2$	(x, y)
-2	$y = \frac{1}{2}(-2) + 2 = 1$	$(-2, 1)$
-1	$y = \frac{1}{2}(-1) + 2 = \frac{3}{2}$	$\left(-1, \frac{3}{2}\right)$
0	$y = \frac{1}{2}(0) + 2 = 2$	$(0, 2)$
1	$y = \frac{1}{2}(1) + 2 = \frac{5}{2}$	$\left(1, \frac{5}{2}\right)$
2	$y = \frac{1}{2}(2) + 2 = 3$	$(2, 3)$

2. x-intercept is $(4, 0)$; **3.** $\sqrt{125} = 5\sqrt{5}$
y-intercept is $(0, 3)$. **4.** $\left(-5, \frac{5}{2}\right)$

A-1

Section 2.2

1. $x = -5$ **2.** $x = -3$ **3.** $x = \dfrac{10}{3}$ **4.** $x = 1$

5. $x = -\dfrac{7}{17}$, excluded values are $-\dfrac{1}{2}$ and $-\dfrac{1}{3}$.

6. $x = \dfrac{1}{3}$ **7.** $m = -\dfrac{2}{3}$ **8.** $y = 4x - 3$ **9.** $x + 3y = 2$

10. Horizontal line: $y = 2$

11.

Parallel lines: equations are written in slope-intercept form.

12. $y = 5x + 3$

Section 2.3

1. 11 and 25   **2.** $C = 2.5x + 3{,}650$ **3.** 45 mi/h

4. $L = 37$ cm, $W = 18$ cm   **5.** 250 ft^2

Section 2.4

1. $\sqrt{-24} = 0 + 2i\sqrt{6}$ **2.**

3. $(3 - 4i) - (2 + 5i) = 1 - 9i$

4. $\dfrac{5}{2} - i$ **5.** $18 + i$

6. $-3 - 4i$ **7.** -1

Section 2.5

1. $(x - 6)(x + 1) = 0$; $x = 6, x = -1$ **2.** $(x - 7)(x + 3) = 0$; $x = 7, x = -3$ **3.** $(x + 5)(x - 5) = 0$; $x = -5, x = 5$

4. $(3x + 2)(4x + 1) = 0$; $x = -\dfrac{2}{3}, x = -\dfrac{1}{4}$ **5.** $x = 0, x = -10,$

$x = -1$ **6.** $x = 4 \pm \sqrt{5}$ **7.** $x = 3 \pm \sqrt{22}$ **8.** $x = -\dfrac{2}{3},$

$x = \dfrac{1}{3}$ **9.** 5 units

Section 2.6

1. $\dfrac{1}{4}$ **2.** 25   **3.** $\{-1\}$ **4.** $x = 0, x = \dfrac{1}{2}, x = -\dfrac{1}{2}$

5. $x = 1$; extraneous solution: $-\dfrac{2}{9}$ **6.** $x = -2$; extraneous

solution: -1 **7.** $x = -1, x = \dfrac{3}{2}$ **8.** $x = -3, 3, -i, i$

9. $x = 2, x = 12$ **10.** $x = -1, 0$ is not a solution.

Section 2.7

1. $[-3, 5]$ **2.** $(-\infty, -2) \cup [3, \infty)$ **3.** $x < 1$ **4.** $x \geq -5$

5. $(2, \infty)$ **6.** $\left[-\dfrac{3}{14}, \infty\right)$ **7.** $6 < x \leq 9$ or $(6, 9]$

8. $\left(-\dfrac{1}{8}, \dfrac{1}{2}\right)$ **9.** $|x - 2| \leq 3$

10.

$-2|k - 4| \leq -6$

$k \leq 1$ or $k \geq 7$; in interval notation, this would be $(-\infty, 1] \cup [7, \infty)$.

Chapter 3

Section 3.1

1. a. Yes **b.** Yes   (Note: If two players had been tied for, say, 4th place, then the name would not have been a function of rank.)

2. $w = f(d)$ **3.** Yes   **4.** $g(5) = 1$ **5.** $m = 8$

6. $y = f(x) = \dfrac{\sqrt[3]{x}}{2}$ **7.** $g(1) = 8$ **8.** $x = 0$ or $x = 2$

9. a. Yes, because each bank account has a single balance at any given time; **b.** No, because several bank account numbers may have the same balance; **c.** No, because the same output may correspond to more than one input.   **10. a.** Yes, letter grade is a function of percent grade; **b.** No, it is not one-to-one. There are 100 different percent numbers we could get but only about five possible letter grades, so there cannot be only one percent number that corresponds to each letter grade.   **11.** Yes

12. No, because it does not pass the horizontal line test.

Section 3.2

1. $\{-5, 0, 5, 10, 15\}$ **2.** $(-\infty, \infty)$ **3.** $\left(-\infty, \dfrac{1}{2}\right) \cup \left(\dfrac{1}{2}, \infty\right)$

4. $\left[-\dfrac{5}{2}, \infty\right)$ **5. a.** Values that are less than or equal to -2, or values that are greater than or equal to -1 and less than 3;

b. $\{x | x \leq -2 \text{ or } -1 \leq x < 3\}$ **c.** $(-\infty, -2] \cup [-1, 3)$

6. Domain = $[1950, 2002]$; Range = $[47{,}000{,}000, 89{,}000{,}000]$

7. Domain: $(-\infty, 2]$; Range: $(-\infty, 0]$

8.

Section 3.3

1. $\dfrac{\$2.84 - \$2.31}{5 \text{ years}} = \dfrac{\$0.53}{5 \text{ years}} = \$0.106$ per year.   **2.** $\dfrac{1}{2}$

3. $a + 7$ **4.** The local maximum appears to occur at $(-1, 28)$, and the local minimum occurs at $(5, -80)$. The function is increasing on $(-\infty, -1) \cup (5, \infty)$ and decreasing on $(-1, 5)$.

Section 3.4

1. a. $(fg)(x) = f(x)g(x) = (x - 1)(x^2 - 1) = x^3 - x^2 - x + 1$

$(f - g)(x) = f(x) - g(x) = (x - 1) - (x^2 - 1) = x - x^2$

b. No, the functions are not the same.

2. A gravitational force is still a force, so $a(G(r))$ makes sense as the acceleration of a planet at a distance r from the Sun (due to gravity), but $G(a(F))$ does not make sense.

3. $f(g(1)) = f(3) = 3$ and $g(f(4)) = g(1) = 3$ **4.** $g(f(2)) = g(5) = 3$

5. a. 8; **b.** 20   **6.** $[-4, 0) \cup (0, \infty)$

7. Possible answer: $g(x) = \sqrt{4 + x^2}$; $h(x) = \dfrac{4}{3-x}$; $f = h \circ g$

Section 3.5

1. $b(t) = h(t) + 10 = -4.9t^2 + 30t + 10$

2.

The graphs of $f(x)$ and $g(x)$ are shown here. The transformation is a horizontal shift. The function is shifted to the left by 2 units.

3.

4. $g(x) = \dfrac{1}{x-1} + 1$

5. a.

b.

6. a. $g(x) = -f(x)$

x	-2	0	2	4
$g(x)$	-5	-10	-15	-20

b. $h(x) = f(-x)$

x	-2	0	2	4
$h(x)$	15	10	5	unknown

7.

Notice: $h(x) = f(-x)$ looks the same as $f(x)$.

8. Even

9.

x	2	4	6	8
$g(x)$	9	12	15	0

10. $g(x) = 3x - 2$

11. $g(x) = f\left(\dfrac{1}{3}x\right)$ so using the square root function we get $g(x) = \sqrt{\dfrac{1}{3}x}$

Section 3.6

1. Using the variable p for passing, $|p - 80| \leq 20$

2. $f(x) = -|x + 2| + 3$ **3.** $x = -1$ or $x = 2$

Section 3.7

1. $h(2) = 6$ **2.** Yes   **3.** Yes   **4.** The domain of function f^{-1} is $(-\infty, -2)$ and the range of function f^{-1} is $(1, \infty)$.

5. a. $f(60) = 50$. In 60 minutes, 50 miles are traveled.

b. $f^{-1}(60) = 70$. To travel 60 miles, it will take 70 minutes.

6. a. 3 **b.** 5.6   **7.** $x = 3y + 5$ **8.** $f^{-1}(x) = (2 - x)^2$; domain of f: $[0, \infty)$; domain of f^{-1}: $(-\infty, 2]$

9.

Chapter 4

Section 4.1

1. $m = \dfrac{4 - 3}{0 - 2} = \dfrac{1}{-2} = -\dfrac{1}{2}$; decreasing because $m < 0$.

2. $m = \dfrac{1{,}868 - 1{,}442}{2{,}012 - 2{,}009} = \dfrac{426}{3} = 142$ people per year

3. $y = -7x + 3$ **4.** $H(x) = 0.5x + 12.5$

5.

6. Possible answers include $(-3, 7)$, $(-6, 9)$, or $(-9, 11)$

7.

8. $(16, 0)$

9. a. $f(x) = 2x$;

b. $g(x) = -\dfrac{1}{2}x$

10. $y = -\dfrac{1}{3}x + 6$

Section 4.2

1. $C(x) = 0.25x + 25{,}000$; The y-intercept is $(0, 25{,}000)$. If the company does not produce a single doughnut, they still incur a cost of $25,000.   **2. a.** 41,100 **b.** 2020   **3.** 21.15 miles

Section 4.3

1. 54° F   **2.** 150.871 billion gallons; extrapolation

Chapter 5

Section 5.1

1. The path passes through the origin and has vertex at $(-4, 7)$, so $(h)x = -\dfrac{7}{16}(x + 4)^2 + 7$. To make the shot, $h(-7.5)$ would need to be about 4 but $h(-7.5) \approx 1.64$; he doesn't make it.

2. $g(x) = x^2 - 6x + 13$ in general form; $g(x) = (x - 3)^2 + 4$ in standard form

3. The domain is all real numbers. The range is $f(x) \geq \dfrac{8}{11}$, or $\left[\dfrac{8}{11}, \infty \right)$. **4.** y-intercept at $(0, 13)$, No x-intercepts
5. 3 seconds; 256 feet; 7 seconds

Section 5.2

1. $f(x)$ is a power function because it can be written as $f(x) = 8x^5$. The other functions are not power functions.
2. As x approaches positive or negative infinity, $f(x)$ decreases without bound: as $x \to \pm\infty, f(x) \to -\infty$ because of the negative coefficient.　　**3.** The degree is 6. The leading term is $-x^6$. The leading coefficient is -1.　　**4.** As $x \to \infty, f(x) \to -\infty$; as $x \to -\infty, f(x) \to -\infty$. It has the shape of an even degree power function with a negative coefficient.　　**5.** The leading term is $0.2x^3$, so it is a degree 3 polynomial. As x approaches positive infinity, $f(x)$ increases without bound; as x approaches negative infinity, $f(x)$ decreases without bound.
6. y-intercept $(0, 0)$; x-intercepts $(0, 0)$, $(-2, 0)$, and $(5, 0)$
7. There are at most 12 x-intercepts and at most 11 turning points.
8. The end behavior indicates an odd-degree polynomial function; there are 3 x-intercepts and 2 turning points, so the degree is odd and at least 3. Because of the end behavior, we know that the lead coefficient must be negative.　　**9.** The x-intercepts are $(2, 0)$, $(-1, 0)$, and $(5, 0)$, the y-intercept is $(0, 2)$, and the graph has at most 2 turning points.

Section 5.3

1. y-intercept $(0, 0)$; x-intercepts $(0, 0)$, $(-5, 0)$, $(2, 0)$, and $(3, 0)$
2. The graph has a zero of -5 with multiplicity 1, a zero of -1 with multiplicity 2, and a zero of 3 with even multiplicity.

3.

4. Because f is a polynomial function and since $f(1)$ is negative and $f(2)$ is positive, there is at least one real zero between $x = 1$ and $x = 2$.

5. $f(x) = -\dfrac{1}{8}(x-2)^3(x+1)^2(x-4)$　　**6.** The minimum occurs at approximately the point $(0, -6.5)$, and the maximum occurs at approximately the point $(3.5, 7)$.

Section 5.4

1. $4x^2 - 8x + 15 - \dfrac{78}{4x + 5}$　　**2.** $3x^3 - 3x^2 + 21x - 150 + \dfrac{1,090}{x + 7}$
3. $3x^2 - 4x + 1$

Section 5.5

1. $f(-3) = -412$　　**2.** The zeros are 2, -2, and -4.
3. There are no rational zeros.　　**4.** The zeros are -4, $\dfrac{1}{2}$, and 1.
5. $f(x) = -\dfrac{1}{2}x^3 + \dfrac{5}{2}x^2 - 2x + 10$　　**6.** There must be 4, 2, or 0 positive real roots and 0 negative real roots. The graph shows that there are 2 positive real zeros and 0 negative real zeros.
7. 3 meters by 4 meters by 7 meters

Section 5.6

1. End behavior: as $x \to \pm\infty$, $f(x) \to 0$; Local behavior: as $x \to 0$, $f(x) \to \infty$ (there are no x- or y-intercepts).

2.

The function and the asymptotes are shifted 3 units right and 4 units down. As $x \to 3, f(x) \to \infty$, and as $x \to \pm\infty, \ f(x) \to -4$. The function is $f(x) = \dfrac{1}{(x-3)^2} - 4$.

3. $\dfrac{12}{11}$　　**4.** The domain is all real numbers except $x = 1$ and $x = 5$.
5. Removable discontinuity at $x = 5$. Vertical asymptotes: $x = 0$, $x = 1$.　　**6.** Vertical asymptotes at $x = 2$ and $x = -3$; horizontal asymptote at $y = 4$.　　**7.** For the transformed reciprocal squared function, we find the rational form.

$$f(x) = \frac{1}{(x-3)^2} - 4$$
$$= \frac{1 - 4(x-3)^2}{(x-3)^2}$$
$$= \frac{1 - 4(x^2 - 6x + 9)}{(x-3)(x-3)}$$
$$= \frac{-4x^2 + 24x - 35}{x^2 - 6x + 9}$$

Because the numerator is the same degree as the denominator we know that as $x \to \pm\infty, \ f(x) \to -4$; so $y = -4$ is the horizontal asymptote. Next, we set the denominator equal to zero, and find that the vertical asymptote is $x = 3$, because as $x \to 3$, $f(x) \to \infty$. We then set the numerator equal to 0 and find the x-intercepts are at $(2.5, 0)$ and $(3.5, 0)$. Finally, we evaluate the function at 0 and find the y-intercept to be at $\left(0, -\dfrac{35}{9} \right)$.

8.
Horizontal asymptote at $y = \dfrac{1}{2}$. Vertical asymptotes at $x = 1$ and $x = 3$. y-intercept at $\left(0, \dfrac{4}{3} \right)$. x-intercepts at $(2, 0)$ and $(-2, 0)$. $(-2, 0)$ is a zero with multiplicity 2, and the graph bounces off the x-axis at this point. $(2, 0)$ is a single zero and the graph crosses the axis at this point.

Section 5.7

1. $f^{-1}(f(x)) = f^{-1}\left(\dfrac{x+5}{3} \right) = 3\left(\dfrac{x+5}{3} \right) - 5 = (x-5) + 5 = x$
and $f(f^{-1}(x)) = f(3x - 5) = \dfrac{(3x-5) + 5}{3} = \dfrac{3x}{3} = x$
2. $f^{-1}(x) = x^3 - 4$　　**3.** $f^{-1}(x) = \sqrt{x - 1}$
4. $f^{-1}(x) = \dfrac{x^2 - 3}{2}, x \geq 0$　　**5.** $f^{-1}(x) = \dfrac{2x + 3}{x - 1}$

Section 5.8

1. $\dfrac{128}{3}$　　**2.** $\dfrac{9}{2}$　　**3.** $x = 20$

Chapter 6

Section 6.1

1. $g(x) = 0.875^x$ and $j(x) = 1095.6^{-2x}$ represent exponential functions.
2. 5.5556   **3.** About 1.548 billion people; by the year 2031, India's population will exceed China's by about 0.001 billion, or 1 million people.   **4.** (0, 129) and (2, 236); $N(t) = 129(1.3526)^t$
5. $f(x) = 2(1.5)^x$ **6.** $f(x) = \sqrt{2}(\sqrt{2})^x$; Answers may vary due to round-off error. the answer should be very close to $1.4142(1.4142)^x$.
7. $y \approx 12 \cdot 1.85^x$ **8.** About \$3,644,675.88   **9.** \$13,693
10. $e^{-0.5} \approx 0.60653$ **11.** \$3,659,823.44   **12.** 3.77E-26(This is calculator notation for the number written as 3.77×10^{-26} in scientific notation. While the output of an exponential function is never zero, this number is so close to zero that for all practical purposes we can accept zero as the answer.)

Section 6.2

1.

The domain is $(-\infty, \infty)$; the range is $(0, \infty)$; the horizontal asymptote is $y = 0$.

2.

The domain is $(-\infty, \infty)$; the range is $(3, \infty)$; the horizontal asymptote is $y = 3$.

3. $x \approx -1.608$

4.

The domain is $(-\infty, \infty)$; the range is $(0, \infty)$; the horizontal asymptote is $y = 0$.

5.

The domain is $(-\infty, \infty)$; the range is $(0, \infty)$; the horizontal asymptote is $y = 0$.

6. $f(x) = -\frac{1}{3}e^x - 2$; the domain is $(-\infty, \infty)$; the range is $(-\infty, 2)$; the horizontal asymptote is $y = 2$.

Section 6.3

1. a. $\log_{10}(1{,}000{,}000) = 6$ is equivalent to $10^6 = 1{,}000{,}000$
b. $\log_5(25) = 2$ is equivalent to $5^2 = 25$ **2. a.** $3^2 = 9$ is equivalent to $\log_3(9) = 2$ **b.** $5^3 = 125$ is equivalent to $\log_5(125) = 3$
c. $2^{-1} = \frac{1}{2}$ is equivalent to $\log_2\left(\frac{1}{2}\right) = -1$
3. $\log_{121}(11) = \frac{1}{2}$ (recalling that $\sqrt{121} = 121^{\frac{1}{2}} = 11$)

4. $\log_2\left(\frac{1}{32}\right) = -5$ **5.** $\log(1{,}000{,}000) = 6$ **6.** $\log(123) \approx 2.0899$
7. The difference in magnitudes was about 3.929.   **8.** It is not possible to take the logarithm of a negative number in the set of real numbers.

Section 6.4

1. $(2, \infty)$ **2.** $(5, \infty)$

3.

The domain is $(0, \infty)$, the range is $(-\infty, \infty)$, and the vertical asymptote is $x = 0$.

4.

The domain is $(-4, \infty)$, the range $(-\infty, \infty)$, and the asymptote $x = -4$.

5.

The domain is $(0, \infty)$, the range is $(-\infty, \infty)$, and the vertical asymptote is $x = 0$.

6.

The domain is $(0, \infty)$, the range is $(-\infty, \infty)$, and the vertical asymptote is $x = 0$.

7.

The domain is $(2, \infty)$, the range is $(-\infty, \infty)$, and the vertical asymptote is $x = 2$.

8.

The domain is $(-\infty, 0)$, the range is $(-\infty, \infty)$, and the vertical asymptote is $x = 0$.
9. $x \approx 3.049$ **10.** $x = 1$
11. $f(x) = 2\ln(x + 3) - 1$

Section 6.5

1. $\log_b(2) + \log_b(2) + \log_b(2) + \log_b(k) = 3\log_b(2) + \log_b(k)$
2. $\log_3(x+3) - \log_3(x-1) - \log_3(x-2)$ **3.** $2\ln(x)$
4. $-2\ln(x)$ **5.** $\log_3(16)$ **6.** $2\log(x) + 3\log(y) - 4\log(z)$
7. $\frac{2}{3}\ln(x)$ **8.** $\frac{1}{2}\ln(x-1) + \ln(2x+1) - \ln(x+3) - \ln(x-3)$

9. $\log\left(\frac{3 \cdot 5}{4 \cdot 6}\right)$; can also be written $\log\left(\frac{5}{8}\right)$ by reducing the fraction
to lowest terms. **10.** $\log\left(\frac{5(x-1)^3\sqrt{x}}{(7x-1)}\right)$

11. $\log\frac{x^{12}(x+5)^4}{(2x+3)^4}$; this answer could also be written $\log\left(\frac{x^3(x+5)}{(2x+3)}\right)^4$.
12. The pH increases by about 0.301. **13.** $\frac{\ln(8)}{\ln(0.5)}$
14. $\frac{\ln(100)}{\ln(5)} \approx \frac{4.6051}{1.6094} = 2.861$

Section 6.6

1. $x = -2$ **2.** $x = -1$ **3.** $x = \frac{1}{2}$
4. The equation has no solution. **5.** $x = \dfrac{\ln(3)}{\ln\left(\frac{2}{3}\right)}$
6. $t = 2\ln\left(\frac{11}{3}\right)$ or $\ln\left(\frac{11}{3}\right)^2$
7. $t = \ln\left(\frac{1}{\sqrt{2}}\right) = -\frac{1}{2}\ln(2)$ **8.** $x = \ln(2)$ **9.** $x = e^4$
10. $x = e^5 - 1$ **11.** $x \approx 9.97$ **12.** $x = 1$ or $x = -1$
13. $t = 703,800,000 \times \frac{\ln(0.8)}{\ln(0.5)}$ years $\approx 226,572,993$ years.

Section 6.7

1. $f(t) = A_0 e^{-0.0000000087t}$ **2.** Less than 230 years; 229.3157 to be exact
3. $f(t) = A_0 e^{\left(\frac{\ln(2)}{3}\right)t}$ **4.** 6.026 hours **5.** 895 cases on day 15
6. Exponential. $y = 2e^{0.5x}$ **7.** $y = 3e^{(\ln 0.5)x}$

Section 6.8

1. a. The exponential regression model that fits these data is
$y = 522.88585984(1.19645256)^x$. **b.** If spending continues at this
rate, the graduate's credit card debt will be $4,499.38 after one year.
2. a. The logarithmic regression model that fits these data is
$y = 141.91242949 + 10.45366573\ln(x)$ **b.** If sales continue at this
rate, about 171,000 games will be sold in the year 2015.
3. a. The logistic regression model that fits these data is
$y = \dfrac{25.65665979}{1 + 6.113686306e^{-0.3852149008x}}$.

b. If the population continues to grow at this rate, there will be
about 25,634 seals in 2020. **c.** To the nearest whole number,
the carrying capacity is 25,657.

Chapter 7

Section 7.1

1. Not a solution
2. The solution to the system is
the ordered pair $(-5, 3)$.

3. $(-2, -5)$ **4.** $(-6, -2)$ **5.** $(10, -4)$ **6.** No solution.
It is an inconsistent system. **7.** The system is dependent so
there are infinite solutions of the form $(x, 2x + 5)$.
8. 700 children, 950 adults

Section 7.2

1. $(1, -1, 1)$ **2.** No solution **3.** Infinite number of
solutions of the form $(x, 4x - 11, -5x + 18)$

Section 7.3

1. $\left(-\frac{1}{2}, \frac{1}{2}\right)$ and $(2, 8)$ **2.** $(-1, 3)$ **3.** $\{(1, 3), (1, -3), (-1, 3), (-1, -3)\}$
4.

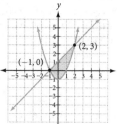

Section 7.4

1. $\dfrac{3}{x-3} - \dfrac{2}{x-2}$ **2.** $\dfrac{6}{x-1} - \dfrac{5}{(x-1)^2}$ **3.** $\dfrac{3}{x-1} + \dfrac{2x-4}{x^2+1}$
4. $\dfrac{x-2}{x^2-2x+3} + \dfrac{2x+1}{(x^2-2x+3)^2}$

Section 7.5

1. $A + B = \begin{bmatrix} 2 & 6 \\ 1 & 0 \\ 1 & -3 \end{bmatrix} + \begin{bmatrix} 3 & -2 \\ 1 & 5 \\ -4 & 3 \end{bmatrix} = \begin{bmatrix} 2+3 & 6+(-2) \\ 1+1 & 0+5 \\ 1+(-4) & -3+3 \end{bmatrix} = \begin{bmatrix} 5 & 4 \\ 2 & 5 \\ -3 & 0 \end{bmatrix}$
2. $-2B = \begin{bmatrix} -8 & -2 \\ -6 & -4 \end{bmatrix}$

Section 7.6

1. $\begin{bmatrix} 4 & -3 & | & 11 \\ 3 & 2 & | & 4 \end{bmatrix}$ **2.** $\begin{matrix} x - y + z = 5 \\ 2x - y + 3z = 1 \\ y + z = -9 \end{matrix}$ **3.** $(2, 1)$
4. $\begin{bmatrix} 1 & -\frac{5}{2} & \frac{5}{2} & | & \frac{17}{2} \\ 0 & 1 & 5 & | & 9 \\ 0 & 0 & 1 & | & 2 \end{bmatrix}$ **5.** $(1, 1, 1)$
6. $150,000 at 7%, $750,000 at 8%, $600,000 at 10%

Section 7.7

1. $AB = \begin{bmatrix} 1 & 4 \\ -1 & -3 \end{bmatrix}\begin{bmatrix} -3 & -4 \\ 1 & 1 \end{bmatrix} = \begin{bmatrix} 1(-3)+4(1) & 1(-4)+4(1) \\ -1(-3)+-3(1) & -1(-4)+-3(1) \end{bmatrix}$
$= \begin{bmatrix} 1 & 0 \\ 0 & 1 \end{bmatrix}$
$BA = \begin{bmatrix} -3 & -4 \\ 1 & 1 \end{bmatrix}\begin{bmatrix} 1 & 4 \\ -1 & -3 \end{bmatrix} = \begin{bmatrix} -3(1)+-4(-1) & -3(4)+-4(-3) \\ 1(1)+1(-1) & 1(4)+1(-3) \end{bmatrix}$
$= \begin{bmatrix} 1 & 0 \\ 0 & 1 \end{bmatrix}$
2. $A^{-1} = \begin{bmatrix} \frac{3}{5} & \frac{1}{5} \\ -\frac{2}{5} & \frac{1}{5} \end{bmatrix}$ **3.** $A^{-1} = \begin{bmatrix} 1 & 1 & 2 \\ 2 & 4 & -3 \\ 3 & 6 & -5 \end{bmatrix}$ **4.** $X = \begin{bmatrix} 4 \\ 38 \\ 58 \end{bmatrix}$

Section 7.8

1. $(3, -7)$ **2.** -10 **3.** $\left(-2, \dfrac{3}{5}, \dfrac{12}{5}\right)$

Chapter 8

Section 8.1

1. $x^2 + \dfrac{y^2}{16} = 1$ **2.** $\dfrac{(x-1)^2}{16} + \dfrac{(y-3)^2}{4} = 1$

3. Center: $(0, 0)$;
vertices: $(\pm 6, 0)$;
co-vertices: $(0, \pm 2)$;
foci: $(\pm 4\sqrt{2}, 0)$

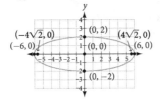

4. Standard form: $\dfrac{x^2}{16} + \dfrac{y^2}{49} = 1$;
center: $(0, 0)$;
vertices: $(0, \pm 7)$;
co-vertices: $(\pm 4, 0)$
foci: $\left(0, \pm\sqrt{33}\right)$

5. Center: $(4, 2)$;
vertices: $(-2, 2)$ and $(10, 2)$;
co-vertices: $\left(4, 2 - 2\sqrt{5}\right)$
and $\left(4, 2 + 2\sqrt{5}\right)$;
foci: $(0, 2)$ and $(8, 2)$

6. $\dfrac{(x-3)^2}{4} + \dfrac{(y+1)^2}{16} = 1$; center: $(3, -1)$; vertices: $(3, -5)$ and $(3, 3)$; co-vertices: $(1, -1)$ and $(5, -1)$; foci: $\left(3, -1 - 2\sqrt{3}\right)$ and $\left(3, -1 + 2\sqrt{3}\right)$ **7. a.** $\dfrac{x^2}{57,600} + \dfrac{y^2}{25,600} = 1$; **b.** The people are standing 358 feet apart.

Section 8.2

1. Vertices: $(\pm 3, 0)$; foci: $\left(\pm\sqrt{34}, 0\right)$ **2.** $\dfrac{y^2}{4} - \dfrac{x^2}{16} = 1$

3. $\dfrac{(y-3)^2}{25} + \dfrac{(x-1)^2}{144} = 1$

4. Vertices: $(\pm 12, 0)$;
co-vertices: $(0, \pm 9)$;
foci: $(\pm 15, 0)$;
asymptotes: $y = \pm\dfrac{3}{4}x$

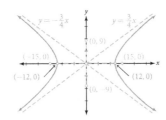

5. Center: $(3, -4)$; vertices: $(3, -14)$ and $(3, 6)$;
co-vertices: $(-5, -4)$ and $(11, -4)$; foci: $\left(3, -4 - 2\sqrt{41}\right)$ and
$\left(3, -4 + 2\sqrt{41}\right)$; asymptotes: $y = \pm\dfrac{5}{4}(x - 3) - 4$

6. The sides of the tower can be modeled by the hyperbolic equation. $\dfrac{x^2}{400} - \dfrac{y^2}{3600} = 1$ or $\dfrac{x^2}{20^2} - \dfrac{y^2}{60^2} = 1$.

Section 8.3

1. Focus: $(-4, 0)$;
directrix: $x = 4$;
endpoints of the latus
rectum: $(-4, \pm 8)$

2. Focus: $(0, 2)$;
directrix: $y = -2$;
endpoints of the latus
rectum: $(\pm 4, 2)$

3. $x^2 = 14y$

4. Vertex: $(8, -1)$;
axis of symmetry: $y = -1$;
focus: $(9, -1)$;
directrix: $x = 7$;
endpoints of the latus rectum:
$(9, -3)$ and $(9, 1)$.

5. Vertex: $(-2, 3)$;
axis of symmetry: $x = -2$;
focus: $(-2, -2)$;
directrix: $y = 8$;
endpoints of the latus
rectum: $(-12, -2)$ and
$(8, -2)$.

6. a. $y^2 = 1,280x$ **b.** The depth of the cooker is 500 mm.

Section 8.4

1. a. Hyperbola   **b.** Ellipse   **2.** $\dfrac{x'^2}{4} + \dfrac{y'^2}{1} = 1$
3. a. Hyperbola   **b.** Ellipse

Section 8.5

1. Ellipse; $e = \dfrac{1}{3}$; $x = -2$

2.

3. $r = \dfrac{1}{1 - \cos\theta}$

4. $4 - 8x + 3x^2 - y^2 = 0$

Section 9.7

1.

Outcome	Probability
Roll of 1	
Roll of 2	
Roll of 3	
Roll of 4	
Roll of 5	
Roll of 6	

2. $\dfrac{2}{3}$ **3.** $\dfrac{7}{13}$ **4.** $\dfrac{2}{13}$

5. $\dfrac{5}{6}$ **6. a.** $\dfrac{1}{91}$ **b.** $\dfrac{5}{91}$ **c.** $\dfrac{86}{91}$

Chapter 9

Section 9.1

1. The first five terms are $\{1, 6, 11, 16, 21\}$.   **2.** The first five terms are $\left\{-2, 2, -\dfrac{3}{2}, 1, -\dfrac{5}{8}\right\}$.   **3.** The first six terms are $\{2, 5, 54, 10, 250, 15\}$.   **4.** $a_n = (-1)^{n+1} 9^n$ **5.** $a_n = -\dfrac{3n}{4n}$

6. $a_n = e^{n-3}$ **7.** $\{2, 5, 11, 23, 47\}$ **8.** $\left\{0, 1, 1, 1, 2, 3, \dfrac{5}{2}, \dfrac{17}{6}\right\}$ **9.** The first five terms are $\left\{1, \dfrac{3}{2}, 4, 15, 72\right\}$.

Section 9.2

1. The sequence is arithmetic. The common difference is -2.
2. The sequence is not arithmetic because $3 - 1 \neq 6 - 3$.
3. $\{1, 6, 11, 16, 21\}$ **4.** $a_2 = 2$ **5.** $a_1 = 25$; $a_n = a_{n-1} + 12$, for $n \geq 2$ **6.** $a_n = 53 - 3n$ **7.** There are 11 terms in the sequence.   **8.** The formula is $T_n = 10 + 4n$, and it will take her 42 minutes.

Section 9.3

1. The sequence is not geometric because $\dfrac{10}{5} \neq \dfrac{15}{10}$.
2. The sequence is geometric. The common ratio is $\dfrac{1}{5}$.
3. $\left\{18, 6, 2, \dfrac{2}{3}, \dfrac{2}{9}\right\}$ **4.** $a_1 = 2$; $a_n = \dfrac{2}{3} a_{n-1}$ for $n \geq 2$
5. $a_6 = 16{,}384$ **6.** $a_n = -(-3)^{n-1}$ **7. a.** $P_n = 293 \cdot 1.026a^n$
b. The number of hits will be about 333.

Section 9.4

1. 38   **2.** 26.4   **3.** 328   **4.** -280 **5.** \$2,025
6. $\approx 2{,}000.00$ **7.** 9,840   **8.** \$275,513.31   **9.** The sum is defined. It is geometric.   **10.** The sum of the infinite series is defined.   **11.** The sum of the infinite series is defined.
12. 3   **13.** The series is not geometric.   **14.** $-\dfrac{3}{11}$
15. \$92,408.18

Section 9.5

1. 7   **2.** There are 60 possible breakfast specials.   **3.** 120
4. 60   **5.** 12   **6.** $P(7, 7) = 5{,}040$ **7.** $P(7, 5) = 2{,}520$
8. $C(10, 3) = 120$ **9.** 64 sundaes   **10.** 840

Section 9.6

1. a. 35 **b.** 330   **2. a.** $x^5 - 5x^4y + 10x^3y^2 - 10x^2y^3 + 5xy^4 - y^5$
b. $8x^3 + 60x^2y + 150xy^2 + 125y^3$ **3.** $-10{,}206x^4y^5$

Odd Answers

CHAPTER 1

Section 1.1

1. Irrational number. The square root of two does not terminate, and it does not repeat a pattern. It cannot be written as a quotient of two integers, so it is irrational.   **3.** The Associative Properties state that the sum or product of multiple numbers can be grouped differently without affecting the result. This is because the same operation is performed (either addition or subtraction), so the terms can be re-ordered.   **5.** -6 **7.** -2 **9.** -9 **11.** 9   **13.** -2 **15.** 4   **17.** 0   **19.** 9   **21.** 25   **23.** -6 **25.** 17   **27.** 4   **29.** -4 **31.** -6 **33.** ± 1 **35.** 2   **37.** 2   **39.** $-14y - 11$ **41.** $-4b + 1$ **43.** $43z - 3$ **45.** $9y + 45$ **47.** $-6b + 6$ **49.** $\frac{16x}{3}$ **51.** $9x$ **53.** $\frac{1}{2}(40 - 10) + 5$ **55.** Irrational number   **57.** $g + 400 - 2(600) = 1200$ **59.** Inverse property of addition   **61.** 68.4   **63.** True   **65.** Irrational   **67.** Rational

Section 1.2

1. No, the two expressions are not the same. An exponent tells how many times you multiply the base. So 2^3 is the same as $2 \times 2 \times 2$, which is 8. 3^2 is the same as 3×3, which is 9.
3. It is a method of writing very small and very large numbers.
5. 81   **7.** 243   **9.** $\frac{1}{16}$ **11.** $\frac{1}{11}$ **13.** 1   **15.** 4^9
17. 12^{40} **19.** $\frac{1}{7^9}$ **21.** 3.14×10^{-5} **23.** 16,000,000,000
25. a^4 **27.** $b^6 c^8$ **29.** $ab^2 d^3$ **31.** m^4 **33.** $\frac{q^5}{p^6}$
35. $\frac{y^{21}}{x^{14}}$ **37.** 25   **39.** $72a^2$ **41.** $\frac{c^3}{b^9}$ **43.** $\frac{y}{81z^6}$
45. 0.00135 m   **47** 1.0995×10^{12} **49.** 0.00000000003397 in.
51. 12,230,590,464 m^{66} **53.** $\frac{a^{14}}{1296}$ **55.** $\frac{n}{a^9 c}$ **57.** $\frac{1}{a^6 b^6 c^6}$
59. 0.000000000000000000000000000000000662606957

Section 1.3

1. When there is no index, it is assumed to be 2 or the square root. The expression would only be equal to the radicand if the index were 1.   **3.** The principal square root is the nonnegative root of the number.   **5.** 16   **7.** 10   **9.** 14   **11.** $7\sqrt{2}$
13. $\frac{9\sqrt{5}}{5}$ **15.** 25   **17.** $\sqrt{2}$ **19.** $2\sqrt{6}$ **21.** $5\sqrt{6}$
23. $6\sqrt{35}$ **25.** $\frac{2}{15}$ **27.** $\frac{6\sqrt{10}}{19}$ **29.** $-\frac{1 + \sqrt{17}}{2}$
31. $7\sqrt[3]{2}$ **33.** $15\sqrt{5}$ **35.** $20x^2$ **37.** $7\sqrt{p}$
39. $18m^2\sqrt{m}$ **41.** $2b\sqrt{a}$ **43.** $\frac{15x}{7}$ **45.** $5y^4\sqrt{2}$
47. $\frac{4\sqrt{7d}}{7d}$ **49.** $\frac{2\sqrt{2} + 2\sqrt{6x}}{1 - 3x}$ **51.** $-w\sqrt{2w}$
53. $\frac{3\sqrt{x} - \sqrt{3x}}{2}$ **55.** $5n^5\sqrt{5}$ **57.** $\frac{9\sqrt{m}}{19m}$ **59.** $\frac{2}{3d}$

61. $\frac{3\sqrt[4]{2x^2}}{2}$ **63.** $6z\sqrt[3]{2}$ **65.** 500 feet   **67** $\frac{-5\sqrt{2} - 6}{7}$
69. $\frac{\sqrt{mnc}}{a^9 cmn}$ **71.** $\frac{2\sqrt{2}x + \sqrt{2}}{4}$ **73.** $\frac{\sqrt{3}}{3}$

Section 1.4

1. The statement is true. In standard form, the polynomial with the highest value exponent is placed first and is the leading term. The degree of a polynomial is the value of the highest exponent, which in standard form is also the exponent of the leading term.
3. Use the distributive property, multiply, combine like terms, and simplify.   **5.** 2   **7.** 8   **9.** 2   **11.** $4x^2 + 3x + 19$
13. $3w^2 + 30w + 21$ **15.** $11b^4 - 9b^3 + 12b^2 - 7b + 8$
17. $24x^2 - 4x - 8$ **19.** $24b^4 - 48b^2 + 24$
21. $99v^2 - 202v + 99$ **23.** $8n^3 - 4n^2 + 72n - 36$
25. $9y^2 - 42y + 49$ **27.** $16p^2 + 72p + 81$
29. $9y^2 - 36y + 36$ **31.** $16c^2 - 1$ **33.** $225n^2 - 36$
35. $-16m^2 + 16$ **37.** $121q^2 - 100$ **39.** $16t^4 + 4t^3 - 32t^2 - t + 7$
41. $y^3 - 6y^2 - y + 18$ **43.** $3p^3 - p^2 - 12p + 10$
45. $a^2 - b^2$ **47.** $16t^2 - 40tu + 25u^2$ **49.** $4t^2 + x^2 + 4t - 5tx - x$
51. $24r^2 + 22rd - 7d^2$ **53.** $32x^2 - 4x - 3m^2$
55. $32t^3 - 100t^2 + 40t + 38$ **57.** $a^4 + 4a^3 c - 16ac^3 - 16c^4$

Section 1.5

1. The terms of a polynomial do not have to have a common factor for the entire polynomial to be factorable. For example, $4x^2$ and $-9y^2$ don't have a common factor, but the whole polynomial is still factorable: $4x^2 - 9y^2 = (2x + 3y)(2x - 3y)$.   **3.** Divide the x term into the sum of two terms, factor each portion of the expression separately, and then factor out the GCF of the entire expression.   **5.** $7m$ **7.** $10m^3$ **9.** y **11.** $(2a - 3)(a + 6)$
13. $(3n - 11)(2n + 1)$ **15.** $(p + 1)(2p - 7)$
17. $(5h + 3)(2h - 3)$ **19.** $(9d - 1)(d - 8)$
21. $(12t + 13)(t - 1)$ **23.** $(4x + 10)(4x - 10)$
25. $(11p + 13)(11p - 13)$ **27.** $(19d + 9)(19d - 9)$
29. $(12b + 5c)(12b - 5c)$ **31.** $(7n + 12)^2$ **33.** $(15y + 4)^2$
35. $(5p - 12)^2$ **37.** $(x + 6)(x^2 - 6x + 36)$
39. $(5a + 7)(25a^2 - 35a + 49)$ **41.** $(4x - 5)(16x^2 + 20x + 25)$
43. $(5r + 12s)(25r^2 - 60rs + 144s^2)$ **45.** $(2c + 3)^{-\frac{1}{4}}(-7c - 15)$
47. $(x + 2)^{-\frac{2}{5}}(19x + 10)$ **49.** $(2z - 9)^{-\frac{3}{2}}(27z - 99)$
51. $(14x - 3)(7x + 9)$ **53.** $(3x + 5)(3x - 5)$
55. $(2x + 5)^2 (2x - 5)^2$ **57.** $(4z^2 + 49a^2)(2z + 7a)(2z - 7a)$
59. $\frac{1}{(4x + 9)(4x - 9)(2x + 3)}$

Section 1.6

1. You can factor the numerator and denominator to see if any of the terms can cancel one another out.
3. True. Multiplication and division do not require finding the LCD because the denominators can be combined through those operations, whereas addition and subtraction require like terms.

B-1

5. $\dfrac{y+5}{y+6}$ **7.** $3b+3$ **9.** $\dfrac{x+4}{2x+2}$ **11.** $\dfrac{a+3}{a-3}$

13. $\dfrac{3n-8}{7n-3}$ **15.** $\dfrac{c-6}{c+6}$ **17.** 1 **19.** $\dfrac{d^2-25}{25d^2-1}$

21. $\dfrac{t+5}{t+3}$ **23.** $\dfrac{6x-5}{6x+5}$ **25.** $\dfrac{p+6}{4p+3}$ **27.** $\dfrac{2d+9}{d+11}$

29. $\dfrac{12b+5}{3b-1}$ **31.** $\dfrac{4y-1}{y+4}$ **33.** $\dfrac{10x+4y}{xy}$ **35.** $\dfrac{9a-7}{a^2-2a-3}$

37. $\dfrac{2y^2-y+9}{y^2-y-2}$ **39.** $\dfrac{5z^2+z+5}{z^2-z-2}$ **41.** $\dfrac{x+2xy+y}{x+xy+y+1}$

43. $\dfrac{2b+7a}{ab^2}$ **45.** $\dfrac{18+ab}{4b}$ **47.** $a-b$ **49.** $\dfrac{3c^2+3c-2}{2c^2+5c+2}$

51. $\dfrac{15x+7}{x-1}$ **53.** $\dfrac{x+9}{x-9}$ **55.** $\dfrac{1}{y+2}$ **57.** 4

Chapter 1 Review Exercises

1. -5 **3.** 53 **5.** $y=24$ **7.** $32m$ **9.** Whole

11. Irrational   **13.** 16 **15.** a^6 **17.** $\dfrac{x^3}{32y^3}$ **19.** a

21. 1.634×10^7 **23.** 14 **25.** $5\sqrt{3}$ **27.** $\dfrac{4\sqrt{2}}{5}$

29. $\dfrac{7\sqrt{2}}{50}$ **31.** $10\sqrt{3}$ **33.** -3 **35.** $3x^3+4x^2+6$

37. $5x^2-x+3$ **39.** $k^2-3k-18$ **41.** x^3+x^2+x+1

43. $3a^2+5ab-2b^2$ **45.** $9p$ **47.** $4a^2$

49. $(4a-3)(2a+9)$ **51.** $(x+5)^2$ **53.** $(2h-3k)^2$

55. $(p+6)(p^2-6p+36)$ **57.** $(4q-3p)(16q^2+12pq+9p^2)$

59. $(p+3)^{\frac{1}{3}}(-5p-24)$ **61.** $\dfrac{x+3}{x-4}$ **63.** $\dfrac{1}{2}$ **65.** $\dfrac{m+2}{m-3}$

67. $\dfrac{6x+10y}{xy}$ **69.** $\dfrac{1}{6}$

Chapter 1 Practice Test

1. Rational   **3.** $x=12$ **5.** $3{,}141{,}500$ **7.** 16

9. 9 **11.** $2x$ **13.** 21 **15.** $\dfrac{3\sqrt{x}}{4}$ **17.** $21\sqrt{6}$

19. $13q^3-4q^2-5q$ **21.** $n^3-6n^2+12n-8$

23. $(4x+9)(4x-9)$ **25.** $(3c-11)(9c^2+33c+121)$

27. $\dfrac{4z-3}{2z-1}$ **29.** $\dfrac{3a+2b}{3b}$

CHAPTER 2

Section 2.1

1. Answers may vary. Yes. It is possible for a point to be on the x-axis or on the y-axis and therefore is considered to NOT be in one of the quadrants.   **3.** The y-intercept is the point where the graph crosses the y-axis.   **5.** The x-intercept is $(2, 0)$ and the y-intercept is $(0, 6)$.   **7.** The x-intercept is $(2, 0)$ and the y-intercept is $(0, -3)$.   **9.** The x-intercept is $(3, 0)$ and the y-intercept is $\left(0, \dfrac{9}{8}\right)$.   **11.** $y=4-2x$ **13.** $y=\dfrac{5-2x}{3}$

15. $y=2x-\dfrac{4}{5}$ **17.** $d=\sqrt{74}$ **19.** $d=\sqrt{36}=6$

21. $d\approx62.97$ **23.** $\left(3, -\dfrac{3}{2}\right)$ **25.** $(2,-1)$ **27.** $(0,0)$ **29.** $y=0$

31. Not collinear   **33.** $(-3, 2), (1, 3), (4, 0)$

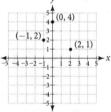

35.

x	-3	0	3	6
y	1	2	3	4

37.

x	-3	0	3
y	0	1.5	3

39.   **41.**

43. $d=8.246$ **45.** $d=5$ **47.** $(-3, 4)$ **49.** $x=0, y=-2$
51. $x=0.75, y=0$ **53.** $x=-1.667, y=0$
55. $15-11.2=3.8$ mi shorter   **57.** 6.042 **59.** Midpoint of each diagonal is the same point $(2, 2)$. Note this is a characteristic of rectangles, but not other quadrilaterals.
61. 37 mi   **63.** 54 ft

Section 2.2

1. It means they have the same slope.   **3.** The exponent of the x variable is 1. It is called a first-degree equation.   **5.** If we insert either value into the equation, they make an expression in the equation undefined (zero in the denominator).   **7.** $x=2$

9. $x=\dfrac{2}{7}$ **11.** $x=6$ **13.** $x=3$ **15.** $x=-14$

17. $x\neq-4; x=-3$ **19.** $x\neq1$; when we solve this we get $x=1$, which is excluded, therefore NO solution   **21.** $x\neq0; x=-\dfrac{5}{2}$

23. $y=-\dfrac{4}{5}x+\dfrac{14}{5}$ **25.** $y=-\dfrac{3}{4}x+2$ **27.** $y=\dfrac{1}{2}x+\dfrac{5}{2}$

29. $y=-3x-5$ **31.** $y=7$ **33.** $y=-4$ **35.** $8x+5y=7$

37. Parallel   **39.** Perpendicular

41. $m = -\frac{9}{7}$ **43.** $m = \frac{3}{2}$ **45.** $m_1 = -\frac{1}{3}, m_2 = 3$; perpendicular
47. $y = 0.245x - 45.662$. Answers may vary. $y_{\min} = -50, y_{\max} = -40$
49. $y = -2.333x + 6.667$. Answers may vary. $y_{\min} = -10, y_{\max} = 10$
51. $y = -\frac{A}{B}x + \frac{C}{B}$ **53.** The slope for $(-1, 1)$ to $(0, 4)$ is 3. The slope for $(-1, 1)$ to $(2, 0)$ is $-\frac{1}{3}$. The slope for $(2, 0)$ to $(3, 3)$ is 3. The slope for $(0, 4)$ to $(3, 3)$ is $-\frac{1}{3}$. Yes they are perpendicular.
55. 30 ft **57.** $57.50 **59.** 220 mi

Section 2.3

1. Answers may vary. Possible answers: We should define in words what our variable is representing. We should declare the variable. A heading. **3.** $2,000 - x$ **5.** $v + 10$
7. Ann: 23; Beth: 46 **9.** $20 + 0.05m$ **11.** 300 min
13. $90 + 40P$ **15.** 6 devices **17.** $50,000 - x$ **19.** 4 hr
21. She traveled for 2 hr at 20 mi/hr, or 40 miles.
23. $5,000 at 8% and $15,000 at 12% **25.** $B = 100 + 0.05x$
27. Plan A **29.** $R = 9$ **31.** $r = \frac{4}{5}$ or 0.8
33. $W = \frac{P - 2L}{2} = \frac{58 - 2(15)}{2} = 14$
35. $f = \frac{pq}{p + q} = \frac{8(13)}{8 + 13} = \frac{104}{21}$ **37.** $m = -\frac{5}{4}$
39. $h = \frac{2A}{b_1 + b_2}$ **41.** Length = 360 ft; width = 160 ft
43. 405 mi **45.** $A = 88$ in.² **47.** 28.7 **49.** $h = \frac{V}{\pi r^2}$
51. $r = \sqrt{\frac{V}{\pi h}}$ **53.** $C = 12\pi$

Section 2.4

1. Add the real parts together and the imaginary parts together.
3. Possible answer: i times i equals 1, which is not imaginary.
5. $-8 + 2i$ **7.** $14 + 7i$ **9.** $-\frac{23}{29} + \frac{15}{29}i$
11.   **13.**

15. $8 - i$ **17.** $-11 + 4i$ **19.** $2 - 5i$ **21.** $6 + 15i$
23. $-16 + 32i$ **25.** $-4 - 7i$ **27.** 25 **29.** $2 - \frac{2}{3}i$
31. $4 - 6i$ **33.** $\frac{2}{5} + \frac{11}{5}i$ **35.** $15i$ **37.** $1 + i\sqrt{3}$
39. 1 **41.** -1 **43.** $128i$ **45.** $\left(\frac{\sqrt{3}}{2} + \frac{1}{2}i\right)^6 = -1$
47. $3i$ **49.** 0 **51.** $5 - 5i$ **53.** $-2i$ **55.** $\frac{9}{2} - \frac{9}{2}i$

Section 2.5

1. It is a second-degree equation (the highest variable exponent is 2). **3.** We want to take advantage of the zero property of multiplication in the fact that if $a \cdot b = 0$ then it must follow that each factor separately offers a solution to the product being zero: $a = 0$ or $b = 0$.

5. One, when no linear term is present (no x term), such as $x^2 = 16$. Two, when the equation is already in the form $(ax + b)^2 = d$.
7. $x = 6, x = 3$ **9.** $x = -\frac{5}{2}, x = -\frac{1}{3}$ **11.** $x = 5, x = -5$
13. $x = -\frac{3}{2}, x = \frac{3}{2}$ **15.** $x = -2,$ **17.** $x = 0, x = -\frac{3}{7}$
19. $x = -6, x = 6$ **21.** $x = 6, x = -4$ **23.** $x = 1, x = -2$
25. $x = -2, x = 11$ **27.** $x = 3 \pm \sqrt{22}$ **29.** $z = \frac{2}{3}, z = -\frac{1}{2}$
31. $x = \frac{3 \pm \sqrt{17}}{4}$ **33.** Not real **35.** One rational
37. Two real; rational **39.** $x = \frac{-1 \pm \sqrt{17}}{2}$
41. $x = \frac{5 \pm \sqrt{13}}{6}$ **43.** $x = \frac{-1 \pm \sqrt{17}}{8}$
45. $x \approx 0.131$ and $x \approx 2.535$ **47.** $x \approx -6.7$ and $x \approx 1.7$
49. $ax^2 + bx + c = 0$
$$x^2 + \frac{b}{a}x = -\frac{c}{a}$$
$$x^2 + \frac{b}{a}x + \frac{b^2}{4a^2} = -\frac{c}{a} + \frac{b}{4a^2}$$
$$\left(x + \frac{b}{2a}\right)^2 = \frac{b^2 - 4ac}{4a^2}$$
$$x + \frac{b}{2a} = \pm\sqrt{\frac{b^2 - 4ac}{4a^2}}$$
$$x = \frac{-b \pm \sqrt{b^2 - 4ac}}{2a}$$
51. $x(x + 10) = 119$; 7 ft. and 17 ft. **53.** Maximum at $x = 70$
55. The quadratic equation would be $(100x - 0.5x^2) - (60x + 300) = 300$. The two values of x are 20 and 60. **57.** 3 feet

Section 2.6

1. This is not a solution to the radical equation, it is a value obtained from squaring both sides and thus changing the signs of an equation which has caused it not to be a solution in the original equation. **3.** He or she is probably trying to enter negative 9, but taking the square root of -9 is not a real number. The negative sign is in front of this, so your friend should be taking the square root of 9, cubing it, and then putting the negative sign in front, resulting in -27. **5.** A rational exponent is a fraction: the denominator of the fraction is the root or index number and the numerator is the power to which it is raised. **7.** $x = 81$ **9.** $x = 17$ **11.** $x = 8, x = 27$
13. $x = -2, 1, -1$ **15.** $y = 0, \frac{3}{2}, -\frac{3}{2}$ **17.** $m = 1, -1$
19. $x = \frac{2}{5}$ **21.** $x = 32$ **23.** $t = \frac{44}{3}$ **25.** $x = 3$
27. $x = -2$ **29.** $x = 4, -\frac{4}{3}$ **31.** $x = -\frac{5}{4}, \frac{7}{4}$
33. $x = 3, -2$ **35.** $x = -5$ **37.** $x = 1, -1, 3, -3$
39. $x = 2, -2$ **41.** $x = 1, 5$ **43.** All real numbers
45. $x = 4, 6, -6, -8$ **47.** 10 in. **49.** 90 kg

Section 2.7

1. When we divide both sides by a negative it changes the sign of both sides so the sense of the inequality sign changes.
3. $(-\infty, \infty)$

5. We start by finding the x-intercept, or where the function $= 0$. Once we have that point, which is $(3, 0)$, we graph to the right the straight line graph $y = x - 3$, and then when we draw it to the left we plot positive y values, taking the absolute value of them.

7. $\left(-\infty, \dfrac{3}{4}\right]$ **9.** $\left[-\dfrac{13}{2}, \infty\right)$ **11.** $(-\infty, 3)$ **13.** $\left(-\infty, -\dfrac{37}{3}\right]$

15. All real numbers $(-\infty, \infty)$ **17.** $\left(-\infty, -\dfrac{10}{3}\right) \cup (4, \infty)$

19. $(-\infty, -4] \cup [8, +\infty)$ **21.** No solution   **23.** $(-5, 11)$

25. $[6, 12]$ **27.** $[-10, 12]$

29. $x > -6$ and $x > -2$ Take the intersection of two sets.
$x > -2, (-2, +\infty)$

31. $x < -3$ or $x \geq 1$ Take the union of the two sets.
$(-\infty, -3) \cup [1, \infty)$

33. $(-\infty, -1) \cup (3, \infty)$ **35.** $[-11, -3]$

37. It is never less than zero. No solution.

39. Where the blue line is above the red line; point of intersection is $x = -3$. $(-\infty, -3)$

41. Where the blue line is above the red line; always. All real numbers. $(-\infty, -\infty)$

43. $(-1, 3)$
45. $(-\infty, 4)$
47. $\{x | x < 6\}$
49. $\{x | -3 \leq x < 5\}$
51. $(-2, 1]$
53. $(-\infty, 4]$

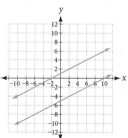

55. Where the blue is below the red; always. All real numbers. $(-\infty, +\infty)$.

57. Where the blue is below the red; $(1, 7)$.

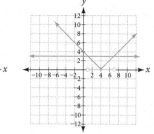

59. $x = 2, -\dfrac{4}{5}$ **61.** $(-7, 5]$

63. $80 \leq T \leq 120; 1,600 \leq 20T \leq 2,400; [1,600, 2,400]$

Chapter 2 Review Exercises

1. x-intercept: $(3, 0)$; y-intercept: $(0, -4)$ **3.** $y = \dfrac{5}{3}x + 4$

5. $\sqrt{72} = 6\sqrt{2}$ **7.** 620.097 **9.** Midpoint is $\left(2, \dfrac{23}{2}\right)$

11.

x	0	3	6
y	-2	2	6

13. $x = 4$
15. $x = \dfrac{12}{7}$
17. No solution
19. $y = \dfrac{1}{6}x + \dfrac{4}{3}$
21. $y = \dfrac{2}{3}x + 6$
23. Females 17, males 56
25. 84 mi
27. $x = -\dfrac{3}{4} \pm \dfrac{i\sqrt{47}}{4}$

29. Horizontal component -2; vertical component -1

31. $7 + 11i$ **33.** $16i$ **35.** $-16 - 30i$ **37.** $-4 - i\sqrt{10}$

39. $x = 7 - 3i$ **41.** $x = -1, -5$ **43.** $x = 0, \dfrac{9}{7}$

45. $x = 10, -2$ **47.** $x = \dfrac{-1 \pm \sqrt{5}}{4}$ **49.** $x = \dfrac{2}{5}, -\dfrac{1}{3}$

51. $x = 5 \pm 2\sqrt{7}$ **53.** $x = 0, 256$ **55.** $x = 0, \pm \sqrt{2}$

57. $x = -2$ **59.** $x = \dfrac{11}{2}, -\dfrac{17}{2}$ **61.** $(-\infty, 4)$

63. $\left[-\dfrac{10}{3}, 2\right]$ **65.** No solution   **67.** $\left(-\dfrac{4}{3}, \dfrac{1}{5}\right)$

69. Where the blue is below the red line; point of intersection is $x = 3.5$. $(3.5, \infty)$

Chapter 2 Practice Test

1. $y = \frac{3}{2}x + 2$

x	0	2	4
y	2	5	8

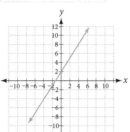

3. $(0, -3)$ $(4, 0)$

5. $(-\infty, 9]$ **7.** $x = -15$ **9.** $x \neq -4, 2; x = -\frac{5}{2}, 1$

11. $x = \frac{3 \pm \sqrt{3}}{2}$ **13.** $(-4, 1)$ **15.** $y = -\frac{5}{9}x - \frac{2}{9}$

17. $y = \frac{5}{2}x - 4$ **19.** $14i$ **21.** $\frac{5}{13} - \frac{14}{13}i$ **23.** $x = 2, -\frac{4}{3}$

25. $x = \frac{1}{2} \pm \frac{\sqrt{2}}{2}$ **27.** 4   **29.** $x = \frac{1}{2}, 2, -2$

CHAPTER 3

Section 3.1

1. A relation is a set of ordered pairs. A function is a special kind of relation in which no two ordered pairs have the same first coordinate.   **3.** When a vertical line intersects the graph of a relation more than once, that indicates that for that input there is more than one output. At any particular input value, there can be only one output if the relation is to be a function.   **5.** When a horizontal line intersects the graph of a function more than once, that indicates that for that output there is more than one input. A function is one-to-one if each output corresponds to only one input.
7. Function   **9.** Function   **11.** Function   **13.** Function
15. Function   **17.** Function   **19.** Function
21. Function   **23.** Function   **25.** Not a function
27. $f(-3) = -11, f(2) = -1, f(-a) = -2a - 5, -f(a) = -2a + 5,$
$f(a + h) = 2a + 2h - 5$ **29.** $f(-3) = \sqrt{5} + 5, f(2) = 5,$
$f(-a) = \sqrt{2 + a} + 5, -f(a) = -\sqrt{2 - a} - 5, f(a + h) = \sqrt{2 - a - h} + 5$ **31.** $f(-3) = 2, f(2) = -2,$
$f(-a) = |-a - 1| - |-a + 1|, -f(a) = -|a - 1| + |a + 1|,$
$f(a + h) = |a + h - 1| - |a + h + 1|$
33. $\frac{g(x) - g(a)}{x - a} = x + a + 2, x \neq a$ **35. a.** $f(-2) = 14$ **b.** $x = 3$
37. a. $f(5) = 10$ **b.** $x = 4$ or -1 **39. a.** $r = 6 - \frac{2}{3}t$
b. $f(-3) = 8$ **c.** $t = 6$ **41.** Not a function   **43.** Function
45. Function   **47.** Function   **49.** Function
51. Function   **53. a.** $f(0) = 1$ **b.** $f(x) = -3, x = -2$ or 2
55. Not a function, not one-to-one   **57.** One-to-one function
59. Function, not one-to-one   **61.** Function   **63.** Function
65. Not a function   **67.** $f(x) = 1, x = 2$
69. $f(-2) = 14; f(-1) = 11; f(0) = 8; f(1) = 5; f(2) = 2$
71. $f(-2) = 4; f(-1) = 4.414; f(0) = 4.732; f(1) = 5; f(2) = 5.236$

73. $f(-2) = \frac{1}{9}; f(-1) = \frac{1}{3}; f(0) = 1; f(1) = 3; f(2) = 9$ **75.** 20
77. The range for this viewing window is $[0, 100]$.

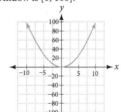

79. The range for this viewing window is $[-0.001, 0.001]$.

81. The range for this viewing window is $[-1,000,000, 1,000,000]$.

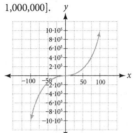

83. The range for this viewing window is $[0, 10]$.

85. The range for this viewing window is $[-0.1, 0.1]$.

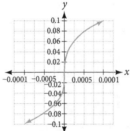

87. The range for this viewing window is $[-100, 100]$.

89. a. $g(5000) = 50$ **b.** The number of cubic yards of dirt required for a garden of 100 square feet is 1.   **91. a.** The height of the rocket above ground after 1 second is 200 ft.
b. The height of the rocket above ground after 2 seconds is 350 ft.

Section 3.2

1. The domain of a function depends upon what values of the independent variable make the function undefined or imaginary.
3. There is no restriction on x for $f(x) = \sqrt[3]{x}$ because you can take the cube root of any real number. So the domain is all real numbers, $(-\infty, \infty)$. When dealing with the set of real numbers, you cannot take the square root of negative numbers. So x-values are restricted for $f(x) = \sqrt{x}$ to nonnegative numbers and the domain is $[0, \infty)$.   **5.** Graph each formula of the piecewise function over its corresponding domain. Use the same scale for the x-axis and y-axis for each graph. Indicate included endpoints with a solid circle and excluded endpoints with an open circle. Use an arrow to indicate $-\infty$ or ∞. Combine the graphs to find the graph of the piecewise function.   **7.** $(-\infty, \infty)$ **9.** $(-\infty, 3]$
11. $(-\infty, \infty)$ **13.** $(-\infty, \infty)$ **15.** $\left(-\infty, -\frac{1}{2}\right) \cup \left(-\frac{1}{2}, \infty\right)$
17. $(-\infty, -11) \cup (-11, 2) \cup (2, \infty)$ **19.** $(-\infty, -3) \cup (-3, 5) \cup (5, \infty)$

21. $(-\infty, 5)$ **23.** $[6, \infty)$ **25.** $(-\infty, -9)\cup(-9, 9)\cup(9, \infty)$
27. Domain: $(2, 8]$, range: $[6, 8)$ **29.** Domain: $[-4, 4]$, range: $[0, 2]$
31. Domain: $[-5, 3)$, range: $[0, 2]$ **33.** Domain: $(-\infty, 1]$, range: $[0, \infty)$
35. Domain: $\left[-6, -\frac{1}{6}\right]\cup\left[\frac{1}{6}, 6\right]$, range: $\left[-6, -\frac{1}{6}\right]\cup\left[\frac{1}{6}, 6\right]$
37. Domain: $[-3, \infty)$, range is $[0, \infty)$
39. Domain: $(-\infty, \infty)$ **41.** Domain: $(-\infty, \infty)$

43. Domain: $(-\infty, \infty)$ **45.** Domain: $(-\infty, \infty)$

47. $f(-3) = 1; f(-2) = 0; f(-1) = 0; f(0) = 0$
49. $f(-1) = -4; f(0) = 6; f(2) = 20; f(4) = 34$
51. $f(-1) = -5; f(0) = 3; f(2) = 3; f(4) = 16$
53. $(-\infty, 1)\cup(1, \infty)$
55.

The viewing window: $[-0.5, -0.1]$ The viewing window: $[0.1, 0.5]$
has a range: $[4, 100]$.   has a range: $[4, 100]$.

57. $[0, 8]$ **59.** Many answers; one function is $f(x) = \dfrac{1}{\sqrt{x-2}}$.
61. a. The fixed cost is $500.   **b.** The cost of making 25 items is
$750.   **c.** The domain is $[0, 100]$ and the range is $[500, 1500]$.

Section 3.3

1. Yes, the average rate of change of all linear functions is constant.
3. The absolute maximum and minimum relate to the entire
graph, whereas the local extrema relate only to a specific region in
an open interval.   **5.** $4(b + 1)$ **7.** 3   **9.** $4x + 2h$
11. $-\dfrac{1}{13(13 + h)}$ **13.** $3h^2 + 9h + 9$ **15.** $4x + 2h - 3$
17. $\dfrac{4}{3}$ **19.** Increasing on $(-\infty, -2.5)\cup(1, \infty)$ and decreasing
on $(-2.5, 1)$ **21.** Increasing on $(-\infty, 1)\cup(3, 4)$ and decreasing
on $(1, 3)\cup(4, \infty)$

23. Local maximum: $(-3, 50)$ and local minimum: $(3, 50)$
25. Absolute maximum at approximately $(7, 150)$ and absolute
minimum at approximately $(-7.5, -220)$
27. a. $-3,000$ people per year   **b.** $-1,250$ people per year
29. -4 **31.** 27   **33.** ≈ -0.167 **35.** Local minimum:
$(3, -22)$, decreasing on $(-\infty, 3)$, increasing on $(3, \infty)$
37. Local minimum: $(-2, -2)$, decreasing on $(-3, -2)$,
increasing on $(-2, \infty)$ **39.** Local maximum: $(-0.5, 6)$, local
minima: $(-3.25, -47)$ and $(2.1, -32)$, decreasing on $(-\infty,$
$-3.25)$ and $(-0.5, 2.1)$, increasing on $(-3.25, -0.5)$ and $(2.1, \infty)$
41. A   **43.** $b = 5$ **45.** ≈ 2.7 gallons per minute
47. ≈ -0.6 milligrams per day

Section 3.4

1. Find the numbers that make the function in the denominator g
equal to zero, and check for any other domain restrictions on f and
g, such as an even-indexed root or zeros in the denominator.
3. Yes, sample answer: Let $f(x) = x + 1$ and $g(x) = x - 1$.
Then $f(g(x)) = f(x - 1) = (x - 1) + 1 = x$ and
$g(f(x)) = g(x + 1) = (x + 1) - 1 = x$ so $f \circ g = g \circ f$.
5. $(f + g)(x) = 2x + 6$; domain: $(-\infty, \infty)$
$(f - g)(x) = 2x^2 + 2x - 6$; domain: $(-\infty, \infty)$
$(fg)(x) = -x^4 - 2x^3 + 6x^2 + 12x$; domain: $(-\infty, \infty)$
$\left(\dfrac{f}{g}\right)(x) = \dfrac{x^2 + 2x}{6 - x^2}$; domain: $(-\infty, -\sqrt{6})\cup(-\sqrt{6}, \sqrt{6})\cup(\sqrt{6}, \infty)$
7. $(f + g)(x) = \dfrac{4x^3 + 8x^2 + 1}{2x}$; domain: $(-\infty, 0)\cup(0, \infty)$
$(f - g)(x) = \dfrac{4x^3 + 8x^2 - 1}{2x}$; domain: $(-\infty, 0)\cup(0, \infty)$
$(fg)(x) = x + 2$; domain: $(-\infty, 0)\cup(0, \infty)$
$\left(\dfrac{f}{g}\right)(x) = 4x^3 + 8x^2$; domain: $(-\infty, 0)\cup(0, \infty)$
9. $(f + g)(x) = 3x^2 + \sqrt{x - 5}$; domain: $[5, \infty)$
$(f - g)(x) = 3x^2 - \sqrt{x - 5}$; domain: $[5, \infty)$
$(fg)(x) = 3x^2\sqrt{x - 5}$; domain: $[5, \infty)$
$\left(\dfrac{f}{g}\right)(x) = \dfrac{3x^2}{\sqrt{x - 5}}$; domain: $(5, \infty)$ **11. a.** $f(g(2)) = 3$
b. $f(g(x)) = 18x^2 - 60x + 51$ **c.** $g(f(x)) = 6x^2 - 2$
d. $(g \circ g)(x) = 9x - 20$ **e.** $(f \circ f)(-2) = 163$
13. $f(g(x)) = \sqrt{x^2 + 3} + 2$; $g(f(x)) = x + 4\sqrt{x} + 7$
15. $f(g(x)) = \dfrac{\sqrt[3]{x + 1}}{x}$; $g(f(x)) = \dfrac{\sqrt[3]{x} + 1}{x}$
17. $f(g(x)) = \dfrac{x}{2}, x \neq 0$; $g(f(x)) = 2x - 4, x \neq 4$
19. $f(g(h(x))) = \dfrac{1}{(x + 3)^2} + 1$
21. a. $(g \circ f)(x) = -\dfrac{3}{\sqrt{2 - 4x}}$ **b.** $\left(-\infty, \dfrac{1}{2}\right)$
23. a. $(0, 2)\cup(2, \infty)$ except $x = -2$ **b.** $(0, \infty)$ **c.** $(0, \infty)$ **25.** $(1, \infty)$
27. Many solutions; one possible answer: $f(x) = x^3$; $g(x) = x - 5$
29. Many solutions; one possible answer: $f(x) = \dfrac{4}{x}$; $g(x) = (x + 2)^2$
31. Many solutions; one possible answer: $f(x) = \sqrt[3]{x}$; $g(x) = \dfrac{1}{2x - 3}$
33. Many solutions; one possible answer: $f(x) = \sqrt[4]{x}$; $g(x) = \dfrac{3x - 2}{x + 5}$
35. Many solutions; one possible answer: $f(x) = \sqrt{x}$; $g(x) = 2x + 6$

37. Many solutions; one possible answer: $f(x) = \sqrt[3]{x}$; $g(x) = x - 1$

39. Many solutions; one possible answer: $f(x) = x^3$; $g(x) = \dfrac{1}{x - 2}$

41. Many solutions; one possible answer: $f(x) = \sqrt{x}$; $g(x) = \dfrac{2x - 1}{3x + 4}$

43. 2   **45.** 5   **47.** 4   **49.** 0   **51.** 2   **53.** 1

55. 4   **57.** 4   **59.** 9   **61.** 4   **63.** 2   **65.** 3

67. 11   **69.** 0   **71.** 7   **73.** $f(g(0)) = 27, g(f(0)) = -94$

75. $f(g(0)) = \dfrac{1}{5}, g(f(0)) = 5$ **77.** $f(g(x)) = 18x^2 + 60x + 51$

79. $g \circ g(x) = 9x + 20$ **81.** $(f \circ g)(x) = 2, (g \circ f)(x) = 2$

83. $(-\infty, \infty)$ **85.** False   **87.** $(f \circ g)(6) = 6; (g \circ f)(6) = 6$

89. $(f \circ g)(11) = 11; (g \circ f)(11) = 11$ **91.** C

93. $A(t) = \pi(25\sqrt{t + 2})^2$ and $A(2) = \pi(25\sqrt{4})^2 = 2,500\pi$

square inches   **95.** $A(5) = 121\pi$ square units

97. a. $N(T(t)) = 575t^2 + 65t - 31.25$ **b.** ≈ 3.38 hours

Section 3.5

1. A horizontal shift results when a constant is added to or subtracted from the input. A vertical shift results when a constant is added to or subtracted from the output.   **3.** A horizontal compression results when a constant greater than 1 multiplies the input. A vertical compression results when a constant between 0 and 1 multiplies the output.   **5.** For a function f, substitute $(-x)$ for (x) in $f(x)$ and simplify. If the resulting function is the same as the original function, $f(-x) = f(x)$, then the function is even. If the resulting function is the opposite of the original function, $f(-x) = -f(x)$, then the original function is odd. If the function is not the same or the opposite, then the function is neither odd nor even.   **7.** $g(x) = |x - 1| - 3$

9. $g(x) = \dfrac{1}{(x + 4)^2} + 2$ **11.** The graph of $f(x + 43)$ is a horizontal shift to the left 43 units of the graph of f.

13. The graph of $f(x - 4)$ is a horizontal shift to the right 4 units of the graph of f.   **15.** The graph of $f(x) + 8$ is a vertical shift up 8 units of the graph of f.   **17.** The graph of $f(x) - 7$ is a vertical shift down 7 units of the graph of f.   **19.** The graph of $f(x + 4) - 1$ is a horizontal shift to the left 4 units and a vertical shift down 1 unit of the graph of f.   **21.** Decreasing on $(-\infty, -3)$ and increasing on $(-3, \infty)$ **23.** Decreasing on $(0, \infty)$

25.

27.

29.

31. $g(x) = f(x - 1)$, $h(x) = f(x) + 1$

33. $f(x) = |x - 3| - 2$

35. $f(x) = \sqrt{x + 3} - 1$

37. $f(x) = (x - 2)^2$

39. $f(x) = |x + 3| - 2$

41. $f(x) = -\sqrt{x}$

43. $f(x) = -(x + 1)^2 + 2$

45. $f(x) = \sqrt{-x} + 1$

47. Even   **49.** Odd

51. Even   **53.** The graph of g is a vertical reflection (across the x-axis) of the graph of f.   **55.** The graph of g is a vertical stretch by a factor of 4 of the graph of f.

57. The graph of g is a horizontal compression by a factor of $\dfrac{1}{5}$ of the graph of f.   **59.** The graph of g is a horizontal stretch by a factor of 3 of the graph of f.   **61.** The graph of g is a horizontal reflection across the y-axis and a vertical stretch by a factor of 3 of the graph of f.   **63.** $g(x) = |-4x|$

65. $g(x) = \dfrac{1}{3(x + 2)^2} - 3$ **67.** $g(x) = \dfrac{1}{2}(x - 5)^2 + 1$

69. This is a parabola shifted to the left 1 unit, stretched vertically by a factor of 4, and shifted down 5 units.

71. This is an absolute value function stretched vertically by a factor of 2, shifted 4 units to the right, reflected across the horizontal axis, and then shifted 3 units up.

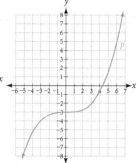

73. This is a cubic function compressed vertically by a factor of $\dfrac{1}{2}$.

75. The graph of the function is stretched horizontally by a factor of 3 and then shifted downward by 3 units.

77. The graph of $f(x) = \sqrt{x}$ is shifted right 4 units and then reflected across the y-axis.

79.

81.

25.

27.

29.

31.

33. range: $[-400, 100]$

35.

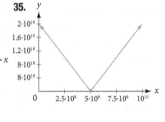

37. There is no value for a that will keep the function from having a y-intercept. The absolute value function always crosses the y-intercept when $x = 0$.

39. $|p - 0.08| \le 0.015$ **41.** $|x - 5.0| \le 0.01$

Section 3.6

1. Isolate the absolute value term so that the equation is of the form $|A| = B$. Form one equation by setting the expression inside the absolute value symbol, A, equal to the expression on the other side of the equation, B. Form a second equation by setting A equal to the opposite of the expression on the other side of the equation, $-B$. Solve each equation for the variable.   **3.** The graph of the absolute value function does not cross the x-axis, so the graph is either completely above or completely below the x-axis.   **5.** The distance from x to 8 can be represented using the absolute value statement: $|x - 8| = 4$.   **7.** $|x - 10| \ge 15$
9. There are no x-intercepts.   **11.** $(-4, 0)$ and $(2, 0)$
13. $(-2, 0), (4, 0),$ and $(0, -4)$ **15.** $(-7, 0), (0, 16), (25, 0)$
17.

19.

21.

23.

SECTION 3.7

1. Each output of a function must have exactly one input for the function to be one-to-one. If any horizontal line crosses the graph of a function more than once, that means that y-values repeat and the function is not one-to-one. If no horizontal line crosses the graph of the function more than once, then no y-values repeat and the function is one-to-one.   **3.** Yes. For example, $f(x) = \frac{1}{x}$ is its own inverse.   **5.** $y = f^{-1}(x)$

7. $f^{-1}(x) = x - 3$ **9.** $f^{-1}(x) = 2 - x$ **11.** $f^{-1}(x) = -\frac{2x}{x - 1}$
13. Domain of $f(x)$: $[-7, \infty)$; $f^{-1}(x) = \sqrt{x} - 7$
15. Domain of $f(x)$: $[0, \infty)$; $f^{-1}(x) = \sqrt{x + 5}$
17. $f(g(x)) = x$ and $g(f(x)) = x$ **19.** One-to-one
21. One-to-one   **23.** Not one-to-one   **25.** 3   **27.** 2
29.   **31.** $[2, 10]$ **33.** 6

35. -4 **37.** 0   **39.** 1
41.

x	1	4	7	12	16
$f^{-1}(x)$	3	6	9	13	14

43. $f^{-1}(x) = (1+x)^{\frac{1}{3}}$

45. $f^{-1}(x) = \frac{5}{9}(x - 32)$

47. $t(d) = \frac{d}{50}; t(180) = \frac{180}{50}$. The time for the car to travel 180 miles is 3.6 hours.

49.

51.

53.

55. $f(x) = |x - 3|$

57. Even   **59.** Odd

61. Even

63. $f(x) = \frac{1}{2}|x + 2| + 1$

65. $f(x) = -3|x - 3| + 3$

67.

69. $f^{-1}(x) = \frac{x - 9}{10}$

71. $f^{-1}(x) = \sqrt{x - 1}$

73. The function is one-to-one.

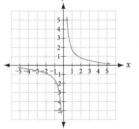

75. 5

Chapter 3 Review Exercises

1. Function   **3.** Not a function   **5.** $f(-3) = -27; f(2) = -2;$
$f(-a) = -2a^2 - 3a; \ -f(a) = 2a^2 - 3a;$
$f(a + h) = -2a^2 - 4ah - 2h^2 + 3a + 3h$

7. One-to-one   **9.** Function   **11.** Function

13.

15. 2   **17.** −1.8 or 1.8

19. $\dfrac{-64 + 80a - 16a^2}{-1 + a}$
$= -16a + 64; a \neq 1$

21. $(-\infty, -2)\cup(-2, 6)\cup(6, \infty)$

23.

25. 31

27. Increasing on $(2, \infty)$, decreasing on $(-\infty, 2)$

29. Increasing on $(-3, 1)$, constant on $(-\infty, -3)$ and $(1, \infty)$

31. Local minimum: $(-2, -3)$; local maximum: $(1, 3)$

33. Absolute maximum: 10

35. $(f \circ g)(x) = 17 - 18x, \ (g \circ f)(x) = -7 - 18x$

37. $(f \circ g)(x) = \sqrt{\frac{1}{x} + 2}; \ (g \circ f)(x) = \frac{1}{\sqrt{x + 2}}$

39. $(f \circ g)(x) = \dfrac{\frac{1}{\frac{1}{x} + 1}}{\frac{1}{x} + 4} = \dfrac{1 + x}{1 + 4x};$ Domain: $\left(-\infty, -\frac{1}{4}\right)\cup\left(-\frac{1}{4}, 0\right)\cup(0, \infty)$

41. $(f \circ g)(x) = \dfrac{1}{\sqrt{x}};$ Domain: $(0, \infty)$

43. Many solutions; one possible answer: $g(x) = \dfrac{2x - 1}{3x + 4}$ and $f(x) = \sqrt{x}$.

45.

47.

Chapter 3 Practice Test

1. Relation is a function   **3.** −16   **5.** The graph is a parabola and the graph fails the horizontal line test.

7. $2a^2 - a$ **9.** $-2(a + b) + 1; b \neq a$ **11.** $\sqrt{2}$

13.

15. Even   **17.** Odd

19. $f^{-1}(x) = \dfrac{x + 5}{3}$

21. $(-\infty, -1.1)$ and $(1.1, \infty)$

23. $(1.1, -0.9)$ **25.** $f(2) = 2$

27. $f(x) = \begin{cases} |x| & \text{if } x \leq 2 \\ 3 & \text{if } x > 2 \end{cases}$

29. $x = 2$ **31.** Yes

33. $f^{-1}(x) = -\dfrac{x - 11}{2}$ or $\dfrac{11 - x}{2}$

CHAPTER 4

Section 4.1

1. Terry starts at an elevation of 3,000 feet and descends 70 feet per second   **3.** $d(t) = 100 - 10t$

5. The point of intersection is (a, a). This is because for the horizontal line, all of the y-coordinates are a and for the vertical line, all of the x-coordinates are a. The point of intersection is on both lines and therefore will have these two characteristics.

7. Yes   **9.** Yes   **11.** No   **13.** Yes   **15.** Increasing
17. Decreasing   **19.** Decreasing   **21.** Increasing
23. Decreasing   **25.** 2   **27.** −2   **29.** $\frac{3}{5}x - 1$
31. $y = 3x - 2$ **33.** $y = -\frac{1}{3}x + \frac{11}{3}$ **35.** $y = -1.5x - 3$

37. Perpendicular   **39.** Parallel   **41.** y-int: $(0, 2)$, x-int: $(2, 0)$
43. y-int: $(0, -5)$, x-int: $\left(\frac{5}{3}, 0\right)$ **45.** y-int: $(0, 4)$, x-int: $(-10, 0)$
47. Line 1: $m = -10$, Line 2: $m = -10$, parallel
49. Line 1: $m = -2$, Line 2: $m = 1$, neither
51. Line 1: $m = -2$, Line 2: $m = -2$, parallel
53. $y = 3x - 3$ **55.** $y = -\frac{1}{3}t + 2$ **57.** 0   **59.** $y = -\frac{5}{4}x + 5$
61. $y = 3x - 1$ **63.** $y = -2.5$ **65.** F   **67.** C   **69.** A

71.

73.

75.

77.

79.

81.

83.

85. $y = 3$
87. $x = -3$
89. Linear, $g(x) = -3x + 5$
91. Linear, $f(x) = 5x - 5$
93. Linear, $g(x) = -\frac{25}{2}x + 6$
95. Linear, $f(x) = 10x - 24$
97. $f(x) = -58x + 17.3$

99.

101. **a.** $a = 11,900$, $b = 1001.1$
b. $q(p) = 1000p - 100$

103.
(graph)

105. $x = -\frac{16}{3}$ **107.** $x = a$
109. $y = \frac{d}{c - a}x - \frac{ad}{c - a}$
111. $y = 100x - 98$
113. $x < \frac{1999}{201}, x > \frac{1999}{201}$
115. 69. $45 per training session

117. The rate of change is 0.1. For every additional minute talked, the monthly charge increases by $0.1 or 10 cents. The initial value is 24. When there are no minutes talked, initially the charge is $24.   **119.** The slope is −400. this means for every year between 1960 and 1989, the population dropped by 400 per year in the city.   **121.** C

Section 4.2

1. Determine the independent variable. This is the variable upon which the output depends.   **3.** To determine the initial value, find the output when the input is equal to zero.   **5.** 6 square units.   **7.** 20.01 square units   **9.** 2,300   **11.** 64,170
13. $P(t) = 2500t + 75,000$ **15.** $(-30, 0)$ 30 years before the start of this model, the town has no citizens. $(0, 75,000)$ Initially, the town had a population of 75,000.   **17.** Ten years after the model began.   **19.** $W(t) = 0.5t + 7.5$ **21.** $(-15, 0)$ The x-intercept is not a plausible set of data for this model because it means the baby weighed 0 pounds 15 months prior to birth. $(0, 7.5)$ the baby weighed 7.5 pounds at birth.   **23.** At age 5.8 months   **25.** $C(t) = 12,025 - 205t$ **27.** $(58.7, 0)$ In 58.7 years, the number of people afflicted with the common cold would be zero $(0, 12,025)$ Initially, 12,025 people were afflicted with the common cold   **29.** 2063   **31.** $y = -2t + 180$
33. In 2070, the company's profits will be zero   **35.** $y = 30t - 300$
37. $(10, 0)$ In the year 1990, the company's profits were zero
39. Hawaii   **41.** During the year 1933   **43.** $105,620
45. **a.** 696 people **b.** 4 years **c.** 174 people per year **d.** 305 people
e. $P(t) = 305 + 174t$ **f.** 2,219 people   **47.** **a.** $C(x) = 0.15x + 10$
b. The flat monthly fee is $10 and there is a $0.15 fee for each additional minute used **c.** $113.05   **49.** **a.** $P(t) = 190t + 4,360$
b. 6,640 moose   **51.** **a.** $R(t) = -2.1t + 16$
b. 5.5 billion cubic feet **c.** During the year 2017
53. More than 133 minutes   **55.** More than $42,857.14 worth of jewelry   **57.** More than $66,666.67 in sales

51. $f(x) = 50 \dfrac{x^2 - x - 2}{x^2 - 25}$ **53.** $f(x) = 7 \dfrac{x^2 + 2x - 24}{x^2 + 9x + 20}$

55. $f(x) = \dfrac{1}{2} \cdot \dfrac{x^2 - 4x + 4}{x + 1}$ **57.** $f(x) = 4 \dfrac{x - 3}{x^2 - x - 12}$

59. $f(x) = -9 \dfrac{x - 2}{x^2 - 9}$ **61.** $f(x) = \dfrac{1}{3} \cdot \dfrac{x^2 + x - 6}{x - 1}$

63. $f(x) = -6 \dfrac{(x - 1)^2}{(x + 3)(x - 2)^2}$

65. Vertical asymptote at $x = 2$; horizontal asymptote at $y = 0$

x	2.01	2.001	2.0001	1.99	1.999
y	100	1,000	10,000	-100	$-1,000$

x	10	100	1,000	10,000	100,000
y	0.125	0.0102	0.001	0.0001	0.00001

67. Vertical asymptote at $x = -4$; horizontal asymptote at $y = 2$

x	-4.1	-4.01	-4.001	-3.99	-3.999
y	82	802	8,002	-798	-7998

x	10	100	1,000	10,000	100,000
y	1.4286	1.9331	1.992	1.9992	1.999992

69. Vertical asymptote at $x = -1$; horizontal asymptote at $y = 1$

x	-0.9	-0.99	-0.999	-1.1	-1.01
y	81	9,801	998,001	121	10,201

x	10	100	1,000	10,000	100,000
y	0.82645	0.9803	0.998	0.9998	

71. $\left(\dfrac{3}{2}, \infty \right)$ **73.** $(-\infty, 1) \cup (4, \infty)$

75. $(2, 4)$ **77.** $(2, 5)$ **79.** $(-1, 1)$ **81.** $C(t) = \dfrac{8 + 2t}{300 + 20t}$

83. After about 6.12 hours   **85.** 2 by 2 by 5 feet

87. radius 2.52 meters

Section 5.7

1. It can be too difficult or impossible to solve for x in terms of y.
3. We will need a restriction on the domain of the answer.   **5.** $f^{-1}(x) = \sqrt{x} + 4$ **7.** $f^{-1}(x) = \sqrt{x + 3} - 1$

9. $f^{-1}(x) = \sqrt{12 - x}$ **11.** $f^{-1}(x) = \pm \sqrt{\dfrac{x - 4}{2}}$

13. $f^{-1}(x) = \sqrt[3]{\dfrac{x - 1}{3}}$ **15.** $f^{-1}(x) = \sqrt[3]{\dfrac{4 - x}{2}}$

17. $f^{-1}(x) = \dfrac{3 - x^2}{4}$, $[0, \infty)$ **19.** $f^{-1}(x) = \dfrac{(x - 5)^2 + 8}{6}$

21. $f^{-1}(x) = (3 - x)^3$ **23.** $f^{-1}(x) = \dfrac{4x + 3}{x}$ **25.** $f^{-1}(x) = \dfrac{7x + 2}{1 - x}$

27. $f^{-1}(x) = \dfrac{2x - 1}{5x + 5}$ **29.** $f^{-1}(x) = \sqrt{x + 3} - 2$

31. $f^{-1}(x) = \sqrt{x - 2}$ **33.** $f^{-1}(x) = \sqrt{x} - 3$

35. $f^{-1}(x) = \sqrt[3]{x - 3}$ **37.** $f^{-1}(x) = \sqrt{x + 4} - 2$

39.   **41.** $[-1, 0) \cup [1, \infty)$

43. $[-3, 0] \cup (4, \infty)$ **45.** $[-\infty, -4] \cdot [-3, 3]$

47. $(-2, 0), (0, 1), (8, 2)$ **49.** $(-13, -1), (-4, 0), (5, 1)$

51. $f^{-1}(x) = \sqrt[3]{\dfrac{x - b}{a}}$ **53.** $f^{-1}(x) = \sqrt{\dfrac{x^2 - b}{a}}$

55. $f^{-1}(x) = \dfrac{cx - b}{a - x}$ **57.** $t(h) = \sqrt{\dfrac{600 - h}{16}}$, 3.54 seconds

59. $r(A) = \sqrt{\dfrac{A}{4\pi}}, \approx 8.92$ in.   **61.** $l(T) = 32.2 \left(\dfrac{T}{2\pi} \right)^2, \approx 3.26$ ft

63. $r(A) = \sqrt{\dfrac{A + 8\pi}{2\pi}} - 2, 3.99$ ft   **65.** $r(V) = \sqrt{\dfrac{V}{10\pi}}, \approx 5.64$ ft

Section 5.8

1. The graph will have the appearance of a power function.
3. No. Multiple variables may jointly vary.   **5.** $y = 5x^2$

7. $y = 10x^3$ **9.** $y = 6x^4$ **11.** $y = \dfrac{18}{x^2}$ **13.** $y = \dfrac{81}{x^4}$

15. $y = \dfrac{20}{\sqrt[3]{x}}$ **17.** $y = 10xzw$ **19.** $y = 10x\sqrt{z}$

21. $y = 4\dfrac{xz}{w}$ **23.** $y = 40\dfrac{xz}{\sqrt{wt^2}}$ **25.** $y = 256$

27. $y = 6$ **29.** $y = 6$ **31.** $y = 27$ **33.** $y = 3$

35. $y = 18$ **37.** $y = 90$ **39.** $y = \dfrac{81}{2}$

41. $y = \dfrac{3}{4}x^2$

43. $y = \dfrac{1}{3}\sqrt{x}$

45. $y = \dfrac{4}{x^2}$

47. ≈ 1.89 years
49. ≈ 0.61 years
51. 3 seconds
53. 48 inches
55. ≈ 49.75 pounds
57. ≈ 33.33 amperes
59. ≈ 2.88 inches

Chapter 5 Review Exercises

1. $f(x) = (x - 2)^2 - 9$;
vertex: $(2, -9)$;
intercepts: $(5, 0), (-1, 0), (0, -5)$

3. $f(x) = \dfrac{3}{25}(x + 2)^2 + 3$

5. 300 meters by 150 meters, the longer side parallel to the river

7. Yes; degree: 5, leading coefficient: 4

9. Yes; degree: 4; leading coefficient: 1

11. As $x \to -\infty, f(x) \to -\infty$, as $x \to \infty, f(x) \to \infty$

13. -3 with multiplicity 2, $-\dfrac{1}{2}$ with multiplicity 1, -1 with multiplicity 3   **15.** 4 with multiplicity 1   **17.** $\dfrac{1}{2}$ with multiplicity 1, 3 with multiplicity 3   **19.** $x^2 + 4$ with remainder 12   **21.** $x^2 - 5x + 20 - \dfrac{61}{x + 3}$

23. $2x^2 - 2x - 3$, so factored form is $(x + 4)(2x^2 - 2x - 3)$

25. $\left\{ -2, 4, -\dfrac{1}{2} \right\}$ **27.** $\left\{ 1, 3, 4, \dfrac{1}{2} \right\}$

29. 2 or 0 positive, 1 negative

31. Intercepts: $(-2, 0), \left(0, -\dfrac{2}{5}\right)$, asymptotes: $x = 5$ and $y = 1$

33. Intercepts: $(3, 0), (-3, 0), \left(0, \dfrac{27}{2}\right)$; asymptotes: $x = 1, -2$ and $y = 3$

35. $y = x - 2$ **37.** $f^{-1}(x) = \sqrt{x} + 2$ **39.** $f^{-1}(x) = \sqrt{x + 11} - 3$

41. $f^{-1}(x) = \dfrac{(x + 3)^2 - 5}{4}, x \geq -3$ **43.** $y = 64$ **45.** $y = 72$

47. ≈ 148.5 pounds

Chapter 5 Practice Test

1. Degree: 5, leading coefficient: -2 **3.** As $x \to -\infty$, $f(x) \to \infty$, as $x \to \infty, f(x) \to \infty$ **5.** $f(x) = 3(x - 2)^2$

7. 3 with multiplicity 3, $\dfrac{1}{3}$ with multiplicity 1, 1 with multiplicity 2

9. $-\dfrac{1}{2}$ with multiplicity 3, 2 with multiplicity 2

11. $x^3 + 2x^2 + 7x + 14 + \dfrac{26}{x - 2}$ **13.** $\left\{ -3, -1, \dfrac{3}{2} \right\}$

15. $1, -2$, and $-\dfrac{3}{2}$ (multiplicity 2)

17. $f(x) = -\dfrac{2}{3}(x - 3)^2(x - 1)(x + 2)$ **19.** 2 or 0 positive, 1 negative

21. $(-3, 0), (1, 0), \left(0, \dfrac{3}{4}\right)$; asymptotes $x = -2, 2$ and $y = 1$

23. $f^{-1}(x) = (x - 4)^2 + 2, x \geq 4$

25. $f^{-1}(x) = \dfrac{x + 3}{3x - 2}$

27. $y = 20$

CHAPTER 6

Section 6.1

1. Linear functions have a constant rate of change. Exponential functions increase based on a percent of the original.
3. When interest is compounded, the percentage of interest earned to principal ends up being greater than the annual percentage rate for the investment account. Thus, the annual percentage rate does not necessarily correspond to the real interest earned, which is the very definition of *nominal*.   **5.** Exponential; the population decreases by a proportional rate.   **7.** Not exponential; the charge decreases by a constant amount each visit, so the statement represents a linear function.   **9.** Forest B
11. After 20 years forest A will have 43 more trees than forest B.
13. Answers will vary. Sample response: For a number of years, the population of forest A will increasingly exceed forest B, but because forest B actually grows at a faster rate, the population will eventually become larger than forest A and will remain that way as long as the population growth models hold. Some factors that

might influence the long-term validity of the exponential growth model are drought, an epidemic that culls the population, and other environmental and biological factors.
15. Exponential growth; the growth factor, 1.06, is greater than 1.
17. Exponential decay; the decay factor, 0.97, is between 0 and 1.
19. $f(x) = 2000(0.1)^x$ **21.** $f(x) = \left(\frac{1}{6}\right)^{-\frac{3}{5}}\left(\frac{1}{6}\right)^{\frac{x}{5}} \approx 2.93(0.699)^x$
23. Linear   **25.** Neither   **27.** Linear   **29.** $10,250
31. $13,268.58   **33.** $P = A(t) \cdot \left(1 + \frac{r}{n}\right)^{-nt}$ **35.** $4,569.10
37. 4%   **39.** Continuous growth; the growth rate is greater than 0.   **41.** Continuous decay; the growth rate is less than 0.
43. $669.42   **45.** $f(-1) = -4$ **47.** $f(-1) \approx -0.2707$
49. $f(3) \approx 483.8146$ **51.** $y = 3 \cdot 5^x$ **53.** $y \approx 18 \cdot 1.025^x$
55. $y \approx 0.2 \cdot 1.95^x$

57. $\text{APY} = \dfrac{A(t) - a}{a} = \dfrac{a\left(1 + \frac{r}{365}\right)^{365(1)} - a}{a}$

$= \dfrac{a\left[\left(1 + \frac{r}{365}\right)^{365} - 1\right]}{a} = \left(1 + \frac{r}{365}\right)^{365} - 1;$

$I(n) = \left(1 + \frac{r}{n}\right)^n - 1$
59. Let f be the exponential decay function $f(x) = a \cdot \left(\frac{1}{b}\right)^x$ such that $b > 1$. Then for some number $n > 0$,
$f(x) = a \cdot \left(\frac{1}{b}\right)^x = a(b^{-1})^x = a((e^n)^{-1})^x = a(e^{-n})^x = a(e)^{-nx}$.

61. 47,622 foxes   **63.** 1.39%; $155,368.09   **65.** $35,838.76
67. $82,247.78; $449.75

Section 6.2

1. An asymptote is a line that the graph of a function approaches, as x either increases or decreases without bound. The horizontal asymptote of an exponential function tells us the limit of the function's values as the independent variable gets either extremely large or extremely small.   **3.** $g(x) = 4(3)^{-x}$; y-intercept: $(0, 4)$; domain: all real numbers; range: all real numbers greater than 0.
5. $g(x) = -10^x + 7$; y-intercept: $(0, 6)$; domain: all real numbers; range: all real numbers less than 7.
7. $g(x) = 2\left(\frac{1}{4}\right)^x$; y-intercept: $(0, 2)$; domain: all real numbers; range: all real numbers greater than 0.

9. y-intercept: $(0, -2)$ **11.**

13. B   **15.** A   **17.** E   **19.** D   **21.** C
23.   **25.**

27. Horizontal asymptote: $h(x) = 3$; domain: all real numbers; range: all real numbers strictly greater than 3.

29. As $x \to \infty, f(x) \to -\infty$; as $x \to -\infty, f(x) \to -1$
31. As $x \to \infty, f(x) \to 2$; as $x \to -\infty, f(x) \to \infty$
33. $f(x) = 4^x - 3$
35. $f(x) = 4^{x-5}$
37. $f(x) = 4^{-x}$
39. $y = -2^x + 3$
41. $y = -2(3)^x + 7$
43. $g(6) \approx 800.\overline{3}$

45. $h(-7) = -58$ **47.** $x \approx -2.953$ **49.** $x \approx -0.222$
51. The graph of $g(x) = \left(\frac{1}{b}\right)^x$ is the reflection about the y-axis of the graph of $f(x) = b^x$; for any real number $b > 0$ and function $f(x) = b^x$, the graph of $\left(\frac{1}{b}\right)^x$ is the reflection about the y-axis, $f(-x)$.
53. The graphs of $g(x)$ and $h(x)$ are the same and are a horizontal shift to the right of the graph of $f(x)$. For any real number n, real number $b > 0$, and function $f(x) = b^x$, the graph of $\left(\frac{1}{b^n}\right)b^x$ is the horizontal shift $f(x - n)$.

Section 6.3

1. A logarithm is an exponent. Specifically, it is the exponent to which a base b is raised to produce a given value. In the expressions given, the base b has the same value. The exponent, y, in the expression b^y can also be written as the logarithm, $\log_b x$, and the value of x is the result of raising b to the power of y.
3. Since the equation of a logarithm is equivalent to an exponential equation, the logarithm can be converted to the exponential equation $b^y = x$, and then properties of exponents can be applied to solve for x.   **5.** The natural logarithm is a special case of the logarithm with base b in that the natural log always has base e. Rather than notating the natural logarithm as $\log_e(x)$, the notation used is $\ln(x)$.
7. $a^c = b$ **9.** $x^y = 64$ **11.** $15^b = a$ **13.** $13^a = 142$
15. $e^n = w$ **17.** $\log_c(k) = d$ **19.** $\log_{19}(y) = x$
21. $\log_n(103) = 4$ **23.** $\log_y\left(\frac{39}{100}\right) = x$ **25.** $\ln(h) = k$
27. $x = \frac{1}{8}$ **29.** $x = 27$ **31.** $x = 3$ **33.** $x = \frac{1}{216}$
35. $x = e^2$ **37.** 32   **39.** 1.06   **41.** 14.125   **43.** $\frac{1}{2}$
45. 4   **47.** -3 **49.** -12 **51.** 0   **53.** 10
55. ≈ 2.708 **57.** ≈ 0.151 **59.** No, the function has no defined value for $x = 0$. To verify, suppose $x = 0$ is in the domain of the function $f(x) = \log(x)$. Then there is some number n such that $n = \log(0)$. Rewriting as an exponential equation gives: $10^n = 0$, which is impossible since no such real number n exists. Therefore, $x = 0$ is not the domain of the function $f(x) = \log(x)$.
61. Yes. Suppose there exists a real number, x such that $\ln(x) = 2$. Rewriting as an exponential equation gives $x = e^2$, which is a real number. To verify, let $x = e^2$. Then, by definition, $\ln(x) = \ln(e^2) = 2$.   **63.** No; $\ln(1) = 0$, so $\dfrac{\ln(e^{1.725})}{\ln(1)}$ is undefined.   **65.** 2

Section 6.4

1. Since the functions are inverses, their graphs are mirror images about the line $y = x$. So for every point (a, b) on the graph of a logarithmic function, there is a corresponding point (b, a) on the graph of its inverse exponential function.   **3.** Shifting the function right or left and reflecting the function about the y-axis will affect its domain.   **5.** No. A horizontal asymptote would suggest a limit on the range, and the range of any logarithmic function in general form is all real numbers.

7. Domain: $\left(-\infty, \frac{1}{2}\right)$; range: $(-\infty, \infty)$

9. Domain: $\left(-\frac{17}{4}, \infty\right)$; range: $(-\infty, \infty)$

11. Domain: $(5, \infty)$; vertical asymptote: $x = 5$

13. Domain: $\left(-\frac{1}{3}, \infty\right)$; vertical asymptote: $x = -\frac{1}{3}$

15. Domain: $(-3, \infty)$; vertical asymptote: $x = -3$

17. Domain: $\left(\frac{3}{7}, \infty\right)$; vertical asymptote: $x = \frac{3}{7}$; end behavior: as $x \to \left(\frac{3}{7}\right)^+, f(x) \to -\infty$ and as $x \to \infty, f(x) \to \infty$

19. Domain: $(-3, \infty)$; vertical asymptote: $x = -3$; end behavior: as $x \to -3+, f(x) \to -\infty$ and as $x \to \infty, f(x) \to \infty$

21. Domain: $(1, \infty)$; range: $(-\infty, \infty)$; vertical asymptote: $x = 1$; x-intercept: $\left(\frac{5}{4}, 0\right)$; y-intercept: DNE

23. Domain: $(-\infty, 0)$; range: $(-\infty, \infty)$; vertical asymptote: $x = 0$; x-intercept: $(-e^2, 0)$; y-intercept: DNE

25. Domain: $(0, \infty)$; range: $(-\infty, \infty)$ vertical asymptote: $x = 0$; x-intercept: $(e^3, 0)$; y-intercept: DNE

27. B   **29.** C   **31.** B   **33.** C

35.   **37.**   **39.** C

41.   **43.**

45.

47. $f(x) = \log_2(-(x-1))$
49. $f(x) = 3\log_4(x+2)$
51. $x = 2$
53. $x \approx 2.303$
55. $x \approx -0.472$

57. The graphs of $f(x) = \log_{\frac{1}{2}}(x)$ and $g(x) = -\log_2(x)$ appear to be the same; conjecture: for any positive base $b \neq 1$, $\log_b(x) = -\log_{\frac{1}{b}}(x)$.

59. Recall that the argument of a logarithmic function must be positive, so we determine where $\frac{x+2}{x-4} > 0$. From the graph of the function $f(x) = \frac{x+2}{x-4} > 0$, note that the graph lies above the x-axis on the interval $(-\infty, -2)$ and again to the right of the vertical asymptote, that is $(4, \infty)$. Therefore, the domain is $(-\infty, -2) \cup (4, \infty)$.

Section 6.5

1. Any root expression can be rewritten as an expression with a rational exponent so that the power rule can be applied, making the logarithm easier to calculate. Thus, $\log_b(x^{\frac{1}{n}}) = \frac{1}{n}\log_b(x)$.

3. $\log_b(2) + \log_b(7) + \log_b(x) + \log_b(y)$

5. $\log_b(13) - \log_b(17)$ **7.** $-k\ln(4)$ **9.** $\ln(7xy)$

11. $\log_b(4)$ **13.** $\log_b(7)$ **15.** $15\log(x) + 13\log(y) - 19\log(z)$

17. $\frac{3}{2}\log(x) - 2\log(y)$ **19.** $\frac{8}{3}\log(x) + \frac{14}{3}\log(y)$ **21.** $\ln(2x^7)$

23. $\log\left(\frac{xz^3}{\sqrt{y}}\right)$ **25.** $\log_7(15) = \frac{\ln(15)}{\ln(7)}$

27. $\log_{11}(5) = \frac{1}{b}$ **29.** $\log_{11}\left(\frac{6}{11}\right) = \frac{a-b}{b}$ or $\frac{a}{b-1}$ **31.** 3

33. ≈ 2.81359 **35.** ≈ 0.93913 **37.** ≈ -2.23266

39. $x = 4$, By the quotient rule:
$$\log_6(x+2) - \log_6(x-3) = \log_6\left(\frac{x+2}{x-3}\right) = 1$$
Rewriting as an exponential equation and solving for x:
$$6^1 = \frac{x+2}{x-3}$$
$$0 = \frac{x+2}{x-3} - 6$$
$$0 = \frac{x+2}{x-3} - \frac{6(x-3)}{(x-3)}$$
$$0 = \frac{x+2-6x+18}{x-3}$$
$$0 = \frac{x-4}{x-3}$$
$$x = 4$$
Checking, we find that $\log_6(4+2) - \log_6(4-3) = \log_6(6) - \log_6(1)$ is defined, so $x = 4$.

41. Let b and n be positive integers greater than 1. Then, by the change-of-base formula, $\log_b(n) = \frac{\log_n(n)}{\log_n(b)} = \frac{1}{\log_n(b)}$.

Section 6.6

1. Determine first if the equation can be rewritten so that each side uses the same base. If so, the exponents can be set equal to each other. If the equation cannot be rewritten so that each side uses the same base, then apply the logarithm to each side and use properties of logarithms to solve.   **3.** The one-to-one property can be used if both sides of the equation can be rewritten as a single logarithm with the same base. If so, the arguments can be set equal to each other, and the resulting equation can be solved algebraically. The one-to-one property cannot be used when each side of the equation cannot be rewritten as a single logarithm with the same base.   **5.** $x = -\frac{1}{3}$ **7.** $n = -1$ **9.** $b = \frac{6}{5}$

11. $x = 10$ **13.** No solution   **15.** $p = \log\left(\frac{17}{8}\right) - 7$

17. $k = -\frac{\ln(38)}{3}$ **19.** $x = \frac{\ln\left(\frac{38}{3}\right) - 8}{9}$ **21.** $x = \ln(12)$

57. $y(x) = 3\sqrt{x^2 + 1}$, $y(x) = -3\sqrt{x^2 + 1}$

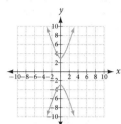

59. $y(x) = 1 + 2\sqrt{x^2 + 4x + 5}$, $y(x) = 1 - 2\sqrt{x^2 + 4x + 5}$

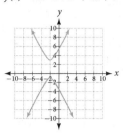

61. $\dfrac{x^2}{25} - \dfrac{y^2}{25} = 1$

63. $\dfrac{x^2}{100} - \dfrac{y^2}{25} = 1$

65. $\dfrac{x^2}{400} - \dfrac{y^2}{225} = 1$

67. $\dfrac{(x-1)^2}{0.25} - \dfrac{y^2}{0.75} = 1$

69. $\dfrac{(x-3)^2}{4} - \dfrac{y^2}{5} = 1$

31.

33.

35.

37.

39.

41.

43.

45. $x^2 = -16y$

47. $(y-2)^2 = 4\sqrt{2}(x-2)$

49. $(y + \sqrt{3})^2 = -4\sqrt{2}(x - \sqrt{2})$

51. $x^2 = y$

53. $(y-2)^2 = \dfrac{1}{4}(x+2)$

55. $(y - \sqrt{3})^2 = 4\sqrt{5}(x + \sqrt{2})$

57. $y^2 = -8x$

59. $(y+1)^2 = 12(x+3)$

61. $(0, 1)$

63. At the point 2.25 feet above the vertex   **65.** 0.5625 feet

67. $x^2 = -125(y - 20)$, height is 7.2 feet   **69.** 0.2304 feet

Section 8.3

1. A parabola is the set of points in the plane that lie equidistant from a fixed point, the focus, and a fixed line, the directrix.

3. The graph will open down.

5. The distance between the focus and directrix will increase.

7. Yes $y = 4(1)x^2$ **9.** Yes $(y-3)^2 = 4(2)(x-2)$

11. $y^2 = \dfrac{1}{8}x$, $V: (0, 0)$; $F: \left(\dfrac{1}{32}, 0\right)$; $d: x = -\dfrac{1}{32}$

13. $x^2 = -\dfrac{1}{4}y$, $V: (0, 0)$; $F: \left(0, -\dfrac{1}{16}\right)$; $d: y = \dfrac{1}{16}$

15. $y^2 = \dfrac{1}{36}x$, $V: (0, 0)$; $F: \left(\dfrac{1}{144}, 0\right)$; $d: x = -\dfrac{1}{144}$

17. $(x-1)^2 = 4(y-1)$, $V: (1, 1)$; $F: (1, 2)$; $d: y = 0$

19. $(y-4)^2 = 2(x+3)$, $V: (-3, 4)$; $F: \left(-\dfrac{5}{2}, 4\right)$; $d: x = -\dfrac{7}{2}$

21. $(x+4)^2 = 24(y+1)$, $V: (-4, -1)$; $F: (-4, 5)$; $d: y = -7$

23. $(y-3)^2 = -12(x+1)$, $V: (-1, 3)$; $F: (-4, 3)$; $d: x = 2$

25. $(x-5)^2 = \dfrac{4}{5}(y+3)$, $V: (5, -3)$; $F: \left(5, -\dfrac{14}{5}\right)$; $d: y = -\dfrac{16}{5}$

27. $(x-2)^2 = -2(y-5)$, $V: (2, 5)$; $F: \left(2, \dfrac{9}{2}\right)$; $d: y = \dfrac{11}{2}$

29. $(y-1)^2 = \dfrac{4}{3}(x-5)$, $V: (5, 1)$; $F: \left(\dfrac{16}{3}, 1\right)$; $d: x = \dfrac{14}{3}$

Section 8.4

1. The xy term causes a rotation of the graph to occur.

3. The conic section is a hyperbola.   **5.** It gives the angle of rotation of the axes in order to eliminate the xy term.

7. $AB = 0$, parabola   **9.** $AB = -4 < 0$, hyperbola

11. $AB = 6 > 0$, ellipse   **13.** $B^2 - 4AC = 0$, parabola

15. $B^2 - 4AC = 0$, parabola   **17.** $B^2 - 4AC = -96 < 0$, ellipse

19. $7x'^2 + 9y'^2 - 4 = 0$ **21.** $3x'^2 + 2x'y' - 5y'^2 + 1 = 0$

23. $\theta = 60°$, $11x'^2 - y'^2 + \sqrt{3}x' + y' - 4 = 0$

25. $\theta = 150°$, $21x'^2 + 9y'^2 + 4x' - 4\sqrt{3}y' - 6 = 0$

27. $\theta \approx 36.9°$, $125x'^2 + 6x' - 42y' + 10 = 0$

29. $\theta = 45°$, $3x'^2 - y'^2 - \sqrt{2}x' + \sqrt{2}y' + 1 = 0$

31. $\frac{\sqrt{2}}{2}(x'+y')=\frac{1}{2}(x'-y')^2$ **33.** $\frac{(x'-y')^2}{8}+\frac{(x'+y')^2}{2}=1$

35. $\frac{(x'+y')^2}{2}-\frac{(x'-y')^2}{2}=1$ **37.** $\frac{\sqrt{3}}{2}x'-\frac{1}{2}y'=$
$$\left(\frac{1}{2}x'+\frac{\sqrt{3}}{2}y'-1\right)^2$$

39. $\theta=45°$ **41.** $\theta=45°$

43. $\theta=30°$ **45.** $\theta=30°$

47. $\theta=37°$ **49.**

51. $\theta=45°$ **53.** $\theta=60°$

55. $\theta\approx36.9°$ **57.** $-4\sqrt{6}<k<4\sqrt{6}$

59. $k=2$

Section 8.5

1. If eccentricity is less than 1, it is an ellipse. If eccentricity is equal to 1, it is a parabola. If eccentricity is greater than 1, it is a hyperbola.   **3.** The directrix will be parallel to the polar axis.   **5.** One of the foci will be located at the origin.   **7.** Parabola with $e=1$ and directrix $\frac{3}{4}$ units below the pole.   **9.** Hyperbola with $e=2$ and directrix $\frac{5}{2}$ units above the pole.   **11.** Parabola with $e=1$ and directrix $\frac{3}{10}$ units to the right of the pole.

13. Ellipse with $e=\frac{2}{7}$ and directrix 2 units to the right of the pole.

15. Hyperbola with $e=\frac{5}{3}$ and directrix $\frac{11}{5}$ units above the pole.

17. Hyperbola with $e=\frac{8}{7}$ and directrix $\frac{7}{8}$ units to the right of the pole.   **19.** $25x^2+16y^2-12y-4=0$

21. $21x^2-4y^2-30x+9=0$ **23.** $64y^2=48x+9$

25. $25x^2-96y^2-110y-25=0$ **27.** $3x^2+4y^2-2x-1=0$

29. $5x^2+9y^2-24x-36=0$

31.   **33.**

35.   **37.**

39.   **41.**

   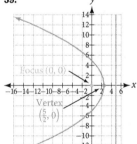

43. $r=\dfrac{4}{5+\cos\theta}$

45. $r=\dfrac{4}{1+2\sin\theta}$ **47.** $r=\dfrac{1}{1+\cos\theta}$ **49.** $r=\dfrac{7}{8-28\cos\theta}$

51. $r=\dfrac{12}{2+3\sin\theta}$ **53.** $r=\dfrac{15}{4-3\cos\theta}$ **55.** $r=\dfrac{3}{3-3\cos\theta}$

57. $r=\pm\dfrac{2}{\sqrt{1+\sin\theta\cos\theta}}$ **59.** $r=\pm\dfrac{2}{4\cos\theta+3\sin\theta}$

Chapter 8 Review Exercises

1. $\frac{x^2}{5^2} + \frac{y^2}{8^2} = 1$; center: $(0, 0)$; vertices: $(5, 0)$, $(-5, 0)$, $(0, 8)$, $(0, -8)$; foci: $\left(0, \sqrt{39}\right), \left(0, -\sqrt{39}\right)$ **3.** $\frac{(x+3)^2}{1^2} + \frac{(y-2)^2}{3^2} = 1$ $(-3, 2)$; $(-2, 2)$, $(-4, 2)$, $(-3, 5)$, $(-3, -1)$; $\left(-3, 2 + 2\sqrt{2}\right)$, $\left(-3, 2 - 2\sqrt{2}\right)$

5. Center: $(0, 0)$; vertices: $(6, 0)$, $(-6, 0)$, $(0, 3)$, $(0, -3)$; foci: $\left(3\sqrt{3}, 0\right), \left(-3\sqrt{3}, 0\right)$ **7.** Center: $(-2, -2)$; vertices: $(2, -2)$, $(-6, -2)$, $(-2, 6)$, $(-2, -10)$; foci: $\left(-2, -2 + 4\sqrt{3}\right)$, $\left(-2, -2 - 4\sqrt{3}\right)$

9. $\frac{x^2}{25} + \frac{y^2}{16} = 1$ **11.** Approximately 35.71 feet

13. $\frac{(y+1)^2}{4^2} - \frac{(x-4)^2}{6^2} = 1$; center: $(4, -1)$; vertices: $(4, 3)$, $(4, -5)$; foci: $\left(4, -1 + 2\sqrt{13}\right), \left(4, -1 - 2\sqrt{13}\right)$

15. $\frac{(x-2)^2}{2^2} - \frac{(y+3)^2}{(2\sqrt{3})^2} = 1$; center: $(2, -3)$; vertices: $(4, -3)$, $(0, -3)$; foci: $(6, -3)$, $(-2, -3)$

17.   **19.**

21. $\frac{(x-5)^2}{1} - \frac{(y-7)^2}{3} = 1$ **23.** $(x+2)^2 = \frac{1}{2}(y-1)$; vertex: $(-2, 1)$; focus: $\left(-2, \frac{9}{8}\right)$; directrix: $y = \frac{7}{8}$

25. $(x+5)^2 = (y+2)$; vertex: $(-5, -2)$; focus: $\left(-5, -\frac{7}{4}\right)$; directrix: $y = -\frac{9}{4}$

27.   **29.**

31. $(x-2)^2 = \left(\frac{1}{2}\right)(y-1)$

33. $B^2 - 4AC = 0$, parabola   **35.** $B^2 - 4AC = -31 < 0$, ellipse

37. $\theta = 45°$, $x'^2 + 3y'^2 - 12 = 0$

39. $\theta = 45°$

41. Hyperbola with $e = 5$ and directrix 2 units to the left of the pole.

43. Ellipse with $e = \frac{3}{4}$ and directrix $\frac{1}{3}$ unit above the pole.

45.   **47.**

49. $r = \frac{3}{1 + \cos\theta}$

Chapter 8 Practice Test

1. $\frac{x^2}{3^2} + \frac{y^2}{2^2} = 1$; center: $(0, 0)$; vertices: $(3, 0)$, $(-3, 0)$, $(0, 2)$, $(0, -2)$; foci: $\left(\sqrt{5}, 0\right), \left(-\sqrt{5}, 0\right)$

3. Center: $(3, 2)$; vertices: $(11, 2)$, $(-5, 2)$, $(3, 8)$, $(3, -4)$; foci: $\left(3 + 2\sqrt{7}, 2\right), \left(3 - 2\sqrt{7}, 2\right)$ **5.** $\frac{(x-1)^2}{36} + \frac{(y-2)^2}{27} = 1$

7. $\frac{x^2}{7^2} - \frac{y^2}{9^2} = 1$; center: $(0, 0)$; vertices $(7, 0)$, $(-7, 0)$; foci: $\left(\sqrt{130}, 0\right), \left(-\sqrt{130}, 0\right)$; asymptotes: $y = \pm\frac{9}{7}x$

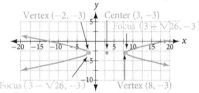

9. Center: $(3, -3)$; vertices: $(8, -3)$, $(-2, -3)$; foci: $\left(3 + \sqrt{26}, -3\right)$, $\left(3 - \sqrt{26}, -3\right)$; asymptotes: $y = \pm\frac{1}{5}(x-3) - 3$

11. $\frac{(y-3)^2}{1} - \frac{(x-1)^2}{8} = 1$ **13.** $(x-2)^2 = \frac{1}{3}(y+1)$; vertex: $(2, -1)$; focus: $\left(2, -\frac{11}{12}\right)$; directrix: $y = -\frac{13}{12}$

15.

17. Approximately 8.48 feet
19. Parabola; $\theta \approx 63.4°$

21. $x'^2 - 4x' + 3y' = 0$

23. Hyperbola with $e = \frac{3}{2}$, and directrix $\frac{5}{6}$ units to the right of the pole.

25.

47.

49.

51.

53. $a_n = 2^{n-2}$

55. $a_1 = 6, a_n = 2a_{n-1} - 5$

57. First five terms:
$$\frac{29}{37}, \frac{152}{111}, \frac{716}{333}, \frac{3188}{999}, \frac{13724}{2997}$$

59. First five terms: 2, 3, 5, 17, 65537

61. $a_{10} = 7{,}257{,}600$

63. First six terms: 0.042, 0.146, 0.875, 2.385, 4.708

65. First four terms: 5.975, 32.765, 185.743, 1057.25, 6023.521

67. If $a_n = -421$ is a term in the sequence, then solving the equation $-421 = -6 - 8n$ for n will yield a non-negative integer. However, if $-421 = -6 - 8n$, then $n = 51.875$ so $a_n = -421$ is not a term in the sequence.

69. $a_1 = 1, a_2 = 0, a_n = a_{n-1} - a_{n-2}$

71. $\dfrac{(n+2)!}{(n-1)!} = \dfrac{(n+2) \cdot (n+1) \cdot (n) \cdot (n-1) \cdot \ldots \cdot 3 \cdot 2 \cdot 1}{(n-1) \cdot \ldots \cdot 3 \cdot 2 \cdot 1}$
$$= n(n+1)(n+2) = n^3 + 3n^2 + 2n$$

CHAPTER 9

Section 9.1

1. A sequence is an ordered list of numbers that can be either finite or infinite in number. When a finite sequence is defined by a formula, its domain is a subset of the non-negative integers. When an infinite sequence is defined by a formula, its domain is all positive or all non-negative integers.   **3.** Yes, both sets go on indefinitely, so they are both infinite sequences.

5. A factorial is the product of a positive integer and all the positive integers below it. An exclamation point is used to indicate the operation. Answers may vary. An example of the benefit of using factorial notation is when indicating the product It is much easier to write than it is to write out
$13 \cdot 12 \cdot 11 \cdot 10 \cdot 9 \cdot 8 \cdot 7 \cdot 6 \cdot 5 \cdot 4 \cdot 3 \cdot 2 \cdot 1$.

7. First four terms: $-8, -\dfrac{16}{3}, -4, -\dfrac{16}{5}$ **9.** First four terms: $2, \dfrac{1}{2}, \dfrac{8}{27}, \dfrac{1}{4}$ **11.** First four terms: $1.25, -5, 20, -80$

13. First four terms: $\dfrac{1}{3}, \dfrac{4}{5}, \dfrac{9}{7}, \dfrac{16}{9}$ **15.** First four terms: $-\dfrac{4}{5}, 4, -20, 100$ **17.** $\dfrac{1}{3}, \dfrac{4}{5}, \dfrac{9}{7}, \dfrac{16}{9}, \dfrac{25}{11}, 31, 44, 59$

19. $-0.6, -3, -15, -20, -375, -80, -9375, -320$

21. $a_n = n^2 + 3$ **23.** $a_n = \dfrac{2^n}{2n}$ or $\dfrac{2^{n-1}}{n}$ **25.** $a_n = \left(-\dfrac{1}{2}\right)^{n-1}$

27. First five terms: $3, -9, 27, -81, 243$ **29.** First five terms: $-1, 1, -9, \dfrac{27}{11}, \dfrac{891}{5}$ **31.** $\dfrac{1}{24}, 1, \dfrac{1}{4}, \dfrac{3}{2}, \dfrac{9}{4}, \dfrac{81}{4}, \dfrac{2187}{8}, \dfrac{531{,}441}{16}$

33. $2, 10, 12, \dfrac{14}{5}, \dfrac{4}{5}, 2, 10, 12$ **35.** $a_1 = -8, a_n = a_{n-1} + n$

37. $a_1 = 35, a_n = a_{n-1} + 3$ **39.** 720   **41.** 665,280

43. First four terms: $1, \dfrac{1}{2}, \dfrac{2}{3}, \dfrac{3}{2}$

45. First four terms: $-1, 2, \dfrac{6}{5}, \dfrac{24}{11}$

Section 9.2

1. A sequence where each successive term of the sequence increases (or decreases) by a constant value.   **3.** We find whether the difference between all consecutive terms is the same. This is the same as saying that the sequence has a common difference.   **5.** Both arithmetic sequences and linear functions have a constant rate of change. They are different because their domains are not the same; linear functions are defined for all real numbers, and arithmetic sequences are defined for natural numbers or a subset of the natural numbers.   **7.** The common difference is $\dfrac{1}{2}$ **9.** The sequence is not arithmetic because $16 - 4 \neq 64 - 16$.   **11.** $0, \dfrac{2}{3}, \dfrac{4}{3}, 2, \dfrac{8}{3}$ **13.** $0, -5, -10, -15, -20$

15. $a_4 = 19$ **17.** $a_6 = 41$ **19.** $a_1 = 2$ **21.** $a_1 = 5$

23. $a_1 = 6$ **25.** $a_{21} = -13.5$ **27.** $-19, -20.4, -21.8, -23.2, -24.6$

29. $a_1 = 17; a_n = a_{n-1} + 9; n \geq 2$ **31.** $a_1 = 12; a_n = a_{n-1} + 5; n \geq 2$

33. $a_1 = 8.9; a_n = a_{n-1} + 1.4; n \geq 2$ **35.** $a_1 = \dfrac{1}{5}; a_n = a_{n-1} + \dfrac{1}{4}; n \geq 2$

37. $a_1 = \dfrac{1}{6}; a_n = a_{n-1} - \dfrac{13}{12}; n \geq 2$ **39.** $a_1 = 4; a_n = a_{n-1} + 7; a_{14} = 95$

41. First five terms: $20, 16, 12, 8, 4$ **43.** $a_n = 1 + 2n$

45. $a_n = -105 + 100n$ **47.** $a_n = 1.8n$ **49.** $a_n = 13.1 + 2.7n$

51. $a_n = \dfrac{1}{3}n - \dfrac{1}{3}$ **53.** There are 10 terms in the sequence.

55. There are 6 terms in the sequence.

57. The graph does not represent an arithmetic sequence.

Index

CPSIA information can be obtained
at www.ICGtesting.com
Printed in the USA
LVHW061753020719
623003LV00010B/355/P

9 781680 920376